Computer Assisted Radiology
Computergestützte Radiologie

Proceedings of the 3rd International Symposium
Vorträge des 3. Internationalen Symposiums

CAR '89 Computer Assisted Radiology

Edited by/Herausgegeben von
H. U. Lemke, M. L. Rhodes, C. C. Jaffe, and R. Felix

Organizer/Organisator: H. U. Lemke

With 357 Figures/Mit 357 Abbildungen

Springer-Verlag
Berlin Heidelberg New York
London Paris Tokyo Hong Kong

Prof. Dr. Heinz U. Lemke

Technische Universität Berlin
Institut für Techn. Informatik
Berlin (West), Federal Republic of Germany

Prof. Dr. Michael L. Rhodes

Multi-Planar Diagnostic Imaging Inc.,
Medical Science Center, Torrance, CA, USA

Prof. Dr. C. C. Jaffe

Yale University, Dept. of Diagnostic Radiology
New Haven, Conn., USA

Prof. Dr. Roland Felix

Freie Universität Berlin
Universitätsklinikum Charlottenburg
Berlin (West), Federal Republic of Germany

ISBN 978-3-642-52313-7 ISBN 978-3-642-52311-3 (eBook)
DOI 10.1007/978-3-642-52311-3

Softcover reprint of the hardcover 1st edition 1989

2068/3020 5 4 3 2 1 0 Printed on acid-free paper

Honorary Chairman:

Heinz Oeser, Berlin (West)

Symposium Committee

Heinz U. Lemke
Technische Universität Berlin (West)

Michael L. Rhodes
MPDI, Medical Science Center, USA

Roland Felix
Freie Universität Berlin (West)

Conrade C. Jaffe
Yale University, USA

Symposium-Organizer

Prof. Dr. Heinz U. Lemke
c/o Institut für Technische Informatik
Technische Universität Berlin
Sekr. CG/FR 3-3
Franklinstr. 28 -29
D-1000 Berlin 10
Tel.: 030 314-73100

Project Management,
Press Coordination and
Editorial Assistant

Bibiana Moser-Illmann
Tel.: 030 216 41 97

CAR'89
in cooperation with:

TU - Technische Universität Berlin

FU - Freie Universität Berlin

acm - Association for Computing Machinery/SIGBIO

Deutsche Röntgengesellschaft

IAPR - International Association for Pattern Recognition

EuroPacs - European Society
Picture Archiving and Communication Systems

IEEE COMPUTER SOCIETY Computational Medicine

IEEE - The Institute of Electrical and Electronic Engineers
IEEE INDUSTRIAL ELECTRONICS SOCIETY

S.I.R.M.N. - Associazione Italiana di Radiologia Medica e Medicina Nucleare

SPIE - The International Society for Optical Engineering

GI - Gesellschaft für Informatik

ACR - American College of Radiology

NVvRd - Nederlandse Vereniging voor Radiodiagnostiek

BIR - The British Institute of Radiology

Deutsche Gesellschaft für Biomedizinische Technik

Radiological Society of Japan

RCR - Royal College of Radiology

Societé Française de Radiologie Médicale

Gesellschaft für Medizinische Radiologie der DDR

IMAC - Image Management and Communication

Oesterreichische Röntgengesellschaft

Schweizerische Gesellschaft für Radiologie und Nuklearmedizin

SCAR - Society for Computer Applications in Radiology

WHO - World Health Organisation

Program Committee

Preface

CAR '89 is a meeting on new methods in radiology and medical disciplines, where digital imaging is an essential part of diagnosis and therapy planning. Such medical activities are increasingly being affected by developments in computer and communication technology.

Many digital imaging or computer assisted medical imaging modalities have been introduced successfully since the 1970's. Digital imaging continues to gain acceptance in the medical community. Its impact is not only felt in diagnostic radiology but also in disciplines such as orthopaedics, radiation oncology, cardiology and surgery, where computer assisted therapy planning is increasingly based on digital imaging modalities.

When radiology takes advantage of the opportunities digital imaging and computer and communication systems can offer, it will play a critical role in the future of medicine. This already is taking place in an increasing number of hospitals in Europe, USA, Canada and Japan. However, there is still a long way to go before it becomes a universal way of providing health care services.

Step by step and in an evolutionary manner, the CAR-symposia are organised to introduce to the medical community current achievements in computer assisted methods for diagnosis and therapy planning. It is realised in coordination with many international medical and technical societies as well as educational institutions and industry.

As a succession to the CAR '85 and CAR '87 symposia, CAR '89 is intended to serve as a forum for discussion between experts from the medical sciences (e.g. radiology, radiotherapy, nuclear medicine, orthopaedics, cardiology, neurology and surgery) and professionals from the computer and physical sciences. The aim is to evaluate the state of the art, (presented in tutorials) and research results (session papers) in the area of Computer Assisted Radiology (CAR).

Computer Assisted Radiology is being discussed with respect to the following topics:

Digital Image Generation

e.g. CT, MRI, DSA, US, DR

Image Management and Communication

e.g. Picture Archiving and Communication Systems (PACS), Digital Optical Recording and Communication, Local and Wide Area Networks, PACS/RIS/HIS Interface, ISDN, Distributed Radiology and Teleconferencing, Standards, Systems Planning and Implementation

Medical Work Stations

e.g. Computer Vision, Computer Graphics, Modelling, Human Computer Interaction

Application Systems for

Computer Assisted Radiological Diagnosis
Computer Assisted Orthopaedics
Computer Assisted Radiation Therapy and
Computer Assisted Surgical Planning.

These four topics are the main tracks of the symposium and each is concluded with an in depth panel discussion.

A special workshop on social aspects of CAR has been prepared, dealing with topics such as

the effect of CAR on the patient and
the effect of CAR on the physician, as well as
the impact of CAR on medical education and
the impact of CAR on the health care system.

Refereed papers towards the above mentioned sessions and workshop have been included in these proceedings.

The international character of CAR '89 is reflected by the large number of countries from where contributions have been received.

The timely publication of the proceedings for CAR '89 by the Springer Verlag is very much appreciated.

Acknowledgement

CAR '89 is a non-profit undertaking and widespread honorary help from many different quarters is essential for its success. As the organizer of CAR '89, it is my task to obtain and coordinate this help. It is therefore my pleasure, to thank and acknowledge a large number of individuals from many different institutions, who have given support to CAR.

One of the first supporters of CAR was our honorary chairman Heinz Oeser, who from the very beginning of the conception of the CAR symposia in 1983, directed some of his infectious enthusiasm towards the acceptance of CAR as a scientific medical method. Who else would (could) have given a paper at CAR '85 entitled "CAR - Flaggschiff der Medizin". Being a professor of the Free University of Berlin, with a long distinguished professional career in radiology, his initial drive for CAR, soon also to be shared by Ogle Burian of the Schering AG, encouraged me to proceed with realising the idea of CAR.

The support of the listed cooperating societies and institutions is also greatly appreciated. Two societies, EuroPACS and SCAR, however, deserve a special mention, as CAR '89 is recognised by them as the event to which to adjust their conferencing and publishing activities. For the future, I hope this bondage becomes even stronger. A special thank you goes to Michael Lüning who introduced to CAR '89, the Gesellschaft für Medizinische Radiologie der DDR, a milestone event, hopefully followed by other Eastern European radiological societies.

Nonetheless difficult was the task of obtaining co-sponsorship of the World Health Organisation. Thank you, Gerald Hanson.

The symposium and program committee, as always, carry the burden of soliciting and refereeing paper and poster contributions, as well as acquiring scientific "work in progress" exhibits. Too many individuals have participated in this process to mention them all, but I am particularly grateful to Roland Felix and his team, for their excellent organisation of the poster session, and Dietrich Meyer-Ebrecht for creating and successfully realising CAR*wipe*.

Professional assistance in project management, organization of the exhibition and press coordination, now already for the 3rd CAR symposium, came from Bibiana Moser-Illmann, Elke Senger and Wolfgang Rogall, respectively. This time, however, for Bibiana Moser-Illmann, it was the hardest CAR ever, as she could not draw on a large team as a back-up. Although some valuable part time assistance was given by

Helga Kallan, Uwe Tronnier, Ulf Jackisch, Michael Henke and Franziska Schweikert, many actions had to be taken single handed by Bibiana Moser-Illmann. The result has been outstanding. I thank also Elke Senger and Wolfgang Rogall for intensively caring as regards industry and press support.

There were moments when I felt that I had taken on too much. To be told, on occasions, that one is on a "Gratwanderung" (engl.: a walk on the edge of a cliff) didn't help either. How much easier life can be, if some of this venture can be shared. Fortunately, there are people prepared to shoulder some of the risk. That is, individuals who would, if something goes wrong with CAR, have some explaining to do in their respective institutions.

I therefore thank in particular, Joachim Rudolf and Lutz Reinke from the staff of the "Senator für Wirtschaft und Arbeit", Berlin, who have, at a very early stage in the planning of CAR '89, secured a basic financial backing. Industrial institutions, which later joined in the exhibiting activity, have also contributed towards the cost of the scientific program and the proceedings.

Support received from those mentioned above, but particularly from the scientific contributors and regular attendees of the CAR meetings, has added to give the CAR symposia a solid basis for the future.

This is now already the 3rd CAR, my wife Jean has managed to live through with me. I have stopped telling her that this is definitely the last one. She would not believe me anyway. But I know, when she sees responses of satisfied attendees of CAR, she gains as much pleasure from the event as I do.

Heinz U. Lemke

Berlin, June 1989

Prelude

Prof. Dr. Heinz Oeser, Berlin

CAR- as an acronym for Computer Assisted Radiology- takes place here in Berlin for the 3rd time.

I have been given the task to open CAR'89 and to welcome you, the participants, most sincerely to this event. CAR demonstrates once again, through discussions and an exhibition, the progress in radiology and nuclear medicine.

Here in Berlin, one frequently likes to demonstrate progress. My statements, however, must therefore be based on issues relating to CAR. I should like to make the following points:

Digital signal processing, using high performance computers, introduced new types of image processing and image producing systems for applications in radiology. Examples are: computer tomography, magnetic resonance imaging, DSA, digital cardiac imaging and positron emission scintigraphy.

The computed reconstructed images allow new insight specifically into the central nervous system, the spinal cord and the joints, showing structural morphological changes which cannot be demonstrated by way of conventional X-ray diagnosis.

Such progress will be demonstrated in lectures for disciplines such as orthopaedics, neurology, cardiology and stereotaxy.

When the radiation source has participated in the functional process, reconstructed images are also mirrors of functions. Amongst other things, such diagnosis of function is possible in relation to circulation of the myocardium and regional areas of the brain.

The MRI spectroscopy allows tissue analysis giving information on biochemical, physiological and pharmacological processes.

The advantage of a computed spatial dose distribution as a precondition for an effective radiation therapy, will be shown in several lectures and has been proven in practice.

The gain and the possibilities of centralised data processing and data storage, augmented with communication to any distance, are also being dealt with in lectures.

CAR offers an overview on what has been achieved and on what can be hoped for.

The search - sometimes already an addiction - for further improvements of imaging modalities will, in my judgement, move towards work on improvements in data acquisition, data processing and image producing systems.

Progress will have an evolutionary effect, when AgBr in X-ray films will be replaced by laser-stimulated digital luminescence, as a converter of the invisible X-rays into visible light. Primarily, this progress promises improved image quality, enlarged image range and a reduction of doses at the time of exposition.

Development of optical components for computers could be one of the next goals.

Such optical computers may try to simulate certain behaviour of our brains. A holographic memory could store holograms as patterns of light, similarly to the holographic plate. In the future, such a computer could not only be a translator in signal processing but could also be a consultant (=advisor) in image diagnostics.

But I will leave these visions of the future to physicists and technicians. As a veteran of radiology, I rather should like to concentrate on aspects which (for the present) are not discussed in the context of CAR. Aspects, however, which one day will have to be accounted for, as they will effect the discipline of radiology and the profile of future radiologists.

Radiology alias Röntgenology has its roots in the discovery of X-rays approximately 95 years ago. With the picture of the skeleton of the hand of his wife, W.C.Röntgen laid the foundation for the medical applications of X-rays.

In the first period, from its tendency, röntgenology was a consequental continuation of photography - to display, to allow nature to image itself through rays.

The bone, containing calcium, appeared rigidly on the X-ray plate, heart and lung followed as clear X-ray visible representations. Organs which were röntgenologically anonymous were made X-ray visible through the application of contrast media. Altogether, the X-ray image was considered as a photograph of the shadow of the object of interest.

The history of röntgenology is being characterised by a steady drive towards an improved image quality, for example, through tubes with a fine focal point or special diaphragm for reducing scattered radiation as well as the attempt to introduce imaging into regions of the body which were anonymous to rays. For example, Pneumoretroperitoneum.

A fundamental change in signal processing took place during the middle of the 1930's. Here in Berlin, Chaoul (at that time my teacher) together with Grossmann, then a designer with the company SANITAS, developed the tomograph for representation of layers of the body. It was then developed further as a tomographic technique because of technical progress in the grounding of high tension cables and high performance X-ray tubes as well as realisations of early conceptions of tomographic techniques by Bocage.

Further development of image diagnostics did not follow the demands for a true-sized representation of organs or sections of the body. The familiar ventro-dorsal viewing direction of the doctor onto and into the patient was abundant as a condition for imaging. Today, given anatomic situations and the goal of representation determine the selection of the layer - the slice... three-dimensional imaging is approaching.

The history of röntgenology can show many relics: the first cathode ray tubes, driven by an electric spark inductor, the Duka, several systems for diaphragm setting, dark red adaptation glasses, the lead apron etc.

The X-ray tube as radiation source is being replaced, as the main object of interest, by the signal processing of the information source. Of topical interest is the discussion around the term FA for imaging procedures in place of radiology - suggested at the membership meeting of the Deutsche Röntgen Gesellschaft on the 14th January, 1989 in Frankfurt. The justification is given as:"Radiological diagnostics based on its content, its systematics and its rules for further education, is the comprehensive discipline in image producing diagnostics, using radiation physical methods".

In his period, the former röntgenologist founded by means of basic publications, image interpretations of form and its individual variants, as well as the development of form and their pathological changes.

Today, the radiologist may be compared with an explorer who, by means of signal generation and its processing, enquires into deviations of our organismus from the norm.

Todays vacancy advertisments underline these changes. The driving license for operating these modern devices is being asked for.

Earlier on, the radiologist was a physician interested in physico-technical aspects and often contributed with features of a hobby mechanic. In the future, technology threatens to overshadow the physician. There is a danger that the radiologist will become as a physician, an assistant to communication technology: seated in front of a PC, he will carry out the interpretation of transmitted images. A course of study of altogether eight semesters for the control and tests of the respective medico-physical technology has just been established at the Technische Fachhochschule in Berlin.

In the acronym of CAR, radiology must remain the key word as a distinguishing characteristic for a speciality of the physician.

In addition to this concern in my introduction, I should like to comment on reproaches made with respect to the usage in our health care system of high performance technology for high performance medicine.

The charges of the media, including in some cases medical journals, against the so called appliance medicine are well known. At times this conflict approaches a form of a religious war, which comes near to breaking point for modern medicine. Such critics often mask the envy-complex of other disciplines. Occasionally, they serve as a promotion for obsolete diagnostic procedures. The earlier believe in progress has reversed into scepsis, possibly even changed into the feeling of a threat from technology.

Perhaps science has missed or at least neglected to present to the public its knowledge in a suitable and understandable manner.

It is not necessary here for me to oppose the programs of political parties, for disciplining or reducing expenditure.

Fault for the expansion of appliance medicine lies partly at the door of the legislator, they have to see and admit that the current fee regulations of the gKv (gesetzliche Krankenversicherung, engl: regulated health insurance) encourages this trend. Today, the apparatus has become to the doctor a question of his existence. The doctors fee is often determined by the apparatus.

Here are a few figures on the extent of the so called Grossgeräte (engl: large scale devices) in our health care system of the FRG.

1988 MRI almost 100

approx. 55% in university clinics and hospitals

CT approx. 600

DSA approx. 400

At this time, approx. 85% of work carried out by the radiologist is being achieved by conventional X-ray techniques, approx. 10% by sonography, approx. 5% by CT.

The share of CT with respect to all X-ray examinations is less than 2%, with respect to costs approximately 7%.

The financial expenditure of the gKv in 1985 was approx. 800 million DM for conventional X-ray examinations, approx. 180 million DM for CT examinations, as opposed to US examinations with altogether 360 million DM.

With respect to the total cost of our health care system, expenditures for imaging systems are annually not even 1% (= 0.6%)

To regard the radiologist who owns MRI, CT or digital techniques has a "Grossindustriellen" (engl: large scale industrialist) is completely unjustified. The doctor today, within the legislated health insurance, is subjected to forced administration.

Only a non-sensical application of imaging procedures, in particular their combination, should be rendered more difficult or halted altogether. As an example: a suspected diagnosed tumor in the breast needs conventional mammography, ultrasound examination and perhaps also MRI. A fine needle biopsy could secure the diagnosis quickly and with great reliability.

I pointed out in my introduction, that CAR represents the progress in radiology, besides NM.

In conclusion I should like to examine and bring attention to this evaluation for the complete spectrum of medicine.

In antiquity, it was the dialogue and exclusively the dialogue, between the physician and the ailing person, which served to recognise the state of a disease. It was a consultation hour. The sense of touch through the palpating hand, was a follow-up as a diagnostic means, today perfected with the ultrasound device. The French physician Laennec founded auscultation approximately 1820, initially it was realised by positioning the ear, then improved by the use of the stethoscope. In the second half of the 19th century, it was the sense of sight which achieved the leading and dominant role in medical diagnostics. The magnifying glass and mirror were positioned before our eyes to allow insight to the body, and the recently discovered X-rays then allowed a view through the body. Aspects which were morphologically tangible and visible, became the content of the teaching of diseases.

The physician, during auscultation and palpation, has a personal contact to his patient, who then senses his sympathy.

The apparatus medicine introduced a distance between physician and patient, the latter felt being left alone - particularly when experiencing radiation therapy in the so called radiation bunkers.

CAR furthers the speechlessness in medicine.

Computer language does not permit communication with the patient.

CAR, therefore, demands reflection on this; in particular, as regards the education of our medical students and this generation in general.

Reflections on the discipline of radiology during the study of medicine are urgently needed; like earlier times we are limping behind future requirements, we give radiation protection a first place - to damp-down the generally produced fright of radiation. We do not recognise the occuring problems in data collection including data processing.

Alone, the immense amount of acronyms and the passion for their propagation threaten to shrink our state of knowledge - except when an appropriate dictionary is on hand.

The value of history taking as well as the clinical examination are often underestimated. The power of information from apparatus medicine, however, is being overestimated.

Particularly in our disciplin, there is a danger that training will be directed exclusively towards the usages of apparatus.

CAR must, not least, lead towards a responsibility for drawing the attention of the disciplin of radiology to possible excesses and outgrowths.

The drive for further continuous technical development comes from the respective industry.

In 1988, the world market in the area of medical electronics totaled to a volume of 11 - 12 billion US dollars. Today, it must be estimated to have doubled, approx. 20 billion US dollars. The radiologist is impressed by the possibilities of a steadily improved communication technology. The exhibition at the ICR in Paris from 3rd to 8th July, 1989 can be compared with an international car show, with representations of the newest apparatus. The development of our disciplin is not anymore driven by the radiologist, we find ourselves in a current of the communication technology.

Soon we will be at the mercy of technico-physical overflowing information systems, where the computer medical specialist (acronym: CMS) monitors the data processing. Sitting in front of a display console, the radiologist carries out image interpretation. Are these thoughts so out of place, when the radiologist as a physician, is only interested in the final product?

With this vision of the future, I should like to close my introduction and thoughts on CAR 1989.

I summerise:

CAR is and remains the flagship of radiology. I introduced this characterisation at the first CAR symposium in 1985, also including the recommendation to take and keep course towards deepening the knowledge within our radiological community and for a good cooperation between radiology, data processing and informatics.

It cannot and must not be avoided to incorporate CAR into the structure of radiology - otherwise CAR will become an industrial fair with exhibitors.

Computer Assisted Radiology has been able to display and allow assessment of morphological, functional and biochemical changes in tissue, which have been hidden until now from X-rays.

The view into the third spatial dimension will be made possible through holography.

The silver mines of X-ray film archives can vanish.

Digital luminescence radiography will make the X-ray film redundant.

Modern communication technology will also demand a high price when applied in radiology. Organisational solutions for effective usage must be developed and realised.

Finally, the most important; also with CAR, only an image will be produced. The decisive information content in the image will be determined by what the physician knows. I should like to refer to the quotation of Goethe: "Erst wenn man weiss, sieht man" (engl:" only when one knows, one can see").

For the time being, the computer does not deliver the diagnosis, but it assists towards a more speedy and improved diagnosis compared to current procedures. The number of false diagnosis, or more appropriately said, false image interpretations, will not be reduced.

This statement should make modest with respect to too high expectations of CAR in the future.

Contents

DIGITAL IMAGE GENERATION

Magnetic Resonance Imaging

Digital Radiography

Digital Subtraction Angiography - DSA

Computer Assisted Orthopaedics

Computer Assisted Stereotaxy

Computer Assisted Radiological Diagnosis

IMAGE MANAGEMENT AND COMMUNICATION

Picture Archiving & Communication Systems - PACS

Communications: PACS/RIS/HIS

MEDICAL WORKSTATIONS

Computer Vision

Computer Graphics

Man Computer Interaction

WORKSHOP - SOCIAL IMPLICATIONS

APPENDIX

Digital Image Generation

Magnetic Resonance Imaging

Three-Dimensional Surface Rendering and Volumetric Rendering in Avascular Osteonecrosis of the HIP

Philipp Lang, Harry K. Genant, Peter Steiger,
Thomas Lindquist, Michael Rhodes, and Roland Felix

Departments of Radiology,
University of California San Francisco, San Francisco, CA 94143,
and University Clinic Rudolf Virchow/Charlottenburg,
Free University Berlin

SUMMARY

Three-dimensional (3-D) surface and volumetric rendering was performed in 11 patients with clinically and radiographically proved avascular necrosis of the hip (AVN). The contours of the femoral head, the acetabulum, and the necrotic zone were simultaneously displayed with use of different colors and transparencies. 3-D MR reconstructions helped to conceptualize the position of the necrotic zone relative to the weight-bearing area.

INTRODUCTION

Magnetic resonance (MR) imaging has been demonstrated to be a very sensitive method in the detection of avascular necrosis of the hip (AVN) (1). MR, however, is not only useful for early diagnosis of AVN. In contrast to conventional radiography and computed tomography, it can also be employed to assess the various pathoanatomic soft tissue changes in the area of involvement (2).

However, although one of the unique features of MR is its direct multiplanar imaging capability, MR images are currently limited by a two-dimensional display of a frequently rather complex three-dimensional structure. In the present study, we developed software techniques to generate diagnostic quality 3-D MR reconstructions. 3-D MR displays were produced in 11 patients with clinically and radiographically proved AVN of the femoral head.

MR IMAGING

MR imaging was performed on a superconductive whole-body system operating at a field strength of 1.5 Tesla (General Electric

Signa, Miwaulkee, WI). Imaging studies included T1-weighted (TR=450 msec, TE=20 msec) and T2-weighted (TR=1500 msec, TE=80 msec) spinecho as well as T2*-weighted gradient-echo images (TR=70 msec, TE=30 msec, flip angle=15^0.

3-D RENDERING

The original data were transferred to an independent imaging system for 2-D and 3-D reconstruction (Model 3200, Dimensional Medicine Inc., Minnetonka, MN). The contour of each individual tissue was computed from the pulse sequence that demonstrated optimal contrast of the particular tissue in question. The different tissues were subsequently co-rendered. A manually or automatically drawn disarticulation contour was introduced in areas of potential overlay such as the acetabular fossa. The data were resampled by linear interpolation to generate a uniform resolution in all three dimensions. 3-D displays were generated either with use of surface rendering or alternatively volumetric rendering techniques (3). The various pathoanatomic structures were simultaneously displayed with different colors and opacities.

RESULTS AND DISCUSSION

The resultant 3-D images helped to understand the extent of the necrosis in three dimensions. The position of the necrotic zone relative to the weight-bearing area was demonstrated (Fig. 1).

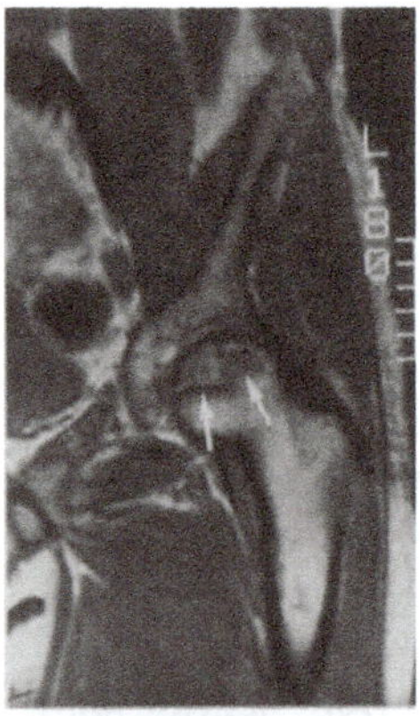

Figure 1A

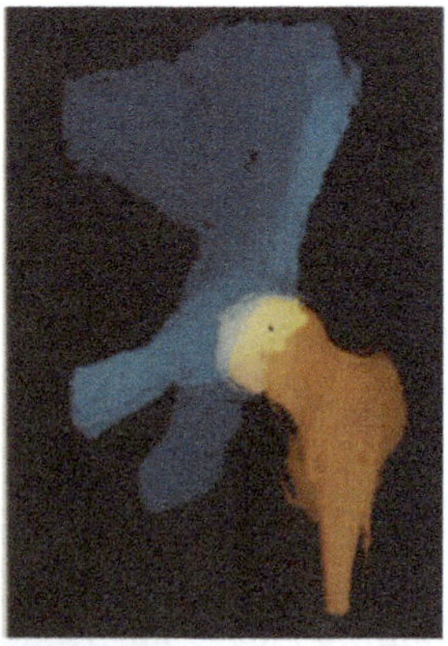

Figure 1B

FIGURE 1:
A. 2-D MR image demonstrates a large area of low signal intensity in the left femoral head (arrows) representing clinically and radiographically proved avascular necrosis.
B. 3-D reconstruction in the same patient shows the extent of the necrosis (yellow). The position of the necrotic zone relative to the weight-bearing surface is demonstrated (blue: hemipelvis; red: femur)

The software and hardware requirements necessary to generate 3-D MR reconstructions in diagnostic quality are extensive at present. However, further improvements of MR imaging and 3-D rendering techniques will help to make 3-D MR displays available on a broader scale. 3-D MR displays in avascular necrosis of the hip are helpful to the surgeon, who, in contrast to the radiologist, is not used to conceptualize a series of two-dimensional images in three dimensions. They may also be useful in preoperative planning of joint preserving surgery.

REFERENCES

1. Mitchell MD, Kundel HL, Steinberg ME, Kressel HY, Alavi A, Axel L. Avascular necrosis of the hip: comparison of MR, CT, and scintigraphy. Radiology 162 (1987) 709-715.

2. Lang Ph, Jergesen HE, Moseley ME, Block JE, Chafetz NI, Genant HK. Avascular necrosis of the femoral head: high-field-strength MR imaging with histologic correlation. Radiology 169 (1988) 517-524.

3. Fishman EK, Drebin B, Magid D, et al. Volumetric rendering techniques: applications for three-dimensional imaging of the hip. Radiology 163 (1987) 737-738.

R e f e r e n c e s

1. McNeil BJ: The value of bone scanning in neoplastic disease. Semin. Nucl. Med. 14(1984):277-286.
2. Batson OV: The vertebral vein system. Amer. J. Roentgenol. Ther. Nucl. Med. 78(1957):195-212.
3. Comarr DR and De Long RP: The role of the vertebral venous system in the metastasis of cancer to the spinal column. Cancer 4(1951):610-618.
4. Smoker WRK, Godersky JC, Knutzon RK, Keyes WD, Norman D, and Bergman W: The role of MR imaging in evaluation of metastatic spinal disease. AJR 149(1987):1241-1248.
5. Sarpel S, Sarpel G, Yu E, Hyder S, Kaufman B, Hindo W, and Ezdinbi E: Early diagnosis of spinal epidural metastasis by magnetic resonance imaging. Cancer 59(1987):1112-1116.
6. Colman IK, Porter BA, Redmond III, Olson DO, Stimac GK, Dunning DM, and Friedl RE: Early diagnosis of spinal metastases by CT and MR studies. J.Comp.Tomogr. 12(1988):423-426.
7. Ludwig H, Tcholakov D, Neuhold A, Fruewald F, Rasoul S, and Fritz E: Magnetic resonance imaging of the spine in multiple myeloma. Lancet 2(1987):364-366.
8. Jung B, Simonsson B, and Sundstrom C: Magnetic resonance imaging in diffuse malignant bone marrow diseases. Acta Radiologica 28(1987):199-205.

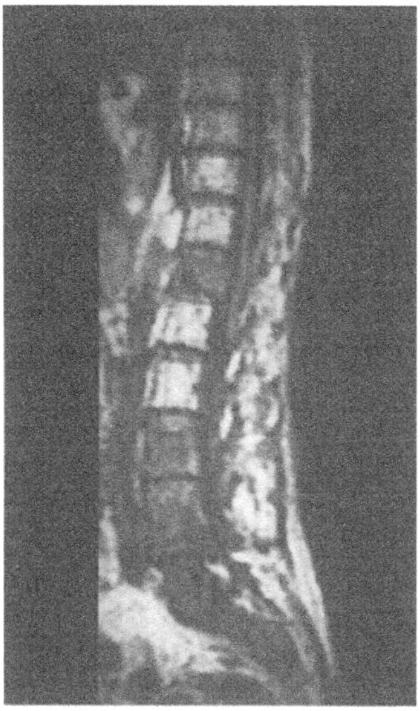

.1 A 56-year-old female 2 years following lt.mastectomy with histologically diagnosed adenocarcinoma. T1 weighted sagittal MR of the lumbo-sacral spine. The L1, L4, L5 vertebrae and the sacrum have weak signals. T2 and proton density (not included) were non significant.

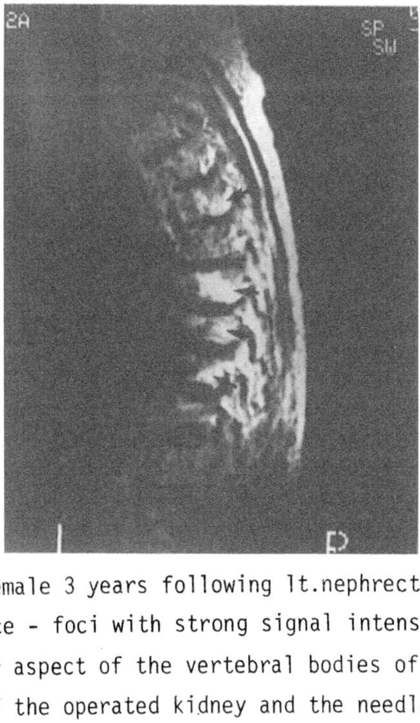

j.2 A 62-year-old female 3 years following lt.nephrectomy. Sagittal MR. On T2 weighted sequence - foci with strong signal intensity are demonstrated in the posterior aspect of the vertebral bodies of D6, D8, D9 and D10 (arrows). The histology of the operated kidney and the needle biopsy from D8 vertebra showed clear cell carcinoma. The T1 weighted sequence and the proton density pulse sequence were unremarkable.

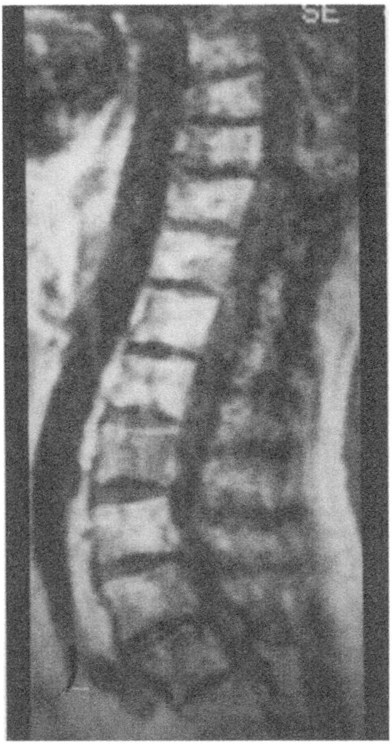

Fig.3 A 52-year-old male with multiple myeloma. T1 weighted sagittal image of th
lower thoracic and lumbar spine. "Mosaic" appearance of the vertebrae.

Comparison of Magnitude and Phase as Observation Techniques in Susceptibility and Perfusion Imaging of the Brain

IR Young, GM Bydder

NMR Unit, Hammersmith Hospital, LONDON W12 ORS, UK

Abstract:

Magnitude and phase data are available from all NMR experiments, including imaging. In the latter method, procedures can be designed for susceptibility mapping and flow measurement wherein the significance of the two components can be quite different, and it is necessary to map both in order to obtain a true picture of what is happening.

Keywords:

MRI, susceptibility, flow measurement, perfusion

Introduction:

All NMR data includes both phase and magnitude data. In imaging, it is the latter which is practically always displayed. Voxel phase in an arbitrary experiment is determined by machine parameters, but the phase difference (between two images which have been acquired with known differences of techniques) results in useful additional data. This has been the basis of susceptibility mapping (seeking the presence of hemorrhagic deposits (1, 2)) and much flow measurement (3, 4). Though it has been reported that phase and magnitude information are both required in assessing what is happening in susceptibility imaging (5), this paper extends the observation to include flow measurement as well.

Methods:

Susceptibility maps are obtained by calculating the phase differences on a voxel by voxel basis between two gradient recalled echo experiments (6) in one of which the data acquisition is delayed by a time t_D relative to the other. The phase difference (φ) between the two relative to a reference point is given by

$$\varphi = \gamma t_D \delta B_o$$

where γ is the gyromagnetic ratio and δB_o is the deviation between the field at the point and that at the reference.

Typically field perturbations due to iron deposits are of the order of 1 ppm, so that the sequences used have a full scale sensitivity of that order, and so main field effects can dominate the appearance of the phase maps. Methods of measuring main field artifacts have been described previously (7).

Magnitude values are determined by sequence design, as usual, with the decay time due to spin-spin relaxation being dependent on T_2* rather than T_2, where T_2* is given by:

$$1/T_2^* = 1/T_2 + \gamma \delta B_o$$

where δB_o is the field deviation across the volume of interest (in this case, a voxel). Note that the two δB_o terms are not normally the same, and we usually identify them differently as δB_{om} for the phase shift component, and δB_{ov} for the magnitude. As can be illustrated, the two do not correlate, and both are needed for proper evaluation of the iron effects.

Flow mapping involves similar problems in most cases, apart from major vessels, with the assumption that there are no partial volume effects. As long as block or laminar flow is assumed, phase can be directly related to flow velocity. The method used involves a pair of spin echo sequences, into one of which a pair of identical gradient field pulses is inserted, one on either side of the inverting rf pulses. If the pulse amplitude is G, its effective duration t_p and the interval between the leading edges of the pulses t_D the phase difference (φ) is given by:

$$\varphi = \gamma t_p t_D G v$$

where v is the flow velocity.

As has been pointed out elsewhere, the situation is much more complex for perfused flow (8) since most of the signal in this case arises from non-moving tissue, and flow patterns and directions can be complex. Resolution of the various components is tedious, and the accuracy of the process is never likely to be high. The analysis of effects is given in reference (8) and will not be repeated here.

Phase changes observed due to blood moving through tissue vary in accordance with TE in most instances, since the T_2 of blood and host tissue is generally quite different. Magnitude change in encoded images arise from dephasing due to macroscopic coherent and quasi-random flows, and from microscopic effects (affecting all signals, including those from nominally static spins). It is this multiplicity of effects that makes their separation so difficult.

Results

Simple experimental results only are given here to indicate the differences in the form of the phase and magnitude signals.

The susceptibility results given first came from head studies, in which phase and magnitude effects were compared over regions of interest of similar size to those used as the voxels in in vivo spectroscopic work. The results were from a sample of pathologies (mainly brain tumours) and the phase effects are given as mean field deviations, while the ratios are those of magnitudes of gradient recalled and spin echo images with identical TE values.

Case No.	Size of Voxel in plane (mm - 10 mm slices)	Field Deviation (ppm)	Range of Magnitude Variation (min/max %)
1	15 x 15	0.30	46%
2	15 x 15	0.30	60%
3	12.5 x 13.5	0.99	17%
4	19.5 x 22.5	0.83	64%
5	10 x 19	0.98	61%

Phase results for flow were derived from an experiment to observe perfusion changes in the arm pre- and post- active exercise. The figures quoted are the differences in phase (relative to a scale of 1) while the IVIM values are the rates of the pre- to post- calculated values.

Case No.	Phase Value	IVIM Value
1	0.04	1.32
2	0.04	0.91
3	0.01	0.80

These results are typical, and it must be born in mind that the very substantial flow sensitization used can lead to significant "motion-like" artifact due to patient involuntary motion, particularly post-exercise. The impact of the artifact is apparent in both phase and magnitude data.

Discussion

The data shows the degree to which phase and magnitude information is complementary rather than repetitive in content. Analysis shows that, in the two instances discussed, this data is derived from different sources and that both sets are necessary in understanding what is happening. While this may seem obvious in connection with flow mapping it is not so clear in susceptibility imaging, but its relevance may be even greater, for the impact on spectroscopic studies of the two pairs of field degradation, represented in their magnitude by δB_{ov} and δB_{om} is significantly different (5).

Acknowledgement

We should like to acknowledge the support of the Medical Research Council in their support for this work.

References:

1. Cox IJ, Bydder GM, Gadian DG, Young IR, Proctor E, Williams SR, Hart I, The effect of magnetic susceptibility variations in NMR imaging and NMR spectroscopy in vivo, J. Magn. Reson. 70 (1) 163-168 (1986)

2. Young IR, Khenia S, Thomas DGT, Davis CH, Gadian DG, Cox IJ, Ross BD, Bydder GM, Clinical magnetic susceptibility mapping of the brain, J. Comp. Asst. Tomogr., 11 (1) 2-6 (1987)

3. Moran PR, A flow velocity zeugmatographic interleave for NMR imaging in humans, Magn. Reson. Imaging, 1 197-203 (1982)

4. Bryant DJ, Bydder GM, Case HA, Collins AG, Cox IJ, Makepeace A, Pennock JM, Use of Phosphorus-31 MR Spectroscopy to Monitor Response to Chemotherapy in Non-Hodgkin Lymphoma, J. Comp. Asst. Tomogr., 12 (5); 770-774 (1988)

5. Young IR, Bydder GM, Khenia S, Collins AG, Assessment of Phase and Amplitude Effects Due to Susceptibility Variations in MR Imaging of th Brain, J. Comp. Asst. Tomogr. (in press) (1989)

6. Hutchison JMS, Edelstein WA, Johnson E, A whole body NMR imaging machine, J. Phys. E, Sci. Instrum., 13 947-955 (1980)

7. Khenia S. Young IR, Image Processing of Susceptibility Maps in Magnetic Resonance Imaging, ed. Lemke HU, Rhodes ML, Jaffee CC, Felix R, Springer-Verlag, Computer assisted Radiology (CAR '87). Berlin p 40-42 (1987)

8. Young IR, Coutts GA, Bryant DJ, Bydder GM, Use of Ultra-Long Echo time Imaging to assist in measuring the velocity of perfused flow in vivo, Magn. Reson. Med. (in press) 1989

Clinical Studies of Head and Neck Tumors by 31p-Spectroscopy

T.Vogl[1], F.Peer[1], V.Reiman[2], S.Holtmann[2], R.Sauter[3], C.Rennschmid[2], J.Lissner[1]

[1] Dept. of Radiology, University of Munich
Klinikum Großhadern, Marchioninistr. 15, D-8000 München 70

[2] Dept. of Head & Neck Surgery, University of Munich

[3] Siemens AG, Medical Engineering Group, Erlangen

Summary

Comparative ^{31}phosphorus magnetic resonance spectroscopy (MRS) and MR imaging (MRI) of 28 patients with masses like sarcoma, carcinoma, meningeoma, lymphoma, adenoma and tuberculosis revealed significant increased concentrations of phosphomonoester, phosphodiester, and inorganic phosphorus in the lesion, whereas the concentration of the phosphocreatine was lower in comparison to muscle or brain tissue. In nearly all masses, pH showed a slight alkaline shift. Existence of necrotic regions detected by MRI was marked by an increase of inorganic phosphorous in the spectra. Tumor growth was characterized by raised concentrations of phosphomonoester. A follow-up study during irradiation of a squamous cell carcinoma revealed a considerable decrease of inorganic phosphate and a subsequent increase of phosphodiester.

Material and methods

28 patients with masses of several histologies underwent comparative investigation by means of magnetic resonance imaging (MRI) and ^{31}phosphorus spectroscopy (MRS). Magnetic resonance imaging was done on a 1.0 Tesla whole-body scanner (Siemens Magnetom). Intravenous injection of the paramagnetic contrast medium Gadolinium-DTPA (0.1 mmol/kg) allowed an exact differentiation of tumor from surrounding soft tissue.
The magnetic resonance spectra were obtained at 1.5 Tesla in a whole-body scanner (Siemens Magnetom) using the ISIS- technique or a 50 mm surface coil for localisation of the measurement volume. The evaluation was made by a least-square-fit procedure, which determines chemical shift, T2-relaxation time and relative area of the peaks caused by the following metabolites: In an in-vivo spectrum of healthy muscle tissue it is possible to differentiate the signals of phosphomonoester (PME), inorganic phosphorous (P_i), phosphodiester (PDE), phosphocreatine (PCr) and nucleoside triphosphate (NTP) (Fig. 1). Since the position of phosphocreatine in the spectra is independent from pH-value, it is used as as reference for the chemical shift.

Results

Results of the investigation of the recurrence of a synovial sarcoma are introduced as an example. The magnetic resonance images revealed a region of

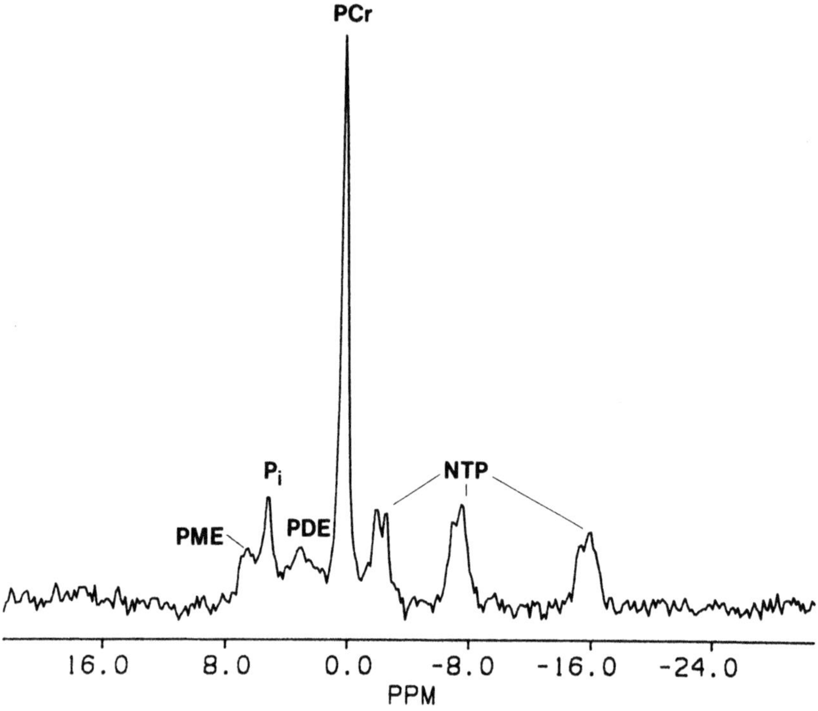

Fig. 1. ^{31}P-Spectra of healthy neck muscle, determined with a 50mm surface coil at 1.5 Tesla.
It is possible to differentiate the resonance of the following metabolites:

- PME= phosphomonoester
- P_i = inorganic phosphate
- PDE= phosphodiester
- PCr= phosphocreatine
- NTP= nucleosid triphosphate

increased signal intensity in the centre of the tumor, as well in the T1- as in the T2-weightened sequence. The substraction image after application of the contrast media shows lower enhancement in the necrotic centre of the tumor, as in the periphery (Fig. 2a). The evaluation of tumor spectrum determined with a surface coil showed some significant changes in comparison to healthy muscle tissue, used as relative value (Fig. 2b). The fit - procedure demonstrated a doubling of the PME- and PDE-concentration, an increase of inorganic phosphate of 35% and a PCr- decrease of 75%. The pH-value was slightly higher (7.21) than muscle pH (7.12).

Furthermore, 9 patients suffering of lymph node metastasis of squamous cell carcinoma underwent magnetic resonance imaging and spectroscopy. Concerning the tumor spectra, the least-square-fit proved some significant differences

Fig. 2. recurrence of a synovial sarcoma near the left orbita

Fig. 2a. MRI, transversal slice orintation, SE T_R/T_E= 500/23 ms, substraction image after application of the contrast media Gadolinium-DTPA. Limited accumulation of Gd-DTPA in the centre of the tumor. Indication of low vascularisation and necrotic areas.

in comparison to muscle tissue. Nearly all cases showed an elevation of phosphomonoester (average + 74%). The phosphodiester were elevated too (average +48%), but not as high as monoester. Similar changes were observed in the concentrations of inorganic phosphorous. Additionally a correlation between the elevation of inorganic phosphorous and the degree of tumor necrosis was suggested by contrast media imaging. Phosphocreatine proved as the most reliable indicator of tumor activity because of it's significant diminition in all tumors.

In a case of squamous cell carcinoma a follow-up study during irradiation therapy was done. Accompanied by considerable shrinking of the tumor, a significant increase of phosphomonoester and a decrease of inorganic phosphorous could be observed. The phosphodiester remained nearly unchanged during the first week and showed an extensive increase after two weeks.

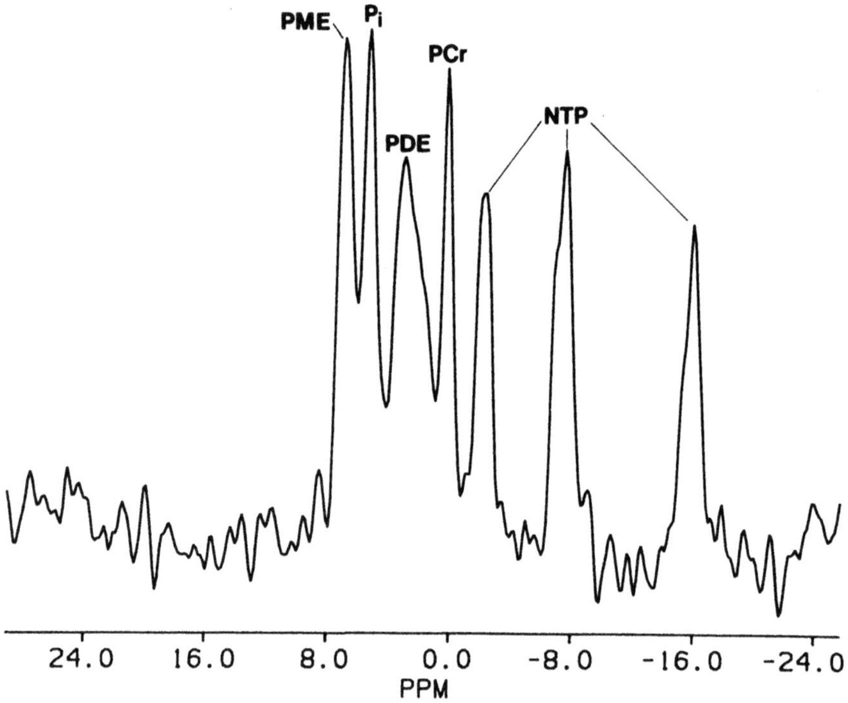

Fig. 2b. ^{31}P-spectra of the tumor

A one week lasting increase was seen in the region of NTP already two days after onset of therapy. The high systematic error of pH-calculation prevents a reliable interpretation of the alterations, observed in the single investigation.

Discussion

The observed differences of tumor spectra in comparison to muscle spectra correlate with the results of in-vivo and in-vitro investigations made by other groups. Tumorous growth is characterized by increased concentrations of phosphoester and inorganic phosphorous and extensive decrease of phosphocreatine. Until now, no reliable connection between tumor histology and ^{31}P-spectra could be detected, whereas most groups report a quite good correlation of spectra and the metabolic state of the tumor, particularly the degree of vascularisation (1). Regions of the tumor most distant from vascular capillaries consist of anoxic necrotic cells. The spectra of such cells exhibit an intense inorganic phosphorous resonance and a very small residual

phosphomonoester peak. The comparison of the evaluated concentrations of inorganic phosphorous and the MR imaging of the respective patients in our study confirm those results.
This fact may be useful in radiation therapy of malignancies, since MR spectroscopy allows the detection of hypoxic cells with low radiosensitivity. Application of radiosensitisers in this cases may improve the effect of the therapy.
Initial expectations to differentiate the tumor histology by ^{31}P-spectroscopy could not be confirmed. The differences within the groups of similar histology are as big as the differences among one other. Errors may be caused by the insufficient definition of the measurement volume. Therefore, the spectra measured upon the tumor consists to some extent of surrounding soft tissue. Our observation of a correlation of PCr-decrease and tumor size may suggest that the phosphocreatine exclusively originates from the muscle, surrounding the tumor.

As a result of this study, we can confirm a significant difference between spectra of tumors and healthy tissue. Even though there is no correlation between histology and spectra visible, it is possible to evaluate vascularisation and metabolic state of tissue by ^{31}P-spectroscopy.Possible clinical application may be the detection of tumor growth, the planning of individual therapy adequate to the metabolic state (1) and therapy control (2,3).

References

1. Lyon R.C., Daly P., Cohen J.S.; Metabolism of Human Cancer Cells In Vitro and In Vivo Monitored by Magnetic Resonance Spectroscopy. Society of Magnetic Resonance in Medicine. Sixth Annual Meeting and Exhibition. Book of Abstracts: 965 (1987)

2. Semmler W.; Monitoring Tumor Response to Chemotherapy in Patients with ^{31}P MR Spectroscopy. Symposium on Positron Emission Tomography and Magnetic Resonance Spectroscopy in Oncology. Heidelberg. Book of Abstracts: 15 (1988)

3. Vogl Th.; Peer F.; Reiman V.; Holtmann S.; Rennschmid C.; Weber H.; Hahn D.; Lissner J.: In vivo ^{31}P Magnetresonanz Spektroskopie und MRI bei Patienten mit oberflächlich gelegenen Tumoren. RöFo, Band 150 (1989) 58-65

Magnetic Susceptibility Effect of Trabecular Bone: Impact on Bone Marrow Imaging

H. Rosenthal[1], K.R. Thulborn[2], D.I. Rosenthal[2], B.R. Rosen[2]

Diagnostische Radiologie I, Medizinische Hochschule Hannover[1]
Department of Radiology, Massachusetts General Hospital[2]
Boston, U.S.A.

Summary

In clinical MR-imaging bone contributes no detectable signal. The obtained signal of the surrounding soft tissue could be decreased in case of a magnetic susceptibility difference between bone and soft tissue. High field spectroscopic studies at 5.88 T have reported such magnetic susceptibility differences between water and bone particles. We have investigated the effects of magnetic susceptibility differences between trabecular bone and water on MR images at 0.6 Tesla. Although susceptibility induced magnetic field inhomogeneities were detected by spectral line broadening using an asymmetric spin-echo technique, our results show only a modest T2* shortening at our field strength. The demonstrated T2* effect can lead to visible signal loss only in case of long dephasing times. In clinical practice long dephasing times only occur in gradient echo imaging with long TE values. In that case low signal of bone marrow is at least in part due to magnetic susceptibility effects.

Introduction

Magnetic resonance imaging is a sensitive method for detection of bone marrow disease. With the growing number of applications a thorough understanding of effects inherent to the method such as chemical shift and magnetic susceptibility and their impact on imaging various tissues is mandatory. Whereas the chemical shift effect of bone marrow has been the subject of numerous studies, little attention has been paid to possible magnetic susceptibility effects of trabecular bone on marrow (1). This study determines the effect of magnetic susceptibility of trabecular bone on adjacent water under imaging conditions at mid field strength. Chemical shift effects from fat and water are avoided by using macerated human vertebral bodies with the intact trabecular network filled with pure water. The single resonance of water is used to measure the change in spectral line width as a direct effect of microscopic of trabecular bone.

Theory

Magnetic susceptibility differences between components of a heterogeneous tissue induce static field inhomogeneities throughout the tissue. In spectroscopic spin echo experiments the 180° refocussing pulse corrects for the effects of all static inhomogeneities at the time of echo (TE) when the maximum signal occurs (HAHN-echo). In imaging experiments, the large gradients used for frequency encoding are designed to dominate any static inhomogeneities. At the time when the gradients are compensated, the bulk of the spins rephase and an echo forms (gradient echo). In the standard SE sequence, the gradient and HAHN-echo coincide, and thus magnetic susceptibility effects are minimized. If the echo time is long however, water can diffuse through regions of different magnetic fields during the course of the spin echo. In this situation the magnetic field variations are no longer static, and irreversible dephasing and hence signal loss will occur. We have used a modification of the asymmetric spin echo technique first introduced by Dixon(2). In this pulse sequence, the gradient echo and HAHN-echo are separated in time by shifting the 180° refocussing pulse from its center at 1/2 TE. The gradient echo formed will then have reduced amplitude, because of incomplete refocussing of static inhomogeneities. By measuring the signal loss as a function of increasing offset between gradient and HAHN-echo with constant TE and TR, the sequence can be used to measure the dispersion of the resonance frequency of the water within and surrounding the bone marrow space in our experiment. This then allows us to directly measure the susceptibility induced broadening of the spectral line (3).

Materials and Methods

Four human cadaveric vertebrae were bisected in coronal plane. The anterior half of the specimen was macerated, defatted and dried, leaving only the inorganic structure and rests of the intervertebral discs. The sample was submersed in water. Air as a possible source of susceptibility effects was completely

removed by degassing in vacuo. The trabecular network was completely permeated by water. No air was detected by thin section CT. Imaging was performed with a standard head coil as RF receiver in a 0.6 T superconducting whole body system. The pulse sequence used a 7.5 mm thick slice selective 90° excitation pulse and a non slice selective 180° refocussing pulse.
The magnetic suceptibility effect was determined by an asymmetric spin echo technique with a fixed TE of 120 ms and TR of 1000 ms. Following an aquisition with a symmetric spin echo, the 180° pulse was shifted towards the excitation pulse in 10 ms increments up to 50 ms. The imaging gradients determining the gradient echo time were kept constant.
Regions of interest were selected in the bulk water outside the specimen and within the trabecular bone of images reconstructed from signal magnitude. The signal intensities were normalized and plotted as a function of dephasing time. For assessment of possible T1 effects of trabecular bone, a series of images with constant TE of 20 ms and varying TR between 125 and 5000 ms were aquired.

Results

Signal intensities of the regions of interest were measured in water within the trabecular bone. No significant signal loss was found for water signal outside the specimen with increasing dephasing time. Signal loss was detected within the trabecular bone. The highest signal loss was obtained for the maximum dephasing time of 100 ms with 43%. Regional signal loss was observed at air/water interfaces. The specimen induced no significant change in T1.

Discussion

A small, but measurable susceptibility effect of trabecular bone on the signal obtained from water is demonstrated. Such a small effect is unlikely to have clinical significance for mid field strength MR imaging, because the dephasing times required for visual perception of the effect are longer than those used in clinical practice. Gradient echo imaging, the method most sensitive to susceptibility effects, uses echo times (= dephasing times) in the range of 10 - 50 ms. Magnetic susceptibility effects of trabecular bone induced a signal loss of only 7-25% for these dephasing times in our phantom.

Davis et al. (4) reported a considerable T2* shortening of water with powdered bone. At a concentration similar to normal trabecular bone T2* was decreased from 102 ms (pure water) to 5 ms, which was interpreted as due to magnetic susceptibility effects. Since these effects are field strength dependent, their results are consistent with our findings. This implies that although the susceptibility effects measured at 0.6 T were small, they may be significantly more important at high field strength systems.

Quantitative chemical shift imaging of bone marrow fat with an asymmetric spin echo technique introduces a dephasing time of 5.3 ms at 0.6 Tesla. This corresponds to a signal loss due to magnetic susceptibility of about 3%. Quantitative measurements are therefore not significantly affected. The magnetic susceptibility effect at mid field strength is probably not strong enough to be used for quantitative assessment of bone itself.

The short T2* of bone marrow must be mainly attributed to mechanisms like chemical shift effects of water and fat proton resonances as well as broad lipid resonances. Only in gradient echo imaging with long TE values a magnetic susceptibility induced signal loss may be detected. Further studies are needed to assess the magnetic susceptibility effects of cortical bone and the influence of higher field strength on bone marrow measurements.

References

1. Wismer, G.L.; Rosen, B.R.; Buxton, R.B.; Stark, D.D.; Brady, T.J.: Chemical shift imaging of bone marrow: preliminary experience. Amer. J. Roentgenol. 145 (1985) 1031-1037

2. Dixon, W.T.: Simple proton spectroscopic imaging. Radiology 153 (1984) 189-194

3. Wismer, G.L.; Buxton, R.B.; Rosen, B.R.; Fisel, C.R.; Oot, R.F.; Brady, T.J.; Davis, K.R.: Susceptibility induced MR line broadening: applications to brain iron mapping. J. Comput. Assist. Tomogr. 12 (1988) 259-265

4. Davis, C.A.; Genant, H.K.; Dunham, J.S.: The effects of bone on proton NMR relaxation times of surrounding liquids. Invest. Radiol. 21 (1986) 472-477

A New Filtering Algorithm for Signal to Noise Ratio Improvement in Magnetic Resonance Imaging

A. De Santis(1), A. Germani(1,2), L. Jetto(3) and R. Passariello(4)

(1) Istituto di Analisi dei Sistemi ed Informatica del CNR, viale Manzoni 30 00185 Rome, Italy. (2) Dipartimento di Ingegneria Elettrica, Università dell'Aquila, 67100 Monteluco (L'Aquila), Italy. (3) Dipartimento di Elettronica ed Automatica Università di Ancona, 60131 Ancona, Italy. (4) Dipartimento di Radiologia, Ospedale "S.M. di Collemaggio", 67100 L'Aquila, Italy.

1. INTRODUCTION. The diagnostic capability of a digital radiological system for NMR, depends on its resolution, that is the number of pixels per unit area, and on the contrast. From a numerical point of view, the problem of the contrast modification is a simple one. On the other hand the resolution improvement due to the increasing of the pixel-density, is relaxed by the effect of the noise, which is always present in any electronic digital device. As a consequence the improvement of the signal to noise ratio (SNR), for a given radiological system, is of great interest. The simplest methods, usually adopted, based on the frequency windowing process, which do not take into account the statistical nature of the image signal, do not give satisfactory response, expecially in the neighbour of the image edges. In [17] a technique to deal with this problem in an NMR system is presented. In [1-9] are proposed more sophisticated methods based on the knowledge of the image autocorrelation function. Other authors [10-14] proposed adaptive filters in order to remove the *a priori* knowledge of the autocorrelation function. The main drawbacks of these last methods are the increased complexity and the high computational cost. The aim of this paper is to present a new filtering algorithm based on a definition of signal model [15] suitable for facing up with the "edge problem". Expecial effort has been devoted to an efficient implementation on a general purpose computer, generally available in a digital radiology laboratory, thus making it sophisticated mathematical subroutines library free.

Here we report briefly the description of the signal model, based on the physical consideration that most of the images are constituted by the union of open disjoint subregions whose interior is enough regular to be well described by a smooth $2-D$ gaussian process. The boundaries of each subregion are the image edges which represent sharp discontinuities in the distributions of the gray level. Of course the image is not differentiable on these dicontinuities so that the filter will have to be suitably adapted to their presence as it is specified in [15]. An image is defined to be *homogeneous* if it does not contain any discontinuity. To this purpose the following hypotheses are assumed:

i) Regularity assumption. The image is modelled by an ensamble of $2-D$ differentiable surfaces. In this case the signal, which represents the gray level, and its spatial derivatives up to a suitably chosen order $\bar{n}$ can be assumed as the state vector.

ii) Stochastic assumption. All the derivatives of order $\bar{n}+1$ of the $2-D$ signal are modelled by means of zero-mean independent gaussian random fields.

Let us indicate with $x(r,s)$ the value of the original image at spatial coordinates (r,s), where the continuous variables r and s denote the vertical and horizontal position respectively, and let us denote with $Z_{i,j}$ the value of any $2-D$ function $Z(r,s)$ in $(i\Delta, j\Delta)$, where Δ is the pixels distance. Because of the smoothness assumption let us define the state as the vector:

$$X_{i,j} = \left[\left. \frac{\partial^n x(r,s)}{\partial r^{n-\alpha} \partial s^{\alpha}} \right|_{r=i\Delta, s=j\Delta}, n = 0, ..., \bar{n}, \alpha = 0, ..., n \right]^T \qquad i,j = 1,2,\ldots M$$

where M is the number of pixels.

By considering a string of L pixels and defining the aggregate state vectors $\bar{X}_i = [X_{i,1}^T \; \ldots \; X_{i,L}^T]^T$, it is possible to derive the following state equation:

$$\bar{X}_{i+1} = [I - F]^{-1} E \bar{X}_i + [I - F]^{-1} Z_i$$

where

$$F = \begin{pmatrix} 0 & \frac{H_1(-\Delta)}{3} & 0 & 0 & \ldots & 0 \\ \frac{H_1(\Delta)}{5} & 0 & \frac{H_1(-\Delta)}{5} & 0 & \ldots & 0 \\ 0 & \frac{H_1(\Delta)}{5} & 0 & \frac{H_1(-\Delta)}{5} & \ldots & 0 \\ \vdots & \ddots & \ddots & \ddots & \ddots & \vdots \\ 0 & \ldots & 0 & \frac{H_1(\Delta)}{5} & 0 & \frac{H_1(-\Delta)}{5} \\ 0 & \ldots & 0 & 0 & \frac{H_1(\Delta)}{3} & 0 \end{pmatrix}$$

$$E = \begin{pmatrix} \frac{H_2(\Delta)}{3} & \frac{H(-\Delta,\Delta)}{3} & 0 & 0 & \ldots & 0 \\ \frac{H(\Delta,\Delta)}{5} & \frac{H_2(\Delta)}{5} & \frac{H(-\Delta,\Delta)}{5} & 0 & \ldots & 0 \\ \vdots & \ddots & \ddots & \ddots & \ddots & \vdots \\ 0 & \ldots & 0 & \frac{H(\Delta,\Delta)}{5} & \frac{H_2(\Delta)}{5} & \frac{H(-\Delta,\Delta)}{5} \\ 0 & \ldots & 0 & 0 & \frac{H(\Delta,\Delta)}{3} & \frac{H_2(\Delta)}{3} \end{pmatrix}$$

$$H_1(\Delta) = e^{A'\Delta}; H_2(\Delta) = e^{A\Delta}; H(\Delta_1, \Delta_2) = H_1(\Delta_1) \cdot H_2(\Delta_2)$$

with A and A' $N \times N$ matrices whose elements $a_{l,m}$ an $a'_{l,m}$ are such that:

$$a_{l,m} = \begin{cases} 1 & \text{if } m = l + 1 + \left[\frac{\sqrt{8l-7}-1}{2}\right] \\ 0 & \text{otherwise} \end{cases}$$

$$a'_{l,m} = \begin{cases} 1 & \text{if } m = l + 2 + \left[\frac{\sqrt{8l-7}-1}{2}\right] \\ 0 & \text{otherwise} \end{cases}$$

and $\{z_i\}$ is a white gaussian sequence with constant covariance matrix Q_z, given by:

$$\begin{pmatrix} \frac{1}{9}S_1 & \frac{1}{15}T & \frac{1}{15}U & 0 & 0 & 0 & \ldots & 0 \\ \frac{1}{15}T^T & \frac{1}{25}S & \frac{1}{25}T & \frac{1}{25}U & 0 & 0 & \ldots & 0 \\ \frac{1}{15}U^T & \frac{1}{25}T^T & \frac{1}{25}S & \frac{1}{25}T & \frac{1}{25}U & 0 & \ldots & 0 \\ 0 & \frac{1}{25}U^T & \frac{1}{25}T^T & \frac{1}{25}S & \frac{1}{25}T & \frac{1}{25}U & \ldots & 0 \\ \vdots & \ddots & \ddots & \ddots & \ddots & \ddots & \ddots & \vdots \\ 0 & \ldots & 0 & \frac{1}{25}U^T & \frac{1}{25}T^T & \frac{1}{25}S & \frac{1}{25}T & \frac{1}{15}U \\ 0 & \ldots & 0 & 0 & \frac{1}{25}U^T & \frac{1}{25}T^T & \frac{1}{25}S & \frac{1}{15}T \\ 0 & \ldots & 0 & 0 & 0 & \frac{1}{15}U^T & \frac{1}{15}T^T & \frac{1}{9}S_3 \end{pmatrix}$$

where

$$S_1 = [Q_2 + H_1(-\Delta)Q_2H_1^T(-\Delta) + 4H_1(-\Delta)QH_1^T(-\Delta)]; \quad S_2 = [4Q_1 + H_1(\Delta)Q_2H_1^T(\Delta)]$$

$$S = S_1 + S_2; \quad S_3 = [4Q_1 + H_1(\Delta)Q_2H^T(\Delta) + Q_2]$$

$$T = [Q_2H_1^T(\Delta) + H_1(-\Delta)Q_2 - 4H_1(-\Delta)Q_1]; U = H_1(-\Delta)Q_rH_s^T(\Delta)$$

and

$$Q_i = \sigma_s^2 \int_0^\delta H_i(\Delta - \tau) \begin{bmatrix} 0 \\ I \end{bmatrix} [0 \quad I] H_i^T(\Delta - \tau)\, d\tau \qquad i = 1,2$$

with σ_s^2 a suitable parameter which determines the signal covariance matrix.

The measurement process is defined by the equation:

$$\bar{Y}_{i+1} = C\bar{X}_{i+1} + V_{i+1}$$

with

$$\bar{Y}_{i+1} = \begin{pmatrix} Y_{i+1,1} \\ \vdots \\ Y_{i+1,L} \end{pmatrix}; C = \begin{pmatrix} C' & 0 & \dots & 0 \\ 0 & C' & \dots & 0 \\ \vdots & & & \\ 0 & 0 & \dots & C' \end{pmatrix}; V_{i+1} = \begin{pmatrix} v_{i+1,1} \\ \vdots \\ v_{i+1,L} \end{pmatrix}$$

$$C' = (1 \quad 0 \quad \dots \quad 0)$$

where $Y_{i,j}$ represents the gray-level at the $(i,j) - th$ pixel and $\{V_i\}$ is a zero mean, white gaussian sequence with covariance equal to $\sigma_o^2 \cdot I$, representing the noise measurement.

The Filter Equation.

The proposed model is in an amenable form for Kalman filtering. Moreover, in order to take into account the fast gray-level variations due to the edges, we propose the following filter structure:

$$\hat{X}_{i+1} = (I - F)^{-1}\hat{X}_i + K(\bar{Y}_{i+1} + Y_{i+1}^e - C(I - F)^{-1}E\hat{X}_i)$$
$$K = \frac{1}{\sigma_o^2}PC^T$$
$$P = \lim_{l \to \infty} P_l$$
$$P_l = (I + \frac{1}{\sigma_o^2}R_{l-1}C^TC)^{-1}R_{l-1}$$

$$R_{l-1} = (I - F)^{-1}EP_{l-1}E^T(I - F^T)^{-1} + (I - F)^{-1}Q_z(I - F^T)^{-1}$$

where Y_i^e is the output of an edge-detector [16] which forces the filter evolution to follows the sudden gray-level variations in presence of the edges.

Let us note that the main computational difficulty is carried on by the inverse matrix operation in the above equations, that has to be performed at every step. The inverse of the matrix $(I - F)$ is given in a closed form [15]. Next procedure gives an easy computational tool for fast inverse operation for the matrix $(I + \frac{1}{\sigma_o^2}R_{l-1}C^TC)$.

The inverse of a partitioned matrix

$$D^{(k)} = \begin{bmatrix} D_{11}^{(k)} & D_{12}^{(k)} \\ D_{21}^{(k)} & D_{22}^{(k)} \end{bmatrix}$$

is a correspondent partitioned matrix

$$B^{(k)} = \begin{bmatrix} B_{11}^{(k)} & B_{12}^{(k)} \\ B_{21}^{(k)} & B_{22}^{(k)} \end{bmatrix}$$

where

$$\begin{aligned} B_{11}^{(k)} &= [D_{11}^{(k)} - D_{12}^{(k)} D_{22}^{(k)-1} D_{21}^{(k)}]^{-1} \\ B_{12}^{(k)} &= -B_{11}^{(k)} D_{12}^{(k)} D_{22}^{(k)-1} \\ B_{21}^{(k)} &= -D_{2}^{(k)-1} D_{21}^{(k)} B_{11}^{(k)} \\ B_{22}^{(k)} &= D_{22}^{(k)-1} - D_{22}^{(k)-1} D_{21}^{(k)} B_{12}^{(k)} \end{aligned}$$

Moreover, for the computation of $B_{11}^{(k)}$, we can iterate the procedure in the following way:

$$B_{11}^{(k)} \triangleq B^{(k+1)} = D^{(k+1)-1} = \begin{bmatrix} D_{11}^{(k+1)} & D_{12}^{(k+1)} \\ D_{21}^{(k+1)} & D_{22}^{(k+1)} \end{bmatrix}^{-1} = \begin{bmatrix} B_{11}^{(k+1)} & B_{12}^{(k+1)} \\ B_{21}^{(k+1)} & B_{22}^{(k+1)} \end{bmatrix}$$

obtaining matrices of reduced order at every step, until $B_{11}^{(k+j)}$ turns to be directly invertible in a closed form. The details will be reported in [18].

Numerical Results.

The proposed filter was tested in [15] showing an improvement of the signal to noise ratio of about 10 decibel.

Tipical 256 × 256 pixels NMR images have been processed. The numerical experiments were performed with $\bar{n} = 1, 2$. Several processing have been obtained for different choices of the ratio $\pi = \sigma_s^2/\sigma_o^2$. Early results showed a significant restoration of the corrupted images, jointly with a satisfactory behaviour in correspondence of the edges.

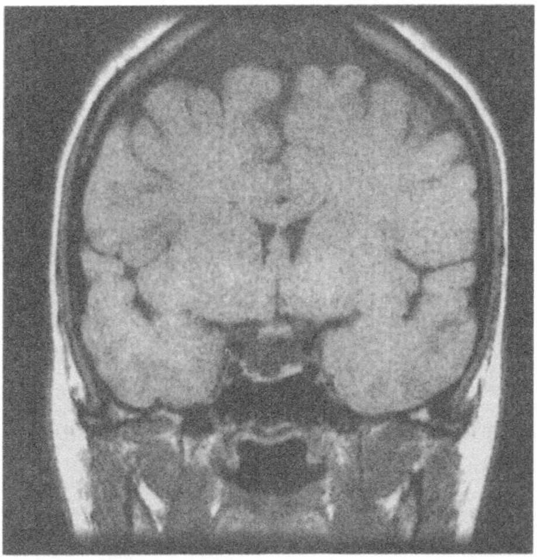

Fig. 1.a Original corrupted image.

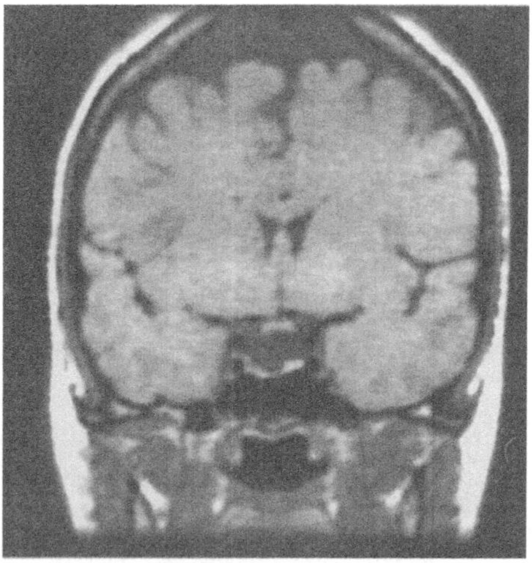

Fig. 1.b Filtered image, $\bar{n} = 2$

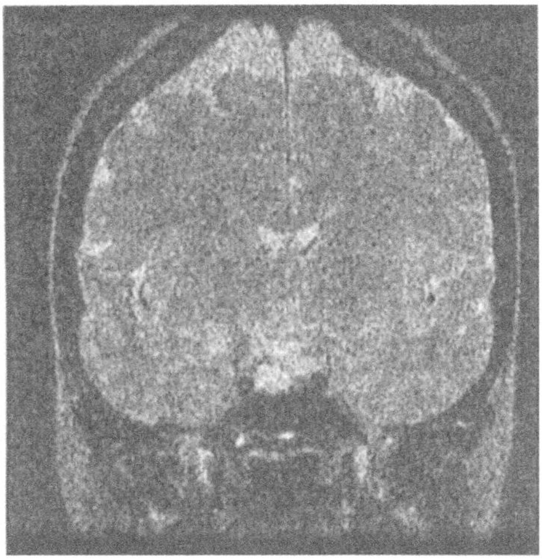

Fig. 2.a Original corrupted image.

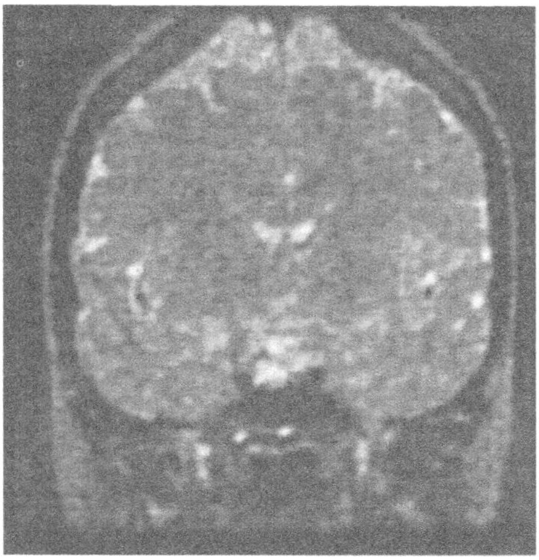

Fig. 2.b Filtered image, $\bar{n} = 2$

As a final remark we can stress the importance of a correct choice of a $\bar{\pi}$ value that, for a given NMR system, depends only on the type of the diagnostic images. This suggest us to provide the diagnostic system of a data base for the best choice of $\bar{\pi}$.

References

1. N.E. NAHI, Role of recursive estimation in statistical image enhancement, *Proc. IEEE* Vol. 60, pp. 872-877, July 1972.
2. N.E. NAHI and T. ASSEFI, Bayesan recursive image estimation. *IEEE Trans. on Comp.* Vol. C-21, pp. 734-738, July 1972.
3. S.R. POWELL and L.M. SILVERMAN, Modeling of two-dimensional covariance functions with applications to image enhancement, *IEEE Trans. Automat. Contr.*, Vol. AC-19, pp. 8-13, Feb. 1974.
4. A.K. JAIN and E. ANGEL, Image restoration, modeling, and reduction of dimensionality, *IEEE Trans. Comput.* Vol. C-23, pp. 470-476, May 1974.
5. J.W. WOODS and C.H. RADEWAN, Kalman filtering in two dimensions, *IEEE Trans Inform. Theory* Vol. IT-23, pp. 473-482, July 1977.
6. A. HABIBI, Two dimensional Bayesian estimate of images, *Proc. IEEE* Vol. 60, pp. 878-883, July 1972.
7. A.K. JAIN, Partial differential equation and finite difference methods in image processing. Part I: image representation *JOTA* Vol. 23, pp. 65-91, Sept 1977.

8. T. KATAYAMA and M. KOSAKA, Smoothing algorithms for two-dimensional image processing, *IEEE Trans. on Syst. Man and Cybern.* Vol. SMC-8, pp. 62-66, Jan. 1978.

9. T. KATAYAMA and M. KOSAKA, Recursive filtering algorithm for a two-dimensional system, *IEEE Trans. Automat. Contr.* Vol. AC-24, pp. 130-132, Feb. 1979.

10. T. KATAYAMA, Restoration of noise images using a two-dimensional linear model, *IEEE on Syst. Man and Cybern.* Vol. SMC-9, pp. 711-717, Nov. 1979.

11. H.R. KESHAWAN and M.D. SRINATH, Enhancement of noisy images using an interpolative model in two dimension, *IEEE on Syst. Man and Cybern.* Vol. SMC-8, pp. 247-258, April 1978.

12. H. KAUFMAN, J.W. WOODS, S. DRAVIDA and A.M. TEKALP, Estimation and identification of two dimensional images, *IEEE Trans. on Automat. Control* Vol. AC-28, pp. 745-756, July 1983.

13. P.E. WELLSTEAD and J.R. CALDAS PINTO, Self-tuning filters and predictors for two-dimensional system. Part I: Algorithms, *Int. J. Control* Vol. 42, pp. 457-478, 1985.

14. P.E. WELLSTEAD and J.R. CALDAS PINTO, Self-tuning filters and predictors for two dimensional systems. Part II: Smoothing application.

15. A. GERMANI and L. JETTO, Image modelling and restoration: a new approach. Circuits Systems and Signal Processing, 1988, in press.

16. J. BIEMOND and J.J. GERBRANDS, An edge-preserving recursive noise-smoothing algoritm for image data, *IEEE Trans on Syst. Man and Cybern.* Vol. SMC-9, pp. 622-627, Octob. 1979.

17. T. BROSNAN, G. WRIGHT, D. NISHIMURA, Q. CAO, A. MAKOVSKY and F.G. SOMMER, Noise Reduction in Magnetic Resonance Imaging, *Magnetic Resonance Medicine*, vol. 8, pp. 394-409, 1988.

18. A. DE SANTIS, A. GERMANI, L. JETTO, An Efficient Numerical Algorithm for Fast Image Restoration. Rapporto Iasi 1989, in press.

MRI 3D Imaging of the Knee Joint Combined with an Image Processing System

G. Adam, K. Bohndorf, M. Drobnitzky, Th. Tolxdorf, R. W. Günther

Department of Diagnostic Radiology and Statistics, Technical University, Aachen FRG

Summary

A prospective study was performed to evaluate the diagnostic accuracy of 3D imaging combined with an image processing system. 60 patients with suspected meniscal tears were examined on a 1.5 Tesla imager and underwent subsequent arthroscopy Using FISP 3D sequences knees were imaged in the axial and sagittal planes. The axial data set was transferred to an image processing system (Kontron, Munich) where secondary reconstructions in every desired plane were performed. MR diagnosis was made on the sagittal images and on the secondary reconstructed data set. MR diagnosis showed an accuracy of 91.6 % compared to arthroscopy.

Introduction

Fast 3D Imaging is not a new concept in MRI of knee joint disorders. However, there are no reports in the literature regarding multiplanar reformatting capability of 3D MR data sets. Therefore, we performed a prospective study using the sophisticated displaying capabilities of an image processing system for evaluation of patients with suspected meniscal tears.

Methods

Sixty patients who had sustained knee trauma underwent MR imaging and subsequent arthroscopy. MR examination was performed on a 1.5 Tesla imager (Siemens, Erlangen) , using a linear polarizing surface coil. After a conventional Spin-Echo scout view knees were imaged by a FISP 3D sequence (a = 40° , TR = 0.03 sec, TE = 12 msec) in sagittal and axial planes.

Axial data sets were transferred to the image processor (Kontron, Munich) Image transfer was intialized from the main console of the MR system. Using the standard software of the image processor, real time reformatting was performed intersecting the original data set at any given position. Results of the reformatting process were documented by photography from the video screen or by transferring data back to the MR system where they were filed together with conventionally acquired MR images. Diagnosis was made from sagittal and reconstructed images and were compared to the arthroscopic findings.

Results

Meniscal tears were defined according to Crues (1) . Degenerative changes were mentioned if a globular or linear signal increase was observed within the meniscal borders. A meniscal tear was diagnosed if the signal increased zone had contact to the meniscal margins or the surrounding hyaline cartilage surfaces. Using the original and reconstructed datas MR diagnosis was true positive in 26 of 60 patients and true negative in 28 of 60 patients. It was false positive in three of 60 patients and false negative in three of 60 patients in comparison to the arthroscopic findings. In one patient the tear was not found on the initial arthroscopy, but MR diagnosis was confirmed on a second session four months later. All lesions diagnosed on sagittal images were diagnosed on reconstructed images too. So MR diagnosis of a meniscal tear had an overall accuracy of 91.6 % in comparison to the arthroscopic findings.

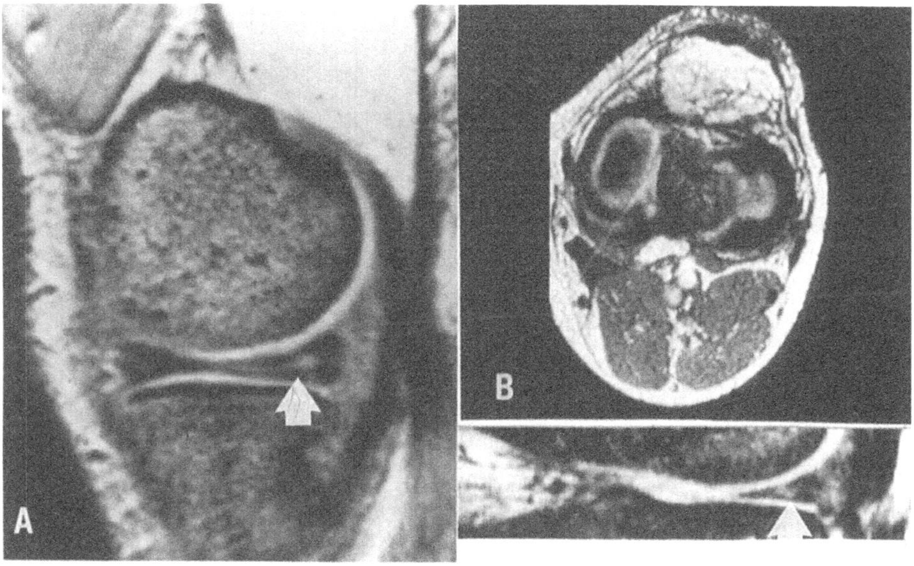

Figure 1 : Typical meniscal tear of the posterior horn in a) sagittal image and b) reconstructed image

Discussion

Except for Tyrell (2), Spritzer (3) and Skalej (4) who worked with 3D imaging, however, there are no reports concerning the reformatting capability of this imaging mode. Once data were transferred to an image processing system reformatting in every desired plane is possible without using the MR computer calculation time. Especially for diagnostic imaging of the meniscus with its curved surfaces and borders the reconstruction process is helpful, because the cutlines can be adapted to the anatomic configuration. Using parasagittal, paraaxial or free drawn cutlines imaging of the posterior and anterior horn as well as of the intermediate part could be improved. Although the meniscal tears were also seen on the original MR data sets, the image processor facilitates reviewing of the bulk of information provided by volume imaging. Instead of looking to " static " images, the image processor allows a dynamic view on the joint by interactive working. Our results demonstrate that 3D Volume imaging in one plane with subseqent reformatting may be most usefull in knee joint imaging for suspected meniscal disorders. In comparison to Spin-Echo imaging it provides the same diagnostic accuracy (1) but in a shorter iamging time.

References

1. Crues J.C., Mink J., Levy T.L. et al
Meniscal tear of the knee accuracy of MR imaging of the knee: first 144 cases.
Radiology 1987; 164 445-448

2. Tyrell R.L., Gluckert K., Pathria M. et al
Fast three dimensional MR iamging of the Knee: Comparison with arthroscopy
Radiology 1988; 166 865-872

3. Spritzer C.E., Vogler J.B., Martinez S. et al
MR iamging of the knee preliminary results with 3d FT GRASS pulse sequence
AJR 1988; 150 597-603

4. Skalej M., Klose U., Küper K.
Optimierte Untersuchungstechniken von Meniskopathien durch kernspintomographische 3D
Imaging bei 1,5 Tesla
Fortschr Röntgenstr 1988; 148,2 183-188

A New Hybrid Method for Analysis of Multiexponential T_2 Relaxation Processes in NMR Images

H.WERN, P.BARTH, K.H.FOLKERTS, M.STAEMMLER, K.GERSONDE

Hauptabteilung Medizintechnik, Fraunhofer-Institut für zerstörungsfreie Prüfverfahren, D-6670 St. Ingbert, Ensheimerstr.48 und Fachrichtung Medizintechnik der Universität des Saarlandes, D-6650 Homburg/Saar (Germany)

Summary

An accurate evaluation of T_2 decay processes obtained from in vitro measurements and in vivo NMR imaging is of main importance for tissue characterization and differentiation. In this paper a hybrid method for the analysis of multiexponential functions is presented. The method is based on the combined use of semilogarithmic regression analysis and of a least squares estimate based on a modified Gauß-Newton-Marquardt algorithm (GNM). The method has been applied to synthetic, in vivo and phantom data in order to evaluate the performance, accuracy and computational requirements. Calculations were performed employing the software system SUNRISE [1].

Introduction

The analysis of a multiexponential decay curve as a sum of monoexponentials is a commonly ill-conditioned problem. Decay curve analysis occurs in various areas, such as chemical kinetics, biomedical engineering, and nuclear science [2-5]. The decomposition of multiexponential T_2 relaxation processes into monoxeponential components is the basis of parameter-selective proton imaging. Parameter-selective imaging allows the characterization and differentiation of tissues [6,7]. Employing a multiple magnetic resonance excitation scheme a set of up to 100 echo images is obtained in a single examination. As a result, a set of NMR parameters, e.g., relaxation times T_{2j}, fractions α_j, and proton densities ρ_j $(1 \leq j \leq 3)$ can be determinated for each pixel by multiexponential analysis methods. The echo images describe the decay of the magnetization for each pixel by 100 data pairs (magnetization,time), given by eq.(1).

$$M(t_i) = \left(M_0 \sum_{j=1}^{n} \alpha_j \cdot e^{-\frac{t_i}{T_{2j}}} \right) + N(t_i) \qquad ;1 \le i \le 100 \qquad (1)$$

M_0 : initial magnetization

n : number of components

T_{2j} : relaxation time of component j

α_j : fraction of component j

N(t) : noise

Due to the excitation scheme the t_i times are equidistant. The term $N(t_i) = n(t_i) + n_0$ describes the noise and contains two different parts: (i) n_0 is a constant noise level originating from the offset of the analog-digital converter and of the additional amplifiers and (ii) $n(t_i)$ represents the noise statistically distributed around zero.

Determination of Noise

In order to eliminate the constant noise level n_0 (eq (1)) the following procedure is performed: In an image matrix the pixels which are outside the object under investigation are used to estimate the mean noise level. If the sum of the first ten data-points of the magnetization decay curve is less than 1.2 times the sum of the last ten points of the decay, the pixel is considered to represent noise. The mean noise level is determined on the basis of 100 of these pixels and subtracted from each decay curve, prior to the application of the analysis methods I and II.

Semilogarithmic Regression Analysis (Method I)

In semilogarithmic regression analysis (SRA) only those points of a magnetization decay curve are used for the determination of the relaxation times, which are greater than 5% of the maximum of the curve (Fig 1). If there only one component exists in the pixel, the curve should be a straight line in the semilogarithmic representation (compare eq.(2)). In the case of the existance of multiple components it is supposed that at least the tail of the curve can be represented as a straight line by semilogarithmic representation (Fig. 2). This assumption is fulfilled if the T_{2j} values differ by a factor greater than three.

$$\ln(M(t_i)) = \ln\left(\left(M_0 \cdot \sum_{j=1}^{n} \alpha_j \cdot e^{-\frac{t_i}{T_{2j}}}\right) + n(t_i)\right) \tag{2}$$

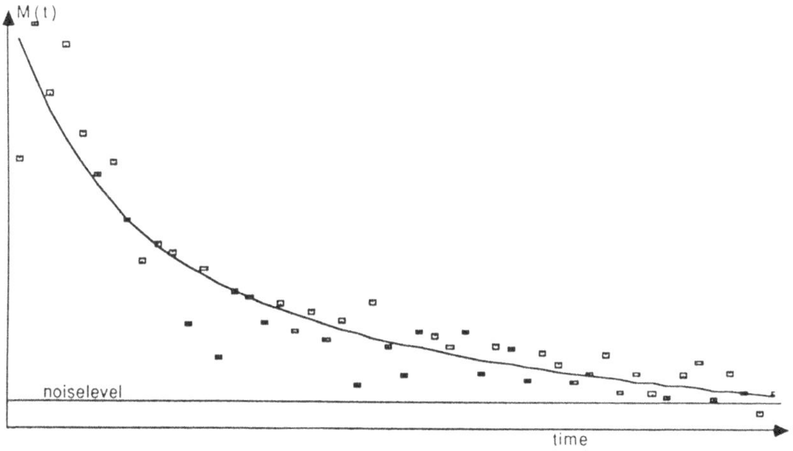

Figure 1: Linear representation of a multiexponential decay curve.

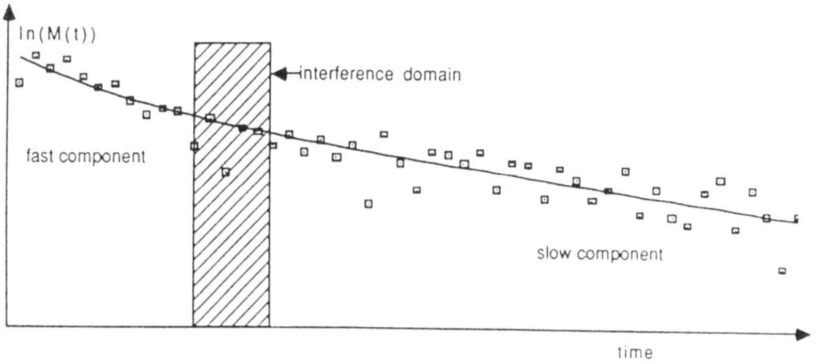

Figure 2: Semilogarithmic representation of a multiexponential decay curve.

The difficulty is now to decide which points belong to the slowest component, that means to find the interference domain between the two components. Therefore, in a first step a regression line is calculated for the last five data-points. Then the regression line and the correlation coefficient are recalculated using stepwise additional data-points. If the correlation coefficient continuously decreases within five successive runs with increasing number of data points, the time domain is defined, where the monoexponential part of the decay curve interferes with the next faster T_{2j-1} component. The actual T_2 value is then determined by regession analysis of data-points excluding the five data-points of the interference domain. With this T_{2j} value a monoexponential function is calculated which is then subtracted from the experimental decay curve. The procedure is repeated with the residual decay curve, if the number of the remaining data-points is greater than six. If the number of remaining data-points is less than six, no further component can be evaluated. Those remaining data-points are therefore included into the set of data-points of the foregoing monoexponential.

Gauß-Newton-Marquardt Algorithm (Method II)

A different approach to determine relaxation times is based on an iterative fit of the magnetization decay given by eq. (1) to the measured data-points of the decay. The accuracy of this fit is controlled by the Gauß-Newton least square method, thus aiming to minimize the nonlinear functional $F(\underline{\alpha},\underline{T}_2)$.

$$F(\alpha,T_2) = \sum_{i=1}^{N} \left[M_i - \sum_{j=1}^{n} \alpha_j \cdot e^{-\frac{t_i}{T_{2j}}} \right]^2 \tag{3}$$

The necessary condition for minimization of the functional is that all partial derivatives with respect to the unknowns α_j, T_{2j} must vanish. This yields a set of 2n nonlinear equations which are difficult to solve. Therefore, the nonlinear equations are linearized. It is assumed that appropriate starting values $\alpha_j^{(0)}, T_{2j}^{(0)}$ are given. With these starting values we expand the fitting function to first order in a Taylor's expansion as a function of the parameters α_j, T_{2j}. For simplicity we arrange the parameters α_j, T_{2j} in a vector $\underline{a}=(\alpha j, T_{2j})^T, (j=1,...,n)$ of the length 2n. Then Taylor's expansion reads as

$$M = M\left(\underline{a}^{(0)}\right) + \sum_{l=1}^{2n} \left[\frac{\partial M_0}{\partial a_l} \delta a_l \right] \qquad , M_0 = M\left(\underline{a}^{(0)}\right) \tag{4}$$

The result is a function which is linear with regard to the parameter increments δa_l. Then, well-known methods of linear least squares can be applied. Thus, a sequence of approximate solutions $x^{(m)}$ is obtained which are improved by iteration [8].This iterative process is called Gauß-Newton(GN) method.

$$\underline{\underline{\gamma}} \cdot \delta\underline{a} = \underline{\beta} \tag{5}$$

For a nonsuitable choice of the startvector $\underline{a}^{(0)}$ the GN method must not necessarily converge to the exact solution. Convergence is achieved if the sum of squares of residuals decreases in each iteration step, thus requiring the existance of a direction of descent $\underline{v}^{(m)}$. Two approaches are used to compute the direction of descent $\underline{v}^{(m)}$.

$$\underline{x}^{(m)} = \underline{x}^{(m-1)} + g \cdot \underline{v}^{(m)} \qquad ,g > 0 \tag{6}$$

The first approch employes the negative gradient of $F(\underline{x}^{(m-1)})$. The negative gradient method fastly converges if the starting values are far from the desired least square minimum. The second approach is the extension of the GN method given by Marquardt [8]. This GNM method properly performs near the minimum and is controlled by the parameter λ. Varying λ to several possible values one descent has to be detected. However, this procedure is time-consuming for image matrices, therefore a hybrid method is employed.

Algorithm of the Hybrid Method

The hybrid method combines the semilogarithmic regression analysis with a modified GNM algorithm to exploit the best features of each method shown in the flow chart (Fig. 3). The regression analysis performs fast and yields the approximate number of relaxation processes and the corresponding T_2 values with fractions α. These results provide better starting values for the modified GNM method rather than randomly distributed ones. The modified GNM method consists of a combination of gradient search and GNM method. The parameter λ is used to guide the relative influence of each method. If λ is very small, a solution similar to the GNM method results (see eq(5)). If λ is very large, the modified GNM method is dominated by the gradient search method which is illustrated as a shaded area in Fig.3. The value for λ should be choosen small enough to take advantage of the GNM method, but large enough to be sure that F rapidly decreases.This is achieved by choosing the s t e e p e s t possible direction of descent.

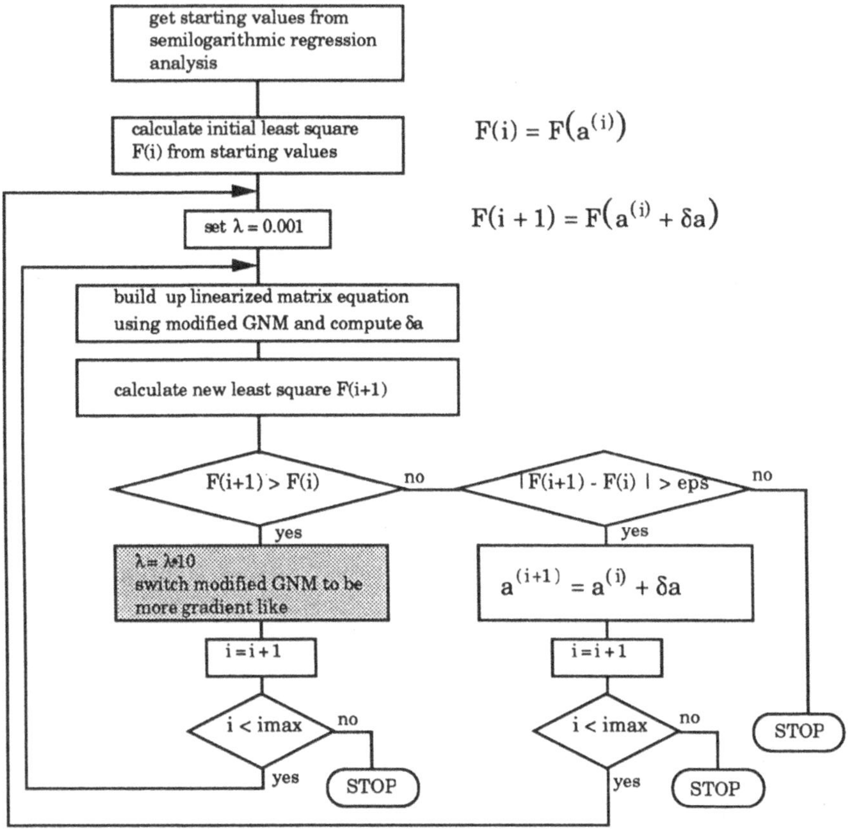

Fig. 3: Flow chart of hybrid method.

The number of possible relaxation times must be specified in advance for the hybrid method. If the actual number of relaxation times is less than the specified one, the hybrid method duplicates relaxation times, but with evenly distributed fractions α. Therefore, the result is corrected for duplicates.

Computational Requirements for the Hybrid Method

Currently the hybrid method is implemented on a system having a computing power of 50 kflops. The SRA is to be about 25 times faster than the modified GNM method. The SRA method needs about 75 ms for the analysis of a decay, whereas the modified GNM method runs for about two seconds. Employing the hybrid method, the number of iterations in the modified GNM method is reduced by a factor of 2 to 3, thus reducing the computational time by about 40 to 60 percent. Nevertheless, the maximum number of iterations has to be limited and the truncation value eps which sets the limit of the difference of two successive root mean squares has to be set properly (usually about 10^{-7}).

Validation by Using Synthetic Data

Several data sets have been simulated and superimposed with Gaussian noise in order to validate the performance, the sensitivity, and the accuracy of all three methods, SRA, modified GNM, and hybrid method. An equidistant time grid of 23 ms was used with a total number of 50 data pairs. The results of these test calculations are summarized in Table 1.

simulated data		regression analysis				modified GNM method			
α	T_2 (ms)	α	T_2(ms)	rel. Error		α	T_2(ms)	rel. Error	
50	100	57,9	109	15,8	9	55,1	99	10,2	1
50	400	42,1	459		14,8	44,9	410		2,5
50	200	46,3	222	7,4	11	51,4	194	2,8	3
50	2000	53,7	2089		4,5	48,6	1939		3
50	400	33,5	301	33	24,8	47	381	6	4,8
50	2000	66,5	1545		22,8	53	1831		8,5
33	100	16,2	74	50,9	26	52,9	100	60,3	0
67	200	83,8	185		7,5	47,1	200		0
50	50	56,1	58	12,2	16	60	52	20	4
50	200	43,9	227		13,5	40	202		1
50	800	11,6	406	76,8	49,3	41,8	772	16,4	3,5
50	2000	88,4	1474		26,3	58,2	1804		9,8

Table 1: Analysis of simulated relaxation data. The amplitudes of exponentials have been normalized and the fractions α are given in %.

In all cases the appropriate number of relaxation processes have been found by either method. The mean relative error for the T_2 values found by regression analysis is about 19% and only about 3% for the modified GNM algorithm indicating the higher accuracy of the latter method. The mean relative error for the fractions is 33% and 19%, respectively. This is a natural consequence of noisy data to which both methods are sensitive. In the present example no attempt was made to smooth the data. The poorer accuracy of the regression analysis is due to the fact that for the largest time values on the time scale only one component is assumed to be left. Whenever this is true, this largest component is better reproduced than the smaller one. On the other hand, if this criterion is not fulfilled, the error drastically increases as illustrated in the last row of Table 1. The modified GNM method, however, produces reliable results. Employing the GNM method usually 6-8 iterations have to be performed to obtain accurate results. With the hybrid method the identical results are obtained but with half the number of iterations.

Validation by Using Phantom Data

In order to check all methods under more realistic conditions, they have been applied to measured phantom data. The phantom consists of two wedge-shaped tanks which are arranged to form a square stone. The tanks are filled with $MnCl_2$ solutions each representing one known T_2 relaxation time. The arrangement allows a complete variation of the fractions α from 0 to 100% but with fixed relaxation times. The results are summarized in Table 2.

method	SRA	GNM
pixels total	6600	6600
1 component	3218	869
2 components	3382	5731
T_{21} [ms]	90 - 130	100 - 120
T_{22}[ms]	800 - 1100	900 - 1100
computing time	8 min	200 min

Table 2: Analysis of measured phantom data.

A component with a fraction of less than 5-10% is omitted. The number of pixels in which two components are detected is thus apparently smaller than the expected one. The linear increase of the fractions is qualitatively reproduced by both methods with a relative error of 35% and 20%, respectively. This confirms results already obtained with the simulated data.

Conclusion

The hybrid method for analyzing multiexponential relaxation processes in NMR images has been proven to yield better results than the semilogarithmic regression analysis. The hybrid method reduces the time consumption to 40-60% compared to the modified GNM method. Nevertheless, the time consumption is still the main bottle neck. Future enhancements rely on improved hardware environments and especially on parallel processor technology. The algorithm is ideally suited for parallel processing as each pixel can be analyzed independently. With the appropriate hardware the improved analysis of NMR images should be possible within a few minutes.

References

1. Staemmler, M.; Brill, R.; Becker, K.; Folkerts, K.H.; Gersonde, K.: SUNRISE-A Software System for Medical Imaging Analysis. **Proceedings of the International Symposium Computer Assisted Radiology CAR 89, Berlin, Springer Verlag, Berlin, (1989).**

2. Golub, G.H.; Pereyra, V.: **SIAM,J.NUMER. ANAL. 10 (1973) 413-432.**

3. Provencher, S.W.: **J.Chem.Phys. 64 (1976) 2772-2777.**

4. Yeramian, E.; Claverie, P.: **Nature 326 (1987) 169-174.**

5. Provencher, S.W.: **Computer Physics Commun. 27 (1982) 213-242.** Provencher,S.W.; Vogel, R.H.: **Progress in Scientific Computing 2 (1983) 304-319, Birkhäuser.**

6. Gersonde, K.; Felsberg, L.; Tolxdorff, T.; Ratzel, D.; Ströbel, B.: Analysis of Multiple T_2 Relaxation Processes in Human Head and Imaging on the Basis of Selective and Assigned T_2 Values. **Magn. Reson. Med. 1, (1984) 463-477.**

7. Gersonde, K.; Tolxdorff, T.; Felsberg, L.: Identification and Characterization of Tissues by T_2-Selective Whole-Body Proton NMR-Imaging. **Magn. Reson. Med. 2, (1985) 390-401.**

8. Schwarz, R.: **Numerische Mathematik, (1988) 313ff, B.G.Teubner.**

9. Bevington, Ph.R.: **Data Reduction and Error Analysis for the Physical Sciences (1969), McGraw-Hill.**

A New Segmentation Algorithm for Knowledge Acquisition in Tissue Characterizing NMR-Imaging

Heinz Handels and Thomas Tolxdorff
Institut für Medizinische Statistik und Dokumentation,
Rheinisch-Westfälische Technische Hochschule Aachen,
D-5100 Aachen, Germany.

Summary

Tissue characterizing NMR-imaging is a new imaging method for differentiation and biochemical characterization of tissue based on multi-dimensional biochemical NMR-parameter information. In contrast thereto NMR-imaging visualizes anatomical structures utilizing only one parameter, the magnetization.

To analyze the relationship between biochemical properties of tissue and NMR-parameter combinations a new segmentation algorithm has been developed by using clustering techniques. The evaluated multi-dimensional parameter cluster are visualized in cluster images showing volume elements with similar biochemical properties. The developed software tool allows an easy and comfortable handling of the complex tissue characteristic NMR-information and supports knowledge acquisition for automatic knowledge based tissue characterization.

Introduction

Based on Carr-Purcell-Meiboom-Gill (CPMG) spin-echo sequences the transverse magnetization decay in each volume element can be recorded (Fig. 1). The measured signals describe spin-spin-relaxation processes, which are correlated to different proton classes [1]. Using non-linear optimization algorithms [2] the transverse magnetization decay can be decomposed in up to three exponential functions (Fig. 2) according to

$$(1)\quad M(t) = \sum_{i=1}^{n} M_{0i}\, e^{-t/T_{2i}}$$

The relaxation times T_{2i} (i=1,..,3) and the intrinsic magnetization at time zero M_{0i}, which are proportional to the spin densities ρ_i, assuming $T_R >> T_1$, characterize different proton classes in one volume element.

$$(2)\quad \alpha_i = M_{0i} / \sum_{j=1}^{n} M_{0j}$$

Equation (2) describes the partial volume, which is assigned to the i-th proton class in one voxel. Employing a combination of CPMG and saturation recovery pulse sequences the T_1 relaxation time can be computed additionally in each volume element.

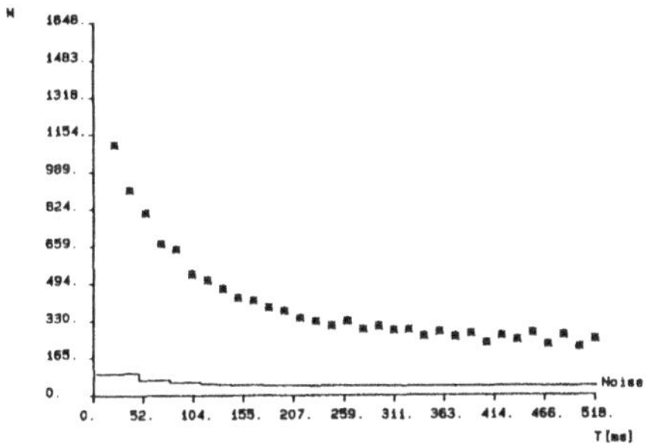

Fig. 1: Measured magnetization decay with 32 echoes

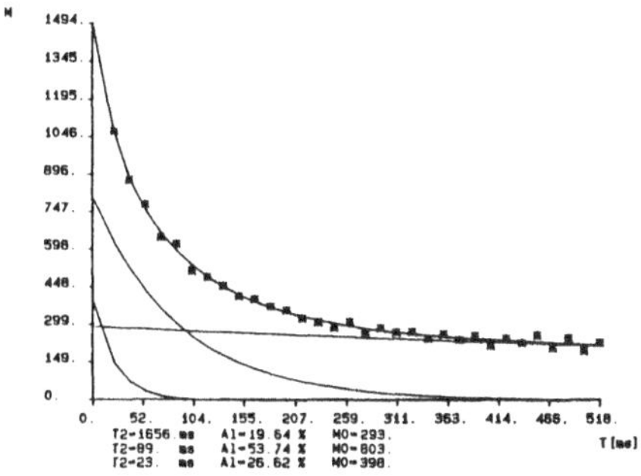

Fig. 2: Decomposition of the magnetization decay into three relaxation components

To support the medical diagnosis the software tool enables the physician to define special regions of interest (ROI). The ROI, which is described by a (nxn)-matrix (default: n=5), can be created easily by positioning a crosshair in the center of the ROI (Fig. 3).

The system computes mean- and standard deviation-vectors of the 1-, 2- and 3-dimensional parameter values of $P \varepsilon \{T_1,T_2,\alpha,\rho\}$ in the considered ROI. Then the following cluster definition is used:

A voxel with $n \varepsilon \{1,2,3\}$ relaxation components and associated NMR-parameters $(P_1,...,P_n)$ belongs to the P-cluster C(P,k), if $\forall$ i=1,...,n

$$(3) \quad | \overline{P}_{ni} - P_i | <= k * \sigma_{ni}$$

$P \varepsilon \{T_1,T_2,\alpha,\rho\}$

$\overline{P}_{ni}$ = i-th component of the n-dimensional mean-vector of the parameter P

σ_{ni} = i-th component of the n-dimensional standard deviation-vector of the parameter P

k = cluster factor

The clustering algorithm is performed by selecting all voxels with a parameter structure similar to the parameter structure of the ROI.

Color Coded Multi-Cluster Images

All P-clusters C(P,k) ($P \varepsilon \{T_1, T_2, \alpha, \rho\}$, k=1,2,3,4,6,8) can be visualized simultaneously in a so called "color coded multi-cluster image". The selection criteria used by the clustering algorithm can be described by the discrete function M_P defined by

$$M_P : [XxY] \dashrightarrow \{1,2,3,4,5,6,7\} \quad \text{with } X=Y=\{0,..,255\},\ x \,\varepsilon\, X,\ y \,\varepsilon\, Y$$

$$(4) \qquad M_P(x,y) = \begin{cases} 1 & \text{, if } P(x,y) \,\varepsilon\, C(P,1) \\ 2 & \text{, if } P(x,y) \,\varepsilon\, C(P,2)\backslash C(P,1) \\ 3 & \text{, if } P(x,y) \,\varepsilon\, C(P,3)\backslash C(P,2) \\ 4 & \text{, if } P(x,y) \,\varepsilon\, C(P,4)\backslash C(P,3) \\ 5 & \text{, if } P(x,y) \,\varepsilon\, C(P,6)\backslash C(P,4) \\ 6 & \text{, if } P(x,y) \,\varepsilon\, C(P,8)\backslash C(P,6) \\ 7 & \text{, if } P(x,y) \not\varepsilon\, C(P,8) \end{cases}$$

The clustering algorithm is performed for the relaxation times T_1 and T_2, the spin density ρ and the partial volume α. Hence, for each of the NMR-parameters a "colour coded multi-cluster image" is generated (Fig. 4). Six different color codes of the heat scale are used to represent six voxel classes evaluated by the selection algorithm. A seventh voxel class is provided to accept those pixels not belonging to any of the six classes mentioned above. This class is visualized using grey values. Color coded multi-cluster images give an overview about the distribution of distances between the parameter mean-vectors of the ROI and the parameter vectors according to each pixel. Furthermore, color coded multi-cluster images are used to find proper parameter cluster sizes for the generation of combined cluster images.

Combined Cluster Images

All voxels of one T_1, T_2, α and ρ-cluster can be displayed in a cluster image using an interactive changeable cluster factor (k=2 by default). The selection can be described by a boolean function

S_P: [XxY] --> {0,1} with $P \varepsilon \{T_1, T_2, \alpha, \rho\}$, $x \varepsilon X$, $y \varepsilon Y$

$$(5) \qquad S_P(x,y) = \begin{cases} 1 & \text{, if } P(x,y) \,\varepsilon\, C(P,k) \\ 0 & \text{, if } P(x,y) \not\varepsilon\, C(P,k) \end{cases}$$

The selected voxels are displayed in overlay mode [3] visualizing anatomical structures (grey values) and biochemical (color values) cluster information in one image (Fig. 5).

To get a specific biochemical characterization of the interested tissue different parameter selection criteria can be combined to visualize all voxels with a multi-dimensional parameter structure similar to the ROI. Therefore, the software system allows to calculate the intersection of generated binary cluster images (Fig. 6).
Based on the voxels of the visualized tissue the calculation of the mean- and standard deviation-vectors of the NMR-parameters T_1, T_2, α and ρ is performed. These tissue characteristic values are displayed and stored in a data base to support automatic tissue characterization.

Conclusion

Cluster analysis of NMR-parameters computed by multi-exponential evaluation is a new method to find regions of similar biochemical properties. With this method the spatial distribution of tissue as for example fat, muscle, grey and white matter of the brain or cerebrospinal fluid as well as pathologic tissue as for example malignant carcinoma of the female breast or astrocytoma in the brain can be visualized. The software tool, which is integrated in the software system RAMSES (RWTH Aachen Magnetic Resonance Software System) [4], enables the physician to use the complex biochemical information evaluated by multi-exponential T_2-analysis in an easy and comfortable way. Furthermore knowledge about the relation between segmented tissue and multi-dimensional NMR-parameter cluster can be stored in a tissue data base to improve medical diagnosis.

References

1. Gersonde, K., Tolxdorff, T., Felsberg, L.: Identification and Characterization of Tissues by T_2-Selective Whole Body Proton NMR-Imaging. Mag. Reson. Med. **1**, 390-401 (1985)
2. Repges, R., Tolxdorff, T: Felsberg, L., Broszio, G. and Gersonde , K.: Decomposition of Multi-Exponential Transverse Magnetization Decays in T_2-Selective Imaging Employing the Subsystem EVALUATE of RAMSES. Proceedings of the International Symposium Computer Assisted Radiology CAR 85 , Springer-Verlag, Berlin, 25-29 (1985).
3. Tolxdorff, T., Felsberg, L., Gersonde, K.: Proton NMR-Imaging: Contrast-Enhanced Images and Combination with Parameter-Selective Images by Overlay Display Technique. Proceedings of the International Symposium Computer Assisted Radiology CAR 87, Springer-Verlag, Berlin, 3-11 (1987).
4. Tolxdorff, T.: Ein neues Software-System (RAMSES) zur Verarbeitung NMR-spektroskopischer Daten in der bildgebenden medizinischen Diagnostik. Medizinische Informatik und Statistik 66, Springer-Verlag, Berlin (1987).

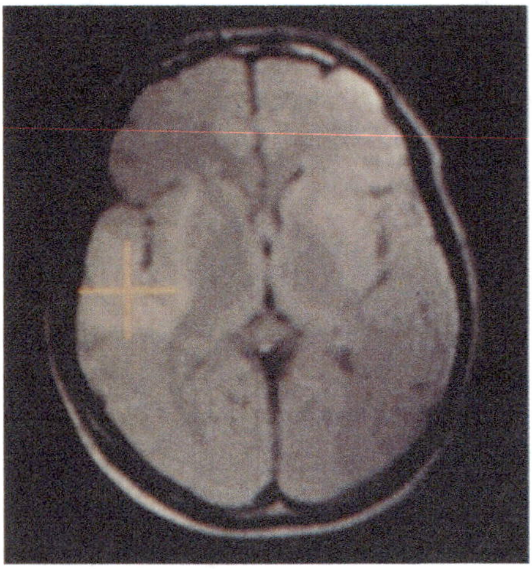

Fig. 3: Creation of a ROI in a supraorbital head slice showing an astrocytoma

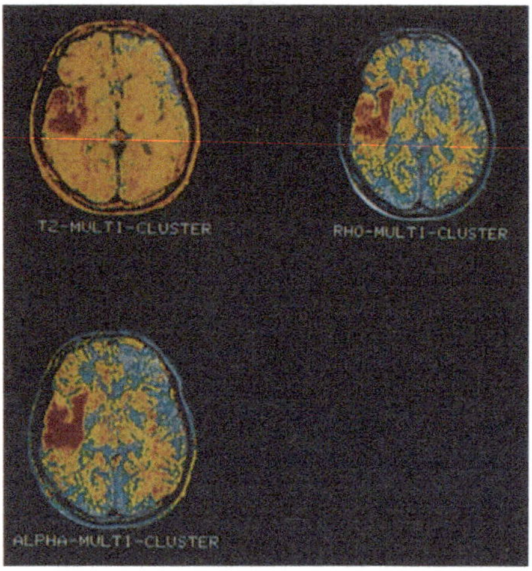

Fig. 4: Color coded multi-cluster images showing T_2-, ρ- and α-clusters with cluster factors k= 1,2,3,4,6,8. T_1 relaxation time has not been measured in this experiment

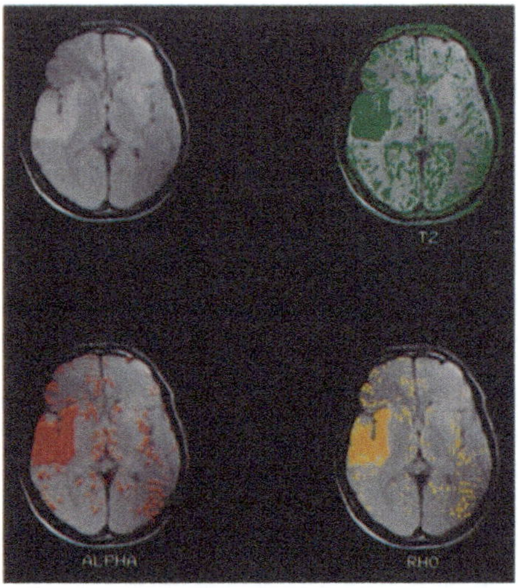

Fig. 5: Cluster images using cluster factor k= 2 for the parameter T_2, α and ρ

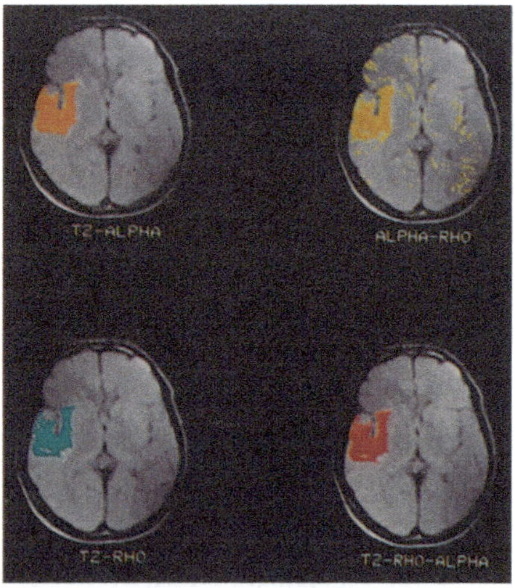

Fig. 6: Combined cluster images for combination of the $T_2 \wedge \alpha$, $\alpha \wedge \rho$, $T_2 \wedge \rho \wedge \alpha$ cluster criteria characterizing an astrocytoma in the head slice

Cardiovascular Cine MRI

G. W. LENZ, T. A. SPRAGGINS, D. A. FINELLI

SIEMENS Medical Systems, Erlangen
Erlangen, FRG

Summary

CINE MRI, using short TR steady-state, free precession sequences, has facilitated the evaluation of cardiovascular function. Three different data acquisition schemes have been implemented and compared: prospective ECG triggering where the ECG-data is used to trigger the measurement sequence; retrospective cardiac gating, where the MR data of the freely running measurement sequence is referenced later to time points in the cardiac cycle; and "in vivo" gating, where a navigation scan provides the necessary cardiac timing information.

Introduction

The advent of gradient-field-echo (GFE) imaging has led to several new techniques in magnetic resonance imaging (MRI) [1-3]. The method to dynamically image and review the same cardiac slice position has been termed Cine Magnetic Resonance Imaging (CINE MRI) [4]. Short repetition times (TR) provide the temporal resolution needed to assess ventricular function, valvular regurgitation or stenosis, and to examine abnormal hemodynamics in congenital heart disease. To correlate the acquired MR data with cardiac motion, three different data acquisition and post-processing schemes have been implemented and compared.

Method

In the first method, prospective ECG-triggering, the same cardiac slice is excited a fixed number of times and the phase-encoding gradient is incremented after a varying dead time at the instant of the next R-wave. This technique allows immediate utilization of the data for cine display without post-processing, but the late portion of diastole, during which atrial contraction is occurring, remains unimaged. The variable dead time at the end of the cardiac cycle disturbs the steady state and leads to increased signal intensity in the first few images of the cine set. When rapidly reviewed in a continuous cine loop, the so-called "lightning artifact" occurs [4]. Depending on the irregularity of the heart beat, TR, and echo time (TE), significant ghosting even from stationary tissue can render the first images suboptimal for definition of cardiac function and blood flow patterns from end-diastole into early systole.

In the second method, retrospective cardiac gating, MR data is acquired continuously, uncorrelated to the ECG-signal, which is merely recorded during the measurement on a personal computer. The data is then retrospectively referenced and interpolated to fixed points in the cardiac cycle. This way, the steady state can be maintained and the entire cardiac cycle is imaged. Up to three cardiac slice positions can be imaged simultaneously by exciting the different levels in an alternate fashion. This leads to an effective TR of 60 msec. Only a linear interpolation scheme can then be successfully used, since higher order interpolation techniques would require more than two points spaced too far apart in time.

The coherent nature of the remnant ghosting artifacts after retrospective data reconstruction identifies respiratory motion as the major contributor to ghosting. Additional retrospective respiratory gating was applied to the data using the recorded respiratory signal. When the imaging experiment was repeated 4-8 times, the Fourier data after the reordering and interpolation could be sorted according to the respiratory level. This made it possible to reconstruct cardiac images at the instant of inspiration and expiration with greatly reduced ghosting artifacts due to respiratory motion. The extension to 3D cardiac imaging allows one to image 3 mm slices contiguously and time resolved with sufficient signal-to-noise ratio (SNR). To reduce the measurement time, the phase encoding table of 126 steps has been split into 3 parts and interleaved, i.e., phase encoding step n, n+42, n+84 (for n = 1-42) have been repeated in that order over one cardiac cycle. This leads to a time resolution of 60 msec for each phase encoding step using an effective TR of 20 msec. For 1 acquisition, the total measurement time was roughly 16 minutes for 16 partitions, depending on the heart rate.

The third method, "in vivo" retrospective gating, is similar but requires no recording of the patient's ECG. Interleaved with the acquisition of the phase-encoded imaging data is that of non-phase-encoded timing data. This timing data can come from either a second echo or another slice; the pulse sequence can be tailored to enhance the motion sensitivity of this second echo. The timing signals are, in general, modulated both by cardiac and respiratory motion; application of a digital filter allows determination of either or both types of motion. From this point, data processing proceeds as in the retrospective gating algorithm above.

Independent of any method a FISP sequence [5] with gradient motion refocussing (GMR) in the frequency and slice-select directions has been used with a flip angle of 40° to 60°.

Results and Discussion

ECG-triggered cine studies may be of diagnostic quality for determination of cardiac function when reliable triggering on the R-wave occurs for the entire measurement, which typically requires 3-5 minutes for completion. The presence of mistriggered data and variable trigger delays severely degrades image quality for the cine study.

Well known artifacts are introduced into the ECG signal during MR imaging, the most important being T-wave enlargement due to Lorentz voltage induced by aortic blood flow, spikes introduced by gradient switching, and large baseline fluctuations due to respiration. These artifacts can cause unreliable R-wave triggering even in normal volunteers. In the patient population with cardiac disease, the combination of pathologically based ECG abnormalities and induced artifacts can render triggered studies useless.

With retrospective cardiac gating, again it is important that the recorded ECG signal be of as high quality as possible. In contrast to the ECG triggered studies, however, processes such as software filtering and extra-systole detection can be applied to the timing data before sorting and interpolation of the imaging data. The quality of retrospectively cardiac gated images was found consistently superior to prospectively gated data. Not only was the lightning artifact eliminated, but the images contained overall less systematic noise and showed improved contrast and edge definition. This effect has to be attributed to the fact that the steady state can be maintained in retrospective cardiac imaging.

The physiologic retrospective gating technique utilizes information obtained from a motion sensitive navigator pulse to construct relative time scale to which imaging raw data is fitted. The primary advantage of this method is that time is saved in placement of ECG electrodes and/or respiratory signal transducers, and that the cost and complexity of the imager can be reduced. Additionally, it may be advantageous to use this method on patients with cardiac arrhythmias. A disadvantage, as compared with the ECG-driven retrospective gating method, is that some time must be devoted to acquiring the timing data, thus either reducing the number of slices or slightly increasing the imaging time. The results with in vivo gating demonstrate that the necessary information on cardiac motion can be extracted from non-phase-encoded projections. The slightly reduced image quality with in vivo gating as compared to retrospective ECG gating indicates that the reliability of the timing information still has to be improved.

Cine GFE techniques have facilitated the development of dynamic cardiac functional analysis using MRI. Of the basic acquisition strategies, prospective ECG triggering or retrospective gating, retrospective schemes produce consistently better image quality and allow more complete evaluation of the cardiac cycle. Retrospective cardiac gated studies were superior to our initial retrospective physiologic gated measurements. Recent use of more sophisticated digital filtering is producing cleaner timing data for both retrospective techniques, with improved separation of cardiac and respiratory timing signals.

References

1. Van der Meulen, P.; Groen, J.P.; Cuppen, J.J.M.: Very fast MR Imaging by field echoes and small angle excitation. Magn. Res. Imag. 3 (1985) 297-299.

2. Frahm, J.; Haase, A.; Matthei, D.: Rapid NMR imaging of dynamic processes using the FLASH technique. Magn. Reson. Med. 3 (1986) 321-327.

3. Sekihara, K.: Steady-state magnetization in rapid NMR imaging using small flip angles and short repetition intervals. IEEE Trans. Med. Imag. MI-6 (1987) 157-164.

4. Herfkens, R.J.; Utz, J.A.; Glover, G.H. et al.: Rapid dynamic NMR imaging of the heart: Initial clinical experience. Society of Magnetic Resonance in Medicine, Fifth Annual Meeting, Montreal (1986) 247.

5. Oppelt, A.; Graumann, R.; Barfuss, H.; Fischer, H. et al.: FISP - a new fast MRI sequence. Electromedica 3 (1986) 15-18.

Multiplanar and Curved-surface Imaging from Volumetrically-acquired MR Data

Rhodes ML, Sanny J, Malhotra D.

MPDI, 2730 Pacific Coast Highway, Torrance, Ca. 90505

SUMMARY

Image planes acquired by MR scanners do not always display the anatomy of interest for optimal diagnostic clarity. For example, curved coronal MR images of the spine, not currently available, are especially desirable because, similar to a myelogram, they provide an opportunity to compare analogous anatomical structures from one level to the next in a single view. This work describes software which generates various curved-surface and oblique views of anatomy reformatted from a single 3D FT scan. This scanning technique allows high-quality reformations not possible with 2D scans since closely spaced slices delivered by 3D FT protocols lead to data of higher isotropy. Sample reformatted images are shown. Presently, two prototypical software packages have been developed, one which runs on an MR scanner's computer system and the other on an independent medical workstation. Each has the capability of producing curved coronals, curved sagittals and obliques. Further development is expected to lead to a clinical tool which would be able to generate any defined curved suface within a given volume.

INTRODUCTION

The clinical applications of three-dimensionally acquired magnetic resonance (3D MR) imaging have become increasingly widespread. The technique has been cited for improved imaging of pituitary microadenomas [1], to usefulness in surgical planning [2], and has shown clinical benefit when used with contrast agents [3]. The advantages of three-dimensional magnetic resonance imaging over conventional 2D multislice imaging are significant:

- 3D MR offers true gapless imaging with minimal interslice crosstalk;
- it avoids the distortion of T1 contrast which occur in the 2D process because of saturation effects;

- o there is no fat/water shift or susceptibility contamination of the slices in the slice selection direction;
- o 3D MR allows for very thin slices which enhance the value of image reformation. This advantage is particularly important for our case.

Since 3D MR involves an extra phase-encoding step, the scanning time is correspondingly increased when compared to 2D imaging. There are two ways to alleviate this situation. The first is through the use of gradient-recalled echo techniques which require very short repetition times. The second is by means of a software diagnostic tool described here. The software will not decrease the scanning time required in 3D MR. Instead, it serves to significantly increase the amount of information that is made available through such a scan by reformation of the image data.

Image reformatting from a series of parallel image planes is well-established in clinical radiology and dates back to the introduction and early use of CT scanners [4]. However, this circumstance is entirely different for magnetic resonance. Before 3D MR acquisition techniques were developed, intervals between MR slices were too large to give clinically useful reformatted views. Once MR 3D scan acquisition techniques become more commonplace, we feel that image reformatting software for MR data will be an essential component of many exams. This will actually improve patient throughput (with likely lower cost) for otherwise time consuming examinations where optimal image orientation at the time of scanning is uncertain.

METHOD AND MATERIALS

Within the volume of acquired data obtained using 3D MR, our software will allow the physician to view any arbitrary planar or curved surface through image reformation. Curved surface reformatted views, particularly coronal images adjusted to follow the curvature of the spine are the most diagnostically useful of the reformatted images this feasibility study has developed. They provide the opportunity to compare analogous anatomical structures from one level to the next in a context that is similar to, but more detailed than an AP myelogram. Because the spine is a curved structure, routine coronal MR images often fail to demonstrate analogous spinal structures at all levels scanned. It is this comparison that often pinpoints the presence of an abnormality to the radiologist. Similar software has been introduced for computed tomography (CT) [5,6] and is currently in widespread use. More recently, Ney and colleague [7] have developed software to merge multiple planar and 3D views of CT data into a single interactive presentation. Curved coronal images designed for MR will be even more diagnostically helpful since MR has the capability of demonstrating intraspinal neural structures to a far greater extent than CT. The goal to reformat curved surface images from 3D acquired MR data is the most compelling clinical justification for our effort.

A series of parallel, regular and closely spaced image sets for volumetrically-acquired data can be reconstructed in an unlimited number of orientations. The most clinically informative and physician preferred orientations for spinal studies have been in the sagittal plane. Though our study has demonstrated the ability to reformat curved surfaces from parallel image data sets of any orientation (see Figs. 1 and 2 for curved coronals from axially oriented planes, and Fig. 3 for sagittally oriented planes), sagittal planes add an extended, more comprehensive diagnostic perspective in its presentation of the spine than a series of axial planes conveying the same anatomy.

DISCUSSION

Figures 1 through 5 are a collection of photographs taken of representative reformatted images produced on the GE Signa and the Philips PICS 2000™ workstation (SunMicrosystem 3/160 and PIXAR image computer). Reformations have been performed on 3D acquired MR data on the cervical spine, the lumbar spine, the knee and the ankle. Such data demonstrate that our image reformation software is neither limited to spinal studies nor is its development handicapped by clinical application to other anatomic structures.

At present, this software has the following capabilities:

- o images along orthogonal planes can be formatted.
- o oblique planes can be generated.
- o curved surface views can be reformatted.
- o image magnifications are possible (Philips PICS 2000 only)

The work described has provided us with the technical foundation to develop a tool for routine clinical use.

ACKNOWLEDGEMENTS

We thank Philips Medical Systems, General Electric Medical Systems and Pixar for their support. Thomas Giordano, Irene Jemcyzk and Robert Applegate of Philips were especially helpful to this program. MR image data was provided by GE. Michael Mantle and colleagues at Pixar were generous with their assistance. SunMicrosystems also participated by aiding MPDI's software group with technical updates.

REFERENCES

1. "3D MRI promoted for imaging of pituitary microadenomas": Radiology Today, November 1988.

2. "Graphics Tools": Computer Graphics Today, December 1988.

3. "Gd-DTPA and Fast Scans extend utility of MRI": Diagnostic Imaging, January 1989.

4. Glenn et al, "Image Generation and Display Techniques for CT Scan Data: The transverse and reconstructed coronal and sagittal planes," Investigative Radiology, 10, 5, September 1975, pp. 403-410).

5. Rhodes ML, Glenn WV, Azzawi YM, "Extracting Oblique planes from Serial Sections," J. Corp. Assist Tomography, 1980, 4: 649-657.

6. Rothman SLG, Dobben GD, Rhodes ML, Glenn WV, Azzawi YM, "Computed Tomographny of the Spine: Curved Coronal Reformations from Serial Sections," Radiology, Vol. 150, No. 1: 185-190.

7. Ney DR, Fishman EK, Magid D, Kuhlman JE, "Interactive Real-Time Multiplanar CT Imaging," Radiology, Vol. 170, No. !, 1989, 275-276.

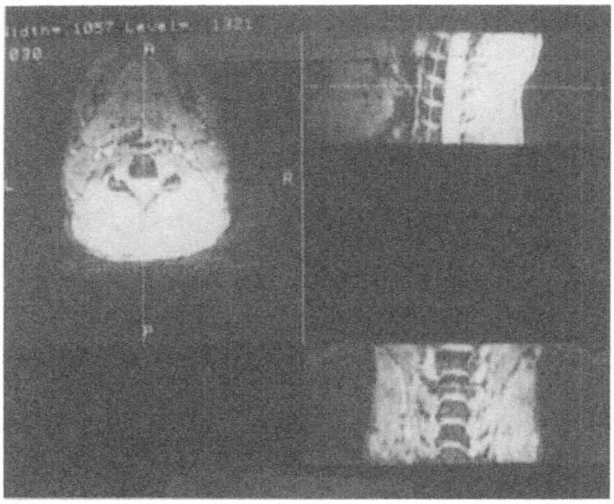

FIGURE 1 Composite presentation of 3D acquired axial image of cervical spine with reformatted sagittal and reformatted curved surface (coronal orientation) view of the cervical vertebral bodies. Feasibility software operational on the GE Signa MR system.

A.

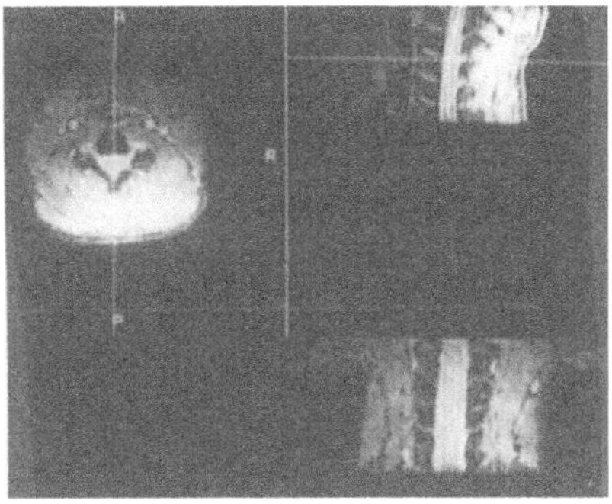

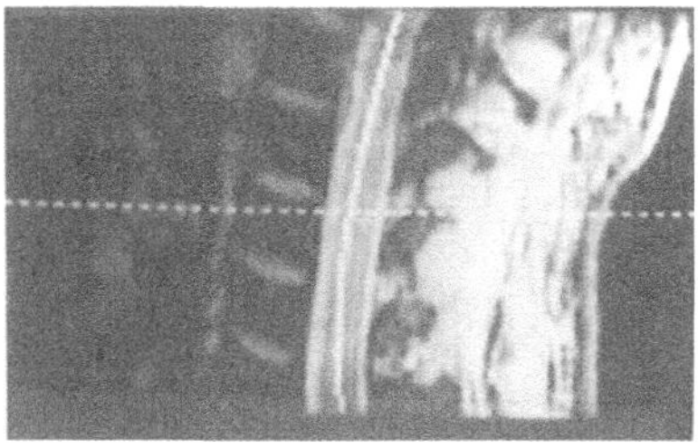

B.

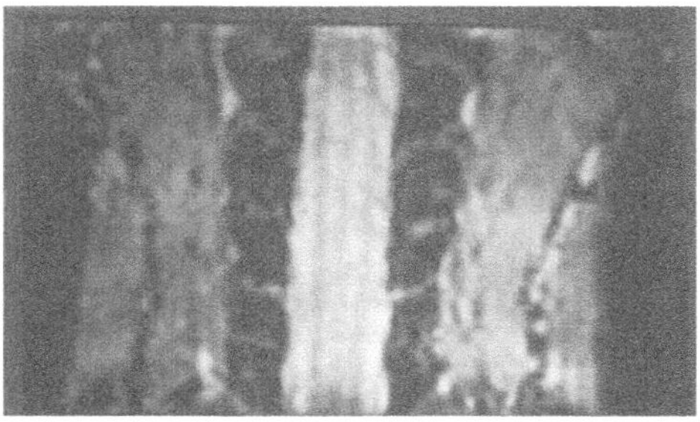

C.

FIGURE 2

a. Composite presentation of axial and reformatted sagittal and curved coronal images of cervical spine. Feasibility software written for GE Signa MR system.

b. Close-up of curve drawn by cubic spline expression for curved coronal view in Fig. 5c.

c. Curved coronal view of cervical spinal cord, notice neuro-structural detail.

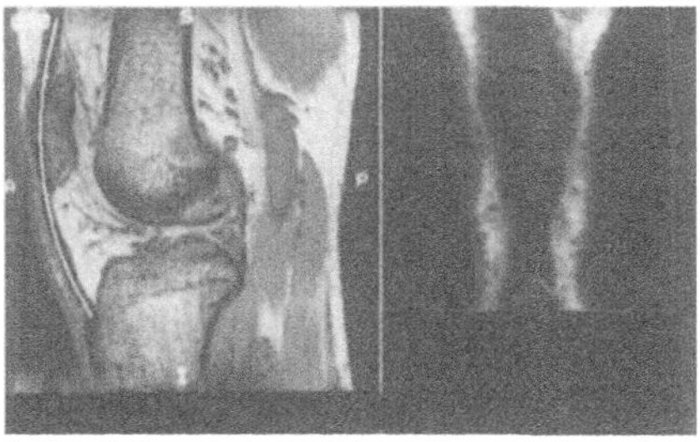

FIGURE 3 The curved surface view shown here illustrates the structure of the patella tendon and the patella ligament of the knee in a single view.

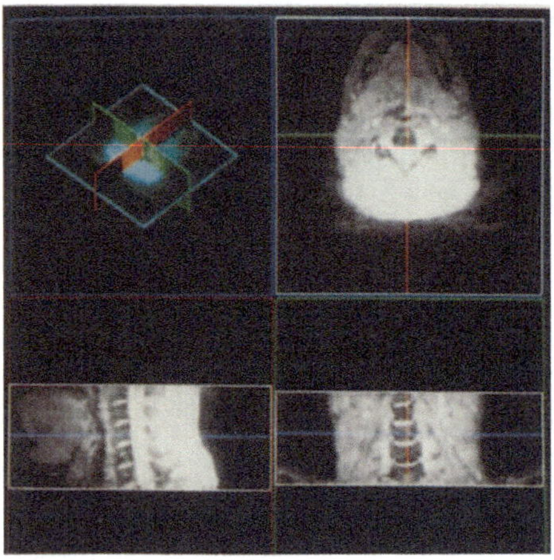

FIGURE 4 Curved surface views reformatted to show cervical spine vertebral body in lower right quadrant, the curve specified on reformatted sagittal (lower left) view. (Sun-based feasibility software)

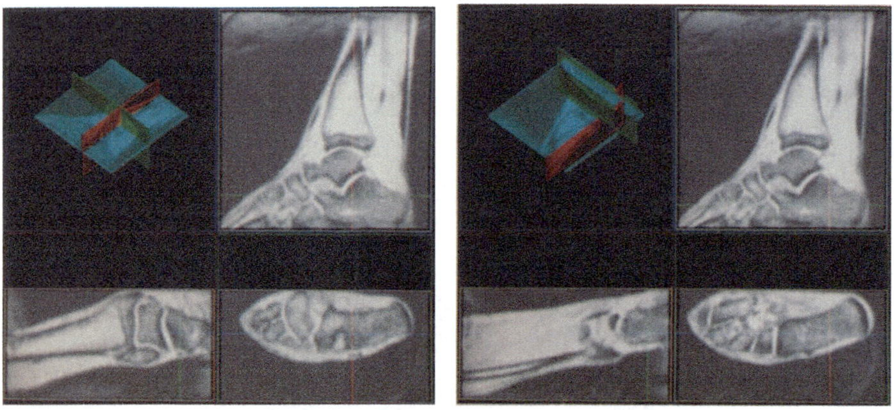

FIGURE 5

a. Three different planes are selected here to review the ankle anatomy. Notice the re-projection of the 3D context presentation in the upper left quadrant.

b. In ths presentation, the 3D context view (upper left) shows a bottom-up orientation; viewing orientation of the context view is user selectable. This feasibility software is based on the Philips PICS 2000 workstation.

First Steps to Quality Assurance for MRI

B. Schnackenburg, D. Schmidt, R. Juran, Ch. Dewey

Humboldt-Universität zu Berlin, Bereich Medizin (Charité),
Institut für Röntgendiagnostik, 1040 Berlin/GDR

Abstract:

This paper will report on the experience and the results in quality assurance gained in a two years work with a 1.5 T-machine (Gyroscan S 15, Philips).

The device is daily used from 8.00 a.m. to 10.00 p.m. Two days per week are reserved for spectroscopy. In imaging there is a share of 55 % for neurological exams and of 45 % for exams of the entire body. Per year there is a total of about 2100 exams.

A good co-operation and a permanent exchange of information between physicians, physicists, medical and technical assistants and service technicians are an essential condition for an efficient quality assurance. Our strategy for an assurance of quality is based on two fundamental ideas.

1. The optimate performance of the exam. This part of quality assurance implies the adaption of the examined region to the MR-System. This requires a sufficient information of the patient on the course of examination and on his chances of an active cooperation (i.e. in breath-holding techniques).

   Another important aspect is the selection of the adequate coil and the correct positioning of the patient.

   In this context it is particular necessary to adapt the measuring system to the specific diagnostic problem. First of all this means to choose the right examination technique with optimal sequence parameters. Characteristic examples may demonstrate the effects of unfavourable sequence parameters on the image quality both in phantoms and in various MR-images got in human examinations. Consequently this part led to organ-specific present procedures which are being improved continously.

2. The other pillar of quality assurance refers to the control of important technical parameters which habe to be kept constant. These parameters are in detail: signal-to-noise ratio, flood-field uniformity, spatial-resolution, spatial-linearity, slice profile, contrast, T1/T2-accuracy and artifacts.

The measurements are carried out with a standard phantom . The evaluation is made both visually and soft-ware guided.

The measurement provides results about signal-to-noise ratio, uniformity and artifacts and takes a time of 10 minutes. The measurement and evaluation of all parameters mentioned above take about 1 hour and are carried out once a week.

Summing up, ist may be said that the trend of fixed measurement results allows a statement on the technical condition of the equipment at any time, thus providing an objective judgement on the existing image quality and the quality to be expected in the near future.

Computer Tomography

3-D Imaging in the Post Surgical "Failed Back" Syndrome

S. James Zinreich, M.D., Henry Wang, M.D., Donlin M. Long, M.D.
Cynthia B. Quinn, R.T. Randy Davis, M.D., Paul C. McAfee, M.D.

Meyer 8-140, Johns Hopkins Hospital, 600 N. Wolfe Street
Baltimore, Maryland 21205, U.S.A.

One hundred consecutive patients with post surgery failed back syndrome (PSFBS) without fusion and 100 patients with PSFBS with fusion were evaluated with direct axial computed tomography (CT), 2-Dimensional (2D) multiplanar and 3-dimensional (3D) imaging. In the patients with and without fusion 3D images were found to best display the following: the surgical procedure and its extent; lateral neural foraminal narrowing; and fractures of the posterior elements. The 3D images enabled optimal evaluation of the integrity of fusion. It was the only modality which afforded an accurate delimitation of fusion: solidity, pseudarthrosis, incorporation of transverse processes and facet joints, and transitional syndrome. In the patients without fusion 3-D provided improved perception of the surgery performed in 31%, it showed 2 additional fractures in 9%, better displayed lateral neural foraminal narrowing in 42% and showed additional posterior element fractures in 11%. 3D uncovered incomplete fusion in 17%, transitional syndrome in 13% and pseudoarthrosis in 6%. Thus the 3D images provided additional information over the direct axial and multiplanar reconstructed images (2D images) in 56/100 patients without fusions, and 76/100 patients with fusion. This imaging modality is now routinely performed in this population at our institution.

3-D CT for Routine Examinations of the Base of the Skull

H. BECKER

Department of Neuroradiology,
Medical School, Hannover, F.R. Germany

Summary

Testing the clinical significance of 3-D imaging of the skull base was the aim of this study. The 3-D images were calculated in 50 patients from the data of axial or coronal CT slices. This method has been used especially in patients with craniofacial anomalies, tumors and fractures. From our experience the 3-D does not result in any fundamentally new aspects in the diagnostical point of view. The relevant information can already be obtained from conventional 2-D images. However, a clearer understanding of the topography of the pathological findings can be achieved. Above all, a better spatial concept of the base of the skull is conveyed using 3-D CT. This simplifies the communication between the radiologist and the attending physician. For surgeons in particular, the preoperative planning can be improved.

High-resolution computed tomography (CT) provides exquisite delination of the skull base and adjacent tissues. The skull base can be examined from the endocranial or exocranial aspects. Each fossa has numerous openings, canals, foramina, and fissures that provide natural pathways for the spread of pathological processes in either direction. The skull base is the anatomical crossroad between the brain and its meninges above, and the orbit, sinuses, and nasopharynx below. CT has been successfully employed in the evaluation of pathological findings of the base of the skull. However, conventional CT-slices yield two-dimensional (2-D) representations only. The third dimension must be imagined conceptually. Secondary sagittal and coronal reconstructions may be helpful in this regard. In recent years, special software programs have been developed with which a threedimensional (3-D) imaging is possible.
The purpose of our investigations was to test the use of the

3-D image procedure with CT, to determine its applicability for the imaging of the complex bony structures of the base of the skull and also its clinical significance.

Methods

Since 1985, we have been working with an experimental 3-D software program on the CT scanner GE 9800. The calculation of the base of the skull to be imaged was elaborate and required one to two hours. Now we have the new software program "Quick 3-D". This program calculates a "baseline image" three-dimensionally within about five minutes. It can then be rotated in angular steps of up to 1 degree around all spatial axes. The rotation intensifies the 3-D impression. This arises in such a manner that structures which are located closer to the examiner are imaged bright. Whereas the regions which are farther away are imaged dark. The 3-D images are calculated from the data of axial or coronal 1.5 or 3 mm slices which have been taken in continous sequence. In most cases, routinely taken CT series were used. The number of slices necessary for the 3-D reconstruction depends on the size of the object to be calculated. In a representative tomogram, the region of interest to be calculated is established. The window level must be selected in such a way that all structures to be imaged show CT absorption values on the Hounsfield scale which are above this level. The simplest way to describe a bony surface is to enter a threshold value. For imaging soft tissue, it can be helpful to use a range within which the structure's CT numbers fall. The calculations are carried out via the computer of the CT system in online operation and can be made during the routine investigations when the quick computer available has free capacity outside the scanning program.

Results

Up to now, 3-D calculations in the region of the skull base have been carried out in 50 patients.
Craniofacial anomalies are conventionally investigated using ordinary radiographs and by conventional computed tomography. Both methods have the major disadvantage of trying to demon-

strate a complex three-dimensional structure, such as the skull, in two dimensions. It is much easier to use 3-D CT in craniofacial malformations prior surgical correction. Radiographs are often inferior to 3-D CT because of superimposition of asymmetric structures or technically unsatisfactory examinations in young children. Axial CT demonstrates reliably the extent of bone deformity but the appreciation of the severity of cranial-facial distortion is usually superior on 3-D CT.

In tumors affecting the skull base (metastases, meningiomas, paragliomas, chordomas, chondromas, chondrosarcomas and primary cholestatomas) a 3-D image of the bone destruction can be obtained from any viewpoint, and also the relationship of the tumor itself to the osseus base can be delineated. This can be most successfully achieved if the tumor tissue is calcified or if marked enhancement occurs after intravenous administration of contrast medium. The large basal vessels of the brain, the falx and the tentorium are then also visualized (Figure 1). 3-D bone images were useful in all cases when skull base erosion was present. The so-called "pseudoforamina" are a special problem of 3-D imaging of the skull base, when thin bone has a density lower than the selected threshold value. Especially "pseudoforamina" around the orbits and antra made the detection or exclusion of bone destruction difficult at these sites. 3-D CT was valuable in severe injuries with multiple fractures, providing a clear demonstration of fracture extent and fragment displacement (Figure 2). The preoperative and postoperative evaluation of maxillofacial trauma is one of many potential applications of this new technology (Figure 3). Many of the spatial relationships not easily appreciated on conventional roentgenograms or conventional 2-D CT are easily conceptualized with the 3-D CT technique. It is surprising to find significant limitations in the interpretation of the 2-D images without a full understanding of the spatial relationships so well depicted on the 3-D images. The 3-D CT provides an easily appreciated comparison of the preoperative condition to the postoperative repair. Thus, postoperative management or the evaluation of postoperative complications as the spatial concepts are more readily assessed in 3-D images. The limitations of 3-D imaging is the decreased resolution for the fine detail in comparison with the 2-D images.

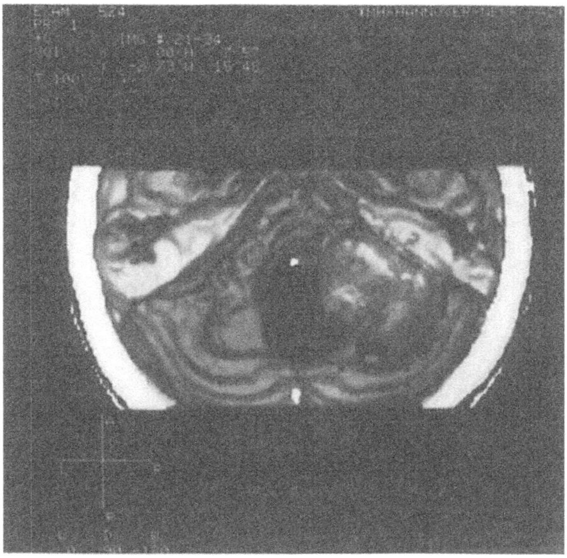

Figure 1: Meningioma of the posterior fossa

The additional diagnostic benefit of 3-D CT is limited. As a rule, clinically relevant diagnostic information can already be obtained from the traditional 2-D images. However, it is important that the topographical anatomy is rendered more readily understandable by 3-D CT. This simplifies the communication between the radiologist and the physician responsible for referral and further treatment. This results in particular in advantage for the planning of the operation, which is the reason for the special interest surgeons take in 3-D CT, since they enable a better spatial conceptualization of the area of surgery. The reconstruction of plastic models of the skull base on the basis of the CT data is a further new possibility to improve the surgical planning, particularly in craniofacial cases.

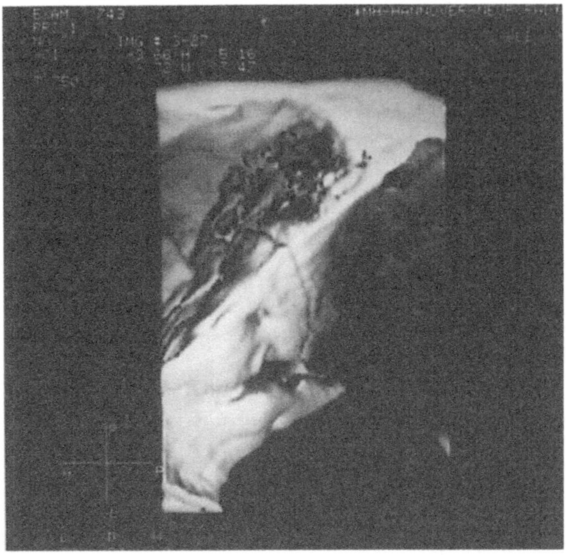

Figure 2: Fracture of the petrous bone

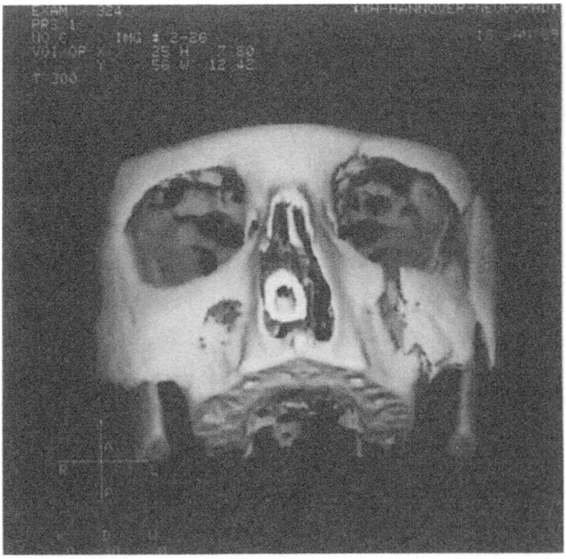

Figure 3: Midfacial fracture on the left

References

1. Becker, H., Grau, H., Hacker, H., Ploder, K.W.: The base of the skull: a comparison of computed and conventional tomography.J. Comput. Assist. Tomogr. 2 (1978) 113-118.

2. Becker, H.: Dreidimensionale kraniale und spinale Computertomographie. Radiologe 28 (1988) 239-242.

3. De Marino, D.P., Steiner, E., Poster, R.B., Katzberg, R.W., Hengerer, A.S., Herman, G.T., Wayne, W.S., Prosser, D.C.: Three-dimensional computed tomography in maxillofacial trauma. Arch. Otolaryngol. Head Neck Surg. 112.

4. Ernsting, M., Zeitler, E., Theissing, J., Imhof, K.: Technik und Ergebnis der Computertomographie der Rhinobasis und der Orbita mit multiplanaren Rekonstruktionen. Fortschr. Röntgenstr. 146 (1987) 376-380.

5. Gillespie, J.E., Adams, J.E., Isherwood, I.: Three-dimensional computed tomographic reformations of sellar and parasellar lesions. Neuroradiology 29 (1987) 30-35.

6. Witte, G., Höltje, W., Tiede, U., Riemer, M.: Die dreidimensionale Darstellung computertomographischer Untersuchungen kraniofazialer Anomalien. Fortschr. Röntgenstr. 144 (1986) 400-405.

3-D-Imaging: A Computing Program for the Future?

Ernsting, Marianne, M.D.

Radiol.Zentrum,Klinikum Nürnberg,Flurstr.17, 8500 Nürnberg

During the last years radiological diagnosis in traumatology and otorhinology has been essentially improved by CT-scanners of the 3rd generation and their further developped computing programmes (9). Prerequisit for this progress was that the osseous skeletal structures could be demonstrated anatomically correct and without artefacts. Necessary therefore were correspondingly thin slices of 1-2 mm and a highly dissolvable algorhythm with a corresponding number of kernels, which enables to show high density differences without overlappings and artefacts.

For traumat logical radiology this results in the advantage that socalled "grey zones", which had been overlapped in conventional radiology or could not be evaluated because of missing details, can be demonstrated. These "grey zones" are the bottom of the orbita, the complete rhino-base (10,11), the structures of the shoulder joint(1), of the scapula (1) or of the rear pelvic ring in proved complete and instable fractures of the pelvic ring.

Additionally we could use a 3-dimensional computing programme, which enables to make 3-dimensional surface-reconstructions out of axial, coronary and semi-coronary scans (2). This means no additional time in the CT-scanner for the patients and no additional exposition to radiation.

First patients with fractures of the rhino-base and intersperse of foreign bodies have been computed 3-dimensionally for a second time. Fractures of the middle face and fractures of the renal base (5,12,13) could be demonstrated and proved surprisingly well. The 3-d-reconstructions show an impressive image of the sort and number of osseous injuries (Fig.1.2,3).

According to the improvement of the computing programme we increasingly also computed traumatological injuries 3-dimensionally for a second time.

HR-CT and the 3-dimensional re-computings show the possibility to demonstrate osseous injuries without overlappings. The 3-d-reconstructions decisively help the surgeon in his preoperative planning (7,8).

Fractures of the scapula and the shoulder-joint (3) can also be clearly demonstrated, as well as injuries in the region of the lower extremities (4).

Summary

3-D-reconstructions give the following additional information:

a) for all traumatologic therapists, who are used to 3-dimensional imaging;
b) for the planning of prostheses in the region of the thigh and at the neck of the femur.
c) for the reconstruction of fractures of the middle face.
d) for the localisation of foreign bodies.

Nevertheless the computing programme has still to be improved and especially the documentation of details in 3-dimensional second computations still needs an improvement.

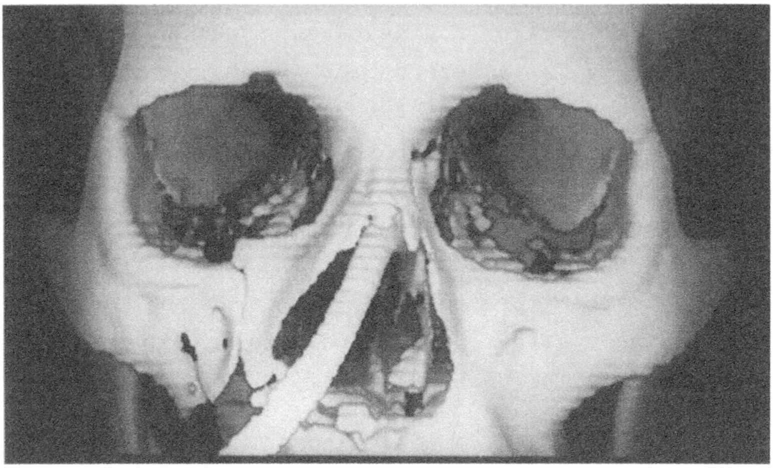

Fig.1:

Fracture of the bottom of the orbita on the right side, several fractures of the maxillary sinus on the right side, partly with dislocations, fractures of the nasal bone and the sieve bone.

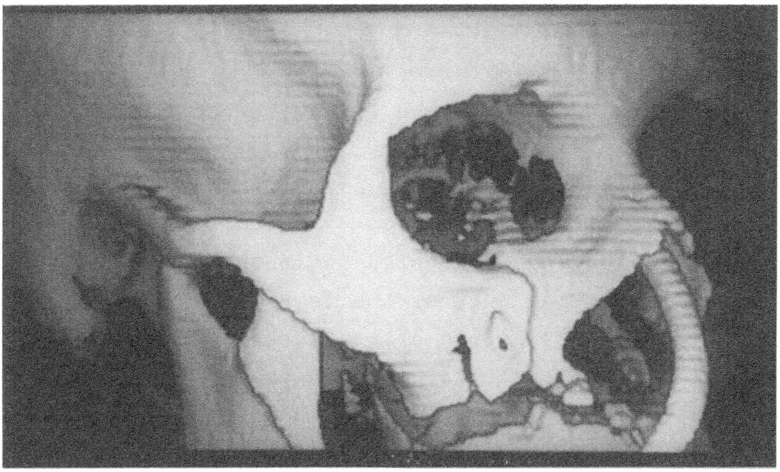

Fig.2:

Additional fracture of the mandibula of the same patient.

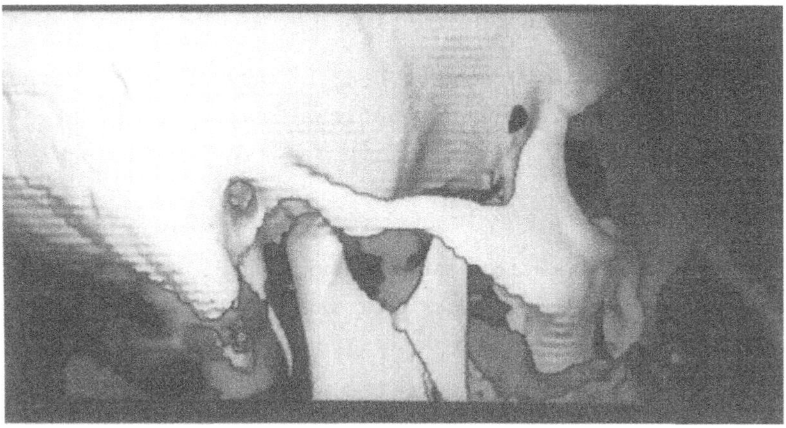

Fig.3:

Clearly demonstrated additional fractures of the upper jaw of the same patient.

References:

1. Castagno, A.A., W.P. Shuman, R.F. Kilcoyne, D.R. Haynor, M.E. Morris, F.A. Matsen, Complexe Fractures of the proximal Humerus: Role of CT in Tretment. Radiology (1987) 165: 759 - 792

2. Chen, L.S., G.T. Herman, C.R. Mayer, R.A. Reynolds, J.K. Udupa, 3 D 83. An easy to use software package for 3-D display from computed tomograms. Proc. SPIE 515, 1984, 309

3. Dihlmann, W., J.Baudich, Computertomographie (CT) der Schulterweichteile, Fortschr.Röntgenstr.147,2 (1987), 147-151

4. Heuchemer, Th., G.Bargon, G.Bauer, W.Mutschler, Vorteile in Diagnose und Einteilung der intraartikulären Kalkaneusfraktur durch die Computertomographie. Fortschr.Röntgenstr. 149, 1 (1988) 8-14

5. Jend, H.-H., W.Crone-Münzebrock, E.Grabbe, Die Computertomographie der Schädelbasisfrakturen. Fortschr.Röntgenstr. 140 (1984) 147-151

6. Kondo, M.,M.Horiuchi, Y.Inuyama,T.Dokiya, T.Tsutsui, Y.Iwata, M.Endo, T.Hashimoto, E.Kunieda, S.Hashimoto, Value of Computed tomography for radiation therapy of tumors of the nasal cavity and paranasal sinuses. Acta Radiol. Oncol.22 (1983) 3-7

7. Marincek, B., B. Porcellini, G. Robotti, Computertomographierte Klassifikation von Acetabulumfrakturen. Radiologe (1984)24: 205-210

8. Pozzi Muchelli, R.S., G. Muner, F. Pozzi Mucelli, M. Pozzi Muchelli, F. Marotti, L. Dalla Palma Three Deminsional Computed Tomography of the Acetabulum. Europ. J. Radiology 6 (1986) 168-177

9. Rosenberg, Z. S., F. Feldmann, R. D. Singson, R. Kane, Ankle Tendons: Evaluation with CT, Radiology (1988) 166: 221-226

10. Som, P.M., CT of the paranasal sinuses. Neuroradiology 27 (1985), 189-201

11. Valvassori, G.E., G.D.Potter, W.N.Hanafee, B.L.Carter, R.A.Buckingham, Radiologie in der Hals-Nasen-Ohren-Heilkunde. Thieme-Verlag, Stuttgart 1984

12. Vannier, M.W., J.K.Marsh, J.O.Warren, Three dimensional CT-reconstruction images for cranio-facial surgical planning. Radiology 150 (1984) 179

13. Witte, G., W.Höltje, U.Thiede, M.Riemer, Die dreidimensionale Darstellung computertomographischer Untersuchungen kraniofazialer Anomalien. Röfo 144, 4 (1986) 400-405

14. Witte, G., F.Böcker, E. Bücheler, K.H.Höhne, Three dimensional reconstruction from digital angiographic sequences. 70th Scientific Assembly of the Radiological Society of North America, Washington 1984

New Algorithm for Metal Artefact Reduction (MAR) in Computed Tomography: Phantom Studies

R. Sokiranski, D. Felsenberg, W. Kalender*

Röntgendiagnostik (Director: Prof. Dr. K.-J. Wolf), Klinikum Steglitz, Freie Universität Berlin, F.R.G.
*Siemens AG, Bereich Medizintechnik Erlangen, F.R.G.

Summary

Depending on their atomic number and structure, metals can produce artefacts which obviously limit CT scan diagnosis. A new metal artefact reduction (MAR) program is capable of clearly depicting interstices in the direct vicinity of metals at intervals of 1 - 1.5 mm. Structures close to the metals which could not be evaluated at all or where only a limited evaluation was possible before became distinctly visible only after MAR. The program is distinguished by quick calculation times, easy operation, minimal individual interaction and good reproducibility. One area where it is especially of value is in the postoperative inspection of metal endoprotheses. MAR makes it possible to observe by computed tomography the transformation processes in the immediate vicinity of metal implants in vivo and under normal conditions.

Introduction

Since computed tomography is a diagnostic procedure offering high resolution and overlapfree images, it was predestined to play a role in the determination of transitional processes in the vicinity of implants. Previously, it was impossible to examine metal implants by CT due to the great degree of artefact formation. The first studies of hip endoprotheses by means of metal artefact reduction were published between 1978 and 1984 by the work group comprising Hinderling, Rüegsegger, Seitz et al. (1-5). Very good MAR programs were already available then, but they could not be widely used because of the technical limitations of the equipment at that time, e.g. long recording times, minimum layer thickness of 5 mm (which resulted in poorer resolution), long computation times due to the nature of the program, and the necessity for large computer storage capacity. A new method of MAR was developed by Kalender in 1986 (6) which, together with technical improvements in the CT equipment, made even routine diagnosis of endoprotheses possible. The goal of this study is to evaluate the efficiency of this new MAR program.

Materials and Methods

A plexiglas water phantom (Figs. 1 and 2) comprising a watertight, reclosable plexiglas cylinder with a diameter of 15 cm (average thigh size) was used in our investigations. Plexiglas disks with a place in their center to accommodate the rods composed of various prothesis alloys were plased inside the phantom. These metal rods were surrounded by a plexiglas tube with an outer diameter of 35 mm. The inner diameter of the tube increased every 12 mm, thus incrementally increasing the distance (d) between the metal rod and the plexiglas tubes in 6 stages where d = 0.25, 0.75, 1.0, 1.5, 2.5 and 3.75 mm, respectively. In order to simulate the absorption conditions present in the softpart structures of the thigh, the entire cylinder was filled with water.

Three types of common prothesis material were studied:

1) a cobalt/chrome/molybdenum alloy,
2) a chrome/nickel/ molybdenum alloy, and
3) a titanium/aluminum alloy.

The interstices were measured before and after reduction of artefacts.

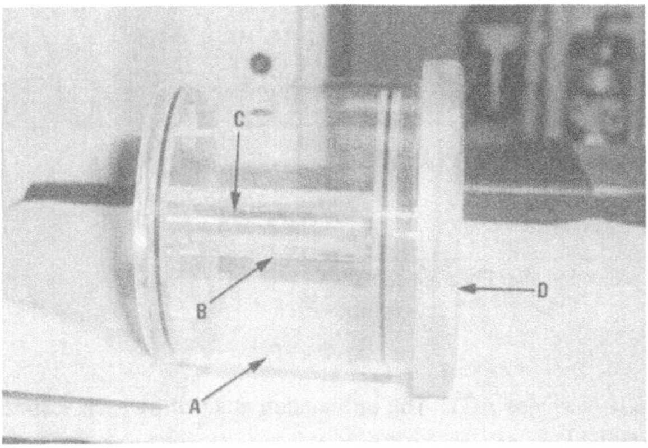

Figure 1

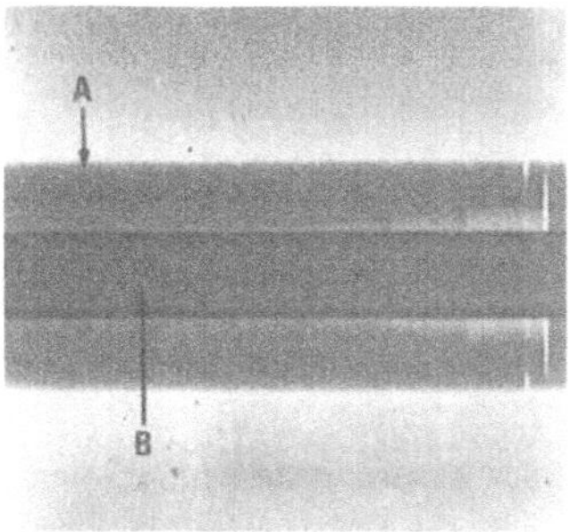

Figure 2

Figure 1: Plexiglas water phantom A) plexiglas cylinder (diameter = 15 cm), wall thickness = 4 mm; B) plexiglas tube; C) interchangeable metal rod (diameter = 13 mm); D) screw-on lid

Figure 2: X-ray of the plexiglas water phantom A) Plexiglas tube with constant outer diameter of 35 mm; inner diameter varies in 6 stages from 13.5 mm to 22.5 mm. B) Central metal rod

Computed Tomographic Examination

1. Fundamentals of metal artifact reduction in CT.

Metals, especially those with high atomic numbers, absorb X-rays almost completely. This X-ray absorption leads to missing projection data, or data gaps, which are the cause of the wellknown, starshaped artefacts in CT images with normal reconstruction algorithms. These disturbances cause an almost total loss of scan information in the immediate vicinity of the metals.

Our MAR program comprises basically the linear interpolation of missing projection data. Completion of these lacking data supplements the measurement data to such a degree that the distracting artifacts can be considerably reduced (Fig. 3). This method was introduced by Lewitt and Bates in 1978 (7).

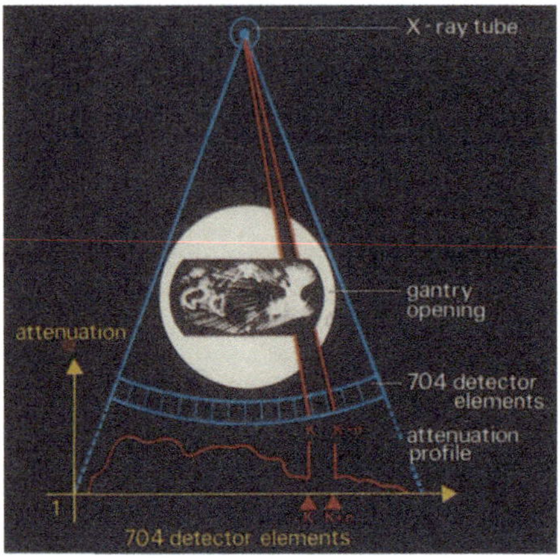

Figure 3

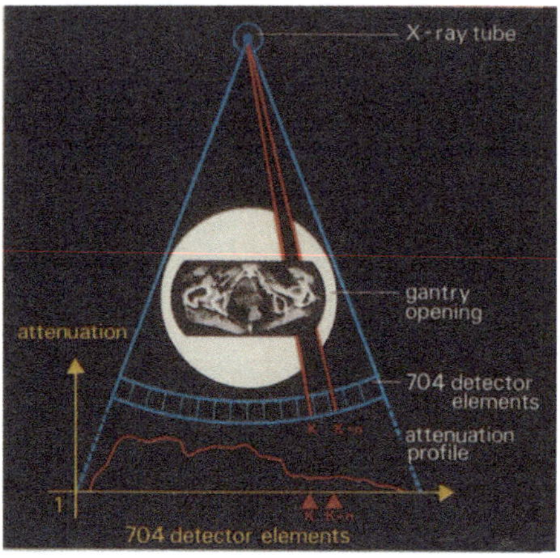

Figure 4

Figure 3: Attenuation profile of an image before MAR

Figure 4: Attenuation profile of the same projection after linear interpolation

2. Performing MAR.

Scans were made with a Somaton DRH (Siemens AG). The information attained by each scan was stored, and the images containing artifacts were reconstructed from this raw data. The area requiring reduction was marked with a light pen. An automatic contour seeker program then identified the exact borders of the implant within the circumscribed area.

By using adjoining projections, information gaps were completed by linear interpolation. A new set of raw data was computed and reconstructed into a now artifact-reduced image.

Reduction of Metal Artefacts in Computed Tomography:

- CT scan is made, raw data is stored

- Images containing metal artefacts are reconstructed

- Area requiring reduction is marked with a light pen

- Automatic contour finding program is implemented

- Missing projection data is linearly interpolated

- An artifact-reduced image is reconstructed from the newly computed projection data

The time required for computation of the data needed to produce the artefact-reduced image was 30 to 45 seconds. A further 30 seconds were needed to reconstruct the image. Thus, a total of 1 to 1.5 minutes were required to convert the artifact-containing picture into an artifact-reduced image.

Results

In all of the aforementioned alloys, the interstitial spaces of 0.25 to 3.75 mm were not visible in the normal CT image due to artefact formation. Table 1 lists the results of interstitial space measurements after reduction of artefacts for each of the alloys.

Table 1: Actual intersticial spaces compared to space measured after MAR

	Interstitial space in mm					
Material	0.25	0.75	1.00	1.5o	2.50	3.75
1. Cr/Co/Mo	0.0	0.0	0.0	0.0	1.0	2.2
2. Cr/Ni/Mo	0.0	0.0	0.0	0.0	1.0	2.2
3. Ti/Al	0.0	0.0	0.0	0.0	1.0	2.2

In the artefact-reduced images, the area circumscribed by the light pen appears as a grey region with a diameter 3 mm larger than the diameter of the metal rod. This means that interstices immediately adjacent to the metal which are less than or equal to 1.5 mm in width cannot be demonstrated. If the actual width of the interstice is greater than 1.5 mm, the demonstrable interstitial space is 1.5 mm smaller than in reality.

Artefact formation is dependent on the atomic number of the implant material. Thus, titan/aluminum alloys produce fewer artifacts than the normal cobalt/chrome/ molybdenum alloys. Even in the latter case, we were able to achieve noticeably improved projection data, whereby the aforementioned 1.5-mm-wide border escapes evaluation here, too (Figs. 5-7).

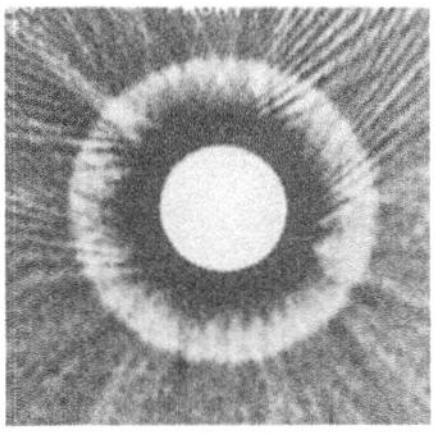

A

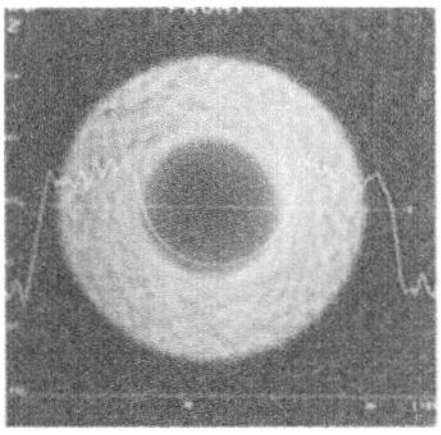

B

Figures 5A and B: Interstitial space = 1.5 mm (Cr/Co/Mo) A) Density measurement impossible without MAR, interstice cannot be demonstrated. B) After MAR, density measurement can be performed without any complications. Interstice not recognizable.

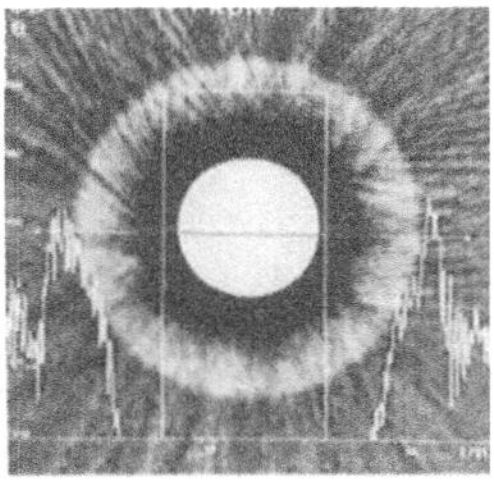

A

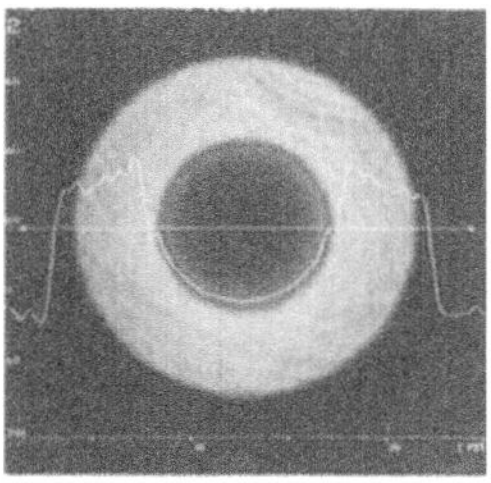

B

Figure 6: Interstitial space = 2.5 mm (Cr/Ni/Mo) A. Before MAR, interstice could not be demonstrated. B. After MAR, a 1-mm-wide space becomes visible.

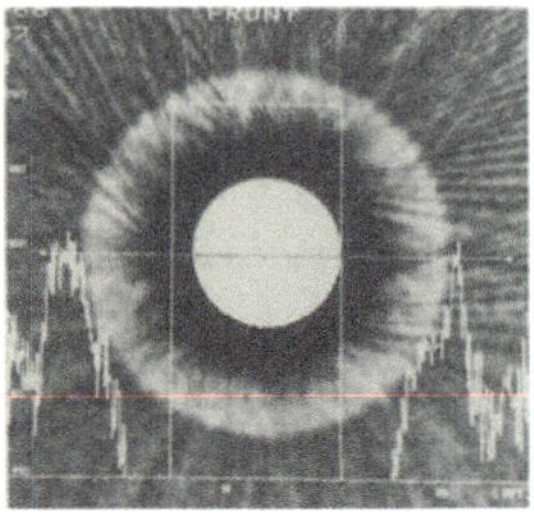

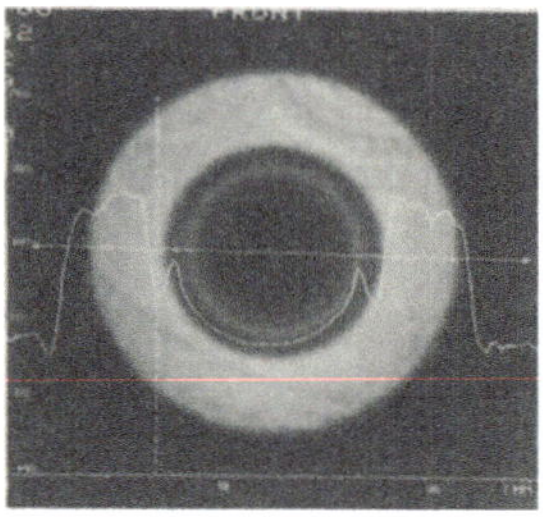

A B

Figure 7: Interstitial space = 3.75 mm (Cr/Ni/Mo) A. Before MAR. B. After MAR; demonstrated space was 2.2 mm wide. Density measurement possible only after MAR.

Discussion

Contrary to previously described MAR programs (3), the program for reduction of CT artifacts that was used by our clinic can be applied under routine conditions due to its quicker computation times, easy use, minimal individual interaction and good reproducibility of results. Technical advances in CT during the past ten years which allow quicker scan times with a resolution of ca. 1 mm have made it possible to achieve a correlation between artefact-reduced CT images and the histological transformation processes in the area of an artificial hip joint (8) (Figs. 8-10).

In addition to its clinical applicability, CT in conjunction with MAR can also be applied in the frame of scientific investigation, offering a new and sensitive method of visualization which will especially have an impact on the further development of prothesis design. With it, one could develop means to more quickly recognize inadequate new types of protheses or find proof to confirm them.

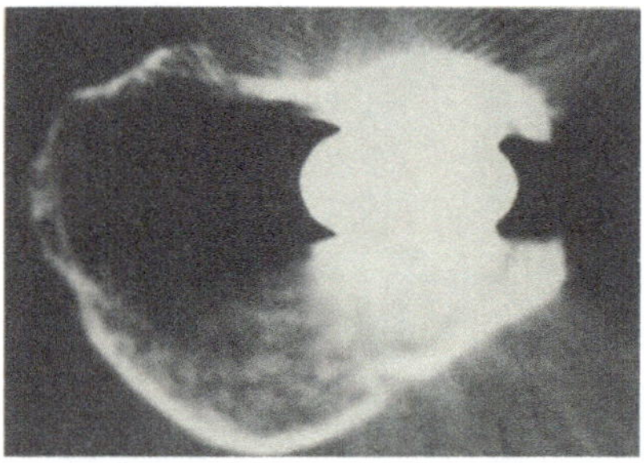

Figure 8: Cemented hip prothesis shaft in CT. Severe artifact formation has impeded evaluation.

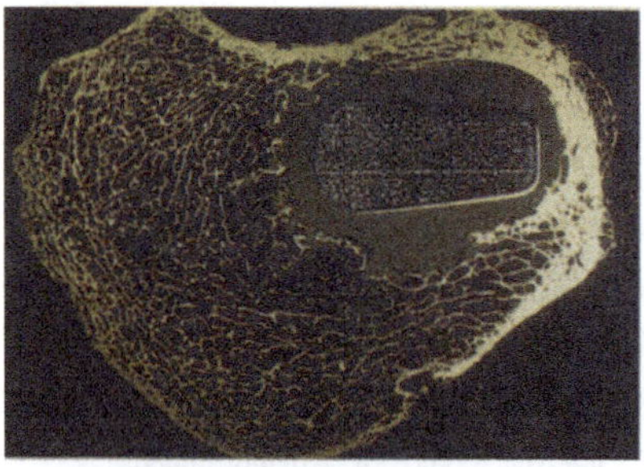

Figure 9: Microradiographic preparation from same CT examination level. Cement and spongiosa have intermeshed well. No signs of loosening. Prothesis implanted 4.5 yrs ago.

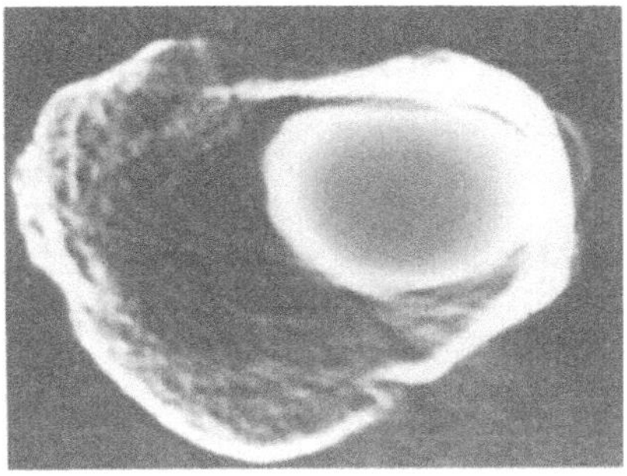

Figure 10: CT picture after MAR. Cement/bone border well delineated. No signs of loosening.

One special area for its use would be in postoperative follow-up of metal endoprotheses. It makes it possible to observe transformation processes in the immediate vicinity of a metal implant under in vivo, routine CT conditions. Late endoprothetic complications like loosening of the prothesis can be examined with the aid of MAR, because CT with MAR allows early detection of interstitial spaces that could be indicative of prothesis loosening and also provides important additional information on the position of the prothesis and distribution of cement. It also makes it possible to quantify changes in bone density resulting from compact degeneration processes due to non-optimal transfer of strength in the area of the shaft (9).
This is confirmed in the first studies involving patients with total endoprotheses. In numerous cases with unclear or even normal X-rays and scintigraphs, loosening and fractures and osteolysis in the region of the implant could thus be diagnosed (10-12). Unlike conventional radiology, which cannot satisfactorily demonstrate early alterations in the prothesis region, metal artifact reduction of CT images provides a speedier and more accurate means of diagnosis.

References
1. Hinderling, T.: Computertomographische Verfahren zur Untersuchung der Gewebe und Zementverteilung in der Umgebung metallischer Implantate. Dissertation, Universität Zürich, 1978.

2. Hinderling, T., Rüegsegger, P., Anliker, M. Dietschi, C.: Computed tomography from hollow projection to in vivo evaluation of artificial hip joints. J. Comput. Assist. Tomogr. 3: 52-75, 1979

3. Rüegsegger, P., Hangartner, H.U., Keller, T., Hinderling, T.: Standardization of computed tomography images by means of material selective beam hardening correction. J. Comput. Assist. Romogr. 2: 184-188, 1978.

4. Seitz, P.: Computertomographische Osteodensitometrie beim metallischen Kunstgelenk. Dissertation, Eidgenössische Technische Hochschule Zürich, 1984.

5. Seitz, P., Rüegsegger, P.: Anchorage of femoral implants visualized by modified computed tomography. Arch. Orthop. Trauma Surg. 100: 261-266, 1982.

6. Kalender, W., Hebel, R., Ebersberger, J.: Reduction of CT artifacts caused by metallic implants. Radiology 164: 576-577, 1987.

7. Lewitt, R.M., Bates, R.H.T.: Image reconstruction from projections. III: Projection completion methods (theory). Optik: 189-204, 1978.

8. Sokiranski, R.: Metallartefaktreduzierung am Beispiel der Hüftendoprothesen. Dissertation, 1989. (in press)

9. Faensen, M., Felsenberg, D., Sokiranski, R.: Quantitative CT-Messung des Knochenmineralgehaltes im Femur bei Hüftgelenkprothesen. First International Symposium on Custom-Made Protheses, Düsseldorf, October 21-22, 1988.

10. Felsenberg, D., Riso, J., Sokiranski, R., Faensen, M., Kalender, W.A., Gross, U.: Metallartefaktunterdrückung in der CT: Erste Ergebnisse bei implantierten Hüftkopfprothesen. Hefte zur Unfallheilkunde 200: 51-52, 1988.

11. Felsenberg, D., Kalender, W.A., Sokiranski, R., Ebersberger, J., Krämer, R.: Reduktion von Metalartefakten in der CT: Klinische Erfahrungen und Ergebnisse. Elektromedica 4: 97-104, 1988.

12. Sokiranski, R., Felsenberg, D., Faensen, M., Kalender, W., Gross, U., Fritz, R.T.: Modifizierter algorithmus zur metallartefaktreduzierten CT-Untersuchung. 4. Treffpunk Medizintechnik, Universitätsklinikum Steglitz, Berlin, 1989.

Classification of Brain Tumours by CT-Image Walsh Spectra

M. Wolf, S. Ziegengeist, F. Bornholdt, S. Michalik, M. Michalik

Klinik und Poliklinik für Neurologie und Psychiatrie des Bereiches Medizin der Humboldt-Universität Berlin
(Charité) DDR-1040 Berlin, Schumannstr. 20/21

For 139 patients with 7 classes of most frequent brain tumours CT image data were taken from squares of 16 x 16 pixels centered in the tumours border, an appropriate set of features was derived from their Walsh spectrum and averaged for several of these square regions. For the analysis of the data and for pattern recognition linear discriminant analysis was used.
We first compared the results with those obtained from a set of features containing image parameters measured on the Evaluscope and anamnestic data. For both sets of features quite similar classification rates of 57% (Walsh data) and 61% (patient and image data) were estimated for a single-step discrimination. Finally a hierarchic classification system based on both sets of features was developed with an estimated classification rate (diagnostic accuracy) of 84 %. Considering the small number of the presently available cases with both histological proof and digitally stored images this is an encouraging result. Recognition by spectral analysis appears to be an adequate concept for brain tumour discrimination and possibly also for other problems in the diagnostics of liver, lung and bones. Image processing may become a substitute for invasive bioptic/histologic methods of tumour diagnostics.

Diagnostic Evaluation of Three-Dimensional CT Reconstruction Methods

M. W. VANNIER, T. K. PILGRAM, C. F. HILDEBOLT, J. L. MARSH, L. A. GILULA
Mallinckrodt Institute of Radiology
Washington University School of Medicine
510 South Kingshighway Blvd.
St. Louis, Missouri 63110
U.S.A.

Summary
Major areas of application of three-dimensional computed tomography (3-D CT) are cranial and skeletal imaging. 3-D CT has an important role in determining the presence and extent of congenital and acquired craniofacial and orthopedic deformities. We compared the diagnostic value of 3-D CT images in the detection and characterization of cranial sutural synostosis and pelvic fractures with planar CT and conventional x-rays. 3-D CT scan processing was performed by both surface reconstruction and volumetric methods. Several trained observers read each scan series in a blinded fashion. The diagnostic quality of the images was determined with receiver operating characteristic (ROC) analysis. The results of this study indicate that 3-D shaded and volumetric reconstructions are superior to conventional modalities for diagnosis of cranial sutural synostosis and equal to conventional modalities for diagnosis of pelvic fracture and associated complications.

Introduction

Three-dimensional computed tomography (3-D CT) has been successful at producing lifelike images of the skull and skeleton. These images are important for communicating information to surgeons regarding congenital or acquired craniofacial and orthopedic deformities. However, the diagnositc utility of these 3D images has not been objectively assessed. We compared the sensitivity and specificity of 3-D CT images in the detection and characterization of cranial sutural synostosis (1) and pelvic fracture with planar CT and conventional x-rays.

Patient Population

Eighty-two (82) subjects were included in the sutural synostosis study. Of these, 53 had synostosis of the metopic suture, 38 of the sagittal suture, 16 of the left lambdoidal, 18 of the right lambdoidal, and 25 of the left and right coronal sutures. Truth was established by review of all modalities, post-operative followups, and especially intra-operative direct observations after reflection of the scalp during cranial recontouring (JLM). All patients had CT scans and 3-D reconstructions (2), while only 41 had conventional x-rays.

Thirty-two (32) hemipelves were studied from 16 patients. There were 17 fractures, 10 of which had fragments in the joint space. Truth was established by review of all modalities. All patients had CT scans and 3-D reconstructions, and 14 had conventional x-rays.

Experimental Procedure

Observers were presented with the images in blinded fashion. No clinical information was provided, and names and birthdates on the images were masked. Diagnoses were made on a five-point rating scale. Because some observers were unfamiliar with some modalities, all images from each modalitiy were rated before moving on to the next modality, to reduce the effect of learning.

Analysis

The primary mode of analysis was Receiver Operating Characteristic (ROC) analysis (3-6). Fitted ROC curves and the areas under them (*Az*) were calculated with the ROCFIT program, and where statistical significance was calculated, the CORROC2 program was used (Metz et al. 1984). In some cases, the observer rating data were degenerate, causing the ROC curve to become vertical or horizontal at some false positive fraction. In these cases, a battery of other diagnostic performance measures were calculated: accuracy, the total proportion of patients correctly diagnosed; sensitivity, the proportion of abnormal patients correctly diagnosed; and specificity, the proportion of normal patients correctly diagnosed.

Results

Cranial Sutural Synostosis

The overall results for both observers suggested that 3D shaded was the most useful imaging modality for diagnosing cranial sutural synostosis, followed closely by 3D volume. CT slice and plain-film were next, and approximately equal in utility. 3D depth was markedly the least useful imaging modality. Figure 1 shows the graphs in which there is the least and greatest differentiation of modalities, and Table 1 the results with all modalities for Observer 2. The results for Observer 1 are similar, but more complex due to the large amount of degenerate data which necessitated calculation of accuracy, sensitivity, and specificity.

Pelvic Fracture

Results for all observers were combined, because the small sample size often resulted in degenerate data for individual observers. There was no clearly superior imaging modality. In general, all modalities were very close in their diagnostic utility, and the ranking of the modalities varied by task. There was a slight tendency for the conventional modalities (CT slice and plain-film) to be superior to the 3D reconstructions, but this tendency was not strong enough to be meaningful, and was almost certainly not statistically significant. Figure 2 and Tables 2 and 3 show the results for the easiest (presence/absence of fracture) and one of the most difficult (presence of fragments in joint space) diagnostic tasks.

The results show differences among imaging modalities for diagnosis of skeletal abnormality and trauma. 3D imaging modalities clearly have an important role in the diagnosis of skeletal abnormality and trauma. 3D shaded and 3D volume images are clearly superior to conventional methods for the diagnosis of cranial sutural synostosis, and appear to be equal to conventional methods for the diagnosis of pelvic fracture and associated complications.

We found that the level of diagnostic task difficulty is important for the quality of results. For cranial sutural synostosis, all tasks except the sagittal suture appeared to have a reasonable level of difficulty, and ROC analysis revealed differences in diagnostic utility. For pelvic fracture and associated complications, several of the tasks appeared either "too hard" or "too easy". All modalities are so useful for determining presence/absence of a fracture that it is difficult to distinguish meaningfully between them ("too easy"). Distinction is also difficult for diagnosing fragments in joint space, in this case because all modalities do so poorly ("too hard" - note that random diagnosis would result in a diagonal line with Az=0.500). The general approach of observers to the rating responses is critical. ROC studies and diagnostic image interpretation procedures do not mesh perfectly, because diagnosis encourages definite responses while ROC analysis requires a certain level of uncertainty to avoid degenerate data. Our experience suggests that a six-category rating system may be better than the more commonly used five-category system (which we used in this study). The absence of a true "middle" category (which tends to be seen as mere equivocation) may encourage observers to use the high-uncertainty responses more often. In addition, where the data are degenerate, lack of a "middle" category aids in the valid calculation of accuracy, sensitivity, and specificity.

References

1. Vannier MW, Hildebolt CF, Marsh JL, Pilgram TK, McAlister WM, Shackelford GD, Offutt CJ, Knapp RH: Determination of the diagnostic value for three-dimensional CT reconstruction in craniosynostosis. Submitted to Radiology.

2. Drebin RA, Carpenter LA, Hanrahan P: Volumetric rendering. Computer Graphics 22:65-74, 1988.

3. Metz CE, Wang DL, Kronman HB: A new approach for listing the significance of differences between ROC curves measured from correlated data. In: Deconink F, ed. Information Processing in Medical Imaging. The Hague: Nijhoff, 432-445, 1984.

4. Hanley JA, McNeil BJ: A method of comparing the areas under receiver operating characteristic curves derived from the same cases. Radiology, 148:839-843, 1983.

5. McNeil BJ, Hanley JA: Statistical approaches to the analysis of receiver operating characteristic (ROC) curves. Med Dec Making, 4:137-150, 1984.

6. Hanley JA, McNeil BJ: The meaning and use of the area under a receiver operating characteristic (ROC) curve. Radiology, 143:29-36, 1982.

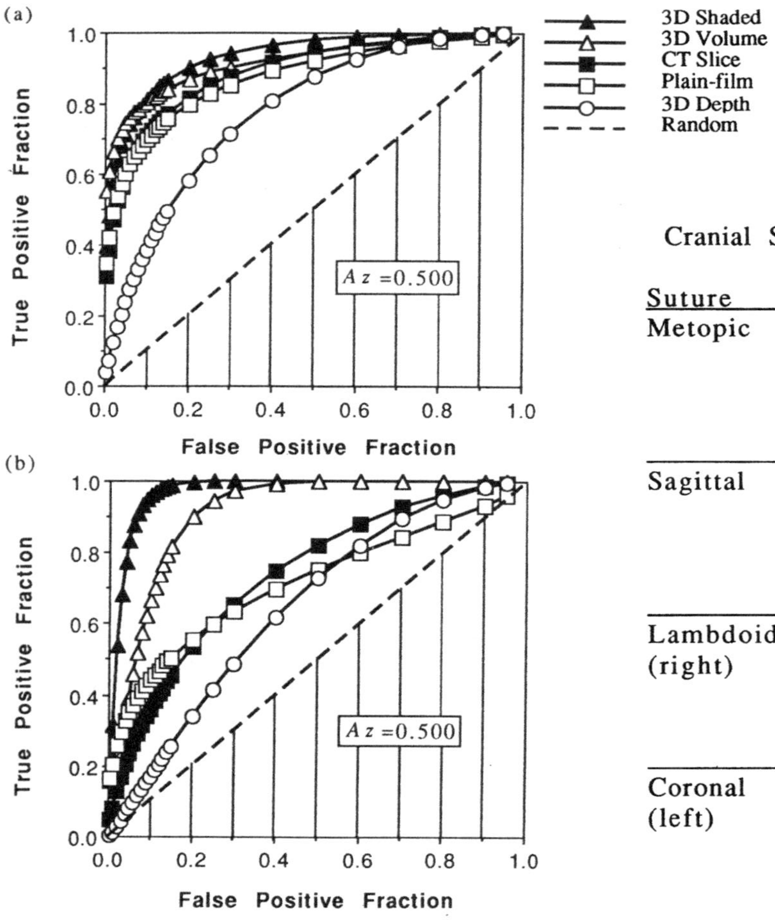

Figure 1. Fitted ROC Curves for Cranium: Sagittal (a) and Right Lambdoidal (b) sutural synostosis for Observer 2.

Table 1
Cranial Sutural Syonostosis Results for Observer 2

Suture	Modality	Az	Mean of Az
Metopic	3D Shaded	0.942	
	3D Volume	0.900	
	CT Slice	0.877	0.818
	Plain-film	0.718	
	3D Depth	0.651	
Sagittal	3D Shaded	0.935	
	3D Volume	0.917	
	CT Slice	0.891	0.880
	Plain-film	0.877	
	3D Depth	0.778	
Lambdoidal (right)	3D Shaded	0.971	
	3D Volume	0.909	
	CT Slice	0.743	0.798
	Plain-film	0.715	
	3D Depth	0.650	
Coronal (left)	3D Volume	0.899	
	3D Shaded	0.893	
	Plain-film	0.800	0.782
	CT Slice	0.667	
	3D Depth	0.649	

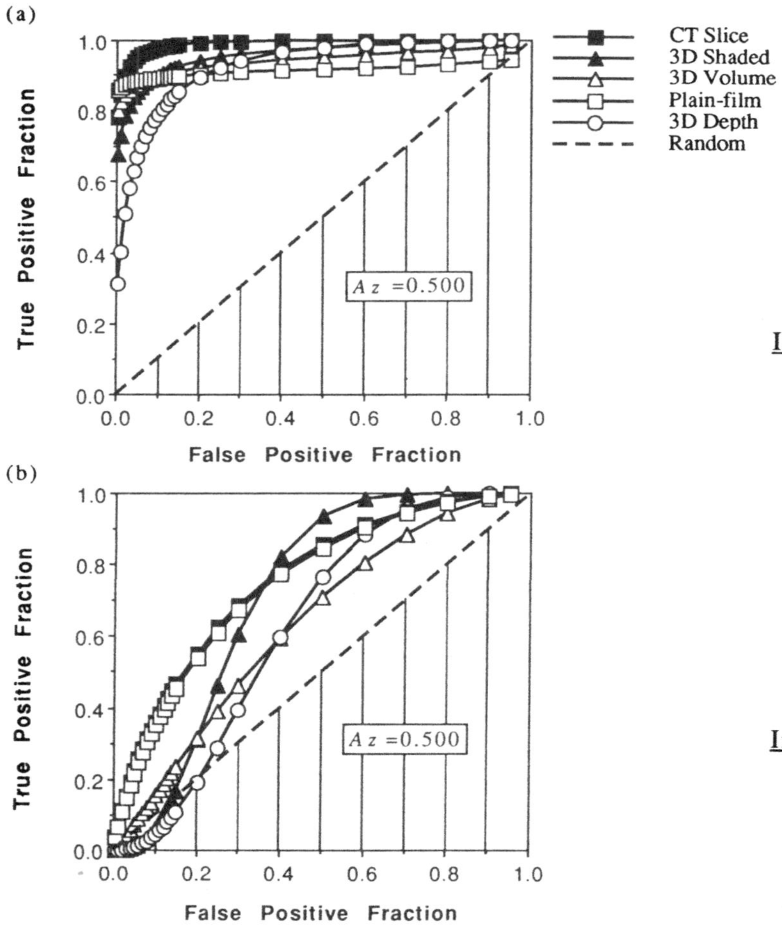

Figure 2. Fitted ROC Curves for Pelvis: Presence/absence of fracture (a) and fragments in the joint space (b) for all observers.

Table 2

Pelvic Fracture (All Observers)

Imaging Modality	Az	Mean of Az
CT Slice	0.990	
3D Shaded	0.961	
3D Volume	0.944	0.948
3D Depth	0.930	
Plain-film	0.915	

Table 3

Pelvic Fracture: Fragments in Joint Space (All Observers)

Imaging Modality	Az	Mean of Az
CT Slice	0.762	
Plain-film	0.754	
3D Shaded	0.721	0.701
3D Depth	0.636	
3D Volume	0.632	

Computer Assisted Techniques for Extracting Quantitative Cardiovascular Data from Ultrafast CT Images

M.J. Lipton[1-2], R. Gould[3], W. W. Holt[1-2], J. Harmon[4], D.N. Levin[1]

1.Department of Radiology, The University of Chicago
2. Department of Medicine, The University of Chicago
3.Department of Radiology, U.C.S.F.
4. Imatron Inc., So. San Francisco, Ca.

INTRODUCTION

Heart disease and particularly coronary artery disease remains the major cause of illness and death in the Western world in those aged 35 - 65 years. Current methods used for diagnosing this terrible disease are unable to identify victims sufficiently early - before symptoms occur - nor can the extent and severity of the disease be measured precisely enough for selecting optimal management. This has recently become more critical with the introduction of modern interventional therapy, specifically, thrombolysm and balloon angioplasty.

The two established modalities most widely used for non-invasive studies of coronary artery disease are radionuclide imaging and echocardiography with or without doppler flow measurements. The sensitivity of exercise thallium imaging is about 85% in good hands but the specificity is less, being approximately 55%, in patients who have symptoms. Furthermore, in asymptomatic patients, it is of very little diagnostic value unless the ECG is abnormal. Despite this fact, many centers perform thousands of these expensive treadmill studies in asymptomatic patients every year. A further well-recognized and significant limitation of radionuclide imaging is its poor spatial resolution. Edge detection is therefore very difficult with this technique and the regions of interest from which counts are obtained must be redefined on each patient. Additional problems occur in obese patients with photon scattering from the subject and poor in-vivo labellings, as well as superimposition of other cardiac chambers. The assessment of regional wall motion is relatively crude and reliable measurements of wall thickness and thickening, therefore, cannot be made. Despite these significant shortcomings, this technique is widely used and relied upon for routine patient management.

Echocardiography is even more widely applied than radionuclide angiography. Echocardiography is a tomographic technique with greater spatial resolution than radionuclide imaging and allows the left ventricular borders to be defined at multiple levels. However, the variability of an adequate acoustic window at multiple imaging planes using two-dimensional echocardiography limits its clinical application for precise measurement of ventricular mass and volumes and, like angiography, this technique requires geometric assumptions.

The need to quantitate the degree of myocardial ischemia and the presence and severity of coronary artery stenosis has been addressed widely in the cardiology literature. The anatomic-geometric approach traditionally applied necessitates invasive coronary arteriography. Numerous algorithms have been developed to predict the severity of stenotic resistance, pressure-flow relationships, or coronary flow reserve. All such methods provide incomplete approximations because coronary artery lesions are complex, multiple and often diffuse and the cumulative haemodynamic effects assumed under standardized conditions do not apply with typical variations of regional vascular perfusion states. Because of all these difficulties, functional assessment based upon resting and maximal myocardial perfusion states has been combined with the anatomic-geometric approach. However, what is needed is a better general population screening procedure and one which could also evaluate patients with symptomatic heart disease for optimal management. The recent revolution in advanced digital imaging technology may change this situation. High speed computed tomography is one of the most recent emerging modalities in this class.

ULTRAFAST CT SCANNER DESIGN AND OPERATION

The first high speed computed tomography scanner was developed at the Mayo Clinic (1). This device, called the Dynamic Spatial Reconstructor (DSR), employs multiple x-ray tubes arranged in a circle around the patient which are fired in rapid sequence. This machine is one of a kind and has primarily been used as a research tool.

More recently, Ultrafast CT (Imatron scanner C-100) is now available. This system is based upon electron beam technology, and, unlike conventional CT, this scanner has no moving parts except for the patient table. During operation, an electron beam is focused to a 1-2mm focal spot size deflected across a series of 4 semicircular tungsten targets beneath the patient, each of which produce a fan beam of x-radiation. Two detector rings lie above the patient so that a pair of CT images are obtained each time any one of the tungsten rings is "swept" by the electron beam; hence as many as eight levels can be scanned without table movement. Several scanning options are currently available including almost simultaneous, multilevel, 50 msec scan acquisition at frame rates of up to 34 images per second (17 at each of two levels). Scans can be acquired at two, four, six or eight levels during a single acquisition. Scanning is usually initiated from the patient's electrocardiogram and commences at a selected percentage of the cardiac cycle from 0-80% of the R-R interval; it can also be triggered manually. The specifications of the system are given in Table I and the scanning capability is summarized in Table II. The x-ray data, once absorbed by the detectors, is converted to digital data . A powerful computer then reconstructs the images and displays them on a cathode ray monitor. Further details of the Imatron C-100 scanner have been described elsewhere (2,3).

ULTRAFAST CARDIAC CT SCANNING PROCEDURES

The basic scanning protocol is usually tailored to the clinical problem. However, in most patients, the following plan is generally followed. No premedications are given and the patient fasts for at least four hours prior to examination. Standard ECG electrodes are placed on the patient's chest for continuous monitoring and for triggering the CT exposures. An 18-gauge, 5cm long intravenous catheter is placed in either an antecubital vein or, when this is not possible, an external jugular vein. The CT table is then positioned to facilitate the acquisition of either short axis or long axis tomograms (4). The mobile scanning platform allows the patient to be precisely tilted and slewed within the wide scanning gantry. Before contrast medium is given, localization scans are obtained and reviewed. The circulation time is assessed by noting deflection time from an ear oximeter following an intravenous bolus injection of 0.2 mg indocyanine green dye. Contrast medium is administered using a flow controlled injector (Medrad Co., Pittsburgh) and for quantitative CT studies a non-ionic agent is essential and injected at between 3-7 ml/second. Approximately 0.5 ml/Kg body weight of contrast medium is administered per injection. The patient is requested to suspend respiration for a few seconds before and during scanning at a comfortable mid or maximal inspiratory level. This level should be constant for all subsequent series. Most cooperative patients manage this quite well. Scanning is usually performed to obtain coverage of all ventricular levels in both the flow and movie acquisition modes (Table II).

CT DATA ANALYSIS

There are two general approaches for obtaining quantitative information from CT images. The first is by means of density measurements (CT numbers) which enable the borders of anatomical structures to be defined objectively. This can be performed for a single image which provides static measurements or it can be done on many images, usually acquired sequentially to provide a movie display and allow measurement of changes in dimensions of cardiac structures during a cardiac cycle. As the slices are multilevel and contiguous, linear, area and volume measurement can be obtained. The second general approach for extracting quantitative data from CT images is to obtain density measurements over longer CT scan intervals (flow mode acquisition); this is used to study flowing blood and quantitate flow.

CARDIAC BORDER DEFINITION AND LEFT VENTRICULAR MASS

It is essential that proper cardiac border (edge) definition and placement be assured for quantitation of left ventricular anatomy and function. A quantitative method of cardiac edge detection for Ultrafast-CT was developed initially for the determination of left ventricular mass by Skioldebrand et. al. (5) and was applied to Ultrafast-CT by Feiring et. al. (6). This technique, which is a direct application of principles originally used for transmission x-ray devices nearly 30 years ago, defines the absolute placement of the cardiac edge at a CT density (Hounsfield scale) halfway between the density of the myocardium and adjacent structures.

This requires evaluation of the CT numbers of the left and right ventricular cavities, the anterior chest wall, lung and adjacent myocardium.

Placement of the endocardial and epicardial surfaces, as a direct and straightforward application of this "full width/half maximum" method noted above, is aided by an operator defined, computer assisted, semi-automated edge definition program included in the standard Ultrafast-CT data analysis software. The operator determines the placement of the cardiac edge by defining the center level of the CT number range (window) that constitute the density of the endocardial or epicardial surfaces. The software then outlines the contour and determines the absolute value of the area within. Minor vagaries in computer aided placement of the cardiac edge can be corrected by the operator using a tract-ball device. Global measurements of left ventricular mass and volumes are performed at the data analysis console or workstation following definition of left ventricular endocardial and epicardial contours at each level from apex to aortic root.

As an initial application of this method of border definition and validation of the quantitative potential of Ultrafast-CT, a study of left ventricular mass was performed in 22 consecutive dogs (6). Short axis tomograms from left ventricular apex to aortic root were acquired in the volume mode during suspended respiration following a slow infusion of iodinated contrast. Following completion of each study the absolute left ventricular mass was then determined by direct measurement on a standard laboratory scale.

Subsequently, end-diastolic scans were evaluated from cardiac apex to base using the border definition method noted above. Calculation of left ventricular mass was made using the modified Simpson's rule and knowledge of the scanning target geometry by subtracting the endocardial volume from the epicardial volume and multiplying by the myocardial mass. Comparison between left ventricular mass in-situ and that calculated by Ultrafast-CT in-vivo is shown on Figure 1. The correlation is excellent (R = 0.99) and the standard error of the estimate on the order of 4.1 gm. Additionally, inter-and intra-observer variability has been assessed and the technique found to be highly reproducible.

Left ventricular mass has also been measured by Ultrafast CT and 2-D echocardiography in a small series of patients and the results correlated (R=0.89), Fig. 2 (7).

CALCULATION OF LEFT AND RIGHT VENTRICULAR VOLUMES

Reiter et. al. (8) performed a study in dogs assessing the determination of left ventricular stroke volume by Ultrafast-CT compared with that derived simultaneously using either a chronically implanted and calibrated aortic electromagnetic flow probe or careful assessment using established thermodilution techniques. Electrocardiographically triggered scans were acquired in the movie mode nearly simultaneously from the cardiac apex through the base in the short axis during intravenous slow injection of iodinated contrast. Stroke volume was then calculated as the difference between end-diastolic (peak of R wave) and end-systolic (smallest cavity) volumes again using a modification of Simpson's Rule.

Figure 3 shows 25 separate determinations of left ventricular stroke volume. These data confirm the precision of stroke volume determinations by Ultrafast-CT (R = 0.99, slope - 1.01, SEE-1.47 ml) even in the presence of acute myocardial infarction. If left ventricular stroke volume can be quantitated by Ultrafast-CT then, by inference, determinations of end-diastolic and end-systolic volume should be, likewise, quantitative. Studies done using casts of canine left ventricles have previously shown the accuracy of static estimates of left ventricular volume by conventional CT (9).

Simultaneous determination of right ventricular volumes is possible with the determinations of left ventricular mass and volumes. As an addition to the studies of left ventricular volume noted above, Reiter et. al. also determined right ventricular stroke volumes (8) by planimetry of the tomographic end-diastolic and end-systolic scans and application of the modified Simpson's Rule. Application of this method then makes no assumptions as to right (or left) ventricular geometry as is commonly used for contrast angiography (10). Calculated right ventricular stroke volume compared absolutely with the left ventricular stroke volumes (Figure 4). As with the left ventricular volume calculations, inter- and intra-observer variability was minimal. The direct measurement of right ventricular (RV) volume by CT is expected based upon studies of RV casts (11).

In conclusion, Ultrafast-CT, can be used to provide highly quantitative data regarding left and right ventricular volumes provided careful attention is paid to proper border definition. Direct application of these data to a normal population has recently been reported (12,13).

INTERVENTIONAL ULTRAFAST CT

Because scan acquisition is so fast and no ECG gating is required, the use of interventions such as exercise stress testing and pharmacological studies is possible in conjunction with Ultrafast CT scanning.

A number of centers are now using bicycle ergometer exercise for quantitative studies of left ventricular ejection fraction and regional wall motion abnormalities (14). Caputo et. al. compared radionuclide exercise studies with Ultrafast CT in one patient subgroup (15). Exercise is feasible with Ultrafast CT and should complement isotope studies in that CT can measure regional wall thickening dynamics (not possible with radionuclide methods) by virtue of its high spatial and temporal resolution as shown in Figure 5a and b. Ultrafast CT procedures can also be successfully performed in conjunction with cardiac pacing electrode interventions.

More recently attention has focussed on studies of the right ventricle with Ultrafast CT. Echocardiography has difficulties with this chamber while CT displays it well and Ultrafast CT can be recommended for most forms of right ventricular dysfunction (16).

Numerous methods are now available for resutering images from different modalities. The most precise method and one which has proven itself to be of direct relevance in the brain and of direct clinical utility to the physician is the "hat on head" approach based on the use of paired sets of image data to construct numeric models of the anatomic surface of the head, which are then translated, rotated and scaled until an optimal match is found (17). This operation defines a coordinate transformation that is then used to map any pixel of one data set onto the other set.

This approach has been successfully applied to the head and brain and allows retrospective fitting of cross-sectioned CT, MRI and PET images from studies in the same patient, from data acquired at different scanning angles and even different planes (18). This technique is now being positively used for lesion localizations in 3D within the brain and allows the neurosurgeon to prepare the surgical approach from the surface using electronic craniotomy rehearsals (19,20). The organ can be rotated by the computer on the TV monitor and the program even allows stereo views to be generated as shown in Figure 6.

The computing power needed to obtain these brain images is considerably less than that anticipated for 3D display of the heart, which is in continuous motion (21,22). We are developing methods which may make this a practical operation using a slice-by-slice or diastolic-versus-systolic frame analysis. A wire-mesh type of display is illustrated for a patient's left ventricle in Figure 7a-d. This approach provides the basis for detailed quantitation of cardiac cavity and myocardial wall dimensions and dynamics and can be performed at rest to follow patient program and response to treatment, or during a variety of interventional procedures, such as cardiac pacing, administration of pharmacological agents and exercise.

The ability to rotate the 3D reconstruction ultimately provides a movie of not only the left ventricle and its walls, but also all four cardiac chambers. Problems clearly remain before this type of display can be expected but the basis for this analysis is already available. Similarly, measurements of blood flow can be studied using time/density measurements and have been reported by Rumberger (23) and Gould (24). It should be possible to combine displays of motion and myocardial perfusion on a slice-by-slice basis using the method of mapping 3-D data sets described above. This could be done for one modality or multiple modalities; for example, a PET metabolic study could be integrated for display with a perfusion study by either PET or Ultrafast -CT. This combined approach may improve diagnostic accuracy.

CONCLUSIONS

Computed tomography represents an optimal approach to x-ray imaging. This conclusion arises from its capacity to solve a fundamental limitation of all forms of projection x-ray imaging, that is, the superimposition of anatomic structures. Because CT is a fully three-dimensional method, this problem is addressed in a manner not subject to the risks, complications and technical limitations of selective angiography, subtraction angiography, conventional tomography, and many other techniques. Therefore, the optimism for Ultrafast CT imaging of the heart is well founded. Apart from the demonstration of anatomic structures in any plane and in real time in movie format, this new generation of CT scanners offers a unique potential for measuring regional myocardial perfusion. This can be evaluated in two ways: by measuring myocardial wall thickening, which is a sensitive indicator of blood flow, and by assessing time-density changes due to the passage of contrast agent through thin slices of myocardium. Neither of these techniques can be performed adequately at present with noninvasive or even invasive techniques. Feasibility studies, however, have demonstrated that this should indeed be possible using Ultrafast CT scanning.

Obviously, further controlled clinical studies are needed to validate the capability of Ultrafast-CT for general clinical purposes, and comparisons will be necessary with established angiographic, echocardiographic, and nuclear medicine techniques. Further studies are also needed in animals as well as patients to determine the sensitivity and specificity of Cine-CT. Should Ultrafast-CT reach its anticipated potential, electron-beam systems of this type may not only replace present third- and fourth-generation CT scanners for general whole body purposes, but also Ultrafast-CT may become a complementary or even the primary screening and diagnostic imaging technique for the heart.

TABLE I
CINE CT SPECIFICATIONS

	Multi-level mode	Single-slice mode
Number of detectors/level:	432	864
Number of levels per scan:	2	1
Scan time:	0.05 sec	0.1 - 1.8 sec (in 0.1 sec steps)
Image matrix:	256^2, 360^2	256^2, 360^2, 512^2
Slice thickness:	8 mm	3 mm, 6 mm, 10 mm

TABLE II
CINE-CT SCANNER CAPABILITIES

50 Millisecond scan time for 2 sections
224 millisecond scan time for 8 sections
VOLUME MODE: Reconstruction in any plane
CINE MODE: 2 simultaneous levels; 17 frames/sec
FLOW MODE: Time vs. density or area profiles
HIGH RESOLUTION MODE: 100 millisecond scan time

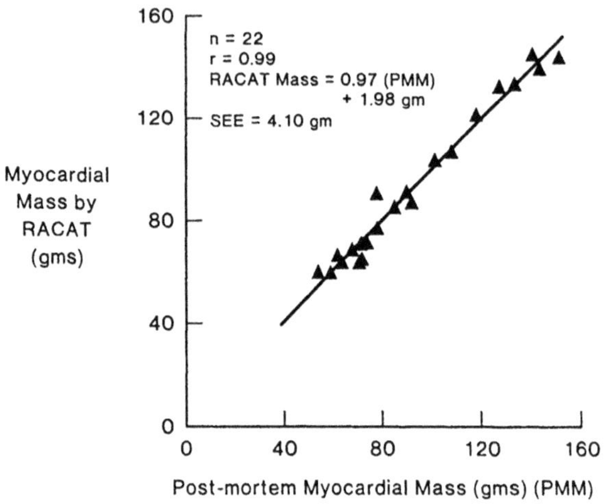

Figure 1 Comparison of absolute postmortem myocardial mass of the canine left ventricle with in vivo determination using ultrafast computer tomography. (Reproduced by permission of the American Heart Association, Reference 6.)

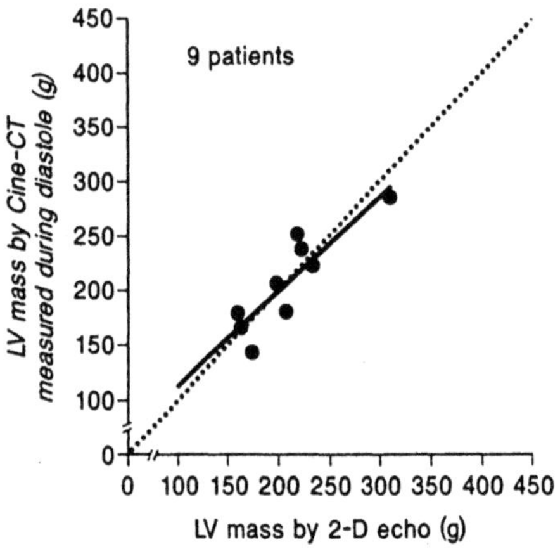

Figure 2 Regression for left ventricular mass measured by Ultrafast CT (systolic images) and calculation by 2-D echo. (Reproduced by permission of Radiology. Reference 7.)

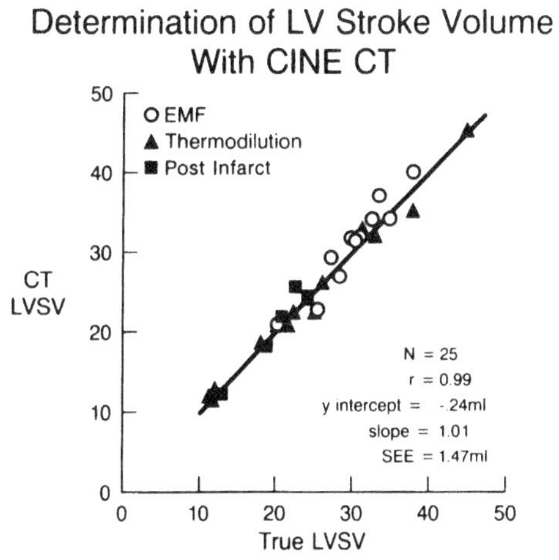

Figure 3 Comparison of true left ventricular stroke volume determined either by electromagnetic flow meter or thermodilution technique in the canine with calculations of left ventricular stroke volume as determined by ultrafast CT. (Reproduced by permission of the American Heart Association, Reference 8)

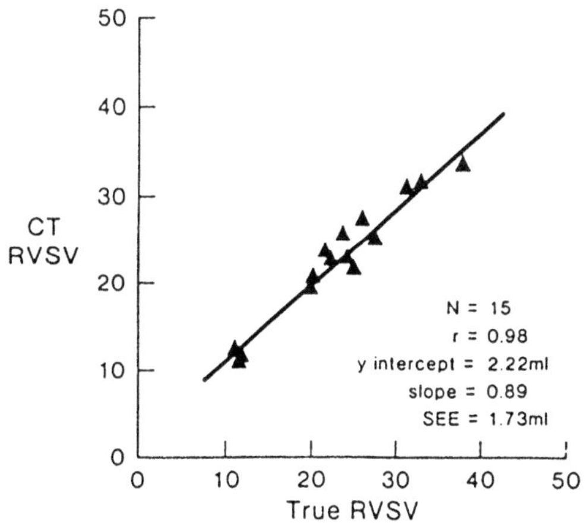

Figure 4 True right ventricular stroke volume compared with CT calculation of right ventricular stroke volume in the dog. (Reproduced by permission of the American Heart Association, Reference 8.)

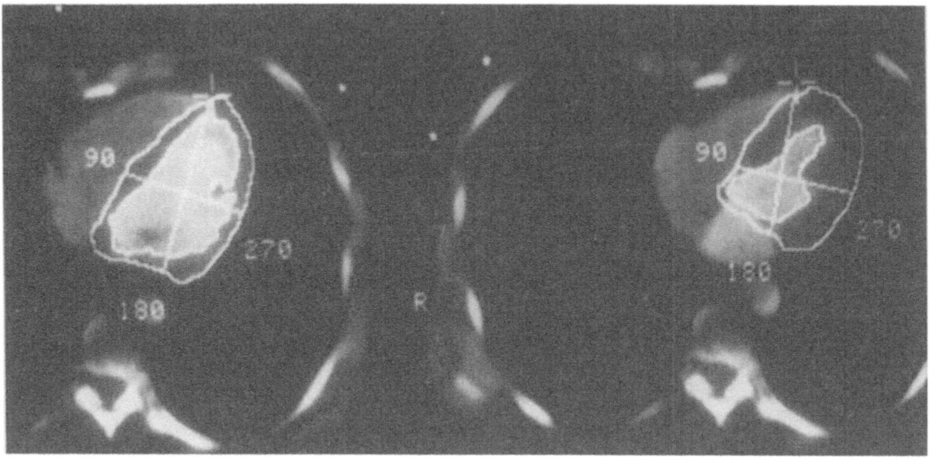

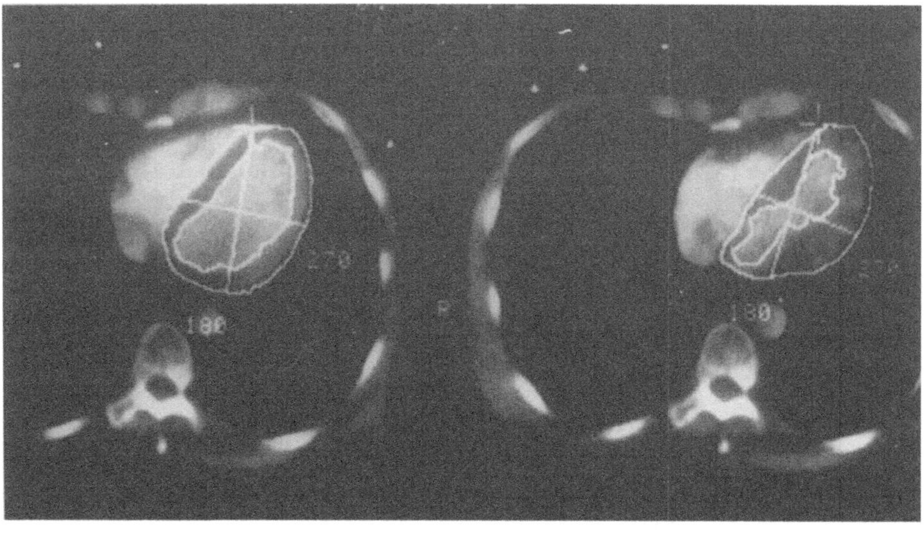

Figure 5a: One level from a multilevel sequence of ultrafast-CT scans depicting normal contraction at rest - the endocardial and epicardial boundaries have been outlined for endiastole and endsystole at rest.

Figure 5b: The same level as shown in 5a above, during exercise showing impaired regional wall contraction in the anterioseptal zones. Note the distortion of the left ventricular cavity compared with the normal appearance as seen at rest in 5a.

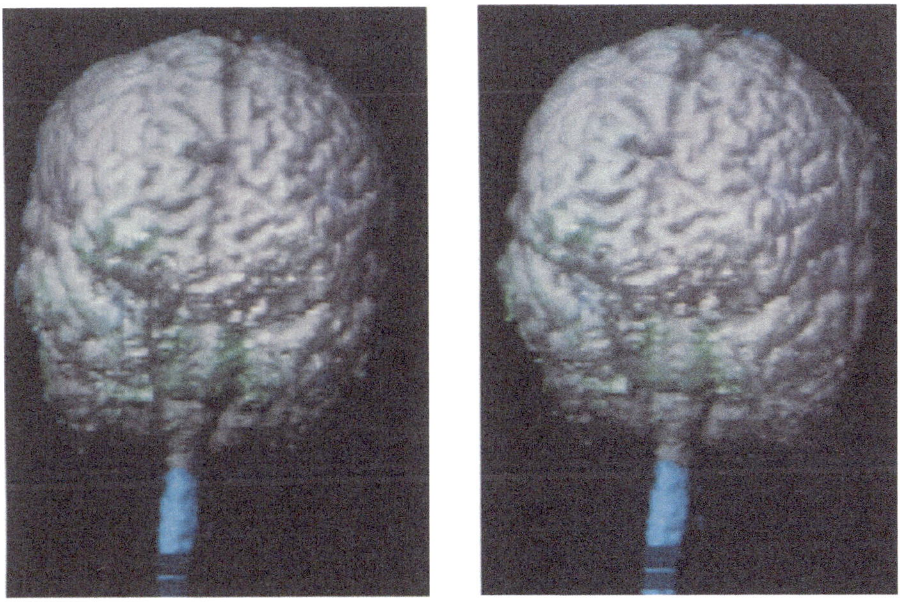

Figure 6: A stereoscopic pair of images of the surface of the brain produced retrospectively from routine MRI and PET studies using a 3-D reconstruction algorithm and TV color coded display monitor.

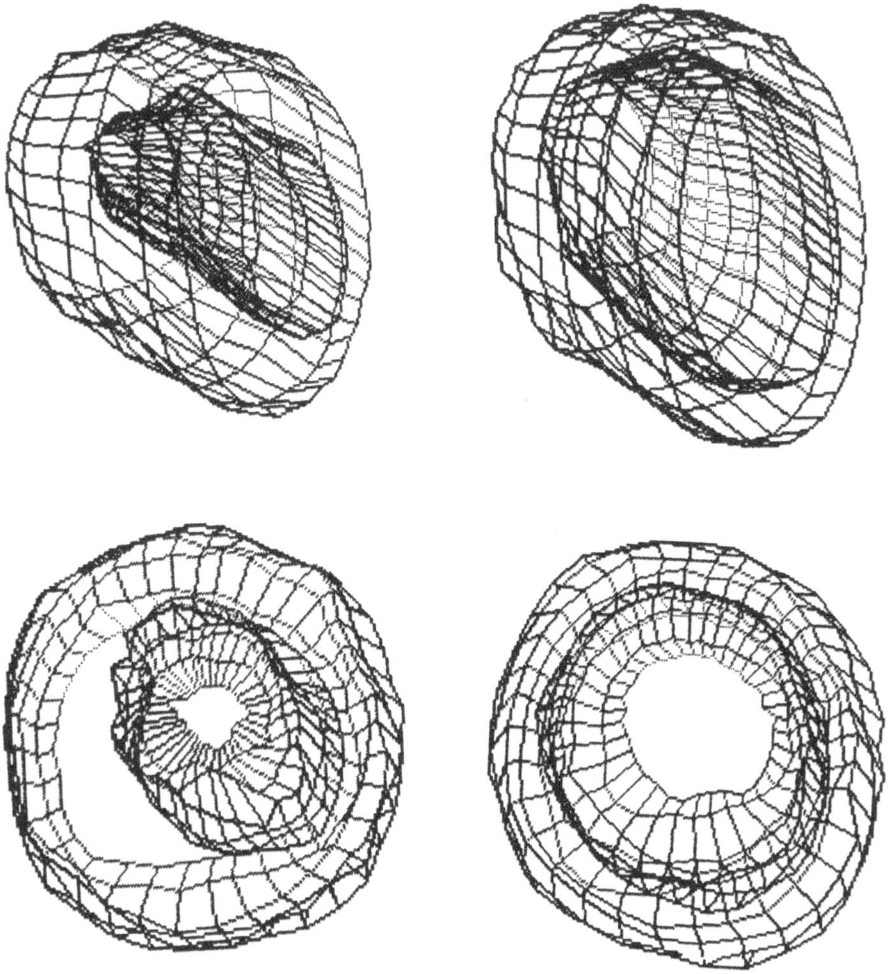

Figure 7a-d: Wire mesh display of left ventricular wall contours created from precise measurements of myocardial boundaries from a routine ultrafast-CT cardiac study. Each contiguous anatomical level is scanned continuously during the cardiac cycle so that all levels imaged during endiastole (7a) and end-systole (7b) can be selected and analyzed. This same data can be rotated on the scanners CRO monitor as shown in figure 7c and 7d.

REFERENCES

1 Ritman EL, Robb RA, Johnson SA, et al: Quantitative imaging of the structure and function of the heart, lungs and circulation. Mayo Clin. Proc., 53:3-11,1978.

2 Boyd DP, Gould RG, Quinn JR, et al: A proposed dynamic cardiac 3D densitometer for early detection and evaluation of heart disease. IEEE Trans. Nuclear Science, 26:2724-2727, 1979.

3 Peschmann KR, Couch JL, Parker DL: New developments in digital x-ray detections. SPIE Proc., 314:5054, 1981.

4 Rees MR, Feiring AJ, Rumberger JA, MacMillan RM, Clark CL: Heart evaluation by cine CT: use of two new oblique views. Radiology, 159: 805, 1986.

5 Skioldebrand CG, Lipton MJ, Mavroudis C, Hayashi TT: Determination of left ventricular mass by computed tomography. Am J Cardiol 49:63-70, 1982.

6 Feiring AJ, Rumberger JA, Skorton DJ, Collins SM, Higgins CB, Lipton MJ, Ell S, Marcus ML: Determination of left ventricular mass in the dog with rapid acquisition cardiac CT scanning. Circulation. 72:1355, 1985.

7 Diethelm L, Simonsen JS, Dery R, Gould R, Schiller NB , Lipton MJ: Measurement of LV mass by ultrafast CT and 2-D echocardiography. Radiology, (in press).

8 Reiter SJ, Rumberger JA, Feiring AJ, Stanford W, Marcus ML: Precision of right and left ventricular stroke volume measurements by rapid acquisition cine computed tomography. Circulation. 74:890-900, 1986.

9 Lipton MJ, Hayashi TT, Boyd D, Carlsson E: Measurement of left ventricular cast volume by computed tomography. Radiology. 127:419-424, 1978.

10 Dodge HT, Hay RE, Sandler H: An angiocardiographic method for directly determining left ventricular stroke volume in man. Circulation. 11:739, 1962.

11 Ringertz HG, Rodgers B, Lipton MJ, Cann C, Carlsson E: Assessment of human right ventricular cast volume by CT and angiocardiography. Invest. Radiol. 20:29-32, 1985.

12 Feiring AJ, Rumberger JA, Reiter SJ, Collins SM, Skorton DJ, Rees M, Marcus ML: Sectional and segmental variability of left ventricular function: Experimental and clinical studies using ultrafast computer tomography. JACC. 12:415-425, 1988.

13 Stark CA, Rumberger JA, Reiter SJ, Marcus ML: Use of cine CT in assessing the severity of aortic regurgitation in patients. Circulation. 74, 4:16, 1986.

14 Lipton MJ, Rumberger JA: Exercise ultrafast computed tomography for the detection of coronary artery disease (Roig et. al): Editorial comment. JACC, in press.

15 Caputo GR, Gould R, Dery R, Diethelm L, Dae M, Botinick E, Lipton MJ: Cine CT evaluation of exercise induced ischemia: A feasibility study. Dyn. Cardiovasc. Imag., in press.

16 Macmillan RM, Rees MR: Measurement of right and left ventricular volumes in humans by cine computed tomography: Comparison to biplane cineangiography. Am. Jour. Card. Imag 2(3): 214-219, 1988.

17 Pelizzari CA, Chen GTY, Spelbring DR, Weichselbaum RR, Chen CT: Accurate three-dimensional resutration of CT, PET and MR images of the brain. JCAT. 13:20-26, 1989.

18 Levin DN, Pelizzar CHA, Chen GTY, Chen C-T, Cooper MD: Retrospective geometric correlation of MR, CT and PET images. Radiol. 169:817-823,1988.

19 Levin DN, Hu X, Tan KK, Galhotra S: Surface of the brain: Three-dimensional MR images created with volume-rendering. Radiol, 1989 (in press).

20 · Levin DN, Herrmann A, Chen GTY, Pelizzari CA, Hu X, Tan KK, et al: Integrated three-dimensional display of MR, CT and PET images of the brain. Radiol 169(P):290, 1988.

21 Hahne KH, Bamans M, Pommert A, Riemer M, Tiede U, Witte G: Dynamic three-dimensional display of the beating heart from four-dimensional MR imaging data. Radiol. 169(P): 472, 1988.

22 Vannier MW, Gutierrez FR, Laschinger JL, Gronemeyer S, Canter CE, Knapp RH: Three-dimensional magnetic resonance imaging of congenital heart disease. Radiographics 8:857-871, 1988.

23 Rumberger JA, Bell MR, Sheedy PF, Stanson AF: In vivo quantitation of intramyocardial blood volume by ultrafast computed tomography. Circulation. 78:II--398, 1988.

24 Gould RG, Lipton MJ, McNamara MT, Sievers RE, Koshold S, Higgins CB: Measurement of regional myocardial blood flow in dogs by ultrafast CT. Invest. Radiol. 23:348-353, 1988.

Transesophageal Echo Computer Tomography ("Echo-CT"): A New Method for Dynamic 3-D Imaging of the Heart

Helmut Wollschläger MD, Andreas M. Zeiher MD, Hans-Peter Klein, Wolfgang Kasper MD, Susanna Wollschläger MD, Anette Geibel MD, Hanjörg Just MD

Medical Clinic, University of Freiburg, F.R.G.

The limited acoustic windows of the chest wall prevent imaging of a sufficient number of parallel sections to permit complete spatial reconstruction of the heart using conventional transthoracic ultrasound. Transesophageal echocardiography provides unrestricted access for ultrasonic imaging of the heart, but is limited by varying transducer positions and thus different imaging planes.

Therefore, we developed a new device for the acquisition of standardized parallel echocardiographic imaging planes covering the entire heart and the great vessels.

This device consists of a conventional transesophageal probe with an additional fully flexible distal portion of multiple semicircular plastic segments (figure 1). After insertion of the probe into the distal esophagus and visualization of a cross sectional view of the left ventricular apex, the lobster-tail-like distal portion is mechanically straightened by means of Bowden cables to form a rigid tube (figure 2). This tube serves as housing for the ultrasonic transducer, which is mounted on a sliding carriage and can be moved freely in axial direction.

The transducer is pulled back from the most distal part of the tube in 0.5 mm increments by means of an ECG gated, computer controlled step motor providing exact parallel and equidistant imaging planes (figure 3). At each plane, one complete cardiac cycle is recorded with ECG triggering delivering images in 30 msec intervals corresponding to the echocardiographic frame rate (figure 4). These images are stored in the correct order in the memory of an image processing computer with 64 MB RAM (figure 5).

This data acquisition process is repeated fully automatically plane by plane covering the whole heart and the origin of the great vessels. By this technique, 128*128*128*8 bit isotropic cubic data sets are generated in 30 msec intervals covering the dynamic cardiac anatomy of one heart cycle (figure 6).

Thereafter, a dedicated software for dynamic 3-D reconstruction allows the computation of any desired tomographic view of the beating heart and the great vessels in real-time (figure 7).

Initial clinical application of transesophageal echo computer tomography demonstrated the applicability of this technique and resulted in high quality computed tomographic echo images of the dynamic cardiac anatomy, providing unique new views of the beating heart.

Figure 1

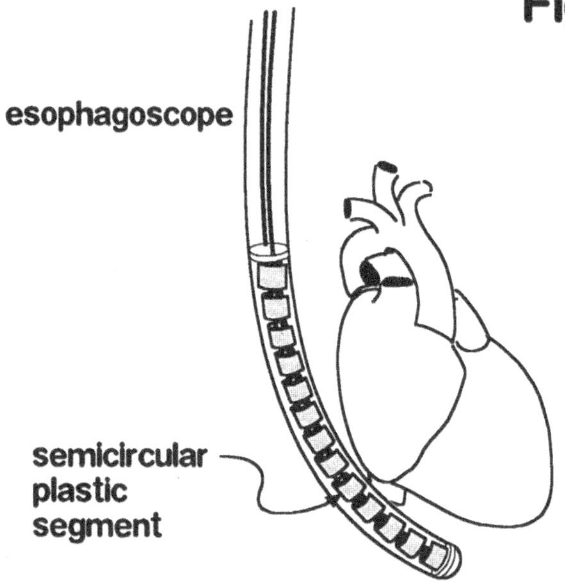

Figure 2

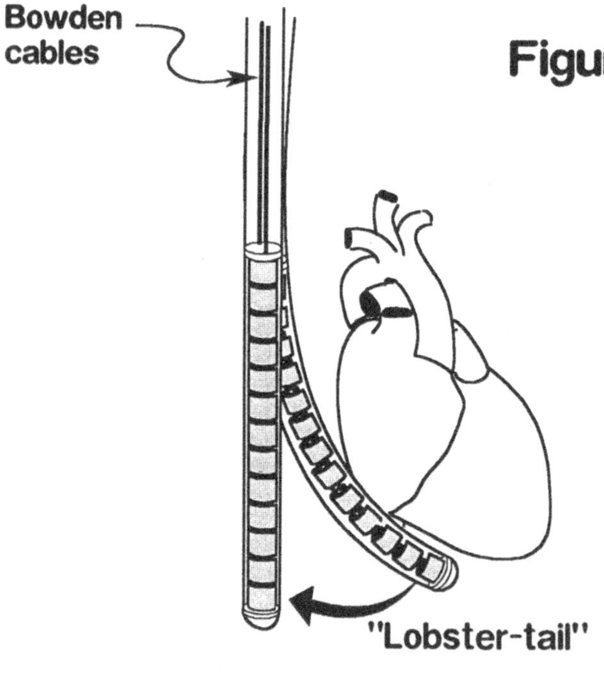

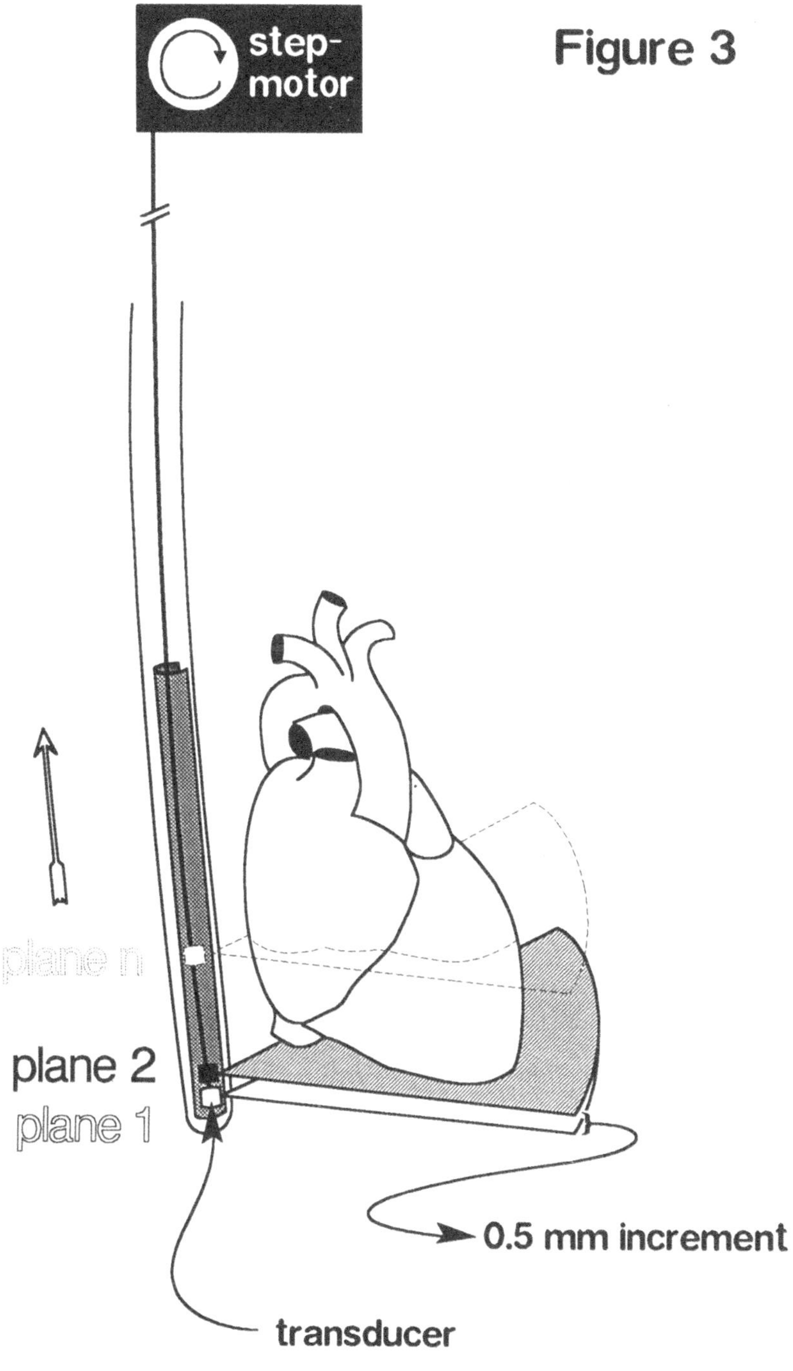

Figure 3
step-motor
plane n
plane 2
plane 1
0.5 mm increment
transducer

Figure 4

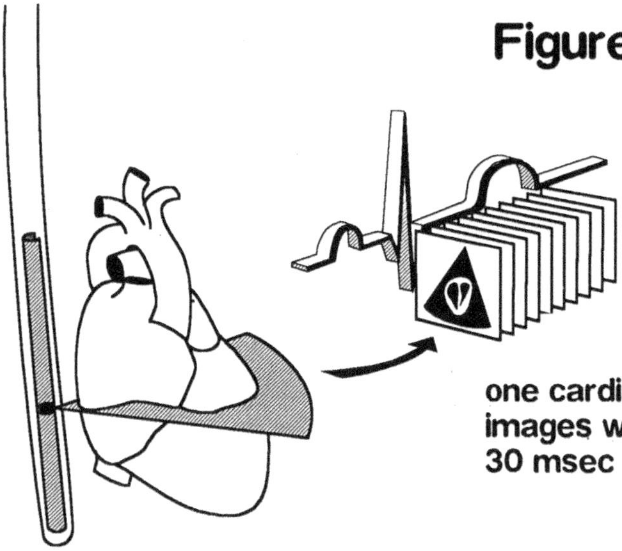

Figure 5

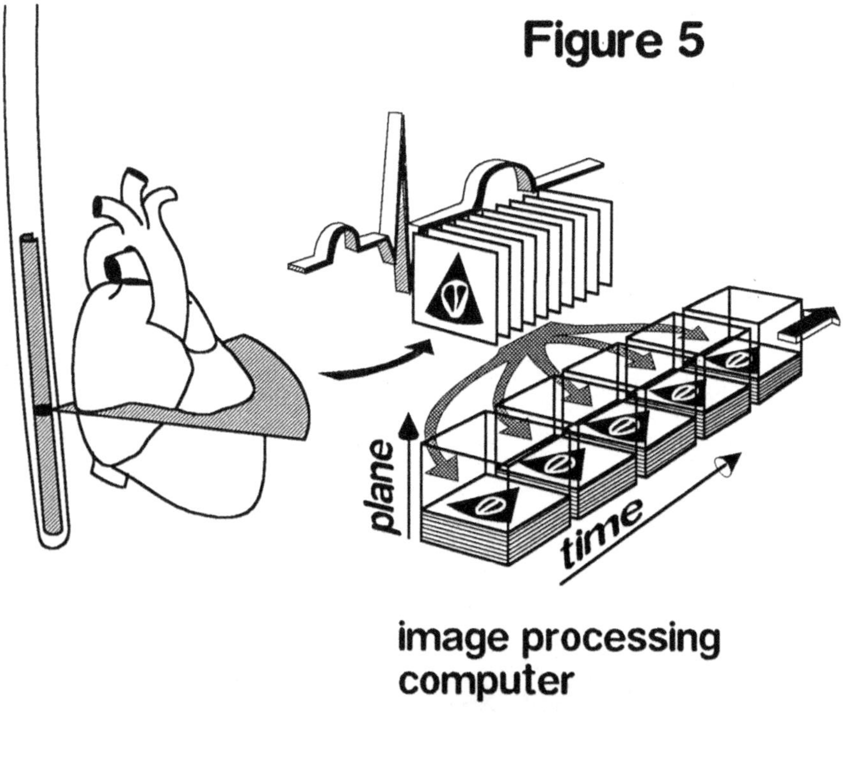

Figure 6

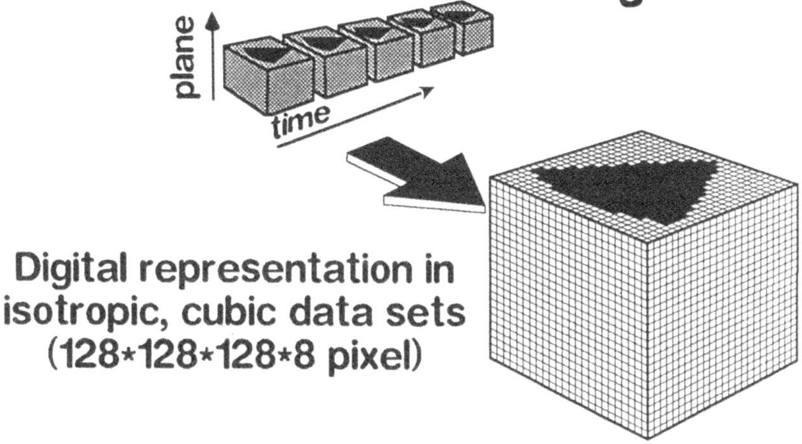

Figure 7

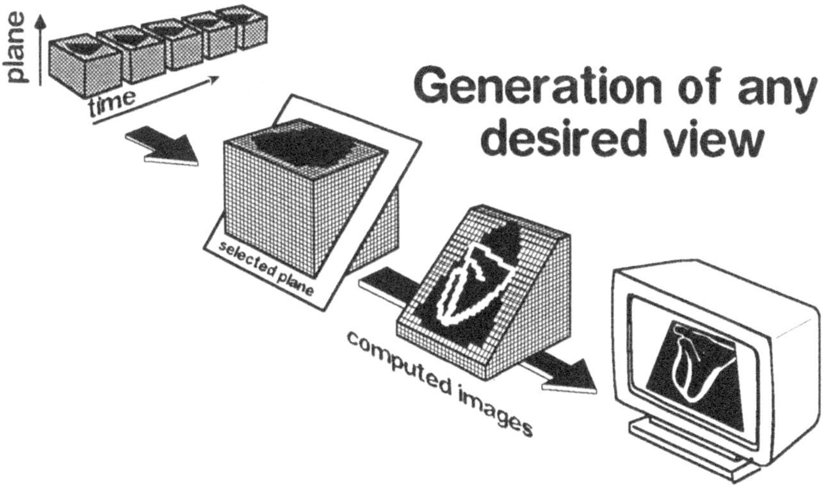

Digital Radiography

Digital Image Intensifier Radiography and Film Screen Systems: Spatial Resolution, Contrast Detail Diagrams, Dose Measurements and Detectability of Pulmonary Nodules

Dölken, W., Th. Krahe, P. Landwehr, K. Lackner

Inst. f. Röntgendiagnostik, Universität Würzburg

Summary

The spatial resolution and the combination of contrast and detail resolution for the DS 1000 sytem are poorer than those for the film screen systems examined. A dose reduction of 76,7% can be achieved using the DS 1000 for chest examinations in respect to the film screen combination used for the conventional systems. No significant differences in detecting pulmonary nodules were evident when comparing the DS 1000 images with the film screen images.

The DS 1000 system (Siemens) is a digital X-ray image intensifier radiographic system used for chest examinations. It provides a 57cm image intensifier (II) with a variable image field diameter (zooming): Z0 = 57cm, Z1 = 47cm, Z2 = 33cm. A 100 mm sheet film camera can be installed at the output screen of the image intensifier. The digitization of the video signal is performed in real time using a 1024x1024 matrix with 256 grey levels. The digital system was compared with the Thoramat and a Potter-Bucky X-ray unit (Siemens) using high definition screens

1. Spatial resolution and contrast detail diagrams

Fig. 1 shows the spatial resolution (LP/mm) of the 100mm photofluorographic images, the DS 1000 monitor and the DS 1000 hardcopies with respect to the image field diameter. The depicted values were obtained from measurements made in the middle of the image field. The spatial resolution was found to be homogen over all the image field. The best results for spatial resolution are those from the 100mm photofluorographic images: 1.5 -2.6 LP/mm. The DS 1000 monitor and hardcopies show lower values: 1.2 - 2.0 LP/mm.

The interaction between spatial and contrast resolution of an imaging system can be evaluated by contrast detail diagrams (CDD). In a CDD, the threshold contrast value for recognizing metal discs in the used phantom (LEEDSTEST T010) is registered vs. the disc diameter. Fig. 2a shows the CDDs of the DS 1000 system for the electron-optical magnification level Z1. A comparison between the DS 1000 monitor and the film screen systems can be found in fig. 2b. The following facilities for digital postprocessing were available during analysis of the digital images: inversion, edge enhancement, gamma correction, windowing.

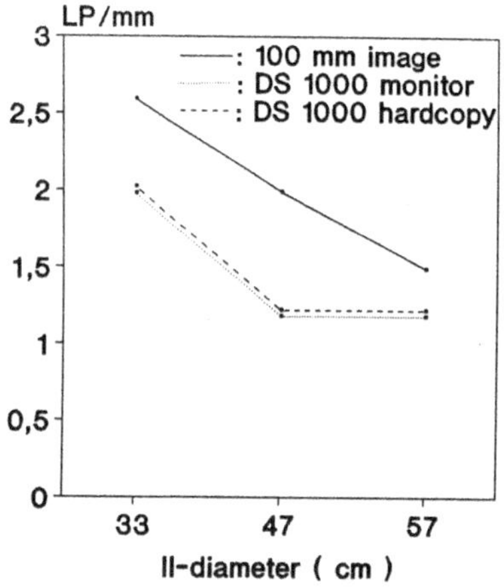

Fig. 1: Spatial resolution for the DS 1000

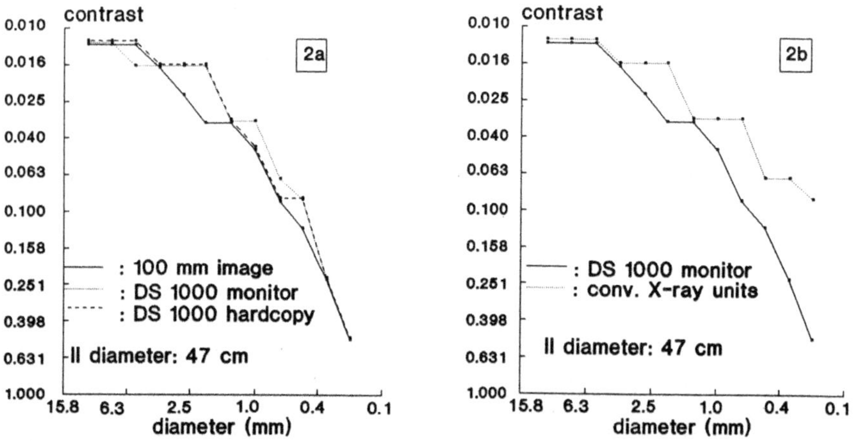

Fig. 2a and 2b: Comparison of contrast detail diagrams

The results for the spatial resolution and the CDDs were obtained by calculating the mean values from three independent observers. There were no marked differences between them. (s.a. [4])

2. Dose measurements

20 patients, ages 17 to 43 years, received a chest radiograph with the Potter-Bucky X-ray unit. 22 patients, ages 19 to 44 years, were examined with the DS 1000 system using Z1. The product of dose vs. area was measured using the Diamentor. This value is independent of the distance between the focus and the measuring unit. The mean dose for a patient was 57.0 +- 26.8 cGy*cm**2 ($\bar{x}$ +- SEM) with the Potter-Bucky X-ray unit. For the DS 1000 we found a mean value of 13.3 +- 9.3 cGy*cm**2. The mean dose reduction was 76.7% for the DS 1000 in respect to the film screen combination used for the conventional systems. The difference was statistically significant ($p < 0.05$) using a modified Wilcoxon test.

3. Detectability of pulmonary nodules

30 patients with known or suspected pulmonary nodules were examined after they were instructed of the examination procedure. Film screen images and DS 1000 images in the posterior-anterior and the left-right projection at 125kV, followed by a tomography of the whole lung (TWL) with a slice thickness of 1cm at 70kV, were taken. The pa chest image was subdivided into separate regions of the lung: supraclavicular region (SCL), retrodiaphragmatic region (RD), paramediastinal region (PM), lateral chest region (LAT) and the rest (SONST). Two independent observers analysed the DS 1000 monitor images and the film screen images. The two observers together viewed the TWL images in order to interprete true positive, false positive and false negative diagnoses. 149 pulmonary nodules were identified using the TWL. 73.9% (110) were also correctly diagnosed by the DS 1000. 21 decisions were false positive, 39 were false negative. 69.1% (104) were correctly identified with the film screen images. There were 25 false positive and 45 false negative decisions. Tab. 1 shows the results for the above defined regions of the lung, fig. 3 those for the whole lung. The comparison between the two observers showed no significant differences for the film screen images.

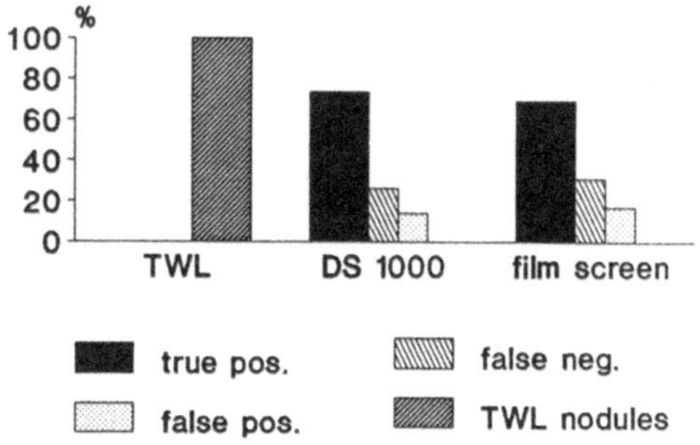

Fig. 3: Results for the total lung

	DS 1000			film screen		
region	N	f.p.	f.n.	N	f.p.	f.n.
SCL	5	1	0	4	3	1
RD	8	0	2	5	1	5
PM	18	2	9	20	7	7
LAT	40	7	9	38	5	11
SONST	39	11	19	37	9	21
sum	110	21	39	104	25	45

Tab. 1: Results for the regions of the lung
(N: true positive, f.p. false positive,
f.n.: false negative)

4. Discussion

The DS 1000 system showed a decrease of the spatial resolution for the digital images and the 100mm photofluorographic images when compared with the film screen images (5 - 7 LP/mm). Looking at the contrast detail diagrams, the poorer spatial resolution of the digital images when compared to the 100mm images can be compensated by the better contrast resolution resulting from the digital postprocessing. Nevertheless the conventional X-ray units show better contrast detail diagram performance. No significant differences between the DS 1000 images and the

film screen images were found when trying to detect pulmonary nodules. This finding includes the results for the whole lung as well as those distinguishing between separate regions of the lung. Other authors found similar results ([1] - [3]).
The digital image intensifier technique should be prefered over the conventional film screen technique when examining pulmonary nodules because of the lower radiation dose for the patient. A comparable diagnostic value for the digital image intensifier system compared with film screen systems may not be expected for interstitial lung diseases, because the spatial resolution found is too small for this problem.

Literature

/1/ Chakraborty, D.P., E.S. Breatneach, M.V. Yester, B. Soto, G.I. Barnes, R.G. Fraser: Digital and Conventional Chest Imaging: A Modified ROC Study of Observer Performance Using Simulated Nodules. Radiology 158 (1986) 35 - 39

/2/ Schuler, M., T. Hübsch, R. Rienmüller, J. Frey, H. Schmidt: DS 1000 - Ein digitales BV-FS-Radiographiesystem in der Thoraxdiagnostik (in german). Digit. Bilddiagn. 8 (1988) 25 - 32

/3/ Sherrier, R.H., C. Chiles, W.E. Wilkinson, G.A. Johnson, C.E. Ravin: Effects of Image Processing on Nodule Detection Rates in Digitized Chest Radiographs. Radiology 166 (1988) 447 - 450

/4/ Dölken, W., T. Krahe, P. Landwehr, K. Lackner: Abbildungseigenschaften der digitalen BV-Radiographie im Vergleich zur konventionellen Röntgentechnik für Thoraxaufnahmen (in german). RÖFO 150 (1989) 86 - 92

Digital Skeletal Radiography with a High Resolution Image Intensifier System

Zwicker, C., Langer, M., Langer, R., Scholz, A.

Department of Radiology, Klinikum Rudolf Virchow, Standort Charlottenburg, Free University of Berlin, 1000 Berlin 19, Spandauer Damm 130

Summary

According to our experiences, Digital Image Intensifier Radiography (DIIR) with small image intensifier input fields of 14 cm and a high resolution matrix in most cases provide suffieient diagnoses of changes in the fine bone structures. Image processing with a free choice of windowing and edge enhancement brings a further improvement of diagnostic validity. Larger image intensifier input fields (28, 40 cm), however, are not able to display alterations of the fine bone structures. Therefore they should only be used in cases of graver alterations, like fractures of the long cortical bones.

Digital imaging systems, like Computed Tomography, Magnetic Resonance Tomography and Digital Substraction Angiography have their undisputed position in radiography. During recent times, different concepts have been deteloped to replace analogeous conventional screen film radiography by digital radiographic systems. Hereby, the Digital Image Intensifier Radiography (DIIR) besides the Digital Luminescence Radiography (DLR) presents a further way to obtain digital images. In this study, the first clinical experiences with DIIR in examining the changes of mainly the arms' and legs' skeletal structures have been evaluated.

All the examinations were performed at the polytron system of Siemens, which allows the alternative application of the Digital Substration Angiography (DSA), the Digital Angiography (DA) and the DIIR. The matrix size was 512^2 and 1024^2 pixel. For the Digital Radiography only the 1024^2 matrix was used. Further, a high resolution television system with a 1024-line monitor and four image intensifier input fields with effective diameters of 40 to 14 cm were at our disposition.
The maximum spatial resolution was shown by the smallest image intensifier input field of 14 cm and 3.4 lp/mm, while the 40 cm image intensifier can only achieve a spatial resolution of 1,4 lp/mm. The minimum tube voltage

was fixed at 40 kV, while the mAs product was regulated automatically and could not be influenced manually. All images were obtained with a small focus (0,3 mm). Hereby, the image intensifier was placed as near to the object as possible. The image processing was achieved by free choice of window position and window size, analogeous to CT, so that the different image properties like soft-tissue or trabecular structure could be emphasized. The weighting of image contrasts was made possible by a continuous edge enhancement achieved by analogeous filtering of local frequency after the mode of an unsharp mask. Only in special cases, amplifying by zooming was used without an improvement of spatial resolution. The documentation was mainly done in the positive mode: dark image on a light background. The negative imaging was only used in single cases. The documentation was obtained by a multi-format camera. The evaluation of all these images was processed with the small intensifier input field of 14 cm, because larger image intensifiers do not provide a sufficient imaging of the fine bone structures. Up to now, 52 patients with mainly either traumatic or degenerative changes of the skeletal systems of the arms and legs have been examined with DIIR. In this study, the results of 34 patients have been evaluated by comparing the image quality and the diagnostic validity of the digital images with the conventional X-ray-images, which were available in all cases.

Four categories of quality have been chosen and three radiologists visually evaluated the image results separately. It became evident that concerning the capability of displaying the fine bone structures as well as the diagnostic validity, DIIR is inferior to the conventionally produced X-ray-images. In this case, we must take into consideration that all these images were processed with the first software generation of polytron and that the image quality has significantly improved after multiple software upgradings. Nevertheless, we can say that in 76% of all cases DIIR provided diagnostically sufficient statements, and in 85% the fine bone structure was displayed in an at least adequately exact quality. (Table 1, 2), (Fig. 1, 2)

Tab. 1 Display of the Fine Bone Structure

	0	+	++	+++
DIIR	15	44	38	3
conv. roe.	0	12	41	47

Tab. 2 Diagnostic Evidence

	0	+	++	+++
DIIR	24	41	35	0
conv. roe.	0	12	53	35

all data in % (n = 34)

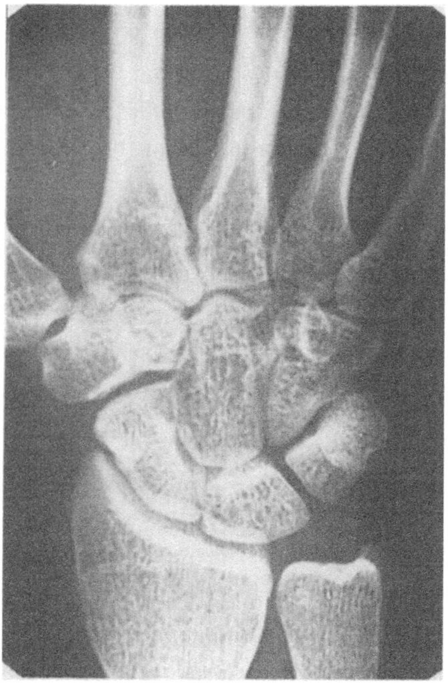

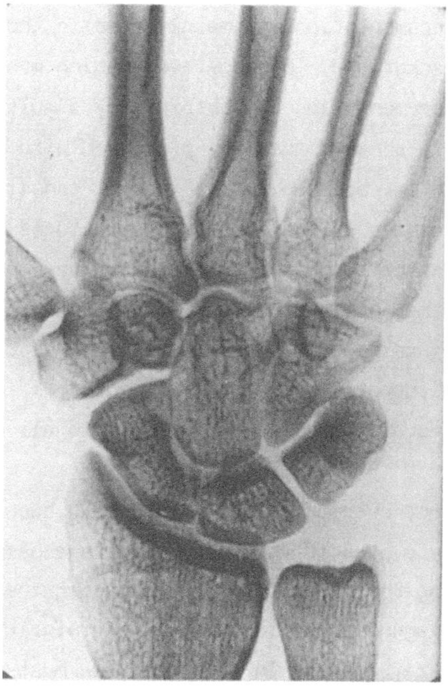

Fig. 1: Normal Carpophalangeal Skeleton (Positive Imaging)

Fig. 2: Normal Carpophalangeal Skeleton (Negative Imaging)

The advantage of Digital Image Intensifier Radiography is its quick image processing under fluoroscopy and thereby the opportunity of adjusting difficult findings optimally.(1,5) The immediate presentation of the image at the monitor improves the examination conditions especially ot traumatized patients. The principal advantages of digital imaging systems are the benefits of image-postprocessing provided by the stressing of different image details demanding not more than a single exposition to Y-rays. (2, 4) Therefore, possible indications for DIIR can be the post-operative control and the checking of conservatively treated fractures in a cast. The disadvantages of digital imaging systems are the generally inferior capacity of resolution of larger image intensifier input fields, like those of the 28- and 40-cm-image intensifier, as well as the fact that still in weakly absorbing skeletal regions like the phalanges only a very low signal-to-noise distance can be achieved. After our experiences, herefore lower voltages than 40 kV must be made possible, while the 10 bit A/D converter of the system itself, according to 1024 grey scales, do not present any limitations to the system's potentialities. (3) The relatively high loss of information, however, produced by image achieving with the multi-format camera is remarkable. An improvement can be brought about by the application of a digital interface in combination with a laser printer within a finite period of time.

At present, diagnostically reliable statements on the fine bone structures and thereafter the assessment of discrete changes are only feasible with small image intensifier input fields (IIIF) like the 14 cm IIIF. Because of their generally worse spatial resolution large IIIFs can only be applied in single and traumatic delicate cases, whereby discrete changes like navicular fractures should only be displayed with small IIIFs.

In the near future, the installation of effective emergency diagnostic units for example will to a sufficient degree be possible by the combination of imaging systems like DIIR, DSA and DA.

Literature:

1. Fujita, H., K. Doi, H. MacMahon, Y. Kume, M.L. Giger, K.R. Hoffman, T. Katafuchi, K. Ohara, H.P. Chan: Basic imaging properties of large image intensifier - TV digital chest radiographic system. Invest. Radiol. 22, 1987, 328-335

2. Hintze, A., G. Jöten: Digitale Radiographie. Fortschr. Röntgenstr. 145, 1986, 91-97

3. Seeley, G., H.D. Fisehr, M.O. Stemski, M. Borgstrom, J. Bjelland, M.P. Capp: Total digital radiology department: Spatial resolution requirements. Amer. J. Radiol. 148, 1987, 421-426

4. Tesic, M.M., R. A. Mattson, G.T. Barnes, R.A. Sones, J.B. Stickney: Digital Radiography of the chest: Design features and considerations for a prototype unit. Radiology 148, 1983, 259-264

5. Zwicker, C., M. Langer, R. Langer, C. Wasmuht, R. Felix: Digitale Bildverstärkerradiographie in der Skelettdiagnostik. Erste klinische Ergebnisse. Digit. Bilddiagn. 8, 1988, 147-151

Detectability Evaluation of a Photostimulable Phosphor Digital Radiographic System

G. S. SHABER, J. M. BOONE, R. SHLANSKY-GOLDBERG

Thomas Jefferson University Hospital, Department of Radiology
Philadelphia, PA 19107

SUMMARY

A digital radiographic system (Matrix Laser Rad) employing a filmless photostimulable phosphor (PSP) was evaluated for spatial resolution and high and low contrast object detectability. Receiver operating characteristic (ROC) curve comparisons were made to film images using equal or lower exposure factors. The low and high contrast test objects for the ROC evaluation were respectively a 5/32 inch diameter hemispheric depression filled with mineral oil and 250-500 micrometer diameter silicon carbide granules, both on a 2 inch thick plexiglas block. These object sizes were shown to be at threshold of detectability for both the film and PSP system at the exposure factors used. The ROC curves were generated from data obtained by radiologists viewing multiple film radiographs and CRT displays of both the digitized film radiographs and PSP images with the test objects in random locations. There were 90 images generated for each system, 45 with test objects and 45 without. Each observer had 15 seconds at a fixed distance to view each image presented at random on either a light view box or CRT display.

The unenhanced filmless PSP radiographs had a reduced detectability threshold for the test objects, however with the use of windowing, level manipulations and appropriate filtering, detectability could be improved for the low contrast objects with the digital system surpassing unenhanced film radiographs. In addition, a dose reduction could be achieved with the PSP filmless system without the loss of significant detectability. Enhanced film radiographs demonstrated the highest detectability characteristics. However, for high contrast objects the PSP system required a significant increase in object size to achieve similar detectability as compared to the film images.

INTRODUCTION

Several recent investigations have focused on the application of digital radiographic techniques to mammography. Digital radiography with its greater dynamic range and separation of detector from display may improve sensitivity and hopefully provide for a concommitant dose reduction in this type of examination. In addition, the application of computer aided diagnosis may further relieve some of the professional manpower requirements to perform mammography. Our recent efforts have been directed towards evaluating the potential application of digital photostimulable phosphor technology to mammography by evaluation of the spatial resolution characteristics and detection capabilities for low and high frequency objects using receiver operating characteristic (ROC) curve analysis (1,2).

The system used in this investigation employs a 2048 x 2048 laser digitizer from both film and reusable photostimulable phosphor (PSP) screens (Matrix Laser Rad). Spatial resolution was determined utilizing a high contrast lead plated line pair phantom and detectability with ROC curve analysis for both low and high frequency objects.

PROCEDURE

The experimental plan employed in this investigation is summarized in table 1.

EXPERIMENTAL PLAN

TEST	PARAMETER		SYSTEM		
			FILM IMAGE	DIGITIZED FILM IMAGE	DIGITIZED "PSP" IMAGE
RESOLUTION	HIGH CONTRAST PHANTOM		★	★	★
DETECTABILITY (ROC ANALYSIS)	LOW FREQUENCY PHANTOM	LOW DOSE 40 mR	★	★	★
		HIGH DOSE 80 mR	★	★	★
	HIGH FREQUENCY PHANTOM	SMALL SIZE 250 MICRONS	★	★	★
		LARGE SIZE 500 MICRONS	--	--	★

Table 1

Spatial resolution test parameters

A high contrast lead plated line pair phantom on 5 cm of plexiglas was used for the test object for spatial resolution evaluation of both conventional film screen systems and PSP screens. A conventional molybdenum targeted, molybdenum filtered x-ray tube with a 0.3 mm focal spot operating at 400 MAS 27 kVp and with a focal image distance of 100 cm was used as the x-ray exposure factors as summarized in Table 2.

Test object	High contrast lead plated line pair phantom on 5 cm of lucite	
X-ray exposure factors	Target	Molybdenum
	Filter	Molybdenum
	Focal Spot	0.3 mm
	FID	100 cm
	MAS	400

Table 2: Spatial resolution test parameters

ROC detectability test parameters.

Test object (signal). So as to arrive at meaningful ROC curves, test objects for both low and high frequency evaluation were selected at the threshold of detectability. The low frequency test object that was evaluated in both systems, conventional film screen and PSP screens, was a 5/32nd inch diameter hemispheric depression filled with mineral oil and in a 5 cm plexiglas block. Initially, the high frequency test object was a 1/4 inch diameter cluster of 6 250-micrometer silicon carbide granules on the 5 cm plexiglas block. However, initial ROC analysis using PSP screens and this size test object revealed that the size of the test object was too small to obtain a meaningful ROC curve. Therefore, the test object size was increased to a 1/4 inch diameter cluster of 6 500-micrometer silicon carbide granules on the 5 cm lucite block and the ROC study was repeated using this size test object.

Test images. 90 3-inch-by-3-inch test images, 45 with the signal at 9 different locations mixed at random with 45 images without the signal were used for this study.

Viewing conditions

Film images. The observer image distance was held constant at 20" with a rigid head holder for both the low frequency and high frequency test objects. Standard viewbox intensities of approximately 140' lamberts were used and the viewing time per image was 15 seconds.

CRT images. The CRT images were displayed on a standard 9" 525 line 2 to 1 interlaced monitor. The distance from the observer to the monitor was adjusted to produce equal apparent image size to the observer as compared to the film images. Naturally, the overall luminosity of the CRT monitor images was less than the film viewbox images. Once again, a rigid head holder was used and viewing time was 15 seconds per image.

Detectability x-ray exposure factors. A molybdenum targeted and filtered x-ray tube was used with a focal spot of 0.3 mm at 27 kVp with a film image distance (FID) of 100 cm. The MAS was chosen to produce threshold detectability of the test object and for the low frequency object two exposure values were used, 40 mR and 80 mR. For the high frequency test obtect, 80 mR.

RESULTS

Using a square response function Figure 1 details spatial resolution characteristics of the three different systems investigated, digitized film, digitized film/screen and PSP screen images. A significant reduction in PSP spatial resolution is confirmed by these results. This reduction in spatial resolution for PSP screens is not specific to the Matrix Laser Rad system. Resolution capability of another PSP screen manufacturer (Fuji) was also investigated and differed only slightly from the results obtained using the Matrix phosphors. There was no more than 0.2 line pair/mm difference in resolution capability for all systems tested.

SPATIAL RESOLUTION RESULTS

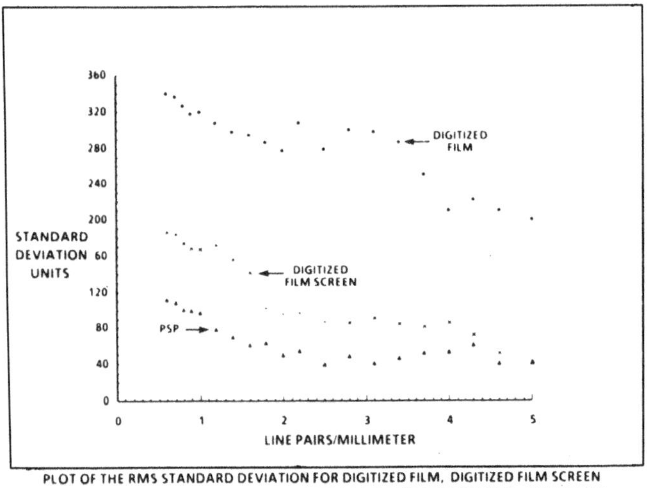

PLOT OF THE RMS STANDARD DEVIATION FOR DIGITIZED FILM, DIGITIZED FILM SCREEN AND PSP IMAGES OF THE LINE PAIR PHANTOM AS A FUNCTION OF SPATIAL FREQUENCY.

Figure 1

Figures 2, 3, 4 and 5 are graphs of four different observer ROC performances, two for the high frequency test object and two for the low frequency test object. The high frequency ROC results reveal that a two-fold increase in test object size was necessary to achieve similar PSP detectability performance as compared to conventional film/screen images. Table 3 summarizes the results of both the spatial resolution and detectability performance of the three different systems investigated.

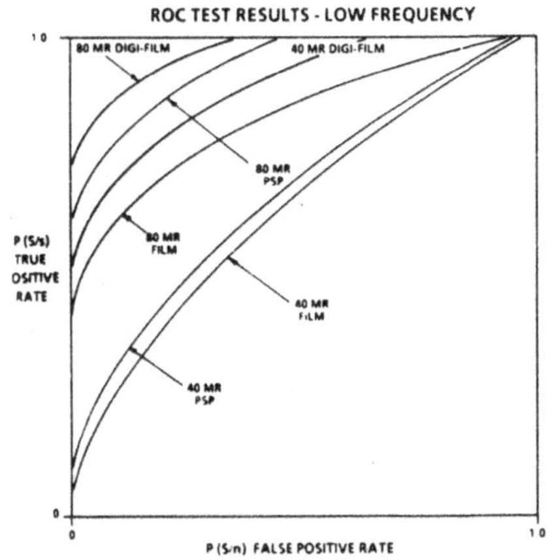

Figure 2: Observer P-1

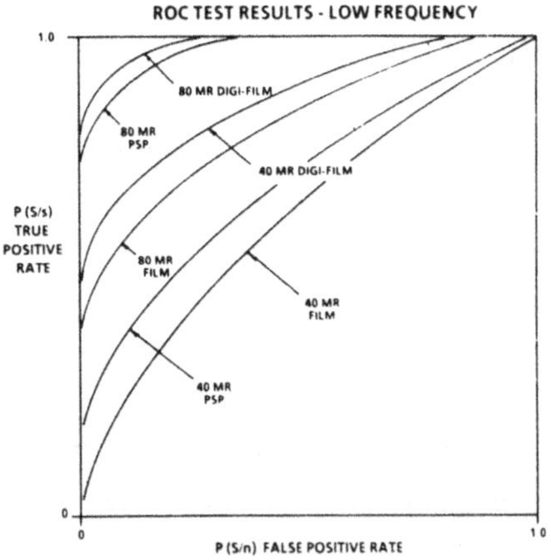

Figure 3: Observer R-1

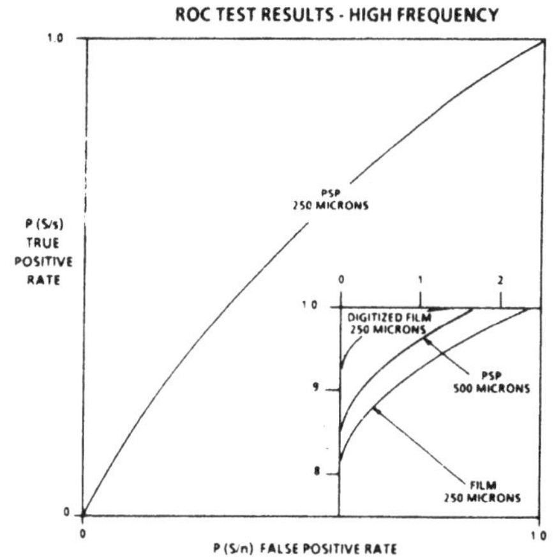

Figure 4: Observer B-1

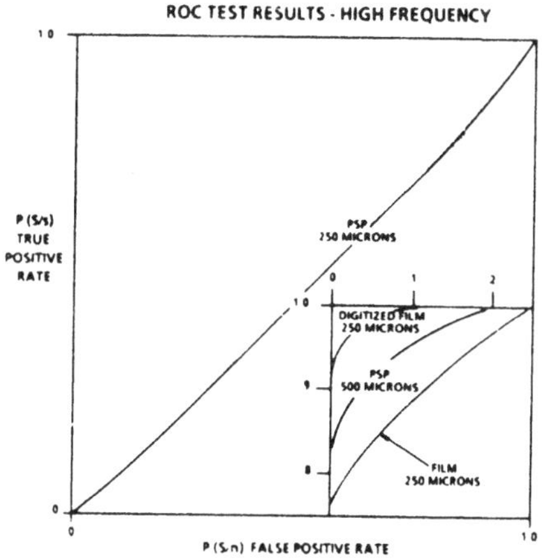

Figure 5: Observer J-1

RESULTS SUMMARY

TEST		RESOLUTION	ROC DETECTABILITY*		
SYSTEM	AIR DOSE	LINE PRS/MM	LOW FREQUENCY	HIGH FREQUENCY 250 MICRONS	HIGH FREQUENCY 500 MICRONS
FILM	80 MR	>5	3	3	--
	40 MR	4	1		
DIGITIZED FILM	80 MR	5	5	5	--
	40 MR	4+	4		
"PSP"	80 MR	2+	4	0	4
	40 MR	1+	2		

*ROC DETECTION SCALE 0-5

Table 3

DISCUSSION

Spatial resolution

There is a modest reduction in the high contrast spatial resolution between the film images and the digitized film images. PSP screen images have a significant reduction in spatial resolution capabilities as compared to film. This decrease in spatial resolution was not specific to the system employed in this investigation but appeared to be universal for all PSP systems. The theoretical limiting spatial resolution factor in our system was the laser dot size of 100 microns. However, it appears that the current state of PSP screen technology is incapable of producing the minimal mammographic spatial resolution requirement of 10 lines/mm.

Detectability

Every effort was made to control for inherent differences in the viewing conditions for CRT and film images. We chose to control for equal image size at the expense of luminosity variation between the light view box and CRT viewing. To maintain the same apparent size of images presented to the eye required differential viewing distances for film system vs. the CRT displayed images. This distance change produced further reduction in the relative luminosity between CRT display and the view box images. We recognize that the decreased luminosity of the CRT display will adversely affect contrast and thus the detectability of low contrast objects, however, it was felt essential that the relative size in the high frequency object detectability study be equal, at the expense of differential luminosity.

With this differential luminosity level as an uncontrolled factor, our results demonstrate that digitized film had the highest low contrast object detectability as the result of window and level manipulation. In addition, the digitized film offered substantial dose reduction over conventional film techniques (50%) with the same detectability characteristics for low contrast objects. In spite of the poor spatial resolution for PSP images, PSP detectability for low contrast objects is better than conventional film techniques. This differential detectability would increase if the luminosity of the CRT display was equal to the viewbox intensities. Therefore, the relative differential detectabilities discovered through the ROC curve analysis would be greater with equal luminosity.

Digitized film images as compared to conventional film images have essentially equal high contrast detectability as determined by receiver operating curve analysis. High contrast object detectability was totally inadequate for the PSP images. Test objects twice the size were necessary to achieve equal detectability for PSP images as compared to film and digitized film images. One observer had a reverse correlation for the 250 micron test object when using PSP techniques indicating that noise was being interpreted as signal more often than not, whereas the signal was totally undetectable.

The results of our investigations reveal that the current state of photostimulable phosphor technology is inadequate for application to mammography due to poor spatial resolution and detection capabilities. The problem most likely lies with the phosphor plate technology with its low signal to noise ratio rather than laser scanning limitations.

REFERENCES

1. Goodenough DJ, Rossmann K, Lusted LB: "Radiographic Applications of Receiver Operating Characteristics (ROC) Curves," Radiology (110), 89-95 (1974).

2. Shaber GS: "Low Dose Breast Electron Radiography (Ionography)", M.C.I., Final Report (1981).

Barium Enema Carried out by Computed Radiography and Conventional Screen-film Systems

KRUG, B.*, STEINBRICH, W.*, LORENZ, R.*, DIETLEIN, M.*, GLATHE, ST.*, FRIEDMANN, G.*, SCHIRRMEISTER, J.**

Radiologisches Institut der Universität zu Köln (Direktor: Prof. Dr. G. Friedmann)* und Institut für Medizinische Dokumentation und Statistik (Direktor: Prof. Dr. P. Bauer)**

Radiologisches Institut der Universität zu Köln
Josef-Stelzmann-Straße 9
D-5000 Köln 41

Summary

In 120 patients double-contrast barium enemas were obtained by both digital luminescent radiography (DLR) and conventional screen-film systems (CSS). The exposure dose used in DLR was 50% of the conventional one. In DLR two differently post processed images were obtained from one X-ray exposure: a display with low spatial frequency enhancement was processed to look like a CSS-radiograph and was opposed to a complementary display with high spatial frequency enhancement. Comparing both methods statistically DLR proved to be diagnostically equivalent to conventional radiography despite the reduction in exposure dose and a slightly diminished image qualitiy. High spatial frequency enhancement did not provide further diagnostical information and should be dispensed with in barium enemas.

According to former investigations it is possible to perform radiological examinations of acceptable image quality by digital lumineszent radiography (DLR) with a diminished exposure dose compared to CSS (1-4,6,7). Because of anatomy barium enemas result in comparatively high gonadial doses, the consistent inset of DLR thus being of radiationhygienical relevance (5). As on the one hand imaging the gastrointestinal tract requires a high spatial resolution and as on the other hand the actual digital spatial resolution of 2,5 lp/mm is distinctly lower than the conventional one, it is questionable, whether the reduction in exposure dose and the lower spatial resolution result in a loss of diagnostical information and whether digital post processing can add to diagnosis.

Methods

In 120 elder patients double-contrast barium enemas (Sireskop 3, Siemens, 125 KV, Iontomat) were obtained by both CR (PCR-SP, Phillips) and CSS (Quanta fast Detail screen-film systems, Dupont) the digital exposure dose being 50% of the conventional one. 65 radiographs in supine position and 55 spot films were compared, the digital exposure following the conventional imaging immediately. In CR two differently post processed images were obtained from one X-ray exposure: a display with low spatial frequency enhancement was filtered to look like a CSS-radiograph and opposed to a complementary display with high spatial frequency enhancement.

The spectrum of differential diagnosis included 58 normal findings, 10 cases of colitis, 16 cases of diverticulosis/itis, 10 polyps/adenoms, 16 colorectal carcinomas, 10 malignant bowel wall infiltrations or extramural processes, respectively.

Definite diagnosis was based on operation in 21 patients, on endoscopy in 31 patients and on clinical data or the clinical course, respectively, in 68 patients.

Conventional barium enemas and digital displays with low and high spatial frequency enhancement were evaluated seperately by 4 independant observers (M. D., St. G., R. L., W. St.) using a questionnaire that includs signs of image quality (f. ex. visibility of mucosal pattern) as well as 6 differential diagnoses. There was a rating score of 3 levels of confidence for the first group of questions and of 5 levels of confidence for the remaining 6 questions.

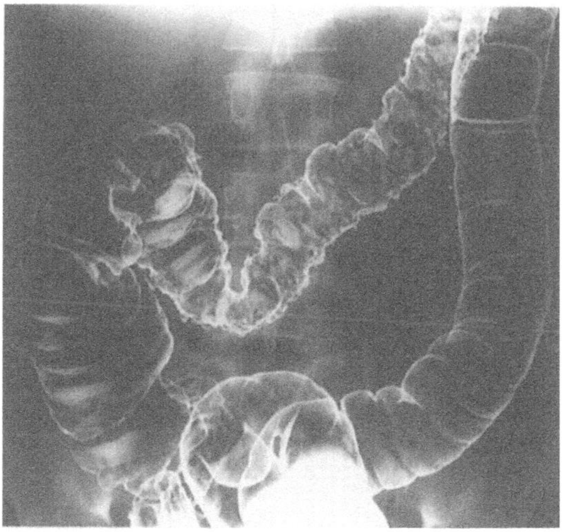

Fig. 1 Digital display with low spatial frequency enhancement processed to look like a conventional radiograph. Barium enema study of a patient with Crohn`s disease.

Results

Analysing the results statistically, in a first step the ranges of all 4 interpreters evaluating one question of the questionaire separated into conventional radiographs and digital displays with low and high spatial frequency enhancement were determined. Then the ranges of conventional versus low spatial frequency enhancement and conventional versus high spatial frequency enhancement were compared by using contingency tables.

Characteristics of image quality like visibility of the mucosal pattern were interpreted with slightly less confidence in the digital displays with low spatial frequency enhancement than by CSS (Tab. 1a, Fig. 1). In contrast there was no difference in diagnostical confidence in establishing the differential diagnoses, demonstrated here by using the example of colorectal carcinoma and malignant bowel wall infiltration (Tab. 2a and 3a). Comparing high spatial frequency enhancement versus CSS and digital images with low spatial frequency enhancement there was always found a lower level of diagnostical confidence with digital displays showing high spatial frequency enhancement (Tab. 1-3b).

LE / CSS	0	1	2	
0	22	46	5	73
1	11	18	8	37
2	1	9	0	10
	34	73	13	120

HE / CSS	0	1	2	
0	10	58	5	73
1	1	29	7	37
2	2	7	1	10
	13	94	13	120

Tab. 1: Range of diagnostical confidence in evaluating mucosal patterns.
Tab. 1a: Low spatial frequency enhancement (LE) versus CSS
Tab. 1b: High spatial frequency enhancement (HE) versus CSS

LE / CSS	0	1	2	3	4	
0	75	10	3	1	2	91
1	5	4	1	1	0	11
2	3	3	1	0	1	8
3	0	0	0	0	0	0
4	1	1	1	0	7	10
	84	18	6	2	10	120

HE / CSS	0	1	2	3	4	
0	68	17	2	0	4	91
1	4	6	1	0	0	11
2	1	4	2	1	0	8
3	0	0	0	0	0	0
4	1	2	1	1	5	10
	74	29	6	2	9	120

Tab. 2: Range of diagnostical confidence in diagnosing colorectal carcinoma.
Tab. 2a: Low spatial frequency enhancement (LE) versus CSS
Tab. 2b: High spatial frequency enhancement (HL) versus CSS

LE / CSS	0	1	2	3	4	
0	91	4	1	1	0	97
1	4	1	1	0	0	6
2	4	1	1	1	1	8
3	2	0	1	1	0	4
4	0	0	2	2	1	5
	101	6	6	5	2	120

HE / CSS	0	1	2	3	4	
0	92	3	1	1	0	97
1	4	1	1	0	0	6
2	2	0	2	3	1	8
3	2	0	1	1	0	4
4	1	0	3	0	1	5
	101	4	8	5	2	120

Tab. 3: Range of diagnostical confidence in diagnosing malignant bowel wall infiltration.
Tab. 3a: Low spatial frequency enhancement (LE) versus CSS
Tab. 3b: High spatial frequency enhancement (HE) versus CSS

Discussion

The main advantages of DLR compared to convenional radiography are the reduction in exposure dose, the possibility of image post processing and the acces to digital archives and hospital communication systems, a digital image quality similar to the conventional one thus already signifying technical progress (1,4,6,7).

Image quality occuring with digital systems is dependant on several factors, the main points being exposure, the signal-to-noise ratio and spatial resolution (2,3,6,7).

The wide digital range of sensitivity allows a marked decrease in exposure dose compared to CSS. However, when decreasing exposure dose, image noise becomes more pronounced, the critical increase in granularity starting at approximately one-eigth of the conventional dose, as Unayma demonstrated casuistically on double-contrast examinations of the stomach (7). The reduction in exposure dose, therefore, must always be a compromise between radiationhygienical interests and image quality (1,7). Some authors suggested the use of varying exposure doses adapted to the diagnostical goal, which seems to be impractical in clinical routine (7). In our opinion a decrease in exposure of 50% compared to CSS is justified in view of the congruous levels of diagnostical confidence in CSS and digital displays with low spatial frequency enhancement.

A number of studies have evaluated spatial resolution requirements in digital chest radiographs indicating that a digital resolution of 2,5 lp/mm is sufficient to detect all pulmonary abnormalities with the exception of fine septal lines (4). As the actual pathology examined in gastrointestinal radiology tends to be relatively large, the resolution requirements for the GI tract are less than for other organ systems: Kastan demonstrated that the resolution necessary for successfully evaluating the colonic mucosa is with 0.8 pixel/mm comparatively low and that more experienced radiologists even detected mucosal details well at lower levels of spatial resolution (3,6). In view of these results the level of diagnostical confidence being equally high with digital displays processed to look like a conventional radiograph as with CSS despite a slight increase in noise and the lesser spatial resolution was to be expected.

Although the use of high spatial frequency enhancement is advantageous in chest radiographs by better demonstrating mediastinal structures and paramediastinal and retrocardiac lung areas, the level of diagnostical confidence was worse with high spatial frequency enhancement than with the digital image processed to look like a conventional radiograph. Furthermore high spatial frequency enhancement did not provide diagnostical information exceeding both other images, so that in consequence digital displays with high spatial frequency enhancement should be dispensed with in barium enemas. The space thus made available should be reserved for the display of a second digital image with low spatial frequency enhancement, consequentely saving film costs and helping to achieve a more practical presentation of barium enemas.

In conclusion image quality of double-contrast barium enemas obtained by DLR using 50% of the conventional exposure dose was evaluated as being slightly lower in the conventionally processed digital display than in CSS. However, the level of diagnostical confidence was similar with both methods, the use of DLR in barium enemas thus being justified. A further increase in digital image quality could be obtained by using the same exposure dose as in CSS, this item not being diagnostically essential. Artefacts by post processing procedures, which were relevant for diagnosis, were not observed· The enhanced digital display was of no diagnostical relevance and should be dispensed with.

References

1. Buchmann, F,: Der Einstieg in die digitale Radiographie. Medizin Technik 107,5 (1987) 1633.

2. Feczko, P.J., Ackermann, L.V., Kastan, D.J., Halpert, R.D.: Digital Radiography of the Gastrointestinal Tract. Gastrointest. Radiol. 13: 191-196 (1988)

3. Kastan, D.J., Ackermann, L.V., Feczko, P.J.: Digital Gastrointestinal Imaging: The Effect of Pixel Size on Detection of Subtle Mucosal Abnormalities. Radiology 162: 853-856 (1987)

4. Krug, B., Neufang, K.F.R., Friedmann, G.: Vergleich der digitalen Lumineszenzradiographie (DLR) mit der konventionellen Thoraxdiagnostik. 3. Frankfurter Gespräche über Digitale Radiographie (in Druck)

5. Lorenz, W.: Strahlenschutz in Klinik und ärztlicher Praxis. Thieme Verlag, Stuttgart 1961

6. Nakano, Y., Togashi, K., Nishimura, K., Itoh, K., Fujisawa, I., Asato, R., Adachi, H., Itoh, H., Torizuka, K.: Stomach and Duodenum: Radiographic Magnification Using Computed Radiograpohy (CR): Radiology 1986; 160: 383-387

7. Unaymama, F.: Upper Gastrointestinal Tract. In: Tateno, Y., Iinuma, T., Takano, M. (Eds.): Computed Radiography. Springer Verlag Tokyo-Berlin-Heidelberg-New York-London-Paris, 1987

Detection of Pulmonary Nodules – A Comparative Study of Low-Dose Digital Luminescent Radiography (DLR) and Conventional Chest Film

K.F.R. NEUFANG, B. KRUG, W. STEINBRICH, R. LORENZ, H. LANFERMANN, W. HEINDEL

Radiologisches Institut der Universität zu Köln
(Direktor: Prof. Dr. G. Friedmann)

Radiologisches Institut der Universität zu Köln
Zentralklinikum Ebene 6, Joseph-Stelzmann-Straße 9
D-5000 Köln 41 (Federal Republic of Germany)

Summary

Biplane low-dose digital luminescent radiography (DLR) and conventional standard radiography (SR) was performed in 17 patients in order to compare their capacities in demonstrating benign and malignant pulmonary nodules. DLR was performed on a PHILIPS PCR-SYSTEM using a 50 % dose reduction compared with the SRs, and a spatial resolution of 2.5 LP/mm. All DLR images were processed and displayed with a low spatial frequencies enhancing (LSFE) filter as well as with a high spatial frequencies enhancing (HSFE) filter. Sagittal and lateral films were arbitrarily divided into 6 areas. Four radiologists were asked to determine independently the presence or absence of nodules in each area of the lung with one out of five levels of confidence. LSFE-DLR images equal SR images in the detection of pulmonary nodules. They tend to be superior to SR images in demonstrating and excluding nodules in the retrocardiac and subdiafragmatic area. HSFE-DLR images are inferior to both SR and LSFE-DLR images.

In contrast to conventional screen-film radiography the detection of the X-ray image is completely separated from processing and display in digital radiographic systems [2,7,15]. Currently, approx. 10-25 % of all radiologic examinations are recorded digitally, e.g. DSA, CT, MRI, ultrasound and nuclear medicine [2,6]. As chest radiographs constitute approx. 40 % of conventional radiographs [9,13], chest imaging has become a major target for the introduction of digital radiography [1,3,6,7,11,12,14,16,18].

Several approaches to digital projection radiography have been developed. Compared with Film Digitization, Image Intensifier, Linear Scanner, Scanning Equalization Radiography [1,3,5-8,10,12,14,15-18], high-resolution storage phosphor imaging, known as Digital Luminescent Radiography (DLR), offers the advantage of direct digital data acquisition without the need for significant modification of current radiographic equipment [5,7,8,14,17].

In this study we attempt (1) to compare the diagnostic value of DLR vs standard film-screen radiography (SR) for demonstration of small benign and malignant pulmonary nodules. Special attention is directed (2) to the question, whether different filtering procedures affect diagnostical accuracy.

Material and Methods

DLR is performed on a PHILIPS Computed Radiography (PCR-SP®) system. Images are obtained on a Bucky stand (source-to-detector distance, 200 cm; 125 kVp; focus 0.6x.0.6 mm^2; automated exposure control with a 50 % dose reduction compared with conventional radiograms). Preliminary evaluation of DLR with chest phantoms and first clinical experience have demonstrated that X-ray exposure could be varied over a wide range with no or only little effect on the perception of anatomical details. A reduction of the system dose by a factor of 2 compared with SR does neither significantly compromize image quality nor does it produce any notable increase of the background noise. Consequently, a reduction of system dose and skin entrance dose by a factor of 2 may be well tolerated in clinical routine [10], resulting in a mean skin entrance dose for p.a. DLR chest radiograms of only 150 µGy.

All SR images are obtained by using a dedicated automated chest unit (SIEMENS THORAMAT®, source-to-detector distance, 200 cm; 125 kVp; focus 0.6x.0.6 mm^2; automated exposure control; SIEMENS TITAN® 2U rare earth-screens).

Biplane Low-Dose DLR and SR images were obtained in 17 patients, the set of digital images being generated immediately following the conventional ones. All DLRs were routinely processed and displayed with a low spatial frequency enhancing (LSFE) filter, which produces images that closely resemble traditional chest films, and a high spatial frequency enhancing (HSFE) filter. The appropriate filter algorithms had been found experimentally before. Both digital images were displayed on one film, so that films had to be cut for further evaluation. As a result, three sets of images (1 with 2 SR films and 2 with each 2 DLR images) were obtained from one patient. For further analysis, sagittal films were arbitrarily divided into 4 quadrants, and lateral films into an upper and a lower part, respectively. The borders of each area were marked on the edges of the films. In total, there were 306 areas to be analysed.

Six radiologists (3 board-certified radiologists, 3 four-year radiology residents) participated in the study. The "Gold Standard" was established by the agreement of two radiologists (K.F.R.N., B.K.) based on the chest radiographs and the results of further radiological investigations (e.g. fluoroscopy, film tomography, computed tomography), clinical data and follow-up. There were 67 areas with one ore more nodules, and 35 areas without nodules. The images were presented randomly to the four readers (W.S., R.L.,H.L.,W.H.), who had to determine without knowlegde of the "Gold Standard" the presence or absence of nodules in each area of the lung with one out of five levels of confidence (1 = definitely absent, 2 = probably absent, 3 = equivocal, 4 = probably present, 5 = definitely present). Furthermore the number of nodules observed had to be rated (no, 1, 2, 3 more than 3 nodules). The results were tabulated, the levels of confidence were calculated, and ROC analysis was performed for each image modality and each area, as well as for the whole group.

Results

Preliminary data analysis shows that DLR equals SR in the detection of pulmonary nodules (Tab.1, Fig.1). There is no difference in the ROC curves nor in the levels of confidence between LSFE-DLR images and SR images, when all areas of the lung are taken together. However, LSFE-DLR images are superior to SR images for demonstration of nodules in the retrocardiac and subdiafragmatic areas of the left lower quadrant. HSFE-DLR images are inferior to both SR and LSFE-DLR images, the level of confidence in their interpretation being significantly lower.

Discussion

DLR is a general-purpose digital radiography system that has several advantages over the conventional radiographic technique [5, 7, 8, 14, 17], virtually eliminating the need for repeats in radiographic examinations [10]. Chest images require high spatial resolution and contrast sensitivity across a wide dynamic range [7]. Using digitized standard radiograms it was found that pixel sizes below 0.8 mm (0.625 line pairs/mm) are not critical for detection of solitary pulmonary nodules. For discrimination of septal lines, however, a pixel size of at least 0.4 mm (1.25 line pairs/mm) is required [11]. These findings correlate with earlier results obtained with a large image intensifier tested for photofluorography of the chest [4, 13], demonstrating that septal lines and reticular interstitial patterns are the most critical low contrast structures in chest radiography rather than pulmonary nodules [4, 13]. As was demonstrated in former studies, their detection is limited in DLR [10] as well as in other digitized chest techniques [11, 12].

In conclusion, DLR equals SR in the detection of pulmonary nodules. Nodules in the retrocardiac and subdiafragmatic areas are occasionally missed on SRs. Due to better penetration of the heart, the mediastinal and soft tissue structures of the upper abdomen, LSFE-DLR images offer advantages exceeding SR images in demonstrating nodules in the critical left retrocardiac area. This may be especially important for single-plane chest radiograms obtained in intensity care units. HSFE-DLR images are not helpful in detecting or excluding the presence of pulmonary nodules. However, they are superior in the demonstration of the mediastinal pleural reflections and of catheters, tubes and pacemakers [3, 10]. LSFE-DLR images, that closely resemble traditional radiograms, may substitute SR images in detecting or excluding pulmonary nodules, thus offering the chance of a further 50 % reduction of radiation dose in chest radiography.

References

1. Chakraborty D.P., E.S. Breatnach, M.V. Yester, et al.: Radiology 158 (1986) 35-39

2. Cox G.C., A.W. Templeton, S.J. Dwyer: Radiol. Clin. North Am. 24 (1986) 37-54
3. Fraser R.G., E. Breatnach, G.T. Barnes: Radiology 148 (1983) 1-5
4. Friedmann G., K.F.R. Neufang, P.E. Peters, et al.: In: Heuck F.H.W., M.W. Donner (Eds.): Radiology Today 2 (p. 311-314). Springer, Berlin Heidelberg New-York Tokyo 1983
5. Fuhrman C.R., D. Gur, B. Good, et al.: Amer. J. Roentgenol. 150 (1988) 1011-1014
6. Goodman L.R., W.D. Foley, C.R. Wilson, et al.: Radiology 158 (1986) 27-33
7. Goodman, L.R., C.R. Wilson, W.D. Foley: Amer. J. Roentgenol. 150 (1988) 1241-1252
8. Hillen W., U. Schiebel, T. Zaengel: Med. Phys. 14 (1987) 744-751
9. Johnson J.L., D.L. Abernathy: Radiology 146 (1983) 851-853
10. Krug B., K.F.R. Neufang, G. Friedmann: In: Riemann R.E., J. Kollath (Hrsg.): 3. Frankfurter Gespräch über Digitale Radiographie. Schnetztor, Konstanz 1989 (in press)
11. Lams P.M., M.L. Cocklin: Radiology 158 (1986) 11-19
12. MacMahon H., C.J. Vyborny, C.E. Metz, et al.: Radiology 158 (1986) 21-26
13. Neufang K.F.R., G. Friedmann, P.E. Peters, et al.: Fortschr. Röntgenstr. 137 (1982) 535-539
14. Sakuma H., K. Takeda, T. Hirano, et al.: Amer. J. Roentgenol. 151 (1988) 27-30
15. Schuler M., M. Rath, D. Hahn, et al.: Digit. Bilddiagn. 5 (1985) 22-39
16. Sherrier R.H., C. Chiles, G.A. Johnson, et al.: Radiology 162 (1987) 645-649
17. Sonoda M., M. Takano, J. Miyahara, et al.: Radiology 148 (1983) 833-838
18. Wandtke J.C., D.P. Plewes, E. Vogelstein: Amer. J. Roentgenol. 145 (1985) 973-978

<u>Table 1</u> Level of confidence vs imaging modality

Pulmonary Nodules	Standard Radiogram	LSFE-DLR	HSFE-DLR
definitively present/absent	306 (75.0 %)	298 (73.0 %)	241 (59.2 %)
probably present/absent	92 (22.5 %)	96 (23.5 %)	132 (32.3 %)
equivocal finding	10 (2.5 %)	14 (3.5 %)	35 (8.6 %)
Statistics	Standard vs HSFE-DLR: $p < 0.0005$, $Chi^2 = 31.74$	Standard vs LSFE-DLR: n.s., $Chi^2 = 0.85$	LSFE-DLR vs HSFE-DLR: $p < 0.0005$, $Chi^2 = 26.70$

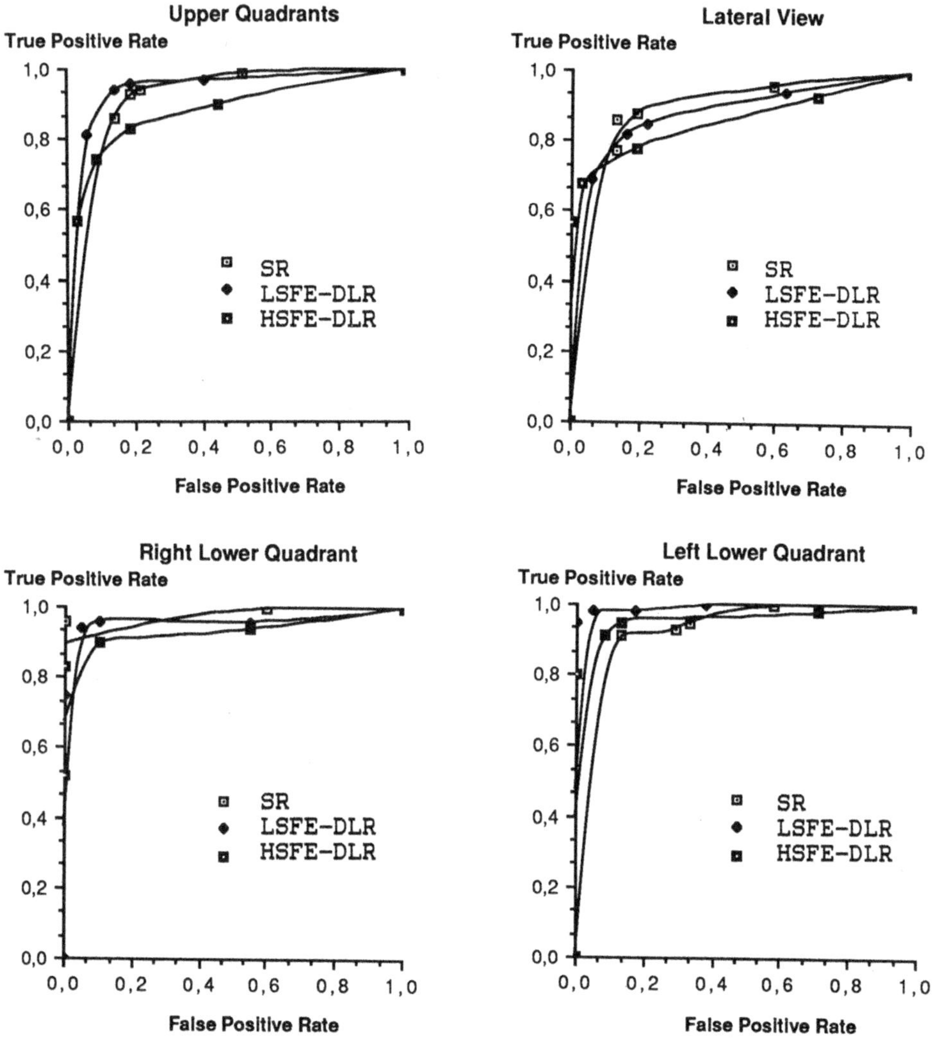

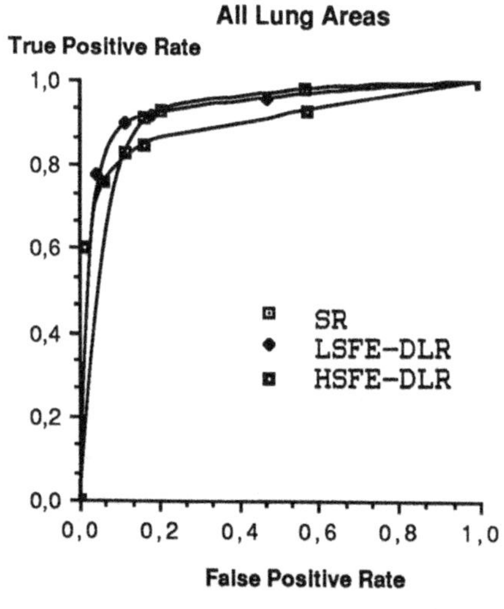

__Figure 1__
ROC curves for SR, LSFE- and HSFE-DLR images

Digital Image Intensifier Radiography in Chest Radiology

I. HARTH, P. HARTH, V. SCHERKAMP, H. E. RIEMAN

Frankfurt on Main University Hospital
Department of Radiology
Theodor-Stern-Kai 7
D-7000 Frankfurt on Main 70

Introduction

In the past few years, efforts have been made to use digital image intensifier radiography as an aid to diagnostic chest imaging. Initially, this technique was but one of several that were being evaluated |1|. Since then, both theoretical |2| and clinical |3, 4| studies have been performed specifically on image intensifier radiography for diagnostic chest imaging, and results are now available.

In Frankfurt on Main, we have carried out a comprehensive clinical study. A total of 803 patients, from our routine material, were examined with both conventional and digital techniques, and the images obtained were compared.

Methods

The digital chest images were produced with a SIEMENS DS 1000 system, which consists of the tube unit, a TS 57 stand, and a memory, the DS 1000, with two TV monitors. The TS 57 stand is equipped with a Sirecon 57 image intensifier (II) with a selection of three input screen diameters (57 cm, 47 cm, and 33 cm). The output phosphor is coupled to a video camera that transmits the video signal. The image is digitized in real time, stored in a semi-conductor memory, and immediately re-converted into an analog image, which can be viewed, on one of the two monitors, as soon as the exposure is finished. The digital image has a 1024 x 1024 matrix and a depth of 8 bit. A disc provides long-term storage for 120 radiographs. Since this archive capacity is limited, we produce additional hard copies using a multi-format camera. The following post-processing facilities are available:

- gamma correction (four levels)
- edge enhancement, by changing the spatial resolution filter; continuously adjustable by means of the Transicon, (2 stage)
- inversion (black/white reversal)
- window and level manipulation.

For our study, we chose the negative, with a Transicon setting of T 30 (stage 1), and a level 2-3 setting on the gamma curve. The radiographic data were:

- focus-film distance (FFD)	183 cm
- kilovoltage	125 kV
- filter	0.3 mm Cu
- grid	not used

We employed an 'air gap technique', i.e. a distance of 15 cm between the patient and the II input screen. The FFD (or rather focus-input phosphor distance) remained unaltered, at 183 cm. This arrangement reduces scatter and obviates the need for a grid. Resolution is marginally improved (Fig.1).

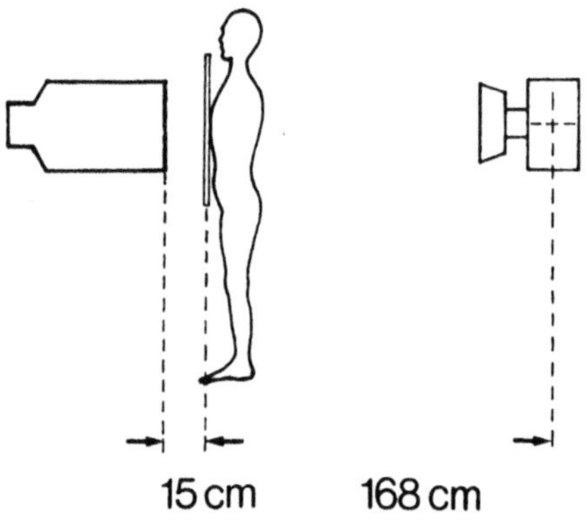

Fig.1 Diagrammatic view of the 'air gap technique' employed by us in digital II radiography

Under these conditions, and with a 57 cm input screen diameter, the patient entrance dose is 0.1 mGy. Lung filters and mediastinal filters were used to prevent overexposure.

The conventional radiographs were obtained using a PHILIPS Vertigraph, with the following technical data:

- focus-film distance (FFD)	200 cm
- kilovoltage	140 kV
- filter	1 mm Al + 0.1 mm Cu
- grid	12/40

The film material used was Trimax XDA, with a Trimax T4 screen (3M). The patient entrance dose measured under these conditions was 0.48 mGy.

The above techniques were used to produce, in parallel, a total of 803 pairs of digital and conventional chest images. Three radiologists with varying degrees of experience were asked to read the hard copies of the digital PA radiographs and the conventional PA films. The images from each pair were not placed side by side for reading. A questionnaire was used to report, first, all the conventional radiographs; the same form was then used for the reporting of all the digital images. Observers were asked to identify and assess normal anatomical as well as abnormal structures in the chest. The normal structures chosen were: trachea, bronchi, paratracheal stripe, paraspinal line, descending aorta, and peripheral pulmonary vessels (4 cm from the chest wall). The detectability ratings for normal structures were 'good', 'partial', or 'poor'. The abnormal structures selected were: homogeneous shadows; nodular opacities up to a diameter of 0.5 cm; nodular opacities with a diameter of more than 0.5 cm; linear opacities; increased interstitial lung markings. Observers were requested to classify these abnormal structures only in terms of 'present' or 'absent'.

Results

There was considerable interobserver variability in the assessment of detectability of normal structures. (These discrepancies will not be discussed in detail in this paper.) Consequently, the mean values of the three results for each rating were calculated. These values are shown in diagrammatical form in Figs. 2 - 4.

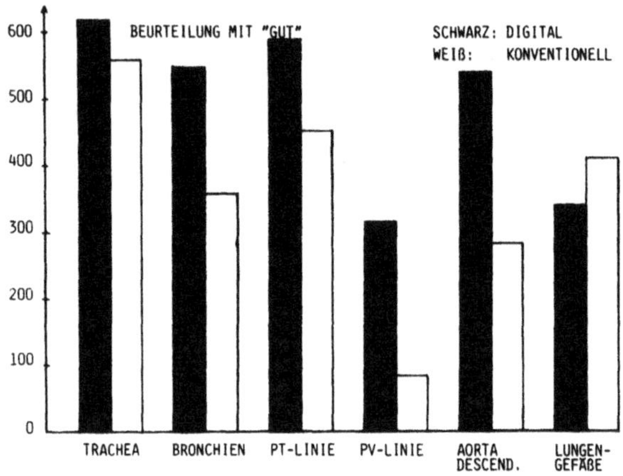

Fig. 2 Number (mean values) of normal structures whose detectability was rated 'good'; (PT stripe = paratracheal stripe; PS line = paraspinal line)

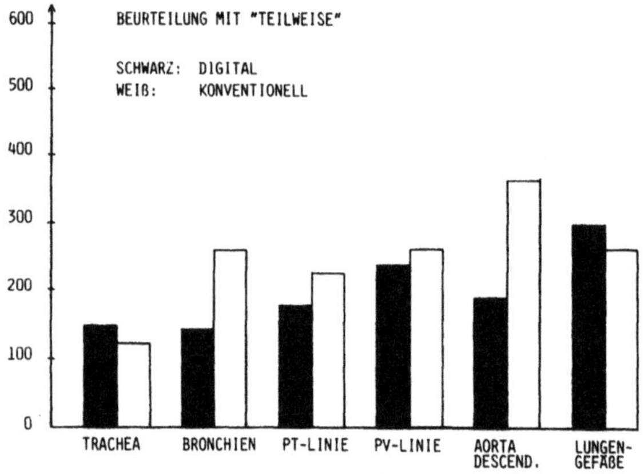

Fig. 3 Number (mean values) of normal structures whose detectability was rated 'partial' ; (PT stripe = paratracheal stripe; PS line = paraspinal line)

In order to form an opinion, it is sufficient to look more closely at the numbers of 'good' ratings. The figures for the other two ratings ('partial', and 'poor' detectability) only provide subsidiary information.

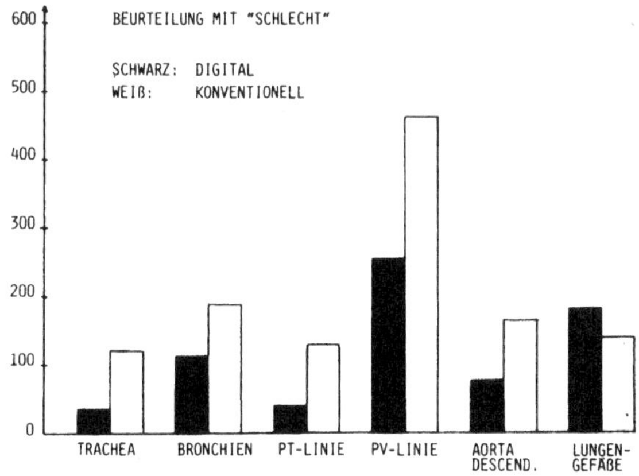

Fig. 4 Number (mean values) of normal structures whose detectability was rated 'poor'; (PT stripe = paratracheal stripe; PS line = paraspinal line)

In the digital radiographs, the mediastinal structures had a noticeably higher percentage of 'good' ratings; this is due mainly to the use of lung filters, which were not employed for the conventional radiographs. It should be borne in mind that, with our film/screen combination, a filter would not appreciably improve image quality.

The result is different for the lung parechyma and the peripheral structures. The detectability of the pulmonary vessels (4 cm from the chest wall) is always rated as better in the conventional radiographs. This is due to the spatial resolutions of the two systems. For the conventional radiographs, we measured 4lp/mm; for the digital image (57 cm II), 1lp/mm. However, the actual results are not as bad as might be suggested by the wide discrepancy between these two values. It should be remembered, firstly, that spatial resolution is not the only factor that determines the visibility of structures; and, secondly, that image interpretation by the human viewer is not a perfectly measurable process.

This fact emerges clearly in the identification of abnormal structures, where the result depends, to a certain extent, on the observer's experience and, even, on his greater or lesser acceptance of the new technique. Figs. 5 and 6 each show 3 'graphs' of the numbers of abnormal structures detected using the two systems. The black line represents the most experienced observer; the dotted line, the viewer with average experience; and the broken line, the least experienced radiologist.

NUMBER OF ABNORMAL STRUCTURES IN THE DIGITAL IMAGE

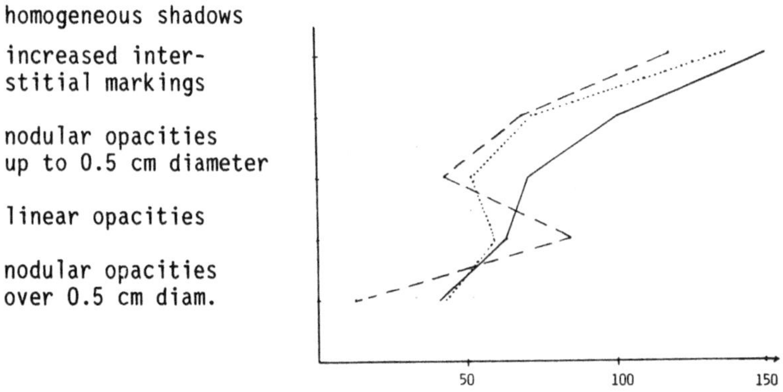

Fig. 5 Numbers of abnormal structures detected by three observers, in the digital image

NUMBER OF ABNORMAL STRUCTURES IN THE CONVENTIONAL IMAGE

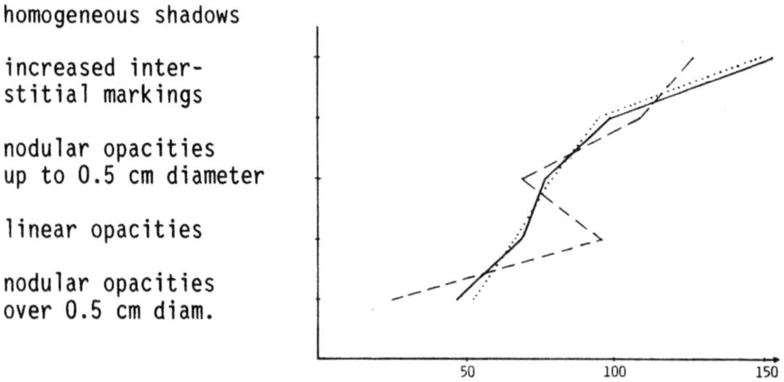

Fig. 6 Numbers of abnormal structures detected by three observers, in the conventional image

For the grosser lesions such as effusions or pleural thickening (homogeneous shadows), detection rates tended to be fairly similar. However, more discreet lesions - nodular opacities with a diameter of less than 0.5 cm - were detected more often in the conventional radiographs. The great discrepancy in the interpretation of the linear opacities appears to be due solely to a certain lack of confidence on the part of some observers. Quite generally, it may be said that the more experienced the radiologist the better he will be at interpreting the digital images.

Summary

In our study, we found one major shortcoming of the digital system, in that, as a result of its poorer spatial resolution, the more discreet structures cannot be reliably identified. However, even in routine imaging work, it is important for the radiologist, and vital for proper patient management, that small round lesions or delicate interstitial lines should be detected and correctly identified.

However, digital radiography also confers the very valuable advantages of speed in obtaining an instant image on the monitor, and of reduction of the dose to the patient. In fact, these aspects are so important that attempts should now be made to improve the resolution of the digital technique.

Literature cited

1. Busch, H.P.; Strauss, L.G.; Georgi, M.: Digitale Bildverstärker-Radiographie - Eine kostengünstige Untersuchungsmethode mit geringer Strahlenbelastung. In: Lemke, H.U., Rhodes, M.L.; Jaffee, C.C.; Felix, R. (Hrsg.): CAR '87 (Proceedings). Berlin, Heidelberg, New York, London, Paris, Tokio: Springer 1987, S. 226-229.

2. Dölken, W.; Krahe, Th.; Landwehr, P.; Lackner, K.: Abbildungseigenschaften der digitalen BV-Radiographie im Vergleich zur konventionellen Röntgentechnik für Thoraxaufnahmen. RöFö 150 (1989) 86-92.

3. Kheddache, S.; Månnson, L.G.; Tylén, U.: Die digitale Lungenröntgenographie mit einem großformatigen Bildverstärker. Elektromedica 55 (1987), 21-27 (Heft 1).

4. Schuler, M.; Hübsch, Th.; Rienmüller, R.; Frey, J.; Schmidt, H.: DS-1000 - Ein digitales BV-FS-Radiographiesystem in der Thoraxdiagnostik. Digit. Bilddiagn. 8 (1988) 25-32.

Digital Multilayer Tomography

C. DÜBER, U. BITTNER, K.-J. KLOSE, M. THELEN

Institut für Klinische Strahlenkunde

Johannes Gutenberg-Universität, Mainz

Siemens AG, UB Med, Erlangen

Introduction

Conventional tomography remains an important radiological method which has not been replaced by computed tomography and magnetic resonance imaging. Digital multilayer tomography is a method combining the principle of tomography and digital imaging techniques.

Principle of Digital Multilayer Tomography

During a linear tomographic sweep a set of projection images is recorded with an image intensifier-TV-system and digitalized . Direct superposition of these images results in a tomogram of layer A. A tomogram of layer B is obtained by shifting the images prior to superposition. With this system reconstruction of tomograms in any desired plane is possible from projection images of a single tomographic sweep (Figure 1).

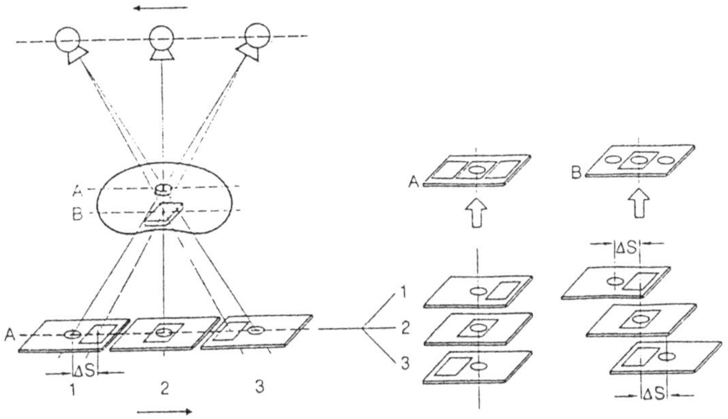

Figure 1: Principle of digital multilayer tomography

Digital Multilayer Tomography System

In the digital multilayer tomography system used in this study a conventional DSA system (Digitron 2,Siemens) was connected to an overtable x-ray unit with tomography facility (Sire-graph,Siemens). A 17 and 25 cm image intensifier was used. The digital images have a 512 over 512 matrix with 10 bit depth. 54 projections are recorded during the linear tomographic movement at a rate of 8 images per second resulting in a total study time of about 7 seconds.

Patients

100 patients with tracheobronchial, pulmonary or skeletal disease were examined. Conventional film tomograms were available for comparison in most patients.

Results

Digital multilayer tomography provides diagnostically useful images. Image quality is comparable with conventional tomography. Advantages of digital tomography include low radiation dose, short study time and digital technology (post processing, archiving, image transfer). Advantages of conventional tomography are high spatial resolution.

Examples of digital and conventional tomograms are presented in Figure 2 and 3.

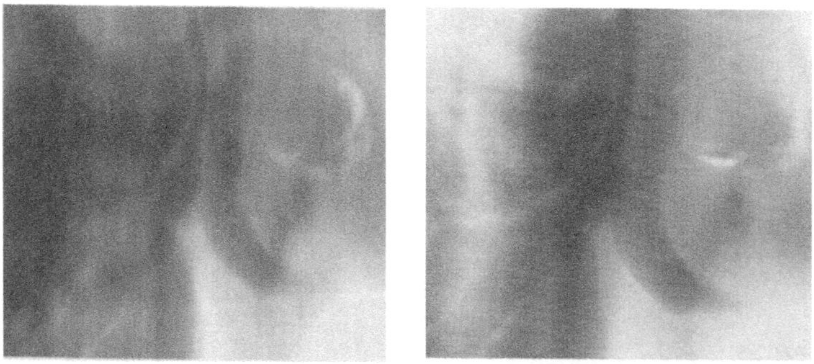

Figure 2: Conventional (A) and digital (B) tomogram:

occlusion of left main bronchus (bronchial carcinoma)

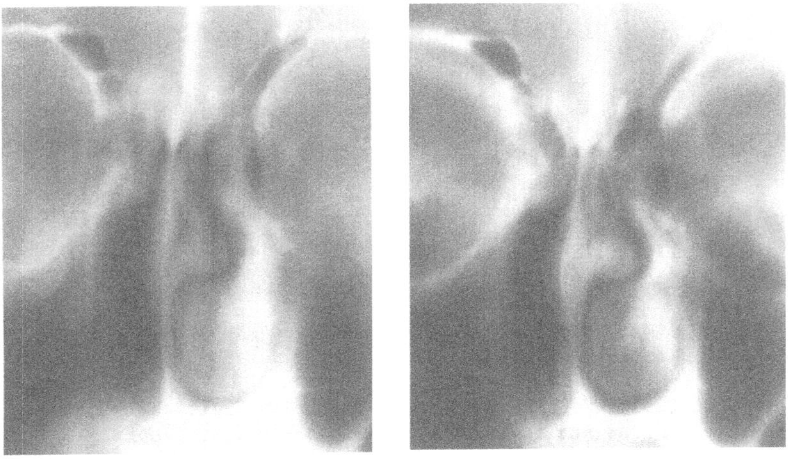

Figure 3: Conventional (A) and digital (B) tomogram: partial resection of left maxillary sinus (sarcoma)

Discussion

In 1935 Ziedses des Plantes (18) described "Serieskopie" - a method for producing an infinite number of tomograms from a set of radiographs. Following this idea various optical and electronical tomography systems have been developed (1,7,10,11, 14,15,16,17). Due to long study times, expensive equippment, or inferior image quality neither system has gained widespread clinical use.

With the advent of digital imaging a number of digital tomography systems have been described (2,4,5,8,9,12,13,).

Digital multilayer tomography is one of the newer approaches to "Serieskopie" (3,6). The method has been implemented in a modified conventional tomography system and DSA unit.

Image quality of digital tomograms is comparable with conventional images. However there is lower spatial resolution due to a 512 over 512 image matrix used in this first clinical study.

Advantages of digital multilayer tomography include short patient study time, reduced radiation dose and facilities of digital imaging.

References

1. Baily, NA, EC Lasser, RL Crepeau: Electrofluoroplanigraphy. Radiology 107, 1973, 669-671

2. Baily, NA, TD Kampp: Digitized longitudinal tomography. Invest Radiol 16, 1981, 126-132

3. Bittner, U, C Düber, R Koch, M Pfeiler: Digitales Vielfachschichten: eine neue Realisierung der klassischen Serioskopie. Electromedica 56, 1988, 16-23

4. Chakraborty, DP, MV Yester, GT Barnes, AV Lakshminarayanan: Self-masking subtraction tomosynthesis. Radiology 150, 1984, 225-229

5. De Vries, N, FJ Miller, MM Wojtowycz, PR Brown, DR Yandow, JA Nelson, RA Kruger: Tomographic digital subtraction angiography Initial clinical studies using tomosynthesis. Radiology, 157, 1985, 239-241

6. Düber, C, K-J Klose, U Bittner, C Gutjahr, M Thelen: Digitale Mehrschichttomographie: klinische Ergebnisse. Electromedica 57, 1989 (im Druck)

7. Dümmling, K: Ein neues Verfahren zum Mehrfachschichten mit Hilfe von Fernsehbildspeichern. Radiologe 9, 1969, 37-40

8. Edholm, PB, G Granlund, H Knutsson, C Petersson: Ectomography. A new radiographic method for reproducing a selected slice of varying thickness. Acta Radiologica 21, 1980, 433-442

9. Friedenberg, RM, JB Lightfoote, SP Wang, MF Smolin: Digital tomography. Description and preliminary clinical results. AJR 144, 1985, 639-643

10. Garrison, JB, DG Grant, WH Guier, RJ Johns: Three-dimensional roentgenography. AJR 105, 1969, 903-908

11. Grant, DG: Tomosynthesis: a three-dimensional radiographic imaging technique. IEE Trans Biomed Eng 19, 1972, 20-28

12. Kruger, RA, M Sedaghati, D Ghosh-Roy, P Liu, JA Nelson, W Kubai, P Del Rio: Tomosynthesis applied to digital subtraction angiography. Radiology 152, 1984, 805-808

13. Maravilla, KR, RC Murry, S Horner: Digital tomosynthesis: technique for electronic reconstructive tomography. AJR, 141, 1983, 497-502

14. Miller, EA, EM McCurry, B Hruska: An infinite number of laminograms from a finite number of radiographs. Radiology 98, 1971, 249-255

15. Nadjmi, M, H Weiss, E Klotz, R Linde: Kurzzeittomosynthese. Erste klinische Erfahrung. Röntgenpraxis 34, 1981, 247-252

16. Richards, AG: Variable depth laminography. Proceeedings of Biomedical Sciences Instrumentation 6, 1969, 194-199

17. Weiss, H, E Klotz, R Linde, G Rabe, U Thiemens: Coded aperture imaging with x-rays (flashing tomosynthesis). Optica acta 24, 1977, 305-325

18. Ziedses des Plantes, BG: Serieskopie. Fortschr Röntgenstr 57, 1938, 605-616

Computed Radiography: Value of Digital Image Processing in Motion Tomography

M. Prokop, A. Quayumi, M. Galanski, S. Reichelt und W. Döhring

Abteilung Diagnostische Radiologie I, Medizinische Hochschule Hannover
Konstanty-Gutschow-Str. 8, D-3000 Hannover 61

Digital image processing is expected to substantially improve the quality of radiographic images. Conventional motion tomography produces composite images of the examined tomographic plane and of unsharp overlying structures. It is therefore an ideal model system to study the capabilities of digital filtering to selectively enhance diagnostic information.
Our installation consisted of a FCR-901 storage phosphor system for digital radiography and a separate Siemens Digiscan digital processing unit.
To study how digital filtering might selectively enhance the image of the tomographic plane, we used decomposable anatomical and geometrical phantoms which enabled us to examine, separately, the isolated tomographic plane and the blurs caused by neighboring structures. By appropriate unsharp mask filtering we were able to enhance the dominant spatial frequencies of the isolated tomographic plane and to suppress simultaneously the low frequency blurs from the neighboring planes. We found the choice of the filter functions to be dependent on the relative contrast of the objects within the tomographic plane and on the motion pattern used.
In the clinical setting, digital filtering alone did not yield additional information. Inappropriate processing could even obscure findings. Appropriate filtering, however, improved the visualization of relevant diagnostic details. Filter functions had to be chosen differently for soft tissue and for bone imaging. Additional information could be gained in conjunction with the properties of the storage phosphor detector system: Digital processing made it possible to fully utilize the wide dynamic range of the detector system and therefore allowed visualization of lesions in conventionally over- or underexposed areas.

Restoration of Digital Radiographic Images by Using Two-Dimensional Antimetric Wave Digital Filters

M. PÄTZOLD, A. TEUNER

Ruhr-Universität Bochum

Lehrstuhl für Nachrichtentechnik

4630 Bochum 1, West Germany

Summary

Every digital radiography system converts a continuous X-ray image into a digital image with restricted resolution of details. Responsible for this degradation effect is the modulation transfer function of the digital radiography system, which has a natural low-pass attenuation characteristic. It is well known that the picture degradation can be efficiently compensated by using linear digital image restoration techniques. But so far nothing had been known about how these techniques can be applied by relying on recursive two-dimensional digital filtering in such a way that the resulting filter satisfies the following important requirements:

- good approximation of the desired frequency response prescribed by the restoration filter,
- satisfying all important stability requirements, (stability under linear and nonlinear conditions, suppression of small and large scale parasitic oscillations, forced-response stability, etc.)
- approximately linear phase in relevant regions of the Nyquist range,
- practicable for real-time applications,
- a modular, regular, and flexible filter structure requiring as little hardware as possible.

In this paper it will be shown that all the requirements listed above can be satisfied by using two-dimensional antimetric wave digital filters designed by a complex network theory approach.

Problem Description

The problem description reads as follows. An X-ray beam penetrates an object in the object plane and engenders an image $f(x_1, x_2)$ in the image plane. A digital radiography system with modulation transfer function $M(\omega_1, \omega_2)$ receives the continuous image $f(x_1, x_2)$ and produces the digital output image $d(k_1, k_2)$ according to

$$d(k_1, k_2) = \mathcal{A}\{\mathcal{D}\{f(x_1, x_2)\}\}, \tag{1}$$

where $\mathcal{A}\{\cdot\}$ denotes an analogue-to-digital converter operator, and $\mathcal{D}\{\cdot\}$ stands for a linear distortion operator that depends on the modulation transfer function $M(\omega_1, \omega_2)$ (see Fig. 1). In addition, the signal $d(k_1, k_2)$ is degraded by noise $n(k_1, k_2)$. The task of the restoration filter $H(\omega_1, \omega_2)$ is the compensation of the degradation effect produced by the modulation transfer function in the presence of noise. Therefore — among linear restoration and enhancement techniques — we have selected the inverse filter, Wiener filter, geometrical mean filter, Wiener-enhancement filter, and the edge-detection filter. A typical characteristic of these filters is that each one (except for the last one) depends on the modulation transfer function of the radiography system. Therefore, a mathematical description of a modulation transfer function belonging to a specific radiography system will first be given.

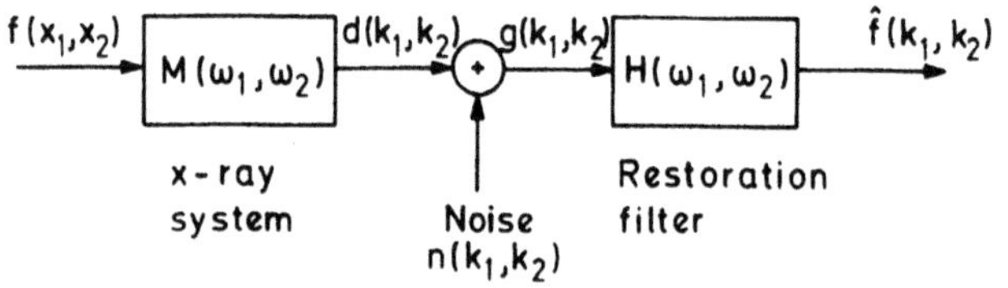

Fig. 1. Schematic image restoration model

Modulation Transfer Function

In co-operation with Siemens [1], we have obtained discrete measured values (see Fig. 2a) which are related to a modulation transfer function of a digital radiography system. As starting point for obtaining an approximate solution for a continuous modulation transfer function in the complete Nyquist range Ω_N according to the sampling matrix $\underline{T} = T_0 \left(\begin{smallmatrix} 1 & 0 \\ 0 & 1 \end{smallmatrix}\right)$, $T_0 > 0$, we have chosen a general function of the type

$$M(\omega_1, \omega_2) = \sum_{i=1}^{I} m_i e^{-(\omega_r/\alpha_i)^2}, \quad (\omega_1, \omega_2) \in \Omega_N, \tag{2}$$

with $\omega_r = \sqrt{\omega_1^2 + \omega_2^2}$ and $m_i, \alpha_i \in R, \forall i = 1, 2, \ldots, I$.

Application of a numerical optimization procedure to (2) has lead to $I = 2$ and

$$m_1 = 0.15, \qquad \alpha_1 = 0.1396\,\pi/T_0, \tag{3a, b}$$

$$m_2 = 0.85, \qquad \alpha_2 = 0.6006\,\pi/T_0. \tag{3c, d}$$

In Fig. 2, the modulation transfer function $M(\omega_1, \omega_2)$ given by (2) in connection with (3) is presented along the ω_1-axis (Fig. 2a) as well as in the complete Nyquist range Ω_N (Fig. 2b).

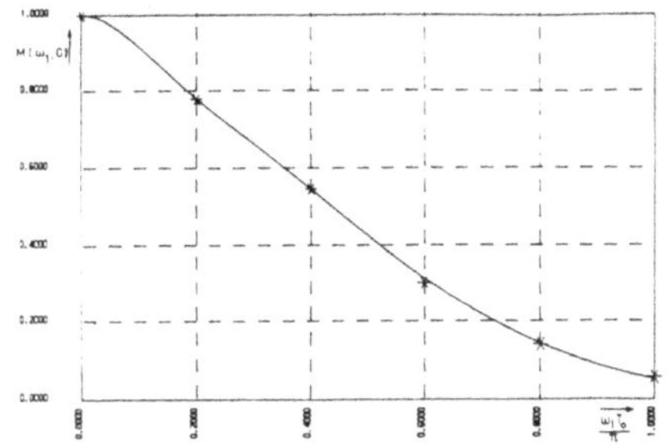

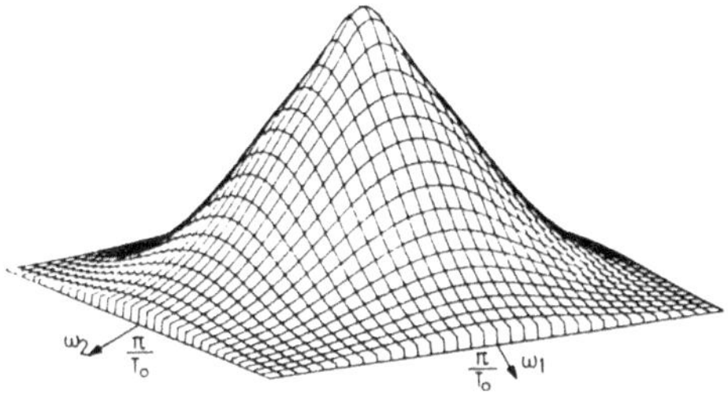

Fig. 2a. Measured (*) and approximated (—) modulation transfer function along the ω_1-axis

Fig. 2b. Modulation transfer function $M(\omega_1, \omega_2)$ in the Nyquist range Ω_N

Design, Optimization, and Realization of Linear Restoration Filters

An approximation of the desired frequency response prescribed by the ideal restoration (enhancement) filter has been obtained by applying an optimization procedure to the coefficients of the transfer function of a two-dimensional antimetric filter. Full details about the design and realization of two-dimensional antimetric filters will not be given here. This class of filters can be obtained from a design method for two-dimensional (but also multi-dimensional) filters having approximately circular (hyperspherical) symmetric loss behavior [2], and can be used as a reference filter for wave digital filters [3]. A complete description of the design and optimization method, as well as of various realization techniques, is intended to be given in [4].

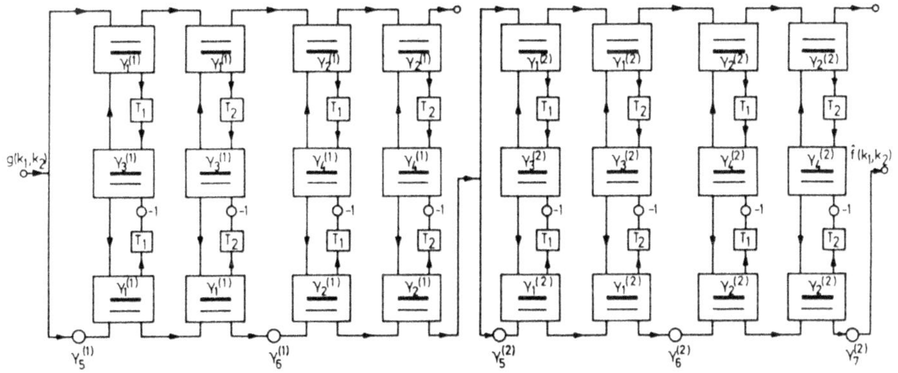

Fig. 3. Resulting wave flow diagram of the restoration and enhancement filters

We have found that an approximate solution of the inverse filter [5], Wiener filter [5], geometrical mean filter [5], Wiener-enhancement filter [6], and the edge-detection filter [7] can be achieved by one and the same filter structure, although necessarily with different filter coefficients. The resulting filter structure is composed of two cascaded two-dimensional antimetric wave digital filters of total degree sixteen (see Fig. 3). This structure can be used as a basis for a hardware realization, so that its installation in the underlying radiography system leads to an improved resolution of details by simultaneously fulfilling the requirements listed in the summary.

Experimental Results

All the optimized restoration and enhancement filters mentioned above have been tested on X-ray images of coronary arteries and hilus obtained from Siemens [1]. These images have been digitized into 512 x 512 x 8-bits. According to a subjective criterion we have found that the best improvement can be achieved by applying the Wiener-enhancement filter [6]. This is (ideally) defined by

$$H(\omega_1, \omega_2) = (1 - \xi) + \xi \cdot T_0^2 \cdot \omega_r^2 \cdot M(\omega_1, \omega_2), \quad \forall (\omega_1, \omega_2) \in \Omega_N, \tag{4}$$

where $\omega_r = \sqrt{\omega_1^2 + \omega_2^2}$, and ξ is the enhancement factor constrained to $0 \le \xi \le 1$.

Fig. 4 shows the ideal Wiener-enhancement filter according to (4) with $\xi = 0.5$, while the actual optimized Wiener-enhancement filter is represented in Fig. 5.

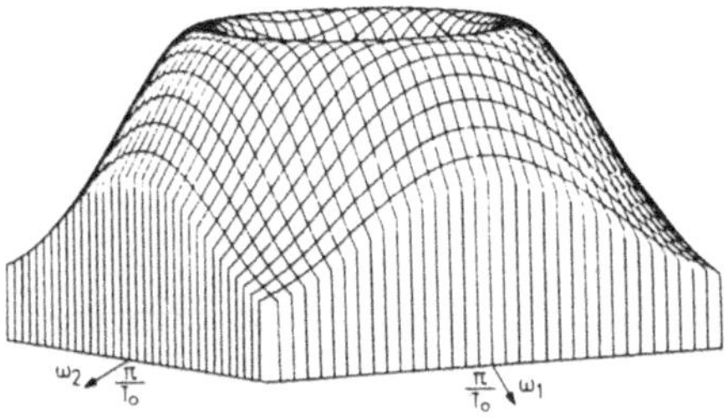

Fig. 4. Ideal Wiener-enhancement filter ($\xi = 0.5$)

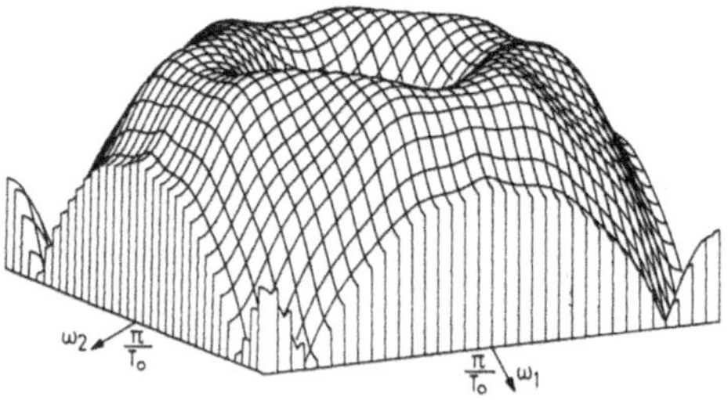

Fig. 5. Optimized Wiener-enhancement filter ($\xi = 0.5$)

The original images of coronary arteries and hilus are shown in Figs. 6 and 7, and the result after processing the original images with the optimized Wiener-enhancement filter can be seen in Figs. 8 and 9, respectively.

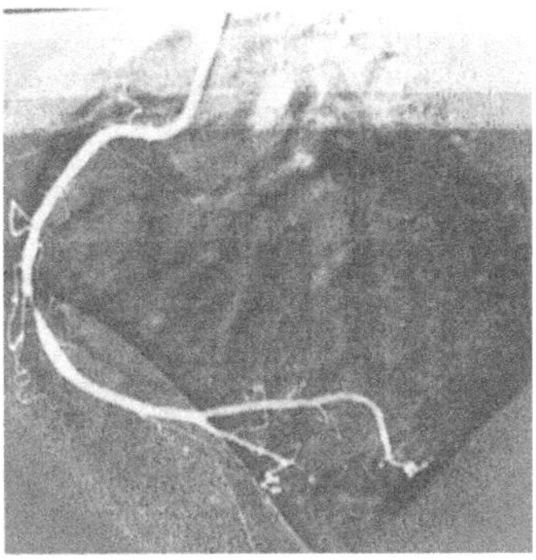

Fig. 6. Original image of coronary arteries

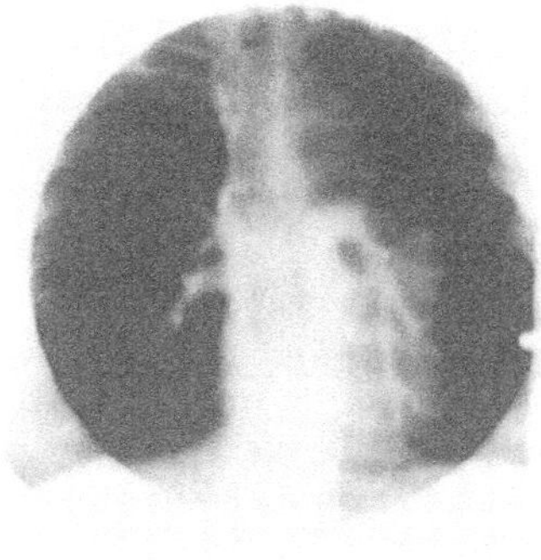

Fig. 7. Original image of hilus

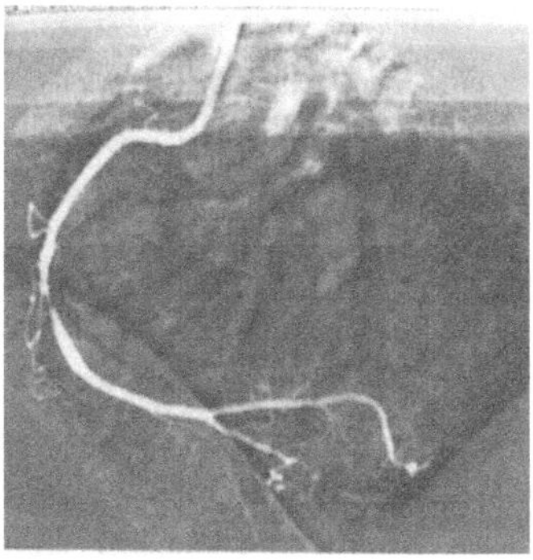

Fig. 8. Filtered image of coronary arteries

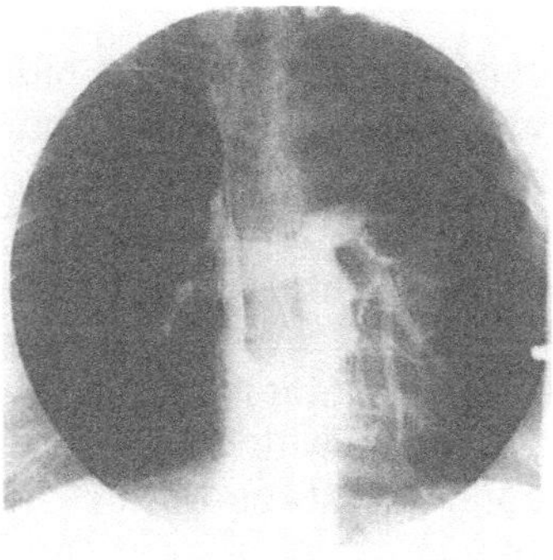

Fig. 9. Filtered image of hilus

References

1. Otto, P.: Private Communication. Erlangen: Siemens AG.
2. Fettweis, A.: Multidimensional digital filters with closed loss behavior designed by complex network theory approach. IEEE Trans. CAS-34 (1987) 338 - 344.
3. Fettweis, A.: Wave digital filters — theory and practice. Proc. IEEE 74 (1986) 270 - 327.
4. Pätzold, M.: Beiträge zur mehrdimensionalen diskreten Signalverarbeitung unter Verwendung antimetrischer Wellendigitalfilter. (Submitted as doct. diss.).
5. Andrews, H.C.; Hunt, B.R.: Digital image restoration. Englewood Cliffs, New Jersey: Prentice-Hall 1977.
6. Fales, C.L.; Huck, O.; McCormick, J.A.; Park, S.K.: Wiener restoration of sampled image data — end-to-end analysis. J. Opt. Soc. Am. A5 (1988) 300 - 314.
7. Modestino, J.W.; Fries, R.W.: Edge detection in noisy images using recursive digital filtering. Computer Graphics and Image Processing 6 (1977) 409 - 433.

Digital Images of Treatment Ports Acquired with Storage Phosphors: Comparison with Conventional Film Technique in a Phantom Model

JOERG W. OESTMANN*, CHRISTOPHER S. WALSH***, STEPHAN REICHELT *, C.C. WANG***, MARY AUSTIN-SEYMOUR***, NOAH C. CHOI***, JOHN E. MUNZENRIDER***, HERMAN D. SUIT***, REGINALD GREENE**

*Abteilung Diagnostische Radiologie I, Medizinische Hochschule Hannover

Departments of Diagnostic Radiology** and Radiation Medicine***,Massachusetts General Hospital, Harvard Medical School, Boston

Summary

We compared conventional radiographs (Cronex 7, DuPont™) and edge and contrast enhanced digital images (TCR 201, Toshiba™) in the detection of soft tissue (paraffin, acrylic) and bone (calcium carbonate) equivalent test objects on a Theratron™ 780 Co60 unit and a Varian Clinac™ 4 MV linear accelerator. Five hundred twenty-eight conventional and 528 digital images were evaluated in a blind and random fashion by six radiation oncologists (1056 observations per observer, 6336 total observations) with a three-day interval between sessions. Receiver operating characteristics (ROC) analysis based on a five-level confidence scale showed no statistically significant improvement in the detection of the test objects with edge and contrast enhanced digital radiographs.

Introduction

Portal films are neccessary to controll, adjust and document the irradiated field in radiotherapy.[1, 2, 3] Their objective is quite different from diagnostic images: Not the detection of pathology is the issue but the recognition of normal anatomic structures that serve as orientation for the irradiated volume. This difference in objective translates into a different approach to post-processing of digital portal films: While edge-enhancement in diagnostic images such as chest radiographs may well decrease the detectability of lesions, because the false positive rate rises due to misinterpretations of normal anatomic structures,[4,5,6] severe edge-and contrast enhancement promises improvements when only the delineation of known anatomic structures is called for. The detection of soft tissue and bone outlines on portal films exposed at megavoltage energies is hampered by low differences in attenuation and subsequently low image contrast.[7] The large focus size and relatively short focus -film-distances add geometric unsharpness.[8] Several techniques have been investigated that optimize the optical density of portal films, most of them starting off with conventional films that are then digitized.[9,10,11,12,13] Digital portal films acquired with storage phosphors are automatically adjusted for optimum density and can undergo edge and contrast enhancement to optimize the display of differences in attenuation. After preliminary clinical images showed some promise and found the approval of staff radiotherapists we designed a phantom study to verify our subjective evaluation.

Methods

Pieces of paraffin, acrylic (to simulate soft tissue:140 HU, 1.5 mm sections, 120 kVp, 9800 scanner, General Electric Corp.™, Milwaukee, WI) and calcium carbonate (to simulate bone:1100 HU) of regular and irregular shape and varying size ranging from 3 mm to 10 mm in maximal sagittal diameter were placed on a grid (manufactured from turkey wire) atop a 9.8 x 29.7 x 29.7 cm acrylic block. The wire grid had 66 sections, in which the test objects were positioned in random order. Conventional radiographs (Cronex 7, DuPont™) and storage phosphor image plates (ST, Fuji™, Stamford, CT,incorporated in the TCR-201, Toshiba ™, Tustin, CA) were exposed in lead-lined cassettes on the Theratron™ 780 Co60 unit with 0.06 minutes and on a Varian Clinac™ 4 MV linear accelerator with 3 (conventional technique) and 2 (digital technique) monitor units . Image data (pixel size 0.2 mm, 1760 x 1760 pixel matrix 10 bits deep) were extracted from the storage phosphors with a helium-neon laser reader and displayed on single emulsion transparencies of 17.5 x 17.5 cm. Edge and contrast enhancement of the digital images was performed on a TCR 201 (Toshiba™, Tustin, CA) with an unsharp mask subtraction technique according to the following formula:

$$I_e = I_o + [F_e (I_o - I_u)]$$

(I_e = Edge enhanced image, I_o = Original image, F_e= Enhancement factor , I_u = spatially averaged unsharp mask). The kernel size of the unsharp mask was selected to obtain low frequency edge enhancement. Maximal enhancement was performed (enhancement factor of 16 on a maximum scale of 16) independent of image density (fig. 1).

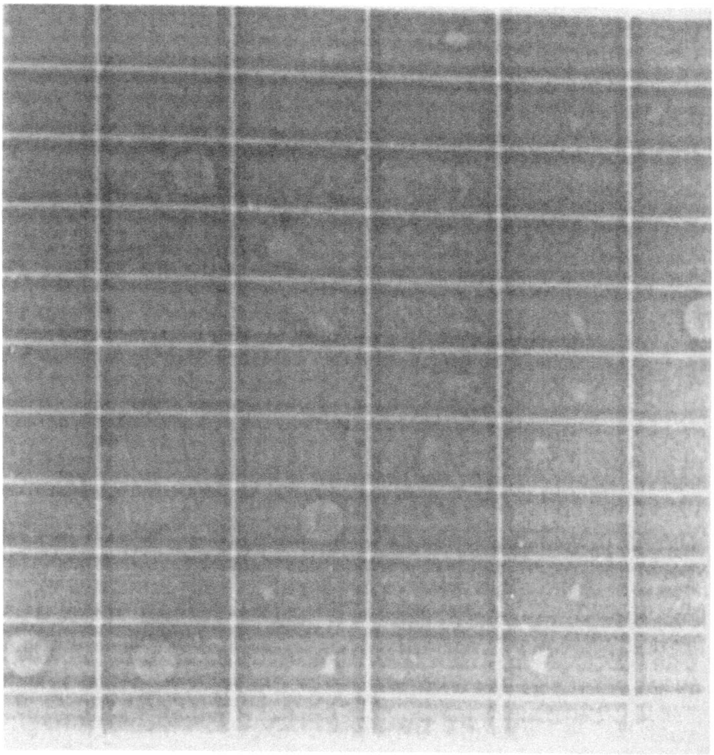

Figure 1:
Test objects of acrylic, paraffin and calcium carbonate superimposed on acrylic block

Five hundred twenty-eight conventional and 528 digital sections (4 x 66/radiotherapy unit) were evaluated in a blind and random fashion by six radiation oncologists (1,056 observations per observer, 6,336 total observations) under standardized viewing conditions with a three-day interval between sessions. They were asked to state wether they saw the test objects within the sections and to rate their confidence according to a five-level scale. Data were analyzed according to receiver operating characteristics (ROC)[14,15] for individual and average[16] performance. Detectability was estimated by the area under the

ROC curve.[17] The significance of differences between the means of the ROC areas of individual readers was tested with Student's paired T-Test.

Results

Receiver operating characteristics (ROC) analysis based on six radio-oncologist readers showed an increased detectability of both "bone" and "soft tissue" test objects with edge enhanced digital images exposed on the cobalt unit(fig. 2). This difference in detectability was not significant, though. Both imaging techniques performed alike on the accelerator unit (fig. 2).

DETECTION OF "SOFT TISSUE" AND "BONE" TEST OBJECTS

ROC - AREA ±SD

	CO-60				4 MV ACCELERATOR			
	BONE		SOFT TISSUE		BONE		SOFT TISSUE	
DIGITAL	0.927	±0.027	0.867	±0.033	0.903	±0.068	0.804	±0.045
CONVENTIONAL	0.878	±0.035	0.812	±0.066	0.922	±0.021	0.808	±0.052

Figure 2:
Mean diagnostic performance of 6 radio-oncologists with conventional and digital edge-enhanced portal films.

Discussion

Post-processing of portal films has been investigated by several groups with a special focus on optimized optical density and contrast enhancement.[9-13, 18] Advantages were seen in subjective comparisons of clinical images in these studies as well as in our preliminary evaluations, which were performed with edge enhanced portal films. We could not fully duplicate these results in observer performance tests that compared conventional and edge enhanced digital portal films. Detectability of both "bone" and "soft tissue" with low frequency edge enhanced digital images and conventional port films showed no significant advantage for the digital technique under the optimized experimental conditions. However, suboptimal optical density of conventional portal films is a frequent shortcoming and was not present in our experimental set-up. In clinical images the advantages of automatic density control and edge enhancement were better appreciated (fig 3a+b). Further investigations with other post-processing algorithms are warranted.

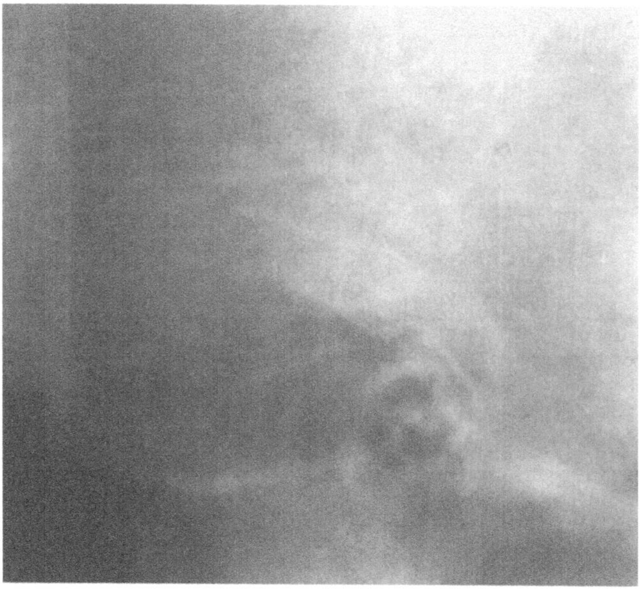

Figure 3a:
Conventional portal film of the skull base.

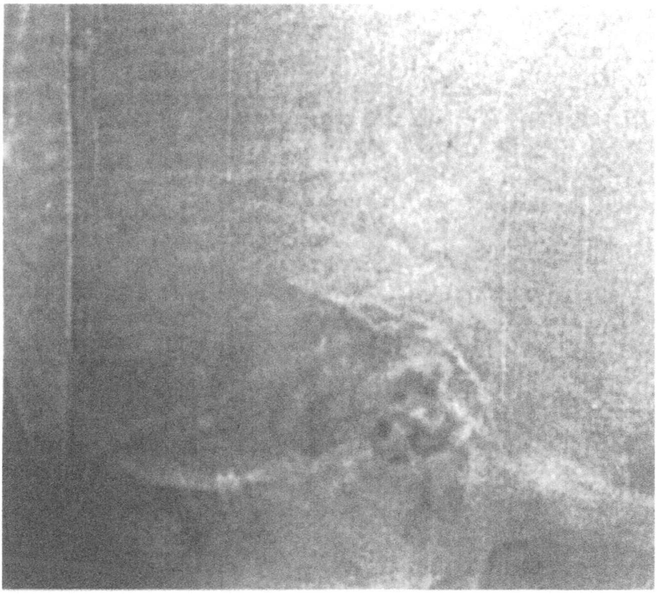

Figure 3b:
Digital portal film of the skull base with severe enhancement of low spatial frequencies.

References

1 MARKS JE, HAUS AG. THE EFFECTS OF IMMOBILIZATION ON LOCALIZATION ERROR IN THE RADIOTHERAPY OF HEAD AND NECK CANCER. CLIN RADIOL 1976; 27:175-177.

2 PEREZ CZ, STANLEY K, GRUNDY G, ET AL. IMPACT OF IRRDAIATION TECHNIQUE AND TUMOR EXTENT IN TUMOR CONTROLL AND SURVIVIAL OF PATIENTS WITH UNRESECTABLE NON-OAT CELL CARCINOMA OF THE LUNG. CANCER 1982;50:1091-1099.

3 WHITE JE, CHEN T, MCCRACKEN J, ET AL. THE INFLUENCE OF RADIATION THERAPY QUALITY CONTROL ON SURVIVAL, RESPONSE AND SITES OF RELAPSE IN OAT-CELL CARCINOMA OF THE LUNG. CANCER 1982;50:1084-1090.

4 OESTMANN JW, BOURGOUIN PM, RUBENS JR, RHEA JT, LLEWELLYN HJ, GREENE R. IMPACT OF POST-PROCESSING ON THE DETECTION OF SIMULATED PULMONARY NODULES WITH DIGITAL RADIOGRAPHY. INVEST RADIOL (IN PRINT)

5 OESTMANN JW, KUSHNER D, BOURGOUIN PM, LLEWELLYN HJ, MOCKBEE BW, GREENE R. SUBTLE LUNG CANCERS: IMPACT OF EDGE ENHANCEMENT AND GRAY SCALE REVERSAL ON DETECTION WITH DIGITIZED CHEST RADIOGRAPHS. RADIOLOGY 1988;167:657-658.

6 FOLEY WD, WILSON CR, SAN DRETTO M. THE EFFECT OF VARYING HIGH PASS FILTER PARAMETERS ON THE DETECTABILITY OF ISOLATED PULMONARY NODULES IN DIGITIZED CHEST IMAGES. INVEST RADIOL 1986; 21:305-310.

7 AMOLS HI, REINSTEIN LE, LAGUEUX B. A QUANTITATIVE ASSESSMENT OF PORTAL FILM CONTRAST AS A FUNCTION OF BEAM ENERGY. MED. PHYS 1986;13:711-716.

8 KRONHOLZ HL, SCHÜTZ J, SCHNEPPER E. KONTRASTREICHE AUFNAHMEN ZUR FELDKONTROLLE UND FELDDOKUMENTATION IN DER STRAHLENTHERAPIE MIT ENERGIEREICHEN PHOTONEN. STRAHLENTHERAPIE UND ONKOLOGIE 1988;164:91-93

9 LEONG J. A DIGITAL IMAGE PROCESSING SYSTEM FOR HIGH ENERGY X-RAY PORTAL FILMS. PHYS MED BIOL 1984;29:1527-1535.

10 MEERTENS H. DIGITAL PROCESSING OF HIGH-ENERGY PHOTON BEAM IMAGES. MED PHYS 1985;12:111-113.

11 PIZER SM, AUSTIN JD, PERRY JR, SAFRIT HD, ZIMMERMAN JB. ADAPTIVE HISTOGRAM EQUALIZATION FOR AUTOMATIC ENHANCEMENT OF MEDICAL IMAGES. PROC SPIE, 1986;626:242-250.

12 REINSTEIN LE, ALQUIST L, AMOLS HI, LAGUEUX B. QUANTITATIVE EVALUATION OF A PORTAL FILM CONTRAST ENHANCEMENT TECHNIQUE. MED PHYS 1987; 15:309-313.

13 SHEROUSE GW, ROSENMAN J, MCMURRY HL, PIZER SM, CHANEY EL. AUTOMATIC DIGITAL CONTRAST ENHANCEMENT OF HIGH-ENERGY RADIOTHERAPY FILMS. INT J RADIAT ONCOL BIOL PHYS 1987;13:801-806.

14 METZ C. BASIC PRINCIPLES OF ROC ANALYSIS. SEMIN NUCL MED 1978; 8:283-298.

15 DORFMAN DD, ALF E. MAXIMUM LIKELIHOOD ESTIMATION OF PARAMETERS OF SIGNAL DETECTION THEORY AND DETERMINATION OF CONFIDENCE INTERVALS - RATING-METHOD DATA. J MATH PSYCHOL 1969; 6:487-496.

16 SWETS JA, PICKETT RM, WHITEHEAD SF, GETTY DJ, SCHNUR JA, SWETS JB, FREEMAN BA. ASSESSMENT OF DIAGNOSTIC TECHNOLOGIES. ADVANCED MEASUREMENT METHODS ARE ILLUSTRATED IN A STUDY OF COMPUTED TOMOGRAPHY OF THE BRAIN. SCIENCE 1979; 205:753-759.

17 HANLEY JA, MCNEIL BJ. THE MEANING AND USE OF THE AREA UNDER A RECEIVER OPERATING CHARACTERISTICS (ROC) CURVE. RADIOLOGY 1982; 143:29-36.

18 POWLIS, WD, BRIKMAN I, SESHADI SB, BLOCH P. PORTAL RADIOGRAPHS: DIGITAL ENHANCEMENT OF CONTRAST. RADIOLOGY 1988;169:839-841.

Noise-Reduction Algorithm for Dual Energy Radiography

W. Bautz[1]), W. Kalender[2])

1)Radiologische Klinik der Universität Tübingen, 7400 Tübingen
2)Siemens AG, UB Med. Technik, 8520 Erlangen

Noise in dual energy radiographs (material density images) obtained by prereconstruction basis material decomposition techniques is negatively correlated. This effect is exploited for noise reduction. The algorithm minimizes noise-related differences between pixels and their local mean value, adhering to the constraint that monoenergetic values, which can be calculated from the density images, have to remain unchanged. Typically, pixel noise is reduced by the factor of 2 - 5. The algorithm also has beneficial effects on motion and sampling artifacts. While quantitative results (e.g. bone density measurements) remain unchanged, edge effects may occur. To suppress these, adaptive procedures have been investigated. Results are presented both for phantom measurements and clinical studies.

Merit Figures for Image Quality Evaluation in a Digital Radiography System and Correlation with Clinical Examination

A. AGNIFILI (1) (2), P. PAVONE (1), A. VIGLIENZONI(3), R. DELFINO (3), R. PASSARIELLO (1)

(1) Department of Radiology, University of L'Aquila, L'Aquila
(2) Department of Electrical Engeneering, University of L'Aquila, L'Aquila
(3) 3M ITALIA Labs, Ferrania, Savona

Summary
A digital radiography system based on storage phosphor panels has been developed in 3M Labs (Ferrania,SV-Italy) and installed for evaluation in a clinical environment at the Department of Radiology of L'Aquila University.
This paper briefly describes this prototype and afterwards deals with a set of merit figures which have been adopted to measure system (sensor and read-out unit) performance. We have tried to relate these merit figures to subjective image quality perception. Radiological examinations have been carried out on an early series of patients. Visual evaluation has been obtained by two experienced radiologists, who have been asked to score the evaluation of normal and pathological structures, and compared with the results of the quantitative analysis.

Introduction

A digital radiography prototipe, developed in 3M Labs at Ferrania (Savona-Italy), was evaluated by carrying out quantitative measurements based upon a set of merit figures. At the moment, the system is installed at the Department of Radiology of L'Aquila University; here the unit has been connected in a PACS configuration with two other digital systems (digital subtraction angiography and digital fluoroscopy) in a separate Section of the Department, with low patients' volume, created for research purposes.
The aim of the experimental installation is the evaluation of the capabilities of Digital Radiology in a clinical setting. After briefly describing the system, image quality measurements will be shown and compared with radiologist subjective evaluation.

System description

The complete system consists of an x-ray sensor (storage phosphor panel) and a read-out unit (laser scanner+ electronics). The scheme of the latter is shown in fig.1. The panel (P) is placed on top of the transport section (TS) which is moving in the y direction. A galvo scanner (GS) makes the laser beam scan the x direction. The laser source is indicated by L while AOM stands for Acousto Optic Modulator used to switch off the laser during the retrace.An elliptical reflecting surface (M) is used to focus the photo-stimulated light onto an optical fibers array (OF)

which drives the light to a filter F and then to a photomultiplier (PM).
The signal from the opto-electric converter is there amplified and filtered (A) and sent to the A/D board of the computer. Panel erasing is carried out by a separate unit based on a high intensity sodium light. The system is also connected to a 3M laser imager to produce hard-copies of the examination after viewing them on a high-resolution monitor (fig.2). The radiological images can be produced with the following digital formats:

- 1024x1024 pixels with 12 bits/pixel;
- 2048x2048 pixels with 12 bits/pixel;
- 4096x4096 pixels with 10 bits/pixel.

Merit figures

Basic merit figures adopted are:

- MFT, modulation transfer function (wich gives informations on spatial resolution);
- Wiener spectrum (on noise power spectrum);
- S/N, signal to noise ratio (on the threshold contrast and noise);
- DQE, detective quantum efficiency (on the influence of noise on information capability).

Calculations are performed taking into account a single direction. To fully characterize the system, measurements in two perpendicular directions are required.

MTF

Modulation Transfer Function is defined according to:

$$\text{MTF (f)} = \frac{\text{DFT (f)reproduced image}}{\text{DFT (f) original image}} \qquad (1)$$

A lead grid with spatial frequencies from 0.05 up to 10 lp/mm is adopted to produce the images.
(1) is normalized to 1 in the case of f=0.
A computer program evaluates the Fourier transform. Owing to systematic (non linearities) and/or accidental (noise, spatial quantization) errors, even harmonics can be non-zero. Two procedures can be meaningfully adopted:

1. MTF values are reliable until (odd harmonics) < (even harmonics)
2. Different spatial frequencies of the bar pattern are considered limiting the analysis to the first harmonic.

The latter procedure is commonly adopted.

Wiener spectrum

It is a frequency analysis of an uniform gray exposure, carried out for both rows and columns and used to detect coating and/or scanning problems.
Here a window with 2^N pixels (N integer up to 9) in the direction under analysis is considered, while in the other one an adequate number of lines is taken to average random noise. Two Fourier transformations are carried out; the first one (after square module) determines a biased power spectrum (according to Parsival theorem), while the second one yields the unbiased spectrum.

S/N

Signal to noise ratio can provide information on sensor uniformity and read-out noise; in particular:

a. spatial fluctuation in the number of absorbed x-ray quanta (image forming fluctuation)

b. spatial fluctuation in sensor absorbtion, associated with sensor random inhomogeneities (space static fluctuation). For this reason this merit figure is fundamental to measure the possibility of detecting low contrast patterns, which are, in turns, very important in a radiographic system.

A stepwedge is used in this case to produce an image showing different gray levels. With the use of a window of fixed dimension mean and standard deviation are calculated for each region; mean signal variation to noise ratio is calculated according to the following formula:

$$(S/N) = \frac{\text{AV. GREY REGION } n - \text{AV. GREY REGION } n\text{-}1}{\text{ST. DEV. REGION } n - \text{ST. DEV. REGION } n\text{-}1}$$

DQE

Detective Quantum Efficiency (DQE) is the ratio between the squares of the output and input S/N.

$$DQE = \frac{(S/N)^2_{out}}{(S/N)^2_{in}}$$

It is possible to interpret the values of DQE as relating to the comparison of spatial frequency fluctuations in the input and output. In terms of the complete Wiener spectra of the input and output noise it becomes possible to get a more complete DQE definition. This is an expression of DQE in terms of spatial fequencies and is one of the most important functions for assessing the performance of an imaging sensor. Such a function is intimately related to the total transfer involved in the imaging process.This can be understood from the formula used to carry out such measurement :

$$DQE(f) = DQE(0) * \frac{MTF(f)}{W(f)}$$

DQE (O) was calculated according to:

$$DQE(0) = \frac{(S/N)^2_{out\text{-->}in}}{(S/N)^2_{in}}$$

All the measurements were carried out on the same panel region and in the same read-out conditions.

Experimental

Two different kinds of phosphor screens were tested, having higher resolution/reduced efficiency, lower resolution/higher speed respectively, corresponding to two different coating weights.
Figs. 3-4 show a comparison between these two sets of panels in terms of the previously mentioned parameters.
In all these cases, pixel size was 100 and operating voltage 45 KVp.

Radiologist subjective evaluation

Qualitative evaluation of digital radiographs and their comparison with conventional x-ray films of the the same patients has been performed by two experienced radiologists. Image quality, spatial resolution and diagnostic content were rated in the chest and abdominal areas.
Positive results were obtained in both areas with digital radiographic images.The radiologists pointed out the high quality of the images obtained (with no re-takes due to erroneous exposure settings) and their utility in detecting low contrast lesions.
Evaluation with different window setting (on the grey level scale) to display mediastinal, bony and parenchimal structures was considered by both radiologists particularly useful and of diagnostic value in difficult anatomical areas. The retrocardiac and costo-phrenic regions were better evaluated on digital films. Also the mediastinal structures could be analized with higher detail.
The radiologists recognise a lower spatial resolution for digital images than for conventional x-rays film when the image format 1024x1024x12 has been used. No significant differences have been observed with the format 2048x2048x12, wich is near optimal.
However, the slightly lower spatial resolution was widely compensated by the gain obtained, in terms of diagnostic information, with the higher contrast resolution of digital system.

REFERENCES

1. Sonoda, M.; Takano, M.; Miyahara, J.; Kato, H.: Computed radiography utilizing scanning laser stimulated luminscence. Radiology 148 (1983) 833-838.
2. Tateno, Y.; Iinuma, T.; Takano M. (eds.) Computed radiography. Tokyo, New York: Springer-Verlag 1988.
3. Brody, W.R. Digital radiography. New York : Raven Press 1984.
4. Fuhrman, C.R.; Gur, D.; Good, B.; Rockette, H.; Cooperstein, L.A.; Feist, J.H.: Storage phosphors radiographs vs conventional films: interpreters' perceptions of diagnostic quality. AJR 150 (1988) 1011-1014.
5. Dainty, J.C.; Shaw, R. Image science. London, New York, S.Francisco: Academic Press 1974.

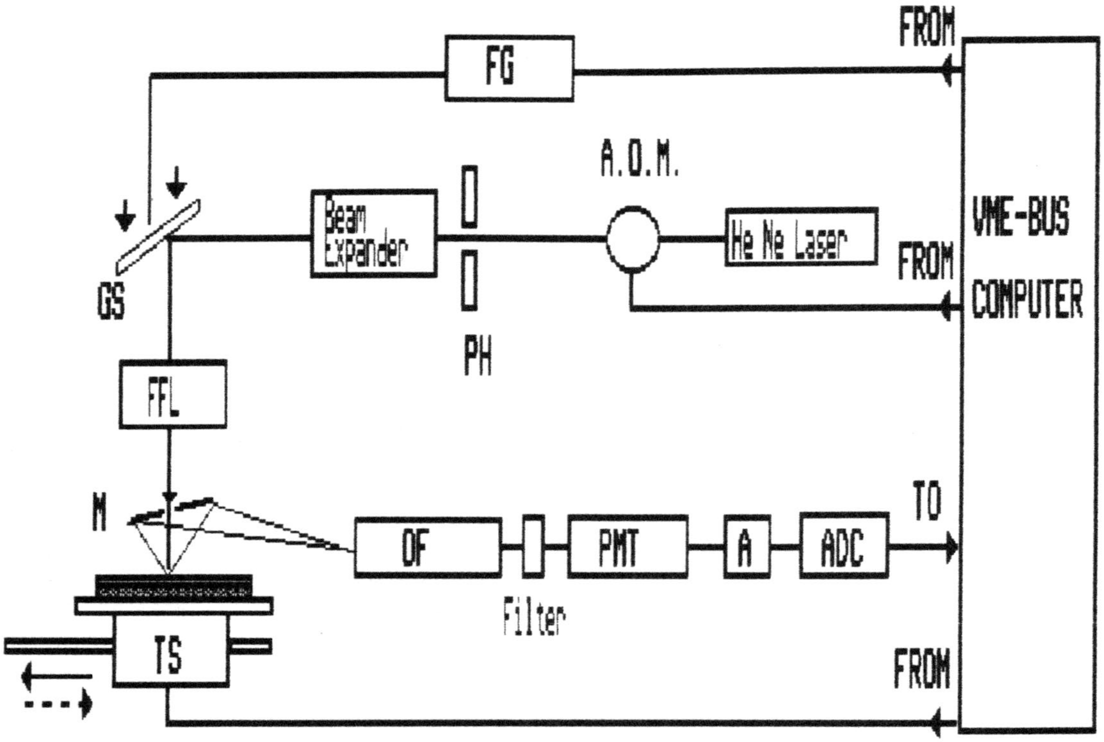

Fig. 1: Internal scheme of the read-out unit

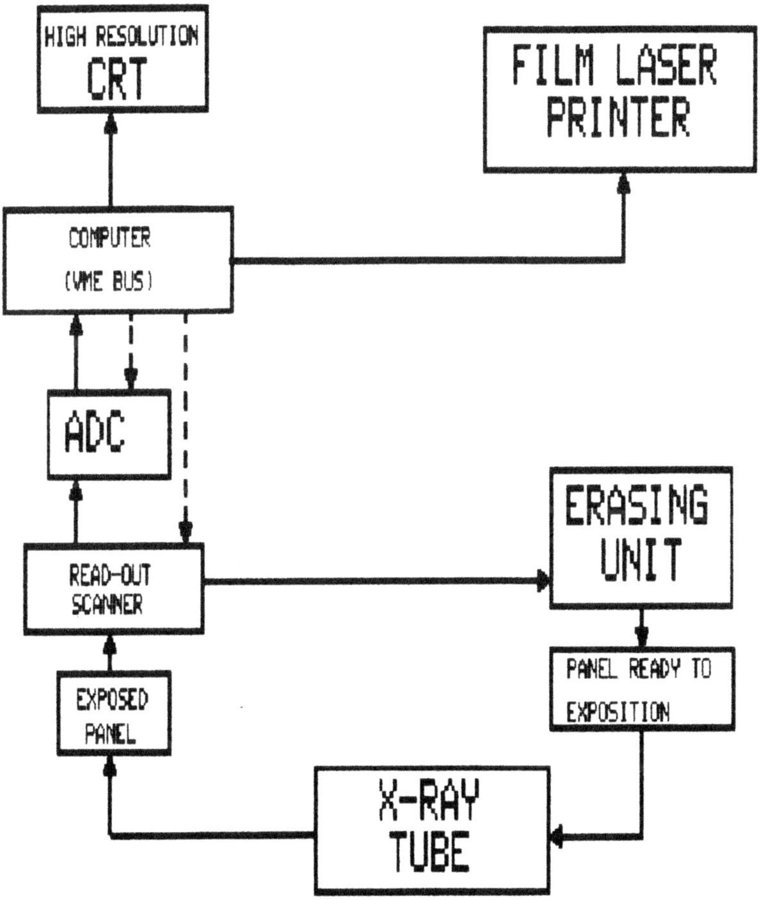

Fig. 2: Logic scheme of the unit

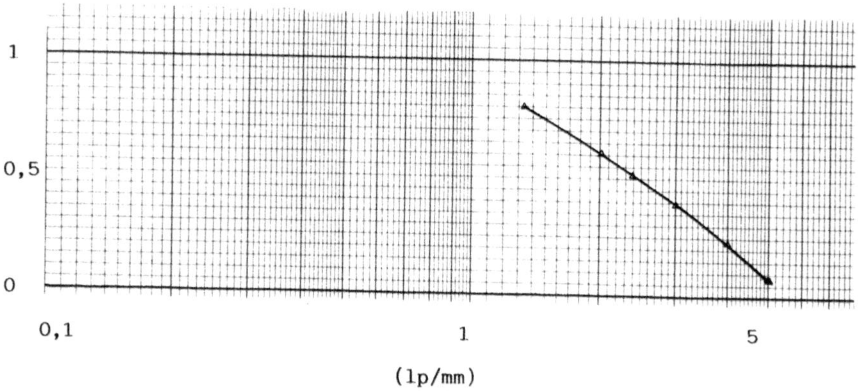

Fig. 3A: Modulation transfer function of the first panel

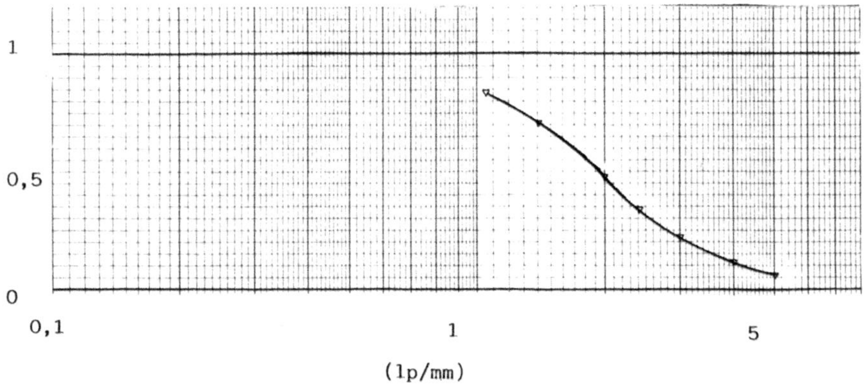

Fig. 3B: Modulation transfer function of the second panel

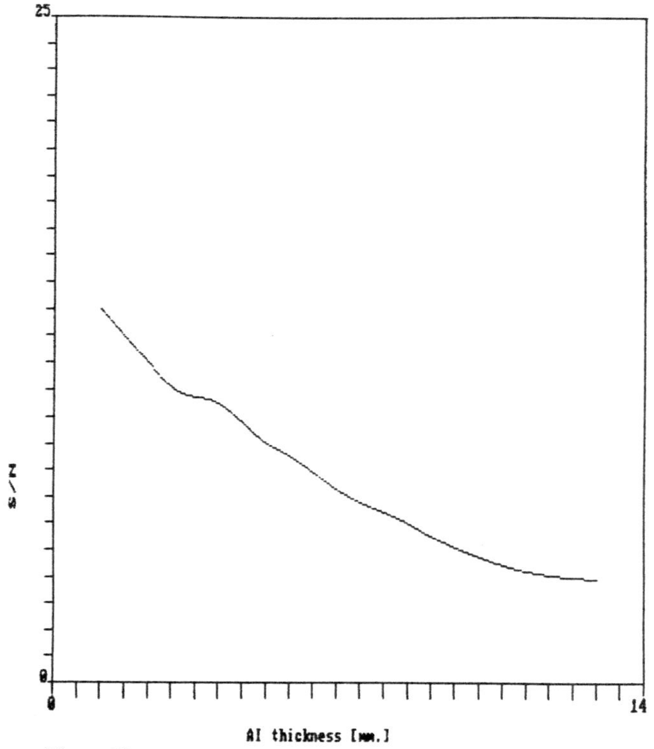

Fig. 4A

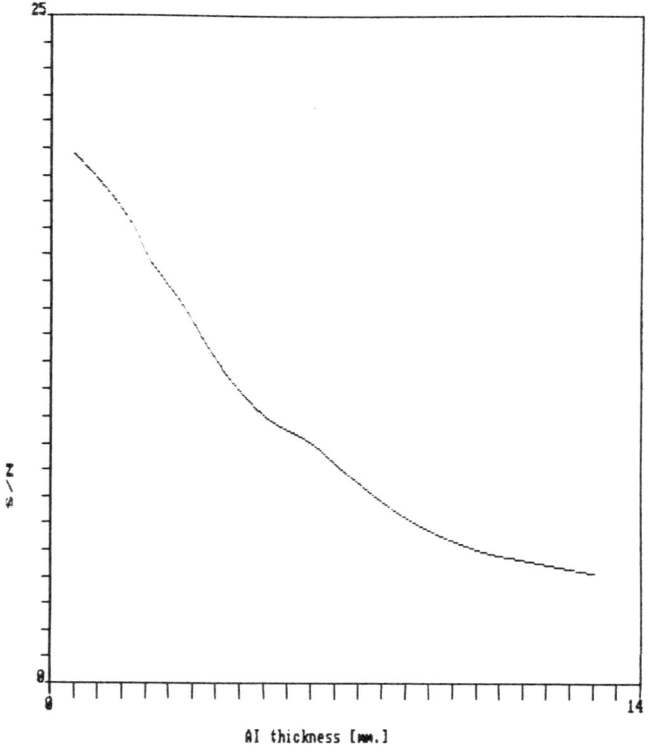

Fig. 4B

Automatic Detection of Clusters of Calcifications in Digital Mammograms

D H DAVIES AND D R DANCE

JOINT DEPARTMENT OF PHYSICS,
INSTITUTE OF CANCER RESEARCH AND THE ROYAL MARSDEN HOSPITAL,
LONDON, SW3 6JJ, U.K.

SUMMARY

The automatic detection of significant clusters of calcifications in digital mammograms has been investigated using a combination of local area thresholding and edge analysis.

Segmentation of any calcification from the background of normal breast structure in any given mammogram was accomplished by using a local area thresholding technique. These objects were then analysed using size, shape and gradient measures to extract clusters of calcifications. The results for a set of 25 mammographic regions, each 5.76 X 5.76 cm^2 in area, show that the computer system can achieve 100% true positive cluster detection with a false positive rate of 12%.

INTRODUCTION

Mammography is an X-ray technique which has been specially developed for taking images of the breast and is able to detect clinically occult early breast cancer. On mammograms approximately 50% of breast carcinomas exhibit microcalcifications (1) and between 60 and 80% of breast carcinomas show calcifications upon histologic analysis. Following the publication of the Forrest report in 1986 (2), the Department of Health and Social Security in the U.K. is setting up a National Breast Screening Programme for all women aged between 50 and 64. The report recommends that single-view mammography, with an interval of three years between consecutive mammograms, is the best method to employ for basic screening in a mass population screening programme. This could result in 1,440,000 screening mammography patients per year and double this number of films to be reported.

The success of the screening programme may be enhanced if an automatic computer method could be employed to pre-read digital mammograms. Ideally, this system will produce two sets of mammograms, one set which can be confidently classified as normal, and the other as abnormal. The latter set of mammograms will require further reading by a radiologist and to aid him in this task all suspicious areas should be labelled. As an initial step in the development of such a system the automatic detection of calcifications is being examined. Other workers in this field, in particular Chan et al (3), have developed a computer system which can achieve a true positive detection rate of 85%.

In this paper we describe our approach to developing an automatic computer system for detecting calcifications in mammograms and present some of the results we have obtained.

IMAGE COLLECTION

In this study digital mammograms were obtained by digitising conventional screen-film mammograms using a Joyce-Loebl scanning micodensitometer with a pixel size of 0.1mm x 0.1mm and 8 bit A/D conversion. The digital images were then transferred to a SUN 3/60G workstation for processing and display. (This workstation is capable of displaying 1152 x 900 pixels and has 256 grey levels).

Twenty five mammograms were selected to be digitised from a training library being developed in the Royal Marsden's X-ray department. These films were a mixture of normal mammograms and mammograms containing clusters of calcifications. From each one a region of interest of size 576 x 576 pixels (approximately one third of the total breast area) was chosen. This size was picked so that both the original region of interest and the processed image could be displayed simultaneously. Of these twenty five regions 16 were normal, i.e. no calcification clusters present, while the remaining 9 regions had 14 clusters of calcifications. The regions were selected to be representative of the complete mammographic image and included both central and peripheral parts of the breast. The clusters were located and labelled by an experienced radiologist.

LOCAL AREA THRESHOLDING

Segmentation of all possible calcifications from the background of normal breast tissue is achieved using a local area thresholding technique based on a point dependent method. The grey level histogram for each subimage is calculated and then classified as being either bimodal or unimodal. If the histogram is bimodal then a threshold is set at the grey level corresponding to the valley of the histogram. However, if the histogram is unimodal then the threshold for that subimage is interpolated from neighbouring subimages.

Each region of interest is subdivided into smaller overlapping subimages. (The overlap of subimages reduces the effect of grey level discontinuities at the boundaries of two different subimages). Inspection of Figure 1 reveals that any quarter of a given subimage is overlapped by four other subimages. Therefore, any given pixel could be segmented from the background by upto five separately determined thresholds. A principle is then applied where a pixel must be segmented from the normal background in a pre-determined number of the overlapping subimages for it to appear in the final threshold image.

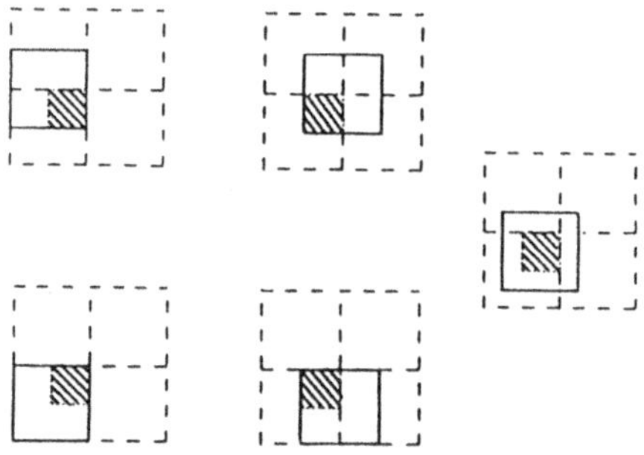

FIGURE 1: The five sub-images used for local area thresholding (solid lines). The region of interest (shown shaded) remains fixed in position whereas the location of the sub-image used to set the threshold is different in each case. Grid lines are also shown (dotted lines).

SIZE AND SHAPE ANALYSIS

In order to reduce the number of detected objects labelled as potential microcalcifications by the thresholding process their digital image characteristics were observed. One of the characteristics analysed was area. Consequently, to eliminate noise and normal breast structures, upper and lower limits are set on the area of detected objects.

To reduce the number of objects further the remainder are analysed according to their shape. Calcifications were found to have relatively small perimeter to area ratios compared with normal breast structures and a shape description based on this ratio was used and an appropriate threshold selected.

GRADIENT ANALYSIS

Calcifications tend to have well defined borders with respect to the surrounding breast tissue. Therefore, to reduce the number of potential calcifications still further, the edge strengths of the remaining objects are analysed. An adaptation of the Roberts Gradient method (4) is used. For each of the remaining objects the mean Roberts gradient value for a pixel edge lying on a given object's perimeter is calculated. Various mean gradient criteria are then set for objects having different shape parameter values.
All items passing these final tests are then labelled as calcifications.

CLUSTERING

Finally, a clustering principle is imposed to locate significant groups of calcifications. If the nearest neighbour calcification is less than a certain distance, and there are more than a pre-selected number linked in this way then a cluster is said to exist. The pre-selected values were chosen according to the clinical expertise of a radiologist.

EXPERIMENTAL RESULTS AND CONCLUSIONS

In order to provide an assessment of the methods described in the previous section, the automatic computer detction system was applied to the twenty-five mammogram regions in our data base.

From our data base of clinical mammogram regions the computer system detected 100% true positive calcification clusters (14/14) with 12% false positives (3/25).

The preliminary results presented in this paper for the detection of calcifications are promising. However, in future work we aim to reduce the number of false positives and examine the results obtained using a larger data set. Also, we intend assessing the affects of increased sampling rates and higher A/D conversions.

Clearly, we have demonstrated the potential for using such a system to detect clusters of calcifications in digital mammograms. With optimisation and further studies we feel confident that the results presented in this publication can be improved upon.

ACKNOWLEDGEMENTS

The Institute of Cancer Research receives support from the Cancer Research Campaign and the Medical Research Council. This work has been funded by CRC/MRC grant number 841 3976 (387/388).

REFERENCES

1. Black J W, and Young B,A Radiological and Pathological Study of the Incidence of Calcifications in Diseases of the Breast and Neoplasms of Other Tissues. Br. J. Radiol. 58: 596- 598, (1965).

2. Forrest P,Breast Cancer Screening. Report to the Health Ministers of England, Wales Scotland and Northern Ireland by a working group chaired by Professor Sir Patrick Forrest. H M S O, (1986).

3. Chan H P, Doi D, Galhotra S, Vyborny C J, Macmahon H and Jokich P M,Image Feature Analysis and Computer Aided Diagnosis in Digital Radiography. 1. Automated Detection of Microcalcifications in Mammography. Med. Phys. 14(4), 538- 548, (1987).

4. Young T Y, Fu K S Handbook of Pattern Recognition and Image Processing. Academic Press, Inc.,pp216-217,(1986).

A Step Towards PACS: The DS 1000 in Traumatology

CH.SCHMIDT*,H.K.DEININGER**

*Bürgerhospital Frankfurt/Main,**Städt.Kliniken Darmstadt

Summary

Medical diagnosis is increasingly being changed by the development of new technologies.Conventional diagnostic procedures are consistently making way for digital methods.Considering the increasing use of digital technologies in administrative and diagnostic areas it only seems natural to begin taking advanture of the benefits which are made possible by this new technology.

A central Picture Archiving and Communication System (**PACS**) seems to be an ideal solution provided that completely using digital technology.Digital image intensifier radiography in Traumatology would provide a number of advantages.The digital workstation presented is capable of demonstrating all clinically relevant changes of the traumatic skeleton,whereby the image quality corresponds approximately to a high-speed film - screen combination.

The development of new technologies (e.g. gamma cameras,computed tomography and magnetic resonance systems) does not only result in a widening of the diagnostic spectrum,but,in addition,has enabled electronics to gain a footing in medical diagnosis.

We do not only find it in the form of automation in conventional x-ray units and in newly developed equipments,but also in image reconstruction,archiving,image communication and in organization. While many features have already become a regular fixture in daily routine work,picture archiving and communication systems (PACS) still encounter difficulties.This is mainly due to the fact that 70 to 80% of all x-ray examinations continue being performed in analog forms,i.e. by direct documentation onto the film.Consequently,this means for a radiology department that it must have two archives:A conventional one for the films and an additional archive for digital data.To remedy this unpractical and uneconomical situation,three possibilities present themselves:

1. Documentation of digital examinations on hardcopy.

2. Digitalization of analog images by means of laser digitizer.
3. Replacing classical radiography by digital procedures (digital imaging intensifier radiography,luminescence radiography).

Today the hardcopy principle is the usual method.This procedure is equally practical and efficient and is only associated with a relatively low loss of information.

Subsequent digitization of analog images has to be dismissed for various reasons.First of all,one does not see why such an expensive and excellent recording medium as the x-ray film should be used only as a kind of "intermediate data storage" and after digital information conversion,be discarded.Secondly a loss of information in digitization,ranging from 30 to 40% approximately, is not acceptable.

In the following,we shall explain the reasons for our deciding on the third method and,in this case,particularly in favor of digital intensifier radiography.

A new diagnostic procedure in traumatology must cope with all problems of the injured patient,and,beyond that,should be able to remove all present difficulties arising from the lack of ability to cooperate on the part of the injured patient.

Furthermore,fast diagnosis by rapid imaging and the possibility of performing all further diagnistic examinations (e.g. angiographies and fluoroscopic examinations) as an emergency,at one workstation,without repositioning the patient,is of vital importance.Fast image transfer from the examination site to the attending services (Surgical casualty ward,operating room) should not be forgotten either.

Since,at the time being,luminescence radiography is no alternative to conventional radiography,we have designed a digital,traumatological workstation, which,in our opinion,largely meets the mentioned requirements.

The **DS 1000** is at the center of our equipment.It is connected to the generator and the central TV control unit of the x-ray installation via an interface.This allows automatic dose control both in fluoroscopy and in radiography.

The DS 1000 is used with a ceiling-mounted C-arm unit.Thanks to the versatile projection directions of the C-arm,it is particularly suitable for all radiodiagnistic examinations,where complete immobilization of the patient is of the utmost importance or,on account of his lack of ability to cooperate,

the highest demands are made on the radiographic system.
The examination unit is equipped with a 25 cm image intensifier whose input size can be reduced to 17 cm for better detail perception.
The objective of our work was to find out whether the digital workstation meets the four criteria which are required of a new method:Does it yield a gain in information?Does it result in less radiation exposure?Are expenses reduced?Are examination results obtained faster?
On account of the radiophysical properties of the image intensifier and the TV tube,a gain in information can be precluded from the beginning.Here,10^7 gray shades and a resolution of 5-12 Lp/mm (depending on the screen-film combination used),as offered by x-ray films,are up against 256 shades of gray and a maximum spatial resolution of 3 Lp/mm as afforded by digital image intensifier radiography.Therefore,first of all it was imperative to find out whether the imaging properties of the digital system are sufficient to represent all those changes of the traumatized skeleton,which are of clinical relevance.

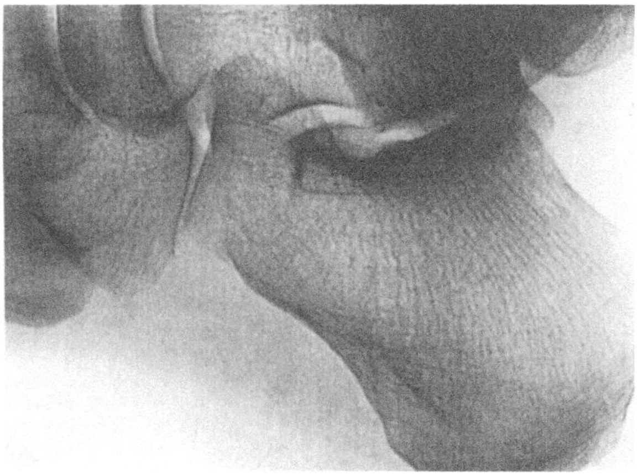

Fig. 1: The maximum resolution capability of our digital examination unit is 3 Lp/mm.This is still clearly below the spatial resolution of conventional x-ray films.Nevertheless,all details of clinical relevance can be detected.

We were fully aware of the fact that such an evaluation would only be possible in close cooperation with traumatologists.Not only the verification of radiological findings was of importance for us.We also wanted to know whether traumatologists - who after all,more frequently work with our radiographs than with the findings - are satisfied with the quality and characteristic of digital exposures.

First of all,we performed a series of phantom exposures.It soon appeared that an acceptable radiographic standard can only be achieved by accurate collimation and by using a scattered radiation grid and special filters. Consequently,old established radiological virtues are particulary effective in digital radiography.In this manner,we not only achieve a reduction in radiation exposure,but also a gain in contrast,definition and signal fidelity.

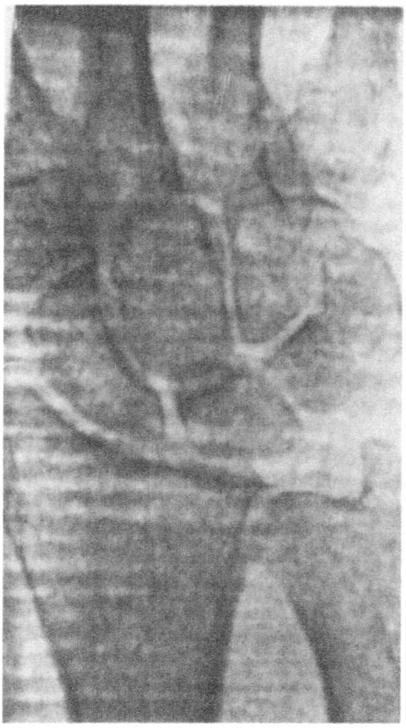

Fig. 2: Follow-up examinations after plaster cast suggest a slight superiority of the digital method.Fragment position and the bony structures can be better evaluated.

By means of image postprocessing,bones and soft tissue can be demarcated very well.Here,it should be said clearly that this Method does not yield any better visualization than that afforded by conventional x-ray film.

So far,we have examined 826 patients,additionally using digital image intensifier technique parallel to conventional diagnosis.

Evaluation of our examinations shows that all visible changes in the radiograph were also visualized digitally.In agreement with conventional radiological diagnostics,we found 369 fractures,whose extent and fragment position could be described accurately.457 patients had either normal findings or the fracture position was evaluated correctly in the control radiograph.Plaster cast checks showed an advantage for digital radiography.In this case,fragment position and bony structures were more clearly visualized.
Summarizing,it may be said that digital image intensifier radiography visualizes all traumatological changes of clinical relevance and,therefore,can be used in traumatology.At the time being,the quality afforded by conventional radiographs cannot yet be reached.

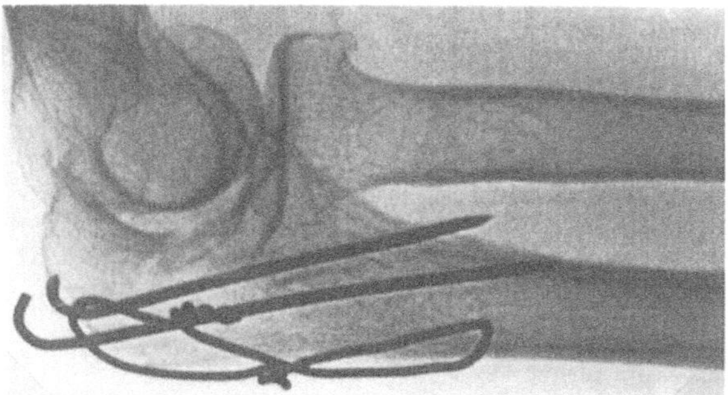

Fig. 3: The radiograph shows a condition after surgical repair of an olecranon fracture.With irritationfree position of the osteosynthesis material,the comminuted zone in the juxta-articular section of the fracture can be clearly recognized

With a marked reduction in radiation exposure (we required only about 1/4 of the usual dose),radiographs are attained immediately after exposure.Therefore this is an extremely fast radiographic procedure.Images can be quickly transferred by video cable to the various work areas.
With digital image intensifier radiography,electronics is not only used in the patient's interest,but also to relieve the medical staff of unnecessary work.

Digital Subtraction Angiography

Studies on Pulmonary Ventilation Using Digital Subtraction Radiography: A Comparison with Ventilation Perfusion Scintigraphy

Lisbeth Denbratt, Bengt Bergman, Bjorn Bake, Ulla Grangard and Lars Bjork

Dept. of Radiology, Chest Diseases and Clinical Physiology, University of Goteborg, Goteborg, Sweden.

Present address: Dept. of Radiology, Faculty of Medicine, Kuwait University, P.O. Box 24923, 13110, Safat, Kuwait.

Summary

Digital subtraciton radiography of the chest is a simple and rapid technique to detect local abnormalities of ventilation. However, as suspected measurements of relative changes in density of the lungs or parts of the lungs did not correlate with perfusion and ventilation measurements using 133-Xenon and scintilation detectors.

Introduction

Digital subtraction radiography has mainly been used in angiography. However the same technique can be used in other investigations where the attenuation of roentgen rays in a part of the body varies as a function of time. Such variations occur in the chest with respiration and in a previous report (1) the experience of this technique mainly in patients with suspected pulmonary embolism was reported. The following is a report of digital subtraciton radiography for studies of ventilation in patients with suspected pulmonary carcinoma. The findings were compared with results of a simple technique for perfusion-ventilation scintigraphy used in our hospital (2).

Material

Twenty-nine patients, 17 men and 12 women varying in age from 30-78 years, were examined. All patients were considered for surgical treatment of suspected carcinoma of the lung. In all patients a preoperative estimate of the distribution of the ventilation in the upper and lower half of both lungs respectively was desired to determine operability.

Methods

For digital subtraction radiography the patients were examined in supine position. The posterior - anterior projection was used.

Both lungs could be included in most patients. Complete breathing cycles during normal breathing and maximum inspiration and exspiration were recorded in a 512 x 512 matrix using a Philips Diagnostic A angiographic stand with a 40 cm imaging intensifier and a DVI II digital unit. The exposures were triggered by the electrocardiogram to occur in late ventricular diastole and two exposures were integrated.

Images of various phases of each breathing cycles were subtracted from each other. In the resulting images, which are composite pictures of the changes in attenuation of roentgen rays in the chest during breathing, any local anomalities in the changes were noted.

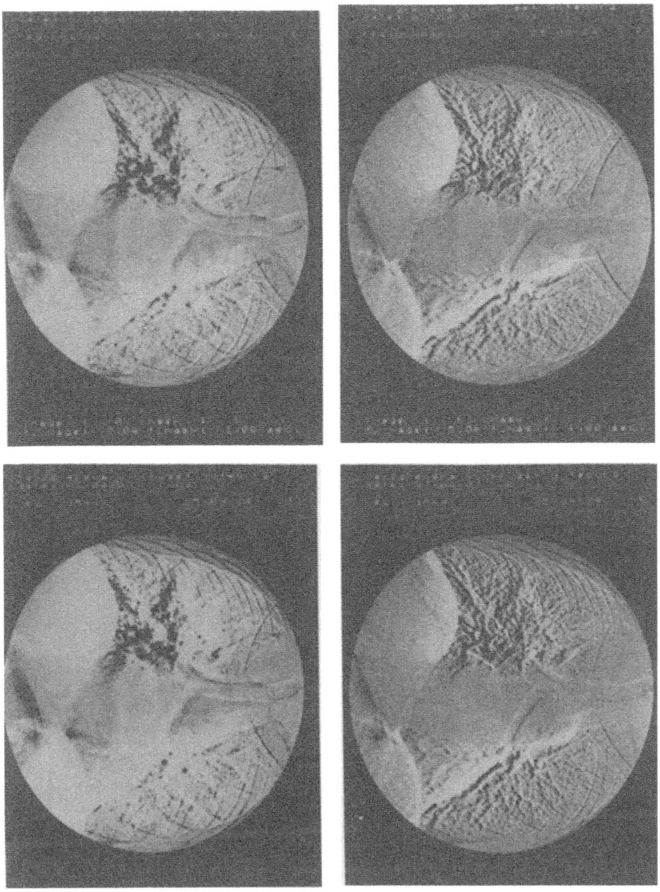

Fig. 1 Digital subtraction radiographs of various phases of respiration. The disturbancies of ventilation caused by carcinoma of the right lung are clearly seen.

In addition an attempt was made to quantify the changes in density in a relative way by expressing the density variations as a function of time.

For comparison with the scintigraphic method the lungs were divided into an upper and a lower half and the maximum density variation in each of these four parts of the two lungs were expressed as percentage of the total sum of density variaitons in both lungs.

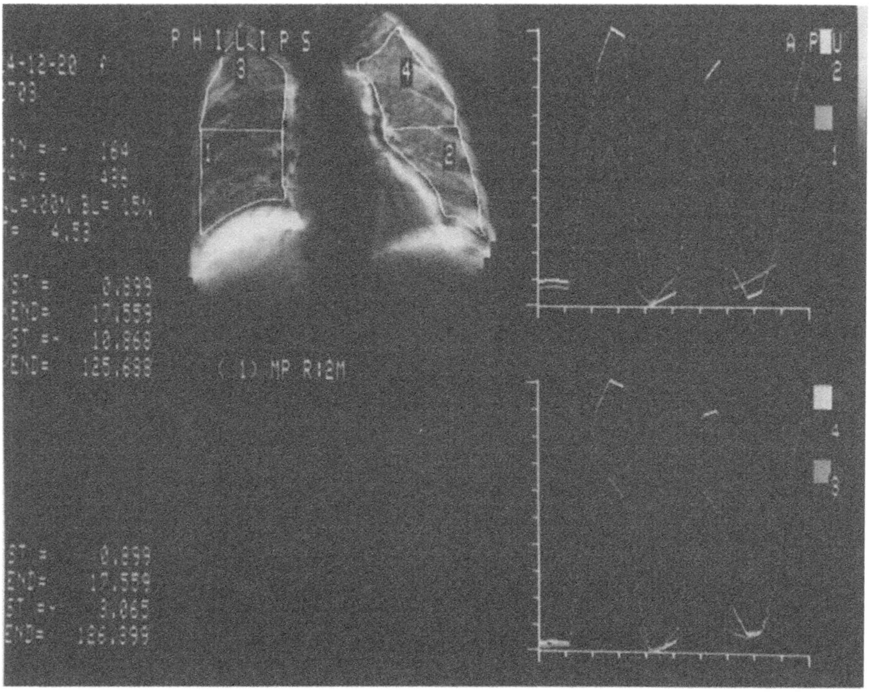

Fig. 2 Relative density variations during breathing in upper and lower half of the lungs.

The scintigraphic examinations were carried out on all patients on the same day and within two hours of the digital subtraction radiography study. For this the technique described by Bake et coll. (2) was used. With this technique radioactive Xenon is injected intravenously and inhaled respectively and the distribution of radioactivity is recorded in the upper and lower half of each lung using four scintilation detectors. In each of the four segments of the lungs the radioactivity is expressed as percent of the total radioactivity.

In addition all patients had conventional chest radiography performed. These images and the images from digital subtraction radiography were examined independently by two observers. The observers measuring the density variations with digital subtraction radiography and the radioactivity distribution in the lungs had no knowledge of each others results.

Results

In eight patient paradoxical changes of the density variations in the affected lung was observed indicating severe disturbances of ventilation from the carcinoma. In these patients no meaningful measurements of density variations to be compared with other patients could be performed. In all these patients the abnormal pattern of ventilation changes appeared very clearly on the digital subtraction radiography images.

In twenty-one patients density measurements from the digital subtraction radiography studies and scintigraphic measurements could be compared.

No clear correlation could be observed when comparing the results of the two methods in upper or lower half of right or left lung or when comparing both whole lungs together. This was true both for comparison between the density variations and scintigraphic ventilation and perfusion studies. At digital subtraction radiography good correlations with the findings at conventional chest radiography was found. The changes in the lungs were in most instances more clearly and easily seen with digital subtraction radiography both with normal breathing and maximal inspiration and exspiration. In all patients the suspected carcinoma could be seen both on standard chest radiograms and on digital subtraction radiography.

Discussion

The lack of correlation between measurements of maximum density variations and measurements of distribution of flow and ventilation using radioactive Xenon was not unexpected.

First there are uncertainties whether the measurements with the two techniques refer to the same volumes of the lungs with both methods. This and the different geometry for recording with the two methods may account for some of the discrepancies between the results.

More important is the fact the density variations are a combination of many changes occuring in the chest during breathing. In addition to the change in airvolume in the lungs, the bloodvolume in the lungs changes. Probably even more importantly the ribcage and the soft tissues change during breathing and these changes contribute to the density changes observed with the digital subtraction radiography method. The digital subtraction method provides a very simple and rapid method to obtain subjective information on local changes of ventilation in the lungs. However, the composite and complex nature of the density changes occuring during breathing makes simple relative measurements of density changes not meaningful without a reference. As proposed earlier the introduciton of a gas as contrast agent which the patient can breath would make more objective measurements of ventilation distribution possible (3). Studies to determine the value of stable Xenon as such a contrast agent are under way.

References

1. Bjork, L.: Studies on pulmonary ventilation using digital subtraction radiography. A preliminary report. Acta Radiol Diagn 27 (1986) 661.

2. Bake, B.; Sixt, R.; Sorenson, S.; Oxhoj, H.: Regional lung function in small cell carcinoma of the lung. Eur J Respir Dis 61 (1980) 50.

3. Bjork, L.; and Bjorkholm, P.J.: Xenon as a contrast agent for imaging of the airways and lungs using digital radiography. Radiology 144 (1982) 475.

Limitations to Quantification of DSA Images

R. LE GOFF, Y.BIZAIS, P.GERLOT CHIRON

The DIMI Project, University Hospital Nantes, France

Introduction

It is important to be able to quantify DSA images in medical diagnosis (densitometric quantification of vessel stenosis, blood flow measurements)[1] This implies to precisely know the intensity to contrast agent amount relationship.
For this purpose, a model of X-rays attenuation through an iodine solution was defined. It requires that no scattered radiations are detected and that the X-rays spectrum is monoenergetic. The model and the validity of such assumptions for a DSA acquisition device is discussed in Section 1.
An extensive experimental study on routine DSA systems was carried out to validate the model. Iodine and water attenuation coefficients were estimated for various radiological (mAs, KV, attenuation material thickness) and geometrical parameters (tube to system-table distance, diaphragm position) (Section 2).
Results are discussed and compared with X-ray attenuation theory. Explanations are proposed for apparent variations of attenuation coefficient (Section 3).
Finally, we discuss the accuracy with which the attenuation coefficient of an iodine solution must be known to permit proper quantification of DSA images.

1 MODELING X-RAY ATTENUATION BY AN IODINE SOLUTION

1.1 Intensity to thickness relationship

Attenuation is a fondamental X-ray property for transmission imaging. A pencil beam propagating in an iodine solution is attenuated according to Beer's law [2] :

$$\phi(x) = \phi(0) \,.\, \exp\left(- \int \mu(E, x)\rho(x).dx \right) \qquad (1)$$

where

$\phi(0)$	is the incident photon fluence (photons/cm2).
$\phi(x)$	is the photon fluence after traveling distance x in the mixture.
μ	is the total mass (energy dependent) attenuation coefficient (cm2/g).
ρ	is the mixture specific mass.

Let m_I be the iodine mass concentration in the solution (mg iodine/ml), and ρ_I be the iodine specific mass. For monoenergetic X-rays, if x_M is the total mixture thickness, logarithmic transformation of Eq.1 leads to Eq.2 :

$$Ln(\phi_1/\phi(0)) = - x_M [\mu_W (1 - m_I/\rho_I) + \mu_I m_I] \qquad (2)$$

where μ_W and μ_I are the water and iodine mass attenuation coefficients.

Eq.(2) holds only for monoenergetic X-rays beams. This is not the case for X-rays produced by Coolidge tubes which cover a continuous energy bandwidth. Moreover, they are emitted within a cone (wide beam geometry) such that scattered radiations are necessarily produced and detected. Nevertheless, the latter are negligible at energies used for diagnosis (about 0.1 Mev) because of anti-scatter grids, and photoelectric absorption is the major attenuation phenomenon [3],[4]. Thus the above modeling still holds in standard experimental conditions.

1.2 The digital output signal

The output signal depends on the behaviour of each chain component downstream the X-rays generator. After attenuation through the mixture, the image intensifier (II) converts the input X-rays fluence into an optical image (O) with a brigthness gain K. The video camera delivers a signal (S) according to : $S = O^\gamma$ where γ is a characteristic parameter equal to 1 for commonly used Saticon tubes. Finally S is digitized into D (ADC). All components are assumed to have a linear response [3].
Consequently the digital output signal (D) is related to the traveled mixture thickness and attenuation coefficients by :

$$D = C - x_M [\mu_W (1 - m_I/\rho_I) + \mu_I m_I] \qquad (3)$$

where

C	is a constant value depending on radiological factors.
$(1 - m_I/\rho_I)$	is the fraction by weight of water.

Eq.3 points out a proportional relationship between output intensity and the crossed mixture thickness x_M, and may be used to estimate the attenuation coefficient of a mixture.

2 PHYSICAL EXPERIMENTS

2.1 Experimental conditions

Experiments were performed on two DSA systems routinely used in our institution. They were chosen because associated digital systems are similar while radiological components (tubes, II) are different. Radiological operating characteristics (KV and mAs) were chosen to cover a wide density range representative of clinical situations. The mixture to be analysed was contained in a glass gauged column and placed on the system table. The column was positioned such that its center coincides with the center of the II field and was at the same distance of both the tube and the II (cf. § 2.3.1).

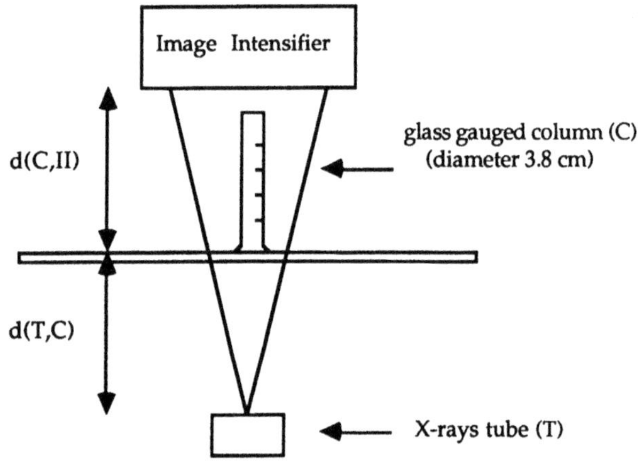

Fig.1 Experimental set up.

2.2 Acquisition conditions

We define series (S_K) as a set of 4 frames acquired in the following conditions:
- fixed radiological settings: KV and mAs, tube to II distance, II diameter (12cm width, zoom position), and diaphragm (lead shutters) position.
- fixed mixture thickness in the column (volume varying from 0 to 200 ml).

Acquisition of series for N mixture thicknesses forms a data set ($\{S_K\}$, $K \in [0,N]$). Each acquisition series lasted 4 s (1 frame/s) to allow the power raise time of the generator to be reached. Only the fourth frame was kept for processing.

An as large as possible region of interest (ROI, 50^2 pixels) was chosen in the part of the image corresponding to mixture intensities. The mean (F_k) and the standard deviation (σ_k) of intensities within the ROI were estimated.

For each data set, the logarithm (f_k) of F_k's was plotted against volumes (Fig 2).

A linear fit was then performed and the linear attenuation coefficient of the mixture was derived from the estimated slope (Eq.3).
In all carried experiments, (σ_k/f_k) was found to be about 0.01 and the regression coefficient of the linear fit was higher than 0.99.

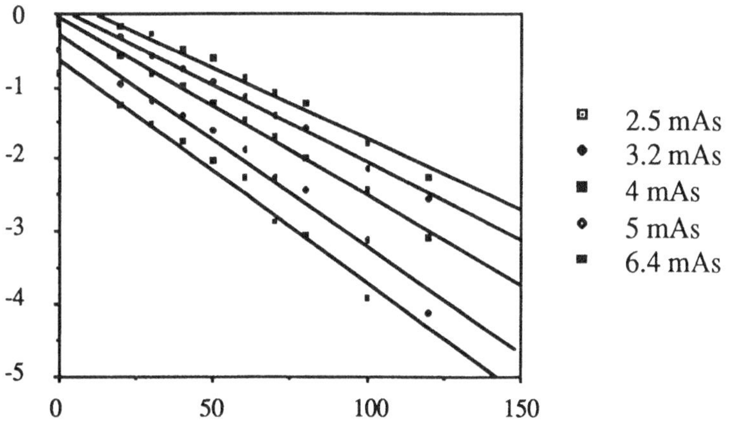

Fig 2: Logarithm of mean intensities .vs. volumes (68KV, diaphragm closed).

2.3. Description of experiments

2.3.1. Variations with tube to column distance : the distance was changed from 20 cm (diagnostic conditions) to 37 cm (maximal distance). Estimated μs for the 2 distances differed by less than 1%. This experiment was carried out for 2 different mAs values with the same result. Consequently, the air layer thickness between the tube and the column does not affect μ, and d(T,C) and d(C,II) were set to 30 cm for the following measurements.

2.3.2. Variations with mAs (diaphragm opened) : data sets were acquired and processed for mAs values equal to 2.5, 3.2, 4.0, 5.0, 6.4, at 68KV. With such values the whole intensity range was covered. Diagrams as in Fig.3, demonstrate that μ decreases when mAs increase. This experiment was repeated and good reproducibility of estimates was observed.

2.3.3 Variations with mAs (diaphragm closed) : the range of mAs was exactly the same as above. The only difference resided in the use of the tube diaphragm which was closed up to surround the ROI to be processed. It limits the photon fluence, which avoids overflow in image background. There is still a decrease of μ when mAs increase, but diaphragm position clearly affects μ (Fig 3) : it is about twice larger than the one estimated in § 2.3.2.

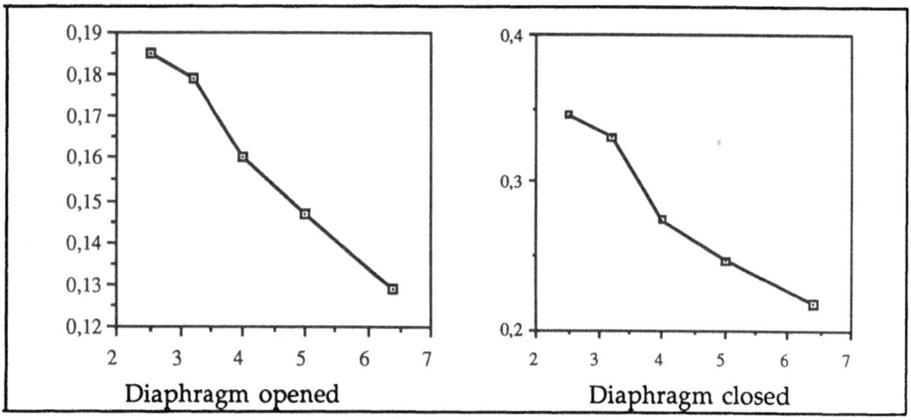

Fig.3: Linear attenuation coefficient .vs. mAs.

2.3.4. Variation with mAs (attenuation material added) : the attenuation material commonly used in DSA performance evaluation is Lucite, because of its attenuation properties similar to those of human tissues. Firstly, two Lucite blocks (15×15x2 cm) were placed directly upon the X-rays tube window. Measurements were performed in the same acquisition conditions as in § 2.3.2.

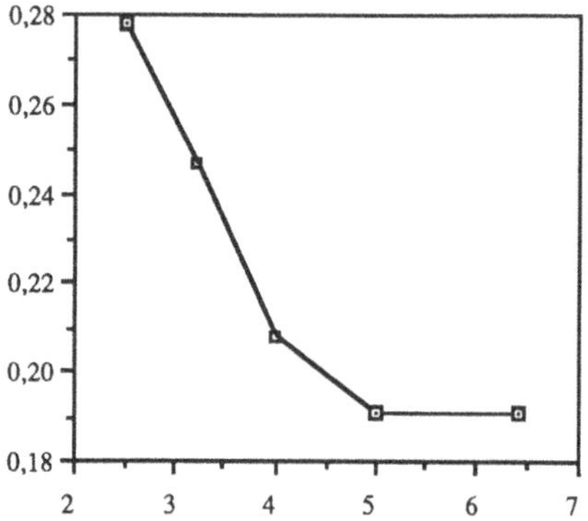

Fig 4: Linear attenuation coefficient .vs. mAs (4 cm Lucite).

A second experiment was realized with 5 Lucite blocks and with higher mAs values (25, 32, 40 mAs) to cover the entire intensity range. Results are consistent with the ones obtained in § 2.3.2.

2.3.5 Variations with operating Voltage : the same experiment as in § 2.3.3 was performed for two voltage values (46 & 60KV). Estimated μs show that attenuation coefficient decreases when energy increases.

3. DISCUSSION.

The described experiments assess the well-known property of the attenuation coefficient : its decrease with X-rays energy for the two tested systems [3].
Nevertheless, it was also established that μ varies with mAs values, which is not explained by the X-rays attenuation theory, and which could be induced by X-rays energy variations with mAs settings. This result was obtained both by varying mAs globally and by varying exposure times at fixed mA settings. The proposed explanation is likely to be true, because this result was obtained with the DSA system equipped with an old generator and not with the newer system.
Most probably, μ values estimated with closed diaphragm are the true ones, since data sets obtained without diaphragms are corrupted by intensity overflows in the II and in the video target.
In summary, the μ values of water can be predicted precisely (less than 4%) when KV and mAs settings are known. Further experiments need to be carried out to study the influence of scatter on apparent μ values [4]. However, because of the efficacy of anti-scatter grids, the latter effect is likely to be negligeable.
The knowledge of apparent μ_I and μ_W values under well-defined radiological conditions is necessary to estimate precisely the amount of iodine crossed by X-rays [5]. This amount can be used to quantify DSA images (blood flow, densitometric quantification). Further experiments are being performed to establish the possibility and the accuracy of such techniques.

<u>References</u>

1. LeFree, M.T.: Digital radiographic assessment of coronary arterial geometric diameter. Proc. SPIE 626 (1986) 334-341.

2. Barrett, H.H.; Swindell, W.: Radiological Imaging. Academic Press 1981.

3. Kruger, R.A.; Riederer, S.J: Basic concepts of digital subtraction angiography. G.K.Hall 1984.

4. Kruger, R.A.; Sorenson, J.A.; Niklason, L.T: Effectiveness of antiscatter grids in DSA. Invest. Radiol. Vol.18 (1983) 293-297.

5. Sandor, T.: High precision quantitative angiography. IEEE Trans. Med. Imaging MI-6 (1987) 258-265.

Improvement of Pulmonary DSA by Measurement of Regional Lung Perfusion

CHRISTINE RICHTER, G. SCHRÖDER, K. RICHTER, H. GERST

Akademie der Wissenschaften der DDR
Zentralinstitut für Herz-Kreislauf-Forschung, Bereich Radiologie
1115 Berlin-Buch, DDR

Summary

A DSA method is being presented linking the diagnostic facilities of pulmonary angiography with the measurement of global and regional lung perfusion by time density curves. The bolus transit is being distinguished systematically within multiple regions of interest over the pulmonary artery, its upper and lower branches, several peripheral lung regions, pulmonary veins, the left atrium and the aortic arc. The method has been developped with DIGITRON 3 (SIEMENS) and can be applied to global pulmonary angiography as well as to selective procedures. Arising from the findings of 105 patients various patterns concerning angiography and regional perfusion could be revealed which characterize pulmonary embolism - including mild types and decreasing stages during therapeutic thrombolysis - different types of pulmonary hypertension, left heart failure, fibrotic and other lung diseases and some congenital heart defects. The method proved worthwhile in particular being applied in ambigous situations of differential diagnosis. In cases of cardiac defects with pulmonary recirculation the left to right shunt can be quantified. Thus a succesfull and rational approach to joint morphologic and densitometric evaluation of pulmonary circulation could be achieved.

Introduction

Beyond the morphologic angiographic information the pulmonary DSA presents favourable conditions for a quantitative evaluation of the pulmonary perfusion. The contrast medium transit can be characterized by parameters of intensity, time and speed concerning the bolus wash in and wash out flow phases. The informations can be achieved from the arterial and venous pulmonary blood flow. Arising from the data of multiple regions of interest (ROI) the regional distribution of lung perfusion is becoming distinguishable. The high density resolution of DSA creates a decesive prerequisite to gain informative time density curves (TDC). Since the ROIs can be positioned with very exact topographic relation to the pulmonary vasculature the perfusion is being evaluated segmentally, subsegmentally resp. over single branches of the pulmonary vessel tree. We report on a method which aims at precise information about regional alterations of lung perfusion in direct context with the morphologic findings of the pulmonary DSA.

Methods

We developped the procedure with the imaging system of Digitron 3, Angioskop resp. Siregraph 2 and Pandoros Optimatic (SIEMENS). Matrix 512 x 512 Pixel, image series of 10 ... 20 sec, image frequency 8/sec, kilovoltage constant.

Global pulmonary DSA by transfemoral right heart catheterisation including pressure measurement, 7 F Pigtail Catheter, 40 ml Visotrast 290 into the right chamber using flow of 23 ml/sec. Imaging in p. a. positioning.

Selective pulmonary DSA was performed by catheterisation of the right resp. left artery main stem, injection of 25 ml Visotrast 290 using flow of 17 ml/sec. Imaging p. a., 30° RAO, 30° LAO.

In postprocessing procedure multiple ROIs are positioned over various lobar and segmental arteries within the hilar and perihilar region, over peripheral areas of the lung coat, the left atrium resp. pulmonary veins near atrium and one ROI covering the aortic arc (fig. 1 and 2). The TDCs are computed and visualized. The curve maximum corresponds with the amount of contrast medium perfusing the ROI. We determine two parameters (fig. 3), the bolus peak time and the mean transit time calculated by the formula

$$t_m = \frac{\int_o D(t)\, dt}{D_{max}}$$

t_m = mean transit time, D = density, t = time

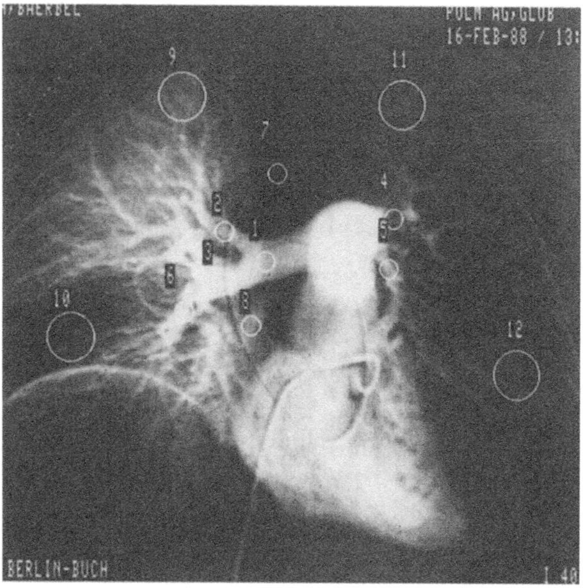

Fig. 1: Multiple regions of interest in global pulmonary DSA (embolism of the left artery and the right lower lobe)

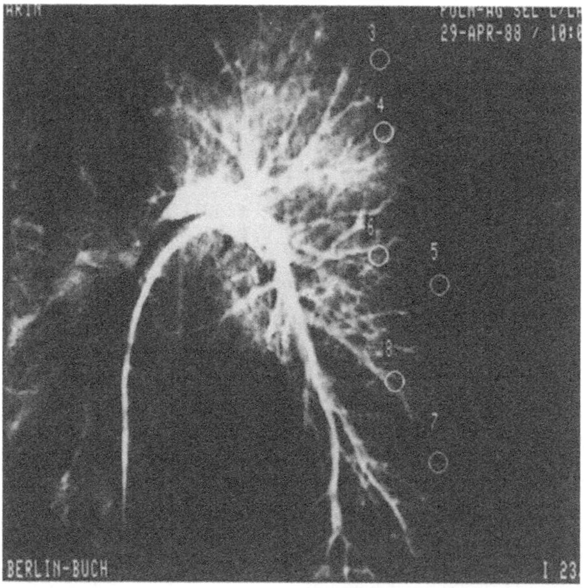

Fig. 2: Multiple regions of interest in selective pulmonary DSA (peripheral embolism in the segments 6 and 9)

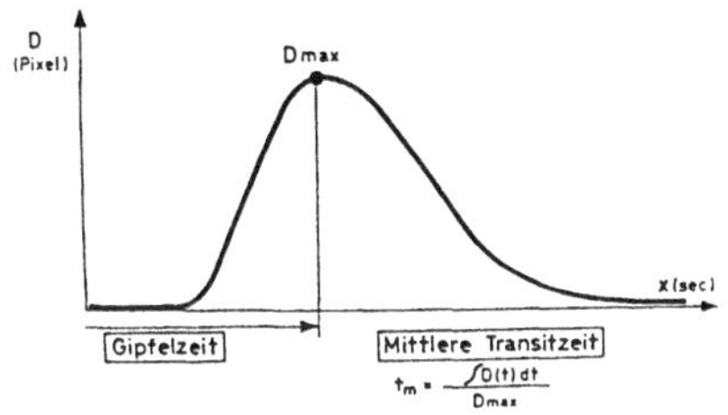

Fig. 3: Time density curve, peak time, density maximum, mean transit time.

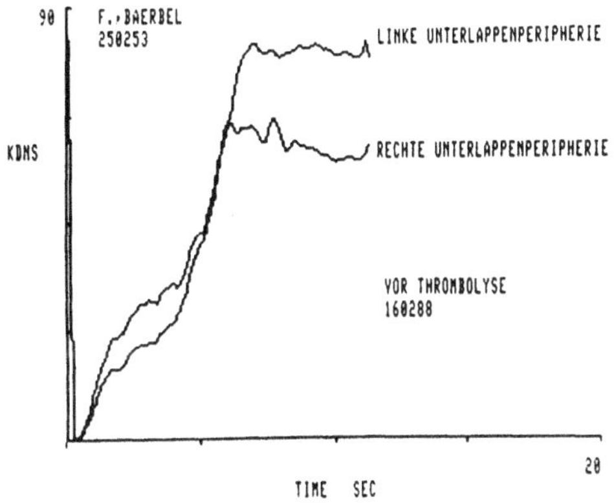

Fig. 4: TDC in massive pulmonary embolism of the left and right lower lobes.

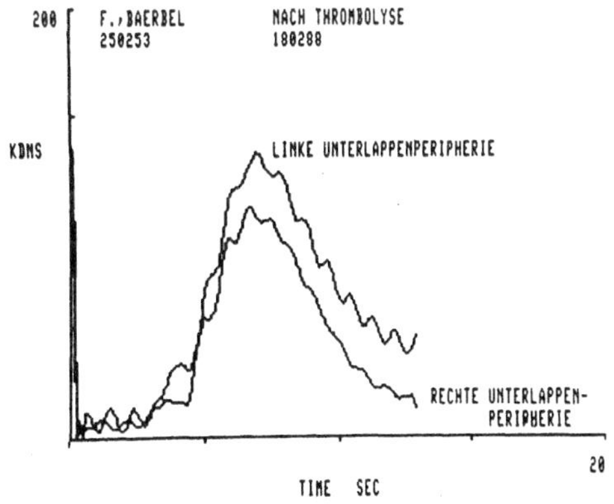

Fig. 5: TDC in the same patient as fig. 4 over the same regions after therapeutic thrombolysis.

The calulated parameters are of a relative nature. The comparative evaluation is being guaranteed by the standardized DSA method, provided that ROIs of identical size and shape are used. If on the occasion of visual (angiographic) image evaluation various regions present different densities these areas may be investigated comparatively too, since the TDC analysis offers further information upon the actual regional circulation.

In the way described we investigated 105 patients suffering from the following diseases: pulmonary embolism, pulmonary precapillary hypertension with different etiology, pulmonary postcapillary hypertension in left heart failure of different pathogenesis, various lung diseases and congenital defects with left to right shunt.

Pulmonary Embolism

Because of the essentially improved visualisation of the parenchymal bolus transit phase the DSA with its highly increased density resolution proved superior to conventional pulmonary angiography. With DSA the restrictive appraisal that pulmonary embolism angiographically would be exclusively detectable down to the occlusion of arteries 2 mm in diameter (1) may be cancelled. The regions peripheral to these small arteries can be judged by TDC (fig. 6). The mean transit time and the bolus peak time increase considerably.

Precapillary Pulmonary Hypertension

In this group of diseases we revealed various patterns of findings. Frequently but not compulsory pulmonary hypertension causes a redistribution of the blood flow in wide areas as known from szintigraphy. This pattern occurs mainly in obstructive bronchopulmonary diseases. In cases of pulmonary hypertension with unknown etiology however patterns prevail which present different perfusion degrees within small lung areas. On fig. 7 an example is depicted with pathological and normal perfusion parameters in neighboring areas of one and the same segment. These findings remind of very similar ones in embolism (fig. 6).

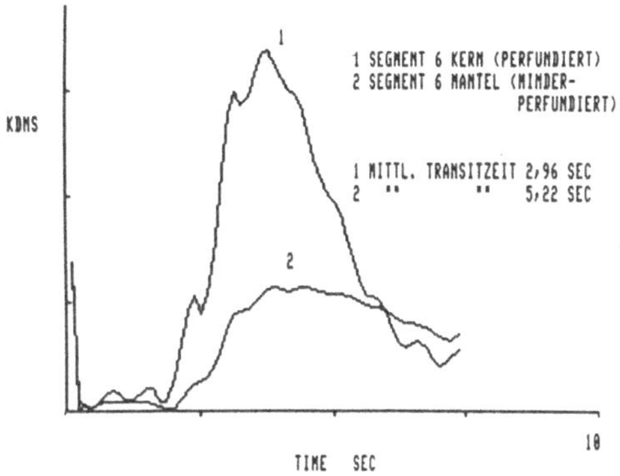

Fig. 6: TDC in peripheral embolism, two regions within segment 6. 1 = normal perfusion area, 2 = decreased perfusion area (the same patient as in fig. 2)

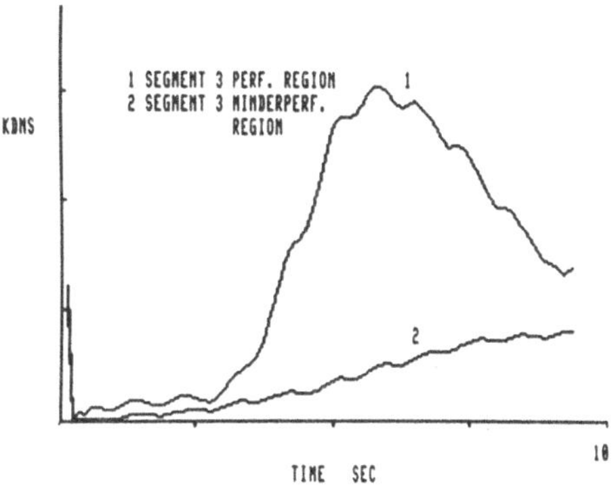

Fig. 7: Pulmonary hypertension of unknown etiology. Areas of normal (1) and diminished perfusion (2) in one and the same segment.

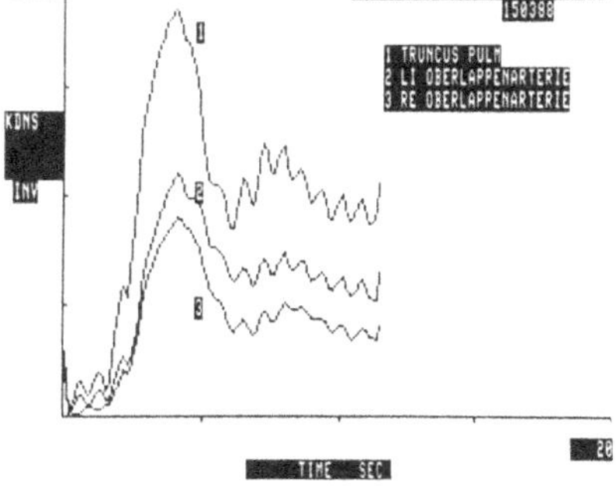

Fig. 8: TDCs in an atrial septal defect. Recirculation wave in the pulm. truncus, left and right upper lobe arteries.

Postcapillary Pulmonary Hypertension

An increased mean transit time of the bolus occurs also in cases of left ventricular failure with increasing left ventricular enddiastolic pressure and increasing pressure in the lung veins. The mean transit time as well as the mean peak time are in particular pathologically alterated in the lower lobes. The results so far obtained are corresponding with those of szintigraphy.

Perfusion Disturbances in Interstitial an Parenchymal Lung Diseases

Is the DSA being performed in patients suffering from lung diseases patterns of pathological regional perfusion are visualized without pulmonary hypertension being present. In such cases the parameters of the TDCs are alterated too but in lower stages than in embolism resp. pulmonary hypertension.

Pulmonary Recirculation in Cardiac Defects with Left to Right Shunt

Another facility to quantify the diagnostic evaluation by the TDC-technique arises in patients with cardiac shunt vitia, (atrial and ventricular septal defects, ductus Botallo or lung vein transposition). The relative shunt volumes can be calculated from the integrated area under the recirculation curve (2) (fig. 8). In comparison with oxymetry we received a rather reliable agreement of the methods (table 1).

Table 1: Quantification of left to right shunt in patients with atrial septal defects and lung vein transposition. Comparison between oxymetry and TDC-Analysis of DSA

Diagnosis	Oxymetry	TDC	QP/QS Oxym.	QP/QS TDC
ASD II	53%	50%	2,00	2,29
ASD II	68%	66%	2,93	2,96
ASD II	71%	67%	3,60	3,04
ASD II	70%	68%	3,30	3,18
ASD I	70%	71%		
Part.Lung Vein Transpos.	57%	61%	2,33	2,57

Conclusions

Finally we consider that the TDC analysis of pulmonary DSA improves the diagnostic angiographic information essentially. The joint morphologic and densitometric evaluation of regional pulmonary circulation proved reliably and worthwhile being applied particulary in certain and ambigous situations of differential diagnosis. The method achieves an increased level in pulmonary angiography and raises the significance and evaluation facilities of DSA imaging.

References

1. Widimský, J.: Pulmonale Hypertonie, S. 294. Georg Thieme Verlag, Stuttgart-New York 1981.
2. Yiannikas, J., u. D. S. Moodie: Assessment of left-right intracardiac shunts using intravenous digital subtraction angiograhy. In: Moodie, D.S., I. Yiannikas: Digital Subtraction Angiography of the Heart and Lungs. S. 23-42. Grune and Stratton. Inc., 1986.

Improvement of Digital Subtraction Angiography Image Quality: High Resolution Processing Coupled with a Real Time Synchronisation Procedure

I.Azancot, X.Favreau, J.P.Melki, A.W.Kedra, G.Soulez, P.Frouge.

Hemodynamic Laboratory and imaging department. Lariboisière hospital. 75475 Paris Cedex 10.

ABSTRACT

The use of DSA (Digital Subtruction Angiography) in pulmonary angiography applications is restricted by the difficulty in: 1/ obtaining a good resolution of the distal branches 2/ avoiding the motion artefacts resoluting of artery and background displacement during the cardiac and respiratory cycles 3/ preserving the dynamic vascular motion 4/ and storing the digital dynamic information on a permanent support. We developed an angiographic system, based upon a high resolution 45 MIPS 32 Moct image processor, interfaced with a 2.75 Goct Wincester drive allowing the real time storage of either 30 frames/s in the 512*512*8 bits matrix or of 7.5 f/s in the 1024*1024*8 bits matrix. The 1024 matrix was found optimal for good distal resolution studies. To preserve the most important dynamic informations on the basis of 7.5 f/s maximal storage rate, we developed a synchronisation device able to recognize on a real time basis, from chronologic delays derived by using both EKG and pulmonary artery pressure curves, the End Diastolic (ED) and the End Systolic (ES) positions, even in the case of heart rates varying from 60 to 120, with a precision of less than 20 ms. The ED and the ES images are stored together with the Mid Systolic (MS) and the Mid Diastolic (MD) image. The visualisation of the sequential ED, MS, ES and MD frames gives the possibility: 1/ of saving the dynamic information, as the most important funtional images can be displayed with a cine loop technique 2/ of performing automatically cardiac and respiratory synchronized subtractions, reducing motion artefacts 3/ of of obtaining functional color coded parametric images 4/ of reducing the amount of data and of allowing the digital and permanent storage of the most important cardiac cycles. The operating procedure that is outlined below allow the examination to be performed rapidly and safely, with substantial decrease of contrast media doses and of X-ray exposure.

METHODS

1. Image acquisition

The digital angiographic system was developed on a high resolution 45 MIDS pipeline processor with 32 MOCT dynamic memory interfaced with a 2.75 Goct capacity and12.5 Moct/s transmission rate. Winchester disk drive, allowing resolutions and real time storage rates: 30 frames/s with the 512*512*8 bits matrix; 7.5 frames/s with the 1024*1024*8 bits matrix. The 1024 matrix was found optimal for good distal resolution studies. To preserve the most important dynamic information on the basis of 7.5 f/s maximal storage rate, we developed a synchronization device able to recognice on a real time basis the END Diastolic (ED) and the END Systolic (ES) positions. From the arterial curve, the computer recognizes the moments corresponding to the opening (ED) and to the closure (ES) of the pulmonary valve. The chronologic delays between these positions and the corresponding EKG R waves are then computed, averaged for 12 basal cardiac cycles, and refered to the RR mean interval. This allows the calculation of the corrected ejection time. After this, the computer is able to work from the simple EKG signal: each RR interval is calculated in real time, and after each R wave, the corresponding ED and ES positions are identified from the above-mentioned calculations. For each cardiac cycle, the computer generates 4 interruption signals, corresponding to the ED, Mid Systolic (MS), ED and Mid Diastolic (MD) positions. For each of these interruption signals, the X-rays generator delivers a pulse with the following characteristics: 6 ms duration, 800 mA and KV varying from 65 to 85 as to obtain a constant X-ray dose of 30 microRoentgen per pulse. This corresponds to a low maximal irradiation of 0.22 mR/s, with a 0.15 mR/s mean value. During each pulse the video camera is on a blanking mode, and after each pulse a slow scan progressive read-out is performed. The visualization of the sequential ED, MS, ES and MD images protect the dynamic information, as it provides a regular sampling between the extreme functional positions.

2. Image processing

The sequential ED, MS, ES and MD images are stored in real time on the fast Winchester drive using the 1024*1024*8 bits matrix. When the patient is able to hold his breath, a single cardiac cycle is recorded before the pulmonary injection and the subtraction is automatically synchronized with the cardiac cycle: for an ED image, an ED mask is used for the subtraction, for the MS image a MS mask and so on. When the patient is not able to maintain apnea for a 10s period, he is allowed to braeth normally during the recording: 3 full resperatory cycles are recorded before the contrast material is injected. During the post-processing period, the cardiac synchronized subtraction images are also synchronized on the respiratory cycle: for the ED image, for instance, the ED mask can be chosen by the computer between the preinjection ED images as to obtain a minimal grey level histogram of the corresponding subtraction frame. It is also possible for the operator to choose very quickly the best. ED mask between the preinjection ED images.

In addition to cardiac and respiratory synchronized images, the software is designed to provide, for each injection, a color coded ED parametric image. A synchronized differential subtraction is first performed. This yields 5 differential images, showing the progression of the contrast medium during the first 6 cardiac cycles: ED2-ED1, ED3-ED2, ED4-ED3, ED5-ED4, ED6-ED5. For each of these 5 differential images, a densitometric study is performed. A color is assigned to each cardiac cycle, each pixel grey level is compared on the consecutive differential images, and the maximal value is allocated to the resulting parametric image with the corresponding color code.

RESULTS AND DISCUSSION

The 1024*1024 matrix was clearly found optimal to obtain a good distal resolution, i.e. beyond the 3d order divisions. With this acquisition format and using intra-arterial injections, the spartial resolution is comparable to that obtained using conventional cut films or 35 mm cine films, while the contrast resolution is better because of the subtraction, filtration and grey level manipulation possibilities. The major problem then remaining with digital subtraction was the motion artefacts due to cardiac and respiratory movements. The synchronization method discribed overcomes most of these problems: 1/A simple EKG trigger is unable to locate the ED and the ES points when the heart rate changes during the examination, and this is often the case. The synchronization device allows the ED and the ES positions (and consequently the MS and MD positions) to be identified in real time. The accuracy of this method was established by comparing the micromanumatric arterial pressure curve and the positions determined by the computer during progressive atrial pacing: even in the case of heart rates varying from 60 to 120 beates/mn, the ED and ES positions are calculated with a precision of less than 20 ms. *The consequence is a functional cardiac synchronization allowing an optimal automatic subtraction.* 2/ In addition to cardiac synchronization, the method discribed above provides a simple and efficient respiratory synchronization when a sufficient level of cooperation is not possible. 3/ The functional parametric color coded images are useful in discriminating unperfused areas from regions where the perfusion is present, but delayed. 4/ This method allows a very efficient functional data compression, as the sequential ED, MS, ES and MD images can be displayed on a cine loop basis, yielding the most important dynamic information (which cannot be obtained with conventional cut films) and reducing the number of images to be stored, without requiring digital compression of the data, thus avoiding any loss of information. During the post-processing phase, it is possible to choose the most important cardiac cycles and to store this reduced amount of data by using 800 Moct laser removable and unerasable digital optical disks at a reasonable cost.

Once these problems has been solved, the digital processing offers several adventages, when compared to conventional analogic films: 1/ The results are obtained in real time. 2/ The angioplastic procedures are significantly facilitated. 3/ The X-ray doses are substantially reduced, by 7 times approximatively in our experience. 4/ The amount of contrast medium is reduced. 5/ The cost of a high-tech digital equipment is compensated by the reduction of X-ray exposure (longer longevity), of the amount of film and contrast medium used, and by the decreased time required to perform the examination and to take the therapeutic decision.

REFERENCES

1/ Azancot I, Kedra A, Friolo S, Mallerque MC, Nittemberg A, Bourthoumieux A, Dibie C, Beaufils P, Slama R. Digital processing of Left Venticilar and Coronary Angiographies. Cardiologia: 33, 563-570 ,1988.

2/ Ergun D.L., Shaw C.G., Kruger R.A., Mistretta C.A., Zarnstorf W.C., Video slow scan technique for mask mode subtraction radiography. Annual meeting of the Radiological Society of North America, Dallas, November 1988.

3/ Hirji M, Gamsu G, Webb W.R., Brito A.C., Kurijama K, Stern R.G., Cox L.. EKG gated digital subtraction angiograhpy in the detection of pulmonary emboli. Radiology 1984; 152: 19-22.

4/ Azancot I, Kedra A, Melki J.P., Soulez G, Favreau X., Azancot S, Blery M, Slama R. Digital Pulmonary angiography: a synchronization procedure for anatomical and functional image optimization. Cardiovasc. Imaging, 1989, in press.

The Improvement of Vessel Quantification with Image Processing Equipment for High Resolution Digital Angiography

K. BARTH, B. EICKER, U. BITTNER, P. MARHOFF

SIEMENS AG
Medical Division, RDEK 211
Henkestr. 127, D-8520 Erlangen

Introduction

In the cardiological clinical application the discussion is still open if digital imaging may replace cine film recording for the assessment of the coronary arteries, especially for the objective quantitative evaluation. This is partly due to previous technical limitations with digital (subtraction) angiography. The limits were low spatial resolution and slow speed of acquisition and replay connected with small digital storage capacity. These restrictions have been overcome by considerable improvements in performance. With a matrix size of 512 by 512 pixels a frame rate of 60 images per second has been realized. Real time storage capacity for this format is more than 20 000 images, which should be sufficient for a day's work in the catheter laboratory. On the other hand, images of the format 1024x1024 can be recorded at a rate of 15 f/s, which is better than five years ago with any smaller matrix. So at least ten images are available for each heart cycle and the selection of an optimal image for stenosis evaluation should always be possible. Only with this matrix size vessel diameters smaller than 1 mm can be differentiated in a reliable way. The spatial resolution in this case is better than 3 LP/mm with a 17 cm image intensifier and is comparable to 35 mm cine film. With any digital image format the superior density resolution proportional to x-ray attenuation is offered, enabling insights beyond the mere spatial resolution limits. And, all these image data are immediately available in reliable digital format and may be duplicated, processed, archived and recalled without loss of information.

General processing concept

A modern DSA system like DIGITRON III® or POLYTRON 1000® contains special digital pipeline processors for handling image data of 0.5 to 2 MBytes per frame in real time. This is a basic requirement for the elementary DSA subtraction operation. These processors in conjunction with semiconductor memory and fast mass storage device (MSD) have been designed in such a general way that they also can be used for more complex image transformations like grey scale modification in a 40ms cycle or for filter operations in the preprocessing step of image quantification. Software libraries are available for controlling the pipeline hardware virtually like a parallel device. A tool box of elementary operations allows the composition of more complex operations. Such elementary operations may be spatial shifts of whole images and simultaneous addition or subtraction with the original or a reference image. Weight factors or truncation of negative pixel values can be included. In such manner MIN or MAX operations between two images can be easily implemented. Real-time lookup tables are further means of fast processing and allow an ATAN-transform for functional imaging and other nonlinear grey scale modifications within 40 ms. In Fig. 1 the realization of a Laplacian filter with the parallel processing concept is explained. Bittner [6] showed theory and implementation of extremely fast nonlinear filtering like median filtering in a similar way.

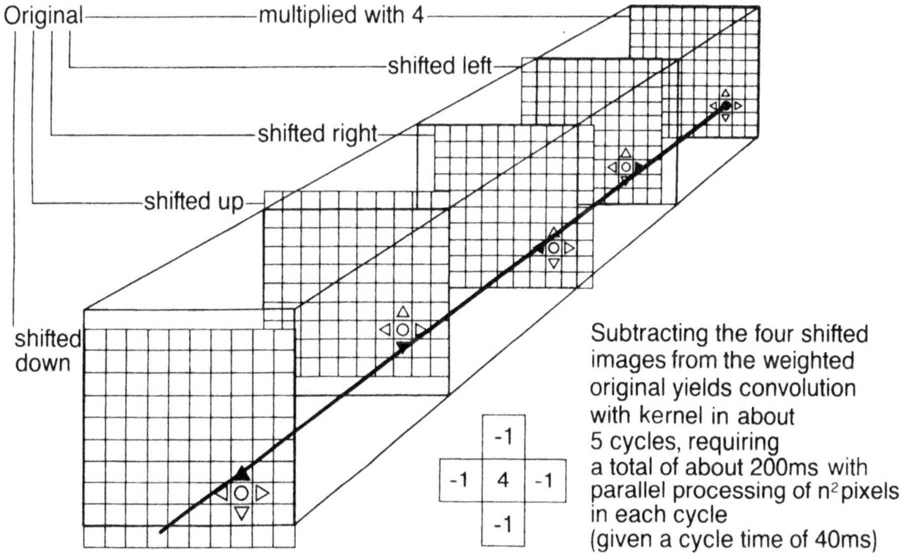

Fig. 1: Laplacian filter for parallel hardware

Measurement technique for vessel quantification

Based on the work of researchers like Sanders, Kirkeeide or Parker [1-4] and in our own investigations we have learnt that the kernel convolution method should be applied for the reliable geometric localization of vessel edges. This convolution or, with appropriate normalization, this correlation is a best-fit matching operation, with a response like a gradient operator but with minimal square error sums. Usually the kernel is shifted over the image in some direction, and in each shift position the correlation factor is calculated. In our concept kernel coefficients are fixed weighting factors and the images are shifted. In each shift position with respect to the original image one of these weights applies to all pixels. In similar fashion any

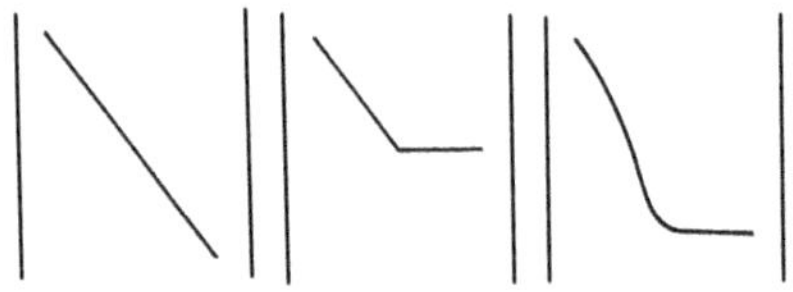

a) linear ramp b) shoulder kernel c) vessel extract

Fig. 2

Fig. 2: Convolution kernels

Fig. 3: Cine film (Houston): Detected vs. true phantom diameters.

Fig. 4: Digital acquisition: Detected vs. true phantom diameters.
a: linear ramp
b: shaped kernels.

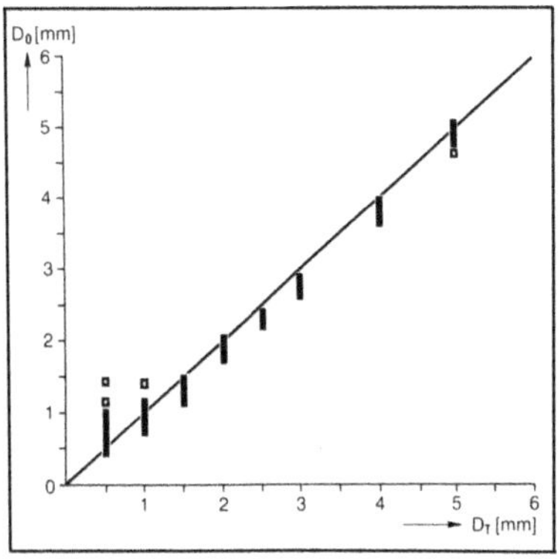

Fig. 3

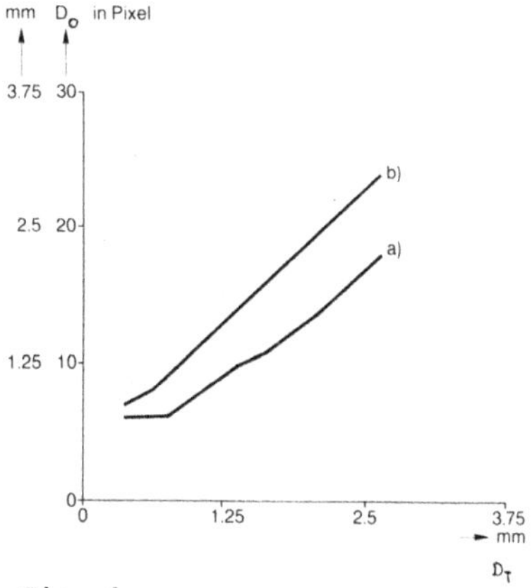

Fig. 4

convolution kernel may be realized (Fig. 2) and applied in main directions like horizontal, vertical and diagonal. We noticed that the response of a shaped kernel (Fig. 2b,c) is more linear than the response of a simple "ramp"-function (Fig. 2a). This is documented in Fig. 4. However, it does not have to be an adapted edge form including the MTF-effect. The abstracted knee-pattern (Fig. 2b) with a given resolution of 3 LP/mm is as good (Fig. 4) as the vessel adaption (Fig. 2c). In Fig. 3 as a comparison data form Kirkeeide [3] are shown, which have been obtained on the basis of cine film digitizations with a "ramp"-kernel.

Shaped kernels are also more noise-tolerant than simple linear templates. Fig. 5 indicates, how for various kernels, a measurement is possible of diameters lower than 1 mm. Vessel phantoms with constant diameter have been evaluated. In the diagram the standard deviation of the measurement in mm and in pixels is plotted over the true vessel size on the abscissa. a1 and a2 are applied kernels of 7 and 9 pixels without shoulder (no "knee"-shape), b1 and b2 are kernels of 9 pixels with shoulders of 3 and 5 pixels. At the resolution limit the diameter is underestimated or overestimated with statistical variance in pixels or in mm. In order to trace the vessel itself, a filter as described in [5], now called "twin"-operator may be used.

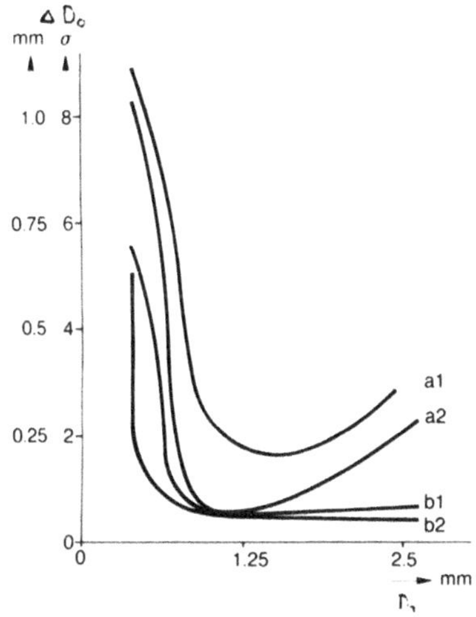

Fig. 5

Fig. 5: Stability of kernels in the presence of noise
a1: ramp with length 7
a2: ramp with length 9
b1: knee, length 9, shoulder 3
b2: knee, length 9, shoulder 5.

Fig. 6: Original AP-angiogram of the left coronary artery.

Fig. 7: Filtered angiogram from Fig. 6. Merely for segmentation the filter would be parametered stronger.

All medical images are unsubtracted, acquired at normal system dose levels (0.1 to 0.2 μGy per frame)

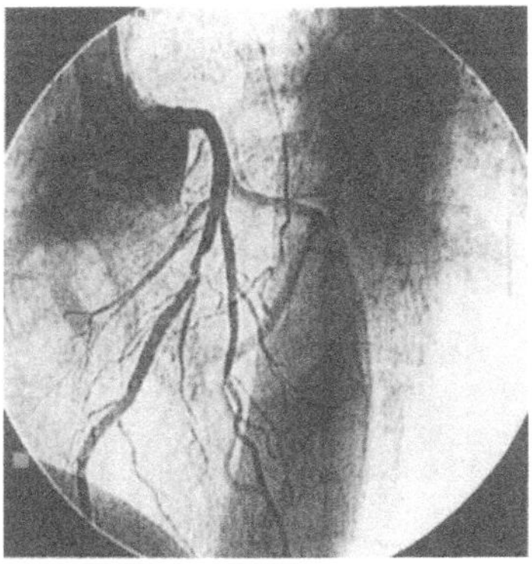

Fig. 6

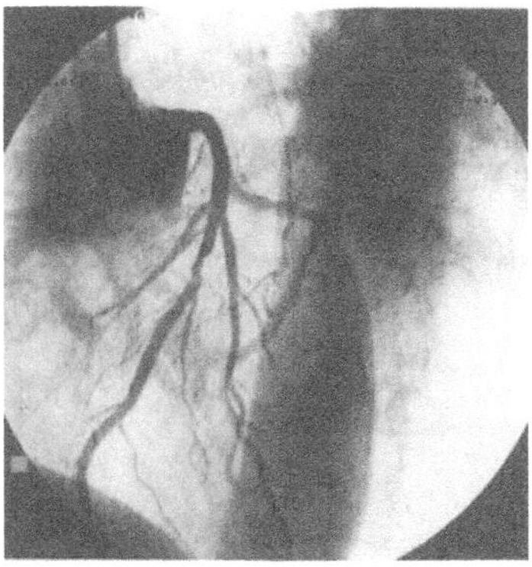

Fig. 7

Automatic object and contour tracing

After preprocessing with the filters of the described principle (Fig. 7), the vessel itself or its borders have to be followed. Dynamic programming, i.e. path finding according to a minimal cost principle has proved to be a powerful tool [4]. With this method, each step (image line) has to be reached in the most cost effective way, which means that the optimal transition from the last line to the actual line is selected under the preposition that the subpaths to all relevant positions in the last line are also optimal. An approximation to this principle can be made with a parallel processing scheme:

For all line-to-line transitions only few alternatives are allowed, e.g. controls 0, -1 and 1 for the straight and the two diagonal connections. The related possible costs can be calculated in parallel e.g. for all odd-to-even line transitions. In this way at the same time a grouping of the whole image in line couples is effected. Now, with a 1024x1024 image 512 couples have to be examined under transition cost aspects, yielding 256 "quadruple costs", ... After 10 steps a line of total costs on paths across the image is generated, which is an approximation to the global optimum. Now a backtracking of the cost-accumulation is performed, down to the quadruples, the couples and finally the line positions. Even across unprocessed images this strict processing scheme generates a path which is close to the complete optimal path. A refinement with more specific weighting of control variables for bigger groupings further improves this partial optimum, but requires more specific processing. Alternately the possible short discontinuities can easily be detected and filled at a postprocessing stage.

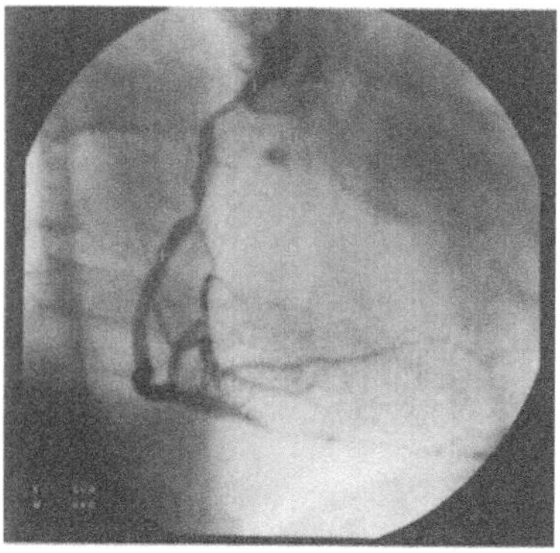

Fig. 8

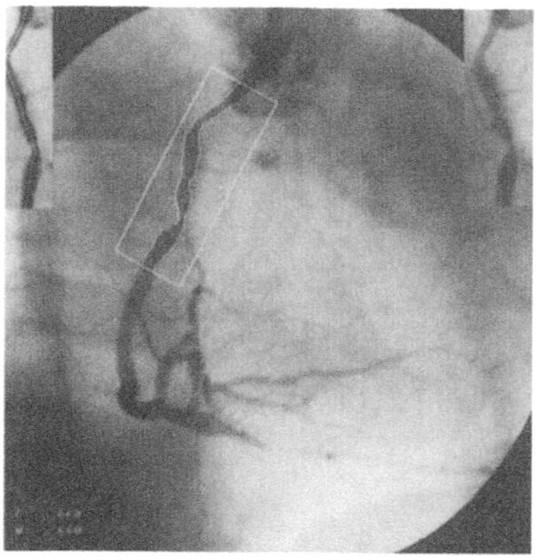

Fig. 9

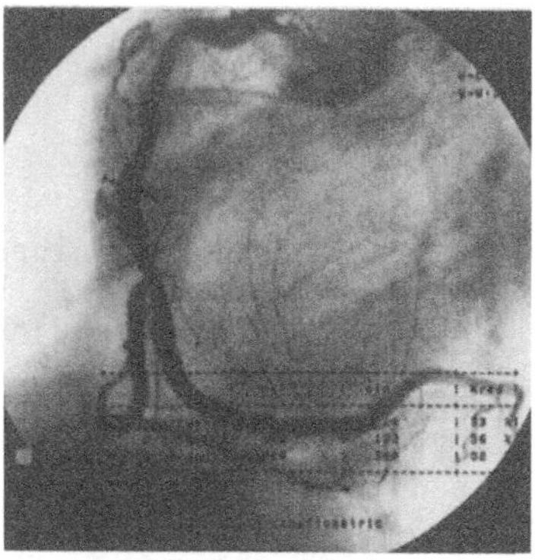

Fig. 10

Fig. 8: Input to vessel tracing of a right coronary artery

Fig. 9: Automatic segmentation of the vessel. Top corners: Pixel extraction, vessel enhancement (left) and centerline generation.

Fig. 10: Stenosis report with automatic "calibre method" on POLYTRON 1000®. Percent stenosis from diameter, area and densitometry.

Conclusion

With these investigations and developments it is attempted to offer object descriptions and contours as the basis for further measurements with a response-time similar to the short access time to DSA-scenes. Such measurements could be 3D-position calculation or stenosis and coronary flow determination with combined geometric and densitometric analysis. With the increased matrix size and a data amount of 2MB per image only parallel processing of the bulk of image data can assure a processing time of few seconds in applications like automatic vessel recognition and segmentation.

Acknowledgement

We would like to thank Prof. Piessens and his group from the University Hospital of Leuven, Belgium, for their cooperation in the clinical cases.

References

1. Sanders WJ, Alderman EL, Harrison DC: Coronary Artery Quantitation Using Image Processing Techniques. IEEE Computers in Cardiology, 15-20, 1979.

2. Barth K, Epple E, Irion KM, Faust U, Decker D: Quantifizierung von Stenosen der Herzkranzgefaesse durch Digitale Bildverarbeitung. Erg. Bd. Biomed. Technik 26, 1981.

3. Kirkeeide RL, Fung P, Smalling RW, Gould KL: Automated evaluation of vessel diameter from arteriograms. IEEE Computers in Cardiology, 215-218, 1982.

4. Pope DL, Parker DL, Gustafson DE, Clayton PD: Dynamic Search Algorithms in Left Ventricular Border Recognition and Analysis of Coronary Arteries. IEEE Computers in Cardiology, 71-75, 1984.

5. Barth K, Koch R, Marhoff P: Fast Automatic Reognition and 3D-Reconstruction of the Coronary Tree from DSA-Projection Pairs. In: Heintzen PH and Buersch JH (eds.): Progress in Digital Angiocardiography (Kiel Symposium 1986), Kluwer Academic Publishers, 193-200, 1988.

6. Bittner U: Fast image processing on parallel hardware. SPIE Vol. 914, Medical Imaging II, 582-590, 1988.

7. Scoblionko DP, Brown BG, Mitten S, Caldwell JH, Kennedy JW, Bolson EL, Dodge HT: A new digital electronic caliper for measurement of coronary arterial stenosis: comparison with visual estimates and computer-assisted measurements. Am J Cardiol 53, 689 - 693, 1984.

Ultrasound

Color-Coded Duplex Sonography: Is this New Method Capable of Replacing More Expensive or Invasive Examinations?

F. Fobbe, H.-C. Koennecke, P. Heidt, M. Dietzel, K.-J. Wolf

FU Berlin, Klinikum Steglitz,
Röntgendiagnostik, Hindenburgdamm 30, 1000 Berlin 45

Abstract:

Color-coded duplex sonography (CCDS) is a further development of duplex sonography. It allows the simultaneous and real-time display of tissue texture and blood motion over the entire scanfield. The aim of our study was to find out to what extent this new method can replace more expensive and/or invasive examinations. 920 examinations were evaluated. For monitoring of the kidney transplants, CCDS seems to be the method of choice to exclude occlusion of renal vessels. The vascularisation of the thyroid gland allows differentiation between the simple nodular goiter and the autonomous adenoma or thyroid carcinoma. For the diagnosis of perpipheral vascular diseases, CCDS simplifies the examination consideralby compared to conventional duplex sonography and can replace i.v. DSA. Testicular torsion can be exluded in cases of acute scrotum by the demonstration of blood flow in the cord vessels.

Color coded duplex sonography is a very promising new noninvasive method which can to some extent replace angiography, scintiscan and needle biopsy in the diagnosis of thyroid, vascular and scrotal diseases and can be used for monitoring kidney transplants.

Measurements of Soft Tissue Deformation Using Echographic Speckle Tracking

M. BERTRAND, J. MEUNIER*, G. MAILLOUX, M. DOUCET and R. PETITCLERC*

Institute of Biomedical Engineering
Ecole Polytechnique de Montréal, Montréal, Canada H3C 3A7

*Institut de Cardiologie, 5000 Bélanger est, Montréal, Canada H1T 1C8

Summary

Measurement of soft tissue motion is of great diagnostic importance, whether it is intrinsic such as with myocardial and skeletal muscle contraction, or extrinsic, such as with externally induced deformation. In the former, tissue motion relates to the dynamics of contractile fiber; in the latter, it relates to passive bio-mechanical properties typically sensed, for example, during palpation examination. In view of this, we propose a new method for estimation of local tissular dynamics, (i.e. rotation, translation and biaxial tissue deformation), by following the motion of ultrasonic speckle through time. The paper shows the usefulness of the speckle tracking method for estimation of fundamental parameters of local myocardial dynamics, namely fiber shortening (equivalent to the local circumferential shortening) and fiber thickening (related to wall thickening). Preliminary results on measurements of contraction thickening of skeletal muscles are presented. Potential for application on bio-mechanical characterization of other soft tissues from externally induced deformation is also suggested.

Introduction

In previous works [1,2], we have discussed the potential of ultrasonic speckle tracking to characterize tissue dynamics based on the assumption that for small changes, there is a direct relationship between speckle pattern motion and the underlying tissue motion. Such motion is computed between successive frames from sequences of ultrasonic images using an optical flow algorithm. In essence, the algorithm tracks the motion of isointensity level through the sequence, the motion being constrained to some form of spatial smoothness [3,4,5].

Method

The optical flow is a vector field which assigns to each point of an image, its apparent two-dimensional motion across the image plane in the interval between two successive frames. In echography, for sufficiently small motion and small region of interest (ROI), the interframe motion of the tissue speckle pattern can be described by a linear 2-D field ($\Delta x,\Delta y$) fitted during the computation of the optical flow [4,5]:

$$\begin{bmatrix} \Delta x \\ \\ \Delta y \end{bmatrix} = \begin{bmatrix} a \\ \\ b \end{bmatrix} + \begin{bmatrix} A\,B & x \\ \\ C\,D & y \end{bmatrix} \qquad (1)$$

Using a sequence of several echographic ROI and the associated interframe optical flow parameters (a,b,A,B,C,D), one can compute various curves that show tissue dynamics as a function of time. To this end, one obtains easily from equation 1 a linear relationship between the position (x_k,y_k) of a point at time t_k (image number "k") and its previous image position (x_{k-1},y_{k-1}) at time t_{k-1} knowing the computed motion $(\Delta x,\Delta y)$ between the two points:

$$\begin{bmatrix} x_k \\ \\ y_k \end{bmatrix} = \mathbf{T}_k + \mathbf{M}_k \begin{bmatrix} x_{k-1} \\ \\ y_{k-1} \end{bmatrix}, \quad \text{where} \quad \mathbf{T}_k = \begin{bmatrix} a_k \\ \\ b_k \end{bmatrix} \quad \text{and} \quad \mathbf{M}_k = \begin{bmatrix} A_k+1 & C_k \\ \\ B_k & D_k+1 \end{bmatrix} \qquad (2)$$

where $\mathbf{T}_k$ is the translation and $\mathbf{M}_k$ is a 2 x 2 matrix that includes rotation and deformation from time t_{k-1} to t_k. This equation can be resolved iteratively $(k>1)$ to get a linear relationship between the position (x_k,y_k) and the initial position (x_o,y_o) at time t_o:

$$\begin{bmatrix} x_k \\ \\ y_k \end{bmatrix} = \mathbf{T}_{ko} + \mathbf{M}_{ko} \begin{bmatrix} x_o \\ \\ y_o \end{bmatrix}, \quad \text{where} \quad \mathbf{M}_{ko} = \prod_{i=1}^{k} \mathbf{M}_i \quad \text{and} \quad \mathbf{T}_{ko} = \mathbf{T}_k + \sum_{j=1}^{k-1} \prod_{i=j+1}^{k} \mathbf{M}_i \mathbf{T}_j \qquad (3)$$

The matrix $\mathbf{M}_{ko}$ is then decomposed into a matrix product $\mathbf{R}_{ko}\mathbf{D}_{ko}$ represented by:

$$\mathbf{R}_{ko} = \begin{bmatrix} \cos\theta & -\sin\theta \\ \\ \sin\theta & \cos\theta \end{bmatrix} \quad \text{and} \quad \mathbf{D}_{ko} = \begin{bmatrix} \alpha & \gamma \\ \\ \gamma & \beta \end{bmatrix} \qquad (4)$$

which respectively describe rotation with the well-known rotation matrix and biaxial deformation with an hermitian matrix. The deformation orientation and strain parameters are obtained from eigenvectors and eigenvalues of $\mathbf{D}_{ko}$.

Results

Myocardial dynamics

We used routine echocardiograms already available at the Institut de Cardiologie de Montréal to test our optical flow method. We considered 150 successive ROI (64*64

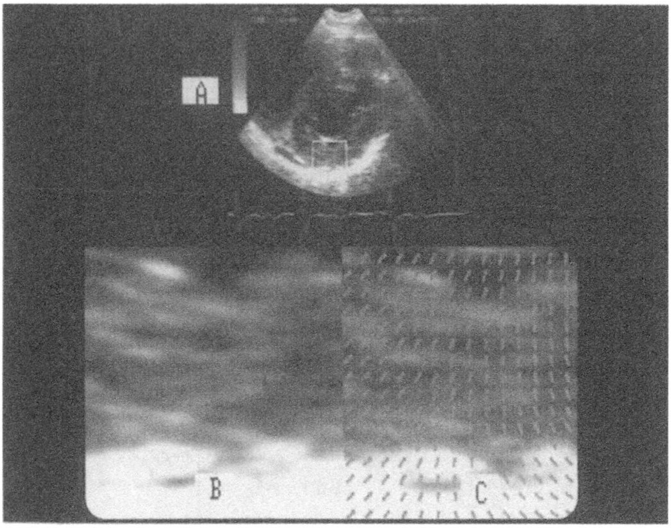

Fig. 1. (A) Example of a short axis echocardiogram showing a posterior wall speckled ROI. (B) and (C): ROI enlargement from two successive scans (1/30 sec.) used for calculation of a velocity field from speckle motions. In (C) the velocity field is shown superimposed on the speckles.

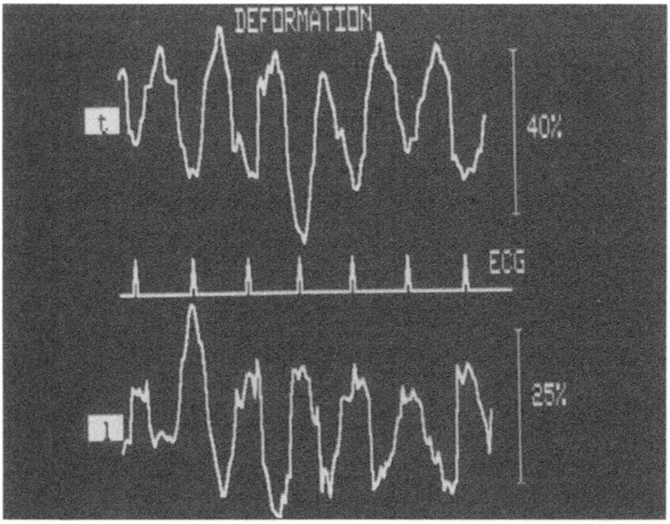

Fig. 2. Deformation parameters as a function of time for the myocardial ROI (Fig. 1). Seven cardiac cycles are represented (ECG). Eigenvalues are labeled "t" and "l" respectively for wall <u>thickness</u> and circumferential <u>length</u> deformation.

pixels) showing a part of the posterior wall (figure 1). The images were lowpass filtered to reduce noise and improve the optical flow performance [5] but without removal of speckle patterns. Also, the parametric curves obtained with the algorithm were bandpass filtered on the cardiac frequency to reject drift component inherent to the integration involved between time t_0 and t_k.

As an example, the biaxial deformation parameters (eigenvalues) are reproduced on figure 2. They characterize the contractility of the myocardium and are therefore the most important variables. As expected, the parameters are in 180° phase opposition and exhibit both cyclic variations and consistent relationship with QRS complex occurrences. Their peak to peak amplitude of 35% and 20% are in agreement with wall thickening and local contraction calculated manually [1].

Skeletal muscle dynamics

Ultrasonic speckle tracking can also be applied to determine local contractility patterns of skeletal muscle. This can be useful in investigating the site and extent of neuromuscular disorders. The feasibility of the method was evaluated using a transverse view (perpendicular to long axis of the fibers) in the mid-plane of the biceps brachii during a short isometric contraction cycle.

Figure 3 shows the muscle deformation curves as a function of time for a 1 cm x 1 cm ROI, as it relates to the applied force. As the force increases, the curves shows similar

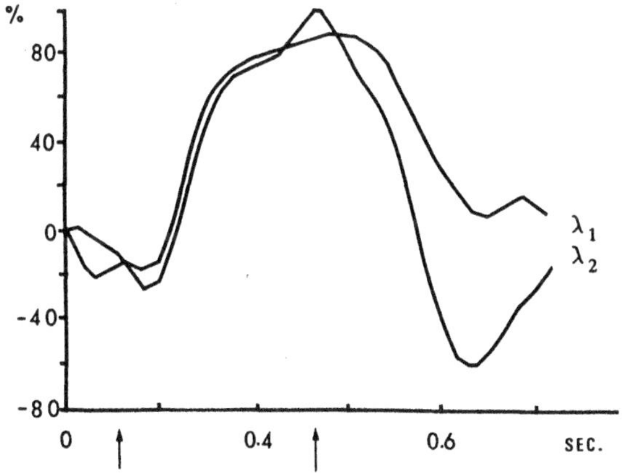

Fig. 3. Results from an ultrasonic speckle motion study for a biceps brachii (1 cm x 1cm ROI) during a brief isometric contraction occurring between arrows. The curves show the time course of the eigenvalues (λ_1,λ_2) of the deformation matrix.

biaxial deformations thereby suggesting symmetric deformation of the fibers in this transverse plane. However, a force/deformation delay is noticeable, presumably due to a particular recruitment behavior.

Discussion

We have presented a new powerful method for the local measurement of contractile performances of muscle using ultrasound speckles as tissue landmarks. Under well controlled experimental conditions, the technique yields reproducible results; sufficiently small deformations occurring between successive time samples are needed to maintain the speckle pattern correlation necessary for adequate speckle tracking.

In addition to the obvious application in cardiology, the method is expected to provide a new look in the study of skeletal muscle contraction, by revealing the force conditions under which a particular region of the tissue shows myogenic activity. Another application in the area of tissue characterization is also being considered. Using an external force, the objective would be to infer the viscoelastic properties of the tissue from its deformation patterns, thereby providing a form of computerized palpation.

Acknowledgments

This work was supported by the Canadian Heart Foundation and Natural Science and Engineering Research Council of Canada. The authors would like to thank G. Ferland and S. DeSerres for their help in the skeletal muscle study.

References

1. Meunier, J.; Bertrand, M.; Mailloux, G.E.; Petitclerc, R.: Assessing local myocardial deformation from speckle tracking in echography. SPIE Proc: Medical Imaging II, Vol. 914 (1988) 20-29.
2. Meunier, J.: Analyse dynamique des textures d'échocardiographies bidimensionnelles du myocarde. Ph.D. thesis, Montréal: Ecole Polytechnique 1989.
3. Mailloux, G.E.; Langlois, F.; Simard, P.Y.; Bertrand, M.: Analysis of heart motion from two-dimensional echocardiograms by velocity field decomposition. Computer in Cardiology, (1987) 441-444.
4. Mailloux, G.E.; Bleau, A.; Bertrand, M.; Petitclerc, R.: Computer analysis of heart motion from two-dimensional echocardiograms. IEEE Trans.Biomed.Eng., BME-34, (1987) 356-364.
5. Mailloux, G.E.; Langlois, F.; Simard, P.Y., Bertrand, M.: Restoration of the Velocity Field of the Heart from Two-Dimensional Echocardiograms. IEEE Trans. Medical Imaging, (1989).

3D Sonographic Reconstruction of Neonatal and Infant HIP

G. BRACCINI°, M. MASSIMETTI°°, O. SALVETTI°°°

° Istituto di Radiologia, Università di Pisa
°° Clinica Pediatrica, Università di Pisa
°°° Istituto di Elaborazione della Informazione del CNR, Pisa

Summary

The preliminary results of a research carried out to investigate possible extensions of two-dimensional hip sonographic examination are presented. A dedicated workstation and software have been designed and partially implemented. By applying suitable analysis and reconstruction algorithms on parallel axial and/or frontal sonographic slices, three-dimensional hip morphology can be obtained in order to perform a more complete anatomical evalu ation and test the actual application of this technique.

Introduction

In the last few years sonography has been added to clinical and radiographic examination as an essential methodology to detect neonatal and infant hip displasia. Hip sonography makes it possible to obtain tomographic views throughout the region of acetabular roof and permits an accurate evaluation of the bony and cartilaginous components of the hip joint. Among the several sonographic techniques, Graf's method has obtained unanimous approval owing to its exact codification. This technique needs a good sonographic scan on a frontal and strictly plane of the acetabular fossa in order to permit angular measurements between the base line and the bony or the cartilage roof lines [1]. These planar measurements are sufficient to detect also slight degrees of displasia; nevertheless, the three-dimensional relation among the anatomical structures can be derived only from a mental reconstruction process based on the understanding of a set of two-dimensional slices.

In this paper we describe a methodology to obtain automatically these relations. By applying image processing techniques, we have developed a hardware/software system able to control the three-dimensional automatic reconstruction of echoes of the chondro-osseous acetabular roof: in this way, we try to determine a geometrical model of the hip joint useful to perform morphometric evaluations.

Materials and Methods

A high-definition EsaoteBiomedica AU-450 ecographic device with 7.5 MHz linear scanner has been used to generate time sequences of images stored on a video tape recorder. A personal computer was used to control image acquisition from VTR, an optical disk mass storage and an image processing board with a video memory of 1024x1024x8 bits. The developed software system was specifically designed for digital image handling and management [2], and the following procedures are available: automatic acquisition of time sequences of sonographic images from VTR; geometric and photometric frames' registration; preprocessing for noise reduction; spatial reconstructions. In particular, the spatial reconstruction procedure is subdivided into two parts. The first one considers the 3D image as a stack of parallel two-dimensio nal slices and computes discrete distance maps of selected echoes according to predefined projective laws [3] ; whereas, the second part, at present under test, defines a geometrical model of the examined object, starting from the previously computed maps.

We have examined 36 normal hips of 18 patients with age in the range from 2 days to 5 months. Parallel scans on frontal and axial planes were obtained and the scanning lengths and the total scanning times for each one were measured. The ecographic acquisition procedure was greatly careful to obtain costant scanning speed and strictly parallel slices. Each sonographic image was digitized to obtain a 256 gray level matrix of

512x512 pixels. The metrics of both the original and processed images were known. The digital image acquisition procedure from VTR selected homogeneous sets of frames from single time sequences in such a way to spatially sample the examined object within the desired precision.

Results and Discussion

In the following, we refer to two frontal and axial scans on the same 4 months old patient. An example of automatic acquisition and time sampling of a sequence of frontal planes is shown in Fig.1.

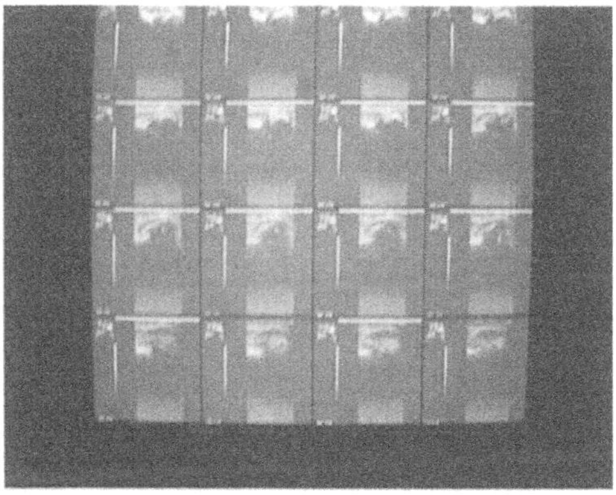

Fig.1. Selected images from a 9 sec VTR registration of a hip joint frontal scan of 40 mm length

The axial planes have been obtained by scanning with a costant speed a distance of 60 mm in a 17 seconds time. The difficult to obtain ecographic parallel scans at costant speed is an actual problem [4]; however, the small dimensions and surface position of the examined anatomical region allow an adequate measuring precision. An interesting rectangular region extracted from a central axial slice is shown in Fig.2a. The selection of the echoes to be 3D reconstructed can be obtained by examining interactively or automatically the graph of the gray levels of chosen rows (Fig.2b).

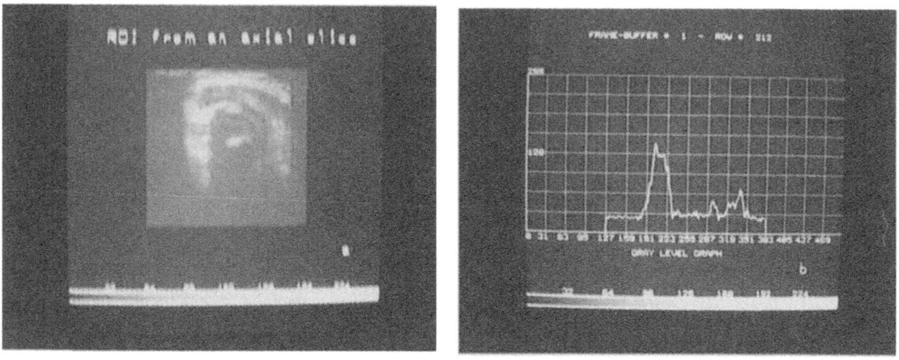

Fig.2. a) ROI from an axial slice and b) the graph of the gray levels of a selected row from the same image as a function of the pixel position

The reconstruction of 3D echo maps has been obtained by processing 34 frames with a time sampling step of 500 msec: in this way the distance between any pair of adjacent planes is about 2 mm. Let's consider a cartesian space having its origin in the center of the hip joint and ZX, ZY and XY planes respectively coincident with sagittal, frontal and axial planes: a reconstruction at 60° along the cranial direction and 180° along the axial direction is shown in Fig.3a; a projection only at 180° along the axial direction is shown in Fig.3b.

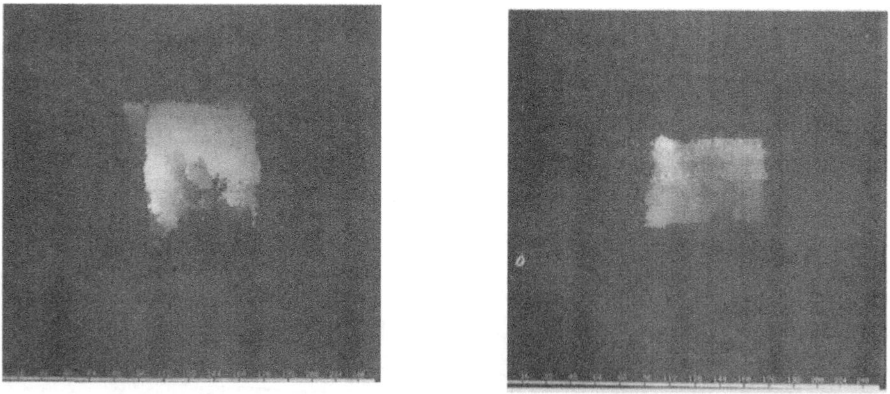

Fig.3. Examples of 3D discrete distance maps reconstruction

Both the computed images needed about 5 minutes to be generated. In particular, Fig.3a corresponds to a projection respect with a point of view placed inside the pelvis at a distance of 150 mm from the hip joint center. The image is displayed with a magnification factor M=2.5. For reconstructions considering selected shapes and echo amplitudes, the image sequence to be processed can be derived by segmenting plane by plane regions of interest.

Conclusions

This paper presents some preliminary results relative to the definition of a geometrical model suitable to perform 3D morphometric analyses on neonatal and infant hip sonographic images. A main concept is to evaluate whether this approach may be considered as a complementary extension of the current methologies. However, it seems to be quite interesting to improve the implemented reconstruction procedure by introducing the possibility of computing pencils around lines and then arbitrary 2D slices.

References

1. Graf, R.: Guide to sonography of the infant hip. G. Thieme, Stuttgart 1987.
2. Salvetti, O.; Braccini, G.; Frassineti A.: A distributed environment for quality control and diagnosis of medicalimages. Mini and Microcomputers and Appl., E. Luque Ed., 1988.
3. Lazzeri, M.; Azzarelli, L.; Chimenti, M.; Salvetti, O.: Computer assisted film dosimetry: 3D dose planning. Proceed. Int. Symp. CAR'87, Springer-Verlag, 1987.
4. Watkin, K.L.: Three-dimensional reconstruction of real-time umtrasonic scans. J. Acoust. Soc. Am. Suppl. 1, Vol. 88, 1988.

Ultrasonic Tissue Characterization of Breast Masses Using a Diagnostic Expert System

A. M. YOUSSEF and Y. M. KADAH

Faculty of Engineering
Biomedical department
University of Cairo, Egypt

Summary

The primary attraction of ultrasound mammography is its ability to image the breast repeatedly with no deletrious effects. In addition palpable breast masses can be conclusively determined to be fluid-filled or solid. Benign and malignant lesions can be further diagnosed. The major limitation is the dependency of technique on the ultrasonographer as well as the complexity of the visual criteriae for differentiation. A real-time ultrasound data acquisition and analysis system has been developed at Badran's hospital of Cairo, employing image textural analysis technique. A diagnostic database is used in the classification of images according to the specific statistical parameters in the database. The overall accuracies for benign cases is 84% by normal echography and 89% by additional tissue analysis methods. The overall accuracies for different malignancies is 82% by normal sonographic methods and is 90.3% by tissue analysis.

Introduction

Using the normal differential diagnostic signs of sonomammography two categories of breast tissue are identified. The first resultii from the echopatterns confined to the tumor namely the boundary echo and shape, as well as the internal echo. The second category is related to the surrounding tissue interaction. Medullary carcinoma demonstrates the highest misdiagnosed tumor. This type of breast cancer has less absorption of ultrasonic energy as confirmed by Calderon {1} who noticed that ultrasound attenuation was highest in the tissue of ductal and scirrhous carcinoma and lowest in medullary carcinoma. As to the ultrasound tissue characterization of tumor tissue, Field and his associates {2} made the observation that connective tissue consisting of fibroblasts and fibrous tissue is one of the factors causing ultrasound attenuation in neoplastic tissue. The ultraound criteriae for differentiating benign from malignant tumors are difficult to interpret. Dependability of diagnosis is highly related to the sonographerś experience {3}. Ultrasound attenuation is usually shown on the echogram by the formation of a retro-tumerous echopatterns as the acoustic middle shadow sign and the complete disappearance of the distal limit of the tumor mass echo are positively correlated with the microscopic documentation of connective tissue content in given breast cancers of different histological types. Connective tissue content is highest in scirrhous and ductal carcinomas lowest in medullary carcinoma and median papillary carcinoma {3}.

Materials and Methods

Seventy malignant and twenty benign cases were included in this study. All these cases were assessed pathologically. The data acquisition system consists of an ultrasound sector 320 Combison (Kretz) scanner, interfaced to a PIP-512 (Matrox) video digitizer card provided with an IBM/AT microcomputer, main storage being 640 KB, hard disk storage of 30 MB. The video digitizer card receives the composite video signal from the sector scanner, digitizes it into 512x512x8 image fram, which may be stored on hard disk. After data acquisition is completed, a set of image textural parameters from first order greylevel distribution, first order gradient distribution and greylevel cooccurence matrix are calculated from the region of interest designated by the ultrasonographer within the mass. Using the most discriminating parameters namely the mean grey level, percentiles, entropy and contrast, domains for malignant and benign regions are defined in three dimensional plots. Further subclassifications of pathologies are possible.

Results

a) Normal breast

The echographic characteristics for a middle aged woman (35 years old) is demonstrated in figure 1. The echopattern tends to be fine and evenly distributed. The breast tissue in young women is usually dense with well developed glandular tissue and decreased subcutaneous as well as interlobular fat. In older women, breast tissue tends to be atrophic with evenly distributed specklings of echoes. This pattern reflects the increased connective tissue and the decreased glandular element. Increased subcutaneous and interlobular fat is also evident in this age group.

b) Benign diseases of the breast

For fibrocystic mastopathy, the echogram demonstrates multiple small fibrocystic areas with homogenous echoes within. Cystic mastopathy is usually anechoic with no internal inclusions and with retrocystic accentuation. For cystosarcoma phylloides the echopattern demonstrates regular boundary and fine homogenous internal echoes, evenly distributed within the tumor area. A typical echographic pattern fo a chronic abscess demonstrates regular boundary with thick wall and heterogenous internal echoes, cystic variations are usually demonstrable. For fat necrosis the tumor boundary is irregular with hypoechogenic texture simulating breast cancer. As for fibroadenomas the characteristic echopattern is uniform sized internal echoes. The delineating boundary is regular and discrete, the differential diagnosis are, cystosarcoma phylloides, benign cyst,benign papilloma, tubular adenoma, and mucous carcinoma. Cystosarcoma phylloides is also simulated by fibroadenomas and mucous carcinomas. Figures 2 through 7 show typical benign lesions together with the corresponding histograms demonstrating altered patterns as compared to the normal breast (see figure 1).

c) Malignant diseases of the breast

A typical pattern for scirrhous or duct carcinoma demonstrates irregular boundary with uneven distribution of internal echoes within the tumor area, evident complete disappearance of the distal limit of the tumor mass is usually demonstrable. For medullary lesions the internal echoes are uneven with disappearance of the distal limit of the tumor mass echo. Mucous carcinoma may simulate a fibroadenoma, cystosarcoma phylloides or metastatic carcinoma. Figures 8 and 9 demonstrate sonograms for malignant lesions with their corresponding histograms. When comparing these histograms with the corresponding benign lesions initiall differentiation between malignant and non-malignant lesions is possible. Table 1 {3} demonstrates the differential diagnostic signs and criteriae for

different benign and malignant lesions. Table 2 demonstrates the textural parameters for the benign and malignant cases. These parameters are the most discriminative. Average values were calculated, further expansion of the database is required to prove that no overlapping of parameter values occur, Table 1 . The discriminative ability is successful for the cases studied as is demonstrable in three dimensional plots of figures 10 and 11.
The parameters of most discriminative ability are, mean greylevel ,percentiles, and cooccurence matrix parameters(entropy and contrast). The diagnostic accuracy rates of ultrasound and tissue analysis technique are demonstrated in table 3, for different stages. The overall accuracies is 84% by conventional ultrasound methods and 89% by additional ultrasound tissue analysis technique. In our study duct carcinoma demonstrates the highest accuracy rate . Diagnostic accuracy rates achieved by various investigators for malignant tumors is 85.2% and 83.1% for benign tumors {3} . The diagnostic accuracy rates for malignant tumors are 79.2%, 90.7%, and 92.3% for stages 1 through three respectively {3}. The diagnostic accuracy for medullary carcinoma is enhanced by tissue textural parameters as is illustrated by figures 10 and 11.

Table 1

Lesion	Boundary	Internal echo	Acoustic mid shadow sign
Scirrhous	Irregular	Nonuniform	present
Papillary	Irregular	Nonuniform	Present in most cases
Medullary	Irregular	Nonuniform	Present or absent
Mucous	Irregular	Nonuniform	Absent
Pagetś	Irregular	Ladder like	Present
Metastatic	Irregular	Uniform	Absent
Cyst	Regular	Anechoic	Absent
Fibroadenoma	Regular	Uniform	Absent
Cystosarcoma	Irregular	Nonuniform	Absent
Fat necrosis	Irregular	Nonuniform	Present
Foreign body	Regular	Anechoic	Present
Abscess	Regular	Complex	Present
Papilloma	Irregular	Ladder like	Absent
Mastopathia	Not clear	Speckeled	Absent

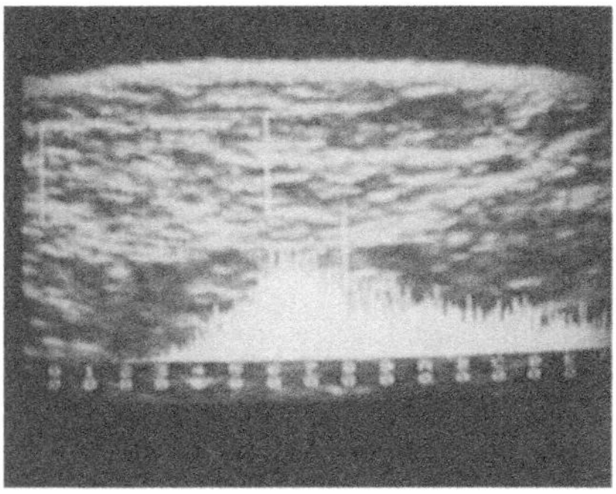

Fig.1. Normal breast echogram with corresponding histogram

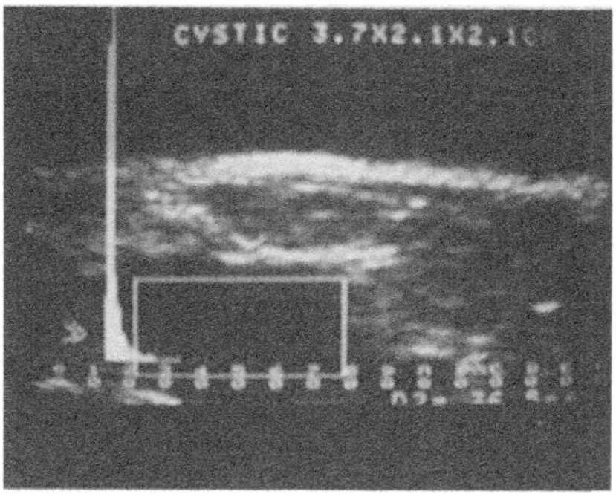

Fig.2. A cystic lesion with corresponding histogram

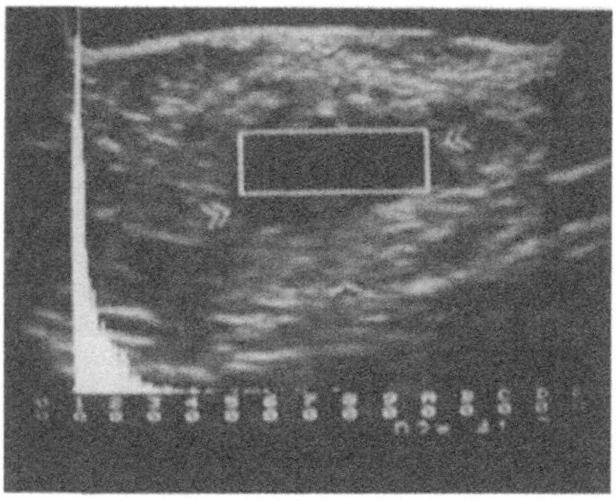

Fig.3. Chronic abscess lesion with corresponding histogram

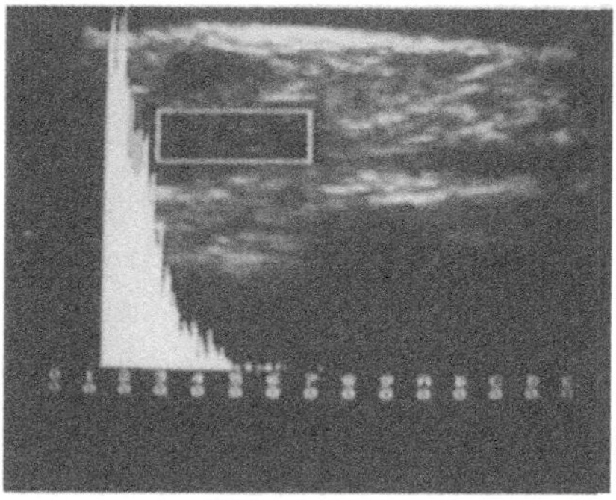

Fig.4. Cystosarcoma philloides sonogram with corresponding histogram

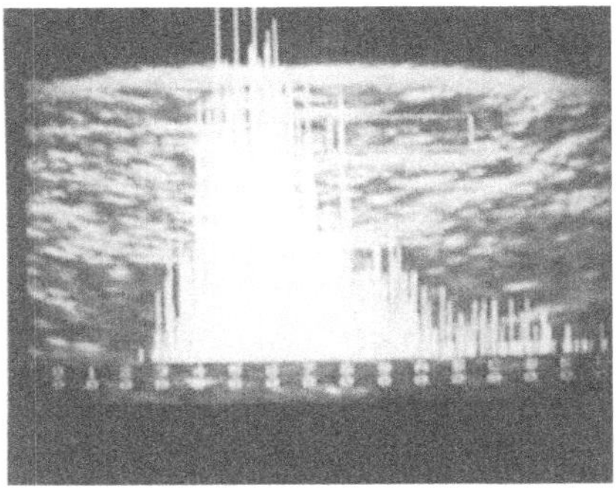

Fig.5. Sonogram for a fibroadenoma with corresponding histogram

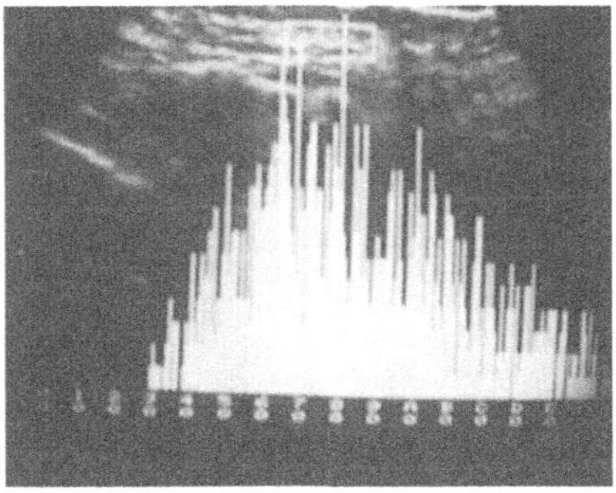

Fig.6. Sonomammogram for a fibrocystic lesion with its histogram

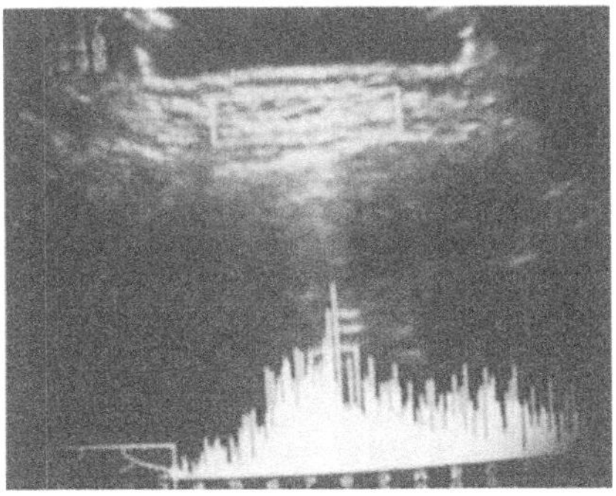

Fig.7. Sonomammogram of fibroadenosis with duct ectasia with its corresponding histogram

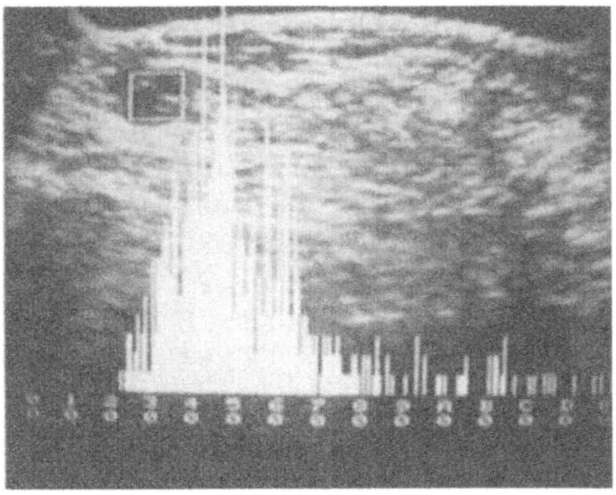

Fig.8. Sonomammogram of duct carcinoma with the corresponding histogram

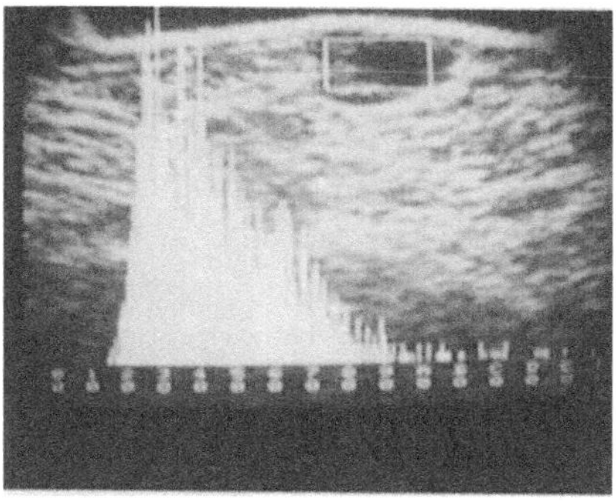

Fig.9. Sonomammogram of medullary carcinoma with the corresponding histogram

Table 2

Different textural parameter values for different pathologies

lab	mgl	percl	absave	absvar	variance	con	asm	ent
L	48.2	29	23.80	809.30	321.700	440.500	0.000512	7.83440
F	143.4	83	80.95	8651.30	2686.800	5331.400	0.000204	8.58047
D	73.7	46	34.40	1584.50	605.020	879.300	0.000978	6.98124
C	25.0	23	3.24	23.91	10.575	12.631	0.083424	3.48133
M	64.8	36	31.90	1497.00	795.000	827.000	0.000420	7.96298
I	122.9	83	56.25	4196.00	1353.700	2450.000	0.000396	7.90645
N	145.5	89	78.60	8007.20	2522.800	5693.300	0.001148	6.82944
A	25.8	16	12.44	371.60	206.860	196.690	0.010816	5.93068

L=Cystosarcoma phylloides, F=Fibrocystic disease, D=Duct carcinoma C=Cyst, M=Medullary, I=Fibroadenoma, N=Normal, A=Abscess, percl=First percentile, absave=Absolute value of gradient ,ent=Entropy absvar=Absolute value of variance, con=Contrast, asm=Angularmoment

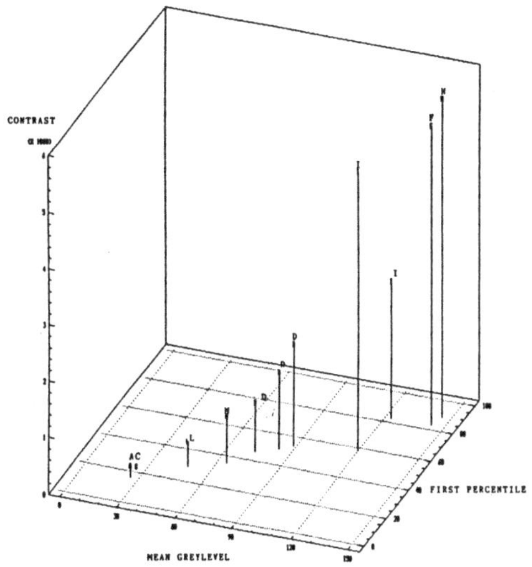

Fig.10. A three dimensional plot for different textural parameters and for different benign and malignant lesions, where A=Chronic abscess,L=Cystosarcoma C=Cyst, M=Medullary carcinoma, D=Duct carcinoma I=Fibroadenoma,F=Fibrocystic, N=Normal

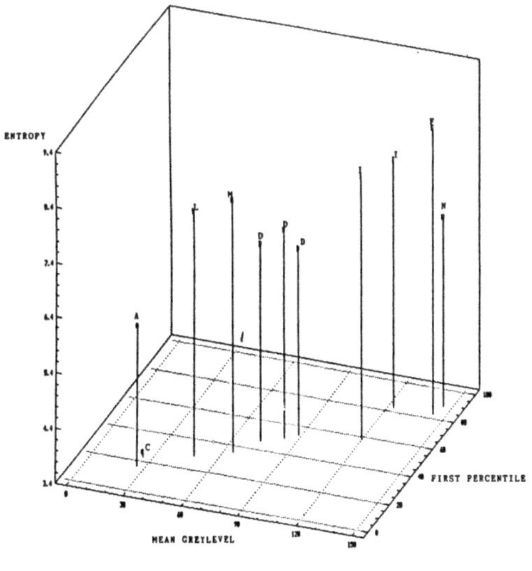

Fig.11. A three dimensional plot for different pathologies at different textural parameters where , A=Chronic abscess L=Cystosarcoma, C=Cyst, M=Medullary carcinoma, D=Duct carcinoma, I=Fibroadenoma, F=Fibrocystic, N=Normal

References

1. Calderon, C.; Vilkomerson, D.; Merzrich, R.; Etzold, K.F. Differences in the attenuation of ultrasound by normal, benign and malignant breast tissue. J. Clinc. Ultrasound 4 (1978) 249.

2. Field, S.; Dunn, F. Correlation of echographic visualization of tissue with biological composition and physiological state. J. Acoust. Soc. Am. 54 (1973) 809.

3. Kobayashi, T.; Present status of differential diagnosis of breast cancer by ultrasonography. Jpn. J. Clin. Oncology 4 (1974) 145.

Application Systems

Computer Assisted Radiation Therapy

Verification of Electron Beam Therapy with Conventional and Storage Phosphor Images

D. Gur*, J.C. Weiser*, M. Deutsch+, R. Schaetzing°

University of Pittsburgh
Departments of Diagnostic Radiology* and Therapeutic Radiology+
Pittsburgh, PA, 15261, U.S.A.

Eastman Kodak Company°
Rochester, NY, 14650, U.S.A.

Summary

Portal verification images were generated by the photon contamination in electron beams produced by a linear accelerator during treatment of patients receiving high-energy electron radiation therapy. Both conventional and storage phosphor methods yielded projection radiographs in which anatomy of the irradiated and surrounding tissue was demonstrated. A preliminary series of 22 cases was evaluated by two radiotherapists and judged subjectively to be of clinical value. Geometric error, or more importantly, the lack thereof, during high-energy electron treatments were easily confirmed with this method. In three cases, the treatment protocol was corrected based on the images obtained.

In electron beam therapy, the limited range of the electrons precludes routine verification using the primary beam. Therefore, in most such instances, a second simulation film or an exposure to photon beams under similar conditions is used for verification. Neither of these procedures is optimal or convenient and, therefore, is not performed routinely during treatment in most instances. Although it could be argued that the need for careful verification of electron beam treatments is diminished by the superficial nature of many of these treatments, a recent increase in combination treatments of high-energy electron ($\geq$ 12 MeV) in conjunction with photons or the use of abutted fields may necessitate adequate documentation of the electron fields.

Photon contamination in electron beams has been shown to originate from a virtual source close to the scattering filters with intensities ranging from 1 to 9 percent of the electron dose at d_{max} depending on the beam energy and SSD [1]. Intensity of the contamination is further modified (somewhat) by the tissue. This intensity of photon contamination in electron beams should allow for portal verification image acquisition during a typical daily treatment. Obtaining images in this circumstance should be convenient and should not require more than the placement of a detector (e.g., screen-film) behind the treated tissue. A preliminary report has indicated that such a procedure is in fact feasible; however, no additional published data on this matter were identified since its publication in 1978 [2]. In this preliminary study, we investigated the feasibility of acquiring both conventional films and storage phosphor images from 22 patients during their daily treatment with high energy (6-20 MeV) electron beams. The main purpose of this study was to: 1) confirm that such images can be acquired in a consistent manner; 2) assess whether or not these images represent an accurate projection of the electron field; and 3) assess whether geometric errors, if any, can be documented, corrected, and the correction verified using a second image. There was no attempt to objectively compare or identify preference between the two detection systems used.

All images (conventional and storage phosphor) were obtained using a clinical treatment system at our institution (i.e., Philips SL/75-20) and were acquired by placing a cassette behind the patient during his/her daily treatment. Conventional images were obtained using DuPont cassettes with lead screens and CRONEX-7 films. For the storage phosphor images, experimental plates (GP) manufactured by Kodak were used in Kodak cassettes with copper screens. There was no added exposure to the patient. We selected the storage phosphor detector for this purpose because of its wide latitude which enables recovery of information in environments where detector exposure may vary by at least an order of magnitude due to both source intensity (energy and treatment protocol dependence) and absorption variability in the body (energy, tissue thickness, and composition dependence). The storage phosphor plates were scanned using a high-contrast-sensitivity (12 bit) laser scanner and were computer processed with an experimental Kodak image processing unit. Images were then laser printed onto film using a Kodak Ektascan laser printer (12 bit).

To confirm that geometric error, or more importantly, the lack thereof, can be documented during electron treatment, our set of images were shown to the treating physician side by side with treatment planning documents (e.g. simulation films, photographs with marked skin).

Conventional and storage phosphor portal images were acquired from 22 patients undergoing high-energy (6-20 MeV) electron treatments in our institution. In all of the cases, anatomy of the exposed and surrounding tissue could be identified. Six of the conventional images were judged to be significantly under- (2) or over-exposed (4), and 16 yielded acceptable images. All storage phosphor images were adjusted prior to digitization for optimal gain and signal levels. Therefore, laser-printed images were all printed in a manner as to optimize film density range (OD to ≅ 1.5).

Nineteen of the patients were judged to be treated as planned. In three cases (14%), the images revealed that the patient's treatment field during the exposure was different than had been planned, and treatment was modified accordingly and verified by a second portal image.

Acknowledgement
This work is supported in part by the Eastman Kodak Company.

References

1. Gur, D.; Bukovitz, A.G.; Serago, C: Photon contamination in 8-20-MeV electron beams from a linear accelerator. Med. Phys. 6 (1979) 145-146.

2. Keller, B.E.: Electron-beam radiographs. Radiology 128 (1978) 830-831.

Presentation and Evaluation of 3D Dose Distributions in Radiotherapy Planning

G.Becker[1], R.Lohrum[1], T.Werner[1], J.Bürkelbach[1], G.Nemeth[2], R.Boesecke[1], W.Schlegel[1], W.J.Lorenz[1]

1.Institute of Radiology and Pathophysiology, German Cancer Research Center, INF 280, 6900 Heidelberg

2.Municipal Oncoradiological Center, Budapest, Hungary

1. Introduction

During the past 25 years, important advances in radiotherapy treatment planning have been made with the aid of computers. Today computers are widely used to calculate 2d radiotherapy plans based on CT scans. Tomorrow, planning systems will be available for 3d treatment plans. These 3d plans will be most useful for the radiotherapy of irregularly shaped target volumes (e. g. larynx), of target volumes that are not parallel to the axis of the patient's body (e. g. oesophagus), or of target volumes in the immediate vicinity of vital structures (e. g. a tumor of the glomus jugular adjacent to the medulla oblongata). Computer systems that are able to calculate spatial dose distributions on the base of CT data have been developed by several research groups (1, 2, 3, 4). After calculating the 3d dose distribution, however, the spatial information has to be presented for evaluation and selection of the optimum 3d plan, a problem which had not been solved in a satisfying manner yet. In today's clinical routine, calculation and visualization are performed in two dimensions - i. e. in the central beam cross section through the target volume. The presentation of the important additional information offered by 3d dose distributions must be clear and distinct as well as numerically unambiguous, so that the radiotherapist can make a quick and safe decision. A system is needed, that shows the physician, which radiotherapy technique will be the best compromise among snugly matching the dose distribution to the contour of the target volume, achieving a homogeneous dose distribution over the target volume, and an optimum protection of the organs at risk. The required device must have the ability to handle the radiotherapy data base efficiently. It should be easy to operate and must be able to display a large amount of data quickly and reliably. All this applies to the evaluation unit which is part of our integrated treatment planning system.

2. Material

2.1. Hardware

Our group has developed an eight cpu-multiprocessor system (Texas Instruments, TMS 320) (5). The local memory (20 MB) can be accessed by all processors simultaneously. (LIT Tim Werner) The system runs with a DEC-MicroVAX II serving as the host computer.

2.2. Software

Time-consuming parts of the evaluation program (slice reconstruction, calculation of the dose-volume histograms, calculation of the 3d projections) have been implemented on the multiprocessor system, whereas user dialogue, administrative parts, file handling, graphics and plots have been realized on the host in Fortran under VMS.

2.3. Data base

The following patient's data are used:

- CT data with a maximum of 64 slices (256 x 256 pixel),
- up to three dose cubes with the same number of slices as the CT data, but with reduced resolution (128 x 128 pixel),
- volumes of interest in form of a binary model

2.4. Paraaortal lymph nodes

The malignomas of the paraaortal lymph nodes comply to all indications for three dimensional treatment planning. The target volume is irregularly shaped, i. e. horseshoe-like with a convex and a concave side, it is not parallel to the axis of the body, and it is situated in the immediate vicinity of vital organs, such as myeloma and kidneys. Patients with indication for irradiation of the paraaortal lymph nodes most commonly are those with malignant tumors of the testicles and those with malignant diseases of the lymphatic system. The problem does not consist in the irradiation of radiosensitive tumors, such as the lymphomas of Hodgkin's disease. They are irradiated using a mantle field technique with a dose of up to 40 Gy within four weeks, a dose the spinal cord tolerates, while shielding radiovulnerable organs such as lung and kidneys with individual metal shielding blocks (6). The remaining radiosensitive tumors such as seminomas are also equally irradiated with a dose of up to 40 Gy. The true problem consists in the irradiation of the less radiosensitive metastases from solid tumors like teratomas, which have to be irradiated with up to 60 Gy (7). Since in this case the tolerance dose of the spinal cord can easily be exceeded, it is therefore of particular importance to protect the myelon. Several authors have experimented with different methods to solve this radiotherapeutic problem.

2.5. Irradiation techniques

In order to demonstrate the capabilities and the limitations of our evaluation program, we have chosen three fundamentally different irradiation techniques out of a host of techniques previously described:

2.5.1. A Y-technique, i.e. one frontal fixed field and two opposing lateral fixed fields. This technique has already been in use for decades in ^{60}Co-teletherapy, but it is still applied in the clinical routine (8). However, we have optimized the lateral fixed fields with the aid of the beam's eye view technique.

kidneys, off the center of the target volume, receives more than 60 percent of the dose, which is the maximum value that appears tolerable (11). For the two other techniques, the smaller load for the kidneys is clearly visible. In case of the MLC plan, the myelon is even outside the 20% dose. The 3d representation allows for a fast survey of the spatial dose distribution under arbitrary angles (Figs. 4 through 9), which shows most clearly that the axis of the target volume is slanted with respect to the axis of the biaxial rotation (Fig. 6). In contrast, the beam's eye view technique allows for the adjustment of the three fixed field to the axis of the target volume (Fig. 4), which also applies to the MLC (Fig. 8). The fact that 80% isodoses pass through the kidneys can be visualized by "cutting and looking into" the kidney (Figs. 4 through 9). By 3d representation, the shape of the dose-fit can be recognized clearly: rectangular in case of the Y-technique (Fig. 5), shaped as a horseshoe for the biaxial rotational irradiation (Fig. 7), and tightly fitting in case of the MLC (Fig 9). Similarly, overdosages and underdosages are shown clearly (Fig. 6,7). For the quantitative analysis, the dose-volume histograms for one volume of interest of the three irradiation techniques are compared (Figs. 2 and 3). In case of the Y-technique, more than 90% of the target volume is covered with a dose of more than 80%, but up to 10 percent of the target volume is underdosed. With the aid of the "combined evaluation", these underdosages can be displayed numerically or graphically for each slice. Only in case of the MLC plan, close to 100% of the target volume is covered with a dose of more than 80 percent (Fig. 2). The sparing of the myelon is clearly visible and impressively substantiated by the dose-voluram (white line Fig.3). The apparently best dose-fit with respect to the protection of the organs at risk and the coverage of the target volume is the MLC plan; however, this will be at the price of the largest inhomogeneities inside the target, as can be seen from the histogram of the dose gradient.

5. Conclusion

It is not subject of this paper to assess three fundamentally differing irradiation techniques of the paraaortal lymph node irradiation. They were used as an example in order to show the general importance in radiotherapy, and especially for the 3d treatment planning, of thoroughly comparing several different plans. The multitude of spatial information cannot be evaluated at a single glance. Equally, the clinical physician needs a system, which graphically and numerically supplies him with a host of information within a short response time. This is the only way for him to safely decide, which irradiation technique is the best compromise. In the future, therefore, it should be required that a medical work station for 3d radiotherapy planning should always provide the possiblity for a parallel evaluation of different radiotherapy plans.

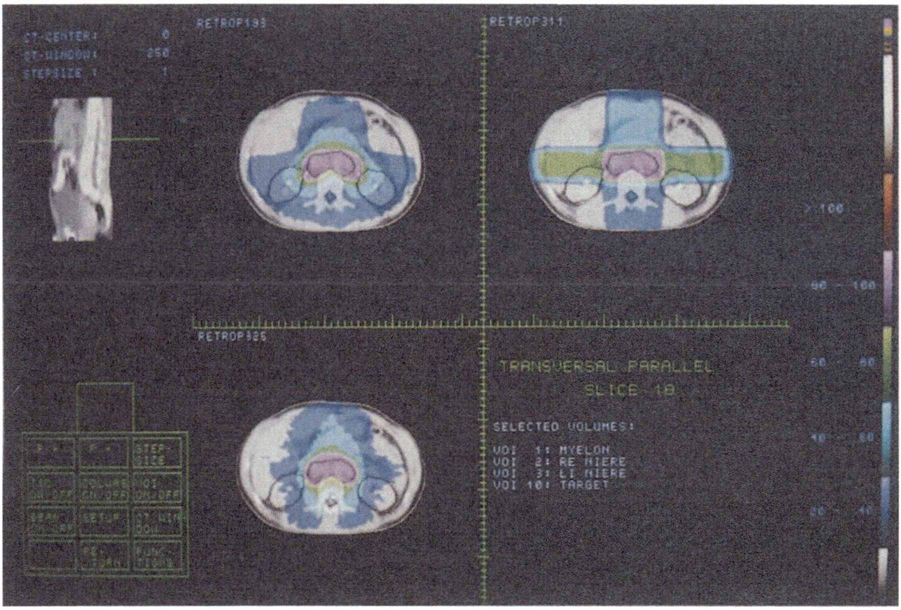

Fig.1: COMPARISON OF THREE IRRADIATION TECHNIQUES
311 Y-TECHNIQUE, 193 BIAXIAL ROTATION, 325 MULTI-LEAF-COLLIMATOR
The dose range 80% to 100% is shown in red, 60% to 80% in green, 40% to 60% in light blue, and 20% to 40% in dark blue. The target volumes and the kidneys are outlined in black. It is shown the transverse section off the center of the target volumes

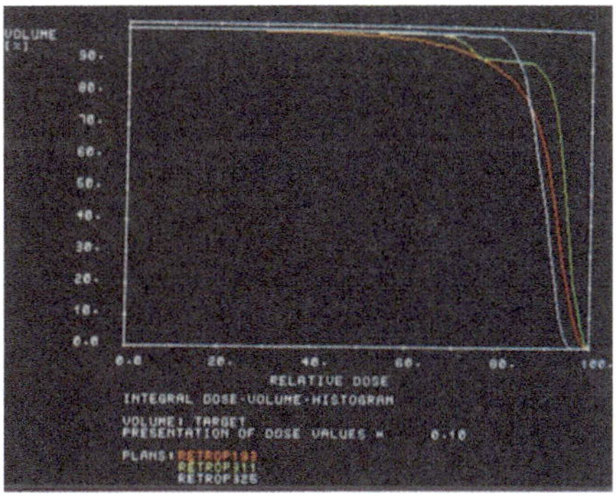

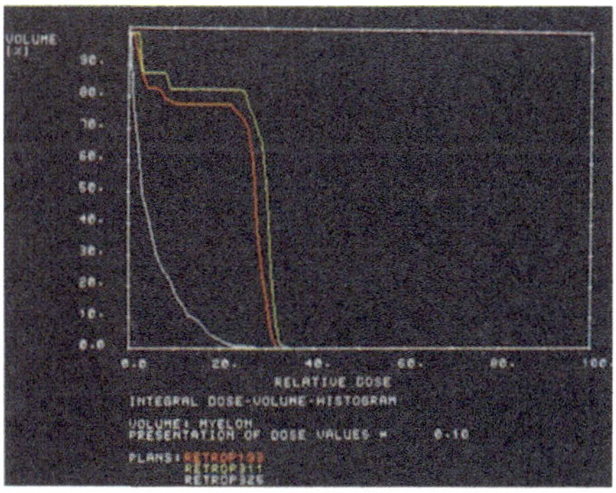

Fig.2: TARGET Fig.3: MYELON
COMPARISON OF DOSE-VOLUME-HISTOGRAMS : 311 = Y-TECHNIQUE green, 193 = BIAXIAL ROTATION red, 325 = MULTI-LEAF-COLLIMATOR white line

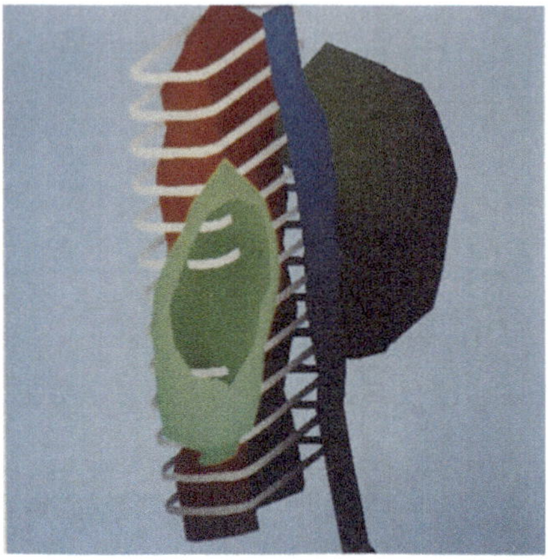

Fig.4: Y-TECHNIQUE Fig.5:

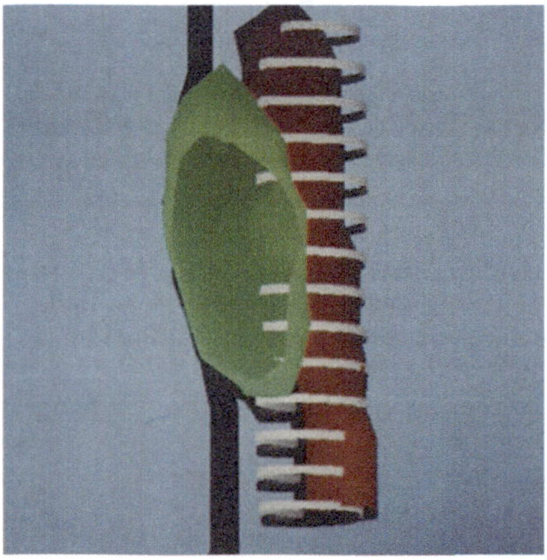

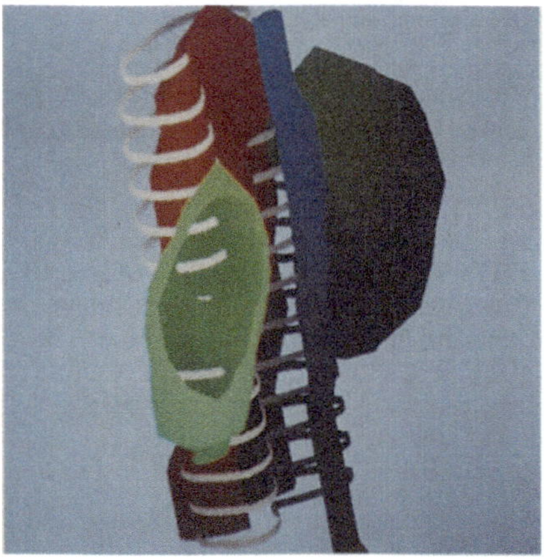

Fig.6: BIAXIAL ROTATION Fig.7:

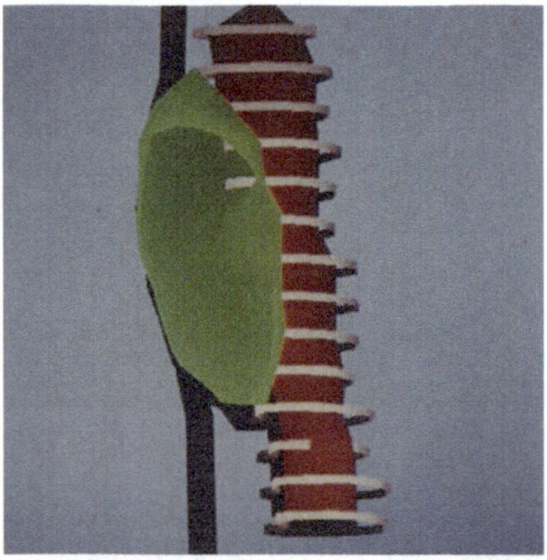

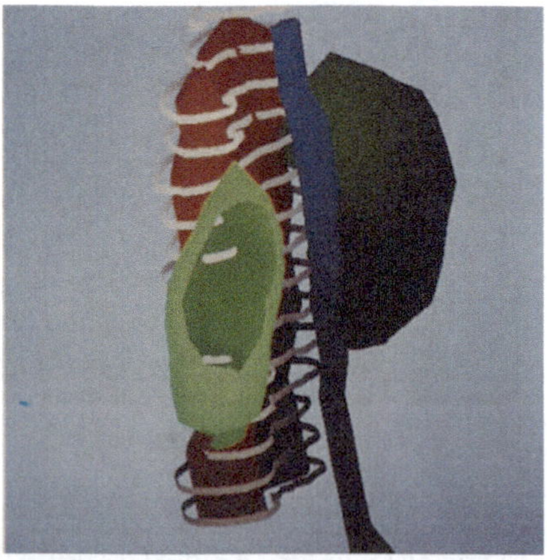

Fig.8: MULTI-LEAF-COLLIMATOR Fig.9:

red = TARGET, blue = MYELON, green = KIDNEY, white = 80 % ISODSES

6. References

1. Goitein M., Abrams M., Powell D., Pollari H., Willis J.: Multidimensional treatment planning: II. beam's eye view, back projection and projection through CT-sections. Int. J. Radiat. Oncol. Biol. Phys. 9, pp. 789-797, 1983

2. Schlegel W., Scharfenberg H., Doll J., Hartmann G., Sturm V., Lorenz W.J.: Three dimensional dose planning using tomographic data. Cunningham J.R. (Ed.), The eighth international conference on the use of computers in radiation therapie, IEEE Computer society press, Silver Spring, pp. 191-195, 1984

3. McShan D.L., Fraas B.A.: 3-D Treatment planning: II. Integration of gray scale images and solid surface grafics. Bruinvis I.A.D. et al. (Ed.), The ninth international conference on the use of computers in radiation therapie, North Holland, Amsterdam, New York, Oxford, Tokyo, pp. 41-44, 1987

4. Mohan R., Barest G., Brewster L.J., Chui C.S., Kutcher G.J., Laughlin J.S., Fuks Z.: A comprehensive three-dimensional radiation treatment planning system. Int. J. Radiat. Oncol. Biol. Phys. 15, pp. 481-495, 1988

5. Werner T., Hermann G., Schlegel W., Lorenz W.J.: Parallel processing in radiotherapie treatment planning. Bruinvis I.A.D. et al. (Ed.), The ninth international conference on the use of computers in radiation therapie, North Holland, Amsterdam, New York, Oxford, Tokyo, pp. 21-24, 1987

6. Todd I.D.H.: Malignant lymphomas (including myeloproliferative disorders). In: The radiotherapy of malignant disease, p. 347. Easson E.C. and Pointon R.C.S. (Eds.), Springer, Berlin, Heidelberg, New York, Tokyo, 1985

7. Nemeth G., Schlegel W.: Radiation therapy of intrathoracic paraaortic lymph node metastases, three-dimensional treatment planning. Acta Oncol. 26, pp. 203-206, 1987

8. Köst S.,Keinert K.: Zur Nierenbelastung bei der Strahlentherapie der paraaortalen Lymphknoten und des Milzbettes. Radiobiol. Radiother. 28, pp. 69-75, 1987

9. Kuttig H., Möller G., Schröder-Babo P.: Untersuchungen über die optimalen Energiewerte von Linearbeschleunigern für die Strahlentherapie. Electromedica 50, pp. 28-32, 1982

10. Boesecke R., Doll J., Bauer B., Schlegel W., Pastyr O., Lorenz W.J.: Treatment planning for conformation therapy using a multi-leaf collimator. Strahlentherapie 164, pp. 151-154, 1988

11. von Fournier D., Nemeth G., Kuttig H.: ale Pendelbestrahlung der aortalen Lymphknoten unter Auslenkung des Nutzstrahlenbündels mit ultraharten Röntgenstrahlen des 42-MeV-Betatrons, Strahlentherapie 146, pp. 43-51, 1973

First Clinical Experience with a Digital Imaging Device for Radiotherapy Dosimetry and Localisation of Treatment Fields

H. LEITNER[1], J. W.WONG[2], K. ARIAN-SCHAD[1], R. BINNS[3]

1 Medical School University Graz, AUSTRIA
2 Medical School University St. Louis, Missouri USA
3 Physics Department University St. Louis, Missouri USA

Summary

A computer based real - time imaging device is used for digitizing verification images as well as for two dimensional dosimetry. The images are sampled from verification films or dosimetry films by a camera connected with an image processor system.
Furthermore we investigated the possibility replacing the film by a scintillator to get the images in real - time.
Concerning verification images (using film or scintillator) a lost of quality is inevitable and cannot be compensated by computer aided contrast enhancement and/or special filters. Dosimetry measurements with film or scintillator demonstrated an accuracy less ± 5 % .

Introduction

Portal films used for verification of treatment fields in radiotherapy are marked by poor contrast. The contrast depends on the energy level delivered from Megavolt - equipment. With an imaging system a selective choice of the gray scale value is possible. The contrast can be enhanced using highbas- and other special filters [3]. Additionaly the film may be replaced by a scintillator to observe the verification image during radiation exposure.
In dosimetry the imaging system is applied for digitation and analysis of dosimetry films and scintillators [2,4].

Material and Methods

Different digital cameras such as a Hitachi KP-140 and a CID2505 with an external cooling system were used in our first clinical tests. The video signals from the camera are digitized by a FG100 plus imaging processor from Imaging

Technology Inc. running at a PC/AT computer. The digitized image is displayed on a RGB monitor. For data storage we use a 20 MB hard - disc and an ETHERNET connection to the hospital data system (VAX cluster).
The image processor (Fig.1.) is equiped with 24 input bits (12 feedback memory bits, 8 data bits and 4 input LUT bits) and 12 output bits (4 output LUT bits and 8 databits). The frame memory has a size of 1024 x 1024 pixels with 12 bit per pixel, the displayable video area is 512 x 512 pixel with 8 bit per pixel. Software for imaging analysis was written by the Forschungszentrum Joanneum - Institut für digitale Biltverarbeitung* using the programm library ITEX 100 from Imaging Technology Incorporated.

Verification

The portal films for verification of treatment fields delivered from megavoltage therapy equipment are known to have two major disadvantages: First the poor contrast caused by the high photon energy and second, the information needed is delayed by film developement.
With our imaging system we are able to enhance the contrast by a selective choice of gray scales, histogram equalization and high frequency filtering.
For immediate information we replaced the film by a fluoroscopic screen using a 45° mirror. Tests with different screens are ongoing.
Our investigations showed that an improved digital verification imaging is assured only by a camera with high sensity and a resolution of at least 1024 x 1024 pixel and 12 bit per pixel.

Dosimetry

To determine the dose distribution of dosimetry - films or scintillators, a correction of camera induced errors concerning linearity and uniformity response is necessary [1]. For these error corrections we were using the input LUT and the feedback memory in real - time mode. With special color output LUT`s we were also able to display the dose

distribution by steps with different colors in real - time. Using a scintillator, an integration time for one image of about 1 - 10 sec is necessary. To reduce the dark current we cooled the camera down to 2° Celsius.
To optimize the efficiancy of the system the spectral sensitivty of the camera and the emission spectrum of the scintillator should be in the same range. We made measurements with different scintillators (BC408, BC470, NE102 and an Kodak intensifying screen Lanex fast f). They were carried out in beam direction with the scintillator vertically placed in a water tank (Fig.2.). To evaluate the dose homogeneity in the horizontal plane measurements were made behind 1 cm plexiglas using a 45° mirror (Fig.3.).
The accuracy achived was ±5% related to ionisation chamber measurements.

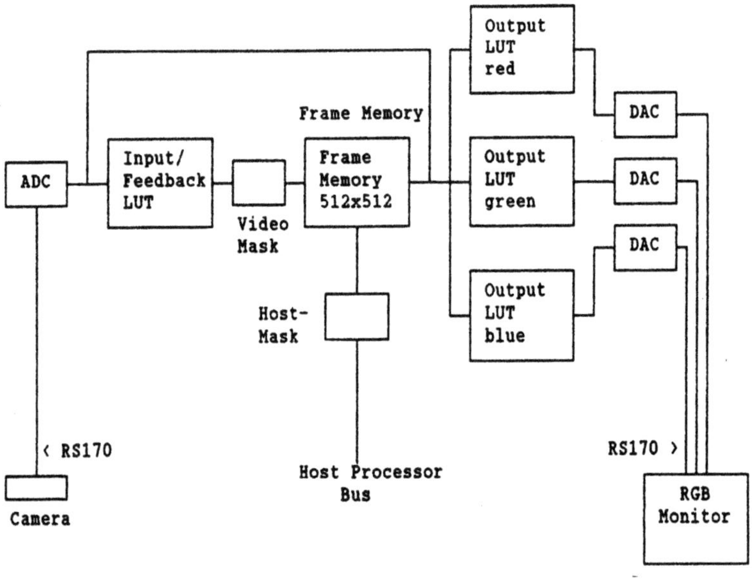

Fig.1. Digital imaging processor FG100 block diagramm.

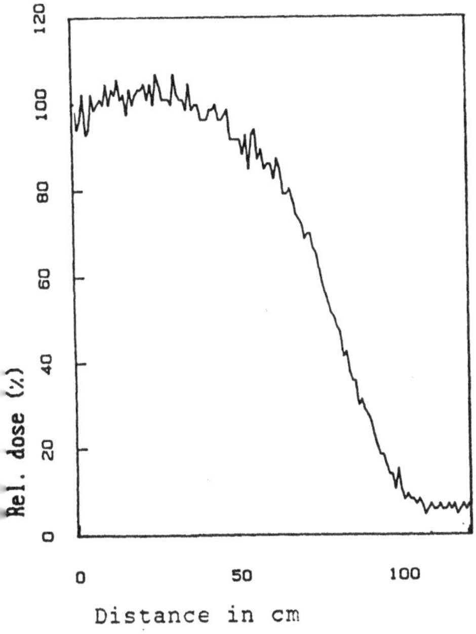

Fig.2. Depth dose distribution of a 20 MeV electron beam, measured in a watertank with a BC408 scintillator.

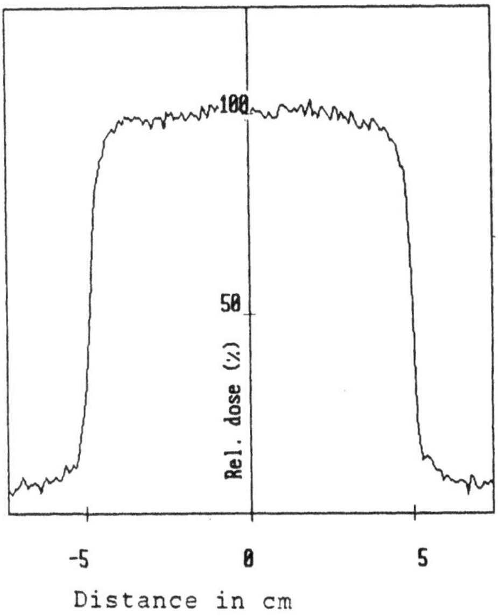

Fig.3. Lateral dose - distribution of a 20 MeV electron beam, measured with a Kodak Lanex fast screen.

* We thank the Austrian Forschungsfonds for supporting this work.

References

1. Ge, W. S.: A two-dimensional scintillation dosimetry system for radiotherapy. Thesis, Washington University St. Louis (1988).

2. Leitner, H.; Wascher, C.: Filmdensitometrie durch digitale Bildverarbeitung. In: H.Bergmann (Hsg.), Medizinische Physik, Wien (1987) 193 - 196

3. Pfluger, T.; Reinfelder, H.E.; Dorschky, K.; Oppelt, A.: Gegenüberstellung verschiedener Filtermethoden bei der digitalen Verarbeitung von Röntgenbildern. Electromedica 3 (1987) 78 - 85.

4. Wong, J. W.; Binns, W.R.; Ge, W.S.; Epstein, J.W.; Klarmann J.; Israel, M.H.: Rapid arealdosimetry using a computer based plastic scintillator - video camera system. In: I.A.D. Bruinvis et al. (eds) The use of computers in Radiation Therapy, Elsevier Science Publishers B.V. North-Holland, (1987) 417 - 420.

Radiation Damage to Vessels Monitored by Tumorfluoroscopy

M. Herbst * und M. Fröder**

*Dpt. of Radiation Oncology, ** Dpt. of Technical Electronics
University Erlangen-Nürnberg, Univ. Str. 27, D-8520 Erlangen

Digital Tumor-Fluoroscopy (DTF) is a new method for the imaging of tumor tissue. A X-ray video chain is expanded by a computed proces sing system with an extension of the grey values up 64 000 (16 Bit) and optimized to detect low iodine contrasts.
Dynamic studies enable the radiooncologist to get direct visual information about the location of brain tumors during radiotherapy simulation. Using DTF it is possible to avoid dislocations of treatment portals especially of excentrically located tumors. Watching the redistribution of contrast medium in the intravasal and extravasal space the characteristics of different brain tumors of 110 patients were defined.
Monitoring blood perfusion in big neck nodes of head & neck carcinomas detailed informations about the perfusion rate in different areas of the node are gained. The perfusion rate of regions of interest within the tumors can be monitored over the whole peried of investigation (15-20 min).These investigations are repeated once per week during treatment for 4-6 weeks. After one to two weeks of irradiation treatment the perfusion rate is increasing. During the following weeks the blood flow decreases and is lowest at the end of treatment. Giving hyperthermia additionally the blood flow rate decreases remarkably.
These findings are of scientific and clinical intrest and will help to answer different questions:

- Is there dose dependency of radiation damage to vessels?
- Does a tumor respond to treatment?
- Are there radioresistent areas in a tumor due to low blood flow?
- Has blood perfuion an impact on the indication for clinical hy perthermia?

Computer Based Workstations in Radiation Oncology

HANS DAHLIN*, PER EKSTRÖM* AND WERNER SCHNEIDER**

*Helax AB, Box 1704, S-751 47 Uppsala, Sweden
**Center for Human Computer Studies, Uppsala University, Sturegatan 9, S-752 23 Uppsala, Sweden

Summary
Computer based workstations are shown to be a prerequisite for the use of the rapidly increasing selection of improved commercially available input and output facilities. Two design strategies are discussed and it is shown that only one of them is suitable for designing the human computer interface in such a way that human work effectivity is preserved in Computer Aided Radiation Oncology.

What is a Computer Based Workstation?

In spite of the amazing progress in the field of computer technology the basic architectural structure of almost all commercially available digital computers has not been changed during the last three decades. Examining the technological progress in the frame of this structure, i.e. the von Neumann model of a digital computer (Fig 1), reveals that until a few years ago most of the development efforts have been devoted to the Processor and Storage components, whereas the Input and Output components were of much lesser interest.

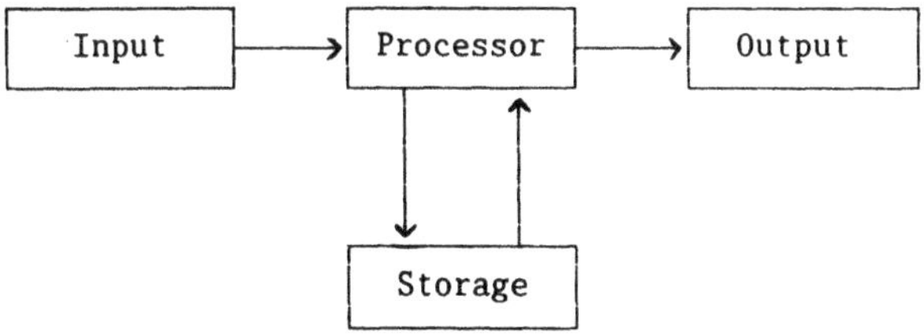

Fig 1
The von Neumann model of a digital computer

During the second half of the sixties the need for graphical presentation of intermediate and final results was important but only a few could afford a real-time display of such information. This was due to the fact that the considerable computing power needed for maintaining even the rather poor graphic terminals of that time was very expensive. During the seventies the problem of being able to afford the computer time needed for interactive work with graphics soon became much less accentuated. Instead the lack of adequate line transmission capacity between the computer and remote graphic terminals became the major problem even though these terminals in general just could produce vector graphics. As a consequence there was a fast increasing need for local computing facilities for the real-time production and display of graphical information in a number of rather well defined application areas such as **computer aided design** in industrial environments and **dose planning systems** in the field of cancer treatment. There arose in other words a need for application specific **graphic workstations.**

The rapidly increasing availability of low price "micro-von Neumann's" in the form of so called PCs during the last few years has dramatically changed the scene. Due to the enormous market potential, the selection of advanced input and output facilities is rapidly improving. Thus computer graphics aided analysis of data using monochrome or colour graphic screens will soon become a tool for everybody at almost any working place. Together with more and more advanced hardware and software tools for word and text processing, electronic mail service, terminal emulation, database handling and file transfer, very versatile and powerful PC-based **general purpose workstations** are now broadly available at a rather decent price.

Design Strategies for Workstations

If human competence and skill are to be preserved in computer assisted work environments, workstations have to be designed specifically for the specific work environment in which the use of computer techniques is envisaged.

There are two strategies for the design of computer based workstations. The first emanates from the aim of adapting an existing way of computer based problem solving to a specific domain of work. This implies in general not only a reorganization of staff and working procedures, it also involves a change in regard to the professional competence and skill required. High priority is given to the ability of handling formal languages (e.g. query languages) and command procedures, Hachimura[2]. Due to the fact that this kind of activities genuinly is "non-human", much effort is made to "humanize" them. Visual programming languages, menu techniques, icons, voice input and other techniques are increasingly introduced in order to achieve an acceptable "user-friendliness".

The second strategy implies an opposite approach, Petterson et.al[4] and Lind et.al[5].
In a first phase the existing problems at a working place are identified and analyzed in relation to the goals to be achieved within the frame of the overall goals of the organization in which the work takes place.
In a second phase different possible ways of solving the problems are investigated involving - in an integrated perspective - alternative staffing enhancement of the competence of the staff, optimal planning of premisses as well as organizational and technical means.
First in the third phase the possibly required computer based tools are specified on the basis of genuine human competence and skill in handling tools out of and in a specific work context.

The Design of an Integrated Computer Based System for Radiation Oncology

As one of the industry members of the Nordic CART project, the Helax AB formed its product development TMS (Treatment Management System) according to the second design strategy described above and recommended by the CART project, Dahlin et.al[1] and Lamm et al[3].

The TMS' products emanate from a number of specifying models.

* **The radiation oncology process** - definition of the different clinical functions.
* **The patient dataflow** - specification of the logical network.
* **The computer application model** - specification of the functions that are suited for computer operations and how they integrate in the information flow.
* **The human computer interface** - technical tools required by different personnel (doctors, physicists and technicians).
* **The ergonomic design** - the ergonomic requirements at different work environments.

The integrated information system, described by Richter et.al[6], for an individual patient is practically realized in TMS. The system provides integrating aids for the whole treatment design process from the stage when the decision on radiation therapy has been made to the initiation of treatment.

Fig 2
TMS-Workstation

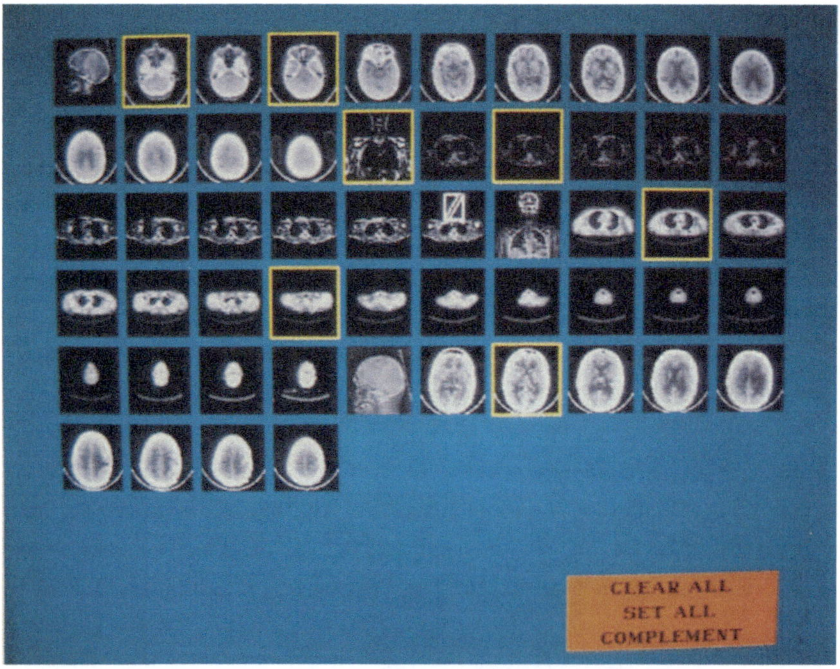

Fig 3
The Index-image gives information of the patient's image directory in the form of image icons (Imics)

The operation of TMS is performed at workstations which are designed in accordance with the second strategy described above. They are connected to a central processor in a cluster environment. The central database holds information of necessary text, graphic and image data for all patients within the oncology process.
The data handling at the different phases of the oncology process is controlled by different kinds of personnel. **The man-machine interface** built-in the TMS workstations is therefore well defined and tested according to CART recommendations.
Two large flicker-free 19" monitors are mounted in an adjustable free hanging box in such a way that the operator gets the feeling of working with one display area only. The functional handling of the TMS is menu driven via a mouse controlled cursor. The same technique is also applied for selecting images and drawings. By pointing and clicking in the appropriate parts (menus or image icons) the operator controls the TMS operation. The fast interaction in TMS depends on the large image memory, which gives real time handling of more than 100 different digital images.

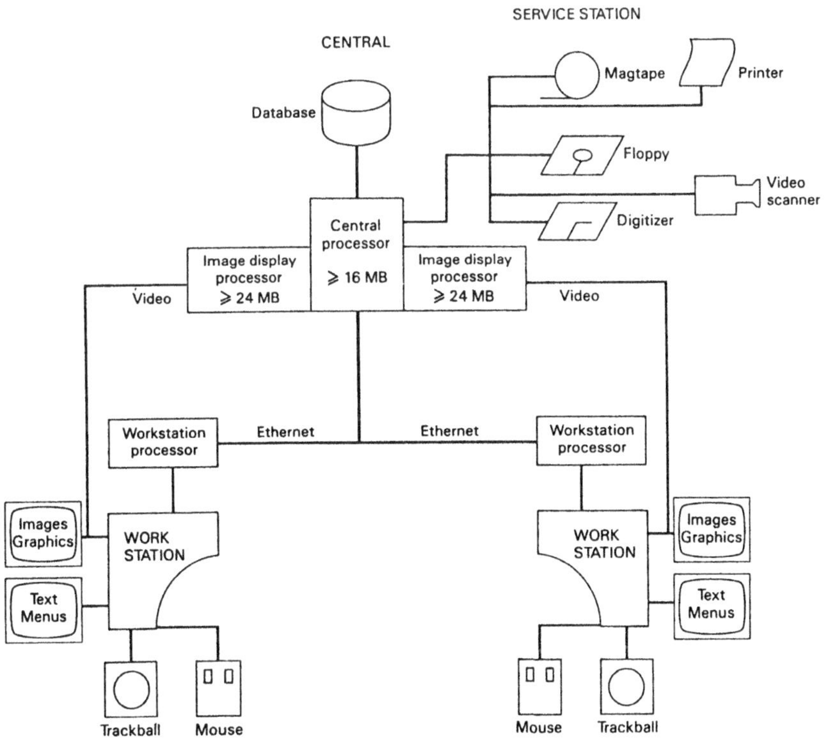

Fig 4
TMS-Configuration

The man-machine concept pronounced in the CART recommendations and further improved at the Center for Human Computer Studies, Uppsala University, has been justified by the TMS in real clinical practice.
The TMS improves the quality of radiation oncology as well as the consistency of the individual patient flow.

REFERENCES

[1]Dahlin, H., Ekström, P. and Lamm, I-L. Digital Imaging for Planning of Radiation Therapy - Practical Considerations. Proc. of the NATO Advanced Study Institutes Program. Pictorial Information Systems in Medicine, Braunlage/Harz, Aug. 1984.

[2]Hachimura, K. Image Query System in Clinical Environment. MEDINFO 86.

[3]Lamm, I-L., Dahlin, H. and Möller, T., 1984. CART - A Nordic Challenge in Medical Computing. Computers in Radiotherapy and Oncology, Session 3. Geneva, March 1984, pp. 81-86.

[4]Petterson, E. and Schneider, W., (eds). Human-Computer Communications in Health Care. Elsevier Science Publishers, Amsterdam, 1985.

[5]Lind, M., Pettersson, E., Sandblad, B. and Schneider, W. Computer Based Workstations in Health Care. Bakker A.R. et.al. (eds), Towards New Hospital Information Systems. Elsevier Science Publishers, Amsterdam, 1988, 235.

[6]Richter, S., Ahnesjö, A., Dahlin, H., Ekström, P., Glimelius, B. and Jung, B. The External Beam Radiotherapy Process. Proc. of the IX Int. Conf. on the Use of Computers in Radiation Therapy, The Hague, 1989, p. 281-.

Improved Brachy-Curietherapy Planning for Brain Tumors Based on a Combination of CT and MRI Parameters

Mohadjer M, Etou A, Birg W, Mundinger F

Department of Stereotaxy and Neuronuclearmedicine
University of Freiburg Medical School
D-7800 Freiburg i. Br., Federal Republic of Germany

Summary

The problem of the tolerance of healthy brain to radiation has hampered the application of conventional, standard fractionated external irradiation of brain tumors. Now, the stereotaxic localization in the so-called radiosurgery and the use of interstitial irradiation can increase the therapeutic spectrum of radiotherapy. Radioactive sources are placed directly in the center of the tumor by means of the CT or MRI stereotaxic method. A beneficial feature of the method is that during the same procedure a serial biopsy is performed to confirm the diagnosis. Since modern neuro-imaging techniques have been incorporated into the stereotaxic method the interstitial curietherapy of inoperable brain tumors has gained in importance.

Introduction

Selective irradiation of inoperable, usually deep-seated benign brain tumors by means of implanted radionuclides has proved to be an effective alternative therapy to external radiation. The objective of the method is to deliver a high local radiation dose to the tumor while protecting the surrounding healthy brain. A so-called necrosis dose is able to kill tumor cells even in non-radiosensitive tumors. The tolerance limit of the brain tissue in conventional external radiotherapy has been given at a maximum of 60 Gy in standard fractionation (3,14). Transgression of this limit can cause acute, early or late reactions with vasogenic edema or necrotic zones in normal brain, independent

of the cumulative dose and the type of fractionation (4,6). The stereotaxic method is ideal for minimizing radiation damage. However, not until computed tomography (CT) and magnetic resonance imaging (MRI) were integrated could the method be carried out with sufficient accuracy.
Both high-energy and low-energy gamma emitters are used for the method. We prefer iridium-192 and iodine-125 (introduced by Mundinger in 1959 respectively in 1979 for treatment of intracranial tumors), which have the following characteristics (5,9,10,13) (Tab. 1):

Physical properties of the nuclides Ir-192 and I-125 used in the radiation therapy of brain tumors

	$T_{½}$	E	K	HVL tissue	HVL Pb	size	ready to use
	d	MeV		mm		mm ø	
Ir-192	74,6	0.31 - 0.61	5.0	60	4,0	0.3	wires in various lengths
I-125	60.2	0.027-0.035	1.4	23	0.025	4.5 x 0.8	Titanium Seeds

$T_{½}$ = physical half life HVL = half-value thickness
E = maximum intensity of radiation in MeV
K = point source dose rate constant in: $\frac{r \cdot cm^2}{n \cdot mc}$ for gamma radiation

Different tissue absorption (a) and dose accumulation (b) of I-125 and Ir-192 (Fig. 1a and 1b):

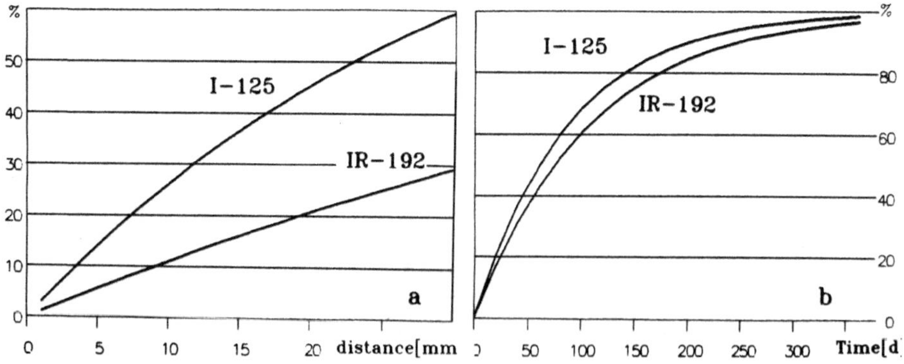

Corresponding to the different characteristics of these emitters, the tissue absorption (Fig. 1a) and cumulative dose (Fig.1 b) also differ.

Method:

The stereotaxic surgical procedure has made it possible to introduce radioactive sources into brain tumors. We use the CT/MRI version (2,8,11) of the Riechert/Mundinger stereotaxic instrument. The operation is usually carried out under local anesthesia. The target point and the tumor volume as well as the dosimetry are determined by means of the CT or MRI examination in the stereotaxic ring. The target device and the holder containing the setting angles that have been calculated by means of a special computer program are secured on the patient ring. A combination probe is advanced through a 2-3 cm long incision and an 8 mm trephination hole up to the predetermined target point. Several tissue specimens are withdrawn with a biopsy forceps having a diameter of 0,8 mm. The histological examination is performed immediately in the smear preparation. At the same time specimens are taken for paraffin embedding, and for special staining procedures. In about 82 % of cases the biological valency and classification of the tumor can be intraoperatively established (1). If indicated, interstitial irradiation can be initiated during the same operation, either temporary by a catheter technique or permanently by direct implantation.

Discussion

The tolerance and reaction of healthy brain is a serious complication associated with conventional radiation therapy of brain tumors. In many cases radiation reactions occur before tumor control has been achieved. Such complications can be reduced by means of CT/MRI stereotaxic implantation of radioactive sources for interstitial irradiation. Emitter selection and dosimetry calculations are based largely on experience. Ostertag (12) was able to prove in animal experiments that necrosis and vasogenic edema can be influenced by the dose and the energy range. Vasogenic edema, however, is also a limiting factor in interstitial irradiation.

The emitters used are iodine-125 and iridium-192. The advantage of the method is that very high doses can be delivered to restricted regions. The cumulative dose required to devitalize the tumor can be achieved either by means of temporary high-dose implants or with permanent low-dose seeds. A high dose delivered over a short time causes a large necrotic zone but can induce extensive vasogenic edema and damage. On the other hand, when radioactive sources are permanently implanted, later correction of location or dose is no longer possible. This fact, along with the proven effectiveness of temporary, afterloading interstial irradiation,are arguments in favor of using the catheter technique. Whereas there are practically no complications associated with the puncture and implantation procedures, the withdrawal of specimens for biopsy has a 1% mortality and a 1 % morbidity rate, largely due to bleeding from pathological tumor vessels (7).

Fig. 2a): metastasis of bronchial carcinoma with isodoses of 100 (distance from the center 13 mm) 25, 15, 5 Gy

2b): 2 month after I-125-implantation, perifocal-reaction around catheter containing the iodine-seed

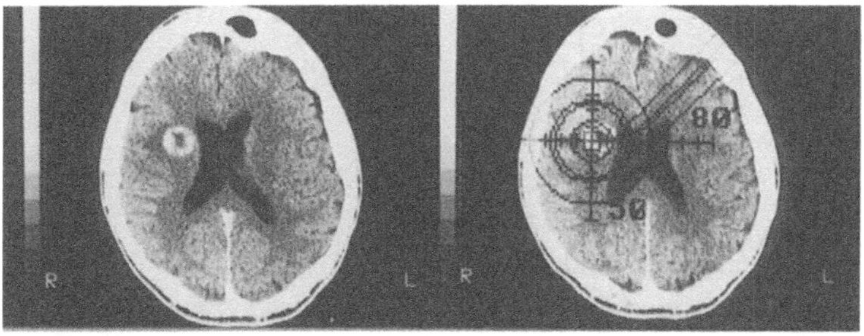

(a)

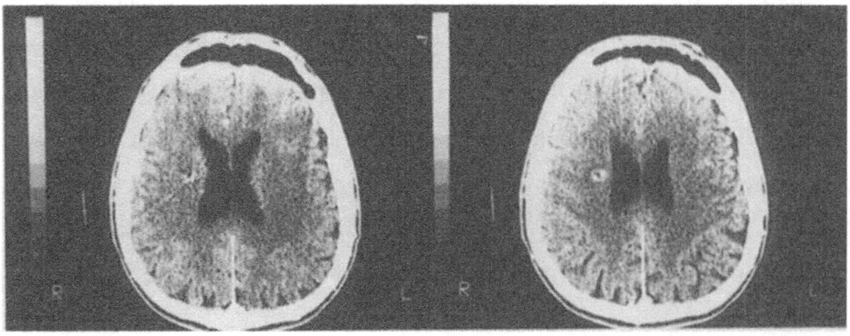

(b)

Literature:

1) Anagnostopoulos J, Kiessling M, Volk B (1988) Validität morphologischer Diagnostik an stereotaktischen Hirntumorbiopsien. In: Bamberg M, Sack H (eds) Therapie primärer Hirntumoren. Zuckschwerdt, München, 270-279

2) Birg W, Mundinger F (1982) The Direct Target Point Determination for Stereotactic Brain Operations from CT Data and the Calculation of Setting Parameters for Polar-Coordinate Stereotactic Devices. Appl Neurophysiol 45: 387-395

3) Burger PC, Mahaley MS, Dudka L, Vogel FS (1979) The morphologic effects of radiation administered therapeutically for intracranial gliomas. Cancer **44**, 1256-1272

4) Fike JR, Sheline GE, Cann CE, Davis RL (1984) Radiation necrosis. Prog. exp. Tumor Res. Vol. 28, Karger-Basel, 136-151

5) Gutin PH, Phillips TL, Wara WM, Leibel SA, Hosobuchi Y Levin VA, Weaver KA, Lamb S (1984) Brachytherapy of recurrent malignant brain tumors with removable high-activity iodine-125 sources. J Neurosurg 60: 61-68

6) Leibel SA, Sheline GE (1987) Radiation therapy for neoplasms of the brain. J. Neurosurg 66(1): 1-22

7) Mohadjer M, Milios E, Mundinger F (1987) Risiken der CT- (MRI)-stereotaktischen intracraniellen Biopsie. Gesamtverb.Deutscher Nervenärzte, Kiel, 28.-30.5

8) Mohadjer M, Mundinger F, Birg W, Friedburg M, Hennig J, Wenz W (1987) MRI-Stereotaxie. In: Lemke HU, Rhodes M, Jaffee CC, Felix R (eds) Computer Assisted Radiology. Springer-Verlag, 369-373

9) Mundinger F (1970) The treatment of brain tumors with interstitially applied radioactive isotopes. In: Yen Wang, Paoletti P (eds) Radionuclide Applications in Neurology and Neurosurgery. Charles C. Thomas, Springfield Ill., 199-265

10) Mundinger F (1988): Stereotactic Biopsy and Implantation of Radionuclides Guided by Computed Tomography or Magnetic Resonance Imaging for Therapy of Brain Tumors. In: Schmidek HH, Sweet WH (eds) Operative Neurosurgical Techniques. Vol.I. Grune & Stratton, Inc., 491-514

11) Mundinger F, Birg W (1984) Stereotactic Biopsy of intracranial Processes.Acta Neurochir.Suppl.33, 219-224

12) Ostertag CB, Weigel K, Warnke P, Lombeck G, Kleihues (1983) Sequential morphological changes in the dog brain after interstitial Iodine-125 irradiation. Neurosurgery 13: 523-528

13) Szikla G, Schlienger M, Blond S, Daumas-Duport C et al (1984 b) Interstitial and combined interstitial and external irradiation of supratrentorial gliomas. Results in 61 cases treated 1973-1981. Acta Neurochir (Suppl) (Wien) 33: 355-362

14) Wigg DR, Koschel K, Hodgson GS (1981) Tolerance of the mature human central nervous system to photon irradiation. British Journ. of Radiology 54:787-798

Computer Assisted Neurology

Computer Assisted Neurology

Gordon Banks
Decision Systems Laboratory
Department of Neurology,
and Intelligent Systems Program
University of Pittsburgh
Pittsburgh, PA 15261

Summary

The field of artificial intelligence (AI) has interested neurologists in two ways: (1) the building of expert systems to assist neurologic and neuroimaging diagnosis and (2) the construction of theoretical cognitive simulations which allow experimentation with cognitive processes. We will survey both traditional AI methods and the new applications of artificial neural networks.

Introduction

The advent of computers has probably affected neurology as much if not more than any other clinical specialty. Computed tomography (CT) and later, magnetic resonance imaging (MRI), have allowed us to visualize the nervous system quickly and safely and have changed forever the way neurology is practiced. Thus, neurologists greet new technology, not with suspicion, but with the expectation of obtaining additional aid in practicing our profession.

The cost and size of computers have both plummeted in recent years and there is no reason to believe the trend will not continue. It is expected that VLSI parallel processors will become more common in the next decade, making new types of computing power available for research and applications. Accompanying the decrease in the cost of computing power has been a diffusion of the technology from large computing centers into laboratories, hospitals, and offices with a consequent increase in accessibility to the average practitioner. It is usual in all branches of computer science that software development lags hardware development and that is also true in the field of medical applications of computers. In this review, I intend to discuss the application of Artificial Intelligence (AI) techniques to neurology. The endeavor in this field can be divided into two branches: (1) conventional AI, and (2) neural networks, known more recently as "the new connectionism".

Conventional AI

Artificial Intelligence as a discipline has the goal of building computer systems which exhibit behavior which would be considered intelligent in a human. One of the earliest applications undertaken by AI researchers was the building of "expert systems", or programs which captured expert knowledge which could be consulted in a manner similar to human experts. The consultation model is well developed in medicine, and medicine is a field which is very knowledge intensive, thus it is natural that AI workers would turn their attention on clinical medicine as an appropriate field to explore expert systems. Expert systems research has largely been concerned with the issues of knowledge representation (how to structure the expert knowledge needed by the program) and methods of performing and controlling inferences (new facts or diagnoses based on the input) made by the system.

Rule-Based Systems

The most familiar type of expert system is the rule-based system. The first significant rule-based medical expert system was MYCIN (Shortliffe, 1976), a program which provided consultation on diagnosis and therapy of infections, including meningitis. MYCIN queried the user for information regarding the case with its questions driven by the goal of first determining the organism and then recommending therapy. MYCIN's knowledge base consisted of several hundred production rules written in LISP. The rules are of the form IF <premise> THEN <conclusion> CERTAINTY <value>.

Rule-based systems have the advantage of a clean, well-defined knowledge base that can be modified and extended by a human expert. Some explanation for the conclusions can also be given by keeping track of the rules that led to them and recapitulating them when asked to explain. The technology for developing such systems is mature and expert system building *shells* are commercially available to allow new systems to be built without starting from scratch. The disadvantages of rule-based systems lie in the difficulty in extending them to large and broad domains. When the knowledge needed for the consultation can be captured in a small number of rules, controlling the inference and the interactions between rules is possible, but as the size of the knowledge base increases, the number of rules needed explodes, and the systems rapidly become intractable.

The general field of neurology is much too broad to allow implementation of a general consultation system using rules, but rule-based systems have been implemented in restricted domains, such as stroke. Hier and coworkers (Hier 1986) have implemented a stroke localization system based on the Michael Reese Stroke Registry (Banks 1983). Spitzer and coworkers (Spitzer 1988) have implemented a suite of rule-based expert systems based on a registry of stroke patients which allow diagnosis of stroke type (hemorrhagic, embolic, thrombotic, etc.) and location.

In the field of neuroradiology, attempts have been made to use rules to guide analysis of objects extracted from CT scans using conventional image analysis techniques such as edge detectors (Kumar 1985). The difficulties with this project lay more with the inability of conventional image analysis tools to extract any but the most gross features than with the rules.

Semantic Networks

Other investigators have used semantic networks to structure the knowledge of expert systems. An early expert system for glaucoma diagnosis called CASNET (Kuliakowski 1982) attempted to connect symptoms such as eye pain with the diagnosis through the intermediate causal pathophysiologic states such as increased intra-ocular pressure. Such as system can be said to have a deeper knowledge of the disease process than simple rules, since the causal pathophysiology is brought into the reasoning. ABEL (Patil 1982) is a program for the diagnosis of acid, base, and electrolyte disorders which contains sufficient knowledge of the underlying pathophysiology to generate quite satisfactory explanations for abnormal patient findings (for example it can link a metabolic alkalosis with vomiting since it understands how that can result in the net loss of stomach acid). Like rule-based systems, causal network expert systems are difficult to employ in large and broad domains. These programs essentially perform searches through the network, and as the size of the network expands, the time required to search them increases exponentially.

Anatomic diagnosis is an essential part of neurology. Given the complexity of the relationships between symptoms and neuroanatomic structures and pathways, clinical neuroanatomy is a natural domain for expert system investigation. Early expert systems of Meyer and Weisman (1973) and Catanzarite (1981) used knowledge bases consisting of simple brain stem and spinal cord sections and a geometric approach to finding lesions which could involve tracts from the user-instantiated symptoms and signs. First (1982) built a program with a knowledge base consisting of all the muscles, peripheral nerves, plexi, and roots. By instantiating muscle weakness or electromyographic (EMG) involvement of a set of muscles, the program could discover all of the possible sites for peripheral lesions which could produce such involvement by tracing the neural connections. Banks and Weimer (1984) constructed a hierarchical network containing over 900 neuroanatomic objects with links to spatial neighbors, vascular supply, and neural connections. Common neurologic symptoms were linked to the objects. This system was a first attempt to provide an anatomic knowledge base for the CADUCEUS program (see below). Hertzberg and coworkers (1987) developed a similar semantic network approach to neurologic localization.

Heuristics

One of the ways to make the search of a large network efficient is through the use of heuristics. Examples of heuristic rules are:

(1) Occam's Razor or the Law of Parsimony.
(2) Proceed from the general to the specific.
(3) Process of elimination.
(4) Common things are common.
(5) Go where the money is.
(6) Don't miss a treatable disease.
(7) Don't leave important data unexplained.

The INTERNIST expert system (Miller, 1982) for providing expert advice with difficult problems in internal medicine is an example of a very large but shallow causal network which employs heuristic rules (especially Occam's Razor) to enhance performance by focussing quickly on a fruitful area for exploration. This system has a knowledge base of thousands of manifestations of disease and hundreds of diseases. It uses a pseudo-probabilistic method plus the heuristics to enable it to make accurate diagnostic suggestions for patients having one or more diseases in its domain. The knowledge base took 15 years to develop. A microcomputer version of INTERNIST, called QMR (Quick Medical Reference) has been implemented.

Eclectic and Hybrid Systems

As expert system development has progressed, systems have appeared which incorporate more sophisticated knowledge structures such as *frames* as well as rules and simple network connections.

INTERNIST suffers from having a relatively shallow knowledge base containing manifestations and diseases with some additional information such as the importance of manifestations and the cost of obtaining them. It has little deep knowledge of pathophysiology and none of anatomy. It also handles information regarding the time course of illness only in a crude and explicit manner (manifestations can be chronic or acute). CADUCEUS (Pople 1985) represents an attempt to build a more general structure which will incorporate frames containing information about pathophysiology, anatomy, and will understand a disease as a process with stages and phases. It is hoped that CADUCEUS will be a rich enough environment to handle issues of treatment and prognosis. Constructing a knowledge base for CADUCEUS will be a formidable task, however, in view of the lengthy period required for the simpler INTERNIST.

The use of anatomic models has been one approach taken by investigators in attempting to build expert systems for neuroradiology (Newell 1986). The neuroradiologist uses his knowledge of the anatomy of the structures that are in the image to interpret the image. In a similar manner, if the program is to interpret anatomically based images, such as CT or MRI, it would be helpful if it had a model of the objects also. In the RAD program (Banks 1987), a neuroanatomy atlas was digitized and stored in the form of octree encoded data sets. Work is underway to use these objects to train a neural network (see below) to recognize the objects from the atlas in order to find their analogs in unknown CT and MRI scans.

Electromyography (EMG) has attracted the attention of several groups of expert system builders. The MUNIN system (Andreassen 1985) employs a Bayesian causal network so that the presence of a disease can be determined probabilistically by tracing the causal links between EMG findings and the disease knowledge base. The Baysean weights are to be determined from actual EMG data gathered in the course of MUNIN's development. An anatomic knowledge base is also part of this system. Jamieson (1987) uses an augmented transition network (ATN) in his Electrodiagnostic Assistant Program which is intended to provide expert advice on performing the EMG examination. The program assures that a sufficiently complete examination is performed to decide among competing hypotheses generated by the EMG request from the clinician and in the course of the data gathering process. Extensive tutorial help is available covering all aspects of the exam. The text for the help screens are composed on the fly by ATNs providing for generation of natural language output. Several other smaller systems have been created for electrophysiologic applications (Gallardo 1987, Raaholt 1987, Villa 1987).

Neural Networks

Artificial neural networks are computational structures consisting of nodes or units (analogous to neurons) and connections between them (analogous to axons and dendrites). Interest in neural networks began more than 30 years ago, but because of limitations of the early theories, they were not successfully able to solve problems as complex as could be handled using conventional AI techniques which use symbolic representations and conventional serial von Neumann style computational algorithms. In a neural network which is programmed to perform a given task, the information is stored and processed in a parallel, distributed manner. Each node does one little piece of the overall computation, communicating with the other nodes and influencing the output states of the system through the connections. Data is given to the network by presenting a "stimulus" pattern to the input nodes. The activation spreads from the input nodes into the system according to the weights on the various connections and finally propagates to the output nodes. The pattern of activation of the output nodes constitutes the "behavior" of the system or the result of the computation. The connection strengths can either be set by the programmer in order to provide the desired type of behavior or the network can be "trained" to exhibit the proper behavior. The training consists of the presentation of known input patterns and observation of the network's output pattern. If the output pattern is incorrect, the network is transformed according to one of several algorithms to adjust the connection strengths such that the probability of a correct response will be greater on succeeding presentations of the same stimulus.

Such a scheme is obviously interesting to neurologists since using these methods, neurologically plausible models can be constructed for performing cognitive tasks in a manner closer to the way that it is thought that biological systems do. Building such models also allows for experimentation with making "lesions" in the network such that cognitive deficits are simulated. If this can be done successfully, it may be possible to experimentally study higher cortical functions in machines without having to be completely dependent on accidents of nature such as strokes and wounds to provide our material. Computer simulations have been constructed for a variety of psychiatric and neurologic deficits, including aphasia (Gigley 1985), anomia (Gordon 1982), dyslexia (Reggia 1986), amnesia (McClelland 1986), mania (Hoffman 1987), depression (Webster 1988), cognitive aging (Salthouse 1988), and epilepsy (Wong 1985).

Arbib and coworkers (1987) have built a neural network model of the toad visual system with its connections to various behavioral modalities, such as prey catching and predator avoidance by actually investigating the histologic structure of the neural connections in the real toad and then recasting them in a computer simulation. When the artificial toad (Rana Computatrix) is presented with various stimuli, the same behavior is observed as that which is seen with identical stimuli presented to real toads.

Expert system builders have also been attracted to neural network methods. Designing conventional AI programs to perform a task requires that we have a good understanding of the principles by which the task is solved. Some tasks that humans do very well, however, such as the recognition of faces, or the identification of features in an x-ray image, may not be easily broken down into logical, symbolic descriptions. The visual system is "hard wired" to perform certain functions and general purpose computers and algorithms can not easily be recruited to do such tasks. Thus AI investigators have been strongly attracted to the use of parallel distributed processing (PDP) or connectionist methods to attempt to solve formerly intractable problems.

Mulsant (1988) has trained a neural network to aid in the diagnosis of demented patients in a geriatric setting. In our laboratory, we have trained a neural network to perform visual field analysis. Using a training set of 25 common visual field defects and a heuristically chosen set of input units, the system was taught to diagnose correctly the visual field deficits in the training set. On average, it took several thousand exposures to each of the items in the training set to satisfactorily train the system. Following training, the system was tested on visual field deficits to which it had never been exposed (but which were among the 25 types that it had learned) taken from a review manual intended for candidates for neuro-ophthalmology board examinations. It was capable of diagnosing correctly all of the unknown fields. Another attractive feature of the artificial neural network model is its ability to generalize. Given a field deficit which is somewhat ambiguous, the output nodes representing the types closest to the input fields will be activated. The networks are also somewhat insensitive to errors in the input data, since the final output is produced by the network's integration over all the input. A disadvantage of the neural network method is that it is difficult for the network to generate explanations for its own behavior, just as it is difficult for a person to explain the process which he uses to discriminate the face of one of his children from that of another. We are currently investigating the extension of this method to the recognition of structures and abnormalities in MRI images. The difficulty here lies in the higher resolution required to analyze such scans properly with the consequent larger number of input units needed. Since these networks are not running on true parallel machines (which are currently quite expensive), but have to be simulated on conventional serial processors, the processing time is an exponential function of the size of the network. Currently, VLSI technology is being applied to creating neural network hardware, so we can expect rapid progress toward the solving of these problems in the future.

References

Andreassen S et al (1985) Knowledge representations for diagnosis and test planning in the domain of electromyography. ECAI-86 proceedings.

Arbib, MA (1987) Levels of modeling of mechanisms of visually guided behavior. Behav & Brain Sci 10:407-465.

Banks G *et al* (1983) The Michael Reese Stroke Registry: a microcomputer-implemented data base. In: Proceedings of the 7th Annual Symposium on Computer Applications in Medical Care (SCAMC). IEEE Computer Society Press, Los Angeles, pp. 724-727.

Banks G & Weimer B (1984) Symbolic coordinate anatomy for neurology (SCAN). J. Med Systems, 8:157-162.

Banks G *et al* (1987) Radiologic automated diagnosis (RAD). Comp Meth Prog Biomed 25:157-168.

Catanzarite VA *et al* (1981) Computer assisted diagnosis and computer consultation in neurology: Preliminary testing of diagnostic accuracy for the Neurologist system. Int. J. Neurosci., 13:43.

First MB *et al* (1982) LOCALIZE: Computer-assisted localization of peripheral nervous system lesions. Comput. Biomed. Res., 15:525-453.

Gallardo R *et al* (1987) Artificial intelligence in the electromyographic diagnosis of cervical roots and brachial plexus lesions. EEG and Clin. Neurophys. 6:S84.

Gigley H (1985) Computational neurolinguistics—What is it all about? International Journal Conference of Artificial Intelligence, Los Angeles.

Gordon B (1982) Confrontational naming: Computational model and disconnection simulation. In: Neural Models of Language Processing, M. Arbib et al, eds. NY: Academic Press.

Hier DB *et al* (1986) Can a patient database help build a stroke diagnostic expert system? Med Inf 11:75-81.
Hoffman R (1987) Computer simulations of neural information processing and the schizophrenia-mania dichotomy. Arch Gen Psychiatry 44 :178-188.
Jamieson P (1987) Symbolic analysis of electromyographic data. EEG and Clin. Neurophys. 6: S84.
Kumar R & Srihari S (1985) An expert system for the interpretation of cranial CT scan images. In: Proceedings of the Expert Systems in Government Symposium, K Karan, ed. McClean VA.
Kuliakowski CA & Weiss HM (1982) Representation of expert knowledge for consultation: the CASNET and EXPERT projects, in: Artificial Intelligence in Medicine, P Szolovits, ed. Boulder, CO, Westview Press, Ch 2.
McClelland J & Rumelhart D (1986) Amnesia and distributed memory. In: Parallel Distributed Processing: Explorations in the Microstructure of Cognition, v. 2. J. McClelland & D. Rumelhart, eds. Cambridge: MIT Press.
Meyer AU & Weisman WK (1973) Computer analysis of the clinical neurological examination. Comput. Biol. Med., 3:111.
Miller RA *et al* (1982) INTERNIST-1, an experimental computer-based diagnostic consultant for general internal medicine. N. Engl. J. Med., 307:468-476.
Mulsant BH & Servan-Schreiber E (1988) A connectionist approach to the diagnosis of dementia. Proceedings of the 12th Annual Symposium on Computer Applications in Medical Care. IEEE Computer Society Press, Los Angeles, pp. 245-250.
Newell JA & Sokolowska E (1986) Model based recognition of CT scan images. In: MEDINFO 86 Proceedings. Amsterdam, Elsevier pp. 619-623.
Patil RS *et al* (1982) Modeling knowledge of the patient in acid-base and electrolyte disorders, in Artificial Intelligence in Medicine, P Szolovits, ed. Boulder, CO, Westview Press, ch 2.
Pople HE (1985) Evolution of an expert system: from Internist to Caduceus. In: Artificial Intelligence in Medicine, I. DeLotto *et al* eds. Amsterdam, Elsevier, pp.179-208.
Raaholt PO *et al* (1987) An expert assistant for electromyography. EEG and Clin. Neurophys. 6:S84.
Reggia JA *et al* (1984) Computer-aided assessment of transient ischemic attacks. Arch. Neurol., 41:1248-1254.
Shortliffe EH (1976) Computer-Based Medical Consultations: MYCIN. Elsevier, NY.
Spitzer K *et al* (1988) Spontaneous subarachnoid hemorrhage: Expert system for appraisal of the prognosis and computer supported decision for therapy. J Neurol 235:335-342.
Villa A *et al*(1987) Are expert-systems an aid for diagnosing neuropathies. EEG and Clin. Neurophys. 6:S84.
Webster C *et al* (1988) DEP: A Depression Emulation Program. Proceedings of the 12th Annual Symposium on Computer Applications in Medical Care. IEEE Computer Society Press, Los Angeles, pp. 287-291.

Automated Identification of Neuroanatomical Structures in CT- and MR-Images

Thomas Schütz, Hans-Joachim Kretschmann, Dieter Müller*

Department of Neuroanatomy
Hannover Medical School
Federal Republic of Germany

*Institute of Computer Science
Hannover University
Federal Republic of Germany

Summary

The computer program AI (Automated Identification) is presented, which is capable of indentifying neuroanatomical structures in cranial CT- and MR-images. The basis for the identification is a data base consisting of 3 dimensional (3d-) reconstructions of neuroanatomical structures derived from serial sections of anatomical specimens. The identification itself is a special case of the consistent labeling problem.

Introduction

The imaging techniques CT and MRI are indispensable methods in modern medical diagnosis. The advances in diagnosis of diseases involving the central nervous system are remarkable.

The first step in interpreting cranial CT- and MR-images is the identification of known neuroanatomical structures. This is done by radiologists, neurosurgeons or neurologists. Their knowledge about the spatial relationship between neuroanatomical structures is one basic requirement for their image interpretation. The ventricular system for example can be delimitated by its contrast to the surrounding grey and white matter in CT- and MR-scans. The caudate nucleus, part of the basal ganglia, can easily be identified if we know its spatial relationship to the ventricular system. The lateral border of the anterior horn of the ventricular system marks the medial edge of the caudate nucleus.

Identifying neuroanatomical structures (i.e. labeling them with their neuroanatomical names) in cranial CT- and MR-images, as described above, can be aided by a computer system. The program AI

---

[1] Supported by Deutsche Forschungsgemeinschaft grant Kr 289/14-2, Kr 289/16-1

(Automated Identification) realizes this approach. A 3 dimensional (3d-) model of neuroanatomical structures of the human brain is the basis for the comparison to the CT- and MR-scans. This model is scaled in order to reduce the influence of interindividual variability.

Spatial relationship and the consistent labeling problem

The main concept of AI is that the strategy of identifying identical structures in two different images by their spatial relationship can be formulated in terms of the consistent labeling problem:

Let O be a set of objects and L a set of labels. R_O is a set of relations on objects. The set R_L consists of relations on the labels. For every relation in R_O there is one and only one corresponding relation in R_L. Let $O' \in O^k$. The problem is to find a $L' \in L^k$ so that for every relation $r_O \in R_O$ which holds on elements of O' the corresponding relation $r_L \in R_L$ holds on the corresponding elements of L' and vice versa.

Now transferred to the problem of spatial relationship let L be the set of names of neuroanatomical structures and O the set of delimited structures in CT- or MR-images. The relations are predicates describing the spatial relationship of the labels and objects. For example "is left of" or "is included in" are such predicates. The statement "the caudate nucleus of the left hemisphere is left of the caudate nucleus of the right hemisphere" holds.

Solutions of the problem like tree search algorithms are well known [1].

AI and NEUREF

CT- and MR-images can show serial sections of the human brain in any plane and direction. Therefore it is necessary to have a neuroanatomical 3d-model for the comparison. The appropriate slice can then be derived by cutting this model with the computer. Creating it is one task of the project NEUREF (NEUroanatomical REFerence system). It consists of three parts:

1. Methods and software for acquisition and processing of data (3d-reconstruction).

2. A data base consisting of 3d-reconstructed neuroanatomical structures derived from serial sections of post mortem human brains.
3. Application software which enables the graphical presentation of neuroanatomical structures and allows a direct comparison with CT- and MR-images (AI).

The data base consists of 97 different neuroanatomical structures like the ventricular system, the parts of the limbic, visual and auditory system, the basal ganglia, cerebellum and cerebral hemispheres. The data is derived from parallel sections of anatomical specimens [2] using triangulation as the method for 3d-reconstruction. AI is part of the application software from NEUREF.

One problem complicating this method is the interindividual variability of neuroanatomical structures. A scaling model is needed for the practical comparison of structures. In AI a modified scaling method according to the proposals of Talairach and Szikla [3],[4] was used (figure 1). This method has succesfully applied in stereotactic procedures. The first step of our modification is to connect the midpoints of anterior commissure (AC) and posterior comissure (PC) to form the bicommissural line. The verticals trough the anterior pole (VAP) and the posterior pole (VPP) of the telencephalon are determined as the last perpendiculars of the bicommissural line touching the anterior or posterior edge of the telencephalon. The axial lines through the superior pole (ASP) of the telencephalon and the inferior pole (AIP) of the temporal lobe are defined as parallel of the bicommissural line. The scaling in axial direction is derived from the extremes of both cerebral hemispheres in the plane of the bicommissural line. This process describes a rectangular boxe surrounding the brain. The positioning of the bicommissural and vertical lines is done manually in the median section of the patient's brain. The axial scaling is performed with the axial section containing the bicommissural line.

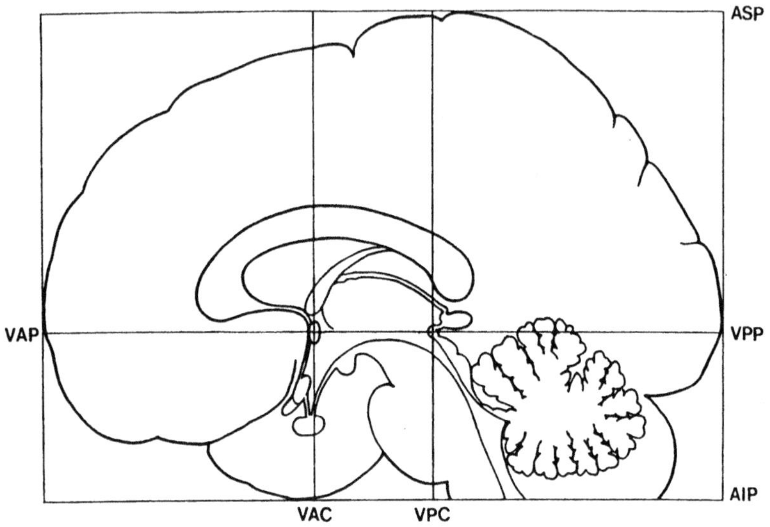

Figure 1:
Modification of the scaling method of Talairach and Szikla.
VAC Vertical line through the anterior commissure
VPC Vertical line through the posterior commissure
VAP Vertical line through the anterior pole of the telencephalon
VPP Vertical line through the posterior pole of the telencephalon
ASP Axial line through the superior pole of the telencephalon
AIP Axial line through the inferior pole of the telencephalon

Implementation

The program AI is implemented on a Silicon Graphics Workstation IRIS 4D/20 G. Its great graphic performance is extensively used by the program. The implementation language is C under UNIX System V.3. The user is guided through the program by a menu. So inexperienced users are familiar with it within a short time.
The limitation of structures in CT- or MR-images is done manually. In certain cases the exchange with an automatic procedure will be possible.

Improvements planed for the future

The interindividual variability of neuroanatomical structures is well known. Therefore it is necessary to adapt the model to a representative number of human brains. Up to now 9 human brains have been examined. We will extend our data within this year. With this statistical spatial model it will be possible to determine the likelihood for the appearence of any neuroanatomical structure within a given voxel. As one consequence it will be possible to measure the variation of patients' brain structures in comparison to the neuroanatomical model.

References

1. Haralick, R.M.; Shapiro, L.G.: The consistent labeling problem: Part 1. IEEE Trans. on Pattern Analysis and Machine Intelligence PAMI-1 (1979) 173-184.

2. Klekamp, J.; Riedel, A.; Herrmann, A.; Kretschmann, H.-J.: A new embedding and sectioning technique (macrovibratome) for macroscopic and morphometric examinations especially of the human brain. J. Hirnforsch. 26 (1985) 33-40.

3. Talairach, J.; Szikla, G.; Tournoux, P.; Prossalentis, A.; Bordas-Ferrer, M.; Covello, L.; Iacob, M.; Mempel, E.: Atlas of stereotaxic anatomy of the telencephalon. Paris: Masson 1967.

4. Talairach, J; Tournoux, P.: Co-planar stereotaxic atlas of the human brain. Stuttgart: Thieme 1988.

The Paradigm of Computer Assisted Radiology and Computer Assisted Neurology

K. Spitzer†, H.S. Stiehl‡

† Neurologische Universitätsklinik Hamburg-Eppendorf, Martinistr.52, D-2000 Hamburg 20
‡ Universität Hamburg, Fachbereich Informatik, Bodenstedtstraße 16, D-2000 Hamburg 50

The domain of clinical neurosciences (neurology, neurosurgery, neuroradiology, neuropathology, neuropharmacology) appears to be an emerging application field for Artificial Intelligence. A large amount of data assembled by clinical examination, laboratory tests and neuroimaging procedures and the demand for fast and reliable information processing challenge the employment of advanced computer programs.
The novel 3-D paradigm [1,2] in clinical disciplines associated with diagnostic brain imaging (CAT, MRI, PET, multi-plane DSA) has been established. Typical clinical tasks in neurosciences, which either directly draw upon three-dimensional imagery or being related to the 3-D paradigm, are:

- ⋆ diagnostic analyses and decision procedures in neuroimaging
- ⋆ investigations of anatomical and functional correlations
- ⋆ planning of neurosurgical and stereotactic interventions
- ⋆ radiation therapy planning

Inferring spatial lesion loci from observable signs and symptoms, investigation of structure-to-function causality, error-prone visual estimation, time-consuming quantification of organ and lesion properties, reasoning and planning in space-time continuum, recovery of spatial configurations and spatio-temporal events from images require intensive computer support, for which existing operational systems are not sufficient yet. These systems lack a variety of useful tools which definitely would have an impact on therapeutic procedures, e.g. accuracy or safety in neurosurgery [3]. A general integrative approach to the design, realization and validation of useful tools in the domain of neurosciences is emerging.
Criteria for the selection of a medical domain for computer assistance have been formulated [4]: existence of demonstrated need for help in the domain, existence of recognized need for help by the physicians, avail-

ability of a core of formalized and readily available knowledge, cooperation with highly motivated collaborators from the domain of expertise, and the possibility of a straightforward introduction of a computer-based tool into daily clinical routine.

While clinical neurosciences with no doubt meet the first three criteria, the last two are directly related to both the necessity for open-minded interdisciplinary basic research and the design, realization, clinical embedding, and validation of future dedicated computer-assisted tools. Tools in the fields of Computer Assisted Radiology (CAR) and Computer Assisted Neurology (CAN) can be applied in:

⋆ topical diagnosis in neurology (association of spatial/functional relationships; localization of lesions within the central and peripheral nervous system; hypotheses formation about the pathogenesis and spatial extent of lesions)

⋆ neuroradiological and neuropathological diagnosis (spatial image sequence processing and understanding; three-dimensional morphometry; recovery of a common cranial reference coordinate system; generation of spatial, iconic or symbolic representations, integration of spatial images from diverse imaging sources)

⋆ neurosurgical treatment planning and simulation (iterative optimization and in vitro simulation based upon a complete and accurate three-dimensional patient representation)

⋆ stereotaxy planning and simulation (minimum risk spatial trajectory simulation)

⋆ radiation planning and simulation (three-dimensional therapy simulation for minimum-error three-dimensional dose distribution planning)

Only a very few laboratory systems do yet exist, which however have to be considered rather far from being applicable in daily clinical routine [5].

Clinical usefulness of such tools depends on the capability of acquisition, formalization, organization, representation, utilization, maintenance, updating and integration of generic domain knowledge as well as patient-specific instances thereof. Availability of expert-level knowledge within a system-supported problem-solving process determines the quality and the acceptance of the system [6]. Moreover, incorporation of knowledge into such workstations will heavily influence the conceptual style and level of interaction which in turn has an effect on the working style of the users in diagnostic and therapeutic strategies.

The role workstations will play in clinical neurosciences can be conceived as an aid for processing, manipulating and visualizing knowledge about three-dimensional structure and function of the human brain. Such workstations have to be able to handle pictorial and domain knowledge. The following requirements have to be met [4]:

- ⋆ capability to support, to amplify and to expand perceptual and cognitive abilities of physicians
- ⋆ technical reliability and expert-level performance
- ⋆ acceptance into the affected community resulting from user-friendly interfaces
- ⋆ benefit of patient care
- ⋆ positive impact on public health care delivery
- ⋆ sense of responsibility for patient data

The major challenge for research toward knowledge-based tools is the development of a coherent knowledge base which is accessible to all representational and computational processes in clinical neurosciences. However, inexactness, incompleteness, and inconsistency generally inheres in medical knowledge which is intrinsically structured along the four representational dimensions of topography, morphology, etiology, and function [7]; divided into seven inter-linked classes (human body structure, functions, diagnostic measures, pathology, etiology, therapies and treatments, concepts and theories of health care systems [8]; and scattered among the interacting three knowledge sources: textbooks (formalized and systemized knowledge), scientific journals (recent research), and patient records (summaries of clinical case histories) [8].

In clinical neurosciences, a common knowledge base should make explicit a diversity of knowledge sources: image formation physics and principles, spatial neuroanatomical morphology, structure-to-function correlations and causalities, neuroanatomical pathways, physiologic/metabolic and sensory/motor/intellectual/psychical function, cytoarchitecture, spatio-temporal alterations due to aging or morphogenetic growth, vascularity, biological variability, pathology, epidemiology, radiation biology, pharmacokinetics, and reference coordinate systems, e.g. for neuroanatomical nomenclature.

A fusion of separately treated research areas regarding explicit representation of generic knowledge about normal and pathologically altered spatio-temporal brain structure and function as well as patient-specific representations could significantly contribute towards an unifying methodological and computational framework for CAR and CAN.

References

1. Stiehl, H.S.: On spatial image sequence understanding. Habilitation research monograph. Berlin: Technische Universität Berlin, Fachbereich Informatik 1987.

2. Lemke, H.U.: The third dimension. In: Lemke, H.U., et al (eds.) Proceedings CAR'85. Berlin: Springer 1985, pp.628-634.

3. Wowra, B.; Schlegel, W.; Kunze, S.: Scopes of computer application in stereotactic neurosurgery. To appear in: Lemke, H.U., et al (eds.) Proceedings CAR'89. Berlin: Springer 1989.

4. Clancey, W.J.; Shortliffe, E.H. (eds.) Readings in medical Artificial Intelligence: the first decade. Reading, Addison Wesley 1984, pp.1-17.

5. Banks, G.: Artificial Intelligence in neurology. To appear in: Lemke, H.U., et al (eds.) Proceedings CAR'89. Berlin: Springer 1989.

6. Lemke, H.U.: Picture archiving and communication systems. In: Chiesa, A., et al (eds.) Proceedings 5th International Symposium on Planning of Radiological Departments. Firenza 1988, pp.198-208.

7. Tsotsos, J.K.: Knowledge organization: its role in representation, decision-making, and explanation schemes for expert systems. Technical Report LCM-TR83-3. Toronto: University of Toronto, Laboratory for Computational Medicine 1983.

8. Gabrieli, E.R.; Reinhardt, M.T.: Computer-based knowledge banks for clinical medicine. J Clin Comp 11 (1983) 195-200.

Scopes of Computer Application in Stereotactic Neurosurgery

BERNDT WOWRA, LOTHAR SCHAD, WOLFGANG SCHLEGEL, STEFAN KUNZE

Clinic of Neurosurgery, University of Heidelberg and Institute of Radiology and Pathophysiology, German Cancer Research Center, Heidelberg, FRG.

Summary

For future scopes of computer applications in stereotactic neurosurgery the development of a composite image stereotaxis is proposed. This system would be frameless stereotaxis working with digital image information from different imaging devices. During operations, image information would be available simultaneously in one composite image with increased pixel depth. Conceptions for image correlation based on neuroanatomical landmark structures and for the establishment of an internal coordinate system are discussed. An intraoperative hardware-software interface controlled by a robot system and a real time digital imaging device would be further system components. Finally, local carrier-mediated chemotherapy could become a possible future application of composite image stereotaxis.

Aim

The following contribution is an attempt to stimulate further CAR applications in stereotactic neurosurgery by reflecting on clinical requirements.

Current status: Integrated Stereotaxis

Since the advent of CT digital image information became available for use in neurosurgery. Spatial information is an inherent property of computer based images. However, image information from different sources (CT, DSA, MRI, or PET) cannot easily be matched spatially to each other. Therefore, image correlation in neurosurgery affords skull fixation with stereotaxic frames. These methods enable now three-dimensional image information from CT, MRI, PET, and DSA to be used *in parallel* for treatment planning and therapy. An integrated stereotaxic system has also been developed in Heidelberg [1,8,9,10,11]. It can be used for diagnostic and therapeutic purposes as well [Fig. 1]. However, there are considerable unsolved problems. It takes several hours to collect image data from different imaging devices. This is very circumstanceous and in addition, very inconvenient for the patients who have to remain rigidly fixed within the base ring during the whole procedure. These experiences and limitations demand for new developments.

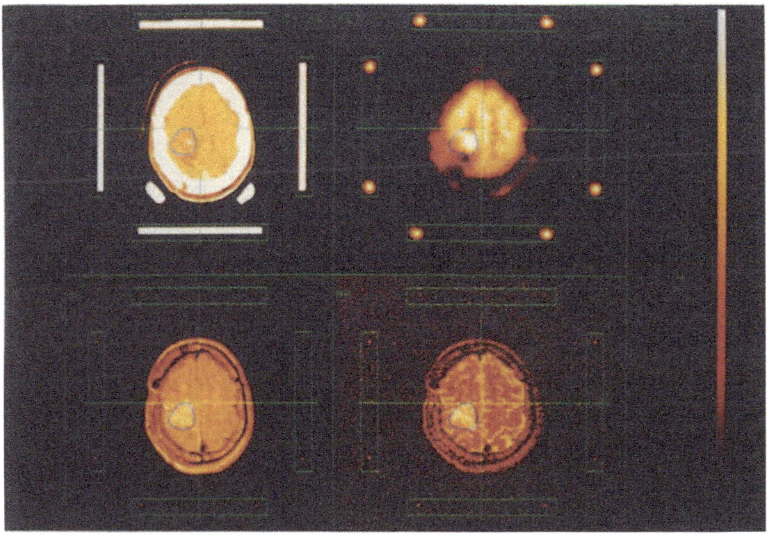

Figure 1: Image correlation with a stereotaxic frame. Definition of the target volume (blue outline) on a set of CT, PET, and MR images. $H_2{}^{15}O$ was used as tracer in PET imaging (upper images CT and PET). For MR imaging (bottom) multislicing double echo sequence TR/TE 1800/35,105 ms was used.

Scopes: Composite Image Stereotaxis

The realization of frameless stereotaxis, a term coined for the system of Roberts [7], should provide precision and high spatial resolution of digital neuroimaging for clinical requirements. This means that the specific information of each imaging method should be made available s i m u l t a n e o u s l y in one composite image with increased pixel depth in order to achieve three-dimensional control of intracranial space. Composite image stereotaxis could benefit from image transfer by PACS and LAN technology.

Landmark structures: Although headholders for scanning free of artifacts would still be necessary, traditional stereotaxic frames would not be required for composite image stereotaxis. Correct spatial information on composite images will be dependent on the identification of landmark structures. These are structures of neuroanatomy suitable to serve as

Table 1: Recognition capacity of CT, MRI, PET, DSA, US for neurostructures

	CT	MRI	PET	DSA	US	score of structures
mid plane	+++	+++	+++	++	++	13
skull	++	-	-	+++	-	5
skull base	+++	-	-	++	-	5
ventricles	+++	+++	++	-	++	10
cortical relief	++	+++	++	++	++	11
vessel architecture	+	++	-	+++	+	7
score of methods	14	11	7	12	7	

- not detectable; + detectable in principle; ++ detectable in sufficient quality; +++ optimal visualization.

fiducial markers for computer processing. They have to be identified on image data files from different sources. When comparing imaging characteristics from different CAR methods, the midplane between both hemispheres, the cerebral ventricles, and the cortical relief present the most readily recognized neurostructures [Table 1].

Selection of neuroimaging methods: When ranking neuroimaging methods by their recognition capacity for relevant neurostructures the most suitable for composite stereotactic imaging apparently are CT, MRI, and DSA [Table 1].

Image correlation procedure: An unequivocal correlation of image information could be achieved with special fitting programs for the above mentioned landmark structures. The reconstruction of the midplane should be comparatively easy by defining a symmetry plane. The cerebral ventricles however, due to their complex three-dimensional structure require fare more sophisticated geometrical analysis. Their detection could probably be facilitated by the homogeneity of the CSF. A surface fitting program could help for analysis and description of individual cortical reliefs.

Internal coordinates: For therapy planning with composite imaging in frameless stereotactic neurosurgery the establishment of an internal coordinate system is essential. One conceivable way to determine the zero point of this rectangular coordinate system could be the identification of the origin of the internal cerebral veins at the dorsal border of the Foramen of Monroe [Fig. 2]. The orientation of the coordinates should be adjusted to the physiological spatial orientation of the brain.

Hardware-software interface: A hardware - software interface should be settled in the operation theatre. This would consist of a workstation with

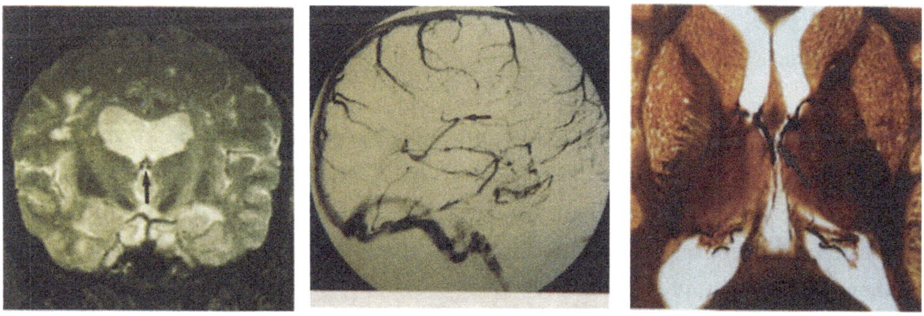

Fig. 2: The origin of the internal cerebral veins [arrows] at the Foramen of Monroe as a possible zero point of an internal coordinate system. Left: coronal section of MRI; centre: sagittal aspect of DSA; right: axial anatomical section.

a high resolution screen, a patient couch with a headholder, and a scanning resp. imaging device. The mechanical system would allow head fixation, correct spatial imaging and the performance of invasive diagnostic and therapeutic procedures. This could be controlled by a robot [3,6,14]. The imaging component of the hardware-software interface should enable the recognition and precise spatial fitting of the landmark structures with the composite image available on the screen display. The imaging method could be either DSA or ultrasound [2]. The latter would require at least two ultrasound probes in order to establisch three-dimensional imaging. Since ultrasound impulses cannot penetrate the skull bone, a small craniotomy would be necessary for each US probe; this however, would be a minor disadvantage because concerned patients are to undergo carniotomy in any case. The advantages of ultrasound e.g. high precision real time imaging, visualisation of brain pulsations, and the lack of a physical burden to the brain tissue could favourable be used.

Future Application: Local Carrier Mediated Chemotherapy

Finally, local carrier mediated chemotherapy for brain tumors [4,5,12,13] as a special application of composite image stereotaxis should be mentioned. Although still at an early developmental stage, it is evident from animal studies that the efficacy this type of local chemotherapy is strongly influenced by variations in tissue texture. Essential informations concerning tissue structure are included in data files of composite images. Hence, it can be expected that these specific information will be available for use in local carrier mediated chemotherapy.

Conclusion

In conclusion from a clinical point of view, essential scopes of computer application in stereotactic neurosurgery will probably be the reconstruction of composite images [Fig. 3], the establishment of internal coordinate systems, and robot technology for precise 3 D - control of therapeutic manipulations.

Abbreviations

CAR	Computer Assisted Radiology
CT	Computed Tomography
DSA	Digital Substraction Angiography
LAN	Local Area Network
MRI	Nuclear Magnetic Resonance Imaging
PACS	Picture Archiving and Communication Systems
PET	Positron Emission Tomography
US	Ultrasound Imaging

Figure 3: 3 D-display of the patients head positions calculated with CT (green) and MR (white) image data. The target volume (red) and 80 % isodose (blue) is indicated by a 'barrel hoop'.

References

1. Bauer, B., Schlegel, W., Scharfenberg, H., Sturm, V., Lorenz, W.J.
Computerized Optimization of Treatment Planning in the Brachytherapy of Brain Tumours.
Proceedings of the 8th International Conference on the Use of Computers in Radiation Therapy (1984) IEE, 395-398 (1984)
2. Berger, M.S.
Ultrasound-guided stereotaxic biopsy using a new apparatus.
J Neurosurg 65: 550-554 (1986)
3. Doll, J., Schlegel, W., Pastyr, O., Sturm, V., Maier-Borst, W.
The Use of an Industrial Robot as a Stereotactic Guidance System.
Proc. Int. Symp. CAR'87 374-378 (1987)
4. Feun, L.G., Lee, Y.Y., Wallace, S., Charnsangavej, C., Savaraj, N.
New drugs and new delivery techniques.
Prog. exp. Tumor Res. 29: 131-139 (1985).
5. Freeman, A.I. and Mayhew, E.
Targeted drug delivery.
Cancer 58: 573-583 (1986)
6. Kwoh, Y.S.
A new CT-aided robotic stereotactic system
70th Scientific Assembly and Annual Meeting, RSNA, Washington D.C. (1984)
7. Roberts, D.W., Strohbein, J.W., Hatch, J.F., Murray, W., et. al.
A frameless stereotaxic integration of computerized imaging and the operating microscope.
J Neurosurg 65:545-549 (1986)
8. Schad, L., Boesecke, R., Schlegel, W., Hartmann, G.H., Sturm, V.
Three dimensional image correlation of CT, MR, and PET studies in radiotherapy treatment planning of brain tumors.
J. Comp. Assist. Tomog. 11(6): 948-954 (1987)
9. Schlegel, W., Scharfenberg, H., Doll, J., Pastyr, O., Sturm, V., et. al.
CT-Images as the Basis of Operation Planning in Stereotactical Neurosurgery.
Proceedings of the 1st International Symposium on Medical Imaging and Image Interpretation ISMIII (1982)
10. Sturm, V., Pastyr, O., Schlegel, W., Scharfenberg, H., Zabel, H.-J.
Stereotactic Computer Tomography with a Modified Riechert-Mundinger Device as the Basis for Integrated Stereotactic Neuroradiological Investigations.
Acta Neurochir. 68, 11-17 (1983)
11. Sturm, V., Schlegel, W., Bauer, B., Wowra, B., Kunze, S., and Lorenz, W.J.
Interstitial Irradiation of Low Grade Gliomas with 125-I.
Treatment Planning, Possibilities and Limitations.
J. Neurooncol. 5: 184. (1987)
12. Wowra, B., Mentrup, E., Zeller, W.J., Stricker, H., Sturm, V.
CT-Kinetik intratumoraler Liposomendepots.
Onkologie 11(2): 81-84. (1988)
13. Wowra, B., Zeller, W.J., Remmele, T., Sturm, V., Stricker, H.
Local methotrexate chemotherapy with polylactide implants for experimental rat glioma.
Regional Cancer Treatment II (1988)
14. Young, R.F.
A robotic system for stereotactic neurosurgery.
In: Lunsford, L.D. (ed.) Modern Stereotactic Neurosurgery
Martinus Nijhoff Publishing, Boston, Dordrecht, Lancaster, pp. 259-266 (1988)

Computer Assisted Cardiology

Digital Image Processing – Clinical Significance and Current Developments

H. Just, H. Wollschläger, U. Solzbach

Medizinische Universitätsklinik - Innere Medizin III
Hugstetterstr. 55
D-7800 Freiburg i. Br.

Imaging techniques are gaining increasing importance in diagnostic and therapeutic applications. Anatomic detail and functional data can be obtained invasively or non-invasively with heretofore unknown precision and have led to a new dimension of "functional" anatomy:

1. Extraordinary insight into the intact organism and organs can non-invasively be obtained, mostly from multiple points of view in multiple planes and tissue quality can be assessed (sonography, computed tomography, nuclear magnetic resonance technique, radioisotope techniques in nuclear medicine).

2. Anatomic structure can be combined with information on tissue metabolism (magnet resonance spectroscopy, positron emmission tomography).

3. Anatomy can be combined with physiological parameters characterizing function (echocardiography, echo-Doppler-techniques, angiography, cine-angiography, fluoroscopy, cine-computed tomography, fast magnetic resonance technique, "echo"-computed tomography).

4. Guidance of needles, catheters, biopsy-instruments for diagnostic or therapeutic applications transcutaneously or within blood vessels or other body activities (echo- or CT-guidance of punctures; fluoroscopy for intravascular or intraluminal catheters for drainage, embolization, recanalization, dilatation, implantation of prosthesis). These techniques have opened the increasing field of transcutaneous, semi-operative procedures, i.e. the interventional catheter-techniques allowing a sizeable reduction of operative, surgical procedures. Basically possible with standard imaging techniques, digital procedures have gratly facilitated the intervention and have remarkably reduced radiation dosage.

What is the role of digital techniques in diagnostic and therapeutic imaging?

Direct images can only be obtained either by transmission and reabsorption (x–ray, ultrasound, infra-red) or reflected radiation. Or emmitted signals from radio-isotopes in nuclear medicine techniques or magnetic resonance signals from activated atoms are used.

Digital techniques are applied in 6 different ways in order to obtain better images or to include additional information:

1. Existing images are "post-processed" (after analog/digital conversion) for contrast enhancement or for quantitative analysis.

2. Existing images are combined with "indicators", such as contour lines, graphic or numerical data explaining or identifying the picture.

3. Primary image information is processed for construction of more complete images and easier handling (filtering, contour completion or -improvement, colour coding, storage for real-time or delayed display (e.g. picture availability for gap filling techniques in fluoroscopy).
4. Primary image information is combined with physiological measurements, likewise after A/D-conversion, e.g. pressure, flow, ECG, volume data for analysis of cardiovascular function. Or the image information in videoformat is made to include technical data characterizing the image production process, e.g. x–ray generator settings, radiation parameters, time signals etc. Also data defining spatial orientation, e.g. position of x–ray stands, or ultrasound transducer can be included.
5. Three-dimensional image information is stored as a digital "data block" from which images can be constructed in any desired plane or viewing angle (CT, NMR, echo-CT).
6. Digital images can be combined with a multiplicity of patient identification data and other information and can then be stored, transmitted, viewed, copied, reduced or enlarged, and manipulated as far as contrast, regional detail analysis, time sequence comparison is concerend (PACS: picture archiving and communication systems).

Examples of clinical application of digital imaging techniques are given in the following:

1. High contrast pulsed fluoroscopy with gap filling:

   This technique is used in interventional cardiology. It allows optimal viewing of minute detail (e.g. intravascular anatomy, guiding catheter, intravascular miniature instruments) with high resolution, high contrast, in real-time motion without flicker, and with greatly reduced radiation dosage. The latter point is of utmost importance from the point of patient exposure and physician radiation load with legthy interventional techniques.

2. The integrated cardiovascular cathetrization laboratry: Digital techniques are used to include physiological measurements from cardiac catheterization, as well as technical data from x-ray generator, tubes, image intensifiers, stands and their resp. position, as well injection pump characteristics, patient and procedure data into the image information data set. This will enable central control and steering as well as documentation, and quality and systems check through a central computer.

   Reproducibility within image sequence, amongst different procedures and over time, and quantitative angiographic analysis, as well as videodensitometric measurements will be possible with such systems.

3. "Echo"-computed tomography:

   Recently one of us (H.W.) has developed a new approach towards a better qualitative and quantitative study of the heart by ultrasound. Reflected ultrasound is emmitted and received from an oesophageal transducer, which, through stepwise motion will scan heart from apex to base. The entire set of image data is stored together with reference coordinates and the ECG-signal. ECG-gated pictures can be obtained. These images can be derived at any desired angle or plane for viewing.

Future developments will aside from better quality images and easier handling of the systems include more information regarding parenchymal function and metabolism; nuclear magnetic resonance spectroscopy and positron emmission tomography will be developed further. However, the digital image processing techniques are already available today. Development will focus on the integration of anatomic structure, metabolic data, physiological measurements, and technical information from image generation and systems function.

It can be expected furthermore, that digital imaging will greatly expand use and success of interventional techniques, the currently most exciting development in medicine.

Overview of Coronary Quantitation Techniques

J.H.C. REIBER

Laboratory for Clinical and Experimental Image Processing,
Thoraxcenter, Erasmus University Rotterdam, the Netherlands.

Summary.

In this presentation an overview of existing quantitation techniques for coronary obstructions is given, either applied off-line from 35 mm cinefilm or on-line by direct digital approaches. To minimize the variabilities in the measurements, standardized procedures for the angiographic data acquisition and analysis should be followed. In addition, extensive (in vivo) phantom as well as clinical studies should be carried out to determine the accuracy and precision of a particular approach.

Finally, it is shown that various techniques are now available for the assessment of the functional significance of an obstruction (coronary flow reserve) during or after cardiac catheterisation.

Introduction

Despite the widespread and long-standing use of coronary angiography in clinical practice, the interpretation of the angiograms has not changed very much since the early days. Although the image quality continues to improve further due to the development of higher quality X-ray sources and image intensifiers, the use of pulsed fluoroscopy, real-time image enhancement and the availability of high quality image digitization, most angiograms are still reviewed visually and therefore subjectively. In the face of an almost exploding field for interventional catheterization procedures, including thrombolysis, balloon dilatation, and other rapidly evolving techniques for transluminal revascularization or recanalization, a more detailed and quantitative analysis of the coronary arteriograms is urgently needed on a routine basis. It is clearly evident that the visual interpretations of coronary obstructions as the basis for scientific studies is no longer acceptable. The spatial and temporal resolution of the angiograms is relatively high, allowing the derivation of both anatomic and functional information on the coronary obstructions. These quantitative data should not only be available off-line from cinefilm, but there is now more and more an urgent need for on-line quantitation during the patient's investigation, thus facilitating and improving clinical decision making. It should be emphasized that the anatomic/geometric and functional/physiologic approaches are complementary, thus using a combination of these approaches is far better than using either one alone. Furthermore, the objective and reproducible analysis will further enhance our understanding about the pathophysiology of coronary artery disease.

Overview quantitation techniques

In general, the quantitative analysis of coronary obstructions is performed from 35mm cinefilm. Great advantages of the off-line cinefilm approach are the high spatial resolution of the cinefilm and that regions-of-interest, e.g. encompassing an arterial segment, can be selected and digitized at a resolution optimally suitable for quantitative analysis through optical or digital zooming; a disadvantage of the film is the non-linear density-log(exposure) transfer function. Since spatial resolution is the most important factor in obtaining a high accuracy and precision in the geometric assessment of arterial dimensions, cinefilm is still the medium of choice for intervention studies, where very often small morphologic changes need to determined with as high a confidentiality as possible.

However, recent developments in digital cardiac imaging systems have been directed towards obtaining such measurements on-line during the catheterization procedure from video digitized images. A few centers have even completely abolished cinefilm. With the present limitations in spatial resolution of these on-line digitized images (512^2), this approach is particularly of interest as a tool for diagnostic and/or therapeutic decision making during the catheterization procedure. On the other hand, the digital systems are characterized by high density resolution and a linear transfer function of the chain from the output of the image intensifier to the brightness levels in the digitized images, making these more suitable for densitometric analysis.

This means that extensive (in vivo) phantom as well as clinical studies need to be carried out to determine the accuracy and precision of these digital approaches as compared to the established off-line cinefilm techniques with optical zooming. It is not clear that higher resolution (1024^2 digital images) will result in improved accuracy and precision. With the use of modern small field-of-view (FOV) image intensifiers (4" and 5" FOV), the present resolution of $512^2 \times 8$ bits or $512^2 \times 10$ bits images at 30 frames/s may even be sufficient to assess the efficacy of intervention studies on-line using the diameter and densitometric cross-sectional area measures mentioned above. Until those points have been resolved, one should be able to use 35 mm cinefilm simultaneously with the digital acquisition without any limitations.

For the assessment of the arterial dimensions, the measurements are usually performed on the original nonsubtracted images. Due to the difficulties in obtaining high quality subtraction images (selection of background image in same phase of cardiac cycle, motion artefacts, etc.) there is no evidence in the literature suggesting any benefit from such approach for the accurate determination of coronary anatomy.

Various groups have been involved in the development and use of computer-based techniques for the assessment of quantitative measurements of coronary arterial dimensions (1). Key elements in these systems are the cine-digitizer for the 35mm cinefilm approaches and the digital camera for the on-line approaches, the software for the semi- or fully-automated contour detection of a catheter segment (calibration) and of the coronary arterial segment (first-, second-derivative approaches or combination), and the software for the assessment of clinically relevant parameters. It is important to know whether the contour data are corrected for pincushion distortion, for the line-spread function of the X-ray system, and for the differential magnification between the catheter segment and the analyzed arterial segment. Furthermore, the question arises how the data from biplane analyses are being handled. Densitometric techniques have been developed for the assessment of absolute and relative cross-sectional area stenosis from a single angiographic view. In these cases data must be presented on calibration and correction procedures, since there are many nonlinear components in the densitometric image analysis chain.

A complete set of quantitative parameters should include obstruction and reference (how defined?) diameters and areas (circular, elliptical or densitometric approach), extent of obstruction, %-diameter and area stenosis, asymmetry of stenosis, area atherosclerotic plaque, the roughness of the arterial segment. By combining all stenosis measurements, the functional pressure-flow effects of the stenosis, as well as coronary flow reserve can be assessed.

Figure 1 shows an example of the automatically detected coronary arterial contours on our new PC-based CLAS-system; the image was digitally magnified by linear interpolation. Processing time, including calibration is less than 1 min.

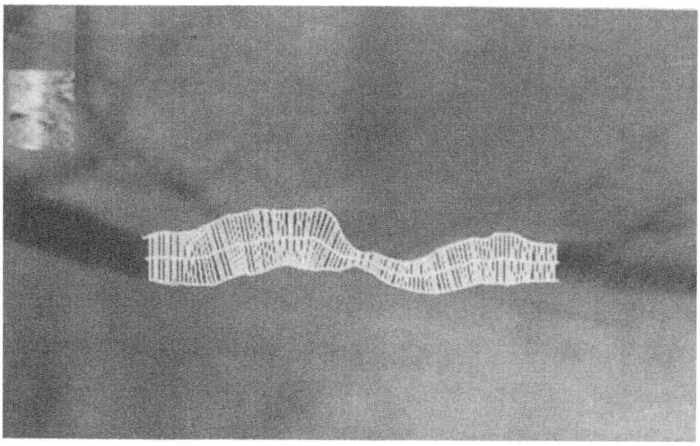

Fig. 1. Automatically detected arterial boundaries on CLAS-system.

Approaches for standardized acquisition and analysis

The majority of the applications of quantitative coronary cineangiography require the comparison of the arterial dimensions either in a control group versus those in a treated group, or pre- versus post-intervention and possibly with the data from some later control-angiogram. The sample size of the number of patients that need to be investigated to demonstrate a certain effect is proportional to the variability of the measurement technique divided by the number of years between the angiograms squared. From a viewpoint of the population size, duration and cost-effectiveness of a study, it is therefore of great importance to minimize the variability of the angiographic data acquisition and computer analysis procedures.

In table I various sources of variation in the angiographic data acquistion procedure have been listed, while Table II shows the corresponding approaches that we have proposed to minimize these variations.

It has been shown that the variabilities in absolute coronary dimensions increase dramatically if these precautions are not followed. On the other hand, when these are followed, the variabilities in absolute dimensions from repeated cine-angiographies and analyses increase only by a factor of 1.5 to 2.2 as compared to those from repeated cinefilm analysis alone (2).

Table I. Sources of variation in angiographic data acquisition.

1. differences in the angles and height levels of the X-ray gantry with respect to the patient at the time of repeated angiography with those used at the time of the pre-intervention study.
2. differences in vasomotor tone of the coronary arteries.
3. variations in the quality of mixing of the contrast agent with the blood.
4. angiographic quality of catheter (image contrast).
5. deviations in the size of the catheter as listed by the manufacturer from its true size.

Table II. Approaches towards standardization in angiographic data acquisition.

1. Accurate on-line or off-line registration of X-ray system settings.
2. Administration of vasodilator immediately before angiographic investigation.
3. Use of nonionic and iso-osmolar contrast media.
   Administration of contrast medium by ECG-triggered injector.
4. Selection of acceptable catheter material (high angiographic image contrast and edge gradient).
5. Measurement of actual size catheter with micrometer following the catheterization procedure.

Validation studies

Validation procedures must include data on the accuracy and precision of the edge detection technique based on (perspex) models with obstruction sizes ranging from 0.5 to 5 mm, as well as on in-vivo models. Furthermore, data on the results from repeated analyses of angiographic studies (reproducibility), and for the densitometric approach data on the variability from different views and under different X-ray system settings must be presented. It is proposed that the validation data are given in terms of the mean difference (accuracy) and standard deviation of the signed differences (precision) between corresponding measurements.

Coronary Flow Reserve

The functional significance of a coronary obstruction can be described very conveniently by the concept of coronary flow reserve (CFR). Basically two approaches have been developed to assess the coronary flow reserve of individual coronary arteries in the clinical setting: 1) by the ratio of maximal to resting coronary blood flow with the maximal flow produced by potent pharmacological coronary vasodilation; 2) predicted from the coronary angiographic data (3).

Following the first approach, three techniques are now available for the assessment of regional coronary flow reserve (CFR) during cardiac catheterization: 1) the pulsed Doppler coronary artery catheter that can measure intracoronary blood flow velocity; 2) radiographic assessment of myocardial perfusion using contrast medium; and 3) an indicator-dilution technique with a platinum-tipped PTCA guidewire using hydrogen as the indicator. Particularly the second approach has received a great deal of attention. Using ECG-gated power injection of a contrast agent at a rate that is presumed to be sufficiently rapid to achieve complete replacement of blood with contrast, a mask-mode subtraction technique was developed by Vogel et al. that allows the determination of myocardial time-density curves before and during coronary vasodilation, before the vascular effects of the contrast medium disturb the ratio between resting and hyperemic coronary blood flow (4).

Conclusions

There is now sufficient evidence from the literature that the visual interpretation of the severity of coronary obstructions is no longer acceptable as the basis for scientific studies. Quantitative coronary angiographic techniques, either applied off-line or on-line, are now becoming available on a large scale, and allow the objective morphologic and physiologic assessment of the severity of the obstructions. The anatomic/geometric and functional/physiologic approaches are complementary, so that a combination of these approaches is far better than using either one alone.

References

1. Reiber JHC. Morphologic and densitometric quantitation of coronary stenoses; an overview of existing quantitation techniques. In: New Developments in Quantitative Coronary Arteriography. JHC Reiber, PW Serruys (Eds.). Kluwer Academic Publishers, Dordrecht, 1988: 34-88.
2. Reiber JHC, Serruys PW, Kooijman CJ, et al. Assessment of short-, medium-, and long-term variations in arterial dimensions from computer-assisted quantitation of coronary cineangiograms. Circulation 71, 1985: 280-288.
3. Gould KL, Kirkeeide RL. Assessment of stenosis severity. In: Reiber JHC, Serruys PW (eds): State of the Art in Quantitative Coronary Arteriography. Martinus Nijhoff Publishers, Dordrecht/Boston/-Lancaster, 1986: 209-228.
4. Vogel RA. The radiographic assessment of coronary blood flow parameters. Circulation 72: 460-465, 1985.

A Pyramidal Approach to Three-Dimensional Reconstruction of a Vascular System from Two Projections

Yongjian Bao

Institut für Technische Informatik
Technische Universität Berlin
FR 3-3, Computer Graphics
Franklinstr 28-29,
D-1000 Berlin 10, FRG

ABSTRACT

A pyramidal approach is developed and presented to treat three-dimensional reconstruction of vascular systems from two orthogonal projections. Through a pyramidal procedure a vascular cross section is refined step by step with improving space resolution. An accurate and very fast reconstruction is attained with this approach. This is especially meaningful for clinical applications and three-dimensional modeling of the coronary arteries.

1. Introduction

In the past decade the application of digital processing techniques in analysis and evaluation of cardiovascular digital or cineangiograms has recieved much attentions, because it can be expected to derive accurate and reproducible quantitative measurements on the severity of the coronary obstructions. Most efforts in this field have been devoted to the development of methods to determine the severity of the coronary arteries in terms of percentage diameter and/or cross-sectional area reduction from single or biplane cardiovascular angiograms [1]. In order to determine the diameter and cross sectional reduction the vascular edges/skeletons and cross sectional area can be acquired by using digital image processing techniques [2] [3].

Recent development in biplane cardiac angiography have made possible techniques to determine the three-dimensional shape and size of coronary arterial segments to a good accuracy. The three-dimensional reconstruction of coronary arterial segments does not only let doctors visualize the morphology of the coronary arterial tree, therefore to determine the severity of coronary obstructions more precisely, but also make an additional number of clinically important parameters available. For example, there is a close relationship between the vascular cross sectional shape and the hemodynamic resistance of an obstruction. The study of vascular three-dimensional morphology and its hemodynamic characteristics may augment our knowledge in searching for the definition of the "critical" stenosis. Finally, a three-dimensional reconstruction of vascular segments is needed to construct an coronary arterial model, which can be very useful in a wide range, from clinical diagnosis to medical teaching.

The long-term objective of our research is to develop a clinical application system based on a medical image workstation, which can aid cardiologists in modeling the three-dimensional structure of coronary arteries. In this paper we present a high speed algorithm to reconstruct the three-dimensional vascular systems.

2. Problem Analysis

X-ray angiograms, DSA (Digital Subtraction Angiography) and film based angiograms, image the coronary arteries after selective injection of an x-ray sensitive contrast agent into one of the main coronary arteries. So far biplane angiograms are widely used clinically, i.e. two orthogonal x-ray projections of the coronary arteries are simutenously obtained. The problem is to reconstruct the three-dimensional coronary systems from the biplane projections. Under the assumption of homogeneous and complete filling of the contrast dye we can define the problem as one to determine the three-dimensional distribution of the dye in the vessels. In an ideal case this is a binary distribution, because the dye is only injected within the vessels, if we do not consider the diffusion across the vascular walls. Because of the fact the absorption of the x-ray phontons is for the greater part due to the dye, we assume our reconstruction to be binary, where for simplicity we define the absorption coefficient of the dye to be unity.

Further the three-dimensional problem can be reduced to a multiple two-dimensional reconstruction by dividing a vascular segment into a stack of parallel cross sections. Here an assumption is implied: the x-rays are transmitted parallel through the vessels. Because of the great ratio of the distance between the x-ray source and the projection plane to the dimension of the space which be occupied with the vessels, this assumption can be held satisfactory.

We focus our attention on the ideal case of parallel rays and homogeneous and complete filling of dye, and state the problem as a reconstruction of a binary matrix from its row and column sums, i.e. in practice the two orthogonal projections.

The problem is thus the estimation of mxn variables from m+n-1 independent equations. In general there is no unique solution to this problem. In order to reduce the ambiguity some restrictions to the object shape [4] [5] or a priori knowledge [6] [7] must be introduced.

In our case a vascular segment is divided into a stack of parallel cross sections. Because of the physiological continuity the spacially adjacent cross sections must resemble each other. Thus a reconstructed neighbouring cross section can be used as a priori knowledge for the currently processed one. To incorporate a resemblance criterion in the reconstruction process a cost matrix with the same size mxn is derived from its neighbour. Through a maximization process the reconstruction problem is optimized according to the neighbour resemblance criterion.

3. Network Flow Approach

As described in section 2, a binary reconstruction with a priori knowledge is treated as a optimization problem. This can be approached by obtaining the maximal flow in a network with minimal cost [7]. The principle of using the network flow approach to solve the reconstruction problem is based on a mathematical isomorphic structure of the two problems [9]. Besides two special nodes, the source and the sink node, there are two groups node in the network. The arcs from the source to the nodes of first group are called source arcs, similarly we can also define sink arcs. The capacities associated with the source and sink arcs describe the two projections and a flow through the arcs between the two groups of nodes gives a reconstructed matrix.

Slump and Gerbrand used this approach to reconstruct the left ventricle [7] and coronary arteries [8]. They have shown that this approach can be very well extended to incorporate the effects of the quantum noise in x-ray imaging [8].

4. A Pyramidal Approach

Although the network approach can satisfactory achieve the reconstruction of vascular cross sections with a priori knowledge, it costs a great computational load, especially in the reconstruction of a large matrix. Therefore we have developed a pyramidal approach to this problem.

The pyramidal approach does not reconstruct a mxn binary matrix directly, as done with all of the other approaches, but instead focuses first on a nonbinary matrix G_1 with low spacial resolution, say $m_1 x n_1$ where $m_1<m$ and $n_1<n$. The matrix G_1 is no longer binary but with k_1 gray levels where

$$k_1 = \frac{n}{n_1} \times \frac{m}{m_1} \qquad (1)$$

The small matrix G_1 can be reconstructed with a modified network approach very quickly. That means the space complex is converted into the gray level complex. In the second step a matrix G_2 with a higher resolution $m_2 x n_2$ and fewer gray levels k_2 than with G_1 is approached by incorporating the matrix G_1 as a priori knowledge. Thus the pyramidal algorithm refines the matrix step by step. Finally an mxn matrix G is achieved.

0 01616 4 0
716161616 5
1012 5161616
0 0 0 11214

a) level 1: $G_1 = 4 \times 6$
$g(i,j) \in \{0,\ldots,16\}$
$1 \leq i \leq 4,\ 1 \leq j \leq 6$

0	0	0	0	1	4	3	4	1	0	0	0
0	0	0	2	4	4	4	4	4	1	0	0
0	1	4	4	4	4	4	4	4	4	1	0
0	4	4	4	4	4	4	4	4	4	4	2
4	4	4	4	4	4	4	4	4	4	4	1
2	2	2	0	0	0	4	4	4	4	4	4
0	0	0	0	0	0	0	2	4	4	4	4
0	0	0	0	0	0	0	0	0	2	4	3

b) level 2: $G_2 = 8 \times 12$
$g(i,j) \in \{0,\ldots,4\}$
$1 \leq i \leq 8,\ 1 \leq j \leq 12$

0	0	0	0	0	0	0	0	0	0	0	1	1	1	1	1	0	0	0	0	0	0	0	0
0	0	0	0	0	0	0	0	0	1	1	1	1	1	1	1	1	0	0	0	0	0	0	0
0	0	0	0	0	0	0	0	1	1	1	1	1	1	1	1	1	1	0	0	0	0	0	0
0	0	0	0	0	0	1	1	1	1	1	1	1	1	1	1	1	1	1	0	0	0	0	0
0	0	0	0	1	1	1	1	1	1	1	1	1	1	1	1	1	1	1	1	0	0	0	0
0	0	0	1	1	1	1	1	1	1	1	1	1	1	1	1	1	1	1	1	1	0	0	0
0	0	1	1	1	1	1	1	1	1	1	1	1	1	1	1	1	1	1	1	1	1	0	0
0	1	1	1	1	1	1	1	1	1	1	1	1	1	1	1	1	1	1	1	1	1	1	0
0	1	1	1	1	1	1	1	1	1	1	1	1	1	1	1	1	1	1	1	1	1	1	0
1	1	1	1	1	1	1	1	1	1	1	1	1	1	1	1	1	1	1	1	1	1	1	0
1	1	1	1	1	1	0	0	0	0	0	1	1	1	1	1	1	1	1	1	1	1	1	1
0	0	0	0	0	0	0	0	0	0	0	0	0	1	1	1	1	1	1	1	1	1	1	1
0	0	0	0	0	0	0	0	0	0	0	0	0	0	1	1	1	1	1	1	1	1	1	1
0	0	0	0	0	0	0	0	0	0	0	0	0	0	0	0	1	1	1	1	1	1	1	1
0	0	0	0	0	0	0	0	0	0	0	0	0	0	0	0	0	0	1	1	1	1	1	1
0	0	0	0	0	0	0	0	0	0	0	0	0	0	0	0	0	0	0	0	1	1	1	0

c) level 3: $G_3 = 16 \times 24$, $g(i,j) \in \{0,1\}$
$1 \leq i \leq 16,\ 1 \leq j \leq 24$
In the last level a 16x24 matrix G_3 is attained.

Fig.1 A practical 3-level pyramidal procedure

Fig.1 gives a practical 3-step pyramidal procedure. In order to reconstruct a 16x24 binary matrix G a small one G_1 with the lower resolution 4x6 is computed in 16 gray levels. In the following steps the matrix G_1 is refined with a greater resolution: a 8x12 matrix G_2 is derived by using G_1 as the initial condition. The 16x24 matrix G_3 is already binary and forms a result to the reconstruction.

In each step a modified network flow alogrithm is empolyed as a core facility of the pyramide. The network flow algorithm is so extended, such that it can also treat nonbinary flow matrices [9].

A flow chart of the pyramidal algorithm is given in Fig.2.

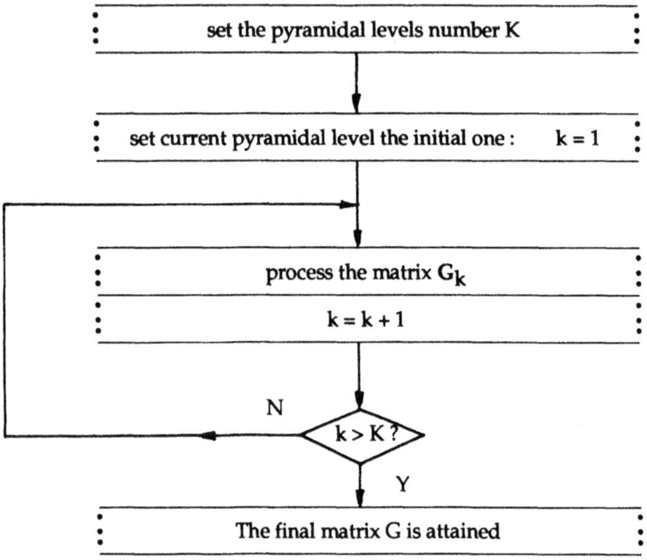

Fig.2 Flow chart of the pyramidal algorithm

5. Results

We have used the pyramidal approach and the network flow approach to respectively reconstruct several simulated vascular cross sections. Both reconstructed the cross sections subject to the resemblance criterion of their reference models. A comparision of the two approaches is performed in terms of the CPU time needed to reconstruct the same cross sections. The results are shown in table I.

Table I. CPU Time Performances of the two Approachs

Matrices	16x24	29x35	26x36
CPU time			
Pyramide	0.663 sec	2.58 sec	1.56sec
Network Flow	10.0 sec	57 sec	63sec

In spite of a few number of holes in some reconstruted cross sections the speed and accuracy performaces of the pyramidal approach are very satisfactory. The holes will be eliminated through improvments of the algorithm.

6. Conclusions

The pyramidal approach to three-dimensional reconstruction of the vascular systems has been presented.

The approach achieved the reconstruction with a good accuracy. As can be seen in table I, it is considerably faster than the network flow approach, so it is very satisfactory for the clinical applications, especilly for three-dimensional modeling of the coronary arteries, by which a large number of vascular cross sections are to be reconstructed. Due to the good structure of the pyramide, either the extended network flow approach or other possible approach can be used as its core facility. Through exploiting the core facility's performace a precise and fast reconstruction is obtained with the pyramidal algorithm.

7. Acknowledgments

This research was supported through the Beschäftigungsplanungsmittel at the Technische Universität Berlin.

The author thanks Mr. Prof. Dr. H.U. Lemke in the Institut für Technische Informatik, Technische Universität Berlin for his support in this research. I would also like to thank Mr. Dr. H. Oswald in the Abteilung Innere Medizin und Kardiologie, Deutsches Herzzentrum Berlin for his helpful discussions.

8. References

[1] Reiber, J.H.C et al.: Transfer Functions of the X-Ray-Cinevideo Chain Applied to Digital Processing of Coronary Cineangiograms, in **Digital Imaging in Cardiovascular Radiology**, International Symposium Kiel, Apr. 1982, 89-104

[2] Thinh V. Nguyen et al.: Computing the Skeleton of Coronary Arteries in Cineangiograms, **Computers and Biomedical Research** Vol.19, 1986, 428-444

[3] Koichi Kitamura et al.: Estimating the 3D Skeletons and Transverse Areas of Coronary Arteries from Biplane Angiograms, **IEEE Trans Med. Imag.** MI-7, Sept. 1988, 173-187

[4] Chang S.K. and Chow C.K. : The Reconstruction of Three-Dimensional Objects from Two Orthogonal Projections and Its Applications to Cardiac Cineangiography, **IEEE Trans Comput**, C-22, 1973, 18-28

[5] Chang S.K. : The Reconstruction of Binary Patterns from Their Projection, **Communications of the ACM**, Vol.14, Jan. 1971, 21-25

[6] Onnasch D.G.W and Heintzen P.H. : A New Approach to Reconstruction of the Left or Right Ventricular Form from Biplane Angiographic Recordings, **Proc. 1976 IEEE Conf. Computers in Cardiology**, 67-73

[7] Slump C.H. and Gerbrand J.J. : A Network Approach to Reconstruction of the Left Ventricle from Two Projections, **Computer Graphics & Image Processing**, Vol.18, 1982, 18-36

[8] Gerbrand J.J. and Slump. C.H. : 3-D Reconstruction of Homogeneous Objects from Two Poisson Distributed Projections, **Pattren Recognition Letters**, Vol.3, 1985, 137-145

[9] Yongjian Bao : Untersuchung und Implementierung der Dreidimensionalen Gefäßsysteme aus Biplanaren Projektionen, Technical Report 89-4, FB 20, TU Berlin, Feb. 1989

A Knowledge-Based System for the Labeling of the Coronary Arteries

Carl Smets, Paul Suetens[1], Andre Oosterlinck, Frans Van de Werf
ESAT-MI2, Department of Electrical Engineering

Department Of Cardiology, UZ Gasthuisberg
Katholieke Universiteit Leuven, Kardinaal Mercierlaan 94
3030 Heverlee, Belgium

Summary

In this paper we will describe a rule-based image analysis expert system for labeling of the coronary arteries on standard RAO and LAO projections. Attention will be focused on the combination of a procedural and declarative programming style providing a general framework for the formalization of most high level coronary anatomical knowledge. The system uses a constraint satisfaction approach to eliminate impossible or unlikely interpretations. Mainly, because most anatomical knowledge can be formalized as constraints on local attributes such as position, grey value, thickness and orientation and as constraints on relations between blood vessel segments such as "left" or "in the same direction". Final interpretation results on clinical coronarograms will be shown.

Introduction

Last years there has been a major increase in the interest of cardiologists for quantitative coronary artery analysis. Although most systems are based on a manual or semi-automatic delineation of the blood vessels, some papers address the usefullness of a fully automatic coronary artery interpretation systems. Most elaborate blood vessel interpretation systems are based on a sequential labeling strategy: First, extract a connected blood vessel tree network. This can be done manually [1],[2], with a simple center line detector [3], or with rather elaborate segmentation and grouping rules [4],[5],[6]. Second, name the most important blood vessel, e.g. the Main Left Coronary Artery or MLCA. Finally, label the consecutive side branches according to their position, direction and greyvalue with the previous labeled blood vessel segments.

The most important criticism is that these approaches will only take into account connectivity relations. For example, once the MLCA is defined in Right Anterior Oblique (RAO) Projection, the connected segment which will run downwards is mostly identified as a part of the Circumflex, while the segment running to the right is identified as the

[1]Paul Suetens is also associated with the National Fund for Scientific Research (NFWO)

Left Anterior Descending (LAD) branch. The main problem is that, due to segmentation errors, a correctly connected blood vessel tree network is not always provided. As a consequence, a sequential labeling strategy starting from the most important blood vessel and making interpretation decisions only relying on connectivity relations is unsufficient for unambiguous blood vessel labeling. Therefore, we propose a general framework for blood vessel labeling where radiological knowledge, general blood vessel knowledge and anatomical knowledge can be easily implemented in. This framework will be discussed in the next sections with a specific focus on the use of coronary anatomical knowledge.

Delineation of blood vessels

In [7] a blood vessel segmentation algorithm based on models which contain common sense knowledge and general blood vessel knowledge has been discussed in detail. The key points can be summarized as follows: First, a maximum intensity detector gives a raw estimate of the center line of a blood vessel. Then, a region growing algorithm hypothesizes a "bar" segment by specifying two parallel lines on each side of the center line. If those lines are not found, if they are not parallel or if they are too far apart, the region is not considered as a part of a blood vessel. In a next stage, we use geometrical and topological knowledge (e.g. bars are not too far apart, bars are in the same direction, vessels are continuous structures) to combine bars into blood vessel segments, defined as blood vessels without crossings nor branches. These blood vessel segments are represented both in a symbolic and iconic way. In the symbolic representation the number of the segment, the begin and end point, the average intensity, the average thickness and the length are stored. This representation provides a fast and easy access to some global characteristics. In the iconic or isomorphic representation we write all the points of the blood vessel skeleton in a label image where the intensity of the pixels correspond to the identification number of the segment. With this representation, there is a direct structural and metric relation to some of the properties of the domain being represented. Answers to questions as "What is the nearest blood vessel?", "How far away is it?" or "What is the orientation?" are implicit to the iconic representation. Figure 1 is an original 1024*1024*10 bit digital coronary angiogram in the RAO view, recorded with a Siemens' Polytron. The delineated blood vessel segments are shown on the right side.

Recognition of the coronary tree using a 2D relational model

The symbolic and iconic descriptions of the delineated blood vessel segments then become the input of the knowledge-based labeling system. In contrast with most other approaches, we consider blood vessel labeling as a constraint satisfaction problem, because it seems that most coronary anatomical knowledge can be formalized as constraints on local attributes such as position, greyvalue, thickness and orientation and as constraints on relations between blood vessel segments. Therefore, we developed a model of the Left Coronary Artery (LCA) in RAO and LAO projection which contains in a rule-based

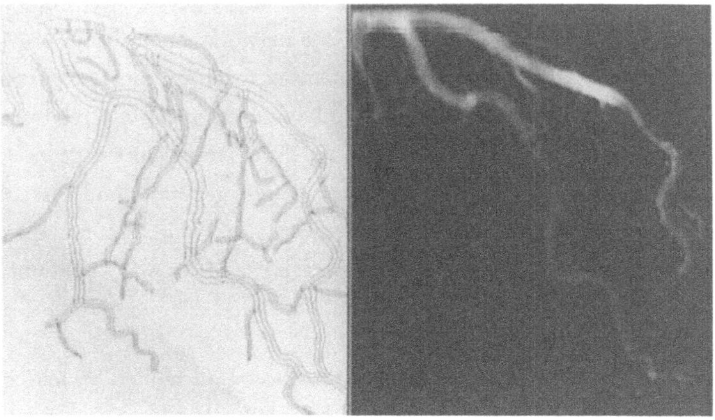

Figure 1: original RAO angiogram with delineated segments

form the coronary anatomical knowledge. For the left coronary circulation, the following interpretations were possible: Main Left Coronary Artery (MLCA), Left Anterior Descending Branch (LAD), Circumflex (C), Atrial Circumflex (ACx), Posterior Descending (PD), Sinus Node Branch (SN), Diagonal Branch (D), Septal Branch (S), Postero-lateral Branch (PB), Obtuse Marginal Branch (OM).

The constraint relaxation approach we adopted can be formulated as follows:
Given is a set V of n variables $\{v_1, v_2, ..., v_n\}$ representing blood vessel segments. Associated with each variable v_i is a domain D of possible interpretations. On some specified subsets of those variables there are constraint relations given that are subsets of the Cartesian product of the domains of the variables involved. The set of solutions is the largest subset of the Cartesian product of all the given variable domains such that each n-tuple in that set satisfies all the given constraint relations. The set of solutions is never empty but a possible solution does not have to be unique, because there might be a considerable overlap between different interpretations. For example, depending on the cardiologist, a postero-lateral branch can sometimes be identified correctly as a Posterior Descending Branch. The main reason for multiple solutions is that even for the medical community the set of possible interpretations is not defined unambiguously.

A set $\{x_1, x_2, ..., x_m\}$ with x_i the interpretation of the ith segment is a possible solution if the next formula assigns to TRUE.

$$Q = P_1(x_1)\&P_2(x_2)\&...\&P_m(x_m)\&P_{12}(x_1, x_2)\&...\&P_{m-1,m}(x_{m-1}, x_m) \quad (1)$$

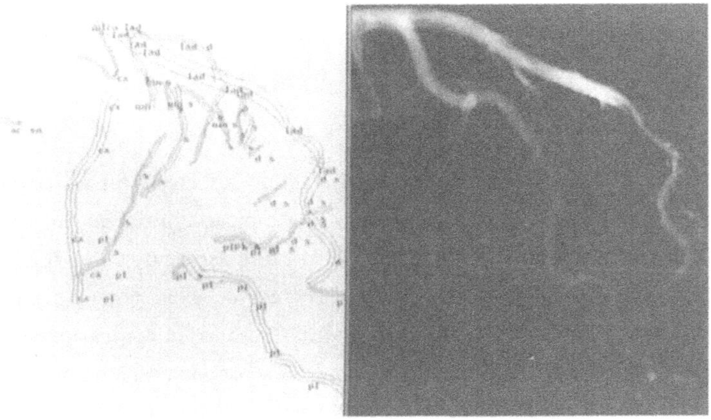

Fig 2: Segments with two interpretations

Here, $P_i(x_i)$ is the boolean truth value for a possible interpretation of the ith segment. In this application, only unary and binary predicates are considered. The unary constraints are related to specific blood vessel attributes such as position, direction, greyvalue and thickness in a specified projection geometry. For example, in RAO projection a septal branch can be identified as a vertical, mostly thin branch in the upper part of the image while the MLCA is a horizontal, thick blood vessel in the upperleft part of the image. These interpretations will be skipped for segments which will not obey at least one of these criteria. The binary constraints are related to specific blood vessel interrelations such as left of, above, thicker than, left connected to and same direction.

Results and discussion

The models discussed in the previous section contribute to the final scene interpretation in the following way: In a first phase, all delineated blood vessel segments are initialized to all possible interpretations. In a next stage unlikely interpretations are skipped according to the unary (attribute) constraints discussed above. These constraints already will rule out a considerable amount of possible interpretations. In general, only 4-5 interpretations remain for each segment. Binary (relational) constraints are not used at this moment because otherways, due to imperfect segmentation, inexact interpretations could relax through the network. Nevertheless, the interpretation result can be improved considerably by hypothesizing likely segments. For example, in most cases the Circumflex in RAO projection can be identified unambiguously, because it is a vertical, thick segment in the leftmost part of the image. By hypothesizing the most likely candidate for the Circumflex, relational constraints will further skip inconsistent interpretations for other segments. In general, only 1-2 interpretations are left, which is an encouraging result if you taken into account that for the RAO projection there is a considerable overlap between the Diagonal and LAD branches. Even cardiologists hardly distinguish these interpretations on a single projection. Blood vessel segments with only one or two interpretations left are depicted in Fig 2.

The main advantages of the constraint satisfaction approach for blood vessel labeling are the following: First, almost all coronary anatomical knowledge can be implemented as unary constraints only depending on the specific segment, or as binary constraints expressed as relations between blood vessel segments. This general framework can be easily modified or updated with new constraints. Second, in contrast with most existing approaches, the labeling does not only depend on a connected blood vessel tree network, where a sequential strategy names the next vessel according to the relations with the previous vessel. Complex relations of all kind can be established between blood vessel segments, thus ruling out unlikely interpretations.

Acknowledgments

The preceding text presented research results of the Belgian National incentive-program for fundamental research in Artificial Intelligence, initiated by the Belgian State - Prime Minister's Office - Science Policy Programming. The scientific responsability is assumed by its authors.

References

1. S Tsuji, H Nakano, "Knowledge-based identification of artery branches in cine-angiograms: an image understanding system which utilizes production-type knowledge," in Computer Science and Technologies, T Kitagawa, ed., North-Holland, 311-321 (1982).

2. HQ Wang, RT Ritchings and ACF Colchester, "Image Understanding System for Carotid Angiograms", Image and Vision Computing, vol5, no2, 79-83 (1987).

3. JL Elion, SE Nissen, "A knowledge-based image processing system for the interpretation of coronary arteriograms", Proc of the SPIE, vol 767, 428-432 (1987).

4. JY Catros, D Mischler, "An artificial intelligence approach for medical picture analysis", Pattern Recognition Letters 8, 123-130 (1988).

5. SA Stansfield, "ANGY: A rule-based Expert System for Automatic Segmentation of Coronary Vessels from Digital Subtracted Angiograms", IEEE Transactions on Pattern Analysis and Machine Intelligence, Vol PAMI-8, no 2, 188-199, (1986).

6. C Smets, G Verbeeck, P Suetens, A Oosterlinck, "Knowledge-based delineation of coronary angiograms", in Signal Processing: Theories and Applications, J.L. Lacoume, ed., 1259-1262, (1988).

7. C Smets, G Verbeeck, P Suetens, A Oosterlinck, "A knowledge-based system for the delineation of blood vessels on subtraction angiograms", Pattern Recognition Letters 8, 113-121, (1988).

Assessment and Analysis of Regional Epicardial Motion from Coronary Cineangiograms

J. MEUNIER, M.G. BOURASSA, M. BERTRAND*, G.E. MAILLOUX*

Institut de Cardiologie, 5000 Bélanger est, Montréal, Canada H1T 1C8

*Institute of Biomedical Engineering
Ecole Polytechnique de Montréal, Montréal, Canada H3C 3A7

Summary

The coronary cineangiograms are used principally to show the coronary circulation (arterial tree morphology, stenosis etc.). However, the coronary arterial tree also provides natural landmarks to appraise the epicardial motion with cardiac contraction. This paper presents a new approach to assess epicardial dynamics, locally, using a motion field description obtained through processing of coronary cineangiographic sequences with an optical flow method. This technique is particularly interesting for the computation of deformation parameters related to local epicardial 2-D extension/compression, rotation and to some extent, wall thickening.

Introduction

In addition to the routine coronary circulation evaluation, cineangiography displays coronary arterial tree motion which is closely related to the epicardial dynamics [1]. By measuring the distance changes between coronary bifurcations, Kong et al. [1] demonstrated the feasibility and diagnostic value of this approach. However, only the one-dimensional changes (linear distance changes between coronary bifurcations) were investigated whereas the epicardium undergoes biaxial deformation.

More recently, computer based methods [2,3,4,5] have been proposed to compute the motion of the arterial tree. The more sophisticated ones use biplane cineangiography with fairly complex algorithm leading to three-dimensional motion estimates that are generally difficult to interpret. However, the operator's assistance is needed to input bifurcation point coordinates.

In our study, an optical flow algorithm, developed in our laboratory and successfully used in echocardiography [6,7], is now used to quantify automatically the interframe motion of the arterial tree. The optical flow is a vector field which describes the two-dimensional motion of each point across the image plane (cineangiogram) in the interval between two successive frames. In our approach, for the small region of interest considered here, the motion of the coronary arteries is described by a linear

2-D field, fitted to the computed optical flow at each step of its iterative computation [6,8]. This allows a decomposition into physically meaningful components such as translation, rotation and deformation [7].

Method

Considering a sequence of several cineangiogram ROI and their associated interframe linear optical flow fields, one easily obtain a linear equation between the position (x_k,y_k) of a point at time t_k and its initial position (x_o,y_o) at time t_o [7]:

$$\begin{bmatrix} x_k \\ y_k \end{bmatrix} = \mathbf{T}_{ko} + \mathbf{M}_{ko} \begin{bmatrix} x_o \\ y_o \end{bmatrix} \tag{1}$$

The vector $\mathbf{T}_{ko}$ is the translation component of the motion while the matrix $\mathbf{M}_{ko}$ is resolved into a matrix product $\mathbf{R}_{ko}\mathbf{D}_{ko}$ represented by:

$$\mathbf{R}_{ko} = \begin{bmatrix} \cos\theta & -\sin\theta \\ \sin\theta & \cos\theta \end{bmatrix} \quad \text{and} \quad \mathbf{D}_{ko} = \begin{bmatrix} \alpha & \gamma \\ \gamma & \beta \end{bmatrix} \tag{2}$$

which respectively account for rotation and biaxial deformation. The deformation matrix $\mathbf{D}_{ko}$ is an hermitian matrix and can be further decomposed to indicate the extent of the 2-D deformation from the eigenvalues λ_1, λ_2 and its orientation ϕ from the orthogonal eigenvectors, while the rotation matrix $\mathbf{R}$ indicates the angular motion (rotation) θ.

Finally another interesting parameter, λ_3, can be deduced from the previous deformation eigenvalues if we assume myocardial volume conservation during the cardiac cycle [3]:

$$\lambda_3 = 1/\lambda_1\lambda_2 \tag{3}$$

We now summarize the physical meaning of these parameters. The 2-D translation vector $\mathbf{T}_{ko}$ is the motion of the local coordinate system origin. This parameter needs to be interpreted with caution since it is origin dependent. The angle θ and ϕ indicate the local twist of the epicardium and the biaxial deformation orientation respectively. The two deformation matrix eigenvalues show the local 2-D extension/compression in the epicardial plane; the last component λ_3, extracted from the previous ones is related to the deformation perpendicular to the epicardial surface, that is wall thickening at the level of the epicardial surface.

Results

We used routine cineangiograms from a data base available at the Institut de Cardiologie de Montréal to test our optical flow method. The images were digitized using a Pie-data system designed for computer analysis of coronary stenosis. We considered 100 successive ROI (64*64 pixels) showing the circumflex branch (figure 1), from an image sequence of RAO 30° arteriography with interframe time interval of 1/45 second. The images were lowpass filtered to reduce noise and improve the optical flow performance [6].

Figure 2 represents the distribution of the deformation principal orientations ϕ as obtained from the computed eigenvectors. It is easily seen that the preferred orientations (maximum occurrence) are somewhere around +60 degrees for the first axis and therefore -30 degrees for the second axis of the biaxial deformation, in agreement with the short and long axis of the heart with the RAO projection. The 90 degrees difference between the two axis, results from a property of the hermitian matrix. Notice that artifactual results would appear when $\lambda_1 \simeq \lambda_2$, a situation where eigenvector angles cannot be identified with confidence.

The biaxial epicardial deformation parameters λ_1, λ_2 and the resulting wall thickening λ_3 are reproduced on figure 3 as a function of time (frame number). Those parameters exhibit cyclic variations and consistent relationship with systolic and diastolic phases of the cardiac cycle. In this example, the first deformation component, λ_1, mainly oriented along the long axis of the heart, shows a peak to peak variation of the order of 10% relative to end-diastole (ED), while its counterpart, λ_2, oriented along the short axis, is less important with 5% changes. The local wall thickening amplitude, derived from the previous eigenvalues, is 15%. These curves are qualitatively very similar to the one previously obtained manually [1,3].

Translation and rotation components were also investigated. The translation of the epicardium as a function of time showed cyclic behavior similar to the ones presented on figure 3 while rotation exhibited a relatively small amplitude of 4 degrees.

Discussion

In this paper, the optical flow tracking of the coronary arterial tree was investigated. Those preliminary results have indicated the capability and potential of the method to characterize locally the time course of the epicardial dynamics. This approach is

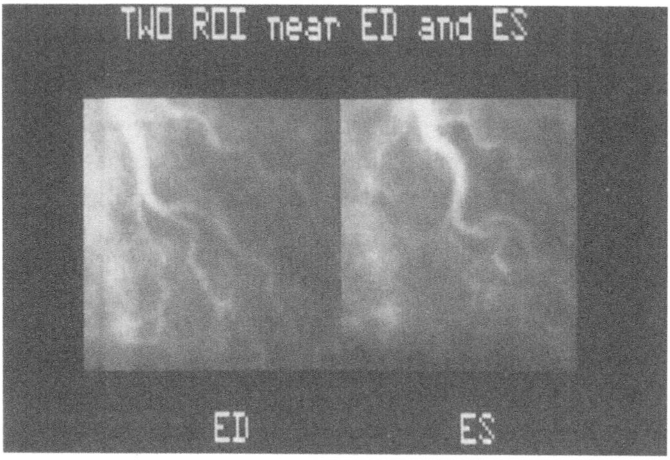

Figure 1. An epicardial region of interest (ROI) that includes a part of the circumflex branch, showing coronary motion from end-diatole (ED) to end-systole (ES). The projection is right anterior-oblique (RAO 30^{0}).

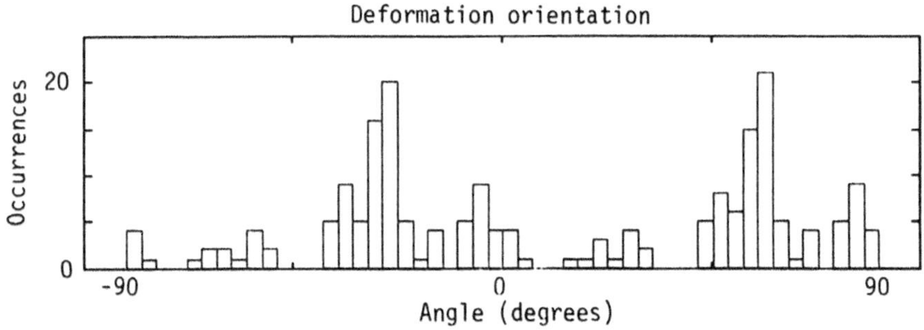

Figure 2. Distribution of deformation orientations ϕ for a sequence of 100 frames. Maximum occurrences are found in agreement with the heart principal axis.

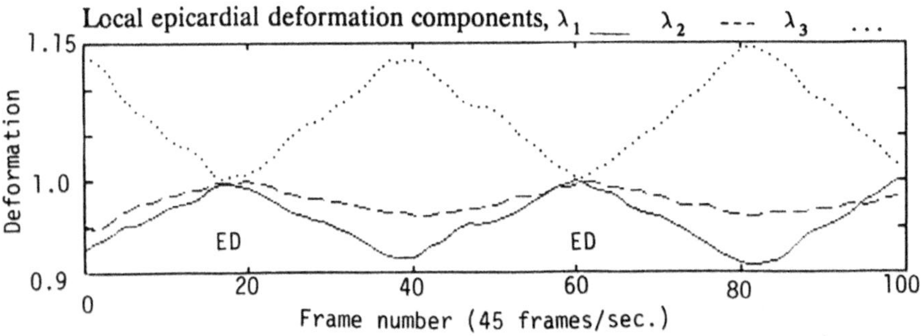

Figure 3. Local epicardial deformation components, λ_1, λ_2 and the resulting local wall thickening λ_3 as a function of frame number (time sampling:45 frames/sec.) (ED = 1).

particularly interesting in the case of deformation parameters which should be able to show normal/abnormal cyclic variations for the local epicardial extension and compression [1]. The method does not require to input bifurcation point coordinates. Finally we believe that this technique is promising for the study of the local biomechanical deformation of epicardium, being complementary to ventriculography which reflects endocardial motions.

Acknowledgments

This work was supported by the Canadian Heart Foundation and Natural Science and Engineering Research Council of Canada. The authors would like to thank Michel Doucet for his help in the present study.

References

1. Kong, Y.; Morris, J.J.; McIntosh, H.D.: Assessment of Regional Myocardial Performance from Biplane Coronary Cineangiograms. Am. J. Cardiology, 27 (1971) 529-537.

2. Potel, M.J.; Rubin, J.M.; Mackay, S.A.: Methods for Evaluating Cardiac Wall Motion in Three Dimensions Using Bifurcation Points of the Coronary Arterial Tree. Invest. Radiol. 18 (1983) 47-57.

3. Kim, H.C.; Min, B.G. et al.: Estimation of Local Cardiac Wall Deformation and Regional Wall Stress from Biplane Coronary Cineangiograms. IEEE Trans. Biomed. Eng. BME-32 (1985) 503-512.

4. Perry, R.A.; Wankling, P.F.; Shiu, M.F.: Measurement of Regional Myocardial Shortening by Digitizing Routine Diagnostic Coronary Arteriograms. Computer in Cardiology (1986) 625-628.

5. Parker, D.L.; Pope, D.L. et al.: Three-Dimensional Reconstruction of Moving Arterial Beds from Digital Subtraction Angiography. Comput. Biomed. Res. 20 (1987) 166-185.

6. Mailloux, G.E.; Langlois, F.; Simard, P.Y., Bertrand, M.: Restoration of the Velocity Field of the Heart from Two-Dimensional Echocardiograms. IEEE Trans. Medical Imaging. (1989).

7. Bertrand, M.; Meunier, J.; Mailloux, G.E. et al.: Measurements of soft tissue deformation using echographic speckle tracking. CAR'89 Computer Assisted Radiology, Springer Verlag (1989).

8. Kanatani, K.I.: Structure and Motion from Optical Flow under Orthographic Projection. Computer Vision, Graphics and Image Processing. 35 (1986) 181-199.

Slice-Wise Three-Dimensional Reconstruction of Myocardial Contrast Perfusion from Simulated Biplane Views

A.C.M.DUMAY, *A.P.ZIJDENBOS, I.M.F.PINTO, *J.J.GERBRANDS, G.J.LAARMAN, P.W.SERRUYS, J.H.C.REIBER

Thoraxcenter, Erasmus University and University Hospital Dijkzigt, Rotterdam, *Department of Electrical Engineering, Delft University of Technology, The Netherlands.

SUMMARY

Slice-wise three-dimensional reconstruction of the myocardial muscle from biplane views is an ill-posed problem. We have developed two algorithms to solve this problem: the segmental reconstruction technique (SRT) and the network programming reconstruction technique (NPRT). Both techniques allow the incorporation of **a priori** information of the myocardial perfusion, for the SRT derived from anatomically related models, and for the NPRT from the adjacent slice. The first slice of the NPRT approach is defined by the SRT. Both methods will be described briefly and have been evaluated on the basis of simulated data.

1. INTRODUCTION

The slice-wise reconstruction problem is defined as $A\underline{x}=\underline{b}$, with $\underline{b}$ representing the logarithmically transformed, mask-mode subtracted density distribution in corresponding biplane scanlines, A the projection matrix and $\underline{x}$ the slice under reconstruction. This system, however, is ill-posed. To limit the number of feasible solutions, **a priori** information on the coronary and left ventricular anatomy as well as the myocardial perfusion must be included in the reconstruction process. The NPRT approach [1] allows the use of the neighboring reconstructed slice as **a priori** perfusion model. For the very first slice the SRT [2] is used. Within the SRT **a priori** information may be incorporated from a selected anatomical model. Both techniques have been evaluated on a segmental basis with similar computer-simulated slices and geometry, with and without the use of **a priori** information. The geometry is assumed to be concentric elliptical, although any other geometry may be used.

2. SEGMENTAL RECONSTRUCTION TECHNIQUE (SRT)

The segmental reconstruction of the average perfusion levels in slices of the myocardial muscle perpendicular to the left ventricular long axis was first proposed by Onnasch et al. [3]. They compared actual density distributions with those of a uniformly distributed model in order to determine which segments in the slice under reconstruction were either normally or abnormally perfused, i.e. a binary outcome. We adopted this idea and extended the technique by determining the segmental reduction in perfusion compared to any model, in a number (≥2) of intensity levels, to be defined by the user. Fig.1 shows an anatomically defined model of a slice selected distal to an obstruction (90% area stenosis) in the Left Circumflex Artery. The relative segmental perfusion levels are indicated, together with the segment separator function (y=mx) and the interval definition (q).

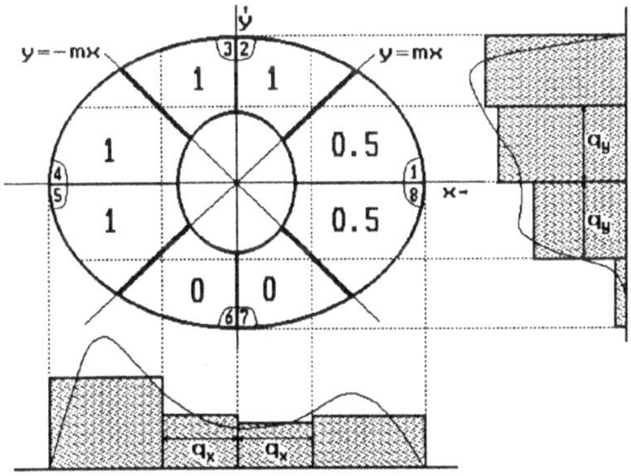

Fig.1 A model and its projections of a slice selected distal to an obstruction (90% area stenosis) in the Left Circumflex Artery, with relative perfusion levels indicated, together with the segment separator function (y=mx) and the interval definitions (q).

The different segments were separated and the projection intervals defined such that a maximal correlation exists between each segment and its projections on the corresponding intervals. The projection data of the model were scaled to the 100% level of the true projections, following which each segment is assigned a perfusion level by an iterative generalized inverse method, including integer computation for grey scale definition. The algorithm has been implemented in the C-programming language on an IBM-compatible PC/AT.

3. NETWORK PROGRAMMING RECONSTRUCTION TECHNIQUE (NPRT)

The reconstruction by NPRT in slices perpendicular to the left ventricular long axis yields a considerably more precise reconstruction of the spatial distribution than the SRT allows. Since each pixel position is uniquely related to its two projections, a capacitated and directed network of nodes and arcs defines the slice under reconstruction [1]. Arcs corresponding to pixel positions not enclosed by the previously reconstructed geometry are excluded from the network. Convex cost-coefficients were used, in stead of the more traditional linear cost-coefficients. Each reconstructed slice is used as a model for the adjacent slice under reconstruction, thus assuming spatial continuity of myocardial perfusion. The very first slice is defined by means of the earlier described SRT.

At present our hardware limits the NPRT to slices of size 32x32 pixels and 8 bits per pixel, although our developments are directed towards a size of 128x128 pixels and 4 bits per pixel. This algorithm has been implemented in the Pascal programming language on the IBM-compatible PC/AT.

4. EXPERIMENTAL RESULTS

The first evaluation of both techniques was performed on computer-generated slices with concentric elliptical geometry. Therefore, an anatomically relevant model of a slice selected distal to a proximal obstruction (90% area stenosis) in the Left Circumflex Artery was disturbed by random noise in each segment, orthogonally projected and reconstructed from its own projections. Both reconstruction techniques were applied to the same projection data. Two situations were distinguished: complete lack of **a priori** information (uniform **a priori** perfusion model), and the availability of **a priori** information by using the anatomical model. The Ti's and the Ri's (i=1,...,8) in the Tables 1, 2 and 3 indicate the mean perfusion levels in segments i of the 'true' slice (T), from which projections are computed, and of the reconstructed slice (R), respectively.

In Tables 1 and 2 the results of the reconstruction for the SRT and NPRT approaches without the use of **a priori** information (uniform **a priori** perfusion model) are presented, whereas in table 3 the data for the NPRT with information included (model from fig.1) are shown. Similar experiments with the SRT yielded correlations better than 0.99. Perfusion values are on a 16 levels grey scale. Biases of corresponding means of input (T) and output (R) segments, and the random errors are computed by the paired Student-test (t-test) for the NPRT-approach. It is clear that the use of **a priori** information reduces both the biases and the random errors.

Table 1 Correlation matrix obtained without using the expected perfusion distribution by the SRT.

	T1	T2	T3	T4	T5	T6	T7	T8
R1	**0.92**	-0.04	-0.10	0.19	-0.26	-0.02	0.00	0.07
R2	-0.03	**0.49**	0.26	-0.33	-0.28	-0.16	0.14	0.00
R3	-0.07	0.24	**0.47**	-0.31	-0.25	0.11	-0.14	-0.01
R4	0.08	-0.29	-0.28	**0.41**	0.30	0.01	0.03	-0.24
R5	-0.30	-0.30	-0.28	0.24	**0.46**	0.00	0.01	0.18
R6	-0.07	-0.44	0.32	-0.06	0.02	**0.81**	0.07	-0.01
R7	-0.02	0.37	-0.38	-0.05	-0.05	0.06	**0.82**	0.03
R8	0.02	-0.05	-0.11	-0.26	0.18	0.00	0.00	**0.92**

Table 2 Mean values, standard deviations, biases of corresponding means and random errors for corresponding segments in 2759 repeated experiments with the NPRT, without use of **a priori** information.

Var	Mean	Std Dev	Var	Mean	Std Dev	Bias	Error
T1	8.65	2.19	R1	9.64	1.74	0.99±0.01	0.91
T2	15.11	1.64	R2	14.14	1.47	-0.96±0.01	0.69
T3	15.09	1.65	R3	15.46	1.01	0.37±0.01	0.87
T4	15.06	1.69	R4	14.96	1.30	-0.09±0.01	0.89
T5	15.16	1.54	R5	14.80	1.37	-0.36±0.01	0.92
T6	0.76	0.91	R6	1.36	0.61	0.59±0.01	0.69
T7	0.74	0.93	R7	1.07	0.55	0.32±0.01	0.64
T8	8.63	2.22	R8	8.94	1.84	0.31±0.02	1.16

Table 3 Mean values, standard deviations, biases of means and random errors for corresponding segments in 2593 repeated experiments with the NPRT, with the use of **a priori** information.

Var	Mean	Std Dev	Var	Mean	Std Dev	Bias	Error
T1	8.69	2.29	R1	8.77	1.71	0.07±0.02	0.98
T2	15.08	1.67	R1	15.07	1.57	-0.01±0.01	0.67
T3	15.12	1.64	R3	15.25	1.36	-0.13±0.01	0.67
T4	15.05	1.76	R4	15.03	1.53	-0.02±0.01	0.77
T5	15.09	1.64	R5	15.19	1.49	0.10±0.01	0.83
T6	0.74	0.90	R6	0.85	0.76	0.11±0.01	0.58
T7	0.72	0.89	R7	0.83	0.68	0.11±0.01	0.52
T8	8.62	2.23	R8	8.97	1.84	0.35±0.02	1.06

The networks were reconstructed such that the dependency on the information in the model was minimal, in order to prevent reconstruction of a priori information.

5. DISCUSSION

In this paper we described briefly both the segmental (SRT) and the network programming reconstruction (NPRT) approaches. Although we have not yet performed the techniques on clinical data, it may be concluded from the evaluations on computer-simulated slices that both techniques allow a reliable slice-wise 3D reconstruction, and that the SRT provides a suitable first slice estimation for the NPRT. However, a priori information should be included with great care, to allow sufficient deviations of the reconstructions from to the model, forced by the densitometric projection data. Biases of means of corresponding segments in the NPRT were proven to be acceptably low if no a **priori** information was used, and even better if this information was used with marginal effectivity.

Further evaluations of both methods are directed towards studies of phantoms and the use of clinical data.

This project is supported by the Dutch Heart Foundation under grantno. NHS86-081.

REFERENCES

[1] Dumay ACM, Minderhoud H, Gerbrands JJ, Zijlstra F, Essed CE, Serruys PW, Reiber JHC. Three-dimensional reconstruction of myocardial contrast perfusion from biplane cineangiograms by means of linear programming techniques. **Intern. J. of Cardiac Imaging** 3: 141-152 (1988).

[2] Zijdenbos AP. Segmental reconstruction of myocardial perfusion from biplane angiograms. **Thesis Delft University of Technology**, December 1988 (in Dutch).

[3] Onnasch DGW, Lindeau J, Rooseboom D, Heintzen PH. Steps towards three-dimensional reconstruction of myocardial perfusion from multiple-view arteriography. **IEEE Comput Cardiol**: 265-268 (1987).

Computer Assisted Orthopaedics

Joint Fluid Quantification Using 3-D Processing of MRI Data

A. Heuck, P. Steiger, D. Stoller, C.-C. Glüer, H.K. Genant

Röntgeninstitut der Technischen Universität,Klinikum r.d. Isar, München, FRG
Department of Radiology, University of California, San Francisco, USA

INTRODUCTION

Joint effusions are frequently associated with a spectrum of articular disorders. In inflammatory arthropathies the amount of joint fluid present may be an indicator for the severity of disease, however correlative studies have not yet been performed (1). Thus the volume of joint fluid may yield an important diagnostic factor and provide an indication of therapeutic response. In the past joint fluid has been quantified only by invasive methods based on arthrocentesis (2).
MRI has recently been shown to be well suited for demonstrating articular effusions (3). When T2-weighted sequences are employed, joint fluid is displayed with high signal intensity due to its long T2 relaxation time and can be easily separated from adjacent articular structures.
Quantitative volumetric studies of joint fluid based on MRI have not yet been reported and 3-D techniques have not yet been applied to volumetric assessment in musculoskeletal disease.
The goal of this study was to develop protocols and to test accuracy and precision of non-invasive joint fluid quantification in the knee using 3-D processing of MRI data.

MATERIAL AND METHODS

MR Image Acquisition

Knees were examined using a superconductive 1.5 Tesla MR imaging system with a saddleshaped extremity coil. For maximum signal and high contrast of joint fluid, heavily T2-weighted SE sequences with TR 2000-2500 msec and TE 120 msec were employed. One knee was imaged in both sagittal and axial planes, all other knees were imaged in the axial plane only. Further imaging parameters included 5mm thick contiguous interleaved slices, a 14 cm FOV, 2 NEX, and a 256x256 acquisition matrix. Syringes filled with varying volumes of saline (10 to 20 cc) were taped on the knee and simultaneously imaged. They acted as a calibration standard and test volume.

Image Analysis and 3-D Data Processing

MRI data were transferred to an independent 3-D imaging system and resampled by linear interpolation. Volume segmentation was performed by thresholding on signal intensity. Image segments that did not form part of the joint fluid, but had identical threshold range, were disarticulated interactively by a radiologist on a slice by slice basis (Fig. 1).

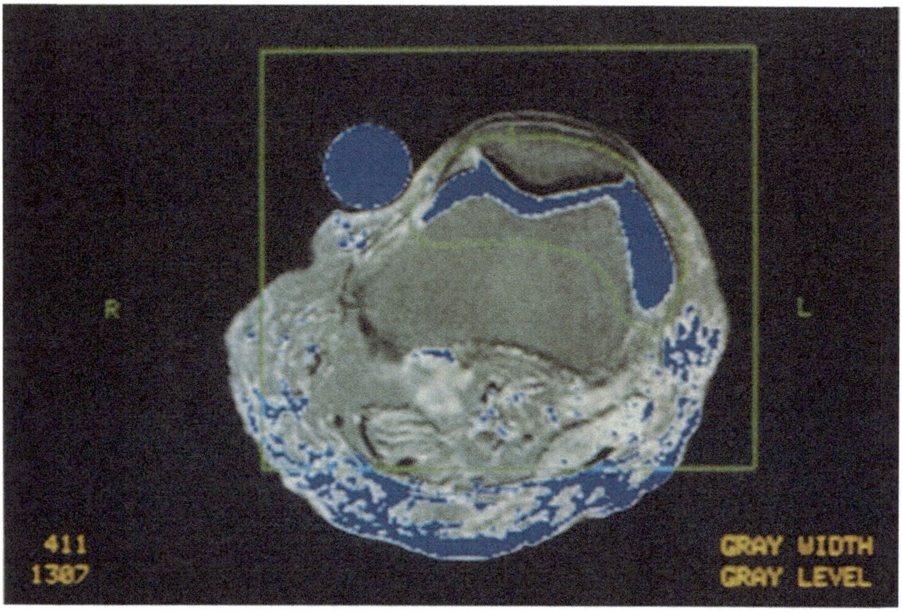

Figure 1 Axial MR slice of a cadaver knee.

Joint fluid and image segments with identical signal intensity are highlighted in blue. Joint fluid is separated by green ROI contour inside the box. Note syringe on medial side of the knee.

Appropriate threshold ranges were determined using the saline of the syringes as a calibration standard. All voxels of the so defined volume were extracted during 3-D processing and the joint fluid volume was quantified in cm^3.

Accuracy Studies on Specimen and Test Volumes

Volume calculations of joint fluid were performed in 5 fresh cadaver knees before and after intra-articular injection of saline, simulating joint effusion. Preinjection volumes were subtracted from post-injection volumes and the differences were compared to the amount of injected fluid. Furthermore, calculated volumes for the syringe content were compared to the true syringe test volumes. The accuracy error was determined as the mean difference between calculated and true values. Error and standard deviation were expressed in per cent.

Precision Studies on Subjects

To determine operator impact on the 3-D volumetric technique, 20 studies of patients with knee joint effusions were analyzed on the 3-D system independently by two radiologists. The precision error was expressed as the mean coefficient of variation (CV) of measurements performed by operator 1 and 2.

RESULTS

Accuracy

In one cadaver knee quantification was performed using sagittal and axial series. In sagittal images the definition of the ROI's for the joint cavity and setting of the signal intensity threshold were confounded by partial volume effects, leading to an 30 % overestimation of joint fluid. Thus sagittal studies were not continued in further investigation. However, a high degree of accuracy could be found using axial images (5 studies). 3-D volumes yielded a mean difference of -2,4 % (SD ± 5,1 %) (Tab. 1).
3-D volume calculations of specified test volumes in syringes that were simultaneously studied showed a mean deviation from true volumes of -0,4 % ± 3,8 %.

Precision

In 20 in-vivo studies, processed independently by two radiologists, their estimated joint fluid volumes indicated a mean coefficient of variation (CV) of 4,5 % ± 2,4 %, representing the precision error (Tab. 1).

TABLE 1. Accuracy and Precision of 3D-Volume Estimates Based on Axial MR-Slices

a. Accuracy	Diff. between estimated and true volumes (mean ± SD)
Joint Fluid	-2.4 % ± 5.1 %
Test Volumes	-0.4 % ± 3.8 %
b. Precision	C.V. for estimates of operator 1 and 2 (mean ± SD)
Joint Fluid	4.8 % ± 2.8 %
Test Volumes	2.9 % ± 2.3 %

DISCUSSION

3-D volume estimates based on MRI data of contiguous axial T2-weighted images revealed good accuracy (4). Although the number of accuracy studies on fresh cadaver knees is limited and saline might be expected to show slightly different T2 signal than does the joint effusion (5), our preliminary data suggest that the technique might be adequate for clinical investigation.
The high degree of inaccuracy for volumes calculated from sagittal images can be attributed to partial volume effects due to the relatively small number of slices (14-15 as compared to 18-26 in axial studies) and their orientation with respect to the anatomy, and the related problems of properly defining ROI's.
Studies on reproducibility (precision) of 3-D processed volumes, performed by two radiologists, revealed a relatively small precision error, suggesting limited operator influence on the outcome of volume estimates, when segmentation is performed by use of a calibration standard.

CONCLUSION

Given these data, we conclude that quantification of joint fluid by MRI using 3-D data processing may provide a promising non-invasive tool for clinical studies when heavily T2-weighted sequences and 5mm thick contiguous axial slices are used.

Literatur

1. Currey HLF, Vernon-Robert B. Examination of synovial fluid. Clin Rheumatic Dis 1976; 2:149-177.

2. Geborek P, Saxne T, Heinegard D, Wolheim FA. Measurement of synovial fluid using albumin dilution upon intraarticular saline injection. J Rheumatol 1988; 15:91-94.

3. Beltran J, Noto AM, Herman LJ, Mosure JC, Burk JM, Christoforidis AJ. Joint effusions: MR imaging. Radiology 1986; 158:133-137.

4. Heuck AF, Steiger P, Stoller DW, Glüer CC, Genant HK. Quantification of joint fluid volume by MR imaging and CT using 3D data processing: a preliminary study. J Comput Assist Tomogr (in press).

5. Baker DG, Schumacher HR Jr, Wolf GL. Nuclear magnetic resonance evaluation of synovial fluid and articular tissues. J Rheumatol 1985; 12:1062-1065.

Acknowledgment:

A.H. was supported by a stipend from the Deutsche Forschungsgemeinschaft.

CAD/CAM/CAO in Total Hip Replacement

G. Aldinger, W. Küsswetter

Orthopädische Universitätsklinik Tübingen
Hoppe-Seyler-Str. 3, D 7400 Tübingen, West Germany

Artificial joints loosen with time. Although the socket fixation is outlined with very encouraging solutions and satisfying results, the stem anchorage still poses an unsolved problem. With the CAD/CAM-technique, we try to approach a solution.

The better a hip stem is fit into the actual stabililzing cavity, the better the expected results. This fact was recognized over 25 years ago. Form filling acrylic cement seemed to solve this problem. The standarized stem received the necessary fit with or better through this filling material. Success was inevitable; the total joint replacement started its triumph. That even such cemented stems loosened over the years, was not due to the form filling function, but, among others, to the unfavorable material characteristics of the acrylic cement.

The once highly praised form filling cement became the scapegoat and cementless anchoring stems were rediscovered, especially for younger patients. The surface of cement was more or less imitated with a porous surface.

The problem of the unsatisfactory fit of standarized stems was answered with numerous, very differently shaped devices in various sizes. The variability of nature, especially the variability of the proximal femoral cavity is far too large to be adequately filled with a realistic and practicable number of confectional stems. Thus far, in order to grant these standardized stems a better fit, the cortical bony cavity had to be drilled, sawed, rasped or shaped. Such a surgical procedure ultimately leads to a weakening of the bearing cortical bone.

The long-termed stability of an artifical joint can only arrise from the bone. Bone stabilizes the artificial joint; bone determines the fate of the implant; bone therefore has priority [1]. Stems should be adapted to bone and not bone to stem. Bone should not be additionally weakened through unnecessary mechanical treatment; on the other hand it should not be too demandingon the stabilizing bone to get the required osseous integration.

Todays modern computer techniques can produce such individual implants. We have been working on this subject since 1981, with the first publication in 1983 [2] and the first human implantation of such fully individual stems in 1985. Since then the idea of the individual prosthesis has convinced - simple as it was: "Stem to bone, not bone to stem". Numerous research groups and projects have been created. What's more, in 1988 even the 1. world congress was held customized implants.

Which problems have to be solved?
It was essential to obtain an exact computer model of the bony cavity, as well as to construct a snug-fitting implantable and if necessary an explantable stem. The surface has to be considered and joint geometry should be adapted to anatomic and/or therapeutic needs.

The question as to material came easier, we used a proven high strength titanium alloy and a tribologically favorable ceramic ball. The production of these titanium stems on computer controlled machines is demanding and expensive. On the whole, one easily confronts financial limits.

A large scale accaompanies our system from the very beginning, from computer tomography until delivery of the device, which also considers the similation of the operative procedure, of the implantation of the stem into bone.

In the beginning, we manufactured each stem twice and tested one in a material testing unit. Computer-assisted testing methods and comparison of data of our meanwhile extensive data bank, make a direct testing only the exception.
Todays computerized imaging procedures enable a good recognition of the outer bony contours, the outer bony shapes. The identification of the bony cavity with the poorly defined transition from the spongious bone, over the trabecular bone to the inner cortex is indeed more difficult (Fig. 1).

A total filling of the bony cavity hinders, on the one hand the implantation and on the other hand the explantation. Implantability, all-round snug fit and explantability therefore demand modification. The computer-assisted design and manufacturing machines, must be supplied with adequate software, the heart of the individual prosthesis [2].

The disk-like surface of our individual stem (Fig. 2) unites the advantages of both recognized and virtually opposing anchoring principles: not only "press-fit" but also "surface enlargement", without having to accept their seperate disadvantages [3]. Through the horizontal load transfering planes, mainly favorable stresses are transfered far proximally, just in the calcar region.

The realisation of such a project demands pragmatic solutions. Our hip patients should not be dependant on a few special centers. In order to guarantee wide-spread medical care, the hospitals and physcians are not envolved in "high-tech". Any surgeon can arrange for the computer tomography to be performed by any radiologist in accordance to our guidelines, after adequate inspection and testing of their units. The "high-tech" remains by the designing engineer and in the production.

Since simple operation techniques normally obtain better widespread results, this device has its own individualy rasp for single use.Any hip surgeon can easily, gently, safely and exactly prepare the bony cavity with such a system.

Thus far over 300 such hips have been performed. The results of the first 100 individual stems are very promising with 96 % good and very good results [4], so now widespread application can can be recommended.

Literature:

[1] Aldinger G., K. Mitzkat (1986), Der Einfluß der Alterung in der Endoprothetik. Z Orthop 124:392-395.

[2] Aldinger G., A. Fischer, B. Kurtz (1983), Computer-aided manufacturing of individual endoprotheses. -Preliminary Communication-. Arch Orthop Traumat 102:31-35.

[3] Aldinger G., A. Fischer, B. Kurtz (1984), Computer-gestützte Herstellung indiduell-anatomischer Endoprothesen. Z Orthop 122:733-736

[4] Aldinger G., The individual cementless stem in THR-Concept and results. Vortrag während des XVII. Weltkongress der Société International de Chirurgie Orthopédique et de Traumatologie vom 16.-21.8. 1987 in München.

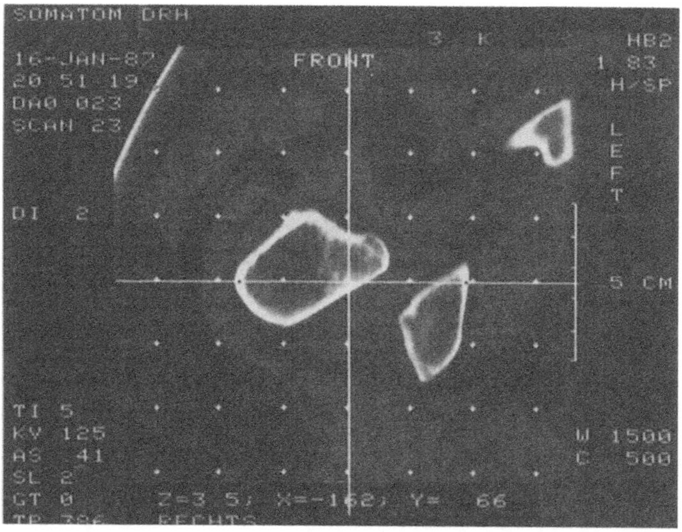

Fig. 1

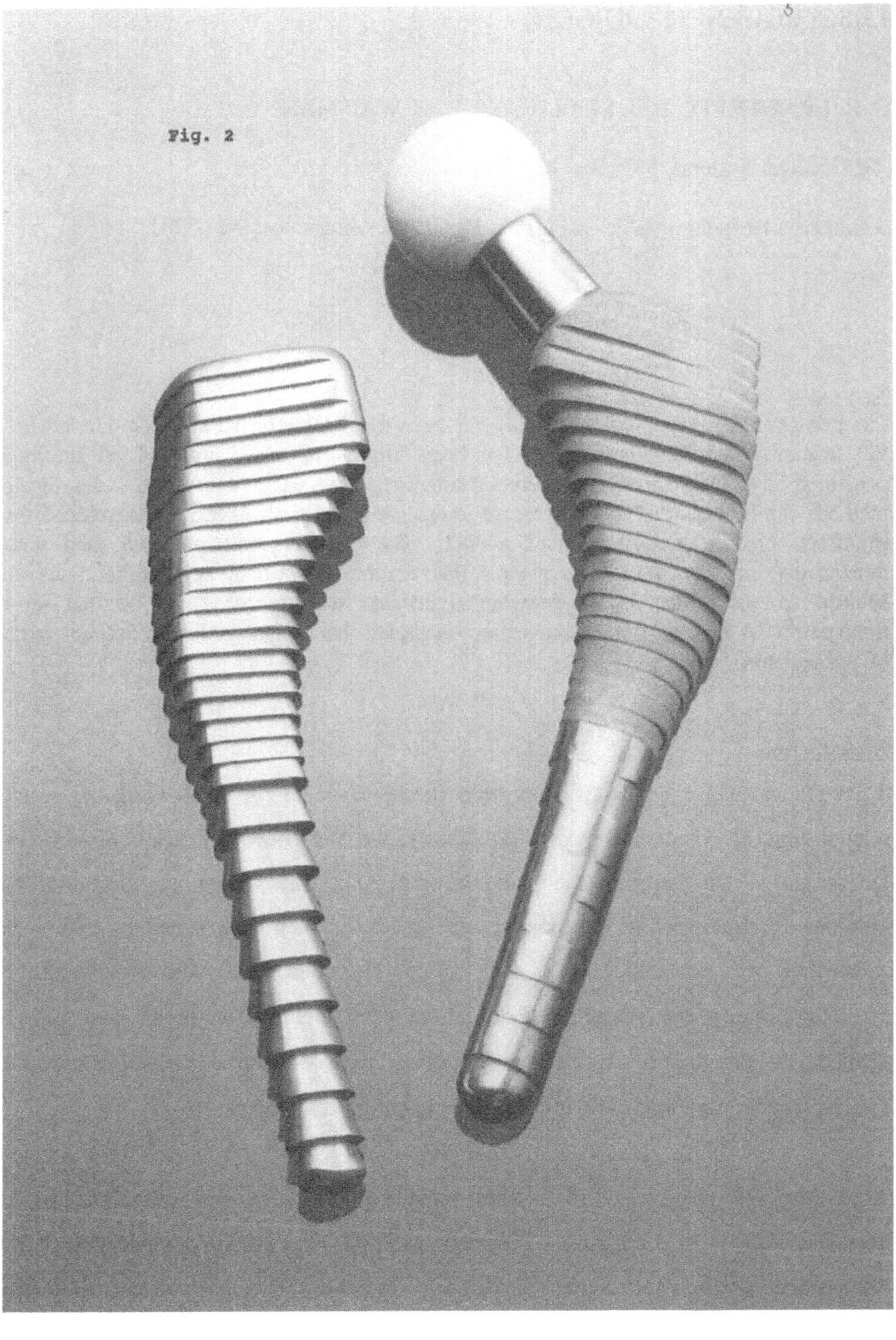

Fig. 2

Three-Dimensional Display of Mathematically Disarticulated Objects

C. R. CRAWFORD[1], R. J. STARSHAK[2], R. C. WAISMAN[2]

[1]GE Medical Systems, P.O. Box 414, Milwaukee, WI 53201

[2]Children's Hospital of Wisconsin, P.O. Box 1997, Milwaukee, WI 53201

Summary

The principal application of 3D display has been the display of the skeleton derived from CT images. The 3D algorithms have been limited to the display of all the bone contained in a user-specified volume-of-interest. This paper shows how the display methods can be extended to disarticulate joints. A set of tools have been developed that, in effect, provide a mathematical scalpel. The primary tools are two- and three-dimensional region growing algorithms that feature anisotropic neighborhood criteria. Multiple passes of the region growing algorithms are applied to the original set of tomograms. A new set of 2D images are generated that contain only the desired portion of the skeleton.

Introduction

Three-dimensional CT (3D-CT) has been shown to be useful in studying the surface relationships of complex anatomy. Previously, we had shown the application of this technology to the study of the pediatric spine [1]. The continuing interest of our orthopedic surgeons led us to study the hip, especially when complicated surgery on the acetabulum was contemplated. In this regard, our colleagues asked if it would be possible to remove the femoral head so that the inner surface of the acetabulum could be studied. For this reason, we began a study of computer techniques that would allow one to disarticulate joints with only minimal interaction from the operator.

Algorithms that display a 3-D surface provide methods to view complex anatomy contained in a set of contiguous CT tomograms [2-7]. The principle application of 3D-CT has been to display the surface relationships of bones. The voxels that correspond to bone can be easily extracted from that axial image with simple thresholding with respect to CT number. In the past, these algorithms have been limited to the display of all the bone that is contained in a user-specified volume-of-interest. It is possible for the user to narrow the volume-of-interest by defining the area to be included on an image-by-image basis. We have found that this amount of user interaction precludes its acceptance in general clinical practice.

This paper will demonstrate how these display methods can be extended to disarticulate joints without extensive user interaction. A set of software tools has been developed that, in effect, provide a mathematical scalpel. These extended algorithms have been applied to disarticulate the hip, knee and ankle. The key functions are two- (2-D) and three-dimensional (3-D) region-growing. We call these programs tools. These tools rely on applying anisotropic neighborhood criteria to the CT images. Multiple passes of these region-growing algorithms are applied to the original set of tomograms. This yields a new set of 2-D images that contains only the desired portion of the skeleton. Typically, only two passes are required and the user has only to supply the location of two seeds from which the growth occurs.

This new set of axial images can then be processed by any 3D surface display algorithm to produce 3D-CT images of the disarticulated structures. In this paper, we use the method of Dividing Cubes [5] to render 3D images. The order in which the tools are applied and the number of applications has been found to depend on the clinical problem and the joint being studied. We have also found that the tools can also be used to segment soft tissue structures from both CT and MRI images.

The tools

3D Region Growing: The user is prompted for a seed (or a number of seeds) that points to a location in a set of slices. The user is also prompted for a criterion that defines a property of interest. In general, the criterion is a range of CT numbers. For the case of bone, the range is replaced with a simple threshold. The seed is then placed on a list. Then a location that initially corresponds to location of the seed is removed from the list. The 3-D region-growing algorithm then checks the values of the voxels that neighbor the removed location to determine if they also satisfy the selection criterion. If they pass the test, the location of the neighbor is added to the list. If they fail the test, growth of the seed in that direction ceases, but the interrogation of neighboring voxels continues. The process is repeated until the list is empty. To prevent infinite loops, a flag is set indicating that the neighbor has been examined when it is placed on the list. Different neighborhood criteria can be used to obtain different results. The routine procedure, however, is to examine voxels that share faces. For objects that possess a thin "skeleton," voxels that share edges and/or points can also be included. In some applications of 3-D region-growing, the neighborhood criterion is anisotropic, that is, it prevents growth along certain axes. It was found that it was common to prevent growth in the distal or proximal directions. A number of options are available after the voxels are marked. The options include: generate the surface corresponding to the marked voxels; replace the marked voxels with a new CT number, e.g., water; or replace all the unmarked voxels that also satisfy the object selection criterion with a new CT number.

Image subsetting: In most cases, the object to be disarticulated is contained in a subset of the original slices. Because the spacing between slices is greater than the pixel size, 3-D region-growing algorithms can easily leak into unwanted anatomy. In order to prevent a region-growing algorithm from leaking into anatomically unconnected regions, only the slices containing the object of interest should be processed. After region-growing has been applied to the subset, the new images are merged back with the complement of subset. A degenerate version of image subsetting is selecting only one image. 3-D region-growing, when applied to only one image, reduces to 2-D region-growing. In typical applications, 2-D region-growing is only required for a small percentage of the original slices.

Manual tracing: Because of the partial volume effect and other related problems, bones can appear to be fused together. This is especially true of portions of joints that are anatomically adjacent like the tibia and fibula near the growth plate. In these situations, only the knowledge of the true anatomy can be used to separate the objects. A manual trace mode has been provided to accomplish this task. The user can outline a region and the computer will change the pixels inside the region to a user-specified value. Alternatively, the user can draw a line that separates two anatomically unconnected regions. In this case, the computer maps the pixels underneath the boundary to some value that will fail the object selection criterion.

Negation: Consider the case of disarticulating the femoral head from the acetabulum. In most applications, we would convert the voxels corresponding to the femur to water and then generate an image of the acetabulum. If we subtract the acetabulum-only slices from the original slices, then only the femur will remain. We call this step negation. The process of negation is really not exactly subtraction. If the pixel values in the original slices do not pass the object selection criterion, like in the case of soft tissue, then the pixels are passed directly to the negated set of slices. This procedure is followed so that a neighborhood of soft tissue is preserved around bones. The soft tissue is required in order to calculate accurate gradient normals [5-7].

Clinical example

A 14-year-old boy injured his right ankle while skate-boarding. A radiograph (Fig. 1) reveals a Salter-Harris IV fracture of the distal tibia with posterior displacement of the distal fragment. A fracture is seen involving the articular surface of the midportion of the tibia. This represents a tri-plane fracture. It is difficult on the basis of the plain radiographs to determine whether this is a two-part or three-part tri-plane fracture. Axial CT (not shown) at the level of the tibial talar joint shows two fracture lines on the articular surface of the tibia. There is a small Tillaux fragment on the posterolateral surface of the distal tibia.

3D-CT (Fig. 2) was performed from the axial CT images. Because the CT number of the cast and bone is similar, the ankle is obscured by the cast. The cast was removed (Fig. 3) by 3-D region-growing with a seed placed in the cast. All the connected voxels were set to water. Because of the layered nature of the cast, edge neighbors were required to pick all of its strands. Because the bones in the ankle were relatively fused, the cast could have also been removed by placing a seed in any bone in the ankle and removing all unmarked voxels after the application of 3-D region-growing. The resulting images clearly show the anterior and posterior components of the tri-plane fracture. The lateral margin remains obscured by the fibula.

The fibula was removed in two sections. Initially, a seed was placed in the proximal shaft of the fibula at the level of the growth plate. 3-D region-growing was performed in the proximal dimension, but growth distally was prevented by image subsetting. The resulting image (Fig. 4) again demonstrates the anterior and posterior portions of the tri-plane fracture. The lateral aspect of the tibia remains obscured by the fibula and a portion of the talus.

Most of the talus was removed by placing a seed at the apex of the tibiotalar surface, applying 3-D region-growing to the talus with growth prevented proximally, and then setting the marked voxels to water. 2D region-growing was required to touch-up artifacts caused by partial volume for the two slices that contained the tibiotalar surface. Restricted-growth 3-D region-growing usually leaves some "floating" voxels behind. Another pass of 3-D region-growing was applied to a seed in the tibia (the unmarked voxels were set to water). The resulting image (Fig. 5) more clearly depict the intra-articular components of the fracture. However, the posterolateral aspect remains partially obscured by the remaining portion of the distal fibula.

The distal fibula was removed by manual tracing in four images to break the surface between the fibula and the tibia. Then 3-D region-growing was applied. The resulting image (Fig. 6) clearly depicts the intra-articular components of this tri-plane fracture. The Tillaux fragment is now clearly seen.

Discussion

Computer disarticulation of 3D-CT images is: feasible with only minimal user interaction; has significant value in surgical planning, especially for procedures on the acetabulum; and its role in evaluating fractures of the acetabulum is promising but, as yet, unstudied. The tools have also been used to extract selected soft tissues from CT and MR slices. We have been able to segment the ventricles from CT slices and the gray matter from MR slices.

<u>References</u>

1. Starshak, R. J.; Crawford, C. R.; Waisman, R. C.; Sty, J. R.: Three dimensional CT of the pediatric spine. Radiology 169(P) (1988) 465.

2. Vannier, M. W.; Marsh, J. L.; Warren, J. O.: Three-dimensional graphics for craniofacial surgical planning and evaluation. Computer Graphics 17 (1983) 263-273.

3. Herman, G. T.; Liu, H. K.: Three-dimensional display of human organs from computed tomograms. Computer graphics and image processing 9 (1979) 1-21.

4. Lorensen, W. E.; Cline, H. E.: Marching Cubes: a high resolution 3D surface construction algorithm. Computer Graphics 21 (1987) 163-169.

5. Cline, H. E.; Lorensen, W. E.; Ludke, S.; Crawford, C. R.; Teeter, B. C.: Two algorithms for the three-dimensional reconstruction of tomograms. Medical Physics 15(3) (1988) 320-327.

6. Weinstein, I. D.: Algorithms for the Display of Surfaces in Three Dimensional Medical Computerized Tomography Data. Marquette University (1986).

7. Santos, E.: Implementation of three-dimensional display algorithms for the display of three-dimensional computerized tomographic images on a large minicomputer. Marquette University (1986).

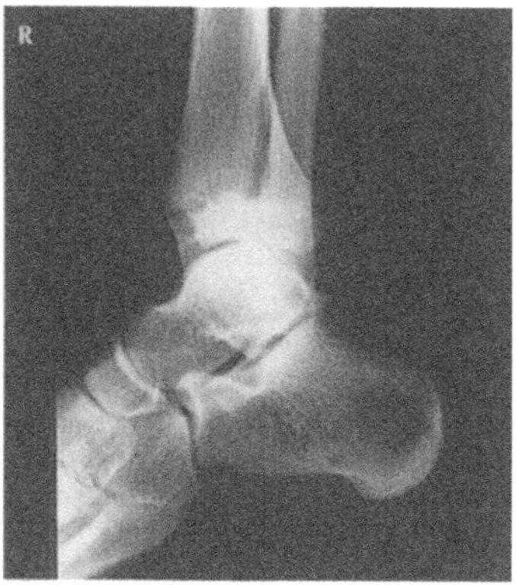

Fig. 1. Radiograph showing a fracture of the distal tibia.

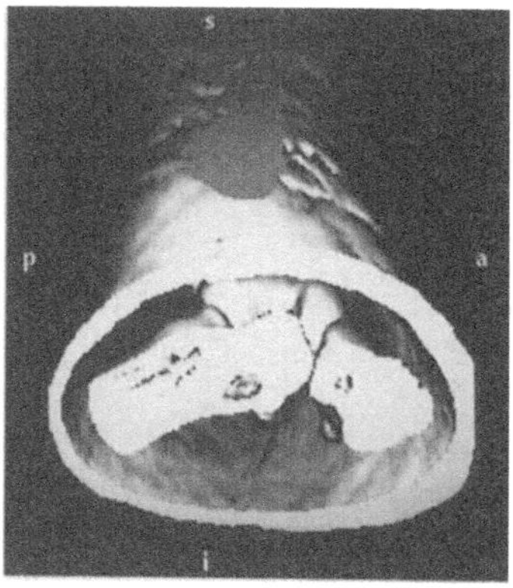

Fig. 2. 3D-CT of the ankle before application of tools.

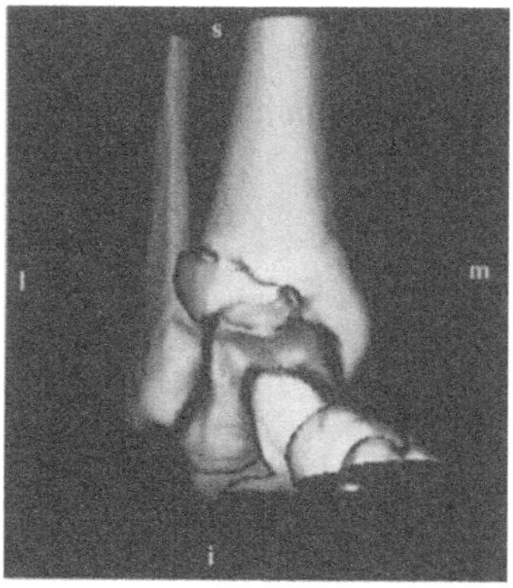

Fig. 3. Image of ankle after cast-removal.

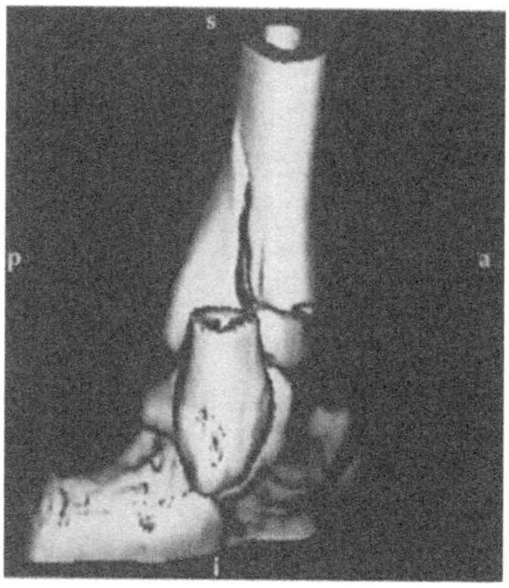

Fig. 4. Lateral view after removal of part of the fibula.

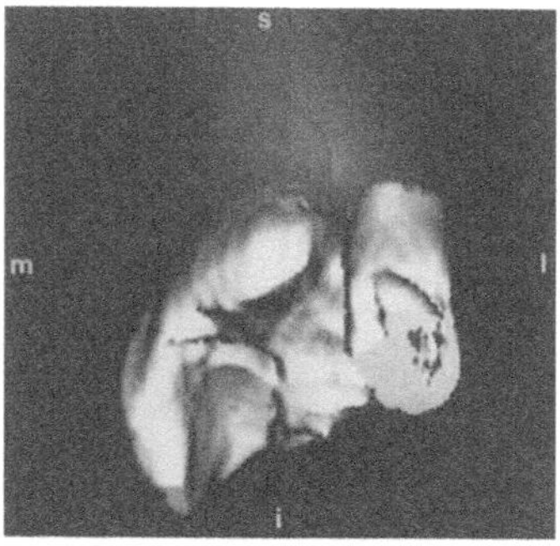

Fig. 5. View of the bottom of the tibia after removal of the talus.

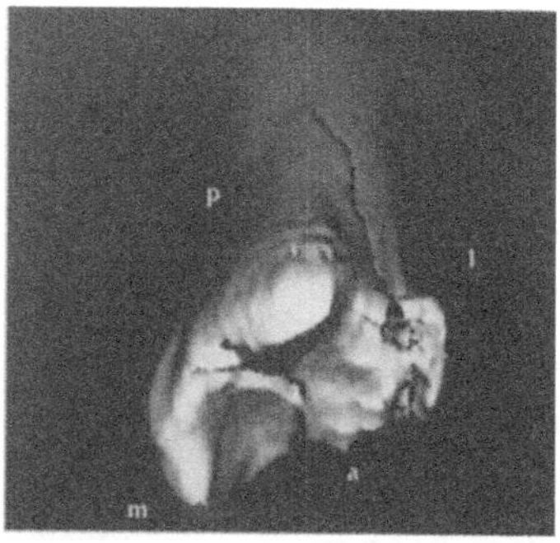

Fig. 6. Image obtained after the distal fibula was removed.

Measurement for the Exact Description of Traumatic Deformities of Individual Vertebrae or Motion Segments of the Thoracic and Lumbar Spine as Basis for a Computer-Assisted Evaluation of X-Rays

H. Daniaux, P. Seykora, E. Beck

Univ.-Klinik für Unfallchirurgie (Head:Univ.-Prof.Dr.E.Beck)
Anichstraße 35, A-6020 Innsbruck, Austria

Summary

Fractures of the thoracic or lumbar spine usually affect only one, seldom two or more vertebral bodies. Similarly, only one or two, seldom more motion segments are normally injured. Traditional measuring methods for scoliosis and kyphosis are not always applied uniformly and so these are increasingly inappropriate for the exact description where the stabilization and fusion of spinal injuries are kept as short as possible. The sagittal index and the kyphotic angle of a vertebra, only provide information about the affected vertebra itself. The adjacent motion segments are not taken into consideration. Therefore, in order to arrive at clear exact description of spinal injuries, we recommend a simple method which permits both a digitalized and a graphic evaluation. In this method, only the corner points of the vertebrae being examined need to be marked in order to obtain the dimensions (SI, KA and the various ULA's) we have recommended.

Injuries of the thoracic and lumbar spine, especially those in the thoraco-lumbar area, most often affect only one vertebra or one or two motion segments (quoted from Junghanns [6]). In contrast to congenital and degenerative malformations, in which arched deformities are predominant, spinal injuries usually involve angular deformities. Most of these are kyphotic rather than scoliotic deviations.
In addition to the restoration of the dynamic and protective functions of the injured spine, the goal of both conservative and operative treatment, from a purely morphological point of view, is the reconstruction of an anatomical spinal axis.

Although in fields such as scoliosis research Cobb's measurement technique is used as an international norm, there are no simple, universally recognized measurement standards for traumatology of the spine. Techniques developed by

Cobb [2], Ferguson [5] and Stagnara [7] are used to quantitatively measure axial deviations but the interpretation of all these techniques can and does vary. The wedge angle (quoted from Debrunner [4]) and Beck's sagittal index [1] are usefull to describe individual vertebrae. Our so-called lower-upper end plate angle (Daniaux [3]), which is formed by the the lower and upper end plates, respectively of the vertebrae immediately adjacent to the injured vertebra, also fails to meet all our requirements.

In order to arrive at a detailed morphological assessment of spinal injuries and a further course of conservative or operative treatment, various dimensions are necessary. These must be derivable from a.p. and lateral X-rays. In chiefly osseous injuries (fractures), the deformation of the injured vertebra from both views, changes in the adjacent motion segments or disc spaces, and axial deviations are of greatest interest. In cases chiefly involving disc and ligament injuries (subluxations and dislocations), the injured motion segment is most important. In order to standarize the follow-up of spinal injuries without taking previous treatment into consideration, we have been using the following dimensions (Fig. 1):

1. Beck's sagittal Index (SI)
2. Kyphotic angle (KA) (quoted from Debrunner)
3. Upper-lower end plate angle (ULA) adapted from Cobb
   a) ULA of the uninjured vertebrae immediately adjacent to the injured vertebra
   b) ULA of the injured and the immediately adjacent cranial uninjured vertebra
   c) ULA of the injured and the immediately adjacent caudal uninjured vertebra

1. SI (Sagittal index)

SI (Sagittaler Index)

This is the quotient of the height of the anterior wall of the injured vertebra over the heigth of its posterior wall. It results in a reproducible and comparable measurement independent of an X-ray enlargement.

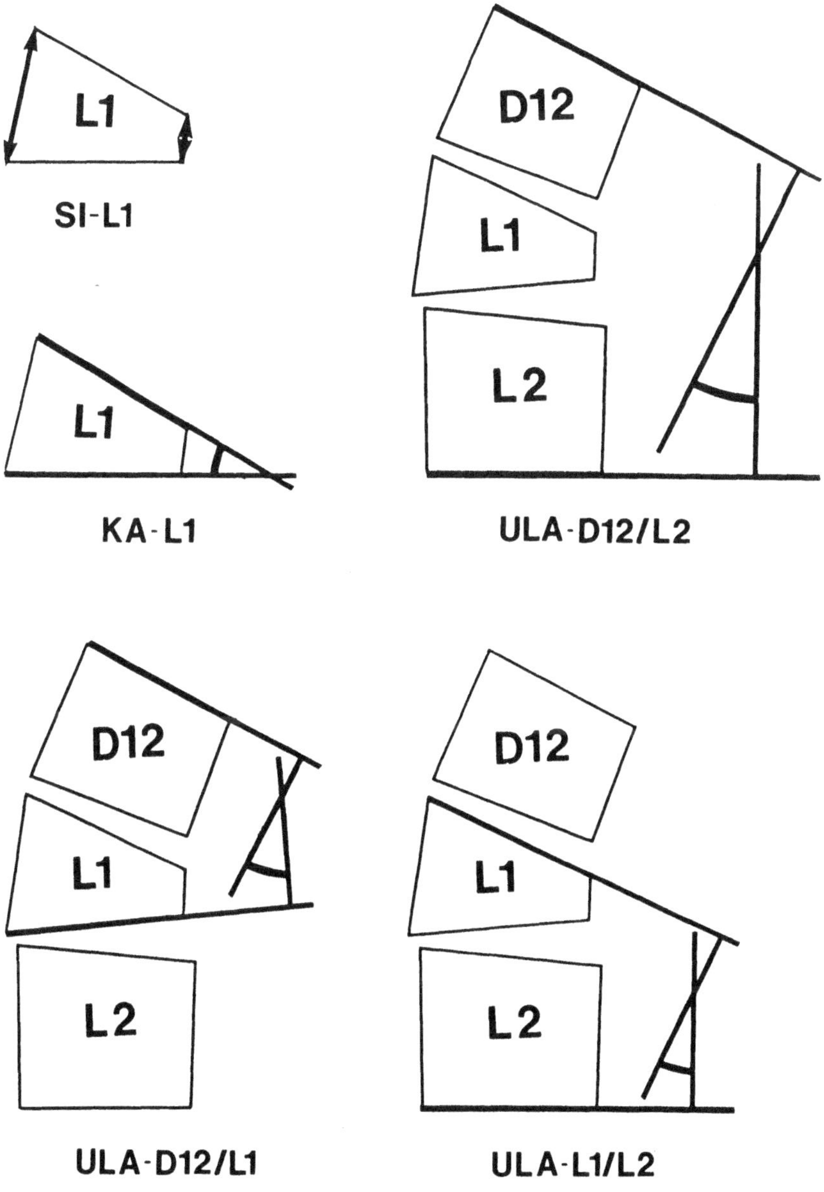

Fig. 1: Measurements (Abbreviations see text)

Example: SI-L1 = 0,3. This figure indicates a wedge-shaped deformation of the L1 vertebra with an anterior wall height reduction to an amount of 30%.

2. KA (Kyphotic angle)

KW (Keilwinkel)

It is derived from the angle between the upper and lower end plates of the injured vertebra. Kyphotic deformations are preceded by a minus sign; lordotic by a plus.

Example: KA-L1 = - 27°. This figure, preceded by minus sign, indicates that the injured vertebra shows a wedge-shaped kyphotic deformation of 27°.

3. ULA (Upper-lower end plate angle)

DGW (Deck-Grundplatten-Winkel)

This is the angle between the upper end plate of a defined cranial vertebra and the lower end plate of a defined caudal vertebra. The use of plus and minus signs is the same as for kyphotic angles.

a) ULA of the uninjured vertebrae immediately adjacent to the injured vertebra

Example: ULA-D12/L2 = - 27°. This figure indicates that the uninjured vertebrae D12 and L2 including the motion segments or disc spaces D12/L1 and L1/L2 as well as the injured vertebra L1 together show a total kyphotic axial deviation of 27°.

b) ULA of the injured and the immediately adjacent cranial uninjured vertebra

Example: ULA-D12/L1 = - 30°. This figure indicates that the uninjured vertebra D12 including the motion segment or disc space D12/L1 and the injured vertebra L1 together show a total kyphotic axial deviation of 30°.

c) ULA of the injured and the immediately adjacent

caudal uninjured vertebra

Example: ULA L1/L2 = - 24°. This figure indicates that the injured vertebra L1 including the motion segment or disc space L1/L2 and the uninjured vertebra L2 together show a total kyphotic axial deviation of 24°.

We take these measurements using digital picture processing. To do so, only the corner points of the vertebrae in question need to be marked on the lateral view X-ray (Fig. 2).

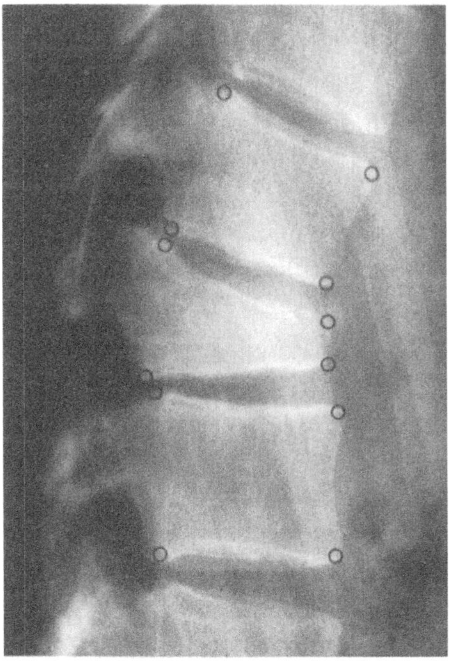

Fig. 2: Unstable compression fracture L1. Note the marked corner points of the vertebrae D12 - L2.

These measurements can also be arrived at by means of simple drawings. In theory, other dimensions such as the angles of individual intervertebral spaces or other indices can be derived using only the corner points of the vertebrae being examined. In practice, however, this method is less than ideal and has therefore not been exploited.
The dimensions described here make it possible to draw morphologically clear conclusions without having to take into consideration the very different classifications of spinal injuries. Vertebrae included in the measurement are especially clearly defined by figures like ULA-D12/L2 = - 27°.

With these techniques it is possible to judge the prognosis not only of the vertebra itself (for example, following conservative treatment consisting of closed reduction and an orthopedic brace) but also of the immediately adjacent vertebral discs or motion segments which are often also affected. The same is true of the operative stabilization methods which are standard procedures today. These fusions, whether dorsal, ventral or combined, are usually kept as short as possible. They involve three or, less frequently, only two vertebrae, and so immobilize only one or two motion segments. In the modern treatment of spinal fractures, longer fusions are used only in exceptional cases or for chain fractures. The measurement technique described here, expanded to include the necessary number of vertebrae, can also be used in these cases.
The SI and the KA quickly and simply deliver dimensions which permit an exact description of the deformation of a single injured vertebra. In combination with the KA, the various ULA's permit both an exact description and a differential analysis of changes after spinal injuries, independent of previous treatment. In our hospital, this type of analysis had led to a definite improvement in our operative stabilization concept.

The dimensions described can, as is the case with Beck's frontal index (FI), also be used in the quantification of scoliotic deformities using an a.p. X-ray on which the corner points of the vertebrae being examined should be marked. In certain cases, other combinations like an upper-upper (UUA) or lower-upper angle (LUA) ect. are possible, and these combinations remain clearly defined thanks to the precisely defined vertebrae involved in the measurement.

Literature:

1. Beck, E. (1971) Röntgenologische Meßmethoden bei Wirbelbrüchen. Hefte zur Unfallheilkunde 108:36

2. Cobb, J.R. (1948) Outline for the Study of Scoliosis. In: The American Academy of orthop. Surg. Instructional course lecture 5:261. J.W. Edwards, Ann Arbor, Michigan

3. Daniaux, H. (1986) Transpedikuläre Reposition und Spongiosaplastik bei Wirbelkörperbrüchen der unteren Brust- und Lendenwirbelsäule. Unfallchirurg 89:197

4. Debrunner, H.U. (1978) Orthopädisches Diagnostikum. Thieme, Stuttgart

5. Ferguson, A.B. (1930) The Study and Treatment of Scoliosis. South Med. J. 23:116

6. Schmorl, G., H. Junghanns (1968) Die gesunde und die kranke Wirbelsäule in Röntgenbild und Klinik. Thieme, Stuttgart

7. Stagnara, P., J. du Peloux, R. Fauchet, B. Faucon (1965) Le plan d'election pour l'examen radiographique de cyphoscoliosis. Rev.Orthop. 51:517

Three-Dimensional Quantification of Established Femoral Head Necrosis

R. WESPE[1], Å. WALLIN[2], K. KLAUE[1], R. GANZ[1], E. SCHNEIDER[2]

[1] Department of Orthopedics, University Hospital, 3010 Bern, Switzerland
[2] M.E.M.-Institute for Biomechanics, University of Bern, Murtenstr. 35, 3008 Bern, Switzerland

Summary
Established necrosis of the femoral head is normally assessed by conventional x-rays or computed tomography (CT) images. The three-dimensional location and extent of this lesion has not been quantified so far. A computer-assisted method is described that allows for visual inspection as well as quantified evaluation of the lesions in all octants of the femoral head. The method is based on the definition of the three-dimensional extent of the lesions and the femoral head from CT images. The preliminary results from six patients agree well with clinical experience. The individual results can be used for planning of surgical therapy.

Introduction

The description of an established femoral head necrosis and the preoperative planning of corrective surgery are usually performed with the aid of conventional radiography, i.e. AP views of the pelvis, true lateral, and tangential views according to Schneider [1] and "faux profil" of Lequèsne [2]. Even with these special radiographic images, visualization of the three-dimensional extent of necrosis in the femoral head is difficult due to the superimposed acetabulum. Computed tomography (CT) provides a better representation of the femoral head necrosis. However, further processing of the CT images is necessary for the three-dimensional image analysis and quantification of the femoral head necrosis. The goal of this paper is to present such a method and to report the results of a pilot study of 6 patients.

Materials and methods

Six patients with variable degrees of necrosis of the femoral head were studied. Etiologies responsible for the femoral head necrosis within the study group included idiopathic diseases (5) and posttraumatic complications (1). Five subjects were male, one was female, with an average age of 46 years (40-55 years). Each patient underwent conventional radiographic analysis [1, 2] as well as a CT study for quantification of the necrotic lesion.

In the standard CT examinations images of the entire femoral head were taken at 2-4 mm intervals. This set of images was transferred to a VAX 11/780 (Digital Equipment Corporation) computer. The osseous structures of the femoral head and neck, as well as all necrotic lesions were defined on each cross section using a graphic terminal and a pointing device. The resulting planar contours (Fig. 1) were used to construct the three-dimensional models for further analysis. In the modelling procedure surface data is generated. This representation is based on a general purpose display software [3] with extensions to cope with the particular problems in processing of CT images [4].

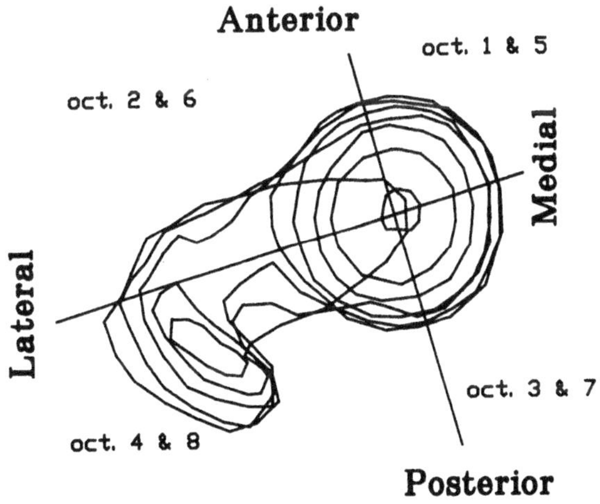

Fig. 1. Collection of planar contours describing the femoral head and neck. The planes for division into octants are indicated.

To compare results of different patients, a local coordinate system related to the geometry of the femur was defined. The three planes (xy, yz, zx) running through the center of the femoral head divide the models into eight areas, known as octants.

The contours describing the femoral head are transformed into a surface. This surface is approximated by a sphere (Fig. 2) using a three-dimensional version of the fitting method developed by Bookstein [5]. The center of this sphere defines the origin of a local coordinate system within the head (Fig. 1).

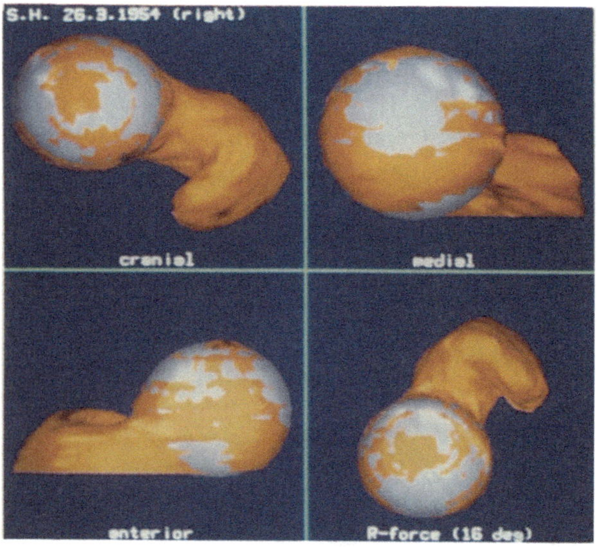

Fig. 2. Four views of the three-dimensional model of the femoral head and neck, and the approximating sphere.

The central plane of the femoral neck was defined by visual inspection. This plane runs vertically from the center of the femoral head toward the greater trochanter and divides the model in an anterior and a posterior part (Fig. 1). A second plane perpendicular to the first through the center of the femoral head divides the model in medial and lateral portions. A final division is made by cutting the model horizontally through the center, thus obtaining cranial and caudal regions.

The result of the modelling procedure is the closed surface of the head, the closed surfaces of eight or less necrotic regions (not all octants have to be occupied), as well as an open surface of the head plus the neck. The volume of each closed and oriented surface is well defined and can be computed by integrating on the surface using Gauss's law:

$$\iiint dxdydz = (\iint x\ dydz + \iint y\ dzdx + \iint z\ dxdy) / 3$$

By means of this formula the volume of the head and the necrotic lesions in each octant are computed. The octants were named according to the following scheme (Fig. 1):

oct. 1: cranial - anterior - medial
oct. 2: cranial - anterior - lateral
oct. 3: cranial - posterior - medial
oct. 4: cranial - posterior - lateral
oct. 5: distal - anterior - medial
oct. 6: distal - anterior - lateral
oct. 7: distal - posterior - medial
oct. 8: distal - posterior - lateral

The surface of the neck is used in the visualization part of the evaluation, where the orthopedic surgeon examines the extension of the necrosis in order to assess the appropriate surgical procedure (Fig. 3). As an added feature it is possible to interactively rotate and move the models using a graphic system (ADAGE 3000) (Fig. 4). Simple simulations of corrective surgical procedures are possible, allowing a surgeon preoperative interactions with the femur [6].

Results

With the three-dimensional image representation we are able to demonstrate the exact position as well as extension of the necrotic lesions in the femoral head (Fig. 3). Table 1 presents the relationship between femoral head volume and volume of the necrotic lesions as well as the surgical procedures which were performed. Table 2 shows the volumes in each octant. The results in table 2 show that the principal part is situated in the cranial-anterior portion (octants 1 and 2), corresponding to reports in the literature [7]. Three of the patients (3, 5, 6) in this series underwent surgery before the three-dimensional analysis took place. From the remaining three patients, patient 2 underwent total hip replacement, and patients 1 and 4 were treated conservatively. This conservative treatment was based on the small volume of the necrotic lesion demonstrated.

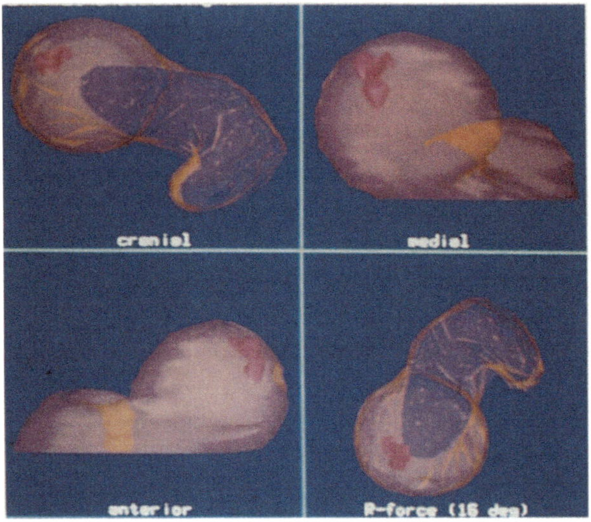

Fig. 3. Four views of three-dimensional model of head and neck displayed transparent. Necrotic lesion visible in dark red.

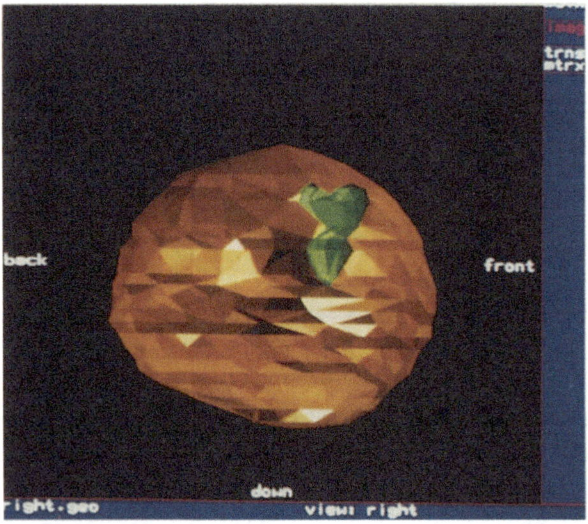

Fig. 4. Three-dimensional model of head and necrotic lesion. Interactive, presurgical planning mode.

Patient	Volume of head (cm^3)	Volume of necrotic lesion (% of volume of head)	Surgical procedure
1	51.1	0.56	conservative
2	72.4	33	total hip replacement (THR)
3	63.1	4.4	THR
4	47.5	3.6	conservative
5	53.0	5.7	intertrochanteric osteotomy (IO) (Flexion 30°)
6	71.3	11.7	IO (Flexion 30°)

Table 1. Volumes of head and necrotic lesions (total)

Pat.	Oct.1 (%)	Oct.2 (%)	Oct.3 (%)	Oct.4 (%)	Oct.5 (%)	Oct.6 (%)	Oct.7 (%)	Oct.8 (%)	Total (%)
1	0.56	0	0	0	0	0	0	0	0.56
2	4.4	7.3	5.4	9.1	1.8	1.5	1.5	2.2	33.2
3	2.4	0.62	0.73	0.67	0	0	0	0	4.4
4	1.7	0.18	1.7	0	0	0	0	0	3.6
5	1.3	0	1.8	0.06	0	0	1.3	1.2	5.7
6	2.8	3.6	1.2	0.37	2.0	1.0	0.72	0	11.7

Table 2. Volume of necrotic lesions in octants, expressed in percent of head volume.

Discussion

The three-dimensional representation permits accurate quantification of the degree of femoral head necrosis as well as the precise localization of the lesion in the femoral head. This enhanced capability facilitates a more precise preoperative planning for intertrochanteric osteotomy. Other authors as Sartoris et al. [8] recommend the preoperative production of a representative model. It is our goal to use three-dimensional representation of femoral head necrosis in order to aid in the decision for osteotomy or total hip replacement. Patient Nr. 2 had a necrotic involvement in excess of 30% of the femoral head; therefore, a total hip replacement rather than osteotomy was performed. In our opinion involvement of greater than 1/3 of the femoral head volume is a relative contraindication to osteotomy [9]. In the future we will attempt to employ three-dimensional image representation and quantification of femoral head osteonecrosis to elucidate the relationship between head volume and necrotic lesion volume which permits decision between osteotomy and total hip replacement. This will improve the quality of preoperative planning.

References

1 Schneider, R.: Die intertrochantere Osteotomie bei Coxarthrose, Springer-Verlag, Berlin, Heidelberg, New York (1979).

2 Lequèsne, M.: Le Faux Profil du Bassin, Revue de Rhumatisme, décembre, pp. 643-652 (1961).

3 MOVIE.BYU, A General Purpose Computer Graphics System, Version 6. Brigham Young University, Provo, Utah (1987).

4 Wallin, Å.; Klaue K.: Three-dimensional In Vivo Modelling and Evaluation of Hip Coverage. Proceedings of SPIE, Vol. 1030, in press (1989).

5 Bookstein, F.L.: Fitting Conic Sections to Scattered Data. Computer Graphics and Image Processing 9:56-71 (1979).

6 Wallin, Å.; Klaue K.; Ganz, R.; Perren, S.M.: Three-dimensional Evaluation of Pathological Joints and Simulation of Corrective Surgical Procedures. Proceedings of NCGA'88, Vol. III, pp. 198-207 (1988).

7 Hungerford, D.S.: Segmental Idiopathic Necrosis of the Femoral Head, Springer-Verlag, Berlin, Heidelberg, New York (1981).

8 Sartoris, D.J.: Computed Tomgraphy with Multiplanar Reformation and Three-Dimensional Image Analysis in the Preoperative Evaluation of Ischemia Necrosis of the Femoral Head. Journ. Rheumatology 13(1):153-163 (1986).

9 Capello, W.N.: Indiana Conservative Hip Arthroplasy, in: Evarts CM, (ed.) Surgery of the Musculo-Skeletal System, Churchill Livingstone, New York, pp. 427-454 (1983).

CAPS - Computer Assisted Prostheses Selection

J.R. Essinger†, M.L. Rhodes†, D.D. Robertson°, J.M. Aubaniac*

†MPDI Medical Research Center, Torrance, CA, USA
°Brigham and Women's Hospital, Boston, MA, USA
*C.H.U. La Timone, Marseille, France

Introduction : Standard prostheses are based on proportionality, assuming that a unique geometry available in several sizes will accommodate the majority of the patients. There are however, multiple different 'standard' geometries on the market.
Choosing the optimal standard prosthesis within the extended choice of available design, relying on the 2D information provided by plane radiographs is somehow cumbersome. Such a selection requires the analysis of the 3D information provided by CT scans.
The purpose of this study was to develop a software capable of helping selection of a design using the 3D geometry of the implants and of patient's bone.

The Software : The software is separated into 4 modules : contour extraction and correction, bone osteotomy, design selection, insertion simulation, and design evaluation.
- Contour extraction : A noise reduction and edge enhancement filter is first applied on each image. Contours are then extracted using a simple edge detection algorithm. The dimension of each contour is then corrected using a 3D reference phantom.
- Bone Osteotomy : The neck osteotomy is performed interactively.
- Design Selection : 3D geometries of implants are stored in a database. In order to reduce the evaluation time, the user may select a limited number of design.
- Insertion simulation : Each selected design is numerically inserted into the osteotomized femur. The insertion is completed when the implant reaches a stable position. Figure 1 illustrates a hip prosthesis successfully inserted by the insertion module.
- Design evaluation : For each tested design, the software provides the user with the following information : display of the insertion position, total and percentage of contact area, location of contact area, total and percentage of filling. The software also evaluates the difficulty of insertion.

Conclusion : This software is still under development, but the early results confirmed that selecting among multiple design optimizes implant-bone fit for off-the-shelf prostheses.

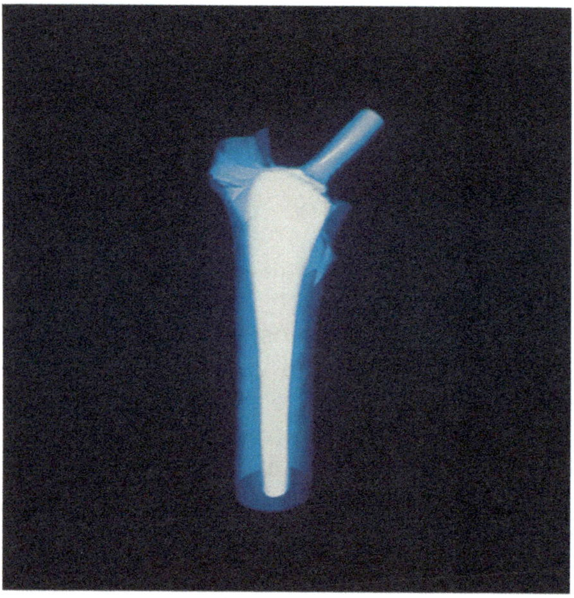

Fig. 1 : Hip prosthesis successfully inserted by the numerical insertion module.

Simulation of Osteotomy Surgery about the Hip Joint

S.B. Murphy*, P.K. Kijewski+, M.B. Millis*, and A. Harless+

Departments of Orthopaedic Surgery and Radiation Therapy, Harvard Medical School and Children's Hospital, Boston

Acetabular dysplasia is a developmental disease that results in an underdeveloped and poorly positioned hip socket. In recent years, new procedures have been designed that separate the hip socket from the surrounding bone and allow the surgeon to move the socket to an improved position. In order to perform corrective osteotomies of the hip joint accurately, the surgeon must have a detailed understanding of the pre-operative hip anatomy and how the proposed surgery will correct the deformity.

By developing computer methods to display anatomy in three dimensions using CT, we recently have been able to analyze pre-operative hip geometry, compare the patient's hip joint to normal data, and then simulate the proposed surgery. This computer simulation involves separation of the hip socket from the surrounding bone, moving the hip socket to the desired position, and then comparing the new hip socket position to normal.

This pre-operative analysis has been performed on over forty hip joints and pre-operative simulation has been performed on twenty-three hip joints. Our current methods of analysis and simulation, and our clinical experience with these techniques will be discussed.

Value of Disarticulation of the Mandible in 3-D Imaging

F.W. ZONNEVELD (1), J.M. VAANDRAGER (2),
M.F. NOORMAN v d DUSSEN (3), J.C. van der MEULEN (2)

(1) Dept. of Radiology, University Hospital Utrecht, Catharijnesingel 101, 3511 GV, Utrecht, the Netherlands
(2) Dept. of Plastic Surgery, Erasmus University, Dr. Molewaterplein 40, 3015 GD Rotterdam, the Netherlands
(3) Dept. of Oral Surgery, University Hospital Utrecht, Catharijnesingel 101, 3511 GV, Utrecht, the Netherlands

Summary

In severe cases of trauma, hemifacial microsomia and Goldenhar and Treacher Collins Syndrome, three-dimensional imaging in general and disarticulation of the mandible in particular provides information which is vital for surgical reconstruction of the mandible and the temporomandibular joint.
Our experience with over 50 cases has shown that disarticulation helps in classifying the preoperative situation and it clarifies postoperative complications such as failed fracture healing or reduction, pseudarthrosis of a reconstructed ramus or fusion of an artificial TMJ. These conclusions will be discussed and illustrated with a number of cases and will hopefully stimulate computer scientists to devise practical methods for the automation of the disarticulation process. We will show a preliminary attempt of the latter as well.

Introduction

Over the past 15 years, the treatment of craniofacial deformities by means of craniofacial surgery has been greatly improved [1]. Immediately after its development during the early 1980's, 3-D imaging was added to the armamentarium of the craniofacial surgeon [2]. While 3-D imaging matured over the past six years by replacement of depth-encoded images by surface-rendered [3] and volume-rendered [4] images of much better quality, it became more and more evident that there is a definite need for the possibility to take morphological entities apart and view them as separate units from any desired viewing direction. This was first demonstrated by disarticulation of the hip joint [5] but it can be applied to any type of joint (we have also experience with separation of vertebrae in the spine) and can be applied to soft tissue (separation of tumor tissue from healthy soft tissue) as well as to bony structures. Disarticulation of the mandible was readily recognized as a necessity for making mandible models [6] to be used, for instance, for manufacturing of subperiosteal implants or for mandibular reconstructive surgery. In this paper we describe our experience with the disarticulation of the mandible, mainly from an imaging point of view, in over 50 cases out of a series of more than 180 cranial 3-D studies.

Materials and Methods

All our CT scans were made with a Philips Tomoscan 350 CT scanner using consecutive 1.5 mm thick slices at a reduced dose level (40 mGy CTDI), a scan time of 4.8 s and a scan cycle of 36 s. The patients were immobilized with adhesive paper tape around the head. No additional immobilization for the mandible was applied. The 3D reconstructions were performed on a Cemax 1500X workstation and the disarticulation was done by combining a preselected CT number range with manually drawn region(s) of interest on each CT scan within the range of the mandible. When necessary, the skull was also disarticulated while ignoring the mandible as well as the cervical spine. Automatic disarticulation was done using dedicated software on a general purpose VAX-750 computer.
The pathologies of our patient population involved hemifacial microsomia (17), Goldenhar Syndrome (3), Treacher Collins Syndrome (9), Plagiocephaly (5), Trauma (7) and miscellaneous deformities (17). In 8 of these 58 cases the mandible and cervical spine were only removed to obtain an unobstructed view of the skull base.

Results

On 3-D reconstructions of the complete skull, the parts of the mandible which are most often obscured by the zygomatic arch are the condylar heads and the coronoid processes; especially in some cases of deformity of the zygomatic arch itself [Fig. 1]. As the heads and coronoid processes themselves are often subject of deformity, the only way to see them in an unobscured way is by the use of disarticulation. As a result, we have found a wide morphological variety within the same grade of deformity of which Fig. 2 gives an example in the case of hemifacial microsomia. In cases of the Treacher Collins syndrome we have frequently observed that the coronoid process has a worn-off appearance giving it a blunt shape in the lateral aspect which, together with the shortness of

Fig.1.
Left lateral aspect of a patient with hemifacial microsomia. The zygomatic arch slants downward thus obscuring the condylar head and the coronoid process which are shown in Fig. 2d after disarticulation of the mandible.

of the condylar neck, reduces the recess between the the coronoid process and the condyle.

In addition to the display of the mandible by itself it is also very useful to show the skull without the mandible; especially if a nonobstructed view of the skullbase from below is desired [Fig. 3.] for comparison of the zygomatic arches and for study of the mandibular fossae. This technique is also extremely useful in cases of cheilognathoschisis and plagiocephaly.

Once the mandible and the skull are available as separate entities, it is possible to recombine them again in one image while displaying them in separate colors (e.g. skull in black and white

Fig.2. Different appearances of the mandibular head and coronoid process in hemifacial microsomia. The mandibular halves are imaged separately. The images on top of each other belong to the same patient showing the affected side on the bottom. The first four cases show the variation within grade II of Pruzansky's classification [7] while the last case is an example of growth disparity between the left and right ramus and corpus. a,b. Absence of the coronoid process with an aplastic condylar head in a 9 year old male. c,d. Aplastic condyle and malformed coronoid process in a 7 year old female (same patient as in Fig. 1.). e,f. Anteriorly displaced hypoplastic condyle in a 6 year old female. g,h. Hypoplasia of both the condyle and the coronoid process in a 6 month old female (same patient as in Fig. 3.). i,j. Growth disparity between the left and right mandibular ramus and corpus in a 15 year old female with a hypoplastic right condylar neck.

a c e g i

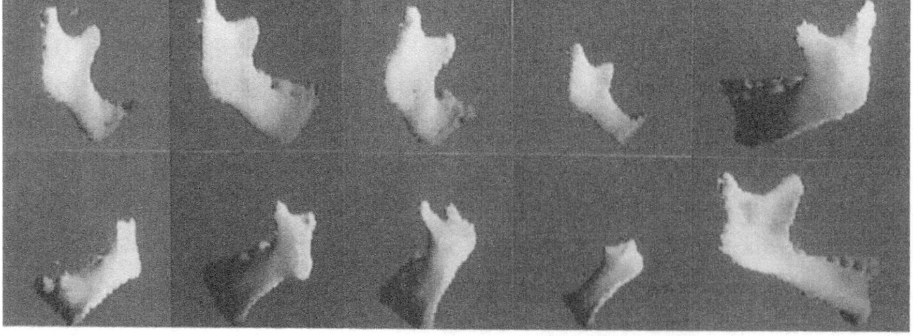

b d f h j

a. b.

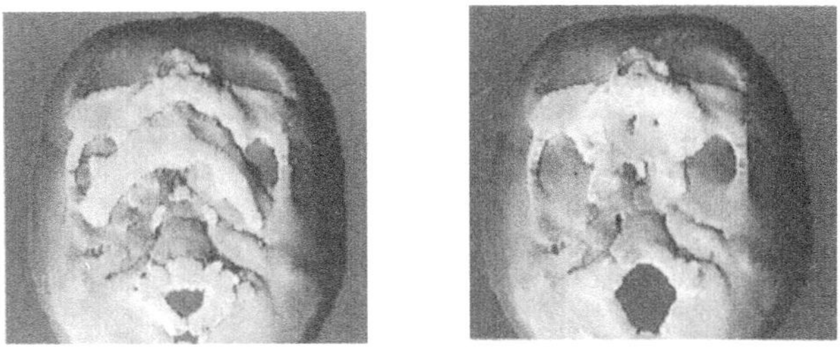

Fig. 3. Comparison of the basal view of the skull without (a) and with (b) removal of the mandible and cervical spine (same patient as in Fig. 2gh). Note the absence of the mandibular fossa and the reduced size of the temporal fossa on the left.

and mandible in heated object scale). This technique is particularly useful in very complicated cases such as trauma or postoperative cases [Fig. 4.]. By using different colors the complete extent of the mandible can be discriminated from the rest of the skull while there is no interference by the complicated structure of the cervical spine.

The time-consuming manual drawing of regions of interest during the disarticulation process can be facilitated by applying automated region growing techniques which recognize all voxels that belong to one entity e.g. the mandible or the skull [Fig. 5.].

Fig. 4. Examples of disarticulation with recombined display in patients with postoperative complications. a-i. Patient who has had a resection of a teratoma in early childhood and subsequent reconstruction of the right ramus with bone resulting in ankylosis. a-c. Overall view showing complexity of the pathology. d-f. Recombined display showing a left hyperplastic coronoid process and condylar aplasia. g-i. Mandible. j-m. Hemifacial microsomia. Status after ramus reconstruction with skull bone and complication by inflammation resulting in ankylosis on the left. m. MPR shows ankylosis and suture within the implant which allowed limited jaw movement. n-o. Hemifacial microsomia after ramus reconstruction with rib bone. Complication due to impossibility to use three fixation screws resulted in instability which consequently developed into pseudarthrosis. The remaining two screws are still visible. Note ankylosis of the remaining part of the graft.

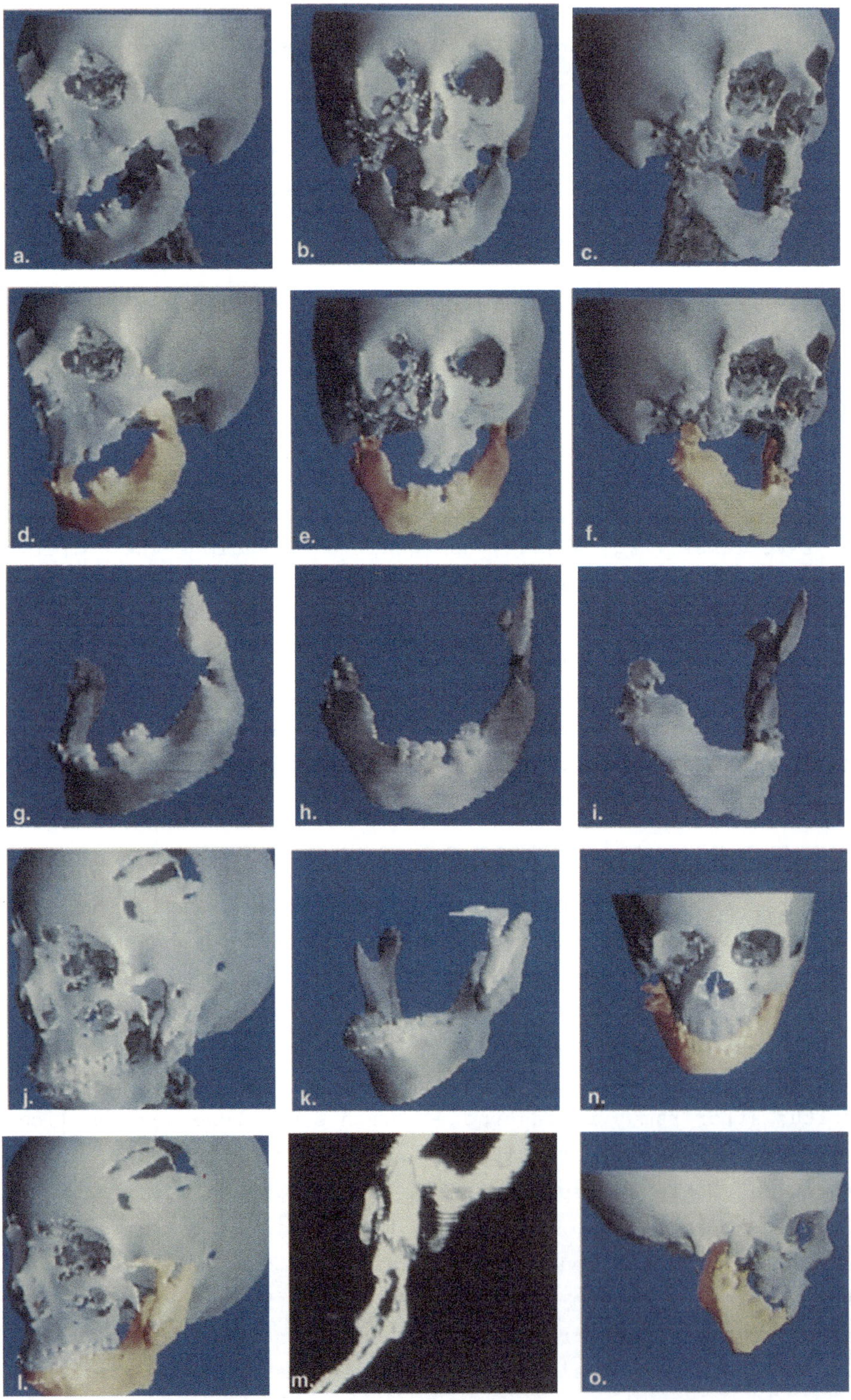
a.
b.
c.
d.
e.
f.
g.
h.
i.
j.
k.
n.
l.
m.
o.

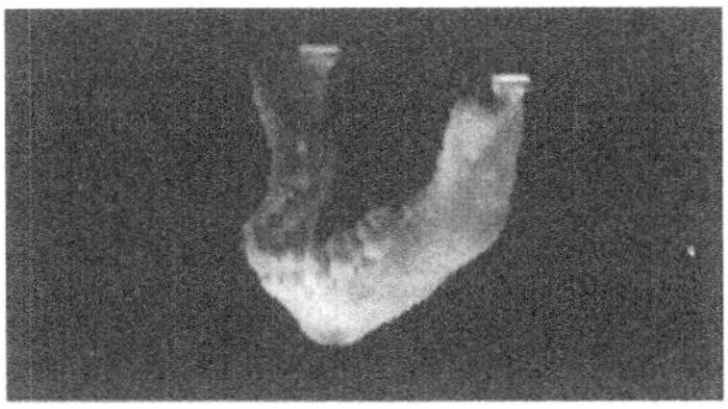

Fig. 5. Example of an automatic disarticulation by means of region growing.
Note the artificial limitation of the region growing at the condylar heads to prevent incorporation of the skull. (Region growing algorithm courtesy of M. Kreuzen, W. Venmans and E. Hol)

Discussion

Disarticulation of the mandible results in better visualization of the deformity of both the mandible as well as the skull base. In combination with soft tissue information on mastication muscle development from the original CT scans we expect that surgical planning is facilitated and a better prognosis prediction is feasible. In complicated cases, recombination display clarifies the relationship between mandible and skull considerably. In automatic disarticulation, the problem of preventing the region growing to cross the temporomandibular joint as well as the problem of connecting fragments belonging to the same entity still requires an elegant solution. Other challenges are the making of accurate models and the extraction of quantitative data.

References

1. Jones, I.T. (ed.): Advances in the treatment of facial deformity. Br. Med. J. [Clin. Res.] 29 (1986) 389-390.

2. Marsh, J.L.; Vannier, M.W.: Comprehensive care for craniofacial deformities. St. Louis, Toronto, Princeton: C.V. Mosby Co. 1985.

3. Zonneveld, F.W., Lobregt, S., Meulen, J.C. van der: 3-D imaging in craniofacial surgery. World J. Surg. (1989) in press.

4. Drebin, R.A., Carpenter, L., Hanrahan, P.: Volume rendering. Comput. Graph. (Proc. SIGGRAPH'88) 22 (Aug. 1988) 65-74.

5. Woolson, S.T.; Fellingham, L.L.; Dev, P.; Vassiliadis, A.: Three dimensional imaging of bone from analysis of computed tomography data. Orthopedics 8 (1985) 1269-1273.

6. Dev, P.; Fellingham, L.L.; Vassiliadis, A.; Woolson, S.T; White, D.N.; Young, S: 3D graphics for interactive surgical simulation and implant design. Proc. SPIE 507 (1984) 52-57.

7. Pruzansky, S.: Not all dwarfed mandibles are alike. In: Bergsma, D. (ed.): Malformation syndromes. Part II. March of dimes birth defects foundation, White Plains, NY, USA, 5 (1969) 120.

Computer Assisted Stereotaxy

Surface Registration for the Segmentation of Implanted Bone Grafts

KLAUS D. TOENNIES, GABOR T. HERMAN, JAYARAM K. UDUPA

Medical Image Processing Group, Department of Radiology
University of Pennsylvania
418 Service Drive, Blockley Hall
Philadelphia, PA 19104, USA

ABSTRACT

Bone grafts implanted inside the human body during a surgical procedure are known to diminish over time. This process can be followed in adult patients using a sequence of three-dimensional representations of the structure from computed tomograms. The segmentation of the bone grafts is done by subtraction of the 3D representations of the structure before surgery from that after surgery. We use anatomical landmarks to compute a compensating transformation for the difference in placement of the patient in the CT-scanner at the different scanning times. Results of preliminary evaluations of the system using a dry skull with bone grafts of known site and volume are presented.

INTRODUCTION

The quantitative comparison of 3D images produced from different modalities or at different time instances requires their registration to compensate for differences in placement of the patient in the scanner. The registration is done by minimizing the distance between corresponding sets of points from the two image sequences. They are derived either from markers (implanted or attached to the patient; see, e.g.,[1]) or from organs which can be segmented in all image sequences[2],[3]. The latter method relies on the fact that organs are contained completely in the image sequence. Since this requirement is hard to meet in our application (the parts of the object which are scanned at different times can be quite different), we use landmarks that can be identified on the surface of the object.

THE ANALYSIS OF BONE GRAFTS

Our aim is to segment and analyze bone grafts implanted in the human body. Our approach is based on the comparison of CT scans of the patient taken prior to and after craniofacial surgery. We assume that the change in position of the patient relative to the scanner at different times can be described by rigid body motion. A registration process produces a set of transformation parameters which are used to correct for the difference in placement of the patient.

The system for the segmentation of bone grafts (see Figure 1) uses discrete surface representations of thresholded skull data from CT images[4]. A high-quality rendition algorithm[5] is used to display the pre-operative representation together with the postoperative representation. We then let the user specify pairs of corresponding landmarks on the displayed surfaces using an interactive pointing device[6]. After each selection of a landmark the program calculates the three-dimensional coordinates in **scanner space** (defined by a coordinate system attached to the scanner). The user may choose a number of views to specify a set of corresponding landmarks.

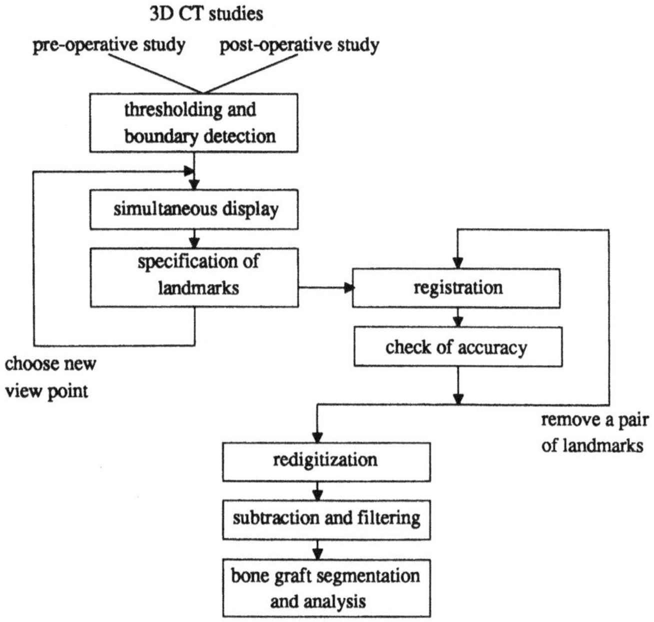

Figure 1: Overall structure of the bone graft analysis system.

Since, ideally, the locations of landmarks in both representations relative to the skull (i.e. in **patient space**) should be identical, their location in scanner space provides sufficient information to map the post-operative representation into the space of the pre-operative representation. Therefore we compute from the landmarks an object reference system in scanner space whose position remains unchanged in patient space. The algorithm uses **principal component analysis**[7], generating an orthogonal object reference system from a set of (at least three) landmarks. The origin of the system is given by the centroid of the landmarks and the axes are those to which the variance of the distances of the sample points reaches its minimum (see Figure 2). To each axis a value is generated which specifies the corresponding variance. The orthogonality of the object reference system ensures that the computation of the transformation will always result in rigid body motion and that errors due to wrong

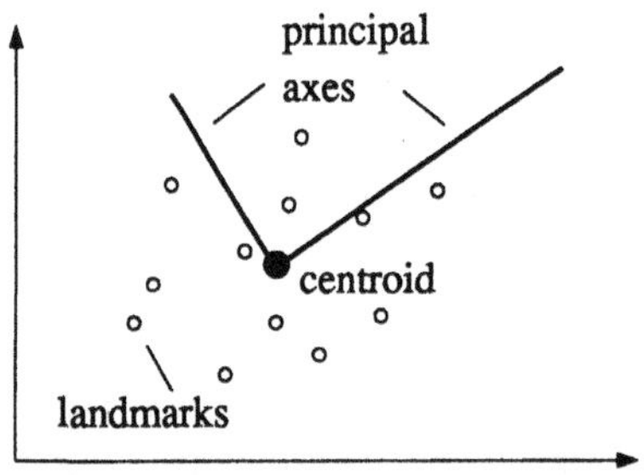

Figure 2: A 2D object reference system

location of landmarks tend to cancel out with an increasing number of landmarks (see [8] for details). Following this registration process redundant information (in the form of more than three pairs of landmarks) is used to check for possible inaccuracies in landmark specification. Pairs of corresponding landmarks are removed if their locations are inconsistent with the other pairs of landmarks and therefore adversely affect the final outcome. The precision of the positioning of landmarks is measured using the variances of the landmarks to the three axes of the object reference system. Since they should be equal for both reference systems, the deviation from this ideal case indicates possible inaccuracies. If for any subset of n-1 pairs of the n pairs of landmarks the deviation is significantly lower than that for the complete set of landmarks, the n-th pair of landmarks may have been positioned inconsistently to the other landmarks. In this case the user should remove that pair prior to the computation of the final object reference systems.

After redigitization of the transformed post-operative representation, the pre-operative study is subtracted from it, revealing the bone graft. Due to redigitization, artifacts in the form of thin surfaces can be expected in the result of the subtraction. These are filtered out using either a median filter or a thinning-thickening filter. Both are context-sensitive and remove thin surfaces while leaving the compact volume of the bone graft basically unchanged.

PRELIMINARY RESULTS

We performed a preliminary evaluation of the described system using a dry skull and bone grafts of known volume, site and origin. The skull was scanned twice, with and without the bone grafts attached to it.

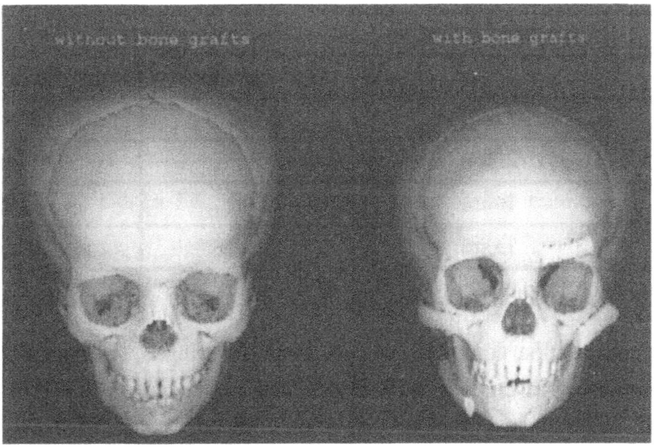

Figure 3: Simultaneous display of the dry skull with and without bone grafts attached to it. Note that the skull without bone grafts (left) is tilted towards the viewer compared to the skull with bone grafts.

A Prototype Medical Workstation for Computer Assisted Stereotactic Neurosurgery

D. Vandermeulen, P. Suetens[1], J. Gybels, A. Oosterlinck, G. Marchal.

Interdisciplinary Research Unit Radiological Imaging (ESAT/MI2 & Dept. of Radiology), Dept. of Neurology and Neurosurgery, Catholic University of Leuven (Belgium).

Summary

We report on the implementation of a display workstation for use in stereotactic neurosurgery planning. The principal features are the simultaneous display of multimodality images with instantaneous transformation of points indicated on one imaging modality image (e.g. CT) to another (e.g MRI or angio) image, three-dimensional viewing and real-time manipulation of a simulated probe trajectory and simulated "microscopic" views of the 3-D CT or MRI data along the trajectory.

Introduction

Computer assisted stereotactic neurosurgery is defined as the flexible, on-line, intelligent and integrated use of all available radiological, patient-specific, data as well as anatomical, generic, models extracted from stereotactic atlases. The prototype system presented in this paper deals only with the radiological data : acquisition, preprocessing, extraction of diagnostic features, and integrated display of all relevant data. All images are acquired using a patient-fixed stereotactic frame with accessory locating devices for CT, MRI and DSA. The digital images are subsequently stored in the workstation for further processing. A first step is the localization of the fiducials on all images. These fiducials enable the transformation of all image data to the stereotactic frame coordinates. This process can be speeded up by having the computer assist as much as possible in the automatic detection as well as the accurate localization of all markers. Furthermore, stereoscopic pairs of angiograms can be made "epipolar". This corrects for any patient movement and facilitates not only the 3D stereoscopic viewing but also the optional module for automatic delineation and reconstruction of the cerebral blood vessels. Other specialized modules extract the outlines of lesions on CT or MRI by simply pointing at their interior. Finally, all data are available for stereotactic planning and confirmation of electrode positioning. Interactive 3D devices simulate the electrode trajectory which

[1]P. Suetens is research associate with the National Fund for Scientific Research (NFWO)

is projected on the stereoscopic set of angiograms. If the automatic reconstruction of the blood vessels has been performed, the vascular tree can be projected from any angle in real time. Hence, an avascular path can be chosen, either manually or automatically, before even opening the skull. Furthermore, by intersecting this simulated electrode with all CT and/or MRI slices, the surgeon also gets information on where this trajectory might pass through vulnerable anatomical regions. Finally, new images orthogonal to the probe trajectory can easily be generated, thus offering additional orientation and confirmation clues to the surgeon. All of these procedures have been implemented on a general purpose workstation with additional graphics accelerators, 3D pointing devices and stereoscopic viewers. In the following sections we will briefly describe the different hardware and software components of the system.

Computer and Imaging equipment

The system consists of an IRIS 2400T 3-D graphics workstation [2], running UNIX with 2MB of RAM memory, a stereoscopic imaging system employing electro-optical shutters in glasses [3], and a three-dimensional cursor with six degrees of freedom [4]. In contrast with the development of PACS consoles with multiple screens, our system uses only a single display with multiple windowing features. Images generated on CT, MRI or DSA are transferred via an Ethernet local network to the workstation.

Instrumentation

All images are acquired using the BRW stereotactic reference frame [5]. It consists of a base ring fixed to the patient's skull. On top of this base ring different accessory rings can be mounted. The existing CT- and MRI-localizers transform CT- resp. MR-coordinates to stereotactic coordinates. We have built an additional Angiograpic Localizer, described in [1]. It has some particular advantages in that the film planes need not have to be parallel to the localizer frame and that absolutely no knowledge of source-target and source-image distances is required. After localization of the lesion, actual surgery is performed using the BRW Arc system mounted on the base ring. It has four angular degrees of freedom which allow a probe to be directed to a given target from almost any direction or through a given point (e.g. burr hole in the skull).

Preprocessing

The 2-D coordinates of the fiducial markers on a single image in a particular sequence are entered manually. The computer then corrects these positions to within submillimeter accuracy. Next, these coordinates are used as initial appproximations to the fiducial

[2]Silicon Graphics, Inc.
[3]STEREOGRAPHICS Corporation
[4]Polhemus Navigation Sciences Division, Mc Donnel Douglas Inc.
[5]Radionics, Inc.

marker positions of contiguous (spatial or temporal) images.
Furthermore, stereoscopic pairs of angiograms can be made "epipolar" using these reference markers. This corrects for any patient movement and facilitates not only the 3D stereoscopic viewing but also the optional module for automatic delineation and reconstruction of the cerebral blood vessels.

Interactive treatment planning

Unlike traditional stereotactic interventions, an additional safety precaution has to be provided in the case of stereotactic biopsies of deeply seated tumor lesions. In order to avoid an haemorrhage due to the accidental puncture of a blood vessel, the surgeon has to know the exact position of the arteries and veins in relationship to the tumor and to the electrode. Since he is usually unable to mentally combine CT/MRI and angiograms, an integrated 3-D image is called for. In the current version of the workstation, two possibilities are offered.

On the one hand, a three-dimensional graphic model of the cerebral vessels, reconstructed from a stereoscopic pair of angiograms [2], can be displayed and viewed from any position in near real-time. A clear three-dimensional perception can be obtained by a combination of both stereoscopy and/or motion parallax. Based on our preliminary experience [3] with three-dimensional projections of graphical objects, we advocate the use of stereoscopy and movement parallax. Movement parallax is to be differentiated from motion parallax in that the former changes the projections according to the viewer's position and expectation, whereas the latter one rotates the object independently from the user's viewpoint. The head tracking is easily accomplished using the three-dimensional position and orientation sensors.

On the other hand, all non-angiographic data can be projected on a stereoscopic pair of angiograms. Points indicated interactively on CT or MR slices can immediately be projected onto the angiograms. Next, using this target point as a pivoting point for a floating line segment, a simulated electrode trajectory can be made to rotate in 3-D. The position of this simulated electrode can be entered most conveniently by using the three-dimensional stylus with six-degrees of freedom. Alternatively, the cross-section of this line with each of the CT or MR slices can be calculated at the same time, providing the surgeon with additional feedback. A last feature that is planned for implementation is reslicing of the original CT or MRI data set orthogonal to the chosen trajectory, resulting in a simulated "microscopic" view of intracerebral tissue at the probe tip.

Discussion

The system described above is still under development and awaits its clinical integration. However, experience at other sites [4] confirms the clinical usefulness of such systems.

Additional features that we plan to implement in the nearby future are the display of the whole time-sequence of DSA images (instead of just a single exposure), and the integration of 3D MR angiographic images. The need for other additional features will certainly emerge once the system will be in clinical use.

Acknowledgments

The authors would like to thank I. Duponchelle and M. Hensmans who have contributed to the implementation and testing of the above procedures. This research is supported by the F.G.W.O. (Belgium) under grant number 3.0009.85.

References

1. Vandermeulen D., Suetens P., Gybels J., Oosterlinck A., "Angiographic Localizer for the BRW stereotactic system", Acta Neurochirurgica, Suppl. 39, pp.15-17, (1987).

2. Smets C., Vandermeulen D., Suetens P., Oosterlinck A., "Knowledge-based system for the 3-D reconstruction and representation of the cerebral blood vessels from a pair of stereoscopic angiograms", to be published in *Medical Imaging III*, Proc. SPIE 1092 (1989).

3. Suetens P., Oosterlinck A., Gybels J., Vandermeulen D., Marchal G., "A pseudo-holographic display system for integrated 3-D medical images", *Medical Imaging I*, Proc. SPIE 767, pp.606-613, (1987).

4. Kall B.A., Kelly P.J., Goerss S.J., "Comprehensive Computer Assisted Data Collection Treatment Planning and Interactive Surgery", *Medical Imaging I*, Proc. SPIE 767, pp.509-514, (1988).

3D-Based Computer Assisted Manufacturing of Individual Alloplastic Implants for Cranial and Maxillofacial Osteoplasties

SCHMITZ, H.-J., *Th.TOLXDORFF, *J.HONSBROK, **Th.FRITZ, **U.GROSS

Clinic for Maxillofacial and Facial Plastic Surgery, Director: Univ.-Prof.Dr.Dr. W.Koberg, Medical Faculty, RWTH Aachen, Pauwelsstraße, D-5100 Aachen, F.R.G.

* Institute for Medical Statistics and Documentation, Director: Prof.Dr. R.Repges, Medical Faculty, RWTH Aachen, Pauwelsstraße, D-5100 Aachen, F.R.G.

** Institute of Pathology, Director: Univ.-Prof.Dr.H.Stein, Steglitz Clinic, Free University of Berlin, Hindenburgdamm 30, D-1000 Berlin, F.R.G.

Introduction:

Improved spatial resolution of CT image data provided a new basis for preoperative planning, manufacturing of anatomical models and custom-made implants. Some 10 years ago, these methods were introduced by TONNER and ENGELBRECHT [14] in Hamburg and BURRI [2] in Ulm. Three-dimensional data files from CT anatomical models of the pelvis were made to allow surgical planning and subsequent manufacturing of custom implants to replace the resected bony structures.

Similar methods were applied for precise preoperative measurement of congenital hip dislocations by MENDES [9].

Lateron these techniques were adopted for diagnosis and treatment of other orthopedic diseases. NERUBAY [10] designed hemipelvic prostheses in 1981; ALDINGER [1] in 1983 and later ROBERTSON [11] in 1985 promoted the manufacturing of CAD/CAM-custom-made total hip-endprostheses.

Initially bone edge detection was performed by photo-optical methods or in time consuming other approaches. These methods, however, have been refined by some authors, e.g. RHODES [12] and CRANHOLM [4] in 1987 - to be processed for the most part automatically.

Necessary tools are image processing, computer graphics, detection of bony or other tissue-edges, CT-data segmentation.

Exciting representations of 3D-models have been given in literature by DEV [6], MEAGHER [8] and by a lot of companies offering workstations that are already commercially available at quite exorbitant expenses. In the mean time in clinical orthopedics several hundreds of individual prostheses have been implanted (ALDINGER [1]). The approach presented here is to point out possible applications of the above mentioned techniques for oral-, dental-, orthodontic-, maxillofacial-, craniofacial- and neurosurgical purposes.

Is customizing of implants necessary ?

From the experience of all surgeons envolved in plastic and reconstrucive surgery of the head, being involved in performing osteoplasties, the inherent difficulties are well known from either homoplasties or alloplasties. Only a small number of preshaped implants is available.

Among polymeres up to now porous implants (Proplast^R PTFE, HTR) have been used either as bulk materials or in granular forms (CRANIN [5]). Ceramic or crystalline materials, e.g. hydroxyapatite as porous blocks or granules have been used as well as glass-ceramics (Ceravital^R) and other bioglasses. Most common is the application of PMMA - polymethylmethacrylate cement (Palacos[R], Sulfix[R]) - which offers the possibility of direct molding into shape on the underlying tissue. There are problems, however, mostly due to unsatisfactory results when symmetrical structures are involved. In the case of the other materials intraoperative shaping is impossible or difficult in ceramics or metals - which, from the materials point of view are better suited to bony implantation beds than methylmethacrylates.

For replacement of the hip joint by total endoprostheses a wide variety of standard off-the-shelf implants is available. For reconstruction of malar bone, orbital floor, frontal bone, and complex structures as mandibula, maxilla, orbital roof together with frontal base standard implants are not available. In most cases direct intraoperative modelling into place is only patchwork of low standard.

Methods:

Due to the very delicate dimensions of facial bony structures in comparison with pelvic or femoral bone - very thin CT cuts are prerequisite. For total hip-arthroplasty about 15 cuts at 2-5 mm intervals are performed in the proximal femur - 5 slices at 10 mm intervals and 5 at 20 mm intervals in the region of femur shaft amounting to about 23-27 cuts for 1 examination. For craniofacial (maxillofacial) purposes up to 70-or 120-140 cuts in 1 or 2 mm intervals may be necessary.

CT data were generated on CAT scans SIEMENS-SOMATON II or DRH with 256x256 or 512x512 matrices. Ten or five second slices were recorded at 1 or 2 mm intervals. Data were stored in an uncompressed mode or in single density mode on 8" floppy discs.

To reconstruct bony defects the CT images were segmented by threshold methods based on Hounsfield Units. The Hounsfield Units were found in the range from -1024 to 1024. For thresholding a suitable interval was selected. The image matrices were transformed in binary matrices according to this Hounsfield interval representing the bony structures of interest.

The contours of the bone structures were detected by a recursive algorithm. The algorithm first traces the boundary, then the inside of the boundary to detect holes using the dualism of inner and outer contours. Especially the algorithm establishes a data-structure (contour tree), which reflected the appearence of holes and branches.

The geometrical model of the bone structure was generated by matching the contour tree elements obtained from adjacent slides. The surface of the bone structures was reconstructed out of two matched contours under the constraint, that the surface of the model was minimized. (CHRISTIAN-SEN [3], FUCHS [7], SHANTZ [13])

Finally the geometric model was interactively corrected to symmetry and the implant was manufactured employing numeric controlled machines.

Results:

a) A 3D-model was generated for planning of dental implants in the maxilla. After 3D-reconstruction a three dimensional anatomical model of the bony alveolar ridge was created. Circular and radial segmentation with subsequent options for measurements of length, areas and volumes provided a useful tool for preoperative quantification of bone available to insert dental root implants.

b) Up to now processing of 3D-CT data for manufacturing maxillo- cranio-facial implants were performed in 11 patients:

6	suffering form severe war injuries
1	temporo-orbital neurofibroma
1	bike accident
1	microphthalmia / anophthalmia
2	edentulous maxillae

Problems encountered are described in typical cases.

G.M., 24 years of age, suffered war injuries with loss of right eye, frontoorbital skull-defect including orbital roof, frontal process of zygoma. Primary wound closure was achieved by split-skin graft directly towards the dura mater resulting in pulsations of the brain in the orbito-frontal defect. 3D-reconstruction is depicted in Fig. 1. The implant was inserted via coronal skin incision after revision of frontal skullbase and covering of ethmoid sinuses by galeaflap. Postoperative X-ray showing individual implants in place (Fig. 2).

B.C. Suffering from loss of zygoma and orbital floor and nose by war injury. 3D-reconstruction is depicted in Fig. 3 showing the osseous defects. By reflexion-techniques along a chosen axis of symmetry the defect is corrected. The implant is shown in Fig. 4.

Conclusion:

The mentioned cases represent only a limited experience. Compared to conventional methods of treatment, customizing of implant fabrication in cranio- maxillofacial surgery and dental implantology represent true landmarks in these fields.

Besides correction for symmetry, the influences of soft tissue have to be concernd to really give exellent results.

There are still problems inherent in 3D-reconstruction facilities and in CAD/CAM manufacturing to real dimenions as well.

Difficulties arose from matching implants to 1-2 mm thin orbital roof as well as 8-10 mm frontal bone. Though a correct symmetrical reconstruction of bone was achieved, the thin skin graft in one case resulted in postoperative deficit of outer contours, thus a postulate of soft tissue segmentation seems mandatory. With regard to soft tissue defects corrections to underlying bone have to bear in mind the outer skin contours as well! Resolution problems and manufacturing problems to delicate cranial dimensions are still imminent.

Futher developement is being done to give relief of inherent obstracles. Thus -in the future- the method of 3D-reconstrution will be undispensable for preoperative planning of dental implants, orthodontic surgery, for surgical treatment of cranial / craniofacial malformations, cranio- maxillofacial disorders necessitating osteoplasties, may they be caused by trauma or resections in tumor treatment and even in aesthetic surgery and plastic and reconstructive surgery in general.

Acknowledgements:

This study was supported by MECRON GmbH, Nunsdorfer Ring 2327, 1000 Berlin 48, R.F.A.

The assistance of Mr. Neubert and Mrs. Dietrich, Fa. Mecron, is gratefully acknowledged as well as the help of Dr.Klose and Mrs. U.Debbing, Clinic for Radiology, RWTH Aachen, for data generation.

References:

1. ALDINGER, G., FISCHER, A., KURTZ, B.: Computer aided manufacturing of individual endoprostheses, Archives of Orthopaedic and Traumatic Surgery 102: 31-35, (1983).
2. BURRI, C., CLAES, L., GERNGROSS, H., MATHYS J.R.: Total internal hemipelvectomy. Archives of Orthopaedic and Traumatic Surgery 94: 219-226, (1979)
3. CHRISTIANSEN, H.N., SEDERBERG, T.W.: Conversion of complex contour lines into polygonal element mosaics. Computer Graphics SIGGRAPH-ACM, Vol. 12, No. 3, Aug. 1978, p. 187-192
4. CRANHOLM, J.W., ROBERTSON, D.D., WALKER, P.S., NELSON, P.C.: "Computer Design of Custom Femoral Stem Prostheses," IEEE Computer Graphics & Applications 7: 26-35, (1987)
5. CRANIN, N.: Polymeric Bone Replacement Materials in Maxillofacial Surgery. Proc. 3. World Congress of Biomaterials, Kyoto, Japan, April 1988
6. DEV, P., FELLINGHAM, L.L., WOOD, S.L., VASSILIADIS, A.: A Medical Graphics System for Diagnosis and Surgical Planning. In: Lemke, H.U., Rhodes, M., Jaffee. C.C., Felix, R. (eds.) Computer Assisted Radiology, Computergestützte Radiologie, Springer Verlag, Berlin, 602-607, (1985)
7. FUCHS, H., KEDEM, Z.M., USELTON, S.P.: Optimal Surfave Reconstruction from Planar Contours. Communication of the ACM, Vol. 20, No. 10, Oct. 1977, p. 693-702
8. MEAGHER, D.J.: Applying Solids Processing Methods to Medical Planning. Proceedings of the National Computer Graphics Conference (NCGA'85), Dallas, April 1985, pp. 101-109, (1985).
9. MENDES, D.G.: Total Hip Arthroplastiy in Congenital Dislocated Hips. Clinical Orthopaedics and Related Research 161: 163-179, (1981).
10. NERUBAY, J., ROBINSTEIN, Z., KATZNELSON, A.M.: Technique of building hemipelvic prosthesis using computer tomography. Proc. Clinical and Biol. Research 99: 147-152, (1981) Alan R. Liss, Inc, NY, NY.
11. ROBERTSON, D.D., NELSON, P., HIRANO, S., GRANHOLM, J., WALKER, P.: Orthopaedic Implant Design from Three-Dimensional Reconstruction of CT Data. Radiology 157: 341-347, (1985).
12. ROHDES, M.L.: An algorithmic approach to controlling search in 3-D image data, ACM SIGGRAPH' 79, Proc. 1979: 134-142, (1979)
13. SHANTZ, M.: Surface definition for branching, contour-defined objects. Computer Graphics SIGGRAPH-ACM, Vol. 15, No. 2, Jul. 1981, p. 242-270.
14. TONNER, H.D., ENGELBRECHT, H.: Ein neues Verfahren zur Herstellung alloplastischer Spezialimplantate für den Beckenteilersatz. Fortschr. Medizin 97, 781-783,1979

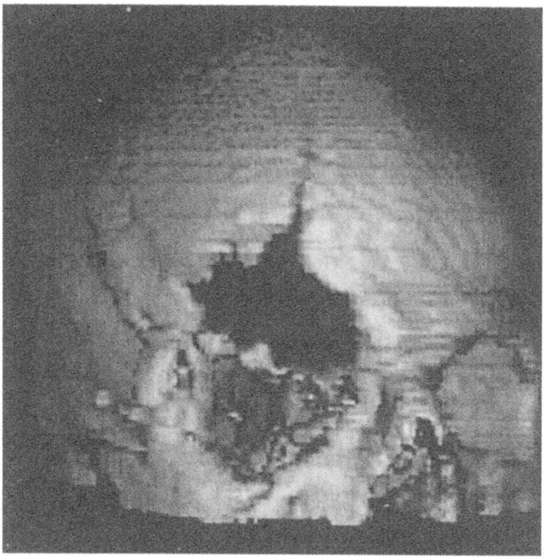

Fig. 1.: 3D-reconstruction of fronto-orbital defect.

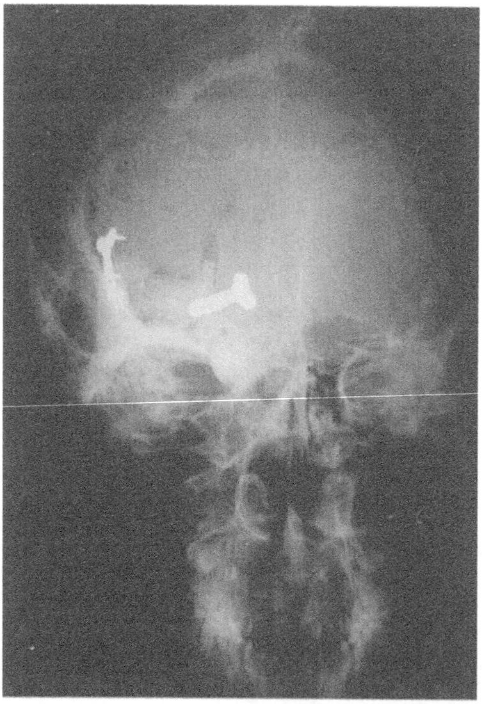

Fig. 2.: X-ray of individual 3D-implant in situ.

Fig. 3.: 3D-reconstruction of zygomatical-maxillary defect with destroyed orbital floor and wall of maxillary sinus on the left.

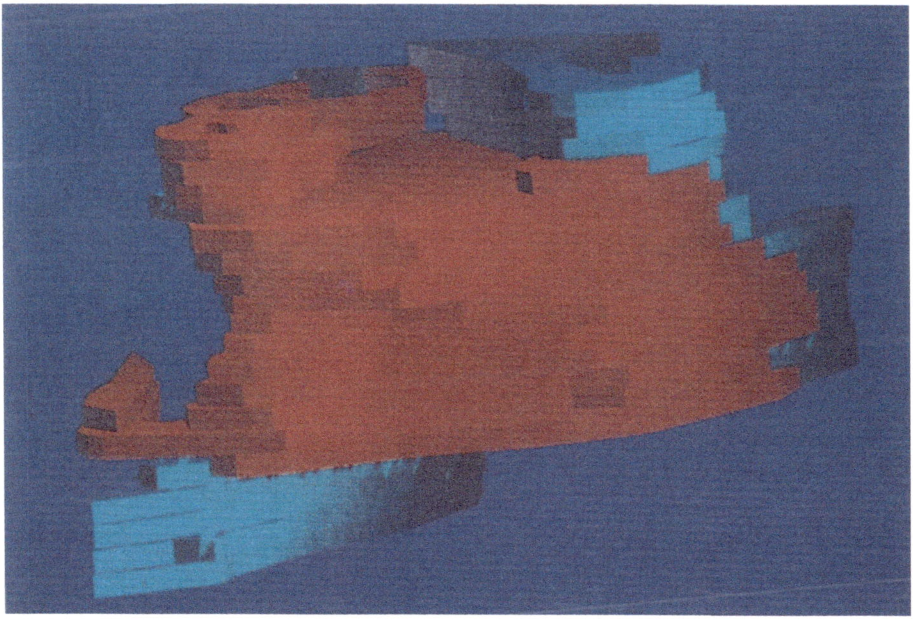

Fig. 4: 3D-simulation of individual implant after correction for symmetry.

3D-CT in Facial Fractures: An Assessment Based on 100 Cases

CHRISTOPHER G. ULLRICH, M.D. AND DAVID C. MATTHEWS, M.D.

Charlotte Memorial Hospital & Medical Center
Charlotte, North Carolina 28232 USA

Summary:

Based on experience with more than 100 facial fracture cases evaluated with surface contour technique 3D-CT images, we conclude that this display method has been excellent for assessing the complex boney relationships of the face. It has been particularly helpful for appreciating the full extent of the facial asymmetries and bone displacements present. Assessment of the mechanism of injury has also been facilitated. It is much less helpful for detecting non-displaced fractures, and associated soft tissues injuries. These 3D-CT images must be interpreted in conjunction with axial, multiplanar (2D) reconstruction and/or direct coronal CT images to provide a complete description of the sustained injuries.

3D-CT in Facial Fractures

The CT images used in this study were obtained with a GE CT9800 (General Electric Co., Milwaukee, Wisconsin, U.S.A.) using the following protocol: 512 x 512 matrix, 3mm image thickness 3mm table incrementation through the face, 2mm table incrementation through the orbit, 2 second scan time, 100mA or 200mA, normal algorithm, and zero gantry angle. Multiplanar (2D) and 3D-CT images were processed using the CEMAX 1500 image processing system (CEMAX Inc., Santa Clara, CA, U.S.A.) If the patient could tolerate it, "direct" coronal CT images of the face were also obtained using 512 x 512 matrix, 3 or 5mm consecutive CT images, 2 second scan time, 100 or 200 mAS, and bone algorithm. In those patients with direct coronal scans, the multiplanar 2D reconstruction images were omitted.

3D-CT surface contour technique images of the boney face were prepared using the standard available software of the CEMAX 1500 system. Images of the whole face were rotated from the AP (zero degree) view about the y axis in several increments from + to - 90°. Rotation about the "x axis" was used if the patient was not in the true AP position. The following "split" 3D-CT images were also prepared; anterior face and mid-face in a coronal plane, and upper orbit, lower orbit and maxilla in the axial plane. These split 3D-CT images are viewed front and back or from above and below as appropriate . Recently, these images have been photographed in lifesize. Orbital volume measurements were obtained in each patient as well. Supplementary 3D-CT images were prepared when associated mandibular, cervical spine or extensive skull injuries were present. Among the more useful "extra" views were upward rotation about the X axis, as well as edited projections which removed overlying bone structures obscuring an area of clinical intent (i.e. removal of the mandible to allow a C1C2 injury to be better seen, etc.).

These surface contour displays have a number of deficiencies. The most vexing problem involves the "drop out" of thin bone structures, resulting in holes where none actually exist. This happens most frequently with the superior, medial and inferior walls of the orbit and the anterior wall of the maxillary sinus. Adjusting the bone threshhold value downward may help but does not entirely eliminate the problem. This maneuver also tends to obscure non-displaced fractures, and may cause the resulting 3D-CT image to appear cluttered or noisey. Direct coronal CT images or multiplanar 2D reconstruction images are extremely helpful for evaluating these thin bone areas and, therefore, are obtained in every case. Surface contour 3D-CT images alone are inadequate for displaying the associated soft tissue injuries, even when composite displays using multiple contours are employed. Misregistration of data due to patient motion occasionally degrades the reconstruction images and cannot be easily corrected. For all of these reasons, careful study of the original axial and coronal CT images will continue to be important for evaluating these patients.

The major benefit of the 3D-CT images is an improved perception of the complex boney relationships of the face. Asymmetry is more easily appreciated. Local depression of fractured bone and its effect on the entire face is clearly depicted. Patterns and mechanisms of injury are more apparent than from the axial and coronal images. The lifesize images simplify measurements of distance and angle for the surgeon. As a result, communication between the radiologist and the surgeon has improved. These images have facilitated both pre-operative planning and the post-operative assessment of our facial fracture patients.

Craniofacial Surgery Planning and Simulation: Current Progress and Problem Areas

Chris McEwan, Richard Robb and Ian Jackson,
Section of Plastic Surgery,
Department of Surgery,
Mayo Clinic and Foundation,
Rochester, Minnesota, 55905.
USA.

Craniofacial surgery is the aesthetic and reconstructive surgery of the cranium, facial skeleton and the associated soft tissues of the head and neck. The principal pathologies involved are congenital, traumatic or neoplastic and may affect the patient either symmetrically, unilaterally or bilaterally. In contrast to maxillofacial surgery (surgery of the jaws and associated soft tissues) where the skull base is assumed to be normal and measurements are made relative to this normal (Cephalometric analysis) the planning of craniofacial surgery relies on aesthetic judgement of the surgeon. The visualisation of the osseous deformity in three dimensions offers an image of greater sophistication than could be achieved by the mental reconstruction of CT slices. However, it is not until measurement of the deformity is offered that true pre surgical planning occurs.

The criticisms offered by Tessier and Hemmy (1986), as they described the state of the art at that time, have largely been addressed as the state of the art has moved on and surgeons committed to computer assisted surgical planning and simulation have made contributions. In their critique they emphasised the importance of the surgeon being involved in the decision making, with respect to data segmentation and display, although methods to gain the involvement of the surgeon were not addressed. It can be seen from a review of the literature since the paper by Tessier and Hemmy that the involvement of surgeons remains peripheral (Donlon et al 1988 and Kaplan 1987) or three dimensional imaging remains a qualitative research tool.

Design Considerations

The recognition of the tasks to be implemented for craniofacial surgery planning took into consideration the practice of craniofacial surgery, as opposed to imitation of the current planning of craniofacial surgery. For the image segmentation and display tasks considerable research and experience are available. The multi-dimensional image analysis and display package [Analyze] (Robb, Stacey, and McEwan, 1987) was used to prototype image segmentation techniques to be combined with a surface shaded display algorithm (Heffernan and Robb 1985). The design requirement was to produce a display with sufficient accuracy, veracity and realism to be acceptable to the astute, not necessarily enthusiastic, physician. The ability to reliably display surgically important anatomic features, such as the foramena for the infraorbital and mental nerves or the mental crest on the mandible. A recent implementation of a ray traced, volume rendering package (Robb and Barillot 1988) with similar interactive speed to the shaded surface display has opened new horizons. Image segmentation can now be visually interactive (a new point of view or segmentation threshold displayed in 3 to 4 seconds using a 280x160x80 volume).

The interaction available to the physician is two tiered, the first tier offers full control of the observers point of view and magnification or minification of the object. The second tier gives access to preference functions and image display variables, such as the ratio of depth and specular reflective effects and direction of light sources etc. In opposition to the detail and realism required is the need for speed in both rendering and display. Ideally the speed of interaction should be real-time, once maximal software optimisation has been performed this is a hardware limitation. Given a particular hardware configuration the parameters that control the speed of execution are the size of the data set, size of the output image and the complexity of the rendering algorithm.

Surgery planning tasks involve the interactive exploration of the data, effectively the rendering and display tasks, interactive at as near real time speed as possible and quantification and reporting of the perceived problem. All of the measurement tasks require reference between screen and data space. The specific measurement tools needed for clinical craniofacial surgery planning are;

linear measurement (point to point in data space),
surface linear measurement (point to point in data space along the surface),
angular measurements (both in data space and projected onto midline sagittal plane),
area measurements
volume measurements (object volumes have readily been available from summation of voxels in the segmented volume however, it is tedious to perform and it has little clinical utility. The measurement of volumes of cavities uses a separate interactive tool using a sampling algorithm),

Documentation of the observations and measurements required an approach that facilitated good record keeping both of the final surgical plan and the intermediate steps in the process. The scan header data is changed to document the sizing and interpolation used to produce the patient data. Observations and measurements can be placed on a notepad at any point in the planning process and then incorporated to produce accurate, coherent and logical reports of the planned procedure.

Surgical simulation requires the provision of direct editing of the volume in a realistic fashion. The resultant pieces must be able to be saved or discarded and manipulated fully to mimic the excision or movement involved in the surgical procedure.

Current Implementation

The current implementation is on an networked imaging workstation using proprietary interface software (SunVIEW: Sun Microsystems, Mountain View California). This software was
chosen despite the limitations of portability for the advantage of speed of development and sophistication of the human computer interface offered. Figure 1 illustrates some of the interface elements used:

1) Multiple windows for graphics (Canvases);
   i) shaded surface display,
   ii) pseudo-shaded contour display,
   iii) ray traced, volume rendered display,
   iv) surface mapping,
2) Control windows containing
   a) system message areas,
   b) interactive input areas
      -single value choice (buttons),
      -variable choice (sliding bar)
      -direct keyboard input areas.

Not shown is the nested nature of menus and sub windows useful to conceal, or reveal, choice lists.

The interaction between the user and the data set, represented as an image on the canvas, is by way of the "mouse". The remote eye hand coordination of "mousing" proved not to be a large obstacle for surgeons, although the problems induced by aliasing in a smooth line drawn on a computer screen constantly irritated. It was also apparent that of the interactive mechanisms, currently commercially available (mouse, trackball, light pen and digitising pad), each had advantages in some area. For example the track ball in controlling rotation or the light pen as a surgical, electronic scalpel, however, the mouse appeared to be the best single device for the scope of interactions required in this project.

It is recognised that there are two groups of people involved in the field of three dimensional imaging; the computer aware and the computer naive. The computer aware have directed the development, but it is the computer naive that will be the main users of any system developed. This prompted two additional areas of special development; appropriate default values to accelerate the new user into useful work and protection of the system and data from inadvertent or inappropriate activity.

Current Usage

Since the project began in June 1987 125 scans (107 patients) have been planned using this system.

52 Congenital conditions
37 Tumours

14 Trauma
4 Degenerative conditions

There has been an evolution from the analysis of retrospective cases to validate the system toward clinically useful pre-surgical planning. Pre-surgical planning and simulation appears to be most useful, in current practice without medico legal pressure, in cases of unique tumour and trauma and the infrequently appearing and mixed congenital case. It also has a role in surgical training, patient education and informed consent, and research, for example into the effects of surgery on facial growth.

Future Prospects

Observation of the development and utility of computerised flight training and simulation and the development of "interactive environments" points to possible future prospects in computer aided surgery planning and simulation. Of more proximate interest is the development of expert systems to by run in conjunction with the surgery planning and simulation package to assist in prediction of soft tissue response and oversee the planning, to ensure possible complications are anticipated.

References

Tessier, P., Hemmy, D. (1986) Three-Dimensional imaging in Medicine: A Critique by Surgeons. Scandinavian Journal of Plastic and Reconstructive Surgery. 20: 3-11.

Donlon, W.C., Young, P., Vassilidas, A. (1988). Three Dimensional Computed Tomography for Maxillofacial Surgery:Report of Cases. Journal of Oral and Maxillofacial Surgery. 46:142-147

Robb, R.A., Stacy, M.C., McEwan, C.N (1987) A networked approach to multi-dimensional biomedical image analysis. Proceedings of the Tenth International conference on Information Processing in Medical Imaging, held in June 22-26 in Utrecht, the Netherlands.

Heffernan, P.B. and Robb, R.A. (1985) A new method for Shaded Surface Display of Biological and Medical Images. IEEE Transactions on Medical Imaging MI-4: 26

Robb, R.A. and Barillot, C. (1988) Interactive Display and Analysis of 3-D Medical Images. IEEE Transactions on Medical Imaging (In Press).

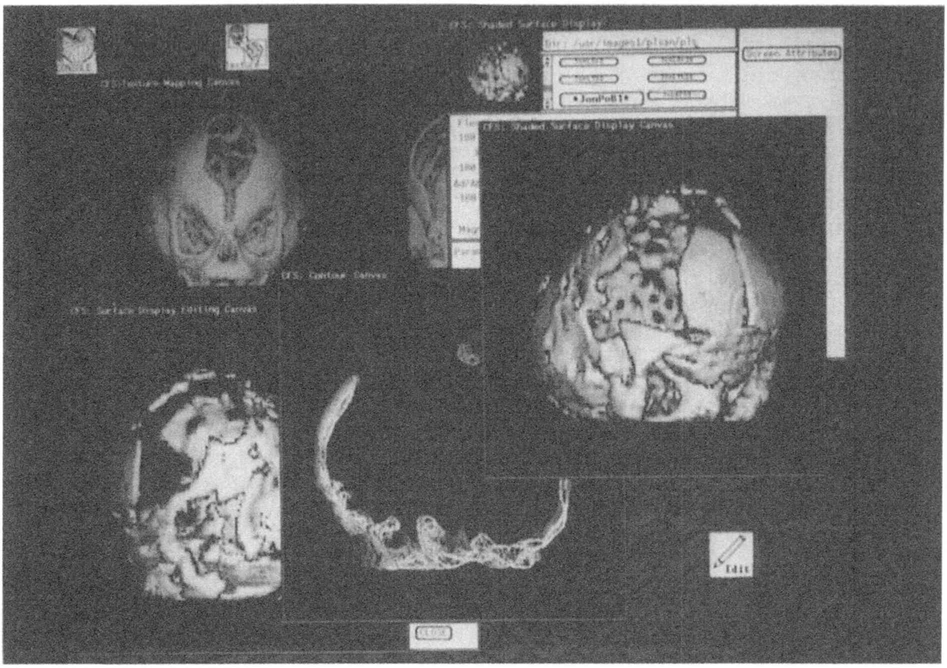

Figure 1

A 3-D Surgical Planning System and its Clinical Applications

Tronnier U.(1), Wolff K.D.(2), Trittmacher S.(3)

(1): TU-Berlin, Fachgebiet Computer Graphics, FR 3-3,
Franklinstraße 28/29, 1000 Berlin 10, W. GERMANY
(2): Klinikum Steglitz Berlin, Klinik für Kiefer- und Gesichtschirurgie
(3): Klinikum Steglitz Berlin, Klinik für Radiologie

ABSTRACT

A 3-D reconstruction system for tomographic CT and MR slice data will be presented. The system is based on an extended cell enumeration model and has the ability for 3-D dissection and display of anatomical structures. It acts as a basic component of a 3-D therapy planning system for oral surgery. The methods will be described and their clinical value will be evaluated by the clinicians involved.

INTRODUCTION

Radiological diagnostic procedures used for oral surgery are generally based on conventional X-ray film and CT data. The introduction of 3-D reconstruction of CT slice data /4,5,8,11/ yields to a consideration of the topological relationships and further allows methods for computer simulated surgical procedures to be implemented /9,12,13/, which have a totally new character because of their third dimension. The functionality of the available systems /1,2,3,15/, however, seems to be rudimentary when compared with the methods available in the CAD area. More complex functions should, therefore, be specified and implemented to answer the question concerning the expected diagnosucal and therapeutical incremental value as compared to the conventional X-ray and surgical planning technique. This is intended by the authors when realizing a 3-D diagnosis and therapy planning system by interdisciplinary cooperation of surgeons, radiologists and computer scientists, and testing it with clinical problems.

METHOD

Figure 1 shows the structure of the pilot implementation. After data conversion with DAM, which allows flexible consideration of varying slice thickness and metric relations, the prepared data sets are conveyed to the central 3-D noninvasive dissection and reconstruction system (NIPRÅS). This module is based on the extended cell enumeration model (XCE model) presented in /10/ and allows arbitrary slice reconstructions as well as immediate 3-D reconstructions by means of multimodal thresholding operators, VOI-definitions, 3-D cutaway functions and interactive 3-D segmentation (all these functions can be executed without any further data reformation). The latter recursively evaluates combinations of the above described operators with a 3-D connected component labeling procedure, allowing 3-D structures of a reconstruction to be uniquely identified and named. This way, hierarchical structuring of the 3-D scene is possible with a set of graph operations. All functions for the dissection process can be based on up to two different slice imaging modalities (e.g. CT and MR as shown in Figure 2), and the use of transparency supports the display of those selected structures. Hence, the structures will be more comprehensive. The display of regions with low contrast and/or arbitrary slice reconstructions will be improved by a histogram equalisation operator. We hope that the consideration of local voxel properties in the context of the virtual line of sight belonging to the ray-casting process will improve the extractability of cavities.

The therapy planning is based on three exemplary cases: STEPS (extracorporal radiation therapy), NEBIS (planning of stereotactic biopsy investigations) and OPS (oral surgical planning). In a first phase OPS should aid the osteosynthetic and readjustment surgical cases. A special problem concerning the implementation is the interactive system response time required for therapy planning and the development of a human computer interface adjusted to the clinicians /14/.

We also use an XCE model for the therapy planning process, although, for example, the necessary affine transformations of segment orientations can be realized faster using boundary representations and high performance graphic workstations. Boundary representations have some disadvantages when simulating material eroding tools. This would complicate the simulation process by several conversion procedures, which generally will introduce some approximation errors. A further reason for using the XCE model is the easy integration of the planning data with the display procedures of the module NIPRÅS resulting in a consistent display of all generated pictures. Near realtime response of the system is realized by consequent use of reference models, which are overlayed with the previously generated 3-D reconstruction, and time stacked computation of pictures with incrementing display quality and accuracy. The interaction methods therefore use z-buffer reliefs of the XCE data. Only the final results will be based on the full XCE model.

In order to gather knowledge about relevant methods of a therapy planning system and their best realization with respect to the graphic interaction phase we decided to construct the methods in a kind of toolbox containing:

1. Evaluation functions for the data material with the module NIPRÅS.

2. Material eroding functions as, for example, sawing, drilling, curetting and milling.

3. Functions for shape definition of implants; for the first time we tried to aid the design of maxillofacial silicone implants and definition of osteosynthetic plates bending.

4. Transformation functions for (re)adjustment surgery. That is: Adjustment of the tooth occlusion, repositioning of bone fragments and positioning of bone segments during readjustment surgery.

Sawing and cutting functions are specified by interactive adjustable reference gridplanes defining the cut direction. Drilling may be specified with trajectories representing the direction of the tool (see Figure 3). The implementaion of curretting and milling functions seems to be much more difficult due to the weak specification possibilities of the tool attachment point and direction by conventional 2-D graphical interaction devices

Material grafts will be designed by the use of bézier-patches and bicubic spline-patches or by mirror operations, which are defined by the use of symmetry points.

To correct bone disorders, a transformation function has been implemented to allow rotation and translation of, for example, bone segments. These transformation functions may also be specified by coincidence points of the involved bone segments or by shifting and rotating a computed local coordinate system of the segment. For this purpose we use a special designed 3-D interactive object manipulation device.

The pilot implementation is based on the memory cube of the PARCUM-II system /6/ in conjunction with a 2 MIPS 68010 UNIX system. Computation times vary from interactive display rates during therapy planning operations and reaches up to 140 min. for a complete 3-D reconstruction of a 256 times 256 times 84 slices dataset.

CLINICAL APPLICATIONS

3-D reconstruction is a valuable tool in certain clinical areas if the instrumental requirements are met. Dislocated fractures of the viscerocranium, especially of the maxillocranium and the frontal base, can be much better detected by surgeons in their topological disorder. The extend of necessary osteosyntheses may become clearer in the preoperative phase. Since the indication for CT data acquisition is often due to inner head lesions and comes from the neurosurgical side, the resulting data material must only be completed at this diagnostic stage for later use in maxillofacial surgery, which is aided by 3-D reconstruction and therapy planning.

In planning ortheodontic surgeries, the 3-D technique has proved to be a valuable tool. In the case of a misproportion of the maxilla and the lower jaw the common question is to decide between a prominence of the lower jaw and a pseudoprogony due to development disturbence of the maxilla. This question may be answered by the most conventional diagnostic techniques, but the 3-D technique shows the extent of the necessary bone adjustment to reach a harmonic patient profile.

Further application ranges are wide area bone defects resulting from tumor sections and/or tumor growth itself, misgrowth or lesions. Several reconstruction alternatives should be disscussed in the preoperative state. The 3-D picture is an aid for selecting the suitable implant and/or transplant. E.g.: Should rip or hip bone be used for a lower jaw reconstruction and how to shape and size an implant for reconstructions of the latteral-midface region after dislocated healed-up malar-complex fractures or defects of the cranial vault.

DISCUSSION

3-D reconstruction results in significantly higher interpretability compared with conventional X-ray films in the case of structural overlays. This is true in the case of the viscerocranium. The frequently used CT is not very userfriendly in the case of multiple strong disordered stuctures. It seems strange that this could be an effect of the high degree of information contained in the CT slices, which must be reconstructed mentally. The 3-D reconstruction of multiple fractures, e.g. the maxillocranium, is therefore an important field in diagnosis and therapy planning. Further indications are congenital cranial anomalies or misproportions of maxilla and lower jaw. The computer simulated planning of osteotomic surgery has been proven a better technique than mental planning based only on 3-D displays of the involved structures. Especially in the area of readjustment osteotomy the possible surgical results may be estimatable. Nevertheless, all methods described here seem to be only of use if a sufficient amount of time is available between acquisition of data and the surgical investigation. Emergency surgeries are doubtless out of the application range, although this field seems to be most interesting. In the field of the realization of a medical workstation for diagnosis and therapy planning /7/ we intend to develop further methods for the system and try to extend it to other therapy areas.

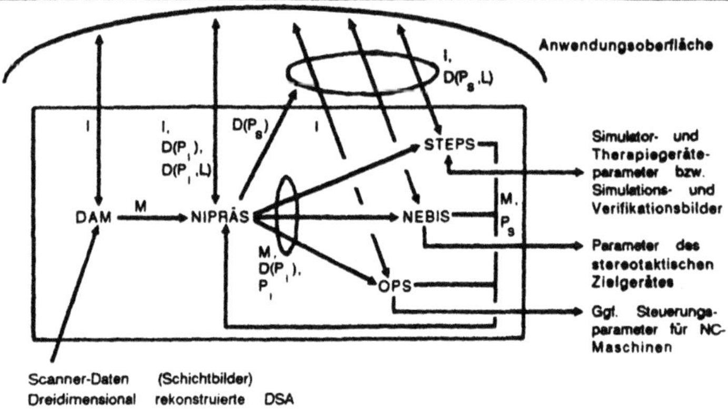

Figure 1: Structure of the pilot implementation. See text.

BIBLIOGRAPHY

1. Anonymous: *CEMAX-1000 Physicians Imaging Console...The New Dimension*; Pub. Contour Medical Systems; California, Mountain View, 1984

2. Goitein M., Abrams M.: "Multi-Dimensional Treatment Planning: I. Delineation of Anatomy" in *International Journal of Radiation Oncology and Biological Physics*; 1983, Vol. 9 Pages 777-787

3. Goitein M., Abrams M., Rowell D., Pollari H., Wiles J.: "Multi-Dimensional Treatment Planning: II. Beam's Eye-View, Back Projection, and Projection through CT Sections" in *International Journal of Radiation Oncology and Biological Physics*; 1983, Vol. 9 Pages 789-797

4. Herman G.T.: "Three-Dimensional Computer Graphics Display In Medicine" in *Pictoral Information Systems*; Pub. K.H. Höhne; NATO ASI Serie F, 1986, Vol. 19

5. Höhne K.H., Riemer M., Tiede U.: "Viewing Operations for 3-D-Tomographic Gray Level Data" in *Proceedings of the International Symposium CAR'87*; Pub. H.U. Lemke, M.L. Rhodes, C.C. Jaffe, R. Felix; Springer-Verlag, 1987, Pages 599-609

6. Jackèl D.: *Eine Rechnerarchitektur zur Visualisierung und Verarbeitung von 'cellular space'-Repräsentationen*; Technische Universität Berlin, Fachbereich Informatik, Dissertation, 1987

7. Lemke H.U., Bösing K., Engelhorn M., Jackèl D., Knobloch B., Scharnweber H., Stiehl H.S., Tönnies K.D.: "3-D computer graphic work station for biomedical information modeling and display" in *Proceedings of the SPIE Medical Imaging*; 1987, Vol. 767, February, 1-6, Pages 586-592

8. Robb R.A.: "A Workstation for Interactive Display and Analysis of Multidimensional Biomedical Images" in *Proceedings of the International Symposium CAR'87*; Pub. H.U. Lemke, M.L. Rhodes, C.C. Jaffe, R. Felix; Springer-Verlag, 1987, Pages 642-656

9. Schellhas K.P., Deeb M. El, Wilkes C.H., Check R.K., Larsen J.W., Heithoff K.B., Fritts H.M.: "Three-Dimensional Computed Tomography in Maxillofacial Surgical Planning" in *Arch Otolaryngol Head Neck Surg*: 1988, Vol. 114, Pages 438-442

10. Stiehl H.S.: *On Spatial Image Sequence Understanding*; Technische Universität Berlin, Fachbereich Informatik, Habilitation, 1987

11. Tönnies K.D.: *3D-Repräsentation der Morphologie von anatomischen Objekten durch Approximation ihrer Oberfläche*; Technische Universität Berlin, Fachbereich Informatik, Dissertation, 1987

12. Totty W.G., Vannier M.W.: "Complex Musculoskeletal Anatomy: Analysis Using Three Dimensional Surface Reconstruction" in *Radiology*: 1984, Vol. 150, Pages 173-177

13. Vannier M.W., Marsh J.L.: "Craniofacial disorders" in *Diagnostic Imaging*; 1983, Vol. 5, Pages 36-43

14. Voigt G. von: *Betrachtung der Mensch-Computer-Interaktion eines bilddatenorientierten medizinischen Arbeitsplatzes für computergestützte Diagnostik und Therapieplanung*; Technische Universität Berlin, Fachbereich Informatik, Dissertation, 1988

15. Wood S.L., Fellingham L.L., Massicotte J.B., Dev P.: "Comparison of image display techniques for solid models in medical imaging" in *Proceedings of the SPIE Third International Conference, PACS III for Medical Applications, Newport Beach, California*; 1985, Vol. 2, Pages 32.12.1-32.12.4

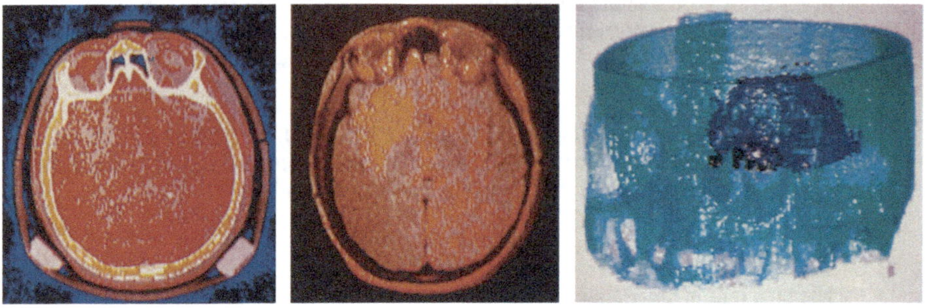

Figure 2: Original slice data and overlayed 3-D reconstruction of CT and MR data. The data have been acquisited, correlated and rectified by the DKFZ Heidelberg.

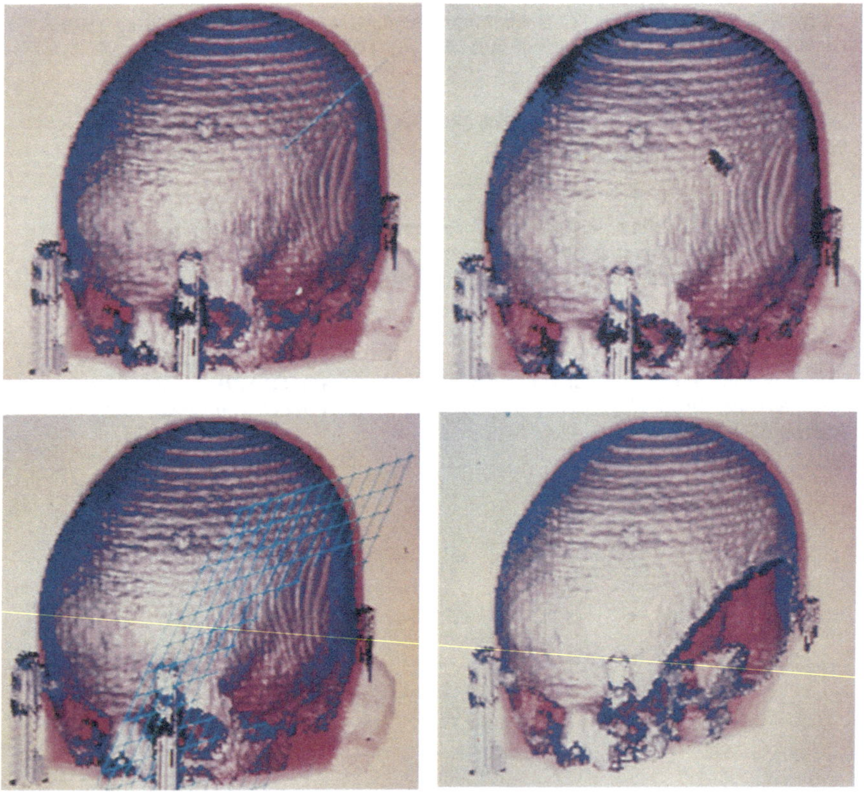

Figure 3: Saw and drill operations.
(a) Specification of the tool working direction.
(b) Final results of the simulation.

Computer-Aided Interactive Surgical Simulation System – Its Clinical Application

M. Fujioka, S. Yokoi, T. Yasuda, J. Toriwaki
Department of Radiology, Dokkyo University School of Medicine and Department of Engineering, Nagoya University
Tochigi and Nagoya JAPAN

Summary

We developed a computer-aided interactive surgical planning and simulation system for skeletal abnormalities based on the three-dimensional(3-D) surface reconstruction CT imaging. We have applied it to several clinical materials either for cranio-facial reconstructive surgeries or orthopedic hip surgeries. We have found that this system is very efficient and we believe that it will be accepted by surgeons as an important tool for more sophisticated and safer surgeries in near future.

Introduction

Since 1986, we have introduced the first version of a computer-aided interactive surgical planning and simulation system(1-4). The system introduced here is the more advanced second version of such system. The last version was obviously limited in its practical usage because only lateral image could be utilized as the reference image for surgical planning just because we intended to start with the simplest surgical procedures which were able to be acheived by rather uncomplicated computer programs. The following improvements have been acheived in the new system:

1) Outlines of bone-blocks can be drawn in any configuration by using selected reference images in any projections on CRT screen.
2) Any of these bone-blocks can be cut and be moved, rotated or reversed on the selected planes of any arbitrary projections on the CRT images.
3) Automatic disarticuration of hip joints can be acheived.
4) Automatic rotatory acetabular osteotomy can be acheived.
5) Overlapping of bone-blocks can be avoided during the

surgical simulation by using special checking function.

6) The intracranial volume can be measured automatically either before or after the surgical simulation.

Experimental trial to clinical materials

1) Cranio-facial reconstructive surgery:

Scaphocephalus is a skull deformity with an increased antero-posterior diameter of the cranium due to premature fusion of the sagittal suture. The reconstructive surgery includes very complicated arrangement of cranial bone-blocks to reform the cranial configuration. The predicted postoperative features are also demonstrated by using this system already included in the previous version(Fig 1).

Acrocephalus is another skull deformity in tower shape associated with exophthalmos which is protrusion of the eye balls seen in a patient with Crouzon's disease. Because of presence of generalized prematuere fusion of the cranial sutures, the patient suffers not only the cosmetic problem but also increased intracranial pressure related to the restriction of the cranial volume for brain growth. Therefore in such case, the reconstructive surgery intends to increase in an amount of the intracranial volume besides the reformation of the cranial configuration more adequately. It was easily done by using this system and approximately 20% increase of intracranial volume was acheived by this planning.

2) Orthopedic hip surgery

Various osteotomy procedures have been introduced to correct the problems mainly due to chronic hip disorders. The computer programs were made for universal usage for the interactive surgical planning in this field.

Clinical trial has been done to utilized some examples.

The fitness between the acetabular socket and the femoral head was tested by using the function of automatic disarticuration in a patient with deformed hip from inadequately treated congenital hip dislocation. The inner surface of the acetabulum was observed after removal of the femoral head and it was concfirmed that the configuration of the acetabular socket was adequate

to hold the femoral head. It was concluded that no surgical treatment might be indicated at this stage and the patient was followed conservatively.
Another patient was planned to have Salter osteotomy in which the mid-portion of the iliac bone on the affected side is cut holizontally and the bone-block is rotated to cover the dislocated femoral head. It was successfully simulated with this system.
The third patient had an exostosis, a bony protrusion, along the medial aspect of the femoral head which prohibits the adequate insersion of the femoral head into the acetabular socket. This was excised and the fitness between newly formed femoral head and the acetabular socket was tested by this method and the effectiveness of the surgery was proved.

Discussion

Most functions which might be required in the various skeletal surgeries such as craniofacial plastic and reconstructive surgeries and various orthopedic osteotomies are included in this new system introduced here. Although the original programs are designed for and can be used in a large-scale general purpose computer(FACOM M382. Fujitu Co. Ltd), we have successfully transfered most of the programs to the new work station system(HP-9000-300, Yokokawa Hewlett-Packard Co. Ltd).
However we still have many problems to be solved for its practical use. For example, the original x,y,z axes of the imaging data are automatically given according to the patient's positioning when the CT examination was done. Since the skeletal images can be displayed in any directions by this method, we need to determine the exact anatomical planes from the certain anatomical landmarks in the original 3-D data. Such images reconstructed according to the certain anatomical planes should become the standard images for any kind of measurements or surgical plannings. This seems to be one of the most important functions of this system and we are now working for it.

References

1. Fujioka M, Nakajima H, Yokoi S, Yasuda T, Toriwaki J: Computer Graphic Simulation of Craniofacial Surgery in Children Radiology 161(P):406, Nov. 1986
2. Yokoi S, Yasuda T, Hashimoto Y, Toriwaki J, Fujioka M, Nakajima H: A Craniofacial Surgical Planning System. Conf. Proc. of NCGA'S Computer Graphics'87 3:152-161,1987
3. Fujioka M, Nakajima H, Yokoi S, Yasuda T, Toriwaki J: Computer-aided Interactive Surgical Simulation for Cranio-facial Anomaly based on the 3-D Surface Reconstruction CT Images Program & Abstracts The 5th Asian-Oceanian Congress of Radiology p281 Sep 21-25, 1987
4. Fujioka M, Yokoi S, Yasuda T, Hashimoto Y, Toriwaki J, Nakajima H: Computer-Aided Interactive Simulation for Craniofacial Anomaly based on the 3-D Surface Reconstruction CT Images. Radiation Medicine 6(5):204-212, Sep-Oct 1988

Figure 1

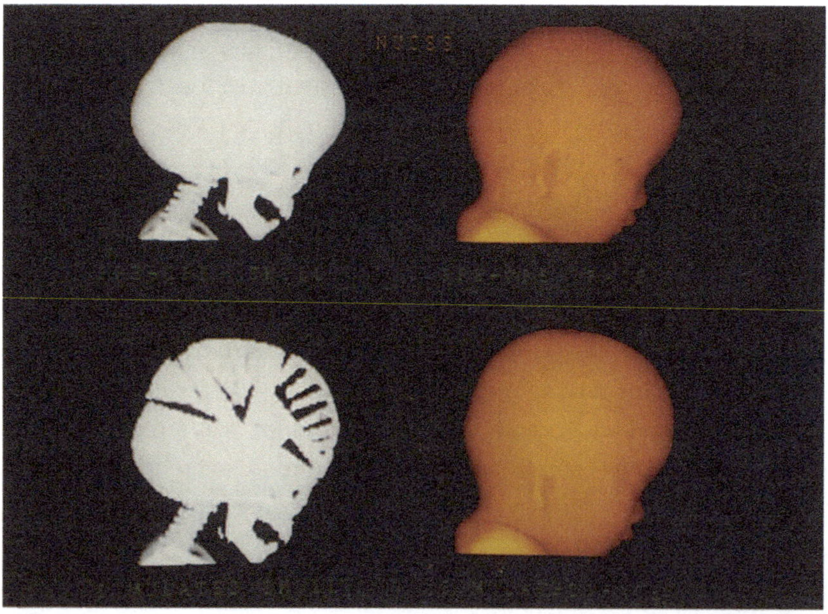

CAS – Computer Assisted Surgery – An Innovative Surgical Technique in Clinical Routine

Mösges, R.;* Schlöndorff, G.;* Klimek, L.;*
Meyer-Ebrecht, D.;** Krybus, W.;** Adams, L.;**
*Dept. of ENT diseases, Plastic Head & Neck Surgery
** LfM Dept. Faculty of Electrical Engineering
University of Aachen

D-5100 Aachen

Summary:

CAS - computer assisted surgery - is an innovative method for intraoperative real-time support of skull base surgeons. More than 50 patients have so far been treated by several surgeons with the assistance of a CAS-system.
Technical items as well as operative techniques are discussed, new developments are presented such as a combination of the CAS-system with the surgical microscope.

Introduction:

The Aachen CAS - development group has first published this new patented method for free hand surgery of the skull base on the occasion of CAR 87. At that time the method was still in an experimental state, a handfull of surgical operations had been performed with the assistance of the CAS-system. We now present the second generation of CAS-systems together with case studies from more than 50 patients and an outlook towards extended applications.
CAS - computer assisted surgery - means real time imaging support to surgeons without additional intraoperative radiation. The surgeon's instrument position within the patient's skull, while performing hypophysoectomy for example, is continously viewed on a CRT-screen. A 3D model of the patient is displayed which is derived from consecutive CT-planes. This model is correlated with the patient in the operative theatre by a 3D measuring device.

Using the same apparatus the position of the surgeon's instrument is determined.

Methods:

The CAS-system has been asssembled from equipment developed by the LfM Institute, Faculty of Electrical Engineering, University of Aachen. The system can process all sorts of digital images such as CT, MR, DSA, PET etc. So far Siemens computer tomographs Somatom 2 and Somatom DR have been the major sources of information. Data is transmitted from the image source via the Aachen University Hospital´s Experimental PACS to a graphical workstation. A three dimensional model of the patient's skull is generated from the CT-planes the voxel size bring 1 mm^3 for surgery of the orbit, 8 mm^3 for surgery of the skull base and the paranasal sinuses. Modelling is done by the surgeon who identifies the reference points. It can be the combined with planning of the surgical procedure, defining the optimal approach and the individual precauntions that have to be taken. The patient model is then transfered to the image workstation in the operative theatre. It will be correlated with the operative situs and the surgeon's instrument, using a newly developed 3D digitizer with a spatial resolution of less than 0.5 mm.

Results:

The system has been used by several surgeons for assistance during some of the most hazardous operations in skull base surgery. It has enabled surgeons to use narrow approaches relying on information from the CAS-system rather than view or touch. It has dramatically reduced uncertainty during surgery equally reducing duration of the operation. Anatomical structures that usually are sacrifized in order to gain better intraoperative view have been preserved in the cases treated with the assistance of the CAS-system. More than 50 patients have so far undergone computer assisted surgery, none of these has suffered from severe adverse events such as blindness, injury of the brain or the meninges.

Discussion:

CAS - computer assisted surgery is a technique reducing uncertainty in the surgical process. It is a method that can dramatically reduce the occurrence of adverse events in skull base surgery, rating as high as 10.5% (1).
Unlike stereotaxy (2) CAS supports free hand surgical operation. The second CAS-generation composed from a 68020 VME Bus workstation an a digital positioning device has proved its feasability in the operative theatre.
For the next generation touchless remote position measuring will be the main objective extending the use of the system towards new applications. Another objective for upcoming CAS-systems is the combination with the operation microscope (3). A special application will be computer assisted radiation using a CAS-system as localisation device.

References:

(1) Draf, W. Micro-Endoscopic Endonasal Surgery of the paranasal 3rd course on endonasal surgery Fulda (1989)

(2) Mundinger,F.; Birge, W.
CT-stereotaxy in the clinical routine
Neurosurg. Rev. 7 19-224 (1984)

(3) Roberts, D.W. et al
A frameless stereotactic integration of computerized tomographic and the operation microscope

Computer Assisted Driving of a Needle into the Brain

S. LAVALLEE[1], P. CINQUIN[1], J. DEMONGEOT[1], A.L. BENABID[2], I. MARQUE[1], M. DJAID[1].

[1] Service d'informatique médicale, TIMB - TIM3, Faculté de Médecine, 38700 La Tronche, FRANCE.
[2] Service de Neurochirurgie, CHRUG A. Michallon, 38700 La Tronche, FRANCE.

Summary
Computer assisted pre-operative planning of a stereotaxic neurosurgical intervention can be combined with the use of a computer controlled positioning arm for the introduction of a tool into brain. We present the methodology and the first clinical results.

1. Introduction

Our general objective is to link up 3D information coming from multi-modalities images with surgical tools (1). A typical instance is Stereotactic Neurosurgery: it requires a high degree of accuracy and repetitive calculations on images. The off-line computation of pre-operative images (CT, NMR, Atlas ...) allows us to select the best trajectory, then the matching of this information with per-operative X-ray images allows us to perform the surgery in a classical operating theatre with the help of a positioning arm. Compared with other approaches (2,3), the proposed method makes use of all the available information, meets accuracy requirements and is implemented outside the CT room. Moreover, it is readily adaptable to other clinical domains.

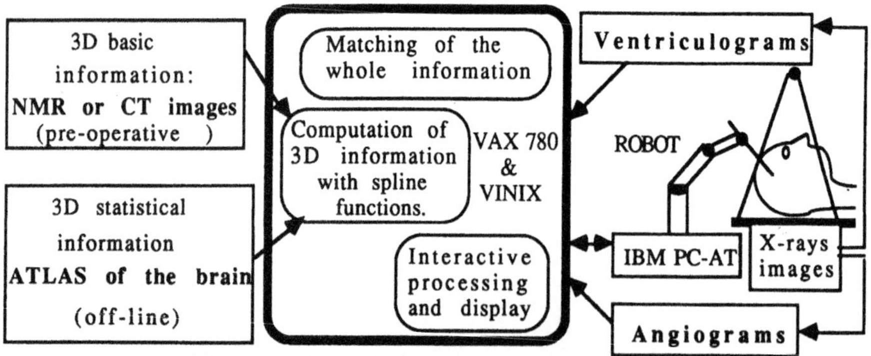

Figure 1: block diagram of the system.

2. Computer Assisted Choice of the Surgical Strategy

The choice of the surgical strategy will be computer assisted with a methodology that we are presently developing. Four steps are necessary. First, a target and a tentative trajectory are defined on CT or NMR based 3D images. We use a continuous modelling of the volume with tricubic spline functions (4), which makes it easy to display images such as the ones presented in fig 2, on which the surgeon can easily select the optimal trajectory of the needle.

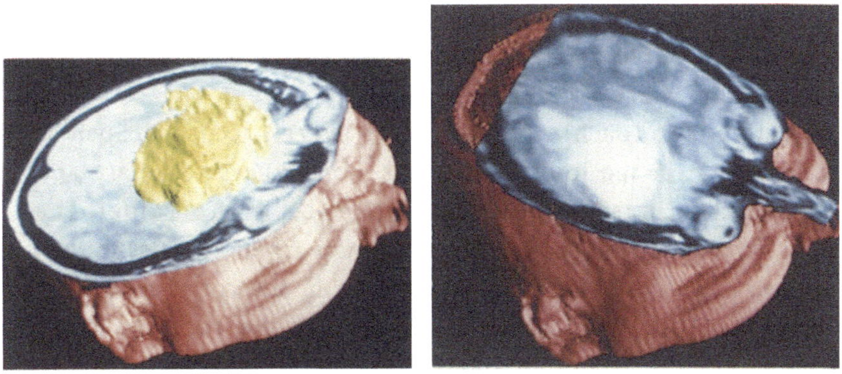

Figure 2: interactive selection of a target on computed NMR sections.

This choice is then correlated with the statistical information contained in a computerized Atlas of the brain, which has to be matched with 3D images of the patient (5,6). The third step consists in matching the whole 3D information with a per-operative system of reference. This matching uses 3D internal structures, such as the ventricular system, that can be seen on both pre-operative and per-operative images (two orthogonal X-ray pictures). Then, angiograms are checked in order to ensure that the selected trajectory does not intersect any vessel. A quick automatic check is indeed possible since it is not necessary to reconstruct the whole 3D vascular tree.

3. Robot Assisted Performance of the Surgical Strategy

We need a computer controlled positioning arm to put a guide in the previously defined location. Once this positioning is achieved, the surgeon introduces the tools (first a drill, and then a probe) through this guide. To carry the guide, we have choosen an A.I.D manipulator (Domène, France), which provides 6 degrees of freedom and a repeatability of +/- 0.2 mm. Using such a robot raises safety and accuracy problems (7).

To reduce as much as possible hardware breakdowns, we have cut down significantly the reduction ratio of each articulation. Several software tests have been added to check the normal behaviour of the robot. For instance, once the robot has reached the final position, as we expect no motion, the variations of the articular variables are tested: any motion first triggers a beep off and then cuts the power off, according to its magnitude. Moreover, with the used motors reductions, the robot is "irreversible": usual human forces cannot move it. Thus if the operator cuts the power off just before the probe is introduced in the guide, the robot does not move and no outcome can result from any mechanical, hardware or software problem.
Our objective is to reach a submillimeter overall accuracy. Such a value is not usual yet in the industrial applications where tasks are specified by coordinates in some external reference frame rather than by learning. So we refered to recent work in advanced robotics to deal with the inaccuracies resulting from the errors in the robot model (8), and we developed our own method to calibrate the robot in the whole system. A modified Denavit-Hartenberg model was designed for our robot. The parameters of this model were estimated with a non-linear least-squared fitting algorithm .
The connexion of the reference system of the robot and of the X-ray device is achieved by the analysis of the position of a calibration cage made of two pairs of parallel plexiglass planes, each of these planes containing 9 opaque metallic balls. The calibration method is inspired by the two-planes method designed for cameras calibration (9). Actually, the robot carries the calibration cage in order to put it on the head of the patient during the acquisition of a pair of reference orthogonal X-ray pictures. The originality of this method is that a unique tool, the calibration cage, is used not only to match the reference system of the robot with the X-ray device, but also to correct the X-ray parallax errors.

4. First clinical results

This methodology was first experimented on a phantom, to test safety and accuracy. Then, clinical experiments were performed, with X-ray based control of the positioning of the tools. Up to now, the matching of pre-operative 3D information with per-operative X-ray images is not yet processed automatically. The following figures show the major steps of the intervention.

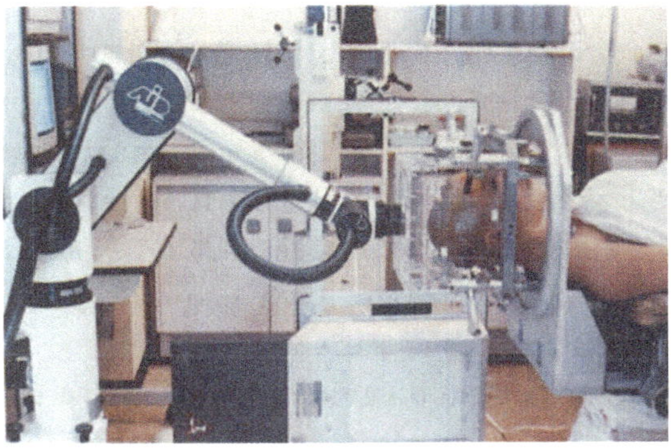

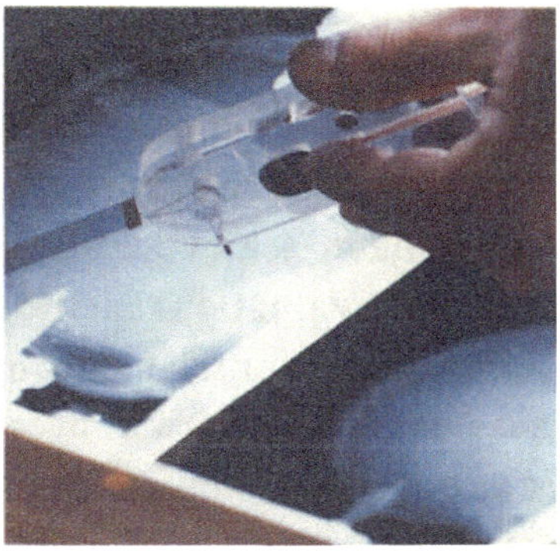

Fig. 3 A, 3 B: The calibration cage with 4x9 landmarks is carried by the robot. A pair of orthogonal X-ray images allow us to find the relationship between the robot, the X-ray device and the anatomical structures.

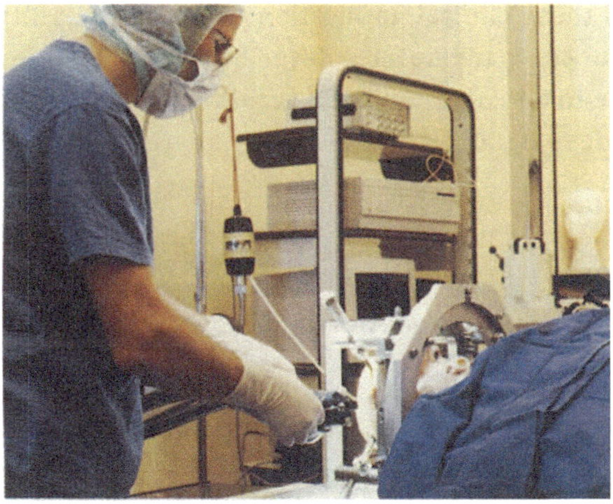

Fig. 3 C: When the manipulator is in the right position, the surgeon pushes a probe through a guide (here a biopsy of a tumour).

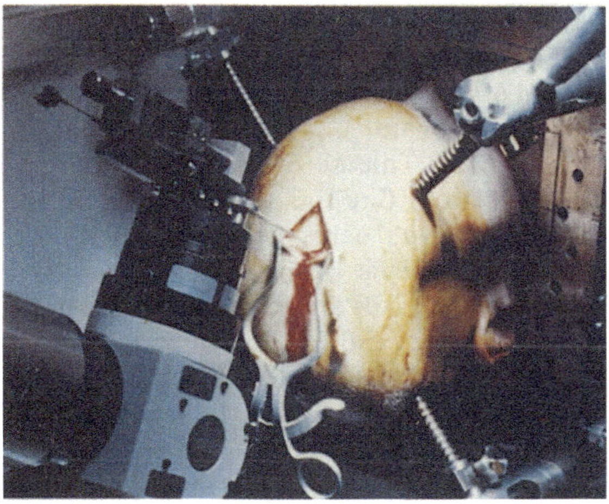

Fig. 3 D: When the probe goes into the brain, the power of the robot is cut off (here a stimulating electrode is put into the thalamus).

5. DISCUSSION.

A method for Computer Assisted Driving of a Needle into Brain has been designed and experimentally and clinically tested. Submillimeter accuracy for the moves of the robot with respect to the X-ray images has already been achieved, but the overall accuracy will be estimated once statistical evaluations of the correlation errors between NMR, CT, Atlas, Angiograms, Ventriculograms and Robot Frame have been obtained. These evaluations are in progress on a series of patients.

We are presently investigating other clinical domains which require the same kind of methodology (spinal repair or dynamic external irradiation for instance). In some cases, the robot will have to be used not only as a positioning arm, but also as an accurate and safe end-effector for complex interventions.

One common problem for all these clinical applications is the availability of the images and of accurate parameters describing the imaging devices. These data will have to be communicated across new generation PACS which still have to be developed.

References:

(1) Cinquin P., Lavallée S., Goret C., Demongeot J., Crouzet G., Peltié P., Computer Assisted Intervertebral Puncture, Proceedings 7th international congress on medical informatics in Europe, Rome,1987, 847-852.

(2) Kwoh Y.S., Hou J., Jonckheere E.A., Hayati S., A Robot with Improved Absolute Positioning Accuracy for CT Guided Stereotactic Brain Surgery, IEEE Trans. on Biomedical Engineering, 35, 2: 153-160 (1988).

(3) Kosugi Y., Watanabe E., et all, An Articulated Neurosurgical Navigation System Using MRI and CT images, IEEE Trans. on Biomedical Engineering, 35, 2: 147-152 (1988).

(4) Cinquin P, Application des fonctions splines au traitement d'images. Thèse d'état es sciences mathématiques. Grenoble , France, Sept. 1987.

(5) Schaltenbrand G., Wahren W., Atlas for stereotaxy of the human brain, 2nd ed., Thième, Stuttgart, 1977.

(6) Garibotto G., Giorgi C., Cerchiari U., 3D Image Processing in Functional Stereotactic Neurosurgery, Proc. SPIE: Applications of digital image processing, Geneva, 1983. 280-287.

(7) Finlay P.A., Advanced Robotics in Medicine and Healthcare, Proc. of the Int. Sympos. and Exposition on Robots, Sydney, Nov.1988. 704-715.

(8) Roth Z. S., Mooring B. W., Ravani B., An overview of Robot Calibration, IEEE J. of Robotics and Automation, 3, 5: 377-385 (1987).

(9) Gremban K. D., Thorpe C.E., Kanade T., Geometric Camera Calibration using Systems of Linear Equations, Proc. IEEE of Int. Conf. on Robotics and Automation, Philadelphia, 1988. 947-951.

Computer Assisted Radiological Diagnosis

Computer Assisted Diagnosis of Brain Tumours in CT-Scan Imaging

J.R. IGLESIAS, J. ESPARZA, K. TÖNNSEN, C. ARUFFO, CH. BRINNEL, K. MAIER-HAUFF.

Neurosurgical Clinic, Laboratory of Neuropathology and Section of Neuroradiology, Universitätsklinikum Rudolf Virchow, Standort Wedding, Augustenburger Platz 1, 1000 Berlin 65, FRG.

SUMMARY

A computerized system for the diagnosis of brain tumours from CT-scans was developed. This system is ideal to learn the radiological characteristics present in each type of brain tumour and their combinations. Furthermore, it indicates the strategy to be followed in establishing a differential diagnosis. The diagnosis of 811 brain tumours was established using this expert system.

INTRODUCTION

Expert and other computerized systems have long been used to diagnose and classify tumours (1, 2). Neuropathologists and neurosurgeons have implemented the use of such systems in the diagnosis and histological grading of brain tumours (3, 4, 5, 6). Due to the proved efficiency of these systems in establishing diagnoses and taking into consideration the great number of cases which must be analyzed daily in a radiological department we have undertaken the development of such a system for neuroradiological diagnosis of brain tumours with CT-scans.

MATERIALS AND METHODS

For the development of this system a pre-existing data base, containing 3,750 brain tumours (7, 8, 9) was used. 811 additional CT-scans of patients with brain tumours were studied to validate the expert system (Tab. 1). A personal computer PC-AT (1MB RAM) with a 40 MB hard disk was used. A discriminance analysis and Bayes' theorem were implemented (10). All variables considered were binary. The radiologic characteristics used by the expert system are observed by the radiologist and it is he who introduces them into the computer. The programming languages used were Turbo-Basic and Turbo-Pascal.

RESULTS

The strategy to be followed when establishing the diagnosis of brain tumours with CT-scans is summarized in Fig. 1. Patients presenting with signs or symptoms of central nervous system involvement such as intracranial hypertension, epilepsy etc, which lead to the possible diagnosis of brain tumours receive an indication for CT-scan. If the CT-scan is negative for the presence of a brain tumour, the patient must be subject to other examinations. If, on the other hand, the characteristics of the CT-scan are suggestive of a brain tumour the first question to be taken into consideration is the age of the patient: is it a child or an adult. The incidence of the different brain tumour types in children and adults is known to be very different (11). Different types of brain tumours vary in their location within the central nervous system. Therefore, the next point to take into consideration would be the predominant localization: frontal, precentral, parietal, temporal, occipital, white matter, basal ganglia, cerebellum, cerebello-pontine angle, brainstem, midline, foramen Monroe, third ventricle, lateral ventricle, fourth ventricle, cisterna chiasmatica, extracerebral, skull base, skull convexity, orbit. Once the localization has been established the form of the lesion should be analyzed since it can be indicative of the tumour type being delt with. Metastasis are, for example, very frequently round. The shapes taken into account are: regular contours (round, triangular, oval, elongated) or irregular contours. The size of the lesion is given by grouping them in smaller or equal to 2 cm or larger then 2 cm. Thereafter, the presence or abscence of multiple lesions should be considered. Multiplicity being decisive for the diagnosis of metastases. Although, for example, lymphomas may also be multiple but in contrast to metastases lymphomas have an irregular contour. Whereas, multiplicity of brain tissue tumours is very rare. The following precontrast CT-scan characteristics should also be taken into account: if the lesion is hypodense or hyperdense and in each case if it is well delimited, spotted or diffuse; and presence of mass displacement. After the application of contrast material, the degree of enhancement (I,II,III) as well as the shape of the post-contrast enhancement must be indicated: spotted, ring, curlly, homogeneous, and "kokadenförmige"; and the presence of perifocal edema. In addition the presence of necrosis, cysts and hydrocephalus should be reported. The sex of the patient being studied should also be registered.

All these data are introducted into a data base and analyzed with a discriminating analysis. The possible diagnoses are expressed in percentages. These indicate the similarity of the combinations of characteristics present in a given tumour with the mathematical function which indicates this diagnosis. If the diagnoses are not very similar and one of them predominates the predominant one should be accepted. If the diagnoses are very similar it is possible to enter into another module in which questions related to the clinical situation of the patient are taken into consideration. This could help in establishing the diagnosis with more precision.

Twelve additional characteristics can be added to the program according to the necessities or experience of any given radiologist.

The patient, the diagnosis and the radiologist who did the study are coded with a number and registered in the data base from where they can be retrived in case of necessity.

The data and characteristics obtained for a given tumour with this expert system can be correlated with those of the histological expert system (12) and with the data base of preoperative clinical symptoms.

DISCUSSION

An expert system for the diagnosis of brain tumours with CT-scans was developed. This system utilizes a maximum of 60 radiological characteristics to reach a diagnosis. The discriminating system establishes a hierarchy of CT-scan characteristics indicating which of them are more important for the establishment of a diagnosis. Hierarchization follows the logical thinking a radiologist would normally use. This structuralization introduces a method to easily learn which combination of characteristics leads to a given differential diagnosis. As a learning method it is particularly appropriate for not very experienced radiologists. The use of this expert system is particularly interesting in the cases of differential diagnoses in which clinical characteristics are also taken into account. This system structurized the characteristics and their combination used in establishing a diagnosis. Thus providing an aid for the radiologist by indicating him or her how sure his decision is. However, the final decision has to be taken by the radiologist. Furthermore, it must be said that there are tumour types which are not included in the data base and can, therefore, not be considered in the differential diagnosis. In the future similar expert systems can be implemented for the analysis of the NMR and PET-scan.

REFERENCES

1. Pfannkuch,F.; Scholman,H.; Iglesias-Rozas,J.;Brinnel,Ch.; Blümcke,S.: "Personal-Computer" in der bioptisch-zytologischen Diagnostik am Beispiel der Schilddrüsen-zytologie. Der Pathologe. 8 (1987) 20-24.

2. Hahn,H.J.; Iglesias,J.R.; Brinnel, Ch.; Pfannkuch,F.: Erfahrung mit einem rechnergestützten "Diagnosepro-gramm", am Beispiel der Histopathologie der Mamma. EDV in Medizin und Biologie. 17 (1986) 53-59.

3. Iglesias,J.R.; Sánchez,M.J.; Sandra,A.; Mohmhaupt,A.: Computer model of archiv and diagnosis of brain tumours based on the WHO classification. EDV in Medizin und Biologie 14 (1983) 40-44.

4. Iglesias,J.R.; Pfannkuch,F.;Aruffo,C.; Kazner,E.; Cervós-Navarro,J.: Histopathological diagnosis of brain tumours with the help of a computer: mathematical fundaments and practical application. Acta neuropathol. (Berl.) 71 (1986) 130-135.

5. Iglesias,J.R.; Aruffo,C.; Esparza,J.; Kazner,E.: Histological grading of brain tumours. EDV in Medizin und Biologie 19 (1988) 38-44.

6. Kroh,H.; Iglesias,J.R.; Matyja,E.; Aruffo,C.; Maier-Hauff,K.; Cervós-Navarro,J.: Computer classification of experimental brain tumours in mice. Exp. Pathol. 35 (1988) 41-46.

7. Iglesias,J.R.; Kazner,E.; Aruffo,C.; Esparza,J.: CT diagnosis of brain tumours: Semiautomatical type-specific diagnosis with the help of a personal computer. Mathematical fundaments and practical applications. In: Computer Assisted Radiology. Proceedings of the International Symposium. Ed.H.U.Lemke; M.L. Rhodes; C.C.Jafee; R.Felix. pp. 443-448. Springer Verlag, Berlin 1985.

8. Iglesias,J.R.; Kazner,E.; Aruffo,C.; Esparza,J.: A model of semiautomatical type-specific CT diagnosis of brain tumours. Mathematical fundaments and practical application. Br. J. Radiology. 59 (1986) 895-900.

9. Iglesias,J.R.; Aruffo,C.; Trempenau,B.; Kazner,E.: An Expert System for the Diagnosis of Brain Tumours. In: Computer Assisted Radiology. Ed. by H.U.Lemke, M.R. Rhodes, C.C.Jafee and R. Felix. pp. 397-401. Springer Verlag, Berlin 1987.

10. Escudero,L.F.: Reconocimiento de patrones. Ed. Paraninfo Sa. Madrid. (1977) 1-688.

11. Zülch,K.J.: Brain Tumours. Their Biology and Pathology. Springer verlag. Berlin 1986.

12. Iglesias-Rozas,J.R.;Brinnel,Ch.; Esparza,J.: "Tumor": Computermodell für das histologische Archiv der Hirntumoren (WHO-Klassifikation). Halbautomatisches Verfahren. Verh. Dtsch. Ges. Path. 70 (1986) 527.

TABLE 1

TUMOURS ANALYZED

Diagnosis	**No. of Tumours**
Angioma	16
Astroblastoma	1
Chordoma	1
Craneopharigeoma	151
Ependymoma	25
Epidermoid Cyst	4
Ganglioglioma	1
Glioblastoma	101
Hemangioblastoma/Hemangiopericytoma	4
Hypophysis Adenoma	22
Hypothalamic Hamartoma	1
Low Grade Astrocytoma	36
Lymphoma	9
Malignant Astrocytoma	59
Malignant Oligodendroglioma	6
Meduloblastoma	16
Meningioma	248
Metastasis	34
Neuroblastoma	1
Neurinoma	43
Neurofibroma	7
Oligodendroglioma	19
Plexus Papilloma	6
Total	**811**

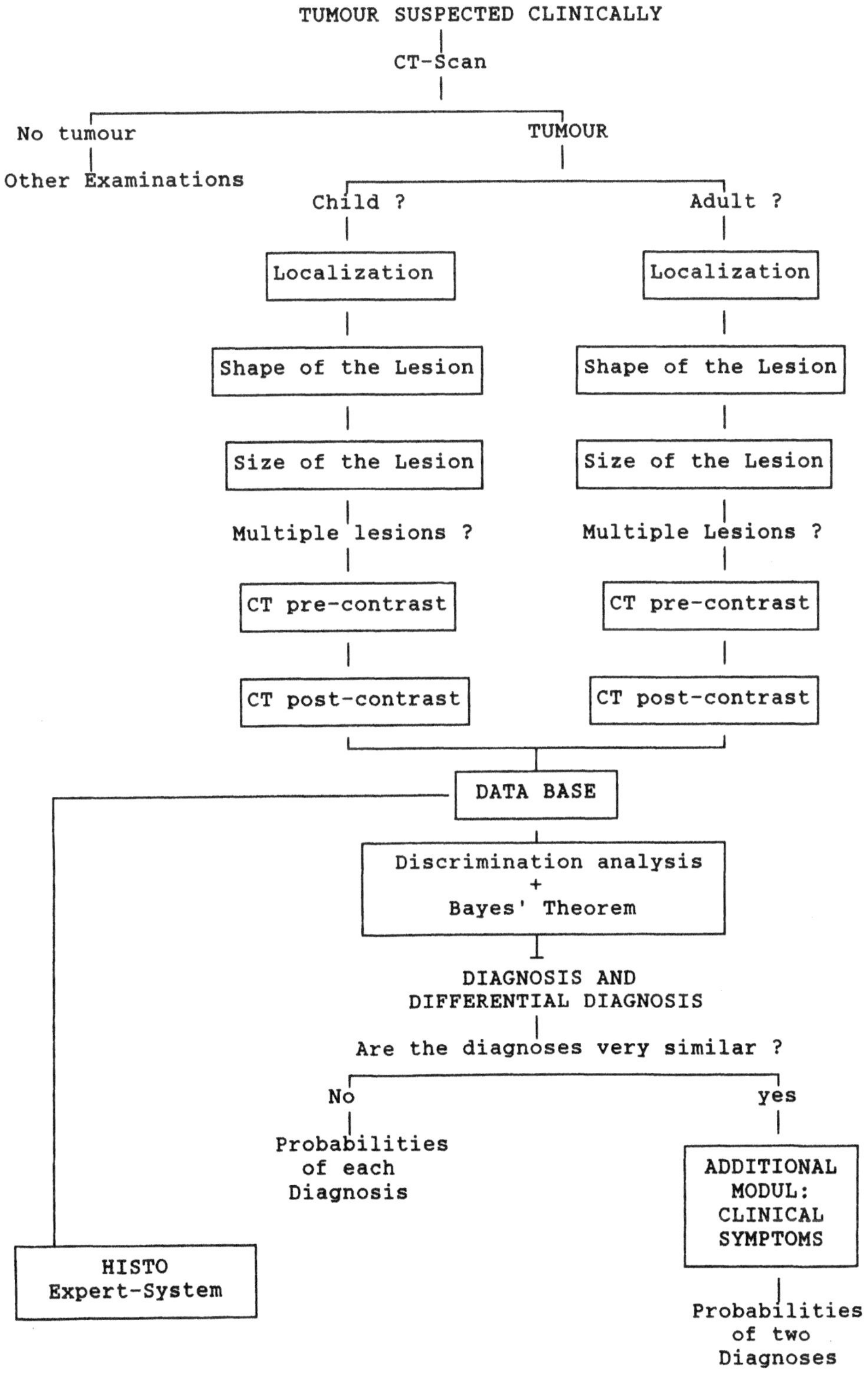
TUMOUR SUSPECTED CLINICALLY
CT-Scan
No tumour
Other Examinations
TUMOUR
Child ?
Adult ?
Localization
Shape of the Lesion
Size of the Lesion
Multiple lesions ?
CT pre-contrast
CT post-contrast
Localization
Shape of the Lesion
Size of the Lesion
Multiple Lesions ?
CT pre-contrast
CT post-contrast
DATA BASE
Discrimination analysis
+
Bayes' Theorem
DIAGNOSIS AND
DIFFERENTIAL DIAGNOSIS
Are the diagnoses very similar ?
No
Probabilities
of each
Diagnosis
yes
ADDITIONAL
MODUL:
CLINICAL
SYMPTOMS
Probabilities
of two
Diagnoses
HISTO
Expert-System

The Design of Acceptable Computer Aids for Radiology

DU BOULAY,G.H.[1], TEATHER,D.[2], TEATHER,B.A.[2], WOODS,A.J.,[2], WILLS,K.M.[2] and PLUMMER,D.[3]

1. Institute of Neurology, Queen Square, London.
2. Medical Systems Research Centre, Leicester Polytechnic, Leicester.
3. Department of Medical Physics, University College Hospital, London.

Summary

The design, development and introduction of computer advisors in medicine represents a significant undertaking. This paper describes the authors' experience of the problems involved in constructing computer advisors in radiology. The approach adopted is based on close collaboration between statisticians, computer scientists and radiologists. The prime determinant of user acceptance is likely to be the system interface. Experience suggests that different forms of interface may be required, depending on both the system's role (training versus routine clinical use) and the clinical expertise of the user. The development of modular systems giving rapid access to statistical information, expert knowledge, textual explanations and reference diagrams and image sets could offer considerable benefits in radiology.

Development of a C.T. Advisor System

A prototype computer system (BRAINS) to aid in the interpretation of C.T. images for the diagnosis of cerebral disease, archiving of disease-related information and training in image interpretation has been developed by the authors [1,2]. The BRAINS system represents one result of a long term collaborative research program involving statisticians, computer scientists and radiologists [3,4,5] at the three institutions.

The system prompts the user to describe 'abnormalities' visible on C.T. images. On the basis of the scan description, advice may be requested on diagnosis and the likely effect on diagnosis of "enhancement" of some tissues by injection of contrast medium. The description is entered by means of menus that form a tree-like structure. "HELP" is available in the form of textual explanations for all questions and options. A reference set of diagrams and images has been established to illustrate the various signs which may appear in the images , and this reference set may be accessed at any point in the dialogue. Images for the current case may be examined at any time, and the normal "windowing" facilities are available. If diagnostic advice is sought, the system is able to indicate those signs which have provided evidence for the disease predicted, together with any that significantly conflict. The full system has been implemented on the Independent Display Console (IC) produced by International General Electric and additionally, in part, on a variety of PCs.

BRAINS : Design Considerations

In constructing the system we have sought to provide more than a simple diagnostic advisor. Thus, the statistical, textual, reference image and diagram databases may be accessed independently of advice. The system may thus serve as an adjunct to the radiologist making his/her own diagnosis.

Diagnostic advice provided by the system is statistically based. The model used to provide disease predictions has been derived by extensive data analysis and the acquisition of medical knowledge regarding which features might differentiate certain disease processes. The need for hard statistical data to calibrate subjective clinical judgements has been noted in both this and other applications [5,6]. We have sought to adopt the knowledge based approach of separation of program and data in that the model specification is stored in text files and may be edited, allowing fine tuning without re-programming of the system.

The need for an efficient means of interaction with the radiologist, requiring only minimal typing skills, led us to design a simple menu-based interface with a standard format for all dialogue screens [2]. We had envisaged that the availability of an extensive range of HELP facilities would provide an interface that was suitable for users with differing levels of computing and radiological expertise. Ease of use and learning to use were paramount considerations in menu design and layout. In this way, we had hoped to minimise the training period required to establish user's confidence.

BRAINS : System Evaluation

A prototype system was first made available at the National Hospital for Nervous Diseases, for preliminary testing and evaluation in 1986. Subsequently, in 1987, the system underwent a major upgrade and was transferred to an "external" (i.e. previously unassociated) site for a rigorous evaluation of acceptability and utility [7]. This external evaluation study involved eight radiologists of differing experience and two radiographers describing the plain C.T. scan studies of up to twenty cases.

The response of external users to the system was mixed. Many were unfamiliar with the concept of formal description of images which underlies the BRAINS system. It is essential that new users are made aware of the concepts underlying the system and the importance of accessing HELP facilities to clarify the use of different descriptors.

In general users were slow to gain confidence in interacting with the system. We had underestimated the extent to which training would be required. It became apparent that users benefitted greatly from person to person training with regular feedback and discussion. Such feedback greatly improved user confidence and performance in obtaining accurate descriptions.

Users reported positively on the educational benefit of the structured approach required to describe images. The training potential of the system was recognised by all users. This may be illustrated by the fact that the most successful user, in terms of entering descriptions and obtaining accurate advice, was a novice radiologist who made extensive use of the HELP facilities and reported most positively on the acceptability of the system.

In contrast, the more experienced radiologists indicated that the benefits of the system in the day-to-day management of patients in a busy radiological department would be limited. The structured menu-based approach, although extremely useful in training, would be too time consuming. These findings confirm those of Kidd [8]. The possibility of an alternative "query-based" interface was discussed at some length. A danger in such an approach is that a referral letter may point the radiologist towards an incorrect initial diagnosis. Querying the system to determine what features might confirm the initial diagnosis may result in the radiologist biasing his/her report and therefore lead to a confirmation of the initial incorrect preconception. The modular nature of the BRAINS system would allow an alternative interface to be designed without the need for re-programming other aspects of the system.

Magnetic Resonance Imaging

In parallel to the evaluation of the BRAINS system, we have examined the feasibility of constructing an advisor system for M.R.I. [9]. Computer advice in magnetic resonance imaging of the head might prove beneficial in two main areas :-

(i) the selection of image acquisition parameters based on clinical presentation,

(ii) image interpretation and the diagnosis of cerebral disease.

Advice on image interpretation and the diagnosis of cerebral disease will be based on a statistical database of cases with confirmed diagnoses. A formal description procedure for M.R.I. has already been established and implemented on a portable lap-top microcomputer. A four year time span is envisaged to establish an adequate database and develop a prototype system.

Radiological Advisors - Discussion

The design, development and introduction of computer advisors in medicine represents a significant undertaking. It is essential that clinical end-users are involved in all stages of the process.

Medical advisors must be capable of adapting to changes in both medical knowledge and user requirements. Modular system design, combined with the separation of program from information/data facilitates continuing system evolution. The assertion that only rule-based knowledge representations can provide explanation facilities is false. Statistical based approaches are not however always appropriate or possible. The prime determinant of user acceptance is likely to be the system interface. Our experience suggests that different forms of interface may be required, depending on both the system's role (training versus routine clinical use) and the clinical expertise of the user.

The development of modular systems giving rapid access to statistical information, expert knowledge, textual explanations and reference diagrams and image sets could offer considerable benefits in radiology. Perhaps the most important benefit offered by such systems is typified by a statement made by radiologists during the BRAINS evaluation : "Although I did not always agree with the approach adopted by the system, or the advice provided, I found the computer sessions very revealing. The system particularly made me consider how I reach decisions". The system may result in a deeper insight into how people work.

References

1. Teather,D., Morton,B.A., du Boulay,G.H., Wills,K.M., Plummer,D. and Innocent,P.R. Computer Assistance for C.T. Scan Interpretation and Cerebral Disease Diagnosis, Statistics in Medicine, 4, 311-315 (1985).

2. Plummer,D., Teather,D., Morton,B.A., Wills,K.M. and du Boulay,G.H. The Development and Assessment of a Computer Aid for use in the Diagnosis of Cerebral Disease, Information Processing in Medical Imaging, Proc. 9th International Conference, Washington D.C., 310-324 (1985).

3. Wills,K.M., du Boulay,G.H. and Teather,D. Initial Findings in the Computer Aided Diagnosis of Cerebral Tumours using C.T. Scan Results, Br.J.Radiol., 54, 948-952 (1981).

4. Morton,B.A., Teather,D, and du Boulay,G.H. Statistical Modelling and Diagnostic Aids, Medical Decision Making, 4,3, 339-348 (1984).

5. du Boulay,G.H., Teather,D., Morton,B.A., Wills,K.M., Innocent,P.R. and Plummer,D. Brains - A Computer Advisor System to Aid in CT Scan Interpretation and Cerebral Disease Diagnosis, Neuroradiology, 29, 196-199 (1987).

6. Leaper,D.J., Horrocks,J.C., Staniland,J.R. and de Dombal,F.T. Computer Assisted Diagnosis of Abdominal Pain using Estimates Provided by Clinicians, Br. Med. Jnl., 4, 350-354 (1972).

7. Teather,D., Teather,B.A., Wills,K.M., du Boulay,G.H., Plummer,D., Isherwood,I. and Gholkar,A. Evaluation of Computer Advisor in the Interpretation of CT Images of the Head, Neurorad., 30, 6, 511-517 (1988).

8. Kidd,A.L. and Cooper,M.B. Man-Machine Interface for an Expert System, Proc. 3rd Technical Conference of the British Computer Society Specialist Group on Expert Systems, 8-18 (1983).

9. du Boulay,G.H., Woods,A.J., Teather,D., Teather,B.A., Wills,K.M. and Plummer,D. Towards an Advisor for MRI, Neurorad., 30, 245-251 (1988).

Computer Characterization of Trabecular Bone Pattern

M.Chimenti[1], D.Trippi[2], R.Bozzi[1], O.Di Munno[3], G.Perri[2]

[1] Istituto di Elaborazione della Informazione (I.E.I.-CNR) Pisa
[2] Istituto di Radiologia, Università di Pisa
[3] Patologia Medica, Università di Pisa

Summary

The paper describes the instruments and the analysis method used in a preliminary study of osteoporosis on conventional X-ray films. Twenty cases have been studied by measuring the integral and the high frequency value of the signals extracted from digitized radiograms; both measurements depend on the bone mineral content and on the trabecular bone pattern.
Further studies will be executed in order to determine the effectiveness of the applied procedures, which allow to obtain indexes of the bone mineral content and of the pattern coarseness in accurately selected measurement sites.

Introduction

The determination of the loss of Bone Mineral Content (BMC) is very important in the diagnosis of several pathological conditions of the skeleton and particularly in the senile osteoporosis.
Radiological methods for the quantitative assessment of the BMC are based on the measurements of the scattered radiation energy (Compton densitometry) or transmitted energy (Quantitative-CT, Dual Photon Absorptiometry or Single Photon Absorptiometry, Film-densitometry). Otherwise morphometric methods for the measurements of the combined cortical thickness and the scoring of the trabecular bone pattern (Singh Index, Calcanear Index, etc.) have been used.
The increased interest for trabecular bone in determining the BMC variations is related to its turn-over which is three to ten times faster than the cortical bone [1]. In order to find some parameters which are able to indicate the variations of the morphology of trabecular patterns we applied to forearm radiograms some procedures of image processing and we compared the results of our determinations to BMC values obtained by traditional Dual Photon Absorptiometry (DPA).
We determined also the coarseness coefficient (C), extracted from the digitized image in different sites of the forearm. The C value represents the web-like structure of the trabecular bone pattern which is normally

constituted by thin and numerous trabeculae. On the contrary in osteoporosis the number of the trabeculae decreases and some trabeculae become thicker modifying the modulation of the signal in the image with respect to normal conditions.

Data acquisition

Twenty women in postmenopausal-age were examined both by DPA and by computer film-densitometry on radiograms of the distal forearm. The radiograms have been digitally processed by using a work-station which consists of a personal computer, an X-Y scanner and a video memory.
The X-ray films have been digitized by a MFA/Digirad scanner [2] with 125μm spatial resolution and 255 levels of photometric resolution.
The video memory has 1MB total size and consists of four frames, each of 512x512 bytes. The processed image is displayed on a RGB monitor with 255 gray levels or pseudocolours.

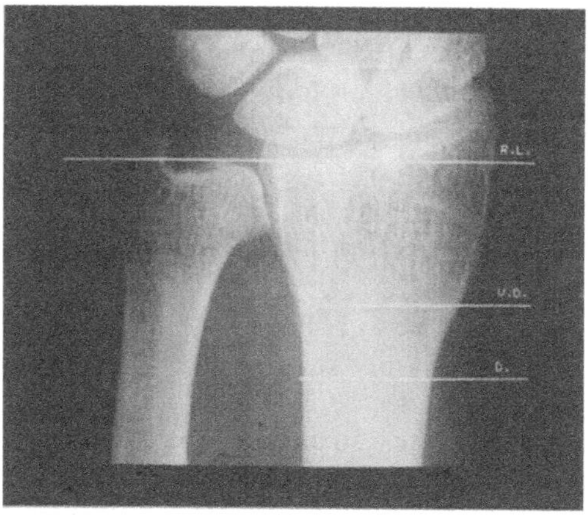

Fig.1. Position of measuring sites in forearm analysis.
RL=reference line; VD=very distal site; D=distal site

The personal computer is supplemented with the image processing software system IPSS, which has been developed at IEI in previous applications of image processing [3]. The actual version of IPSS is composed of about 100 software modules for image acquisition, enhancement and display, measurement and data archiving.
Each processed film contains the image of the portion of skeleton to be analyzed together with an image of an Al step-wedge. In the film digitization procedure the mean optical density of each step is first measured and the pixel values of the image to be analyzed are converted

in mmAl equivalent thickness values. This way normalized data are obtained from different X-ray films and variations of the film optical density due to exposure or dark-room processing fluctuations are compensated.
The digital image is displayed and two measurement sites, according to the DPA practice, are interactively selected.
We considered two transverse scans of the radius: a distal scan at 1/10 of the total length of the radius from the radio-ulnar articular surface, and a very distal scan, 10mm distal from the former (see fig.1). A monodimensional analysis of the two rows of the matrix obtained from the image acquisition has been performed. Each row defines the equivalent heigth f(x) of the tissues along the direction of the transverse scan. The value of f depends on the attenuation of the radiation beam interacting with soft tissues and bone tissues in the patient forearm. In order to evaluate bone tissues only, at least in a first approximation, digitized data have been corrected in the following way. An interval (x_1-x_2) (ROI interval) is first selected, where the extremes x_1 and x_2 correspond to the edges of the radius on the digitized image, then the area value A is computed according to the equation:

$$A = \sum_{x_i=x_1}^{x_2} f(x_1) - P \qquad (1)$$

where P is the area of the pedestal defined by the horizontal axis and the straight line passing through the points $f(x_1)$ and $f(x_2)$ (see fig.2).
The obtained values in distal and very distal sites have been compared with the BMC values given by the DPA on the same nominal sites: in fact BMC values depend only on bone tissues. The correlation coefficients between the two sets of values show the effectiveness of the procedures (see fig.3 and 4).
The trabecular component of bone tissues was evaluated by measuring the high frequency rms signal in the ROI interval on the selected rows (see fig.5); this way the coarseness C is obtained which permits to quantify in an acceptable way the web-like pattern determined by the trabecular component [4]. C is computed according to the equation:

$$C = \frac{1}{x_2 - x_1} \sqrt{\sum_{x=x_1}^{x_2} (f(x_i) - g(x_i))^2} \qquad (2)$$

where g(x) is the low frequency component of the digitized signal, obtained by means of the convolution of f(x) with a constant value step mask.

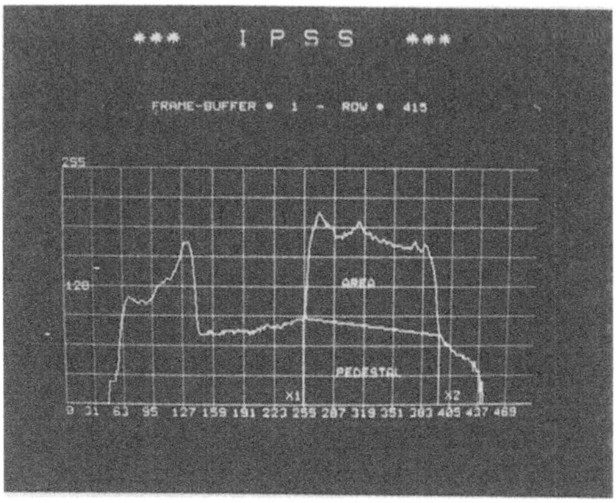

Fig.2.Measurement of the area under the row profile. The measured area is the difference between the total area of the graph from x_1 and x_2 and the pedestal area. Horizontal axis is quoted in 1/8 mm; vertical axis in mmAl, projected onto the 0-255 gray levels interval.

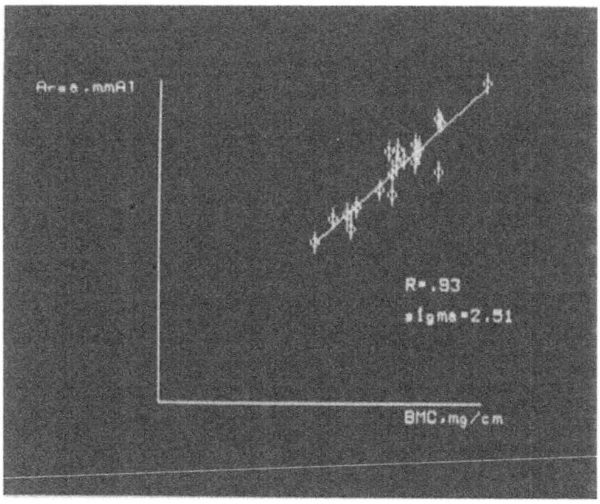

Fig.3.Scatter diagram of measured area vs BMC in distal site. Area values range from 26 to 53 mmAlxmm; BMC values range from 551 to 1170 mg/cm. The correlation coefficient is r=93 and the standard deviation is σ=2.51

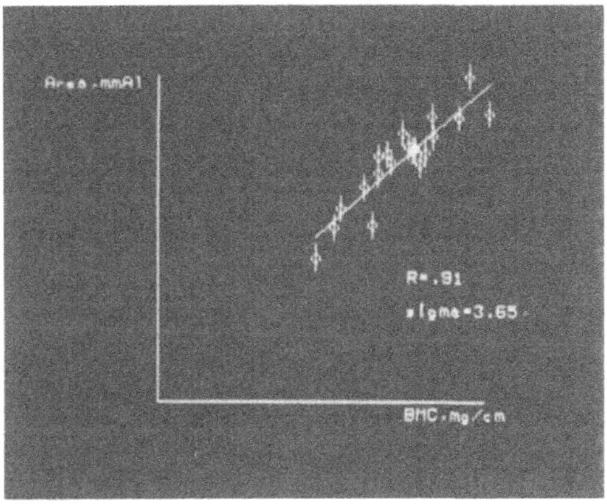

Fig.4.Scatter diagram of measured area vs BMC in very distal site.
Area values range from 27 to 62 mmAlxmm; BMC values range from 542 to 1145 mg/cm. The correlation coefficient is r=91 and the standard deviation is σ=3.65

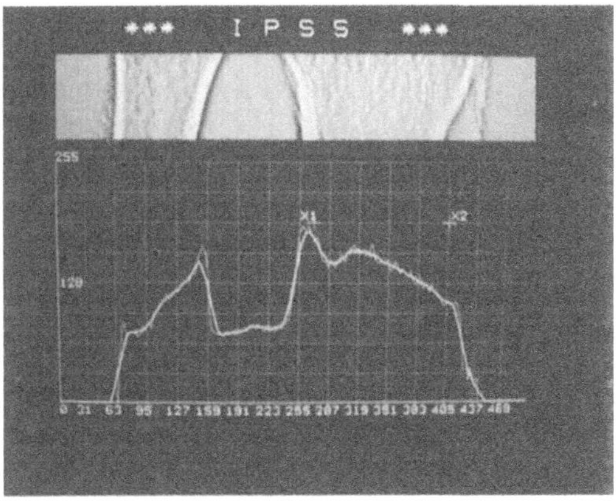

Fig.5.Measurement of coarseness C in distal site.
The original signal is represented by the dark curve. The smoothed curve used to compute C is represented by the white curve; x_1 and x_2 define the measurement interval. In the upper side of the picture the high frequency components of the digitized image are shown.

Conclusions

The use of procedures of signal analysis and image processing on off-line digitized X-ray films provides the radiologist with a further versatile instrument in research activities dedicated to the diagnostic quantifications and to the standardization of parameters.
The work-station used for our experiment presents two very important pecularities: a high resolution device for image digitization, which allows to exploit the high spatial resolution of the X-ray films, and a software system which permits to develope efficiently programs and procedures for the image processing.
The scanning resolution allows to define the coordinates with an error of .5mm. The repositioning errors in longitudinal studies of the same patients will be very low.
Up to now the obtained measurements on twenty cases agree with reported observations in different methods. The C values increase from the distal site to the very distal site determination; moreover the variations of the values with respect to the mean value, are matched to the observed pathological aspects.
The study is continuing on a larger set of patients in order to verify if the values of C or C/A ratios and the variations of such parameters in the axial direction of the examined bone can provide further information for the diagnosis.

References

1. Snyder, W.S.; Cook, M.J.; Nasset, E.S.: Report of the task group on reference man ICRP Pubblication 23. New York, Pergamon Press 1975.

2. Azzarelli, L.; Chimenti, M.; Lazzeri, M.; Salvetti, O.: Interactive integrated modular system for digital radiology. In "Computer Assisted Radiology" Procs. Car'85, U.H.Lemke, M.L.Rhodes, C.C.Jaffee, R.Felix eds., Berlin Springer (1985) 320-328.

3. Azzarelli, L.; Chimenti, M.; Salvetti, O.: Distributed systems for image processing in digital radiology. In "Planning considerations in diagnostic imaging and radiation therapy", A.Chiesa, R.Gasparotti, RR.Maroldi eds., Florence, (1988) 318-321

4. Trouerbach, W.T.; Grashuis, J.L.; Zwamborn, A.W.; Clermonts, E.C.; Schouten, H.J.: Microdensitometric analysis of bone structures in X-ray images. Skeletal Radiology (1987) 16, 190-195

Artificial Intelligence in Chest Radiology: Computer-aided Detection and Quantitation of Disease

H. MAC MAHON, K. DOI, M.L. GIGER, S. KATSURAGAWA, N. NAKAMORI, Y. SASAKI.

Kurt Rossmann Laboratories for Radiologic Image Research,
Department of Radiology,
The University of Chicago, Chicago, Illinois, U.S.A.

Summary

A unique advantage of digital imaging is the facility to apply computer techniques, not only to enhance the images for viewing by the radiologist, but also to analyze the contents objectively. Though automated image analysis for computer-aided diagnosis (CAD) is still in its infancy, the results of preliminary studies clearly indicate a major potential for the future.

Several facets of CAD are presently being developed at the University of Chicago for application in chest radiology as well as in mammography and vascular imaging. To date, we have focused on three types of chest pathology, namely, interstitial lung disease, pulmonary nodules and cardiac abnormalities. The ultimate goal of these endeavors is to provide a system for comprehensive automated image analysis, the results of which could be accepted or modified at the discretion of the radiologist.

Methods

The general approach has been as follows: (1) Selection of conventional analog radiographs containing examples of specific pathology as well as normal controls. (2) Digitization of radiographs using drum scanner or laser scanner. (3) Transfer of digital data to VAX 11/750 or MicroVax computers. (4) Development and application of CAD techniques including background suppression and feature extraction methods, to identify, localize, and classify abnormalities. (5) Evaluation of CAD programs by comparing computer-aided and unaided radiologists' performance. (6) Development of optimal output formats to facilitate use of CAD in clinical practice.

Computer-aided diagnosis of interstitial lung disease:

Diagnosis of interstitial lung disease by radiologists is notoriously unreliable due to inter and intra observer variations. Therefore, a computer technique which could detect and classify abnormal lung textures would be useful to reduce subjectivity and improve diagnostic accuracy. The ability to provide quantitative measures of pulmonary infiltrate would also be important, especially in the pneumoconioses.

The computerized method that we have developed uses physical measures derived from an analysis of the power spectrum of the lung texture [1]. The basic scheme is illustrated in Figure 1. Selected conventional chest radiographs are digitized with a 0.1mm pixel size. Regions of interest (ROIs) of 6.4 x 6.4mm are then selected from the posterior intercostal spaces. A program to select ROIs automatically is being developed in our laboratories, though semiautomated selection is used at present. The contents of each ROI is analyzed, and correction for non-uniform background (background trend) is made. A 2-D Fourier transform is performed and the data are filtered using the human visual system response filter. From this, two basic parameters are derived, namely the root mean square (RMS) pixel variation, and the first moment of the power spectrum. These measures describe the magnitude and coarseness, respectively, of the interstitial pattern.

When the RMS values are plotted against the first moment of the power spectrum, distinctive trends are seen depending on the nature and severity of the interstitial abnormalities (Fig.2). Abnormal lungs with interstitial infiltrate are characterized by higher RMS pixel variations and/or lower frequency components depending on the character of the infiltrate. As shown, the computed results correspond well with the visually perceived findings.

To evaluate the accuracy of this technique compared to unaided radiologists, we applied it to 20 abnormal lungs with mild interstitial infiltrates and 47 normals that had been previously selected for an independent ROC study. The computed results compared favorably with the average results of 12 radiologists in distinguishing normals from abnormals.

In a more recent study, the program was applied to the ILO standard pneumoconiosis films. Again a good correlation between the ILO gradings and the computed results were obtained, provided that appropriately representative areas of the radiographs were used [2]. Standardizing radiographic density and contrast through digital acquisition could further improve the accuracy and reproducibility of this technique.

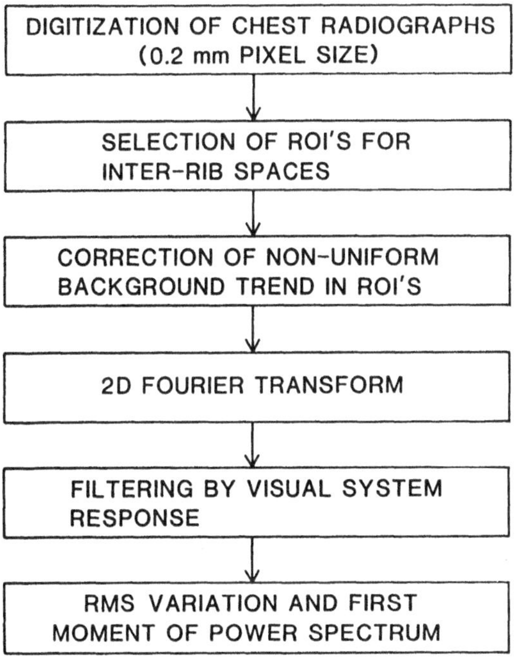

Figure 1. Outline of automated scheme for detection and classification of interstitial lung disease.

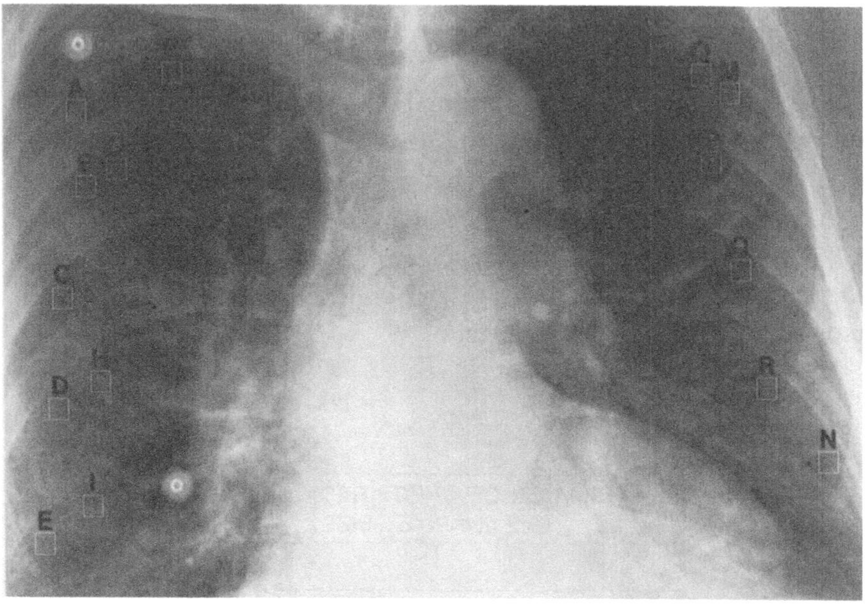

Figure 2a. Chest radiograph of a patient with bilateral lower lobe reticulonodular interstitial infiltrates. The location of each ROI is indicated by a letter.

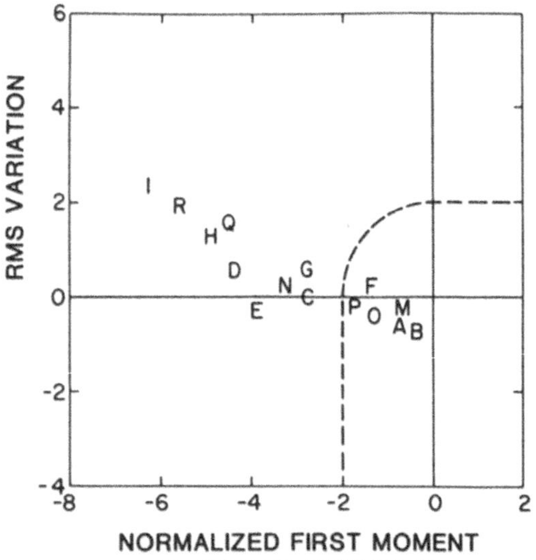

Figure 2b. Plotted values for each ROI indicate that abnormal (lower lobe) areas of the chest exhibit higher RMS variations and lower frequency components compared to normal (upper lobe) areas. Dashed line indicates upper bound for normal lung values (2 standard deviations), based on 100 normal cases.

Computer-aided detection of pulmonary nodules:

According to some estimates, radiologists fail to diagnose pulmonary nodules in up to 30% of positive cases. In one large lung cancer screening program, 90% of peripheral lung cancers were visible in retrospect on earlier radiographs [3]. The objective of our nodule detection program is to direct the radiologist's attention to suspicious densities in radiographs which may represent pulmonary nodules. The general scheme of the nodule detection program is outlined in Figure 3. Our approach differs from previously described nodule detection schemes in that a "background suppression" method is used to minimize structured noise prior to application of feature extraction techniques [4].

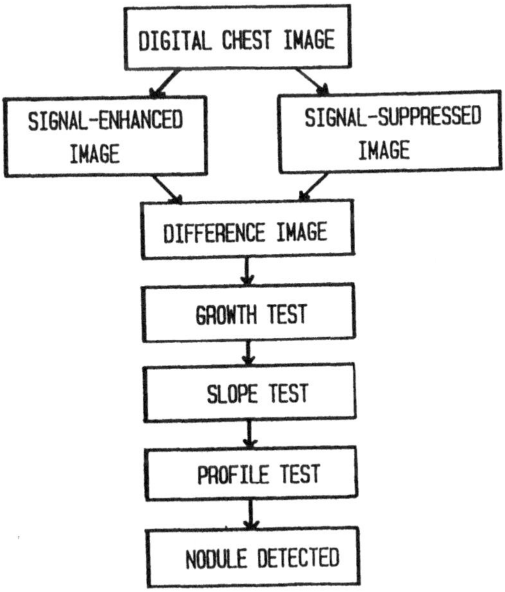

Figure 3. Outline for scheme for automated nodule detection.

First, chest radiographs are digitized with a 0.1 mm pixel size and spatial averaging is performed to reduce the matrix to 512 x 512, corresponding to an effective pixel size of 0.6 mm. From this initial image "signal enhanced" and "signal suppressed" images are produced by filtering. From these two, a "difference image" is obtained in which background structured noise has been minimized. Briefly, this image is subjected to:

(1) A "growth test" which examines the degree of circularity of possible nodules over a range of grey level thresholds (True nodules tend to retain their circular shape at varying thresholds)

(2) A "slope test" which examines the ratio of the change in effective nodule diameter to the change in threshold level (pseudonodules tend to change more rapidly and sporadically than real nodules with thresholding).

(3) A "profile test" is applied in the perihilar regions to examine the contrast of possible nodules relative to their size (real nodules tend to have lower contrast than end-on vascular structures, though both may appear circular).

Each of these feature extraction techniques helps to eliminate false positive detections.

At present, false positive detections are undesirably high with an average of 7.5 per case for a nodule detection accuracy of 70%. However, it should be noted that these data are based on tests including some extremely difficult and subtle nodules. In a preliminary observer test involving 60 clinical cases (30 normal and 30 with nodules of varying subtlety), the performance of the radiologists was slightly improved by using the computer output even with the existing false positive rate. Our present goal is to increase the accuracy of the program further both by reducing false positive detections and by increasing sensitivity for real nodules.

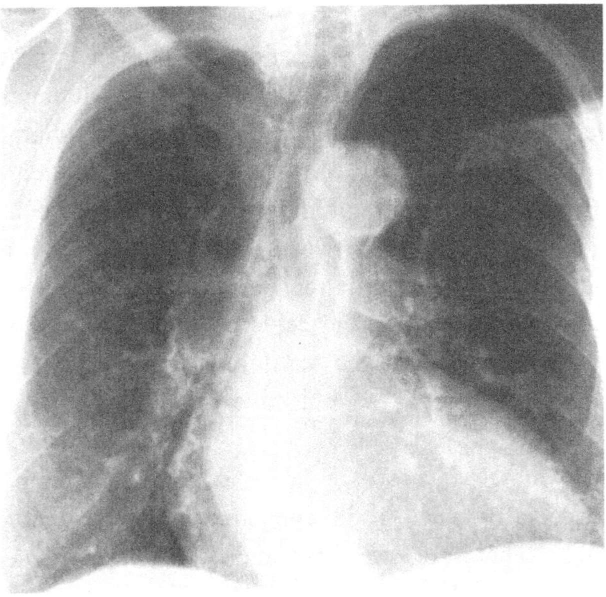

Figure 4a. Chest radiograph of a 78 year old woman with small right upper lobe lung cancer, partially obscured by ribs and clavicle.

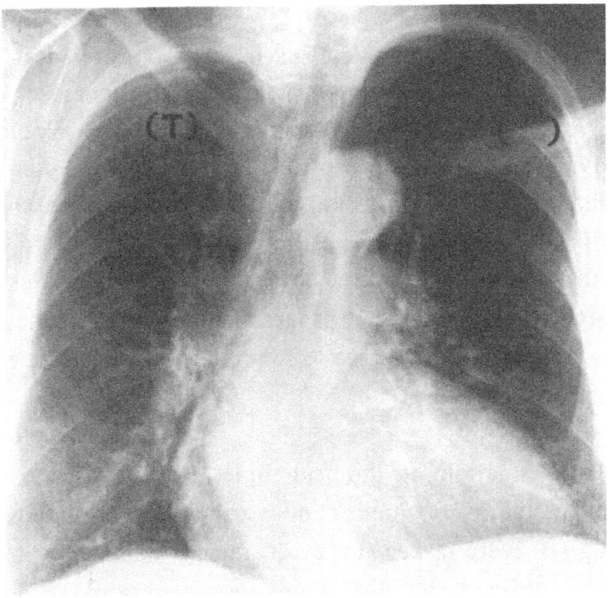

Figure 4b. CAD results showing successful detection of tumor (T) with two false-positive detections in the left lung. Computer display is simulated here in interests of clarity for illustration.

Detection of Cardiac Disease:

Cardiomegaly is one of the more common and important abnormalities diagnosed on chest radiographs. Although a grossly enlarged heart is an obvious finding, mild cardiac enlargement or a modest change in heart size can easily be overlooked in clinical practice. Even when the radiologist's attention is focused on the heart, confident distinction between normal and abnormal can be difficult, and interobserver variations are common. Though complex cardiac measurements have been devised, they are not widely used, and despite its limitations, the cardiothoracic ratio is still widely employed by radiologists as an objective criterion for cardiac size.

The projected frontal area of the heart is potentially a more accurate parameter, especially when it is compared to the overall dimensions of the thorax. Therefore, we have developed a program which can automatically define the cardiac and thoracic margins, and provide objective measurements and ratios of cardiac and thoracic dimensions. The general scheme of this technique is shown in Figure 5.

Digitized chest radiographs were used. The initial large matrix of 3500 x 3500 was reduced to 128 x 128 to decrease computation time. The algorithm used to detect cardiac boundary points is based on an analysis of signature profiles and their first derivatives [5]. Sample contours for the projected cardiac shadow were based on the estimates of four experienced radiologists. The heart contours as drawn by the radiologists were best approximated by using a shift-variant cosine function. The program automatically determines the cardiac area relative to the area of the thorax and also estimates the cardiothoracic ratio (Fig.6).

As an initial evaluation of the technique, it was applied to 46 chest radiographs, including examples of various heart shapes and sizes. Ten radiologists were asked to evaluate the accuracy of the cardiac silhouette as defined by the computer program. The computer result was rated as acceptable to excellent in the majority of cases.

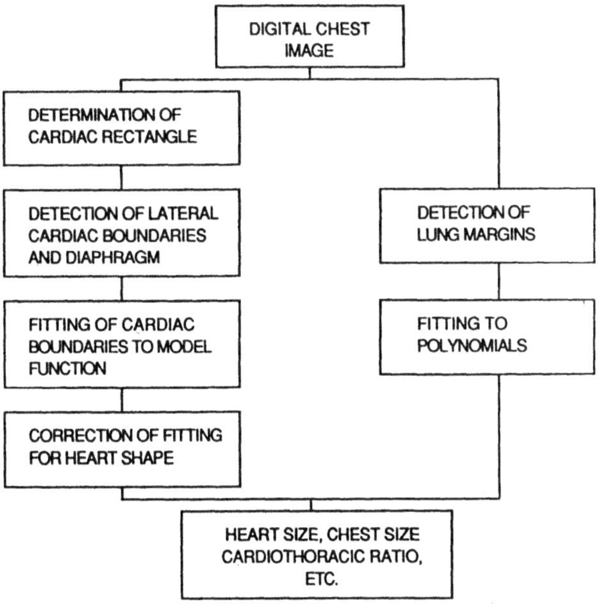

Figure 5. Scheme for automated determination of cardiac size.

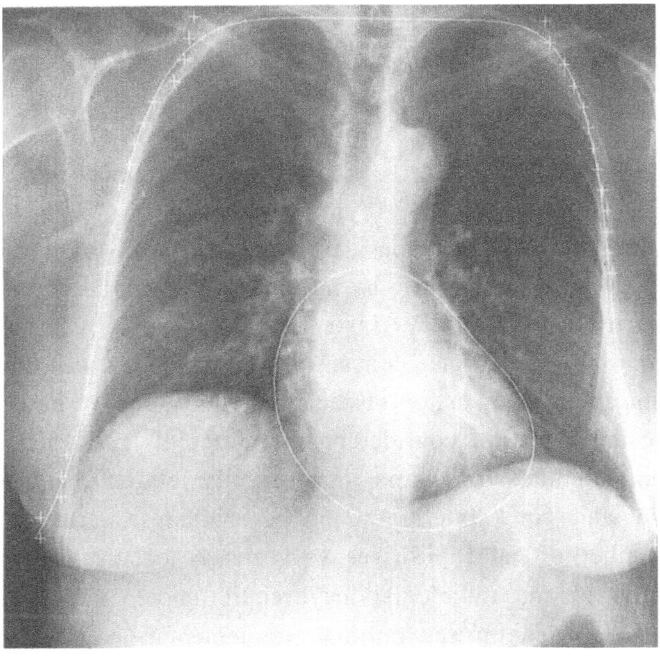

Figure 6. Computer derived cardiac and pulmonary borders.

Discussion

The limited evaluations described here illustrate the potential for using CAD to improve diagnostic accuracy and reduce variability in radiological interpretation. If CAD is to be used routinely, it will be necessary for the system to achieve good sensitivity while keeping false positive results to a minimum. It is not yet clear how many false positives radiologists will be willing to tolerate in this type of system. However, the preliminary results shown here as well as other CAD evaluations suggest that radiologists' diagnostic accuracy can be improved even while computer generated false positives remain relatively high [6,7].

A further question concerns how best to evaluate the clinical utility of CAD. Though ROC analysis has come to be regarded as the gold standard in this area, it has certain practical limitations. In an ROC test, the radiologists attention is sharply focused and concentration tends to be high due to the high likelihood of subtle pathology in the test cases. The kind of detection errors that are common in clinical practice are less likely to occur in this situation. Thus, it may be difficult to demonstrate the real value of CAD programs using ROC tests alone.

In the future, CAD programs could be applied routinely to all digital radiographs prior to viewing by the radiologist, though he or she might choose to view and consider the CAD results only in selected cases. If CAD programs can be made highly accurate, however, they might merit use in every instance. Although some of these algorithms are very time consuming at present, we do not see computation time as a major obstacle in the future. These programs can be designed to run considerably more rapidly, and also could be run in the minutes or hours that usually elapse between image acquisition and final interpretation. Nonetheless, assuming that a certain number of false positives will always occur, analysis of the CAD results by the radiologist may require more time than unaided interpretation. Therefore, optimal implementation to allow rapid review of CAD output displays will be critically important in order to gain acceptance of CAD among experienced radiologists.

Acknowledgement:

This work was supported by USPHS Grant CA 24806 and a Whitaker Foundation Research Grant.

References:

1. Katsuragawa S, Doi K, MacMahon H: Image feature analysis and computer-aided diagnosis in digital radiography. Detection and characterization of interstitial lung disease in digital chest radiographs. Med Phys 15: 311-319, 1988.

2. Katsuragawa S, Doi K, MacMahon H, Nakamori N, Sasaki Y, Fennessy JJ: Quantitative analysis of lung texture in the ILO standard radiographs for pneumoconiosis. Radiology 169 (P): 369, 1988.

3. Muhm JR, Miller WE, Fontana RS, Sanderson DR, Uhlenhopp MA: Lung cancer detected during a screening program using four-month chest radiographs. Radiology 148: 609-615, 1983.

4. Giger ML, Doi K, MacMahon H: Image feature analysis and computer-aided diagnosis in digital radiography. 3. Automated detection of nodules in peripheral lung fields. Med Phys 15 (2): 158-166, 1988.

5. Nakamori N, Doi K, Sabeti V, Chan HP, Katsuragawa S, MacMahon H: Computerized analysis of the heart and lungs in digital chest radiographs. Radiology 169 (P) 160, 1988.

6. Chan HP, Doi K, Vyborny CJ, Schmidt RA, Metz CE, MacMahon H, Lam KL, Wu Y: Effects of computer-aided detection on radiologist performance: Pilot study on detection of mammographic microcalcifications. Radiology 169 (P): 159, 1988.

7. Giger ML, Doi K, MacMahon H, Metz CE, Yin FF: Computer-aided human detection of pulmonary nodules in digital chest images. Radiology 169 (P): 444, 1988.

Computer Assisted Diagnosis of Bone Tumors

K.H.ROTTE, E.PERLICK , G.KUNDT[x]

Central Institute for Cancer Research of the Academy of Sciences of the GDR, Berlin-Buch

[x]Dept. for Medical Statistics, Faculty of Medicine of the Wilhelm-Pieck University, Rostock

Summary

Computer assisted approaches are helpful in diagnostic decision making. The relevant X-ray symptoms of bone tumors were determined by re-evaluation of the films of 374 patients. Using the BAYES´s equation the diagnosis and differential diagnosis were performed by a simple personal computer dialogue. As a result we obtained an accuracy of 79 % for 5 different bone tumors and tumor-like lesions.This result was better than the empiric medical accuracy of an observer study. This approach is simple to handle and can improve the diagnosis of bone tumors.

Introduction

Mathematical approaches can be helpful in special fields of radiodiagnosis. The radiologist has to decide which roentgenographic findings are most likely responsible for the diagnosis of determined diseases. He also has to determine how confident he can be about his diagnosis in each case. In the diagnosis of bone lesions often a great number of items are involved which interact very variably,making it difficult to take a diagnostic decision. Bone tumors are relatively uncommon and have an incidence of approximately 1 % of all malignant tumors. Many radiologists are of little experiences in the diagnostics of bone tumors.This present study was performed to develope an approach for improvement of the diagnostic accuracy,to create a method, which can be used by personal computer (PC) and which also may be applied in smaller radiologic units. The aim of this study is to reduce an iatric caused delay in classifying bone lesions.

Material and method

In a retrospective re-evaluation of the X-rays of 374 patients

with histologically proved bone tumors the incidence of different signs and symptoms was determined and lined up to a symptom index.For further calculations we used the most frequent and relevant symptoms as age, localization,site to the bone axis,shape of destruction, margins,tumor matrix, cortical alterations, periosteal reactions and soft tissue involvement.Our study was limited on 5 of the most common bone lesions like osteosarcoma,chondrosarcoma,Ewing´s sarcoma, giant cell tumors and osteomyelitis, which represent more than 60 % of the bone lesions in our material.
BAYES´s theorem was chosen as mathematical approach.In connection of a-priori probabilities with limited probability of one or several symptoms we are able to calculate the probability of the disease D, if the symptoms S are present.The programme is in FORTRAN. A personal computer PC 1715 of ROBOTRON (GDR) was used.Based on a computer menu the present symptoms were put in, then calculations were started. We distinguished 4 safety degrees: highly suspicious with a probability of more than 99 %, probable with 75 - 99 %, possible with 10 - 75 %, and unlikely with less than 10 % probability. The diagnosis with the highest probability was taken as the correct diagnosis.For the practical use we favour this verbal assessment because there is more scope for the interpretation of the results and it does not give a wrong safety due to the frequency of percentages, which can be less precise in the range of lower probabilities.

Results

To determine the accuracy and the value of this approach we did a prospective study only,that means that these cases are not included in the datas , we used for the symptom index.

Tabl.1

Results of diagnostic calculations in bone tumors

Diagnosis	n	correct	%
osteosarcoma	15	12	80
chondrosarcoma	6	5	83
Ewing´s sarc.	8	5	63
giant cell tumor	4	4	100
osteomyelitis	10	8	80
total	43	34	79

The result of this computer study was compared with the results of an observer study, performed in the same cases by 5 radiologists. In the observer study the accuracy differed between 68 % (experient radiologists) and 44 % (radiologists with little experience in the diagnostics of bone tumors). A further improvement of the computer results seems to be possible.

Discussion

Computer assisted diagnostics has been proved as an improvement of the diagnostic accuracy of bone tumors, compared to the medical diagnosis done by radiologists.Our results are similar to the results of Lodwick et al.(3) and of Virtama et al. (5) with an accuracy of 78% respectively of 82 %. A higher accuracy of 96,7 % is reported only by Fotter et al.(1) who included the most common paediatric bone lesions in their study.

Our model can be performed by a PC in a simple way and can support the diagnosis of radiologists, particulary of those with little experience in bone diagnostics. This approach can not replace the histomorphologic diagnosis.The aim of this study was to develope a simple supporting method for the radiodiagnosis of bone tumors and tumor-like lesions. An earlier tumor diagnosis may shorten the time of the reference to a specialist or special hospital.

The use of a computer assisted diagnostic model requires a precise analysis of X-rays to obtain the relevant datas for the decision process. This simple evaluation programme by use of a PC makes the application of a mathematical approach easier in smaller radiologic departments and may be better accepted by radiologists than previous expensive and labour intensive diagnostic models (4).

References

1. Fotter,R.,Gell,G.,Melzer,G.,Kopp,W.,Lehnert,M.,Weybora,W.: Computergestützte Diagnostik von Knochentumoren und tumorähnlichen Skeletterkrankungen:Kritische Bewertung des klinischen Einsatzes. Fortschr. Röntgenstr. 148 (1988) 498 - 504

2. Lodwick,G.S., Hun,C.L., Smith,W.E., Keller,R.F.,Robertson, E.D. : Computer diagnosis of primary bone tumors. Radiology 80 (1963) 273 - 275

3. Lodwick,G.S., Reichertz,P.: Computergestützte Diagnostik von Tumoren und tumorähnlichen Veränderungen des Knochens Das begrenzte BAYES´sche Konzept. Röntgenblätter 22 (1969) 162 - 166

4. Rotte, K.H.: Computergestützte Röntgendiagnostik am Beispiel peripherer Lungenprozesse. Akademie Verlag Berlin, 1977

5. Virtama,P., Katevuo,K., Mäkela,P., Mäkinen,E.O.: Computer aided diagnosis of bone tumors. Acta Radiol. Diagn. 20 (1979) 70 - 74

STRES: An Expert System for Strokes

D. I. KOUNTANIS
E. C. KOUNTANIS*

Computer Science Department
Western Michigan University, Kalamazoo,
*Kalamazoo Neurology, Kalamazoo, MI

Summary

This paper describes an Expert System for diagnosis and treatment of strokes. The Expert System is intended to be used by Physicians in emergency rooms, so a major effort was directed towards real time responses by the system. The system has a network architecture that includes various knowledge base modules, a scheduler, a communications protocol, a message manager and a number of inference subsystems. It can be used as a problem solving mechanism, as a tutoring facility and finally as a learning tool. The system is still in the development stage.

Introduction

The diagnosis problem using binary tests has been shown to be NP-complete problem [2]. This means that there is no procedure that solves the problem in polynomial time. The proof is based on a graph representation of the problem [1] and the use of vertex covers [4] to obtain a diagnosis that explains a set of observed symptoms and signs. Peeters et. al. [7] have proposed an expert system for diagnosis of strokes based on set coverings. They have presented two algorithms, one sequential and one parallel, to design a system that works on real time for emergency physicians. On the other hand, Kountanis et. al. [9] have proposed an integrated approach to the control of expert systems that is used in the implementation of the STRES. The integration utilizes the power of Petri Nets to represent parallel processes and also to unify the various knowledge base representations used in STRES. The architecture of STRES closely follows the one presented in [10] and also in [8]. The inference engine uses a heuristic search similar to the one presented in [5] and utilizing patterns on the representations of the knowledge base described in [6].

STRES Architecture and Module Definition

The diagram in figure 1 illustrates the architecture that is used in the STRES system. The definition of the various system modules is as follows:

Network of Knowledge Bases: Contains a communications subsystem, a scheduler, a controller, and a message handler. In addition, this module contains the various knowledge bases.

Anatomy Expert: This module contains information about the anatomy of the brain, i.e. a brain map with the functions associated with each locality. There are 54 different localities that have been considered..

Localizer: Given a set of disorders the localizer determines the part(s) of the brain that have been involved in the stroke.

Stroke Expert: Contains knowledge about the cause-effect relationships in the occurrences of strokes.

Treatment Expert: When the diagnosis of a stroke has been determined and its locality has been defined, this part of the system contains the knowledge to suggest a therapy to be followed.

Explanation: When the system reaches a diagnosis, the explanation module provides the inference path that has been followed in the reasoning process.

Prognosis Expert: What is the future outlook for the patient? This expert system provides a prognosis for the patient.

Inference Engine: This part is the workhouse of the system. It uses the knowledge bases and emulates the expert's reasoning. It also analyzes situations, hypothesizes diagnoses or localities and checks their validity.

User: The user wants to use the system to find answers to given problems.

Tutor: The user wants to enhance the knowledge of the system.

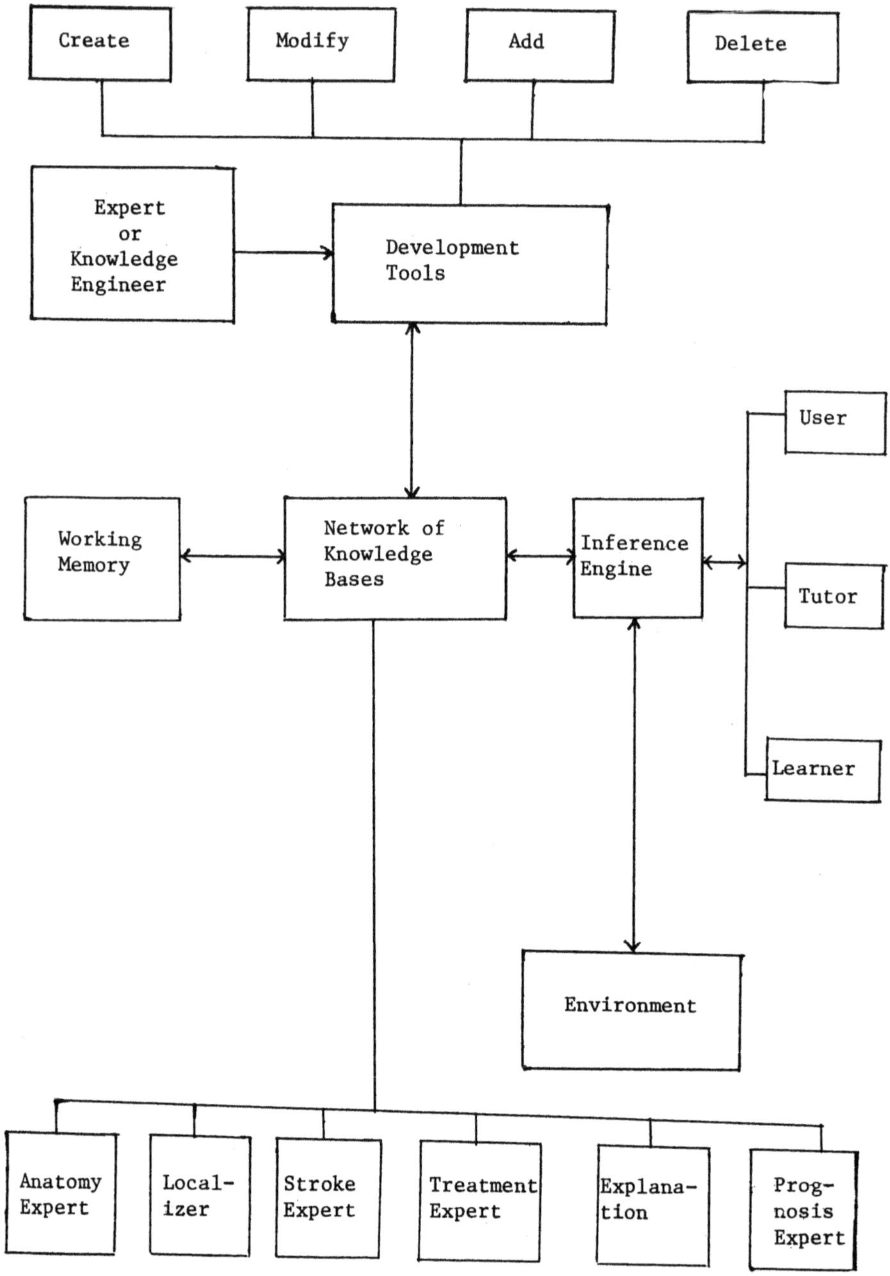

Figure 1. STRES Architecture

Learner: The user wants to learn from the expertise of the system, i.e. the system acts as a teaching device.

Development Tools: This part is the knowledge acquisition subsystem and it is vital in the creation of the expert systems. The tools provided by this subsystem include an editor and modules to create, delete, add, and modify information in the knowledge bases.

Working Memory: Here the knowledge bases are modified by the inference engine. Here also information is manipulated by the inference engine in pursuit of a solution.

The Diagnosis Process

A diagnosis is obtained through values on a set of specified parameters. The inference engine of STRES uses these values to search a table that represents cause-effect relationships in order to obtain the most probable diagnosis. Baye's Theorem is used to choose the most likely diagnosis. The formula used is

$$P(D_i/S) = \frac{P(D_i) \cdot P(S/D_i)}{\sum_{j=1}^{n} (P(D_j) \cdot P(S/D_j)}$$

where:

$$P(S/D_i) = P(S_1/D_i) \cdot P(S_2/D_i) \cdots P(S_n/D_i),$$

D_i is one of n disjoint diagnoses,

$P\ (D_i)$ is the priori probability that the patient has disease D_i before any evidence has been collected,

$P(S/D_i)$ is the conditional probability when S symptoms have been observed.

Conclusion: The parallel process capabilities of STRES obtained through its architecture and its integration through Petri Nets promises a real time stroke diagnosis that is very vital to emergency room physicians.

[1] Behzad, M., Chartrand, G., and Lesniak-Foster, L., Graphs and Digraphs, Wadsworth International Group, 1979.

[2] Garey, R., and Johnson D., Computer Intractability: A Guide to the Theory of NP-Completeness, Freeman and Company, 1979

[3] Greenwell, M., Knowledge Engineering for Expert Systems, Ellis Horwood Limited, 1988

[4] Kohavi, Z., Switching Theory and Finite Automata Theory, McGraw-Hill Book Company, 1978

[5] Kountanis, D., Heuristic Search in Biomedical Images on an Image Analysis System, The 6th Scandinavian Conference on Image Analysis, 1989 (to appear).

[6] Kountanis, D., Muller, B., and Rashidi, R., The Genus of Graphs and the Strong 2-Cell Imbeddings, Capital City Conference on Combinatorics and Theoretical Computer Science, 1989 (submitted).

[7] Peeters, M., Evans, M., and Hier Daniel, A Comparison of the Generalized Set Covering Algorithm With a Decision Tree Algorithm for the Diagnosis of Stroke, Proceedings of Expert Systems for Advanced Manufacturing Technology, Dearborn, Michigan, 1987.

[8] Quinlan, R. (editor), Applications of Expert Systems, Turing Institute Press, 1987. (Based on the Proceedings of the Second Australian Conference).

[9] Rashidi, R., and Kountanis, D. Integrated Techniques for Control of Flexible Manufacturing Systems, Expert Systems Solutions in Manufacturing, Detroit, Michigan, 1989.

[10] Wolfgram, D., Dear, T., and Galbraith, C. Expert Systems for the Technical Professional, John Wiley & Sons, 1987.

Odile: An Analytic Approach to Automated Interpretation of CAT-Scan Images of the Cervical Spine

Zollner G., Wackenheim A., Lachmann F.

Service de Radiologie I, Pavillon Clovis Vincent, Hospices Civils de Strasbourg, B.P. 426, 67091

Summary:

ODILE *(Outil de Diagnostic Informatisé pour La recherche et l'Enseignement en radiologie)* represents an analytic and modular approach to computer assisted diagnosis of CT-scan images. It continues the work begun with ORFEVRE on the cervical spine within the concept of a combination of low and intermediate image treatment procedures with an expert-system approach for high level analysis. It integrates three modules:

- an image processing module, based on the software VISILOG,
- a rule based expert system and
- an image data base using the ORACLE environment.

Parameter extraction is accomplished by automated procedures, combining standard low and intermediate level image treatment routines and specially developed procedures. Knowledge is transferred from the image processing module to the expert system by parameter exchange using a structured file. After numerical - symbolic conversion the parameters are analyzed by the expert system.

The rules of the expert system take into account the clinical context of the patient.

The associated image data-base is strongly parametrized in terms of radiologic parameters and signs and queries can be made for these parameters which in their majority are still inaccessible to automated extraction.

Individual analysis is presented for the cervical canal, the differential diagnosis of lytic lesions and the evaluation of vertebral symmetry.

Introduction:

Automated diagnosis and feature extraction from radiologic images is still in its infancy and until now there has been little flow from research studies to commercially available procedures in CT and NMR: a rare example is automated density measurement of the vertebral spongiosa. This underlines the fact that human vision still outperforms computer vision in Radiology and even for automated extraction of quantitative features there is a lack of generally appliable procedures. A framework has been published by Stiehl (Stie87). Automated diagnosis relies heavily upon feature extraction - corresponding to the extraction of signs and parameters in the radiologist's sense and therefore has still more problems: the inference from roentgen signs and parameters to diagnosis must be resolved. If probabilities are lacking and experts judgements are intervening, the expert systems approach is reasonable (for CT:Rada 83) It allows to develop and test the knowledge base even if a complete set of feature extraction procedures is not available.

Materials and methods.

Within the concept of a unified approach for image analysis, expert system and data base, the UNIX environment has been chosen. It provides compatibility between developments of the different partners and provides the facilities (NFS with ETHERNET) for data exchange between the two computers harbouring the different components of the system:

- a SUN 3/140 workstation (4 MB RAM) and a Matrox card (512*512 pixels resolution) for
  with the image analysis software VISILOG from noesis (Jouy, F) and an expert system, which for small developments is a c-based engine and for the whole rule-base the expert system COGNITIF from COGNITEC (Paris, F).

The images are taken by video camera from photographic support, exceptionally the long way to read CT-tapes is engaged.

Parameters are transferred from Visilog to the expert system by G-files: structured files, where the parameters of labelled regions (defined by low and intermediate image analysis procedures) are written. These parameters associated to regions (which can be anatomic regions as the vertebral body) are then converted to symbolic parameters for further analysis by the expert system (which provides symbolic high level analysis).

- an IBM PC-RT workstation (16 MB RAM) and a MEGAPEL card (1024*1024 pixels) with its proprietary UNIX environment AIX for the SQL-image data base which is now being ported to ORACLE. Developments with neuronal network techniques are done on both workstations.

The image are transferred from the SUN workstation to the IBM workstation on ETHERNET by NFS.

Concept of ODILE

A modular approach has been chosen:

- Automated feature extraction is provided by an image analysis module, with a workbench of standard image analysis procedures. Within the image analysis module individual prodedures are clickable with a window oriented userinterface. Automated sequences of analysis routines can be tested, written and implanted. Within the image analysis module automatized procedures have been developed for the identification of anatomic structures like vertebral body, vertebral canal, uncus, transverese foramen and posterior arc. These identification routines are clickable from the user panel or can be integrated into independent applications.

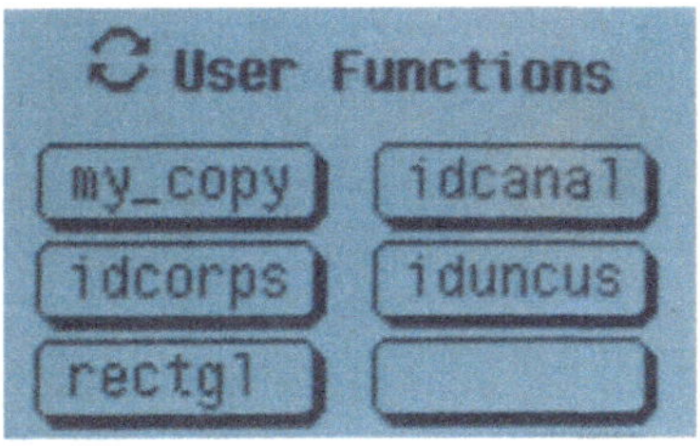

Fig. 1: *"mouse" clickable identification routines for the spinal canal, the vertebral body and the uncus.*

Identification algorithms are developed for bony structures or holes in bony structures and can therefore use thesholding and region labeling based on identification of first big structures or labelling of holes, combined with techniques of morphologic mathematics. Representative axial slices A, B and C as defined with ORFEVRE are to be chosen. Example: for identification of the vertebral body the representative slice B is chosen. The vertebral body is the first big bony structure if one is looking from in front. "Big" is defined by the number of erosions necessary to eliminate bony structures in front of the vertebral body (hyoid bone, calcified thyroid cartilage): fixed to five erosions by heuristic trials. After identification of anatomic structures, parameters are measured: Feret diameter, orientation, first point, inertia moments) These parameters are compared with normal values.

There may be problems: if the Feret diameter of the vertebral body is normal, too big or not too small (>half of normal) structure alanysis procedures (hole detection) are engaged. If the Feret diamenter measured with this algorithm is too small (< half of normal) there are three possibilities: 1. the vertebral body is really so small (congenital), 2. there is a huge osteolysis with disapearence of almost all bony substance of the vertebral body or 3. the vertebral body is fragmented and the bony piece measured was the first of several fragments. To solve this problem, a closing with 4 dilatations is initiated. If the diameter of the closed structure is almost the same as the original body, the structure identified was really the vertebral body or is the last bony piece of an huge osteolysis and further form analysis (cortical bone contour?) can differentiate the two. If the diameter of the closed structure is bigger (>10 %) than the original one, the vertebral body was fragmented by fracture or tumour and further analysis is necessary. The identification of the vertebral body in a simple case is shown in Fig. 2. Parameters are extracted automatically.

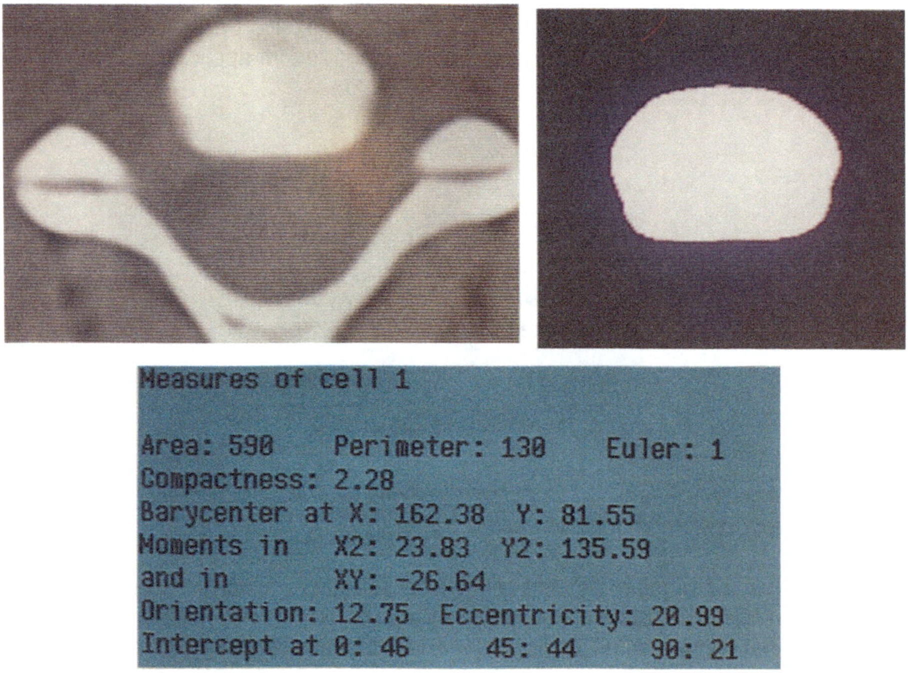

Fig. 2: *Automated identification of the vertebral body*
a. original image
b. vertebral body identified
c. automated parameter extraction fed into the expert system

An algorithm for the identification of a vertebral metastasis has been developed: first the vertebral body is identified. Then the holes are detected and a detailed analysis of the holes is made: in this case the diagnosis is based on the fact that there is one big lysis of very irregular shape (number of convexity and concavity) and not at least the clinical context.

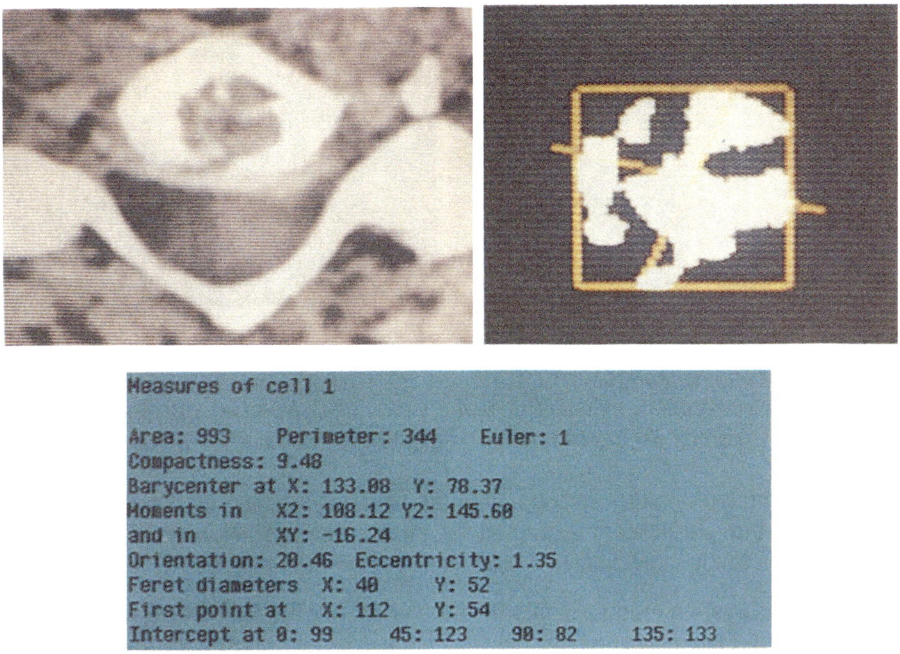

Fig. 3: *Identification of metastasis*
a. original image
b. osteolysis identified
c. parameter extraction

This is to be compared with the pathognomonic image of a hemangioma which is distinguished by a great number of holes in the vertebral body without contour change of the vertebral body. Demonstration of a high number (>10) of holes of the vertebral body with a diameter of 1 to 4 mm detects the hemangioma.

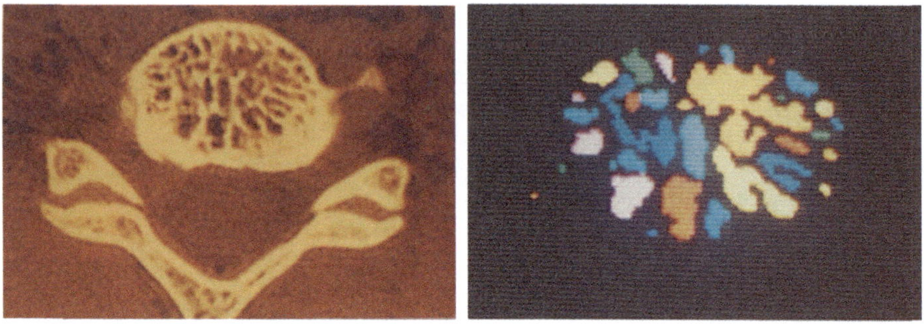

Fig. 4: Hemangioma
a: original image
b. vertebral body with (33) lesions identified. Number and diameter of the lesions distinguish the pathognomonic image of the hemangioma.

The image analysis module is coupled to the expert system which provides symbolic analysis of the parameters extracted by image analysis. Numerical to symbolic conversion is made individually for each item of the parameter file.

- The expert system module is rule-based and the original ORFEVRE rule-base has been modified from a sector-oriented to a more anatomically oriented system. It uses anatomical concepts as defined by the identification routines for the vertebral body.

  The underlying radiological concept is based on the verification or falsification of normality. Without entering into philosophical discussions, we assume that normality can be proved by a limited set of rules: the rules of normality. If the system cannot conlude to normality, there must be a pathologic lesion which is to be identified.

Example: analysis of the cervical canal (bony structures):

The analysis proceeds with three steps:

1. Identification of the canal
2. Measurements
3. Identification of pathologic lesions

1. the canal is identified by the following algorithm (taking the representative slice A):

   a. after thresholding the holes of the vertebra are identified,
   b. three erosions eleminate small holes,
   c. the canal is the first hole seen from 180 degree.

2. Standard measurements include:

   Feret diameter, surface, number of concavity and convexity, inertia moments, symmetry.

The a.p. Feret diameter serves for a crude evaluation of vertebral stenosis.
If the a.p. feret diameter of the canal is < 10 mm then there is stenosis of the canal. But these standard parameteres are not sufficient to characterize pathologic lesions, especially osteophytes. Therefore a special algorithm (taking a concept from ORFEVRE) has been developed: assuming that the vertebra is aligned with respect to the y-axis,

- the image of the canal is subtracted from the image of its Feret rectangle.
- the remaining (median and anterior) structure is identified and analyzed separately.

The remaining analysis is rule-based:

- if the diameter a.p. (Feret) of this region is > 2 mm there is a posterior osteophyte in median position and its diameter has to be indicated.
- if the a.p. Feret diameter of the cervical canal is < 10 mm and there is no osteophyte, the the patient has congenital stenosis of the canal.
- if the a.p. Feret diameter of the cervical canal is < 10 mm and there is an osteophyte, then the patient has congenital stenosis and an osteophyte.
- if the a.p. Feret diameter of the cervical canal is > 10 mm and there is an osteophyte, then the patient has no congenital stenosis and an osteophyte.

Fig. 5 shows the detection of a posterior osteophyte with an algorithm using Ferets rectangle and subtraction. The osteophyte is the bony structure impinging on the canal and its diameters are measured. The residual diameter of the canal (important for judging the clinical importance of the lesion) can also be measured automatically.

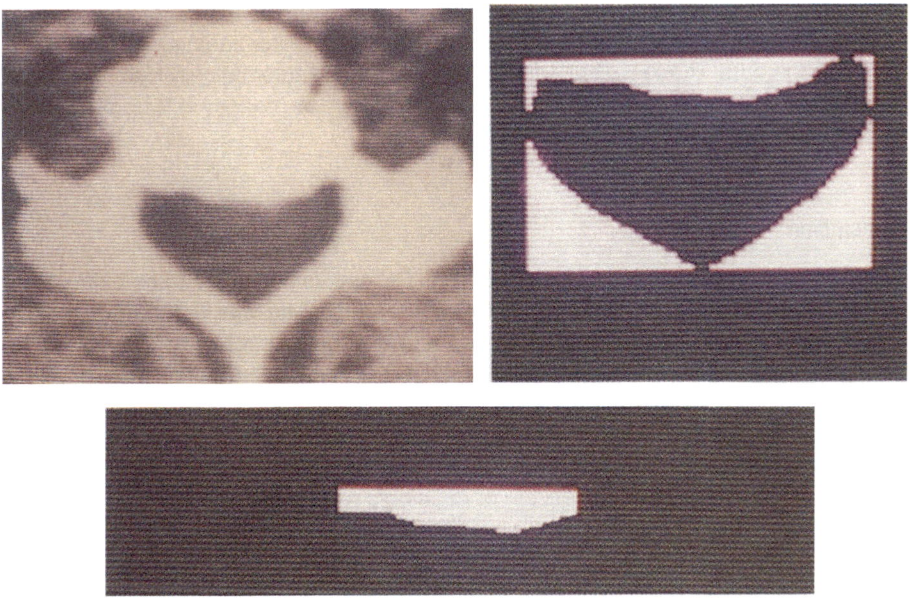

Fig. 5: Osteophyte
a. original image
b. spinal canal identified automatically with Ferets rectangel superposed.
c. Osteophyte isolated by subtraction and identification of the median region.

Comparison with human observer performance:

- median osteophytes,

which are readily identified by the raiologist are also identified by the system under the following conditions:

- the slice a representative A-slice. If the spinal canal is open, the identification procedure fails.
- the vertebra is not too much turned. If the vertebra is not a.P. oriented, the system allows to calculate and to correct the orientation. For time economic reasons this correction is not implanted.

- hemangioma

the image of a hemangioma is pathognomonic and the (rare) cases can be identified with the proposed algorithm.

- metastasis

the automated identification of lysis of the vertebral body functions well if there are isolated holes, but for more complicated lesions, which are not holes, the algorithm fails.

The image date-base:

An image data base associated to the system relies upon a strong parametrization of the images in terms of anatomic structures (which can in part be automatically identified and measured) and lesion. The lesion presents signs with anatomic location, dimensions, form and contours. These characteristics can be extracted automatically for osteophytes, hemangioma and lytic lesions of the vertebral body, but the majority of lesions need still an individual analysis by the user.

Conclusion:

ODILE continues the work begun with ORFEVRE and its concept can be extended to other imaging procedures and anatomic regions. Combining image treatment module and expert system creates an analogy to the radiologists practice of looking first for roentgen signs (which with Odiles analytic tools is still a long sequential task) and then interpreting the signs on a higher symbolic stage. A corollary of the work is to show the application of automated identification and parametrization tasks for anatomic structures and lesions with a user friendly interface. Following the politics of small steps this seems one way to go along for the extension of radiologic workstations.

References

(MARI88) Mariot P., Vesoul P.: Les techniques des systèmes en analyse d'images médicales, application à ORFEVRE. Intelligence Artificielle et Imagerie Médicale, ed. Wackenheim A., Springer, Paris, 1988

(RADA83) Rada R., Ackermann LV: An expert system for interpretation of computerized axial tomogramms of the brain. Proc. of the Oakland conference on Artificial Intelligence, 233-240, 1983

(STIE87) Stiehl H.S., Jackel D.: On a framework for processing and visualizing spatial images. Computer Assisted Radiology, CAR '87, ed. Lemke H.U., Rhodes M.L., Jaffe C.C., Felix R., Springer Berlin Heidelberg, New York, 1987

(VES087) Vesoul P., Mariot P., Haton J.P., Ghedati F., Tombre K., Wackenheim A., Dosch J.C., Zollner G.: The expert System ORFEVRE (Outil de Reconnaissance des Formes et d'Expertise sur les Vertèbres pour le Radiologiste). Computer Aided Neuroradiology, ed. Calabro A., Leonardi M., CIC Edizioni Internationali, Roma, 1987

(WACK88) Wackenheim A., Zollner G., Vesoul P.: Système expert et Intellegence Artificielle, develloppement actuel du système ORFEVRE: Radiologie, J. Cepur, 8/4, 241-245, 1988

Computer Assisted Evaluation of Bone Scintigrams

J. Markwardt, H.-J. Sprenger, G. Richter(*), U. Plaga,
G. Nauber, G. Schmidt

Academy of Sciences of the G.D.R.

Central Institute for Cancer Research
Nuclear Medical Dept.
Berlin
and (*) Central Institute for Astrophysics
Potsdam

SUMMARY
A method for the automatic identification of bones in bone scintigrams and the evaluation of their pulse contents is described. An individual identification and quantification of the more than 100 bones of the skeleton is avoided by a general solution: the bones in the scintigram are matched with the bones of a standard skeleton. Sets of bone masks are used for the identification of bones and the quantification of regional pulse contents. It is possible to investigate an image of differences between two scintigrams. The normalization of pulses in the presentation of bones and the elimination of background in the scintigram are inavoidable prerequisites.
The automatic quantitative evaluation of bone scintigrams leads to an improvement of reproducibility, accuracy, and data presentation for the health care of cancer patients. Percentual deviations to normal and changes in time are localized and quantified in primary diagnostics, follow up and therapy control.

1. INTRODUCTION

The detection of abnormal accumulation of 99mTc-methylene diphosphonate (MDP) at any site in the skeleton is performed by an experienced physician on the basis of visual interpretation of bone scintigrams. This procedure leads to an early diagnosis of e.g. bone metastasis. The basis of this interpretation is the detection of increased activity compared with the surrounding of the same bone or the neighbouring bones, with the bones of the other side or of a previous image. The deviation must exceed 20 % (CONDON et al. [1], PITT et al. [2]) before it becomes visible. A comparison to normal is general practice of the physician. It is performed on the basis of its experience.
The quantification of the result is requested for the follow up and the therapy control, but it is not yet solved in a convincing manner (ANGER et al. [3], CITRIN et al.[4]). By a computer assisted procedure these problems are to be solved to get percentual deviations of the identified bones to normal or to the previous investigation resp. the other side of the skeleton of the same patient. The procedure should run automatically up to the type written report.

2. MATERIAL, METHODS

Digitized whole body scintigrams of the bone are made by a whole body scintigraphic, microprocessor controlled device, developed and constructed in this department (MARKWARDT et al. [5]). The matrix has a size of 128 x 310 addresses, with a pixel size of approximately 6 mm. The data are transferred on line to an image processing computer (MIPRON, KONTRON Instruments, FRG) with a pipeline array processor of up to 20 MOPS and an

image memory of 8 MBytes. The scintigram archive consists of 10 MByte removable disks.
The time for the investigation of one patient is 20 minutes. Within this time the procedure must be completed. To each image belongs a set of patient's and technical data.
The procedure consists of the following parts:

a. Rectification: matching with a standard skeleton
b. Normalization of bone pulse content
c. Elimination of background
d. Comparison:
   - left to right
   - with normal
   - with a previous scintigram
e. Localization of deviations and estimation
f. Writing the report
g. Image and data handling

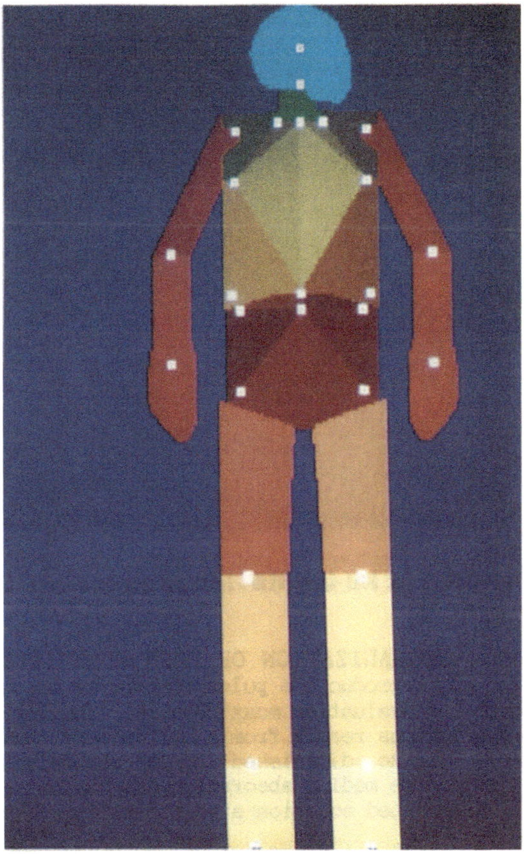

FIG. 1
MASKS AND MARKS for the transform of bone scintigrams

2.1. RECTIFICATION

The bones of the scintigram are matched with the standard scintigram in size, location, and shape by a transform. Since it is impossible to transform it as a whole, the skeleton is disintegrated in triangular regions (FIG. 1), which are transformed. A linear two- dimensional transform consisting of geometrical translation, rotation, and streching is performed (SPRENGER et al. [6]).
The size and location of these regions are described by landmarks, which are positioned at major joints and some reproducible contour points (FIG. 2). The position of the landmarks is determined by different procedures: filters to enhance special structural elements, contour detection, maximum search etc., sometimes followed by an approximation to get mathematical descriptions, to use the complete information of not clearly recognizable regions as the forearm in some cases.
The other basis for the determination of the marks is the knowledge of the proportions of the skeleton. So the length of the spinal column is related to the length of the femur (standard deviation: < 10 %). This knowledge was essential for abbreviating the search procedures and to test for plausibility to avoid errors e.g. in cases with accumulation of activity in metastasis in long bones. The results of ablative surgery in the extremities are taken into account.
Compared with the primary scintigram, the result of the rectification shows a skeleton, moved in a lot of joints and with bones of a normalized size irrespective of age with the same grey levels as before (FIG. 2).

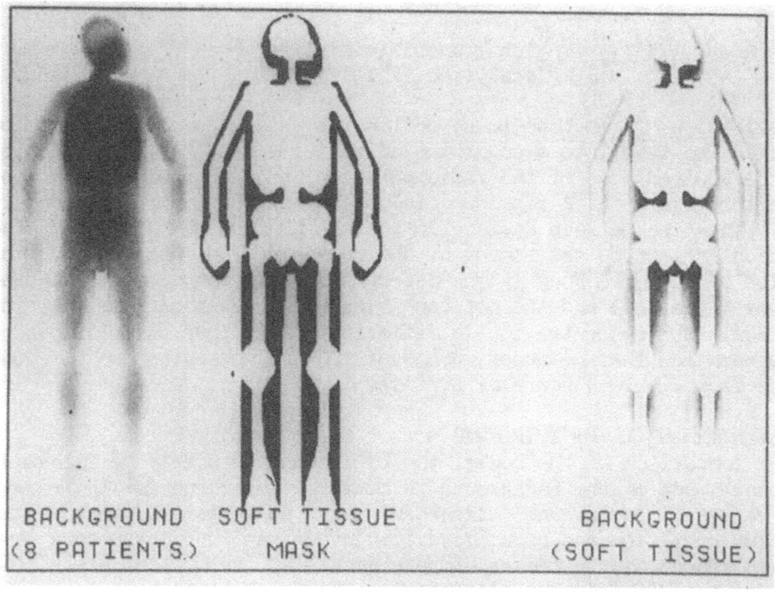

FIG. 3
ADAPTATION OF THE BACKGROUND left: background image, middle: soft tissue mask, right: soft tissue background

2.5. LOCALIZATION OF DEVIATIONS

By masking the skeleton with different masks the identification of bones and the numerical values for the changes are gathered. The parameters of interest are size, dimensions, and percentual changes. The following sets of masks will be used:

- total skeleton,
- skeleton parts: skull, spinal column, pelvis, chest, and extremities (FIG. 2 right side),
- single bones,
- parts of single bones, and
- background regions.

2.6. WRITING THE REPORT

The data are transferred to a dBase file, which is accepted by a special text writing program. The report is written in the form of a letter and informs on the observed changes left to right, to normal or in time.

2.7. IMAGE AND DATA HANDLING

Data files of the patients are created in dBASE. All images are stored on disks for the possibilitie to compare the patient images with previous ones.

3. RESULTS

This study is a work in progress. Preliminary results show the practicability of the procedure. The program runs for approximately 15 minutes. A rough estimate shows, that the rectification procedure fails in its automatic version in < 10 % of patients, esp. when anatomical abnormalities are present (e.g. state after hemipelvectomia) or in insufficiently performed scintigrams (time after injection to short, poor MDP uptake by the bones, extremely positioned extremities). In the evaluation of patients data at now only limited experience is available. All visible changes are presented (FIG. 4-7). Minor differences up to 5 % caused by inaccurate matching are eliminated. The comparison with normal images of different age groups is a valuable tool for non oncological investigations (joints, systemic diseases).

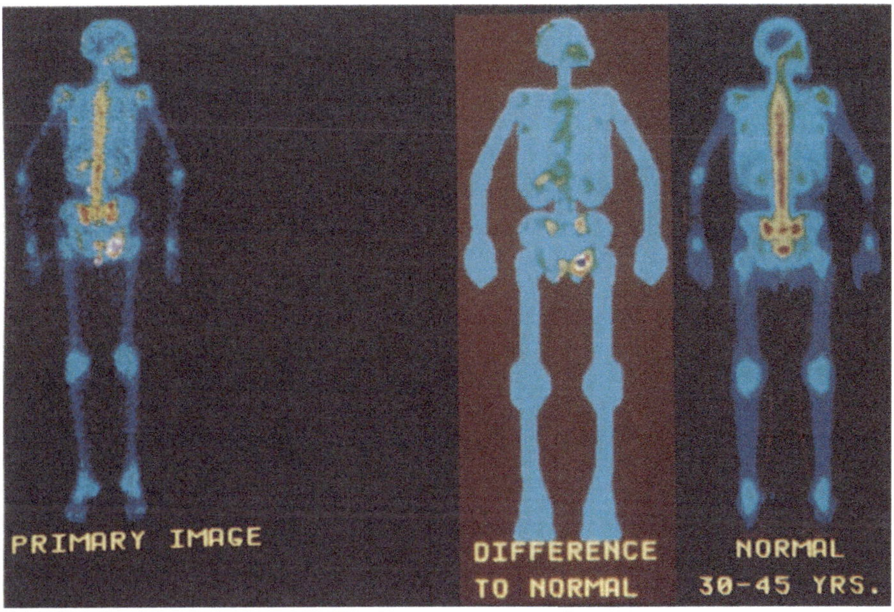

FIG. 4
PRIMARY BONE TUMOR in the right pelvis. Presented are the tumor and a reactively increased uptake in some joints by changed mobility of the patient and the kidneys (Comparison with normal bone scintigram of 10 normal patients after normalization and rectification)

4. DISCUSSION

After numerous attempts to quantify data from bone scintigrams by ROI-techniques, this is a procedure, in which no interaction is requiered. This became possible by the use of a fast image processing computer and adequate algorithms to handle the problem. The major advantage of the described method is the direct comparison of the scintigrams, which makes it much easier to fulfil the task than by searching and determinating each single bone.

By the described method it was first time possible in bone scintigraphy to run completely automatic a procedure which finds and describes regions of impaired function in the complex image of the skeleton. Cases of failure are indicated. A surprisingly low number of landmarks were necessary for the transform.

Precautions are to be taken in the scintigraphic procedure, comparable to those in microscopy image evaluation to get scintigrams of adequate quality. The method will be used in the primary diagnostics, follow up of cancer patients and therapy control in up to 3000 patients per year.

It may be essential, to adapt the absorption correction of quanta and the background elimination to local specificities. It is planned to develop a knowledge based system on the basis of the data to qualify the report over a mere data sheet of a scintigram.

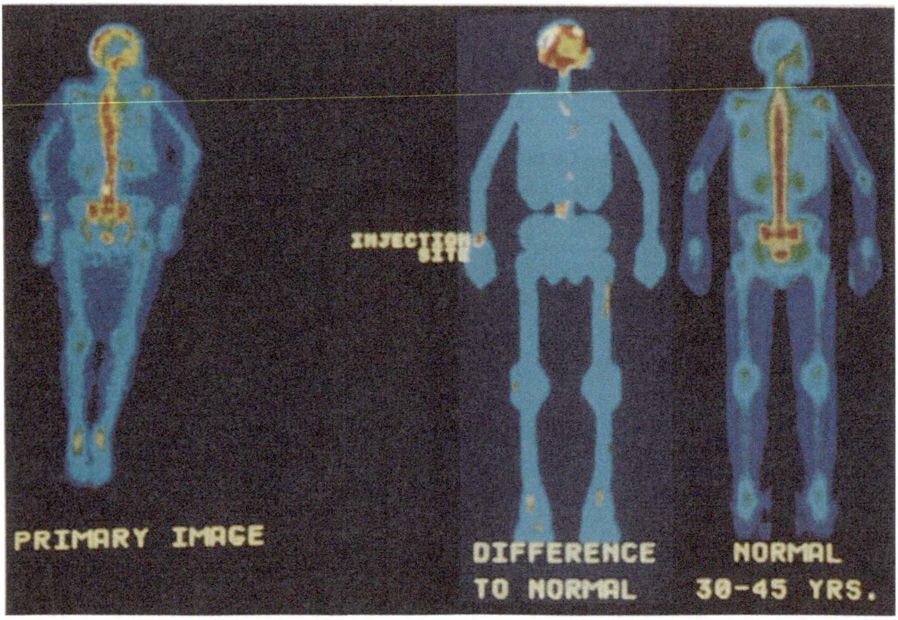

FIG. 5
Patient with METASTASES FROM MAMMARY CANCER compared with normal image: changes in the skull, the spinal column, left and right femur and the arthrotic joints of lower limbs.

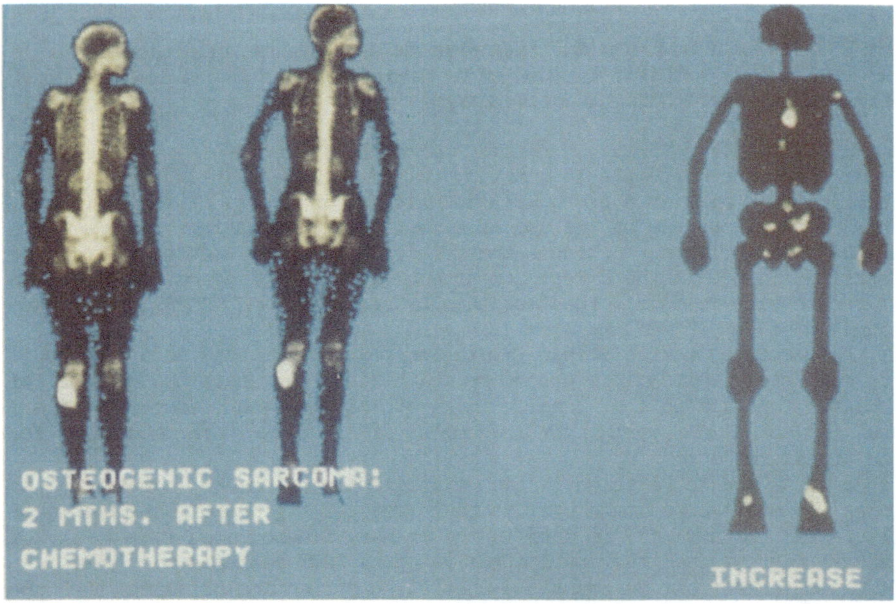

FIG. 6
Comparison of two images of the same patient with an OSTEOGENIC SARCOMA prior and after unsuccessful chemotherapy: increase by a metastasis in spinal column and by changed mobility in some joints. No change of tumor in the left tibia (in this image: no increase)

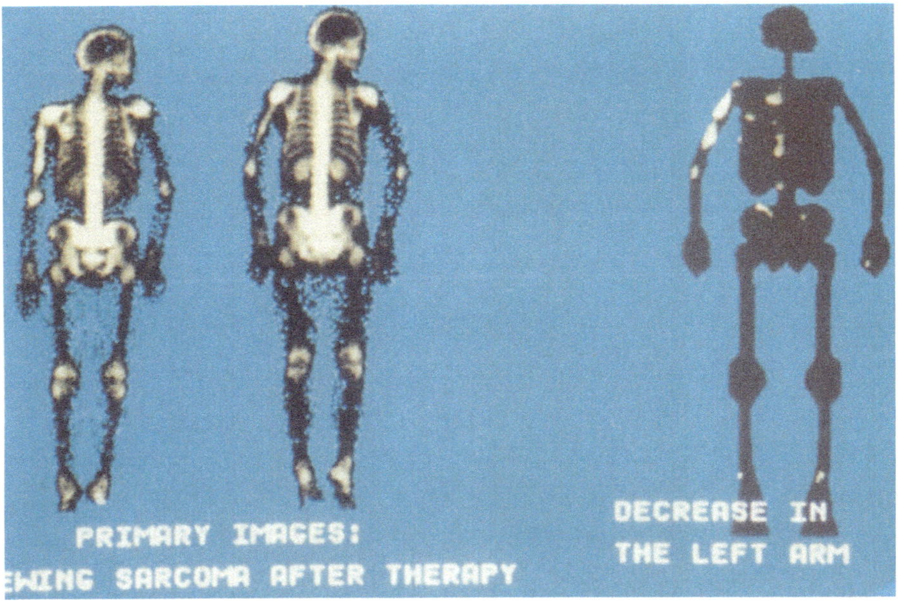

FIG. 7
Patient with EWING-Sarcoma before and after successful chemotherapy. Decrease of activity in the left arm tumor and other locations.

5. CONCLUSION

- The determination of pulse contents in every bone of the bone scintigram is possible.
- The comparison of each bone with this bone of the opposit side or of a previous scintigram reveals essential diagnostic information.
- By comparison with a "normal" scintigram of the same age group, the "normal pattern" in the experience of the physician may be replaced.
- The level of detectability of changes is decreased to below 20 % .

6. REFERENCES

1. CONDON, B. R.; BUCHANAN, R.; GARVIE, N. W.: Assessment of progression of the secondary bone lesions following cancer of the breast or the prostate using serial radionuclide imaging. Brit. J. Radiol. 54 (1981) 18-23.

2. PITT, W. R.; SHARP, P. F: Comparison of quantitative and visual detection of new focal bone lesions. J.Nucl.Med. 26 (1985) 230 -236.

3. ANGER, K.: Quantitative Ganzkoerper-Skelettszintigrafie, Kinetik und klinische Bedeutung. Habilitationsschrift, Universitaet Tuebingen (1975).

4. CITRIN, D.L.; HOUGEN, C.; ZWEIBEL, W.: The use of serial bone scans in assessing response of bone metastases to systemic treatment. Cancer 47 (1981) 680-685.

5. MARKWARDT, J.; PLAGA, U.; LAUERMANN, K.; NAUBER, G.; ANDERT, K.: Ganzkoerperszintigrafie - Bildbearbeitung in einer geraetetechnischen Loesung. Bild und Ton 39 (1986) 89-92.

6. SPRENGER, H.-J.; Richter, G.: Geometrische Entzerrung durch lokale lineare Koordinatentransformation. Bild und Ton (1989 in press).

Image Management and Communication

Picture Archiving & Communication Systems

Storage, Retrievel, and Transmission of Diagnostic Quality Digital Images

Joseph N. Gitlin

Director and Advisor, Vortech Data, Inc.
Reston, Virginia
Associate Professor of Radiology, The Johns Hopkins Medical Institutions
Baltimore, Maryland

Summary

Recent advances in technology have resulted in new options for handling digital data that will be critical to the implementation of PACS. Of particular note are the related improvements in the areas of storage, retrieval, and transmission of diagnostic quality images.

Background

Since the beginning of this decade when the concept of PACS was formulated, several major technical problems have been recognized as critical to its implementation in diagnostic radiology practice. Among these are the related functions of storage, retrieval, and transmission of diagnostic quality images. For the purposes of the paper, "diagnostic quality" has been defined as the full spatial and contrast resolution produced by digital imaging modalities such as computed tomography (CT) and magnetic resonance imaging (MR) and approximately six million bytes of data resulting from the digitization of radiographs produced by conventional x-ray procedures.

Storage and Retrieval

Initial efforts to resolve the problems associated with digital image storage were limited to widely used magnetic tape and relatively expensive magnetic disk systems. Traditional image storage methods have been continued through the conversion of digital to analog images on multi-format film and the manual filing of patients' jackets. As the resolution of digital imaging modalities increased so did the number of tapes at those institutions that opted for this medium with its inherent serial search limitations when it becomes necessary to retrieve the images. While magnetic disks have been well suited for temporary storage of high volume digital images, high costs preclude archival applications where the need for retrieval is relatively low.

Recent improvements in the practice of high density optical media that offer a cost-effective solution to the storage and retrieval requirements of diagnostic imaging have encouraged the development of complete systems with data capacities of several terabytes. The architecture of such a system will be discussed with emphasis on efficiency and utility utility as compared to conventional practices.

Transmission

The original interest in transmitting diagnostic images for interpretation was associated with the need for specialized medical care by large numbers of people in relatively remote areas where a qualified radiologists was not available. "Teleradiology" was investigated as a potential solution to the problem using relatively low resolution film digitizers to transmit images over existing telephone lines to medical centers where trained specialists viewed the images on CRT displays.

The early results were generally favorable and many companies entered the market to supply teleradiology systems to medical care providers who served distant population groups. Among the several problems with such systems were the reliance on existing telephone lines that were limited to relatively slow data transmission rates. The solution that most system suppliers adopted included compression algorithms that reduced the volume of data, the transmission time, and the associated costs. Data reduction ratios of 16:1 were frequently achieved; however, such algorithms are accompanied by the inability to reproduce the original image and often introduce artifacts that may affect the radiologist's diagnostic interpretation. This is particularly true when a case requires image enhancement techniques for primary diagnosis.

Today, medical imaging has several options among transmission media that provide high speed data rates at relatively low cost and obviate the need for destructive compression techniques. High speed data rates are particularly critical in accommodating the increased spatial and contrast resolution of established digital imaging modalities and the introduction of phosphor plate technology and image intensifying system designed to replace conventional film procedures. The potential benefits and costs of available and emerging high speed data transmission media will be presented with a view toward the provision of timely medical care to underserved areas.

PACS and Medical Selection Imaging

J. L. Lamarque - J. M. Bruel - H. Lestienne - P. Lopez - M. Gareil

Department d'Imagerie Medicale - Hospital Lapeyronie -
555, Poute de Ganges - F-34059 Montpellier Cedex

Modern medical imaging gives an increasingly significant place to digital images. On average, 30 % of the production of an imaging department is at present in digital format.

PACS, still at an experimental level, is designed to allow an easy manipulation of the digital image (storage, visualisation, transfer, archiving). It is therefore unnecessary to retranscribe images on a film support avoiding loss of information or increase in cost.

However, one of todays technological and financial limits in the PACS development, is the great quantity of numeric data that has to be processed. Indeed, we may consider two image classes, on one hand images produced by modern technological modalities (CT, MRI) with a matrix usually not exceeding 512 x 512 and where a large number is processed for only one examination and on the other hand, images produced matrices (probably more than 2000 x 2000). Selection is of course necessary where data is to numereous to be managed.

The original option experimented in the Univerity Hospital of Montpellier is that of a medical selection of one or few indispensable images of the patient. These images are chosen after interpretation of the radiologic examination, and only these images will be transmitted and archived.

Nine months of daily training on a PACS console with a view to a medical selection of the indispensable images and in close collaboration with therapist clinicians, allow to present the first practical results on this work hypothesis.

In particular, this experiment allows to appreciate by the clinicians, of a data selection depending on the physicians responsible for the imaging and on the present technical possibilites of the PACS faced with a restricted number of images it can manage.

archival procedure. An *overview image* is then computed in order to sum up all the images of the acquisition on a single additional image.
- the third step is the archiving procedure itself: a specific archiving protocol was defined allowing the creation of the Patient, Study, Acquisition and Image entities in the Copernique central data base.

This archiving procedure has also been implemented on the SIRENE VAX system. As this computer has been used for many years for research purposes, a number of programs exist to load various medical image data from magnetic tapes (MRI, DSA, etc...). As long as the other imaging equipments of the hospital are not connected to the network, this provides a convenient way to load multi-modality image data into the SIRENE system.

Retrieval facilities

• On the central data base

A centralized menu-based *retrieval application* has been developed in the application machine of the SERVER 32 in order to process the retrieval requests arising from the workstations. The search criteria can be entered via menus allowing the selection of patients and examinations: the currently used criteria are basic administrative criteria such as patient name, date of birth, sex, imaging modality, study number, study date, etc... Once the study has been selected, images are transferred to the station by means of user transparent File Transfer Procedures.

• On the workstations

Every SIRENE stations use the retrieval capabilities provided by the central retrieval application. The implementations have been achieved in slightly different ways for the PC based stations (communicating over X25 virtual circuits) and for the VAX and Faxicolor workstations (connected in high speed SCSI mode). With regards to simplicity, a reduced File Transfer Protocol is run in the PC environment.

2. System performances

2.1 Archiving from the CT scanner

The following time table gives details about the archiving procedure: CT to SM90 gateway transfer, data reformatting, overview image computation and image transfer and storage into the data base.

	CT -> SM90	reformatting	overview image	archiving	total duration (minutes)
20 images 256	4	3	6	10	25
20 images 512	17	12	18	25	72

2.2 Retrieval (a 512 512 1 byte image was considered)

The image retrieval durations concerning the various Sirene workstations are given below:

	Actual Transfer (seconds)
Faxicolor (SCSI)	12
Vax (SCSI)	25
PC (X25)	90

3. Discussion

3.1 Interfacing of medical imaging sources

Since the beginning of the project, the direct interfacing of medical image sources has been considered has a major objective. The reason for this choice is probably that the people who started the project were mainly therapists and researchers, working in domains requiring high accuracy of image data (neurosurgery, radiation therapy,..). This initial choice turned out to be a very difficult challenge, due to the heterogeneity of hardware amongst the equipments to interface.

In 1985, the ACR-NEMA recommendations provided a theoretically general solution to this problem but in practice, such a solution could not be used for two main reasons. Firstly, the equipments to be connected did not provide the ACR-NEMA interface, and would not in the near future. In addition, the architecture of the SERVER 32 is not suitable to be connected in ACR-NEMA mode because of the separation of image and alpha-numerical data on separate machines, whereas ACR-NEMA recommands a unique data link between the equipments.

In fact, only the CT scanner could be interfaced up to now; the CGR Magniscan (MRI) could have been easily connected but unfortunately, it was decided to replace it by a GE-CGR MRMAX system at the beginning of 1989.

So the problem of the MRI and DSA (DG200 GE-CGR) interfacing remains complete. Solutions are being looked for in collaboration with GE.

3.2 Software architecture

The question whether the archiving of medical images within PACS must be centralized or distributed has not anywhere received a definitive answer. The respective advantages and drawbacks of both approaches are well known.

Nevertheless some choices with regards to the software architecture within SIRENE may be criticized. Especially, the retrieval application user interface is entirely driven by the server: such an approach is rational for telematic applications, which are accessible from cheap terminal like the "Minitel" but is less suitable for the management of the user interface to a medical image Data Base. In fact, this orientation was chosen just at the beginning of the project and it turned out that it could hardly be changed, although the evolution of the medical workstations towards user-friendly windowing systems was clearly perceived.

Nevertheless, evolutions of SIRENE are possible towards the implementation of retrieval functions within the stations, by means of a different distribution of the application level tasks: it seems judicious to leave the station the entire responsibility of the user-interface. Shortened response times and an improved user-friendliness could be expected. Furthermore, the application itself could be run in the station to release the server from tasks it has not been optimized to deal with.

3.3 Performances

technical aspects

The tables given in section 2 give an idea of the SIRENE system performances. A difference has to be done between the global system response time and actual image transfer durations. Benchmarks have been performed to measure the transmission time between two CCUs of the network. The actual data rate provided on SCSI image channels is about 2 Mbytes per second.

functional aspects

The **archiving procedure** form the CT scanner (or the Vax) is fairly long: nevertheless, it is automatic, and it does not involve the scanner once the data have been transmitted to the gateway. The image decompression stage would better be performed after the image transmission, on the server, prior to the image storage in the physical data machine. Unfortunately, the present server architecture can't afford such processes.

The computation of the "overview image" is also a relatively slow operation but one can consider worthy this small time wasting during the archiving procedure because the presence of this additional image subsequently very much helps the selection of pertinent images.

At the present time, the **retrieval** of images is not possible using SCSI image channels on the PC workstations. The reason for that is an historical one. Initially, PC stations were supposed to be low cost workstations allowing only the retrieval of compressed images (20 to 40 Kbytes) able to be transmitted over 64 kBits/s links in a few seconds. The availability of powerful PC image workstations with rich image processing capabilities has significantly modified the initial approach; a high throughput interface would be very interesting because today, full resolution images can be manipulated and processed on these stations.

After the interactive selection of a study, the retrieval of images from the VAX may either be performed in an image by image mode, or automatically for further local reviewing (visualization in a multi-modality context, 3-D processing, radiation therapy planning,..). The global transmission delays are not negligible; they are mostly due to the multi-user environment and to the data transfers over the Vax Unibus from the SCSI controller board and to the magnetic disk.

Faxicolor is the most integrated SIRENE workstation. The multi-processor architecture of the SM90 allows high data rates on both SCSI and X25 links providing excellent communication performances.

4. Conclusion and further developments

General conclusions about the SIRENE experience can't be drawn today. A first feedback from users experiences is expected in the next months. Further developments are scheduled, regarding optical disks extensions on the central database, and the development of Data Base Administration capabilities.

Acknowledgments

We gratefully acknowledge the participation of all the persons of the CCETT SIRENE group, especially R. Renoulin, G. Babonneau, J.C. Launay and Y. Troude. We also appreciate the participation and support of Copernique and General Electric Companies. SIRENE has been granted by the Regional Council of Brittany and the Ministry of Research and Technology.

References

1. Horii, S.C.; Schimpf, J.H.; Maguire, G.Q.; Zeleznik, M.P.; Noz, M.E.: Broadband coaxial cable image viewing and processing for radiology.SPIE PACS (1983) 247-257.
2. Meyer-Ebrecht, D.:On the architecture for pictorial information systems.NATO ASI series, Pictorial information systems in medicine, HK Höhne (ed) Springer Verlag Berlin Heidelberg New-York Tokyo (1986) .
3. Stockbridge, C.; Ravin, C.E.: Phased Implementation of ATT PACS at Duke University Medical Center. SPIE PACS (1986) 570-573.
4. Jost, R.G.; Hill, R.L.; Blaine, G.L.; Cox, J.R.:PACS experience as a motivation for a campus-wide picture network. SPIE PACS (1986) 549-556
5. Creasy, J.L.; Leondorf, D.D.;Hemminger, B.M.:Initial experience with a prototype storage system at the University of North Carolina. SPIE PACS (1986) 631-636.
6. Hemminger, B.M.; Thompson, B.G.; Pizer, S.M.; Stancil, P.: The ongoing implementation of a prototype medical communication system at the University of North Carolina. SPIE PACS (1986) 574-580.
7. Templeton, A.W.; Cox, G.G.; Dwyer III, S.J.: Digital Image Management Networks: Current status. Radiology 169 (1988) 193-199.
8. Taira, R.K.; Mankovich, N.J.; Huang, H.K.: One-Year Experience with a PACS module in Pediatric Radiology: System Viewpoint. SPIE PACS Vol 914 (1988) 1046-1056.

PACS in Practice: From Software Link to On-Line Communications

L.J.Th.O. van Erning Ph.D., M.J.J. Knots M.Sc., W. Guyt B.Sc., J.H.J. Ruijs M.D.

Section for Physics and Informatics
Institute of Diagnostic Radiology
St. Radboud University Hospital Nijmegen
P.O. box 9101 6500 HB The Netherlands

Abstract.

Initiated by clinical questions a bottom-up development is used for investigating a gradual introduction of a Picture Archiving and Communication System (PACS) in a diagnostic radiology department. On-line communications between a Diagnostic Reporting Console (DRC) and a Computed Tomography (CT) apparatus form a second step after a preparation phase in which a software link was realized. The software link consists of a software package with conversion modules for DSA-, CT-, and MRI-images of various manufacturers. Image-formats can be converted into one file format, i.e. the format of a general MAThematical FUNctions (MATFUN) software package. From this a conversion to the Standard Product Interconnect (SPI) format can easily be realized.

The first results will be presented of a clinical PACS project in which on-line communications were used in daily routine. Specific groups of CT-investigations were selected in order to enable the radiologists to reserve parts of their time to diagnose investigations in a "digital" as well as a "conventional" manner.

Parallel to this clinical evaluation of the DRC, the connection of a Prototype Digital Archive (PDA) to the DRC is studied. The PDA has been developed to determine which data has to be stored in a digital archive and which requirements users have in accessing these data. An overview of all the activities of the department (organization study) will be presented based on activity diagrams. The PDA shows (by means of various functions) the possibilities of a digital archive, especially it's support in scientific research. The latter will be demonstrated by means of a clinical project in which CT-images are processed in an "interactive" as well as an "automatic" way in order to quantify clinical results.

Introduction.

The St. Radboud University Hospital has 1000 beds and handles about 426,000 patients per year (26,000 inpatients). The St. Radboud Hospital Information System (RAHIS) runs on two IBM 4381 computer systems and a VAX 11/785. A number of RAHIS elements are used by the department of diagnostic radiology. This department has 24 different conventional radiology rooms and a separate digital division with 2 CT-, 2 DSA- and 4 US-units. MRI procedures are performed on an external facility. As it became clear that diagnostic imaging techniques are to become more and more based on digital computerization [1], projects were started to investigate to what extent the existing computer infrastructure might support an at first local PACS development. Longterm PACS objectives are:

- resolving existing clinical questions for image processing, image retrieval and image access.
- an on-line connection of all imaging modalities to a centrally located system with image server.
- integration of still separated systems towards a Radiology Information System (RIS) containing a PACS and extensive functional connections with the Hospital Information System (HIS).

Because of the department's role in the diagnostic process, only PACS related projects that result from clinical questions are started, i.e. a bottom-up approach in which the picture system, the archive system and the communication system are considered as more or less separate items.

The **picture system** concerns projects that actually process the images, e.g. to increase their diagnostic output [2,3,4,5] or calculate specific parameters from them. Also aspects of human visual perception will have to be considered when new display techniques are involved e.g. comparison of standard and reversed video images at the same time.

Within the actual PACS set-up the earlier and separately developed processing techniques will be integrated. The picture system functions as a framework into which the various developments may be incorporated according to the ACR-NEMA standard [6]. So applications are invoked from the same user-interface. To the radiologist it should look like an integrated concept.

With respect to the **archive system** a project was started to develop a small scale prototype of a digital archive to determine what information has to be stored in a digital archive and which requirements users have for accessing data [7,8]. At the moment the department still stores images on film from all modalities for 10 years and archives CT-images on magnetic tape from all routine procedures for 1.5 years. In the near future the PDA and a DRC will be interfaced for further research in the field of archive systems.

The **communication system** has not been a separate subject of investigation. However, the realized software link may be considered as a first approach of a system to transfer images between diagnostic data-acquisition modalities and computer systems at various departments within as well as outside the Hospital. The first on-line communications based on Ethernet within the department of diagnostic radiology have been realized between a Somatom DR3 CT-scanner and a DRC in connection with an image server. At the DRC the images are converted to the SPI format [9,10,11]. At this point the profits of previous work come out. Due to the possibility to convert images from various modalities via the software link, the hardware connection at one node (CT) opens up image data transfer from all other imaging modalities.

Picture system projects: general perceptual aspects.

Human visual perception is the last part in the imaging chain that goes from radiation source to eye and brain [12]. Within the brain an image of the outside world is formed based on what is perceived. Aspects of the image, like mean luminance, contrast range, movement, steepness of contours and forms of structures influence perception. Before optimal display strategies within PACS can be realized a number of perceptual aspects have to be considered. For example, in quantitative CT-measurements the drawing of a ROI border in an interactive way is depending on the actual display window setting [13], in DSA the pixel shift amount to reduce movement artifacts may be influenced by the overall image quality in stead of only the ROI they are applied to [4] and in the diagnosis of conventional X-ray films Mach-bands may influence the results [5,14].

The inversion of contrast is used in daily clinical routine of DSA and US investigations. The perceptual implications have been described by various authors [15,16,17]. On various PACS workstations contrast inversion may even be applied within image parts. As a consequence the judgement of detail size (and eventually the severeness of corresponding pathology) may be

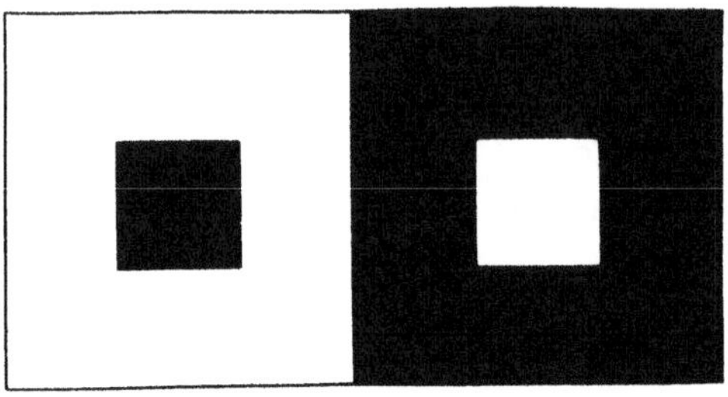

Figure 1: The "irradiation" or "classical" size illusion. The light and dark square have physically the same size.

over- or underestimated. In Figure 1 an extreme example is given of the "irradiation" or "classical" size illusion. Based on exactly the same physical dimensions subjective judgments may result in different values [18].

In a clinical trial, images from a routine DSA procedure are used for automatic pixel shift calculation in order to reduce movement artifacts [4]. The use of DRCs enables radiologists to link images in such a way that within a specific ROI the reversed video display of the corresponding image part is displayed. The effects for clinical diagnosis are subject of current research.

Picture system projects: Texture parameter determination.

The PACS software-link was used in a number of interdepartmental projects [19,20]. The results initiated clinical questions to combine the applied algorithms for automatic determination of ROIs in dual energy CT images of the lumbar vertebrae.

Dual energy CT-scanning is used to determine Bone Mineral Content (BMC) in patients suspected to suffer from osteoporosis. In scientific research as well as in clinical routine an automatic procedure is needed because calculations in a larger series take too much time due to the inevitable human interaction. Therefor processing in batch mode is essential. An important advantage of automatic outlining is expected because only one single technique with objective criteria is used. ROIs drawn by radiologists have a very subjective characteristic.

The diagnosis of osteoporosis is mostly based on the quantification of the BMC indicated by the average X-ray absorption value. Texture analysis may lead to a more accurate and better

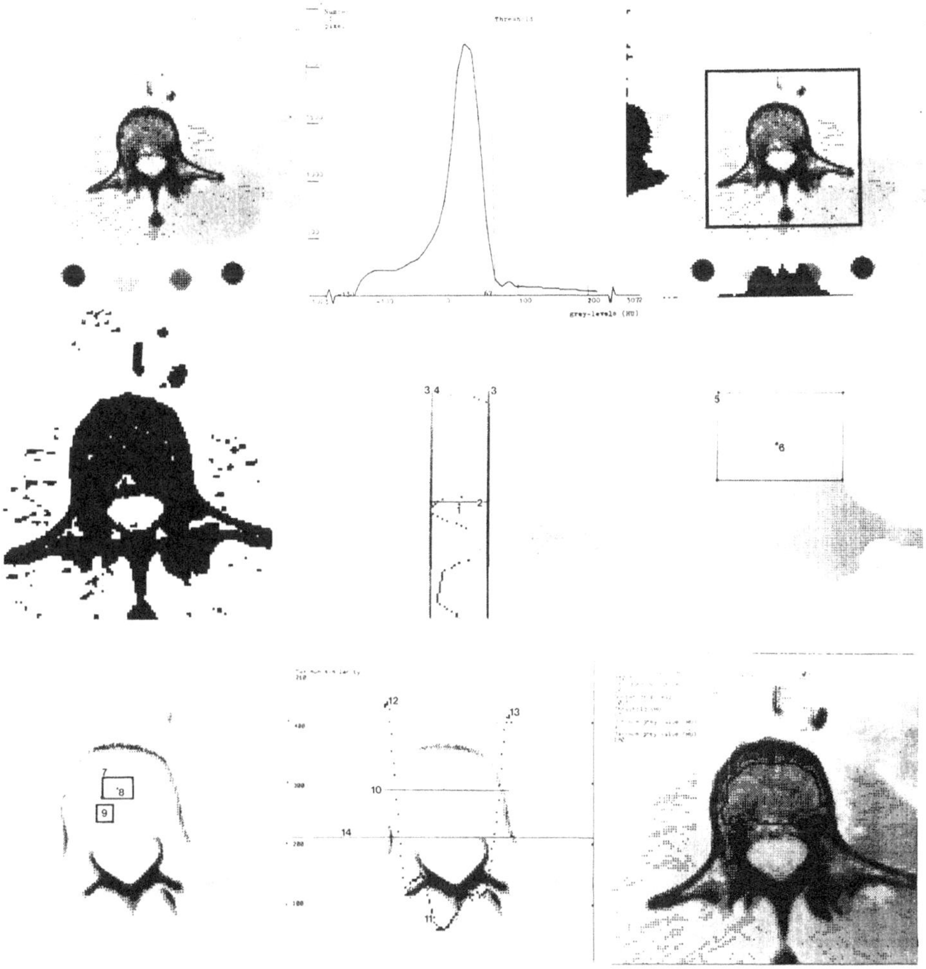

Figure 2: Sequential processing of a CT image to determine a vertebral ROI. (a) The original CT image. (b) Determination of a threshold for locating bony tissue by using the grey-level histogram of the CT image. (c) Finding a 256^2 area containing the vertebra. (d) Thresholding the 256^2 area. (e) Locating the vertebral foramen. 1: The centre of gravity of the binary object. 2: The calculated top of the vertebral foramen. 3: A small strip for tracing the vertebral foramen. 4: The distance to the left side of the small strip represents the number of black pixels. This black pixel count can be used for detecting gaps inside the vertebra. (f) Locating the cancellous bone and positioning of it's reference point. 5: Small window containing the cancellous bone. 6: Reference point. (g) A local window and the reference window. 7: The reference window. 8: The reference point. 9: A local window. (h) Calculation of the maximum similarity. 10: For each point on this line the similarity is calculated. 11: The calculated similarities. 12: Maximum 1. 13: Maximum 2. 14: Maximum similarity = 50% * (Max 1 + Max 2) / 2. (i) Irregular ROI.

method that improves the predictability of osteoporosis.

Following the "software-link"-datapath the CT image files (Siemens Somatom DR3) are converted from RT11 files to RSX 11-M on a system level and thereafter from CT image format to MATFUN format [21] on an application level. On the second system image processing is performed.

In Figure 2 various steps are indicated, i.e. region growing [22], morphological transformations [20,23], blobcoloring [22], calculation of the "Mahalanobis distance" [24] and contour outlining [13].

To calculate additional texture parameters [25] a second software-link datapath is used: conversion of the resulting ROI in MATFUN format to the image file format of pathologic anatomy. From the ROI 19 texture features like geometry, density and texture (Runlength and Markovian) are extracted [23].

An archive system project: Prototype Digital Archive (PDA).

The first step in the development of a Prototype Digital Archive (PDA) was an organization study performed according to a variant of the Information System work and Analysis of Change (ISAC) method [7,8,26,27,28]. It resulted in an overview of the personnel involved, the activities at the department, the information flow between these activities and an overview of the problems in the current situation, the goals and the corresponding needs for changes.

The department consists of a central section and five subdepartments. Four of them have their own archive of which the central one functions as the base archive.

Because of the PACS project the digital subdepartment obtained special attention. The activities of this department are depicted in activity diagrams. An information analysis has been performed according to a variant of the Nijssen Information Analysis Method (NIAM) [26,27]. This resulted in a definition of the structure of a database and its related set of tables.

The implementation of the PDA was done on a PC using the Application Programming Interface (API) for the Structured Query Language (SQL) [29]. API application programs are coded in (turbo) Pascal and contain embedded SQL statements. With SQL various questions can be asked about the data in the database.

One of the programs supports scientific research on a certain clinical picture (i.e. "Paget's disease").

Pointers to a file are generated to display images. These pointers can be used to address the images stored on an external device. Currently various possibilities are being investigated to interface the PDA with the DRC for displaying the images.

A communication system project: Evaluation of a Diagnostic Reporting Console.

In order to evaluate the use of a DRC (type DRC-20) in a clinical environment, an on-line communication has been realized between the DRC and a CT-scanner. The display console of the DRC is connected to a host computer with an image server. The host is connected, via Ethernet, to the CT-scanner (see Fig. 3).

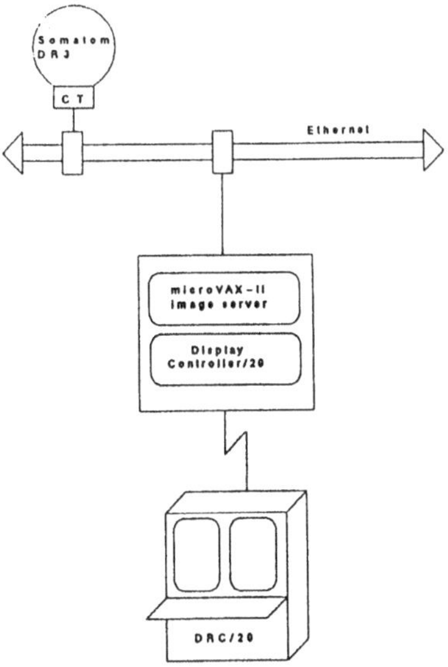

Figure 3: Configuration of a PACS pilot study: connection of a CT via a local area network to a Diagnostic Reporting Console.

For the evaluation of the system various aspects are considered:

- Organizational aspects: time scheduling, development of a protocol for sending images from the CT to the DRC, training of personnel.
- Technical aspects: transfer time, amount of space used.
- Medical aspects: diagnosing of digital images, user-interface of the DRC.

Predefined sets of CT-images are send, via the local area

network to the DRC. The procedure is running in parallel with the actually standardized clinical routine, although it costs extra time for the radiologists and radiographers.

First experiences.

1 Training in using the DRC.

An important point is the training of personnel in using new technology. To evaluate the console the radiologists must learn to use a mouse for interacting with the console and they must get used to diagnosing images on a monitor. Often these users have little or no experience with the user-interface of nowadays personal computers so they need some time to actually get acquaintanced to the console.

A number of lessons have been made in the form of self training sessions with the console. In this way the users can experience the various possibilities of the console. The lessons are also used for education of students. Some concepts can easily be explained by means of interacting with the console, e.g. the concept of window leveling.

2 Time scheduling.

Because of the experimental character of this project it is running in parallel with the normal clinical routine. This implies that it requires extra time, which is, however, not always available. Therefore it is difficult to make a good time schedule to enable the radiologists and radiographers to perform their normal routine firstly. Before starting the project on site it is important to discuss the new way of working first and to explain the objectives in order to get full cooperation.

3 Identification of a patient.

Storing images according to the ACR/NEMA standard [6] implies that different records are available for name, identification number, birthday and sex. However, on the CT-scanner involved only two fields are available, so the radiographers were asked to record the patient identification in a standard way. It turns out to be difficult to realize a situation in which the annotation data are always entered in a consequent way. Often people tend to do it in their "own" way.

4 Amount of storage capacity.

At the present configuration 500 MBytes of external memory are available i.e. about 1000 CT-images (512 x 512). This capacity is only sufficient for short time storage because examinations of some patients, including special reconstructions, appear to consist of more than 100 images. For long time storage one has to look for larger storage media, like optical disks (Write Once Read Many (WORM) times).

5 User interface of the DRC-20.

The DRC-20 is still in a test phase, which means that not all functions considered are implemented yet. So far parts of the user-interface have been tested in a clinical environment at a number of sites world wide.

It is not easy to introduce new technologies, such as a diagnostic reporting console, into a clinical environment. Many obstacles must be overcome. The motivation and training of the medical personnel on every level must not be underestimated. Often people can only be motivated if advantages, new features or extra possibilities are offered in comparison with the current situation. This is not always possible in a pilot study, so it is important to keep everybody informed about the objectives of the projects and to start projects on a small scale in order to keep the extra workload as small as possible. In doing so experiences of users can be taken into account for future developments through which a better acceptance can be expected.

REFERENCES

1. de Valk JPJ, Bakker AR, Bijl K, Heijser W, Boekee DE, Reijns GL. PACS reviewed: Possible and coming soon? Proceedings Medinfo '86 Fifth World Conference on Medical Informatics, Salamon R, Blum B, Jörgensen M, (eds) 1986; 990-995.
2. van Erning LJThO, Bongers PAG, Rothuizen P. PACS in practice: software linking as a first step. Lemke HU et. al. (eds), Computer Assisted Radiology 1987; 496-507.
3. van Erning LJThO, Thijssen MAO, Karssemeijer N, Guijt W. PACS in praçtice: The status of the PACS-project at the St. Radboud University Hospital in Nijmegen, the Netherlands. Part A: Introduction and the Picture System. Med. Info. 1988; 13:255-264.
4. Bongers PAG, van Erning LJThO. Movement artefact reduction in digital subtraction angiography. Int. report Dep. Diagnostic Radiology St. Radboud University Hospital Nijmegen & Technical University Eindhoven, the Netherlands 1986.
5. Lemmens JAM. Analyse van de radiologische veranderingen rond de totale heupprothese volgens Charnley-Müller. Thesis, University of Nijmegen 1987; 24-26.
6. ACR-NEMA Digital Imaging and Communication Standard, Publ. 300 NEMA, 2101 L NW, Suit 300, Washington DC 20037, 1985.
7. Knots MJJ, Bouman HD. A Digital Image Archive: Information Analysis and Development, Masters Thesis no. 70, Department

of Information Systems, University of Nijmegen 1988.
8. Bouman HD, Knots MJJ, van Erning LJThO, Brinkkemper S. PACS in practice: The status of the PACS-project at the St. Radboud University Hospital in Nijmegen, the Netherlands. Part B: A Digital Image Archive: Information Analysis and Development. Medical Informatics 1988; 13:265-278.
9. SPI Standard Product Interconnect for Compatibility of Digital imaging. Int. report Siemens & Philips.
10. Tesche G, SPI: A PACS interface specification. Abstracts 6th EuroPACS Meeting, Utrecht/Leiden, Netherlands, April 1988.
11. Rienhoff O, Greinacher CFC. A General PACS-RIS Interface. Lecture Notes in Medical Informatics 1988; 37.
12. James AE Jr., Goddard J, Price RR, Jones Th, Powis RL. Advances in instrument design and image recording. Radiologic Clinics of North America 1980; 18:3-20.
13. Keller JM, Edwards F, Rundle R. Automatic Outlining of Regions on CT scans. Comp. Ass. Tomography 1981; 5:240-245
14. Lane EJ, Proto AV, Phillips ThW. Mach Bands and Density Perception. Radiology 1976; 121:9-17.
15. Sanders RC. Comparison between black and white backgrounds for Ultrasonic images. Clin J. Ultrasound 1980; 8:413-415.
16. Sommer FG, Orphanoudakis SC, Taylor KJW. Preliminary evaluation of display format effects on perceptibility in a low contrast ultrasound test object. Med.Phys. 1981;8:155-157.
17. Thickman DI, Ziskin MC, Goldenberg NJ. Effect of Display Format on Detectability. Ultrasound Med. 1983; 2:117-121.
18. van Erning LJThO, Gerrits HJM, Eijkman EGJ. Apparent Size and Receptive Field Properties. Vis. Res. 1988; 28:407-418.
19. Rothuizen P, van Erning LJThO, Huiskens R. The accuracy of criteria for automatic 3-D graphics reconstruction of bone from computed tomography. Selected Proceedings of the Fifth Meeting of the European Society of Biomechanics 1986.
20. Karssemeijer N, van Erning LJThO, Eijkman EGJ. Recognition of organs in CT-image sequences: a model guided approach. Computers and Biomedical Research 1988; 21:434-448.
21. Handleiding voor de ervaren gebruiker van MATFUN. Internal report University of Nijmegen, the Netherlands.
22. Ballard DH, Brown CM. Computer Vision. Dep. of Computer Science, University of Rochester, New York, 1982; 149-155.
23. Pahlplatz MMM, Katzko MW, Hesselmans GHFM, Oud PS, Vooys GP. Two Methods for Analyzing Pleural Smears for the Presence of Abnormalities. Pattern Recognition Letters 1986; 4.
24. Mailloux GE, Bertrand M, Stampfler R, Ethier S. Local Histogram Information Content of Ultrasound B-Mode Echographic Texture. Ultras. in Med. & Biol. 1985; 11:743-750.
25. Klotz E, Henscke F, Kalender W. Morphometrishe Analyse Hochauflösender CT-bilder der Wirbelsaule. Proceedings of the German Association of Medical Physics Meeting 1986.
26. Falkenberg ED. Data Bases and Information Systems II, Lecture Notes University of Nijmegen 1987.
27. Verheyn GMA, van Bekkum J. NIAM, An Information Analysis Method. Olle, TW, Sol, HG and Verrijn Stuart, AA (eds) Information Systems Design Methodologies: A comparative Review. Proc. IFIP WG 8.1 Working Conference 1982; 537-590.
28. Lundeberg M, Goldkuhl G, Nilsson A. Information Systems Development - A Systematic Approach 1981.
29. Turbo Pascal, Application Programming Interface for SQL, user's guide. Queensland Inform. Technology Pty. Ltd. 1986.

Review of Experience with PACS Cost Analysis Model

JOHN VANDEN BRINK
Technology Marketing Group, Des Plaines, IL

JOZEF CYWINSKI, PH.D.
Corsan Engineering, New York, NY

Based on interviews with nearly 1500 radiologists and radiology administrators over the last four years, Technology Marketing Group has documented a strong perceived need for PACS technology in order to achieve space efficient archiving support, immediate communication among physicians, and ready accessibility to images. The chief obstacles to acquisition of the technology as cited by interview respondents are the expense of present technology, cost justification and funding.

However, when TMG asked radiologists and administrators what the aggregate annual costs are of present methods of image management, most who respond provide estimates which vary widely, defy rational explanation and/or are based on high profile cost data such as film or file room clerks. In many cases, respondents indicate they just don't know.[1]

Obviously, if cost justification is an issue in the acquisition of expensive new PACS technology, a first step has to be to reliably determine present costs. If the new PACS technology is less expensive than present costs, it can be cost justified. If it is more expensive, then it must be determined how much more expensive and whether the extra expense can be justified based on other values such as improved patient care, reduced hospital stay, etc.

To address the issue of cost analysis, Technology Marketing Group approached several manufacturers[2] to seek funding for expansion and test of an existing computer model[3,4] to evaluate present costs of manual image management, archiving and communication in a given institution or clinic. (See Exhibit 1) This analysis of present costs is called Phase I.

PHASE I:

PRESENT COST ANALYSIS (manual PACS)

- INPUTS
- COMPUTATIONAL SECTION (PROCESSOR)
- MATRIX OF DEFAULT VALUES
- OUTPUT: PRESENT COSTS
- EXECUTIVE SUMMARY

Exhibit 1: PACS COST ANALYSIS MODEL

A second phase development of the program extends the model to compare the costs of PACS acquisition by component or system provided by manufacturers to costs of present methods (Phase I). This Phase II of the model also extrapolates the analyses into the future to measure, on a present value basis, the cost impact of acquiring the component or system.

One objective of the model development was to design a simplified input data sheet usable by any institution or clinic. The input sheet designed requires readily available data from department operations so that the cost analyses can be easily implemented. (See Exhibits 2 & 3)

8-88

FILM MANAGEMENT COST ANALYSIS MODEL

INPUT DATA

TMG
TECHNOLOGY MARKETING GROUP
850 LEE STREET, DES PLAINES, IL 60016
(312) 297 1404

DATE (m m d d y y)

IMAGING INSTITUTION

STREET ADDRESS

CITY, STATE, ZIP

PREPARED BY: PHONE#

REFERRED BY: PHONE#

DATA FOR PERIOD BEGINNING: (m m d d y y) TO: (m m d d y y)

LOCATION: URBAN ☐ RURAL ☐

DATA IS: ACTUAL ☐ PROJECTED ☐

FACILITY TYPE: FEDERAL GOVERNMENT ☐ STATE GOVERNMENT ☐ NON-AFFILIATE HOSPITAL ☐ MULTIHOSPITAL GROUP ☐ PRIVATE PRACTICE ☐

HOSPITAL:
- # OF BEDS
- AVG PATIENT STAY DAYS
- AVG ANNUAL CENSUS %
- AVG COST / PATIENT-DAY
- SATELLITE IMAGING FACILITY ☐ yes ☐ no

TOTAL # OF IMAGING PROCEDURES @ ALL SITES

NUMBER OF FILM HANDLING DEVICES (CAPITAL EQUIPMENT):
- ALTERNATORS
- MULTIFORMAT CAMERAS
- LASER CAMERAS
- FILM PROCESSORS
- FILM COPIERS
- FILM MULTILOADERS
- SILVER RECOV. DEVICES

IMAGING FACILITY:
- # OF RADIOL EXAM ROOMS
- # OF CT SCANNERS
- # OF MRI SCANNERS
- # OF NUC MED IMAGERS
- # OF U-SOUND IMAGERS
- # OF DSA SYSTEMS

RELEVANT ANNUAL OPERATING EXPENSES (DOLLARS/YR):
- TECHNOLOGIST LABOR
- CLERICAL LABOR
- DEPARTMENTAL OVERHEAD
- TOTAL FILM EXPENSES
- ALL OTHER CONSUMABLES
- ALL EQUIPMENT SERVICE
- FILM HANDLING DEVICE SERVICE
- ALL EQUIPMENT LEASES
- FILM HANDLING DEVICE LEASES
- FILM ARCHIVE CONTRACT

STAFF IN RADIOLOGY:
- # OF FULL TIME MD STAFF
- # OF RADIOL MD RESIDENTS
- # OF FTE RADIOL TECHNOL
- # OF FTE RADIOL CLERKS

FILM STORAGE SPACE:

	SPACE SQ. FT	Cost $ /sq.ft
SHORT TERM		
INTERMEDIATE		
ARCHIVE		

RADIOLOGY INFORMATION SYSTEM: NONE ☐ TERMINAL TO HIS ☐ STAND ALONE ☐ STAND ALONE INTERFACED TO HIS ☐ OTHER ☐

Exhibit 2

FILM USE: film - based exams

EXAMINATION TYPE	EXAM CODE	TECH CHARGE PER EXAM	NUMBER OF EXAMS/YEAR	FILM USED PER EXAM					
				FILM CODE	NR.OF SHEETS	FILM CODE	NR.OF SHEETS	FILM CODE	NR.OF SHEETS
CHEST 1 VIEW	CH 1								
CHEST 2 VIEW	CH 2								
ALL OTHER CHEST-RIBS	CHR								
UPPER EXTREM-SHOULDER	SHD								
LOWER EXTREM-PELVIS	PEL								
SPINE-NECK	SPN								
HEAD	HED								
UPPER G.I.	UGI								
BE / COLON	BEC								
ABDOMEN / KUB	KUB								
ALL OTHER ABDOMEN	AAO								
I V P	I V P								
G.U.TRACT, ALL OTHER	GUT								
ORAL CHOLECYSTOGRAM	OCH								
GALLBLADDER, ALL OTHER	GLB								
MISCELLANEOUS SPECIALS	MSP								
MAMMOGRAPHY	MAM								
ARTERIOGRAPHY	ART								
PORTABLE CHEST	PCH								
ALL OTHER PORTABLES	POR								
MYELOGRAPHY	MYL								
OTHER	ORX								

FILM USE: digital - based exams

CT SCAN: HEAD	CTH								
CT SCAN: BODY	CTB								
DIGITAL X-RAY	DIG								
MAGNETIC RESONANCE	NMR								
NUCLEAR MED: BODY SCAN	NUB								
NUCLEAR MED: CARD SCAN	NUC								
NUCLEAR MED: OTHER	NUO								
ULTRASOUND: CARDIAC	USC								
ULTRASOUND: GENERAL	USG								
DIGITAL SUBTRACT ANGIO	DSA								
OTHER DIGITAL	ORD								

FILM PURCHASE COST:

FILM CODE	A	B	C	D	E	F
BRAND/TYPE						
SIZE (INCHES)	14X17	11X14	10X12	8X10		
PRICE $/100 SHEETS						

Exhibit 3

The flow chart of the Phase I model is demonstrated in Exhibit 4, and illustrates how the data input of the radiology department is organized for analysis and presentation by the model.

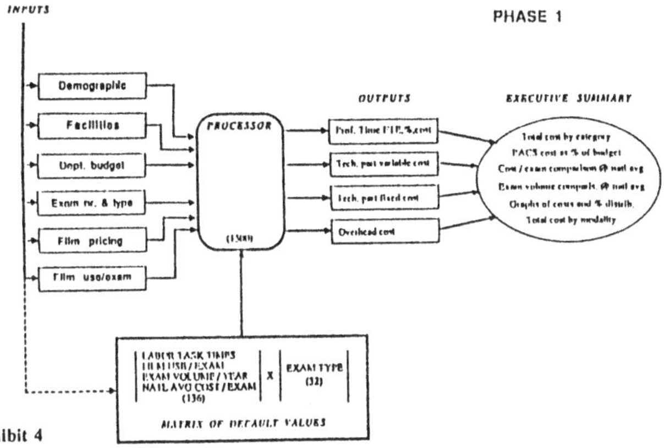

Exhibit 4

A sample of the second page from the executive summary is shown in Exhibit 5 and provides summary data by modality, comparison to data from the TMG tracking study and breakdown of professional and technical costs by major category.

NAH #2

FILM MANAGEMENT COST ANALYSIS MODEL

SUMMARY PAGE 2 OF 4

Listed by modality the costs of present film handling system are as follows:

Professional Costs:	FILM-BASED X-RAY	ALL CT	MRI	ULTRA SOUND	NUCL. MED.	DSA & other dig	TOTALS
Radiology MDs	$411.699	$58,702	$23.999	$32.620	$41,145	$10,368	$578,533
Other Depts MDs	$366,503	$45,392	$17,987	$28.875	$34,818	$6.930	$500.504
Total Professional Cost	$778,202	$104,093	$41.985	$61.495	$75,963	$17,298	$1,079,037
Average Profes. Cost/Exam	$9.77	$10.35	$14.23	$9.32	$8.88	$17.16	$9.92
Technical Costs:							
Film & Materials	$358.753	$77,094	$32.004	$26,536	$16,796	$6,139	$517,323
Radiol Tech & Cler Labor	$312.106	$41,976	$13.978	$24,764	$32,104	$5,521	$430,450
Fixed Technical Cost	$47,443	$9,958	$2,921	$6,536	$8,473	$998	$76,328
Overhead	$193,935	$24,494	$7,185	$16,076	$20,840	$2,455	$264,986
Total Technical Cost	$912,237	$153,522	$56,089	$73,912	$78,213	$15,114	$1,289,087
Average Tech. Cost/Exam	$11.46	$15.27	$19.01	$11.20	$9.14	$14.99	$11.85

The total cost of present methods of film management per examination (including professional and technical components) and the annual procedure volume at your imaging department compare to the national averages (determined by TMG PACS Tracking Study of 1986) as follows:

FILM MANAGEMENT $$/EXAMINATION

	YOUR FACILITY	PERCEIVED NATL AVG FOR 400+ BED HOSP.
FILM-BASED XRAY	$21.23	$15.20
C T SCANS	$25.62	$43.08
MRI	$33.25	#NA
ULTRASOUND	$20.52	$18.47
NUCLEAR MEDICINE	$18.02	$27.40
DSA/OTHER DIGITAL	$32.15	$96.80
ALL EXAMS	$21.77	$20.25

ANNUAL PROCEDURE VOLUME

	YOUR FACILITY	REPORTED NATL AVG FOR 400+ BED HOSP.
FILM-BASED XRAY	79,620	79,620
C T SCANS	10,056	10,056
MRI	2,950	2,950
ULTRASOUND	6,600	6,600
NUCLEAR MEDICINE	8,556	8,556
DSA/OTHER DIGITAL	1,008	1,008
ALL EXAMS	108,790	108,790

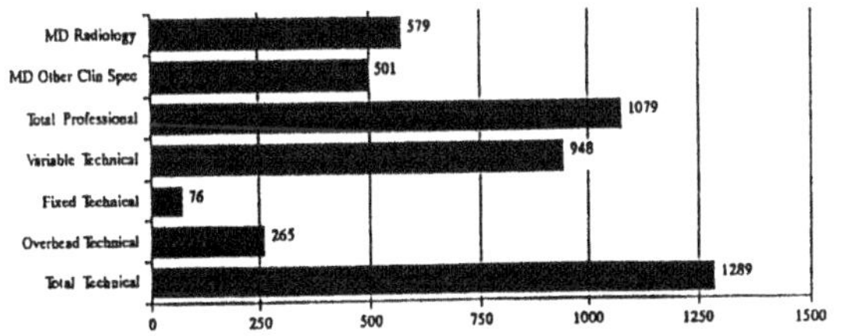

DISTRIBUTION OF ANNUAL COSTS OF FILM MANAGEMENT (in $1000's).

10/12/88

NATL AVERAGE HOSPITAL

Exhibit 5

Validation of the model has been made through detailed evaluations at 24 hospitals of digital modalities in the previous work cited by Cywinski and Gray,[3,4] evaluation of film usage for x-ray procedures based on computer modeling by a major film manufacturer (who wishes to remain anonymous); and data from the TMG tracking studies. Additionally, independent detailed analyses of the algorithms, default values and assumptions have been performed by two equipment suppliers to the diagnostic imaging market who have proceeded to development of the second phase of the model which is specific to their equipment.

The Phase I model (cost analysis of present methods) availability was announced in a TMG news release dated 12 September 1988.[5] As previously noted, two of the sponsors have proceeded with the second phase development of the model to compare the cost of installation of their equipment to present costs using manual methods.

In the process of developing and testing the model, several findings of interest are noted:

- o Nearly half of the present costs of present methods can be attributed to non-productive time spent by radiologists and referring physicians waiting for film, each other, searching out cases that have been misplaced, misfiled or lost, or going to and from the file room or remote sites. This suggests that a major justification for the acquisition of PACS technology will be to free physicians from these actions which will give them more time to dedicate to patient care. Depending on how this time is valued, these savings are likely to be a major element of cost justification for PACS since it affects the efficiency of the whole health care system including patient stay.

o Film costs constitute a little over one fifth of present costs. A common misperception is that PACS can only be justified if it virtually replaces film. In addition to the low likelihood that film will be replaced at any time soon, the savings achieved, even if all film is replaced, will likely be inadequate to cost justify PACS. (See Exhibit 6)

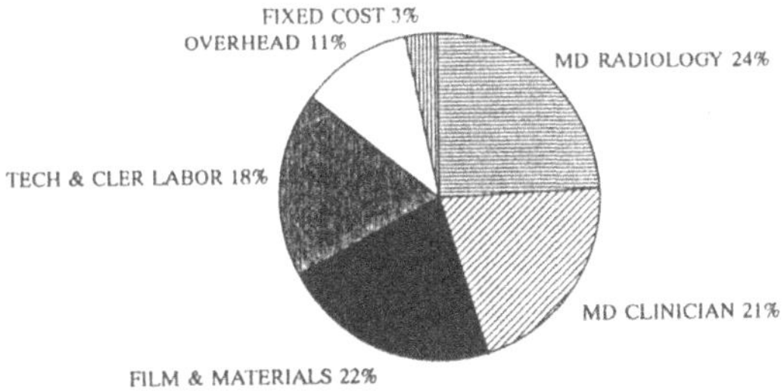

Exhibit 6: COSTS IN AVG. US HOSPITAL
(400+ BED @ '87 TMG SURVEY)

o There are many who believe that PACS will reduce patient stay. TMG tracking study estimates range from 6% by hospital administrators to 13% by radiologists. Assuming that hospital stay is reduced, the value of this savings is a function of whether or not more patients are served per unit time as a result or whether the facilities lay idle. If more patients can be served per unit time, the increased revenues will likely more than justify PACS costs apart from any other consideration.

o Of key interest to both radiologists and administrators is how PACS technology, through providing immediate and broad accessibility to data, will give the institution a competitive advantage over other institutions. While this issue is not directly linked to the cost analysis of present methods, it can become a major rationale for justification of the acquisition of PACS technology, particularly by the administration of the institution.

o The average variable cost per examination of present methods manual of film management is $17.92. The fixed cost of present methods, regardless of the size of the institution, is $181,000 and is expressed in terms of the following formula and chart derived from detailed computer modeling at 14 imaging departments:

$$C = 0.01792\ E + 180.863$$

Where: C = Total cost of manual film management in $1,000's
E = Number of all examinations/year

(See Exhibit 7)

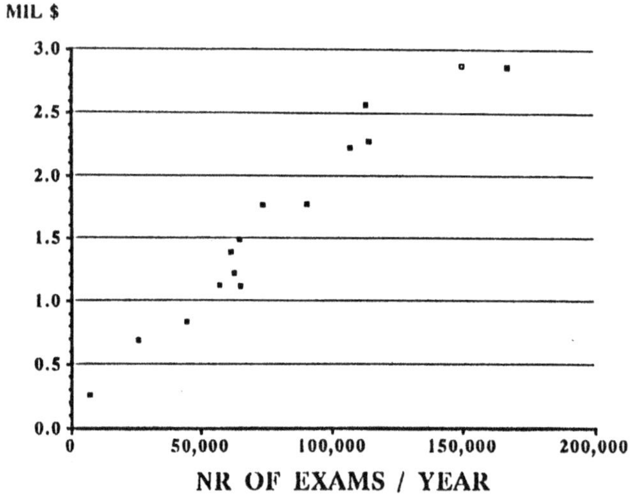

Exhibit 7: COST vs. EXAM. VOLUME

o Costs by exam type are relatively similar with DSA costing the most at $27/exam and nuclear medicine costing the least at about $17/exam. (See Exhibit 8)

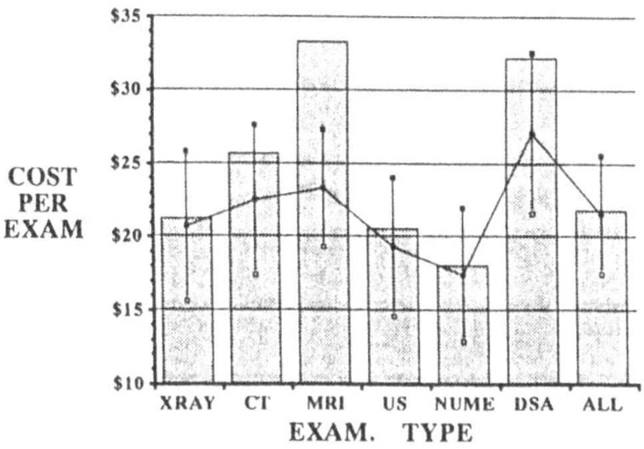

Exhibit 8: AVG. COST / EXAMINATION
14 IMAGING CENTERS -*line &SD;*
US AVERAGE 400+BED HOSPITAL-*columns*

o In cost analyses (Phase II) of a proposed Commview installation at a major U.S. university, cost of the technical component only of manual PACS (not including radiologists and referring physician time) is virtually equivalent to the costs of the PACS technology over the first five years. Significant savings begin to occur in the sixth year. (See Exhibit 9)

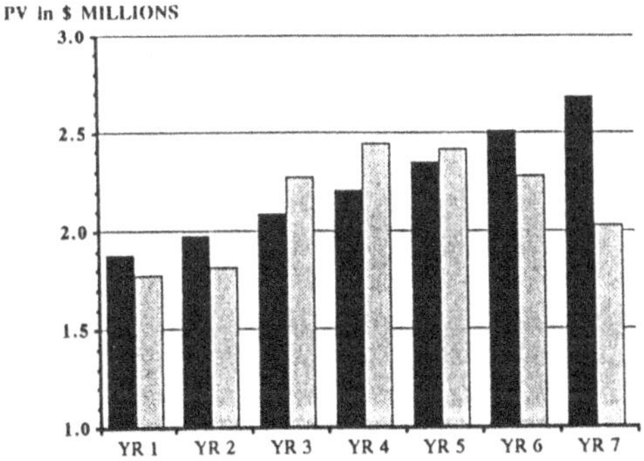

Exhibit 9: TECHNICAL COST OF PACS
MANUAL PACS - black; COMMVIEW - shaded columns.

o During that same period of time in the same institution, the time of the M.D. radiologist for patient care increases significantly as a result of the CommView installation. (See Exhibit 10)

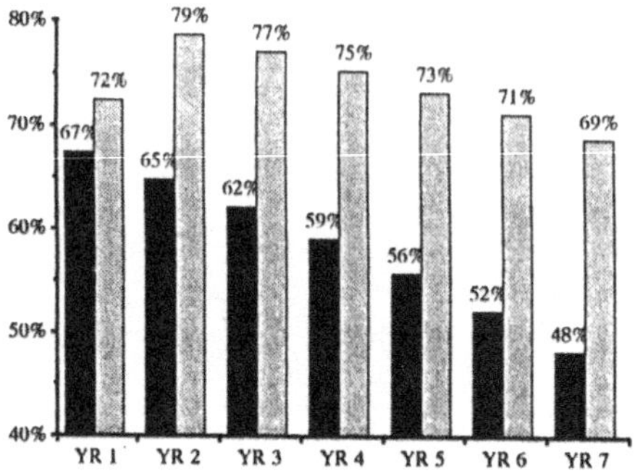

Exhibit 10: TIME OF MD RADIOLOGIST
AVAILABLE FOR PATIENT CARE AFTER FILM MGMT TASKS
(MANUAL PACS - black; COMMVIEW - shaded columns).

A detailed sample print-out of the cost analysis, as well as the simplified input data sheet, are available from Technology Marketing Group.

Summary:

A computer model for cost analysis of present methods of image management, archiving, and communication, has been developed, validated and made available to the medical community. Analyses of the data derived to date suggests that physician time in non-procedural related activity is one of the largest cost components of present methods of image management and that film constitutes one fifth of present costs of present image management methods. The cost analyses of present methods using the model, provides a basis for evaluating investment in PACS technology in terms of components, subsystems or systems. The model can be used to demonstrate cost justification for PACS technology in terms of reduced costs or to provide a basis for assessing the investment required to achieve potential benefits of PACS such as reduced hospital stay, physician bonding, increased referrals and/or improved patient care.

REFERENCES

1. Vanden Brink J., Cywinski J., and Szerlag C. H.,: "Cost Analysis of Present Methods of Image Management" in: SPIE Proc Conf on Med Imag; v.767,pp.758-764, 1987
2. Agfa-Matrix, AT&T, GE, IBM, Philips plus a film manufacturer who wishes to remain anonymous.
3. Cywinski J., Gray M.,: "Computer Model for Estimating Cost of Manual PACS"-IEEE 1984 International Symposium on Medical Images and Icons-ISMII '84
4. Cywinski J., Gray M., and Cywinski L. M.,: "Cost-Benefit Implications of Automated PACS for Digital Modalities" in SPIE '85 Proc 3 Conf PACS vol.536
5. Technology Marketing Group; 950 Lee St. Des Plaines, IL 60016; News Release Sept 12, 1988

Functional Integration of a Radiological Department: A Pilot Project

R. PASSARIELLO (1), A. AGNIFILI (1) (2), P. PAVONE (1), G. SALCITO (3), A. DE SANTIS (3).

(1) Department of Radiology, University of L'Aquila, L'Aquila
(2) Department of Electrical Engeneering, University of L'Aquila, L'Aquila
(3) CITEC, Rome

Summary

In order to evaluate the feasibility of a totally digital radiology department in the new 1500 beds hospital (wich is going to operate in 1992), a pilot project in a low volume setting has been started.
The Digital Radiology Section includes four different system: a digital subtraction angiography unit, a digital fluoroscopy unit, a computed radiography prototype and a digitizer for conventional films. Functional integration of these units has been obtained with a link in a local area network through broadband coaxial cable (Ethernet LAN).
The central computer manages the long-term optical archive and the film laser printer. A large library of image processing procedures is available on the central processing unit for enhancement and restoration of radiological images. Some graphic workstations are connected through the LAN to the file server, allowing fast retrieval and processing of the images. Patients' data are acquired through a radiological information system (RIS) interfaced to the LAN and are stored together with the images.
The interfacing of the single units and the development of the PACS has required close cooperation with the involved manufacturers.
This paper describes the main features of the system and the problems encountered in its development.

A global look at PACS configuration

The Digital Radiology Department at L'Aquila University is equipped with four digital radiographic units:

- a digital subtraction angiography (DSA) system;
- a digital fluoroscopy unit;
- a computed radiography (CR) prototype, based on storage phosphor panels;
- a digitizer for conventional films.

These units are connected together through an Ethernet LAN implemented with broadband coaxial cable. An high performance computer (IBM 6150) acts as network-server and controls image flow to three graphic workstations, used by the radiologists to write medical reports.
Long term image storage is achieved on a juke-box of optical disks with two platters, while hard copies are obtained with a film laser printer.
A unit for patients' data acquisition and examinations'

booking is also connected to the LAN.
The complete scheme of the PACS configuration is shown in fig.1.
In the next section, we are going to summarize the main features of the units. Then we are going to define the amount of data running on the PACS and we'll discuss problems encountered in network development with adopted solutions.

DSA unit

The DSA unit produces diagnostic images in 512x512x8 bits format, but upgrading to 1024x1024x8 bits is forecasted in the near future. The system can acquire up to 4 ECG gated images, but everyday load is 1-2 images per second.
The unit is equipped with a small mass storage device, wich is used during image acquisition and post-procesing.
Apart from conventional image processing procedures, the system allows tridimensional reconstruction of diseased arterias with stenosis index calculation and bi- and tridimensional evaluation of regional contractility. Cardiac chambers reconstruction ia also performed by the software. Selected images (up to 6/patient) are stored through the LAN in the central archive and are sent to the laser printer.

Digital Fluoroscopy

The digital fluoroscopy unit allows acquisition of up 32 frames with an image format of 1024x1024x8 bits. Fluoroscopic images can be freezed and dispalyed on the monitor and are stored , during the acquisition, in the RAM memory of the system. No mass memory devices are present so, at the end of each examination, the radiologist selects the images (up to 6/patient) of clinical value to be stored in the optical archive.
The fluoroscopic unit is particularly useful in dynamic studies of the gastrointestinal and genito-urinary tracts. The studies performed with this unit are more detailed, since a large number of images can be obtained, without increasing the examination cost. Some conventional radiographs of the examined area are also taken during the examination in order to provide images at higher spatial resolution.

Computed Radiography

We are testing a CR prototype developed at 3M Labs (Ferrania, Savona - Italy). This high performance unit is based on storage phosphor panels and is composed by a read-out unit and a VME-bus computer, equipped with a fast A/D board and a 20 Mb image memory.
The system works with a general purpose R and F unit and acquired images can be displayed on a high resolution monitor. The diagnostic images can be produced in the following formats : 1024x1024 pixels with 12 bits/pixel, 2048x2048 pixels with 12 bits/pixel or 4096x4096 pixels with 10 bits/pixel. Owing to the amount of data contained in a single image, the radiologist has to select carefully the images to be stored in the central archive.
The lower spatial resolution (vs conventional films) is widely compensated by the higher contrast resolution and the ability to perform image processing.

Film Digitizer

A digitizer for conventional films is also connected to the LAN allowing the storage in digital format in the central

archive of existing conventional radiographs provided by the patients. The system is based on a digital video camera (512x512 pixels) connected to an original graphic processor ("EIDES COMPUTER" developed by CITEC, Rome) equipped with a 8 Mb RAM and uses an IBM AT host.
The software available on the computer allows graphic and statistic functions and is quite user-friendly.

Data Flow and Memory Requirements

The Digital Radiology Department has research purposes and a low patients' volume. So, for each day we perform 6 angiography, 10 fluoroscopy, 15 computed radiography and we digitize up to 30 film radiographs.
Memory requirements of the images produced from different digital units are summarized in the Table 1 :

TYPE	MATRIX	RESOLUTION	AMOUNT OF DATA
DSA	512 x 512 1024 x 1024	8 bits/pixel 8 bits/pixel	256 Kb 1 Mb
FLUOROSCOPY	1024 x 1024	8 bits/pixel	1 Mb
CR	1024 x 1024 2048 x 2048 4096 x 1096	12 bits/pixel 12 bits/pixel 10 bits/pixel	1.5 Mb 6 Mb 20 Mb
DIGITIZER	512 x 512	10 bits/pixel	320 Kb

Knowing the number of examinations performed per day (on different units) and the maximum number of images per patient stored in the central archive, we have estimated that the total amount of memory for image storage is 0.36Gb/day. Table 2 details the above calculations.

UNIT	(A)	(B)	(C)	(D)	(E)
DSA	6	1 Mb	6	36	36 Mb
FLUOROSCOPY	10	1 Mb	6	60	60 Mb
CR (*)	15	6 Mb	3	45	270 Mb
DIGITIZER	30	320 Kb	1	30	1 Mb
TOTAL					367 Mb (0.36Gb)

LEGEND:
(A) examinations/day; (B) memory requirements/image;
(C) images stored/examination; (D) images stored/day;
(E) total memory request of the single unit.
(*) The 2048x2048x12 bits format has been considered for the CR prototype.

Since we would like to have "on-line" the diagnostic images produced in one year, we should have a central archive of 100-120 Gb (i.e. 0.36 Gb/day x 300 working days/year = 108 Gb). Owing to the amount of memory required we decided to realize the system central archive with an optical WORM disks' juke-box with two platters and a total capacity of 132 Gb.

System Development : problems and solutions

The main problems faced in the system development have been : the different digital formats of the images produced by the four units and the hardware differences between the same four radiological systems.

In effect, two units (DSA and dig. fluoroscopy) are based on Digital PDP11 computers, while the CR prototype uses an high performance VME-bus computer (based on MC68020 processor) and the digitizer is interfaced with an IBM AT computer. These hardware differences and the total lack from the market of commercially available interfaces to the new IBM Token Ring LAN have forced us to adopt Ethernet. Since the Ethernet's data transmission rate is lower than that of the IBM Token Ring (10 Mbits/s vs 16 Mbits/s), we are going, in one or two years, to change the adopted network to increase the "response-time" of the system. This parameter can be defined as the time nedeed to retrieve an image in the central archive and to transfer it in the graphic workstation local memory.

The design of the graphic workstation plays a key role in the project of the PACS configuration : we have planned to have three graphic workstations with multiple high resolutions monitors (1024x1280 pixels). The selected unit is based on the "EIDES COMPUTER" graphic processor equipped with a 8 Mb RAM, a video memory of 1024x2048 pixels and 2D and 3D image processing firmware and software. The central library of graphical procedures can also be accessed from these units. The available software allows the radiologist to perform easily the processing and the enhancement of radiological images.

The big RAM memory (8 Mb) spares the user too frequent access to the central archive and minimizes the "wait-time" at the terminal when a new image, already selected, has to be displayed .

With respect to the image format, we have decided to use the ACR-NEMA standard to obtain that all radiological images be stored and processed in the same way from the central processing unit.

The interfacing of the single units with the Ethernet and the development of the needed software has required close cooperation with the involved manufacturers.

Results and Conclusions

At the moment, the system is in an early stage of testing and we can't show its performances. We expect the actual data

transmission rate to be about 200 Kbits/s in front of the value of 10 Mbits/s, wich is the Ethernet's theoretical maximum. To increase data transfer rate we have also planned to use some data compression and decompression algorithms. The radiologist, selecting the images to be stored in the long-term optical archive, has to decide about the use of a conservative or non conservative algorithm. An improvement of the software is forecasted in the future to allow the use of both algorithms in different areas of the same radiological image.
We are even planning the connection to the PACS configuration of the CT unit and the MRI unit, wich are located in a separate block of the Hospital.
In order to provide immediate evaluation of the images by the clinicians, we are considering connecting to the PACS some remote workstation placed in the main Sections of the Hospital (e.g. the intensive care unit, the operatory rooms, etc...).
System completion is planned for the early 1990.

REFERENCES

1. Mun, S.K.; Benson, H.: Fast paced progress in computers ensures the feasibility of PACS. Diagnostic Imaging 4 (1987) 56-62.
2. Lemke, H.U.: Picture archiving and communication system. In Plannig Considerations in Diagnostic Imaging and Radiation Therapy. Florence : Ed. Clas 1988.
3. Cannavo, M.J.: Low risk strategy for PACS calls for modular phase-in. Diagnostic Imaging 7 (1988) 135-145.
4. Cannavo, M.: Compressed data transmit faster without degrading image quality. Diagnostic Imaging 8 (1988) 85-89.
5. Chandler, R.: PACS - Applications in the modern Radiology Department. Administrative Radiology (June 1987).

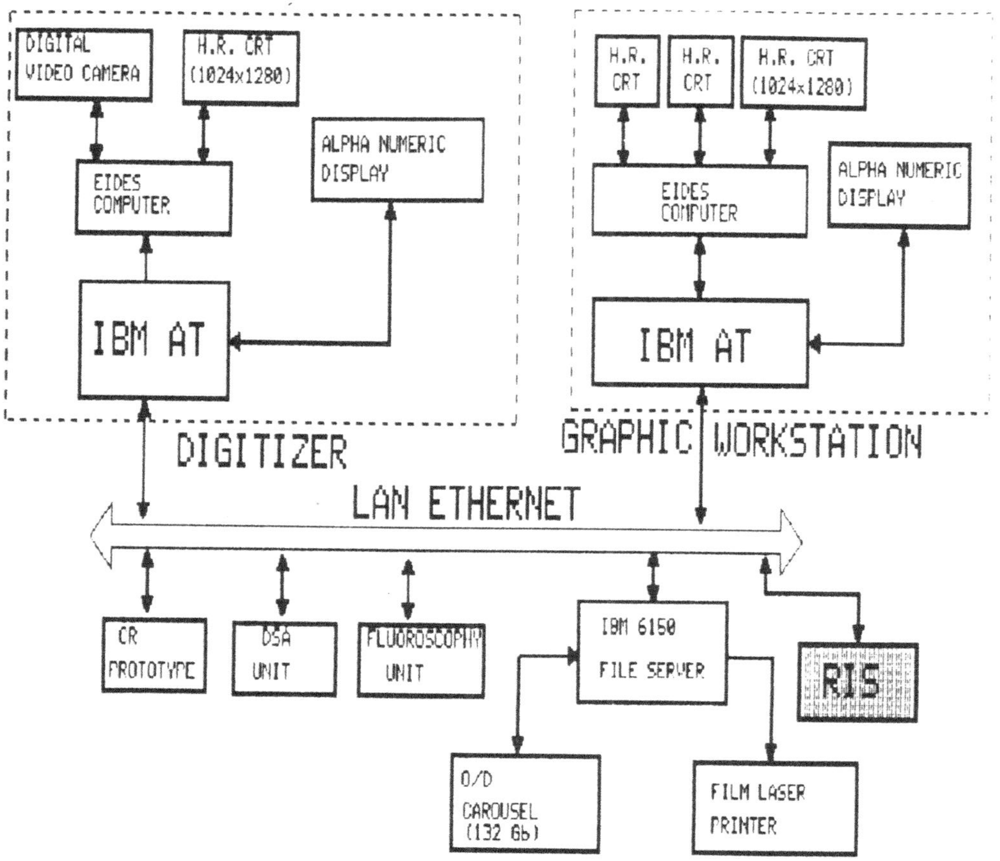

Fig. 1: Global scheme of the PACS configuration

A High Resolution CCD Scanner for X-Ray Films

U.Kliegis, Th. Kortmann, W.Schwesig, R. Neumann, R. Mittelstädt, H.Weigel, H.Wendhausen[+]

MEK Medizinelektronik GmbH, Düppelstr. 71, D-2300 Kiel 1, FRG
[+]Radiologische Klinik, Klinikum der CAU, Arnold - Heller - Str.9, D - 2300 Kiel 1, FRG

Introduction

Digital archiving and communication is the keyword in all areas where electronic imaging devices deliver large amounts of data that have to be stored and processed to be usable for evaluation purposes. With digital radiography, a method is introduced that at least equals the properties of optimally processed conventional x - ray film with respect to spatial and density resolution. The digital format inherently leads to the application of image processing tools like contrast optimization, edge enhancement, structural analysis and other ways of modification or quantification of the information contained in the original data. All this is done to improve the informativity of the image. Digital archiving and transporting the data electronically in networks via coaxial cable allows to display the pictures simultaneously on display screens in different locations, remote from the archive itself where the data remains inert. Thus the problem of films lost somewhere in a hospital environment is solved as well.

Electronic and conventional archives

With the introduction of digital radiography in any institution, however, an ambiguity arises as in all radiological departments a conventional archive exists that can comprise as many as some million single film sheets which rather demand for the continuation of the traditional way of handling them. Additionally, referring doctors and other hospitals send in more conventional films. It can not be expected that the whole medical community will change to digital imaging in one swift sweep. Thus, institutions employing the new digital techniques have to be prepared to import conventional radiographs into their system.

Interfacing conventional and electronic archives

A device for translating image data to the digital format with the best possible resolution is necessary. The data produced by this device should be equivalent to the data of a congruent digital radiography of the same object. The device should be compatible to the digital imaging camera or scanner producing digital radiography data. For practical purposes, using such a film scanner should be as simple as possible, no special user intervention like calibrating, range determination or other tasks should be necessary (2-5). And the device should be fast, as a large area of film will have to be scanned each day.

Apart from the application as an input device for Picture Archiving and Communication Systems (PACS) such a scanner could form a stand alone system in combination with an image processing computer for information enhancement tasks that could not be performed on the original film.

With these applications in mind we decided to develop a high resolution scanner dedicated to transform conventional x - ray films into electronic data sets.

Implementation of a digitizing scanner

We chose to employ a Charge Coupled Device (CCD) light sensitive line array to measure the light passing through the film. The light is produced by fluorescent tube located directly behind the film. It is filtered in a narrow width short wavelength band. The film is moved linearly across the plane connecting the light source and the CCD array. The line scan camera converts the received translucent light strip to a voltage transient proportional to the light intensity profile projected onto the linear array by a low distortion lens. Each element of the CCD array contributes a discrete part to the output signal that represents a serialized analogon to the parallelly received density dependent light profile. The output signal of the CCD array is synchronized by an external clock that switches the charges produced in the single elements like in a peristaltic transport towards the output where they are converted to a voltage. So each element's signal can be exactly sampled, held and converted to a 12 bit digital word. This raw output signal has to be corrected since each receiving element has its individual specific sensitivity as well as its own offset. Thus, a correction of the raw output signal is performed by a normalisation of all samples to result in a uniform sensitivity as well as a uniform offset.

Characteristics of the scanner

The density range covered by the CCD device could be extended to three optical density units, equaling three orders of magnitude of light intensity. The modulation transfer function (MTF) delivers a value of 60 % at five line pairs per millimeter on the film. At two line pairs per millimeter the MTF is 96%. These values have been measured with a 4096 element CCD array depicting a line of 400 mm on the film. A 6000 element array will be used in the future, thus increasing the resolution headroom.

Apart from the optical properties of the lens the spatial resolution mainly depends on the mechanical stability of the whole device. Any vibration or flutter immediately reduces the spatial resolution due to blurring effects. Special emphasis was put on a solid antitorsion construction of the mechanical frame of the scanner.

The characteristics of this system fulfill all demands to a high resolution x - ray film scanner since the parameters surpass the optical properties of x - ray film.

A number of special precautions had to be taken to avoid any alteration of the resulting data set caused by non - ideal properties of the sensor. These become apparent as non - matching MTFs in horizontal and vertical directions, if no measures are taken to correct certain effects of rest charges, shadow effects and more. All these tasks are performed as a pipelined process sequence by an array of transputers. It must be noted here, that without such a postprocessing of the sensor signal the characteristics of the scanner would be about one order of magnitude worse than they are in our system.

On the first glance it may seem to be some sort of an overshoot to equip a rather simple device - as a scanner is - with a parallel computing engine. But if one thinks of implementing features like real time data correction and on-line data reduction in the peripheral device itself, this decision makes sense. As it can be foreseen that the I/O - channels and computing power of any central PACS controller will be burdened with many tasks synchronously, a decentralized dedicated preprocessing of the image data is the optimal solution.

Interfacing the scanner to PACS

The scanner must communicate with any outside devices via some medium. Typically, two types of interface are used to connect system components in PACS networks: A serial connection like the Ethernet[R] LAN interface or a fast parallel I/O channel as it is implemented in the ACR/NEMA standard 300 (1). The implementation of the physical and the network layer - according to the ACR/NEMA and the ISO-OSI standard -,however, is not as essential as the implementation of the transport protocol which means that all devices in a network speak the same language. So, the implementation of the transport protocol has to be dependent on the environment in which the scanner will be used. A very simple interface provides the data in word parallel form, that can be used for a direct interconnect to a dedicated image processing computer in a closed configuration. Such an interface works at the highest speed since no command language overhead has to be regarded. The universality of such an interface, however, is very restricted. On the other end of the spectrum, a fully implemented ISO-OSI or ACR/NEMA standard will result in processing a lot of administrative information relating only very loosely to the relevant data. This applies especially to the recognition of the whole addressing and command structure of the communication standard. But only a subset of all commands has to be understood by a scanner, all others can be classified as unknown. Here again the advantages of the transputer array show up: one processor can be dedicated to control the interface to the outside world, independently from the choice of the physical and the network implementation.

We decided to implement the most important ACR/NEMA commands since most probably this will be the common nominator of all relevant future PACS components. The scanner responds to the *GET_REQUEST*, the *MOVE_REQUEST*, the *CANCEL_REQUEST* and issues all related responses. The *FRAME DESCRIPTOR WORD*, the *PACKET DESCRIPTOR WORD* and all other data describing the format as well as the contents of the image itself are coded according to the standard.

The scanner in clinical practice

The scanner has proved to be an instrument well applicable in clinical routine. Its resolution allows to process conventional x - ray films as if they have been produced by a digital radiography system. Thus, all methods of image enhancement become accessible for the evaluation of conventional films, even allowing to compare enhanced DR-generated images to previously made documents. The spectrum of useful applications of such a postprocessing, however, cannot yet be determined now.

References:

1. ACR - NEMA Standards Publication N. 300 - 1985, Washington 1985

2. Harth et al., Erste Erfahrungen mit einem Filmdigitalisierer, Proc. CAR '87, Ed. H.U.Lemke, M.L. Rhodes, C.C. Jaffee, and R. Felix, Berlin 1987

3. W.H.McFee, Radiographic Film Digitizing Devices: A Comparison between Laser Scanners and Video Scanners, Proc. 74th Scientific Mtg, p.79, Chicago 1988

4. Y.Ishimitsu, Frequency Characteristics of the Laser Film Digitizer, Proc. 74th Scientific Mtg, p.279, Chicago 1988

5. L.J.Th.van Erning, PACS in Practice: Clinical Experiences with a Bottom-up PACS Approach, Proc. 74th Scientific Mtg, p.280, Chicago 1988

Evaluation of Diagnostic Image Quality Using the FEASIBLE Software Package

F.P. Ottes(1), J.P.J. de Valk(1), F.H. Barneveld Binkhuysen(2), H.M. Kroon(3), G.W. Seeley(4).

(1) BAZIS Central Development and Support Group Hospital Information System, P.O. Box 901, 2300 AX, Leiden, Netherlands
(2) Department of Radiology, Utrecht University Hospital, Utrecht, Netherlands
(3) Department of Diagnostic Radiology, Leyden University Hospital, Leiden, Netherlands
(4) Health Sciences Centre Radiology Department & Optical Sciences Centre, University of Arizona, Tucson, USA

Summary

A software package has been developed to assist the performance of ROC evaluation studies concerning the diagnostic accuracy of radiological imaging devices and techniques. This package, called FEASIBLE, runs on a standard PC and guidelines an observer performance study during its phases: the set-up of the whole experiment, the execution of the psychophysical sessions and the calculation and plotting of the ROC curves.
This paper gives a concise overview of the history, the structure and the features of the package. The applications of the program in cooperation with two neighboring university hospitals are summarized. Future improvements are discussed.

Introduction

The present advancement in digital technologies constitutes a technology push behind the development of new digital tools for radiological imaging. Before these new devices can be used to support the daily diagnostic procedures in a hospital their diagnostic accuracy must be investigated.
The method commonly used to evaluate diagnostic imaging systems is the Observer Performance study [1]. This method compares the diagnostic quality of the procedure using the newly developed equipment, which is under evaluation, with that of the conventional method, which could be replaced.

Because of the essential role of the radiologist in making the diagnosis he/she takes part as the observer in psychophysical

sessions. To avoid the influence of subjectivity a certainty scale is used.
To collect a statistically optimal set of data about half of the images in the used series contains pathology and the others are normal cases. To eliminate the influence of foreknowledge the normal and pathological images are presented in a random sequence.
The data are statistically analyzed using the Receiver Operating Characteristic (ROC) technique, which involves a complicated calculation to estimate the ROC curve.

The execution of an ROC study could be supported by software. For example by means of a program images could be presented in a random succession and the collected data could be reordered in a structural arrangement, automatically. The ROC curve could be calculated at once after the data collection.
Within the IMAGIS project, as carried out by BAZIS, the need for software to support the Observer Performance studies was recognized. In 1986 it was decided to develop a program specially tailored for the evaluation of medical image systems. The package was called FEASIBLE, which is an acronym for Feature Evaluation And System Inspection By Logged Experiments. An early version of the program was used to carry out a pilot study on the evaluation of a data-compression algorithm. The experiences with the program during this and subsequent studies have been used as feedback to improve the package continuously. The current version (1.2), has been completed in the middle of 1988 [2,3].

The FEASIBLE package

The main program consists of three modules, matching the three phases that can be distinguished in each observer performance study.
The DESIGN module can be used to set-up the experiment. The questions, the certainty scales, the instructions for the observer, and the information about the images that will be used throughout the experiment is entered into the program´s database.
The EXECUTION module can serve as a guide line to conduct the

psycho-physical sessions. The data entered during the design phase are used to generate the randomized image presentation sequence for each session. At the beginning of each session the instructions are displayed to the observer and some identifying or demographic questions are asked. During the actual session for each image shown one or more relevant diagnostic questions are presented, and the response data are stored in the database.

The STATISTICS module calculates the ROC curves and the related accuracy indices, Area Under Curve and d´(e) [4], for each observer, and if requested for the pooled data from all observers. This module can also produce various tabulated overviews of the observer responses.

A separate module called ROCGRAPH can be used to plot the calculated ROC curves.

One of the principles behind the design of FEASIBLE has been that it should be usable within the medical environment. For example, since it runs on any IBM compatible PC, it can be used at any location in a hospital. The graphical functions require the commonly used Color Graphics Adapter (CGA) card. Medical specialists are prone to called away at any time for patient care, so special measures have been taken that a session can be interrupted at any time while the data are stored, and that the interrupted session can be continued. The program has been designed in such a way, e.g. the user interface is menu oriented, that it can be used by skilled experimenters as well as by those not yet trained in the performance of ROC studies. Several control points have been built in. For example, before the following module may be used, it is first checked whether the previous module has been completed correctly.

Parts of the software are based on previous work done at the University of Arizona [5] and the University of Chicago.

The FEASIBLE package is distributed non-commercially by BAZIS.

Application

Until now FEASIBLE has been used in three different types of ROC studies.

1. In cooperation with the Leyden University Hospital and the Delft University of Technology a pilot study was undertaken to investigate the influence of data compression and reconstruction on diagnostic image quality using a series of CT images of the upper abdomen [6].
2. Within the scope of the Dutch PACS project a research program jointly carried out by the Utrecht University Hospital, BAZIS and Philips Medical System, the diagnostic accuracy of a prototype PACS has been evaluated. The total chain of digitizing, storage, display equipment was evaluated. First an early version of the PACS, including a video digitizer was investigated with a series of chest phantom x-rays [7] and a series of clinical mammographies [8]. Based on the results of these studies the video digitizer was replaced by a laser scanner. This new PACS configuration is under evaluation using a set of head CT images and the same series used previously.
3. The influence on diagnostic image quality of the compound process of digitizing and reprinting radiological images has been investigated by an ROC study at the Leyden University Hospital [9].

Improvements

During the eight studies, which were conducted using FEASIBLE until now, several improvements have been requested by the researchers involved. For example, the first version only provided image presentation sequences which contained a mixture of original (unprocessed) and processed images. This proved adequate for an evaluation of an image data compression technique, where both the original and the compressed/reconstructed images could be displayed on one and the same monitor. However, in other ROC studies, e.g. an evaluation of a PACS subsystem, where analog films viewed on a lightbox are diagnostically compared with digital images displayed on a monitor, it is not practical to mix the images from both series in one psychophysical session. An extra option was added for the generation of sessions with only one type of images.

Currently, the program is being further developed. The next version will among other things be easier to use, be less hardware dependent and have more graphical functions.

References

1. Metz, C.E., ROC methodology in radiological imaging. Invest. Radiology, 21 (1986), 720.

2. de Valk, J.P.J., Ottes, F.P., Seeley, G.W., FEASIBLE (I)- The PC software package for Feature Extraction And System Inspection By Logged Experiment, Medical Informatics, 13, 4 (1988), 331.

3. Ottes, F.P., de Valk, J.P.J., Winter, L.H.L., FEASIBLE (II) - Its use in actual PACS evaluation study, Medical Informatics, 13, 4 (1988), 337.

4. Swets, J.A., and Pickett, R.M., Evaluation of Diagnostic Systems Methods from System Detection Theory (Academic Press, New York).

5. Seeley, G.W., Borgstrom, M., Mazzeo, J. (1982) A general interactive computer program for running signal detection experiments. Behaviors Research Methods & Instrumentation 14, 6 (1982), 555.

6. de Valk, J.P.J., Kroon, H.M., Seeley, G.W., Algra, P.R., Noordveld, R.B., Rompelman, O., Bakker, A.R., First results using a PACS component evaluation chain based on psychophysics: image compression, Journal of Medical Imaging, 1, (1987) 325.

7. Winter, L.H.L., Butler, R.B., Becking, W.B., Klessens, P.L.M., Ottes, F.P., ter Haar Romeny, B.M., de Valk, J.P.J., Diagnostic image quality of video-digitized chest images: a phantom study, Journal of Medical Imaging, (in print).

8. Klessens, P.L., Barneveld Binkhuysen, F.M., Ottes, F.P., Winter, L.H.L., Willemse, A.P.P., de Valk, J.P.J., Diagnostic evaluation of a PACS subsystem using mammographies: an ROC analysis, Medical Informatics, 13, 4, (1988) 323.

9. Lemmers, H.E.A.S.J., Schultze Kool, L.J., Ottes, F.P., Mankovich, N.J., Kroon, H.M., de Valk, J.P.J., Comparison of conventional and equalized (AMBER) radiographs before and after digitization using a 3M chest phantom, Medical Imaging III conference (in print).

to be equal for the situations with and without PACS, because extra digital image manipulations might counteract the time gained by faster image retrieval.
Costs for the personnel in order to learn how to control PACS could not be estimated.

- The personnel related overhead costs calculated per FTE include costs for the restaurant, managers, administration, telephone, etc.
- The power consumption by PACS equipment was taken into account, but the costs for extra air conditioning could not be obtained.
- The renovation costs for creating a digital reading room in the UUH were multiplied with the expected total number of reading rooms (15) in a university hospital. Other costs related to PACS installation, like loss of revenue while installation occurs, were estimated at 40 % of the rebuilding costs (personal communications, Seattle).

3. Options for PACS introduction

Introduction of PACS at a university hospital always includes (a part of) the radiology department. In the rest of the hospital installation may occur in several ways:

Option 1: Hospital-wide without film usage

This option considers a rather hypothetical situation. Currently the image quality of PACS prototypes is not enough for the primary diagnosis. The required advancement in technology for realisation of a hospital-wide PACS, without film consumption, has not been reached so far. This method of PACS introduction however, would be most advantageous to physicians and radiologists. The termination of film consumption implies a large financial benefit. Old images would have to be digitized if the patient visits the Radiology department again or could be destroyed if they have been stored for the legally established period without being requested.

Option 2: Hospital-wide with film usage

Same as option 1, but with film is still in use. This situation is likely to occur over the next few years. This option implies that costs of film, chemicals, floor space, etc. will remain.

Option 3: Only digital modalities

Only the modalities which directly produce digital data (CT, MRI, US) are connected to PACS. According to Greinacher [2] approximately 25 % of the radiodiagnostic procedures is performed digital already and this percentage is expected to grow in future. At the university hospitals studied 11 % of the total surface of film was consumed by digital acquired images (such as CT, MRI and DSA) or images which are expected to become acquired digital directly within a few years (e.g. US).

Option 4: Only a few departments

If selection is based on the highest expected profit of fast image availability the Intensive Care Unit, Reanimation and the Casualty Departments will have most benefits. These departments, responsible for 14 % of the procedures, are less dependent on the archival of film because at the Intensive Care Unit for instance, a thorax X-ray of almost each patient is made every day.

4. Results

The cost estimates for these options are shown in Table 1.

5. Conclusion

The calculations are based on the figures of the clinical evaluation study concerning a prototype PACS with 1024 x 1024 resolution monitors. The accuracy of these monitors is not sufficient for primary diagnosis [3]. This is one of the reasons why film will not be abolished in the near future. According to our estimations partial introduction of PACS at a few wards in the hospital (Option 4) or for digital acquisition equipment only (Option 3), show the lowest costs of PACS. The costs for installation in the total hospital with (Option 2) or without film usage (Option 1) seem to be very high at the moment.

6. Discussion

The costs and benefits have only been calculated for the first year of introduction, based on the figures of the radiology departments and PACS equipment over the year 1987.

Effects of PACS introduction which are not quantifiable at the moment were not taken into account. These aspects however might have great influence on the costs and benefits of PACS. The trends of hardware prices in the forthcoming years, for instance, may be important because hardware prices in general will decrease [4]. Currently a software package for these long-term calculations, called CAPACITY [5], is being developed by BAZIS.

Another benefit of PACS might be a shorter length of stay [1]. The shortened time to undertake clinical actions, shown by Desimone [6], makes a reduction in length of stay likely. Unfortunately this could not be proven, probably caused by the short mean length of stay (4 days) at the Intensive Care where the evaluation study was performed.

If these aspect are taken into account PACS might become cost effective sooner for all options of installation.

7. Acknowledgments

We are very grateful to the financial managers of the Radiology Departments of the UUH, L. Neeleman, and of the Leiden UH, C.Th. Ruygrok and P.H. Elsakkers. We also thank L. Johnson, K. Mitchell, M. Nemerever and J. Russell of the Washington UH in Seattle, USA, for providing us with their preliminary report.

8. References

1. Andriessen, J.H.T.H.; ter Haar Romeny B.M.; Barneveld Binkhuysen F.H; van der Horst-Bruinsma I.E.. "Savings and costs of a Picture Archiving and Communication System in the University Hospital Utrecht." Proc. SPIE, Vol. 1093, Medical Imaging III 1989, Newport Beach.
2. Greinacher, C.F.C.. "Möglichkeiten und Grenzen eines PACSystems." Digitale bildgebende Verfahren, Integrierte digitale Radiographie. Berlin 1986, pp.894-899.
3. Hayrapetian, A.,Motioka C., Huang, H.K., et al." ROC comparison of analog film to 512, 1024 and 2048 line digital displays." Proc. SPIE, Medical Imaging III, 1092-60, 1989, Newport Beach (in print).
4. Bakker, A.R.; Stut jr., W.J.J.;de Valk, J.P.J.; Reijns, G.L.. "PACS costs: modelling and simulation." Proc. SPIE, Vol.914, Medical Imaging II 1988, Newport Beach.
5. Bakker, A.R., B.M. van Poppel, I.E. van der Horst-Bruinsma, et al. " Cost modeling of PACS". Proc. SPIE, Medical Imaging III, 1989, Newport Beach (in print).
6. DeSimone, D.N.; Kundel, H.L.; Arenson, R.L.; Seshadri, S.B.;et al. "Effect of a digital imaging network on physician behavior in an intensive care unit." Radiology, Vol. 169, No.1, October 1988, pp. 41-44.

Table 1. Costs and savings of 1st year PACS introduction (based on data of the year 1987).

	Option 1 (kDFL)	Option 2 (kDFL)	Option 3 (kDFL)	Option 4 (kDFL)
EQUIPMENT				
alternators (#=20)	-251	0	-28	-35
developing equipm.(#=24)	-149	0	-16	-21
PACS (# workstations=40)	5939	5239	653	831
installation (PACS)	375	375	41	53
renovation (PACS)	938	938	103	131
MATERIAL				
film	-1350	0	-149	-189
chemicals	-59	0	-6	-8
profit silver regain	40	40	4	6
filmfolders, etc.	-105	0	-12	-15
DOR (PACS)	700	700	77	98
electricity (PACS)	50	50	6	7
FLOORSPACE				
archive	-175	0	-19	-25
PERSONNEL				
photographer, technician (#=4)	-89	0	-10	-12
technologists (#=56)	-266	0	-29	-37
filmarchive	-288	0	-32	-40
digit. operator (PACS, #=8)	288	288	58	58
operators (PACS, #=3)	147	147	29	29
physicians (PACS, #=21)	-79	0	-9	-11
OVERHEADCOSTS				
personnel	4	12	0	0
TOTAL	5670	7789	663	819

Legend Table 1.
- = savings caused by PACS;
kDFL = 1000 dutch guilders (1 DFL = 0.47 $).
Option 1: PACS installed at the total hospital without film usage; Option 2: Same as Option 1, but with film usage;
Option 3: Only the digital modalities are connected with PACS;
Option 4: Only 3 wards are connected with PACS.

PACS Design Issues: Modelling and Simulation

W. J. J. Stut jr. [1], M. R. van Steen [2], L. P. J. Groenewegen [3], A. R. Bakker [1], B. M. ter Haar Romeny [4]

[1] BAZIS, Central Development and Support Group Hospital Information System, P. O. Box 901, 2300 AX Leiden, Netherlands

[2] TNO Institute of Applied Computer Science, P. O. Box 214, 2600 AE Delft, Netherlands

[3] Department of Computer Science, Leiden University, P. O. Box 9512, 2300 RA Leiden, Netherlands.

[4] Department of Radiology, Utrecht University Hospital, Catharijnesingel 101, 3511 GV Utrecht, Netherlands.

Summary

Modelling PACS requires detailed insight into both diagnostic working methods and hardware characteristics. This paper discusses the background and potential benefits of modelling techniques which may be used to specify and structure a PACS.

Introduction

BAZIS has been actively involved in the PACS field for quite some time now. In 1984, its IMAGIS (Image Information System) project [1] was initiated. Since the construction of image information systems appears to be extremely difficult in practice, BAZIS decided to support their design by computer modelling and simulation. In conjunction with real-world experiments, a performance model is used to predict, analyse and assist in solving performance problems.

Although the results obtained by applying our current simulation package MIRACLES have shown to be useful [2], the construction of models as well as the simulation studies themselves, can be further improved. Major aspects of concern are the application of widely accepted current software engineering concepts and graphical interfacing. Based on our practical experiences obtained we will develop an improved simulation model. This new model is to be used both to do theoretical research into PACS performance issues, and to simulate concrete PAC systems in existing clinical environments.

To these aims, it is necessary that the model and its implementation are accurate, easy to modify, efficient, reliable and

understandable. These requirements are more easily met when using formal modelling techniques (which benefit consistency, completeness, and avoidance of ambiguous specifications), combined with application of basic software engineering principles like data abstraction, information hiding and modularity (see e.g. [3] for more information about these notions).
This paper will discuss some existing formal modelling techniques which may serve as a basis for PACS modelling.

Data modelling

Before constructing a new model, thorough insight into the structure of the problem area is a necessary prerequisite. To this aim, we start with identifying and structuring the data, facts and real-world components (called objects) which play a role when considering PACS in a hospital. Examples of objects are radiologists, technologists, devices, and images.
In the area of conceptual database design [4] so-called data models are used to express the data's logical structure and use. Early models, like primitive file models, as well as the hierarchic, network and relational models, were computer oriented. More recent, so-called semantic data models, provide higher level concepts by defining a new, more abstract level of system description. They allow database designers to think of data in ways that correlate more directly to how data arise in the world [5].
The Entity-Relationship (ER) model is considered to be one of the first true semantic data models to appear in the literature [6]. Applications are represented as networks in which objects are nodes and relationships are edges. However, aggregation is the only form of data abstraction which is directly supported by the original ER model.
Further research has led to more powerful data abstraction concepts, such as generalization, by which similar object types are abstracted into a single, higher-level object type. The resulting subtype/supertype relationships (also called ISA-relationships) are a powerful tool to express a system's hierarchical structure. Moreover, by using so-called inheritance, properties of the general object can be passed on to the more specific objects. We intend to adopt a high-level semantic data modelling

technique as the first step in constructing a PACS simulation model.

As the information that initially interests us, is nonquantitative, and rather, of a structural nature, a visual representation will help to understand the structure of the problem area. One way to represent the data model, is to use higraphs [7], a formalized visualization technique for representing set-theoretical relations. The building blocks of a higraph are (rounded) rectangles enclosing so-called blobs, each blob denoting a certain set. The ISA-relations and inheritance of properties can be visualized quite naturally by using a higraph's subset representation, as is illustrated in Figure 1.

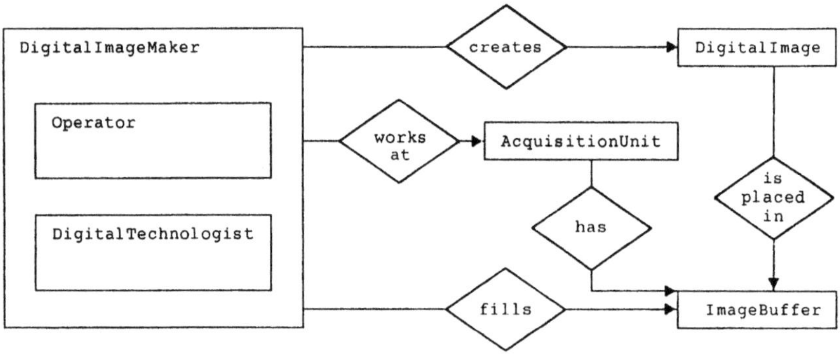

Fig. 1. A simplified example of a higraph-based semantic data model referring to the creation of digital images. The overlapping activities of the DigitalTechnologist and the Operator are abstracted into the higher-level DigitalImageMaker. The properties of the DigitalImageMaker are inherited by the DigitalTechnologist and the Operator). For example, both Operator and DigitalTechnologist inherit the "works at" property.

Semantic data models on their own are not appropriate (nor intended) to make a simulation model of PACSs since they focus on the structure of objects, i.e. their static properties and relations. They do not provide facilities to model dynamics of objects (i.e. their behaviours) or communication between objects. Moreover, they do not allow for the notion of time to be incorporated. However, the intended use of semantic data models when making a simulation model of PACSs seems to be justified by their high-level expressive power. In this way, an abstract description of PACSs based on semantic data models might serve as a first step in deriving an accurate simulation model.

Modelling dynamics

In the literature several formal techniques for modelling a system's dynamics can be found. Net-based formalisms such as Petri nets [8] are frequently used for modelling dynamic properties, but by nature, often lack sufficient support of a modular set-up of models. Recently, the PARADIGM formalism has been introduced for modelling parallel systems in a formal way, supporting modularity and hierarchical organization of models [9]. PARADIGM describes a parallel phenomenon as a vector of communicating decision processes. A decision process is represented by means of a directed graph as is shown in Figure 2.

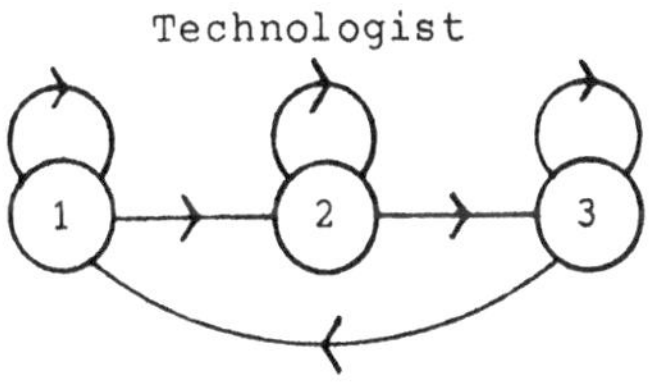

Meaning of states:

1. Waiting for next patient
2. Doing examination procedure
3. Sending patient back to ward

Fig. 2. A simplified example of a PARADIGM decision process. Each state of a decision process is mapped to one node in the graph, whereas a transition from one state to another is represented as a directed edge between the corresponding nodes. If processes communicate, this communication will have an effect on the permissible behaviours. In PARADIGM, this restriction of behaviour is made explicit by removing some edges (i.e. disallowing the corresponding transitions) and some nodes (i.e. prohibiting that the process is in the corresponding states).

Modelling a parallel phenomenon with PARADIGM consists of three steps. In the first step, starting from a given description, the parallel phenomenon is divided into a number of sequential components. Secondly, the communication between these sequential components is also divided into parts. For the control of each part a different manager is created. This manager is also modelled as a decision process. Finally, all sequential modules are put together such that their dependencies reflect their mutual influence, exactly as follows from the description of the parallel phenomenon.

The advantage of PARADIGM is, that the resulting model is a highly modular map of the phenomenon studied, of its sequential components, and of their communication. Unlike Petri nets, the components are easy to distinguish. Moreover, it is easy to adapt the model, to add or refine modules, or to add processes in order to control more complex communication.

Work in progress

We are currently investigating the applicability of both semantic data models and PARADIGM to PACS modelling. Although the ideas as discussed in this paper seem promising, both their practical usefulness and mathematical basis have to be further explored.

When the high-level specification of the model has been completed, a simulation program will be built. Although there are many ways to implement a model, object-oriented programming (in SIMULA) seems to fit in well with the correspondingly object-oriented approach of PARADIGM.

References

1. de Valk, J.P.J.; Stut jr., W.J.J.; Lodder, H.; Bakker, A.R.; ter Haar Romeny, B.M.: "The IMAGIS projects: past, present and future." In Proceedings Medical Imaging II Conference, Schneider, R.H.; Dwyer III, S.J. (eds.), SPIE Vol.914, 1988.
2. Stut jr., W.J.J.; de Valk, J.P.J.; ter Haar Romeny, B.M.; Bakker, A.R.: "Modelling and simulation within the Dutch PACS project." In Proceedings Medical Imaging III Conference, Schneider, R.H.; Dwyer III, S.J. (eds.), SPIE 1989.
3. Sommerville, I.: "Software engineering." 2nd ed, International Computer Science Series, Addison-Wesley Publishing Company 1985.
4. Brodie, M.L.; Mylopoulos, J.; Schmidt, J.W. (eds.): "On Conceptual Modelling." New York, Springer-Verlag 1984.
5. Hull, R.; King, R.: "Semantic database modeling: survey, applications, and research issues." ACM Computing Surveys, Vol.19, No.3, September 1987, pp. 201-260.
6. Chen, P.P.: "The entity-relationship model - Toward a unified view of data." ACM Transactions on Database Systems 1, 1, March 1976, pp. 9-36.
7. Harel, D.: "On visual formalisms." Communications of the ACM, Vol.31, No.5, May 1988, pp. 514-530.
8. Peterson, J.L.: "Petri net theory and the modeling of systems." Prentice-Hall, Englewood Cliffs, N.J. 1981.
9. Groenewegen, L.P.J.; Oosting, G.; van Steen, M.R.: "Modelling parallel phenomena." In Proceedings 2nd European Simulation Congress, Antwerp, Belgium, Vansteenkiste et al (eds), September 1986, pp. 45-51.

An Experimental Evaluation of the Impact of a PAC System on the Operations of Radiology Department

W. UKOVICH*, D. BRAVAR**, R. CUTTIN ZERNICH***, P. GIRIBONA**, M. M. MAFFESSANTI***, F. STACUL***, L. DALLA PALMA***

* Dipartimento di Elettrotecnica, Elettronica ed Informatica, Università di Trieste, Italy.
** Biomedical Evaluation Centre, Research Area, Trieste, Italy.
*** Istituto di Radiologia, Università di Trieste, Ospedale di Cattinara, Trieste, Italy.

Abstract

The process undergone by images and reports from their acquisition and formulation up to their final archiving is analyzed and evaluated. The manual operations carried out for CT examinations are compared with the ones performed using a PAC System on the basis of experimental measures. It is seen that PACS mainly affects operations related to image and report distribution, but not their acquisition and formulation.

Introduction

Picture Archiving and Communication Systems (PACS) are gaining an increasing attention as effective and convenient tools to support several operations in the management of a Radiology Department. The importance of PACS is due to the number of radiological studies performed by digital modalities: presently they cover 20-30% of all the investigations, and this share is expected to grow in the immediate future [1].

As a consequence, there is a need for a comprehensive evaluation of such systems. The present paper is devoted to analyze and evaluate the aspects connected to the process undergone by images and reports from their availability and formulation up to their final archiving. To such an aim an experimental investigation has been performed, comparing the usual operations carried out for CT examinations with the corresponding ones performed using a PAC System.

It should be stressed that the present study is based on field tests, and not on ad hoc experiments, in order to compare actual procedures in a real environment. This feature, especially for what concernes PACS, partially distinguishes our approach from the one of similar and more detailed studies, such as for instance [2] and [3].

Material and Methods

Since September, 1988 a PAC System (CommView by AT&T and Philips) is in operation in the Radiology Department of the University Hospital of Trieste; it is used to manage all the images acquired by a CT unit, a NMR unit, a Digital Angiographic equipment, and part of the images acquired through two US units.

The present configuration of the system consists of a Data Management System (DMS), an Acquisition Module (AM), and a Display Workstation (DW). It is supposed to be expanded in the immediate future by acquiring another DW to be located within the same Department, two

Consulting Workstations (CW) in other Departments of the same hospital, another DW in the Radiology Department of another hospital about 6 km away (connected via optic fiber), one Film Laser Scanner and one Optical Juke-Box (89 plates).

Presently, the routine reporting activity of brain and abdomen CT examinations is carried out on PACS. This stimulated the investigation on the possible differencies and relative merits of the ordinary (manual) and automated (PACS) operations.

The operational study excluded emergency cases due to their peculiar management, and has been carried out in two phases: first the elementary operations have been identified and analyzed; then they have been monitored and evaluated in quantitative terms with respect to their time requirements. The latter analysis has been carried out from two different points of view: on one side the times elapsed among subsequent relevant events, such as films made available for delivering and actual delivering, have been measured; on the other side the actual time required to perform some specific activities, such as retrieving images for reporting, has been recorded.

The motivation for the latter distinction between timing of events and duration of activities is originated from the fact that cases are ordinarily processed in batches, i.e. groups of elements that can proceed through a new operation only after all of them have completed the previous one. So from the point of view of the system productivity the throughput time is relevant, that is the time elapsed from the moment of the beginning of the first operation up to the moment of the completion of the last one; conversely, from the point of view of the resource use (equipment, workforce, etc.) the duration of the activities involving such resources is important.

Results

Concerning the manual procedure seven relevant events have been identified:

0) end of images acquisition;
1) images available for reporting (at this moment the batch of cases is formed);
2) beginning of the reporting session;
3) end of the reporting session (batch available for typing);
4) report and images available for delivering;
5) report and images retrieved by the requesting department;
6) report and images stored in the film archive (for inpatients only).

The times elapsed between such events have been recorded on the basis of ten "tracing" cases that have been monitored (Tab. I).

Tab. I.

Events	Times	
	average	range
0 — 1	25 min	6 min — 1 h
1 — 2	2 h 20 min	20 min — 6h
2 — 3	1 h 45 min	1 h — 3 h 30 min
3 — 4	11 h 40 min	10 h — 14 h 30 min
4 — 5	4 h 40 min	3 h — 8 h 30 min
5 — 6	27 days	13 days — 67 days

The two last rows of Tab. I refer only to inpatients.

For the automatic procedure the present configuration of the system allows only to monitor the first five events (up to 4); actual delivering of images and reports through PACS will be effective only after some other departments will acquire CWs. The corresponding timings for the first five events have not been recorded, since they are only affected by the way the Department is managed, and not in any way by the use of PACS, with the only exception of the 0 — 1 interval (less than 5 seconds), and the 2 — 3 interval (the reporting session), which deserves some further insigh.

Within the reporting session, the sequence of elementary activities related to each case consists of:

a) image retrieval (on the view box or on the CRT screen);
b) analysis and reporting;
c) image folding (for the manual procedure - for the automatic it is included in a).

The average times recorded for these activities in the manual and in the automatic procedures are shown in Tab. II.

Tab. II.

Activity	Times	
	manual	automatic
a	13 sec	39 sec
b	2 min	2 min 10 sec
c	7 sec	—

Discussion

The results of the previous section suggest the following remarks:

- in the case of PACS the events don't necessary follow a rigid chronological sequence, as in the manual case, and this offers new margins for efficiency improvements with a different management of operations;
- time 0 —1 in the present manual procedure could be reduced down to 3min using a laser imager; however, it is quite irrelevant with respect to the other times;
- time 0 — 5 in the manual procedure show a nearly daily pattern, with minor deviations; this has to be attributed partly to the batch arrangement of the operations and mostly to the Department operations;
- time 5 — 6 exhibits a remarkable dispersion; however, the actual relevance of this figure could be controversial;
- in the automatic procedure times 0 — 1 and 4 — 6 make no sense any more, since images are immediately available to anyone connected to the systems; lack of space doesn't allow to analyze different scenarios into which image accessibility could be implemented in practice;

- the analysis and reporting times experienced for the single cases are exceptionally short in both procedures because all cases have been documented by the same radiologist reporting on them;
- reporting times are comparable for the manual and automatic procedure, with a slight advantage for the former one, for each of the three activities considered; it should be mentioned that using alternators instead of manual unfolding wouldn't affect significantly the operations (from 20 to 12 sec);
- no case considered in the reporting procedure on PACS required image processing (zooming, window level modifications, etc.); using such facilities would severely affect the automatic procedure.

Conclusions

The results show that for this class of operations the most relevant improvements in productivity (in terms of throughput times of cases) and use of resources (particularly for qualified workforce) could be rather achieved from an adequate management of the operations than from simply transferring them to a new technology. Nevertheless, the mere availability of such technologies may trigger a rationalization of the existing procedures.

PACS is expected to produce more effective improvements in some operations which have not been contemplated here, such as image archive management as far as the resource use is considered, and image retrieval as far as both resouce use and productivity is concerned. Those are the next issues that will be studied in our project for the assessment of the impact of PACS on the operations of a Radiology Department.

Parallel to the present research, an investigation is carried out on the clinical and diagnostic effectiveness of PACS. The first results are presented in [4]. It is evident that diagnostic and operational effectiveness are both qualities that will decide the actual implementability and acceptance of PACS in the immediate future.

References

1. Lemke H.U.: Picture Archiving and Communication Systems, In: Chiesa A., Gasparotti R., and Maroldi R. (eds.): Planning Considerations in Diagnostic Imaging and Radiation Therapy. CLAS International, Brescia, Italy, 1988, pp. 198-208.
2. Mun S.K., Benson H., Welsh C., Elliott L.P., Davros W.: Baseline Study of Radiology Services for the Purpose of PACS Evaluation. In: Schneider R.H., Dwyer III S.J. (eds.): Medical Imaging II: Image Data Management and Display, Proc. of SPIE Vol. 914 (Part B), pp.978-987.
3. ter Haar Romeny B.M., van der Wielen J.M.M., Achterberg A.J., Barneveld Binkhuysen F.H., Zuiderveld K.J., Andriessen J.H.T.H, Bakker A.R.: PACS Efficiency: a Detailed Quantitative Study of the Distribution Process of Films in a Clinical Environment in the Utrecht University Hospital, presented at the SPIE Medical Imaging III Conference, Newport Beach, Ca., USA, 1989.
4. Stacul F., Bravar D., Cuttin Zernich R., Giribona P., Magnaldi S., Pozzi-Mucelli R.S., Ukovich W., Dalla Palma L.: Clinical Assessment of a PAC System: Effectiveness of a Diagnostic Evaluation on the CRT Screen. This conference.

PACS for Radiotherapy

INGER-LENA LAMM* and **TORGIL MÖLLER#**

*Department of Radiation Physics, #Department of Oncology, University Hospital, S-221 85 Lund, Sweden

Summary
PACS developments and tests have been dedicated mostly to the field of diagnostic radiology, with emphasis on the requirements of the diagnosticians as "human viewers". A second generation PACS is now appearing, supporting image management and integrating the use of image and non-image data. The PACS concept covers the whole organization of the "PACS clinic", and the new tools available will affect the ways of working with images on all application levels. Some of the specific PACS aspects of the radiotherapy workstation will be discussed, as well as computer assistance to use "all knowledge available"; computer vision tools, as well as a futuristic radiotherapist's workstation.

1 INTRODUCTION

The concept of PACS, Picture Archiving and Communication Systems, is definitely not "**just a product, but a way of thinking**". A second generation of PACS is now appearing (1), "which integrates image generation, display and analysis, storage (both short term and archival), production of hardcopy, communication between all of the subsystems and between systems, and management of both the images and the system." The scope of these second generation systems is thus much wider than that of the first generation, implying integration with other information systems, such as Hospital Information Systems, HIS, and Radiology Information Systems, RIS; not only to support image management but also to integrate the use of image and non-image data (1, 2). To distinguish the latter type of system functionality, new acronymes such as IMACS - Image Management And Communication Systems - have been created (the translation Information Management And Communication Systems fits as well); or why not use MACS - Medical Archiving and Communication Systems - for short?

PACS developments and tests have been dedicated mostly to the field of diagnostic radiology, where today the largest amount of images are created. Emphasis has been on the requirements of the diagnosticians as "human viewers", but other users of images must not be forgotten. Image analysis functions are for instance expected to play an increasingly important role in future PACS.

2 INTRODUCTION OF INFORMATION TECHNOLOGY IN MEDICINE

PACS components exist and are used increasingly in clinical routine, dominantly as stand-alone subsystems. It is of the utmost importance to have an overall idea today of the future "PACS clinics", and it is imperative that the user community is encouraged to participate from the beginning in defining requirements both for the PACS and for the subsystems.

The introduction of information technology in the clinical environment will always affect the organization of a department; new routines, new tasks, who does what, new distribution of influence etc. New technology often works as a catalyst, bringing organization problems to the fore. Common incentives for introducing new technologies are expected increases in quality and flexibility, decreases of operational costs and more direct control of management. Large effects should not be expected until some time after the introduction of new systems; the users first have to learn to do things correctly, after that they will be ready to start doing the correct things, i.e. exploiting the possibilities of the system and thus pushing new developments.

3 THE RADIOTHERAPY INFORMATION SYSTEM TODAY

In an integrated information system, fulfilling the clinical requirements of radiotherapy, all relevant data should be available at the different steps in the radiotherapy procedure, and the system should also include appropriate tools for handling these image and non-image data. Such a system belongs to the class of second generation PACS, and is schematically illustrated below.

In order to find the "optimal" way of delivering radiotherapy to an individual patient, we must use all knowledge available; i.e. knowledge about host factors and tumour factors for the specific

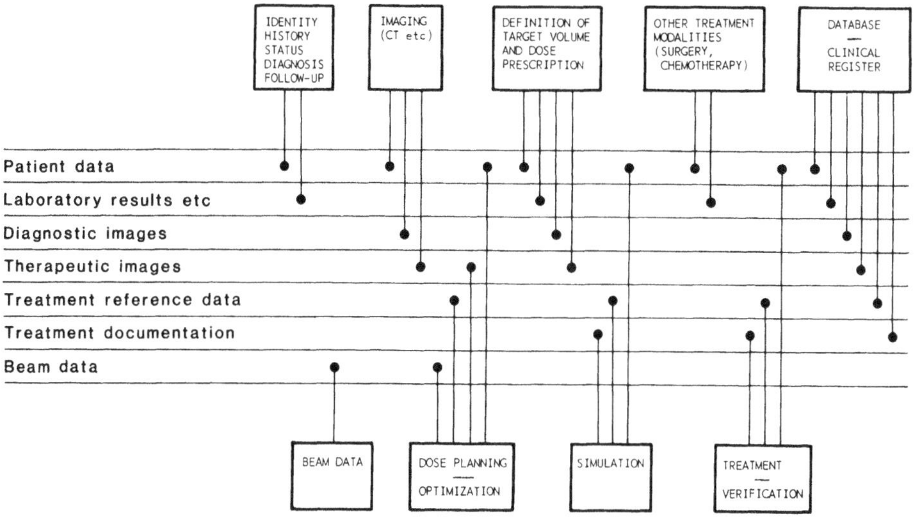

Representation of the flow of information within a radiotherapy department. The different types of data are represented as horizontal logical network channels, the boxes represent the steps of the radiotherapy procedure, and each dot marks information exchange as input and/or output to the network. The data flow in the two image channels is vastly greater than in the other channels. IMAGING stands for all types of image acquisition. (Reproduced from ICRU 42 (6).)

patient as well as knowledge about patients treated earlier, given as both image and non-image information. PACS aspects of the radiotherapy workstation (RTWS) are commented on below.

Images are used abundantly in radiotherapy procedures:

o in treatment planning to define tumour and target volumes and to define the optimal irradiation technique for the patient

o in treatment delivery to verify the equivalence between planned and given treatment;

o in follow-up to evaluate and record treatment response.

Tools should be available to

o integrate all types of image information in 3D, projections as well as sections;

o handle/analyze/present patient information, image (in 3D) and non-image;

- o define treatment beams/sources in relation to the patient in 3D;
- o evaluate/optimize intended/given radiotherapy: present "patient + treatment + effect" in 3D;
- o verify given treatment;
- o perform Quality Assurance of the treatment procedure.

Concerning **selective retrieval and response time** in using images, the RTWS must allow interactive access to all the images for the active patient during planning, i.e. instantaneous (<1-2 sec) display of whatever image is asked for. The emphasis in RT on 3D presentation and image integration is strong. The RT procedure of today makes it possible to use some simple type of prefetching of image data for the active patients, both from central archives and from local image sources such as a dedicated radiotherapy CT-scanner and a conventional simulator. The prefetching of patient data could be initiated well in advance of a planning session, and performed during non-busy periods. Thus, neither response time nor selective retrieval are problems for this application, as long as the active images can be temporarily stored locally in an interactive RTWS.

A way of enhancing the response time of a system, as well as reducing storage requirements, is to use **data compression** techniques. The method of compression must be related to the purpose of doing it (3); lossy irreversible compression might be used when the purpose is simply to view the image. Suffice it here to state that only the reversible compression techniques, preserving the image content, are acceptable for quantitative use of the data and when image processing is to be performed. From the radiotherapist's point of view it is further to be noted that a data reduction, where specific images are selected entirely from a diagnostic point of view, is not acceptable; a full 3D representation of the patient requires not only e.g. the CT slices where the tumour is shown, but also images (information) covering the total volume to be treated.

Storage of RT-images should be distributed to the RT-department. The RTWS allows the user to pick anatomical images - either as "raw unprocessed images" or with some markings from processing e.g. in the diagnostic department, and produces a patient representation consisting of organs, target volume etc, a density volume for dose distribution calculations, a beam/source representation, a treatment effect volume and various computer generated images. The RT images are of interest primarily in the RT department and should be used extensively for treatment verification, quality assurance of the treatment procedure and for follow-up studies.

The **image analysis/processing** part of the RTWS is in its infancy, and will be considered briefly in section 4.

To summarize the present status, we have today the basic tools to "handle" 3D image information, but we do not have the tools available to really help in managing all the information and extracting the relevent parts of it in an intelligent way.

4 FUTURE DEVELOPMENTS

In a future PACS system, the user will be able to pick up information at different workstations. These workstations should be "identical", i.e. they should present the same types of information in the same way, and the tools should be recognized as performing specific functions in the same way. It is further essential that the user gets confirmation from the system that the correct information has been used, be it image or non-image. The tools should ideally help the user to use his/her own competence in a better way. To design many of these tools we must have an extensive model of the radiotherapy procedure, remembering that radiotherapy is one of several cancer treatment modalities.

The management of a patient could be described schematically in three steps: i) definition of the specific treatment (based on cancer care programs); ii) planning and delivery of treatment (according to the program); iii) evaluation and follow-up (results achieved according to predefined endpoints in the program).

4.1 Computer Vision in radiotherapy

The tools required to plan, verify and evaluate radiotherapy have been presented in section 3, where it was stated that the tools of today are inadequate. In the future, tools based on computer vision methods to extract and manage relevant information might be the answer.

Computer vision is a fairly new discipline, combining digital image processing, image analysis, image understanding, neurophysiology, cognitive psychology etc., and it is one of the most challenging areas of artificial intelligence. Experimental systems have already shown the feasibility of using CV methods in clinical radiology (4). The radiotherapy applications are in their infancy, but developments have started.

The development of image understanding procedures will be important in many other areas, such as retrieval from image data and use of symbolic image descriptions in knowledgebased systems. Medical images are "difficult", and we are definitely some way from a system "that can automatically produce a good symbolic representation of the shape of a nonrigid, poorly defined structure in the presence of other structures", as stated in (5).

4.2 The radiotherapist workstation

We have been playing with a vision of a futuristic radiotherapis t workstation, in terms of functional user requirements. This workstation should not only replace the piles of patient records forms to be filled in etc, but in addition truly support the use in the daily routine.

With the radiotherapist workstation, it should be quick and easy to fetch images for viewing purposes, and to compare images e.g. to see if a tumour is changing in size. The patient record should be available simultaneously; for quick perusal - humans are very good at flipping through pages to find the relevant piece of information - and for thorough reading. Retrieval of similar cases should be possible, with "similar" defined by the user. Laboratory test should be presented as graphs if desired, e.g. with the requested test values given as functions of time. Standard template images (in 3D) should be available, where the therapist could draw a tumour and get a suggestion for TNM classification. Knowledgebased support systems should help in managing treatments, determining therapy, suggesting preliminary reference irradiation techniques etc, and also in presenting pertinent reference literature. Scribbles should be made on a memo pad and also directly on images, and the user should be able to interact by pointing at "things" and by using his/her voice. The workstation "should behave as expected", and be very forgiving towards user abuse and misbehaviour. It should further be possible to use ordinary analog paper as well while working, requiring a horisontal surface of at least part of the workstation.

PACS (or IMACS) is the necessary invisible giant behind this formidable workstation, integrating information, handling archives, communications, retrievals, data security etc. Such a workstation would be used not only by radiotherapists, but by all oncologists and many other specialists as well, after suitable adaptions.

REFERENCES:

1 G.Q. Maguire, M.E.Noz, A. Bakker, K. Bijl, H. Didden and J.P.J. de Valk. Introduction to PACS for Those Interested in Image Processing (IPMI 1987).In: Information Processing in Medical Imaging, De Graaf C.N. and Viergever M.A. (eds), Plenum Press, New York

2 J.P.J. de Valk, A.R. Bakker, K. Bijl, W. Heijser, D.E. Boekee and G.L. Reijns. PACS Reviewed: Possible and Coming Soon? In: MEDINFO 86, R. Salamon, B. Blum and M. Jorgensen (eds), Elsevier Science Publishers B.V. (North-Holland)

3 A. Todd-Pokropek. Image data compression techniques: a survey. In: Mathematics and Computer Science in Medical Imaging, M.A. Viergever and A. Todd-Pokropek (eds), NATO ASI Series F:39, Springer-Verlag, Berlin, 1988

4 H.S. Stiehl. On Spatial Image Sequence Understanding. Habilitation Monograph. Technische Universität Berlin, 1987

5 J. Fox and N. Walker. Knowledge based interpretation of medical images. In: Mathematics and Computer Science in Medical Imaging, M.A. Viergever and A. Todd-Pokropek (eds), NATO ASI Series F:39, Springer-Verlag, Berlin, 1988

6 Use of Computers in External Beam Radiotherapy Procedures with High-Energy Photons and Electrons.ICRU Report 42 (International Commmission on Radiation Units and Measurements), Bethesda, Maryland, 198

What are the ACR/NEMA Standard and the SPI Specification?

M. HERFORTH, K. MÜLLER, C.F.C. GREINACHER

Siemens Medical Group, Henkestraße 127, D - 8520 Erlangen

Abstract:

The ACR/NEMA committee where radiologists (American College of Radiology) and manufacturers (National Electrical Manufacturers Association) are represented, published the ACR-NEMA Standard publication no. 300-1985 "Digital Imaging and Communications" in October 1985.This standards publication defines a protocol for a point-to-point communication, regardless of source format and device manufacturer. It was neither intended to specify a PACS nor a network standard.

Therefore in addition to the ACR-NEMA Standard "Digital Imaging and Communications" the SPI (Standard Product Interconnect) was specified by Siemens and Philips and was finished in December 1987. This specification is strictly based on the ACR/NEMA standard. The goal of SPI is to provide compatibility between diagnostic imaging modalities from different vendors and a PACS. SPI supports multiple application oriented services and permits a PACS to grow with minimum redesign to higher performance level and functionality.

1. Introduction

Imaging systems which produce digital images are being used to an increasing extent in medical diagnostics. More than 50 % of the images in a modern radiological department are generated digital today. Understandably, the wish was born for a digital picture archiving and communication system (PACS) instead of the analogue film documentation only. To integrate different manufacturers imaging modalities into a PACS,i.e. to provide an open system architecture, an uniform digital imaging standard is necessary.

2. The ACR-NEMA Standard:

In 1983 the American College of Radiology (ACR) and the National Electrical Manufacturers` Association (NEMA) formed a joint committee to develop a standard to "promote communication of digital image information, regardless of source format or device manufacturer ..."[1]. The resulting standard, published in 1985, forms the basis of most commercial developments in the area of Picture Archiving and Communications Systems (PACS).

The authors of the ACR-NEMA Standard limited themselves to a very specific goal: to develop a point-to-point communication protocol that would permit two pieces of imaging equipment to exchange digital image data. The standard was explicitly **not** intended to specify a network protocol, nor was it intended to provide extensive database manipulation functions.

The standard describes the necessary image and command structures for imaging equipment (IE[1]) to successfully communicate with each other through a standard interface defined in hard-and software. The interface is designed to permit data signalling rates of up to 8 Mbytes per second.

If an imaging equipment has to be interfaced to a network, a so-called network interface unit (NIU) is necessary to transform the individual network protocol into the ACR-NEMA standard interface (Fig. 1).

[1] IE = a PACS entity that produces and handles images

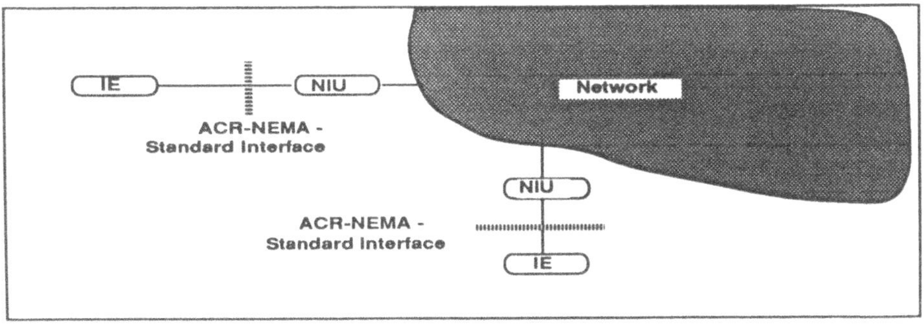

Fig. 1: The ACR-NEMA standard interface in a network environment

However, there are **no** specifications for:

- the communication protocol within PACS
- the physical network
- the management of stored images
- the storage format on off-line exchange storage media e.g. digital optical disks

In practice, this means that images send from the IE via the ACR-NEMA standard interface would disappear into an unspecified region.

Following the organization of the ISO-OSI Reference Model, seven layers are defined from "Application/ Presentation" down to the " Physical " layer :

6/7	"Application / Presentation" defines theapplication oriented logical data structures for the communication commands, image data and data accompanying images.
5	"Session" defines the protocol for establishing and disconnecting logical connections to other units.
4/3	"Transport / Network" converts messages into blocks of up to 2048 words, controls data flow and performs addressing mechanism.
2	"Data Link" is responsible for correct data transfer. It recognizes, recovers and recovers transfer errors. Data packaging is in the form of so-called "frames".
1	"Physical" defines the electrical signal characteristics at the ACR-NEMA interface.

To communicate any information across the ACR-NEMA standard interface the information is structured into packages which are called messages. A message normally comprises the command group and the data group.

Concept

The concept developed in this paper aims to supply physicians who don't work in institutions of the large scale medical health care, like local physicians, with facilities to interchange images over long distances fastly, to support communication involving the images, and to archive and to manipulate them.

Concerning the communication the concept provides for transmission of digital images as well as supplementary information. This integration is in accordance to a transformation of the folder, which holds all the relevant documents of a patient, into an appropriate data structure [2]. This can include diagnosis reports, overviews and directories of existing images, ECG-data or other laboratory data, just to give some examples. To support verbal and textual communication existing standardized kinds of communication, like telephone and telefax, will be integrated.

In addition to the transmission of relevant image information, tools for image manipulation will be provided or made accessible. These tools aim to unburden the physician and to support him finding the diagnosis. The spectrum of the image manipulation tools covers a wide area, reaching from simple tools like contrast enhancement, edge emphasis or distance measurement up to ambitious tasks like tools for tissue differentiation [3], segmentation, threedimensional reconstruction of interesting structures [4] or the use of expert systems [5].

Because the outlined image processing methods require different levels of intelligence, the concept does not provide the computation of all the tasks locally but to process complex tasks needing very much computational power at proper locations like a computing centre. To allow a close adaption to marginal economical conditions different performance classes for the local equipment are provided. Whether an application will be computed locally depends on the installed computer power. This enables the user to configurate a system satisfying his specific requirements without losing functional flexibility.

The outlined forms of communication and image processing require an information interchange that has to suffice a set of marginal conditions. To obtain an acceptable transmission duration it must be sufficiently fast. Futhermore it must be guaranteed, that any physician can participate in the system disregarding his location. Moreover the transmission medium to be employed shall be useable at an economically justifyable expense. Beyond this the confidentiality of the highly sensible patient data has to be preserved. The arrangement of the information interchange with this partially diverging marginal conditions offers the use of public networks, because they are existing all over and are commonly accessible. The problems of identity security of the participants arising from the transmission of highly confidential patient data can be solved by public key crypto systems [6].

However the capacity of nowadays available transmission networks (DATEX, telephone) is not acceptable for image transfer [7]. By introducing ISDN a capacity will be accessible, which seems to be practicably usable for the transmission of such an amount of data. Thus the transmission of an image consisting of 512*512 pixel and a greylevel information of 12 bit will take about one minute at a transmission rate of 64 Kbit/s. Using proper compression algorithms [8] this time can be further reduced. Finally ISDN offers an easy way to integrate verbal communication by providing multiple effective channels simultaneously [9].

The outlined forms of information transmission and processing are efficiently supplemented by archiving the manipulated data. For reasons of organisation and data security each participant should build up and manage his own archive. An access shall be realized in a way that a user will contact the appropriate archive proving his identity whereupon the image will be transfered. To supply such an archive, different performance classes are provided, respecting the extent as well as the speed, which can be dynamically adapted to the particular demand [10]. It is possible to use optical discs or magnetic tape cassettes as media for archiving. Such a tape can store more then 1000 images of the size mentioned above.

A serious problem for transmission as well as for archiving of the relevant images is the enormous amount of data related to an image. This problem can be reduced applying appropriate data compression algorithms. Medical image data are so redundant, that it is possible to reduce their data volume by information preserving methods down to the half. Such methods are important if the image will further be analysed in a way that all details may become relevant. Substantially higher compression factors are achievable using data approximating methods by which the original image can be reconstructed from the compressed data in an approximate way only. These methods can be used in applications where the loss of precision is tolerable. Special algorithms for substantial acceleration of the image presentation task can be applied by first presenting an image of relative low resolution which will be refined during the following image transmission until the desired detail fidelity is reached. In addition to a saving of archive capacity the data compression affects transmission. It leads to a reduction of the transmission duration accompanying a shortening of waiting time and an improvement of interaction facilities.

The explained concept bases on digitized images. Many known techniques of image acquisition, like ultrasound, CT or MRI, are already producing such images. X-ray images made with conventional technique can also be digitized using scanning devices. In this field however a tendency to produce digital image data directly can be observed following the current expansion of digital radiography [11].

Summary

A variety of application fields is imaginable for a system developed according to the outlined concept. In the simplest case the image transport presently done by mail or patient can be accelerated by electronic transmission. Thereby the availability of the images would be increased reducing costs and stress for the patient due to avoiding multiple exposures. The provided communication facilities can be used for consiliary purposes. The spectrum of these consultation activities reaches from an informal dialogue between colleagues up to an institutionalized consultation with experts of a central service. An application of image communication is also conceivable supplementing radiological stand-by services in a way sending images to be reported in combination with the medical history and further patient data to the corresponding working place. The indicated facilities of image processing can be used to support reporting since they unburden the physician from routine work and contribute to the certainty of the diagnosis.

References

1. Greinacher, C.F.C.: Systemarchitektur und Bausteine eines digitalen Bildinformationssystems für die Radiologie. Proceedings CAR'87, Springer Verlag Berlin Heidelberg New York 1987: 544-557.
2. ACR-NEMA Standards Publication No. 300 - 1985. Digital Imaging and Communication. Washington 1985.
3. Ameling, W.: Digitale Bildverarbeitung und Echokardiographie: 3D-Rekonstruktion und Texturanalyse. Fortschritte der Echokardiographie, hrsg. R.Erbel, J.Meyer, R.Brenneke, Springer Verlag Berlin Heidelberg 1985: 3-15.
4. Schneider, W.; Klein, M.; Jensch, P.; Krasny, R.; Ameling, W.; Günther, R.: Volumenorientierte Objektrepräsentation zur Operationsplanung: Proceedings Fifth International Symposium on Surface Topography and Body Deformity; Wien 1988 (in print)
5. Bunke, H.: Modellgesteuerte Bildanalyse. B.G. Teubner Stuttgart: 1985
6. Pfitzmann, A.; Pfitzmann, B.; Waidner, M.: Datenschutz garantierende offene Kommunikationsnetze. Informatik-Spektrum (1988) 11: 118-142.
7. Gabler, H.: Paketvermittlung im IDN. Proceedings Kommunikation in verteilten Systemen II, Springer Verlag Berlin Heidelberg New York Tokyo 1985: 124-144.
8. Jain, A.K.: Image Data Compression: A Review. Proceedings of the IEEE, Vol 69, No. 3, March 1981: 349-389.
9. Rosenbrock, K.H.: Die von der deutschen Bundespost angebotenen Dienstmerkmale zu Beginn des ISDN-Serienbetriebs. Proceedings telematica '88, Verlag Reinhard Fischer München: 37-55.
10. Didden, H.W.; Bijl, K.; De Valk, J.P.J.; Bakker, A.R.: Top-Down Design of a PACS, by Means of the Simulation-Based Performance Prediction Tool Miracles. Proceedings CAR'87, Springer Verlag Berlin Heidelberg New York 1987: 563-573.
11. Hintze, A.; Jötten, G.: Digitale Radiographie. Fortschr. Röntgenstr. 145, 1 (1986): 91-97.

PACS, Step by Step

Drs Robert G.A. Langenhuysen

EUDINICS BV
P.O. Box 42
6850 AA Huissen
Netherlands
Tel:int + 31 85 254573
Fax:int + 31 85 254745

Summary
The complexity of a full-functioning PACS calls for a step by step introduction of this method. The reasons behind the necessity of this approach are discussed and a few examples are given on where to start with PACS.

PACS (Picture Archiving and Communication System) for a long time have been the dream of many futuristic researchers in the medical world. Many clinicians and researchers realize the many shortcomings in the current system of medical image handling. Mostly there is only one image available as making and storing duplicates is very expensive. Consequentually, this iamge can only be at one place at the same time. In the modern multidisciplinary clinical environment where teamwork is more and more a daily routine, a patient is treated by several clinician simultaneously. These clinicians however are located at different places of the building and often even part time working in different hospitals. The result is that when complete information availaibility is desired the treatment of the patient is done in a serial (sequential) mode. This means that the images have to be brought from department to department and obviously much time and money is wasted. This 'foot ware' system can be diminished by simply not making the images available to the clinicians. Doing so would permit a more parallel way of diagnosing, however without all the details at hand. Recently, a publication showed that in a clinical environment situations can exist that as much as 25% of the images are not available for disagnostic purposes due to many causes. These causes can be as clear as the image being at the collegues place, in transit or even simply lost. Alternatively one can imagine a parallel mode where the members of the team gather and review the images simultaneously. Obviuosly a very expensive way in solving this problem. All this is at the basis of all past dreams about PACS.

This dream is turning in reality now by virtue of the availability of fast and economical computers. Both of these aspects are required to make PACS economical and practically realizable. With the current prices of the equipment we will see a vast number of hospitals implement PACS.

Some of these implementations will start with Teleradiolgy. This is one of the many ways to get the feet wet in the world of PACS. Teleradiology and PACS are brother and sister of each other. Often they can use the same type of equipment and some vendors offer Teleradiology equipment which can be easily upgraded to a PACS system or become part of a PACS system. A Teleradiolgy system can perfectly solve the need for fast transmission of pictures over virtually unlimited distances. The world is spanned with a fairly reliable and economic electronic communication system called the telephone network. With the availibility of spread spectrum modems data rates over telephone lines have been multiplied in the recent years. A modem (MODulator/DEModulator) is an essential device for turning digital data (like digitized images in a computer) in analogue data (which can be handled by telephone lines) and vice versa. Fast modems are essential for fast transmission of images. Unlike the ordinary computer files we encounter in the office environment, image files contain huge amounts of information. The information content of a digitized image is hundreds or thousend times more than is contained in ordinary office files. These fast modem make the implementation of Teleradiolgy a reality together with the techniques of datacompression. Data compression is a clever way in reducing the electronic size of an image by mathematical routines such that no or only a little difference between the original and compressed image is observeable. The smaller the electronic size, the faster the image is transmitted. Fast transmission of medical images are required in environments like the ICU, the First Aid and the O.R. It will be obvious that for the clinician on call, the availability of a Teleradiology system will enable him to study an image at a distance so the need for him or her to go to the hospital can be minimized. In nowadays trafic certainly an aspect to consider seriously. The electronic size (and with that also the time it takes to transmit an image by any electronic medium) is also heavyly influenced with the resolution of the image. In PACS terms the word resolution is often misused. One often hears that an image has a resolution of 512 x 512 or 1024 x 1024 whereas one means the total number of datapoints in that image.

Theoretically resolution means someting like number of distinguishable point over a unite distance. In the following we will misuse the word resolution as is used in most PACS discussions.

An other approach to start with PACS is the implementation of a mini PACS. This has the advantage that much more experience can be obtained with a much more PACS like system than a Teleradiology system. In many cases we see also an economic advantage. If for instance the images are from digital origin, the need for making a photographic reproduction of the electronic image as presented on e.g. a CT, an MRI or on a Sonographic system can be large reduced or even totally removed. Much in this respect is dependant on how the organisation of the department is. In departments where only the report is sent to the referring physician, there is no need for making a high quality photographic reproduction of the image. The cost of a photographic reproduction of the CT images can be easily as much as DM 10.- per patient. To this, also the cost for transportation, and archiving must be added. This on its own can easily be as much as DM 20.- for these images during the lifetime of 10 years. With numbers like this one rapidly approaches budgets where a mini PACS system is affordable. Such systems not only have a few review consoles with high resolution screen for viewing multiple images at resolutions of up to 2500 x 2000 but also contain an electronic network and a mass storage device.

The network allows transmission speeds which are thousand times faster than ordinary telephone lines would allow. When many images are to be transmitted over networks, speed is becoming a critical factor. The larger the speed is the better the system will function. Unfortunately higher speeds means higher costs. In the beginning of a PACS situation, high speeds are not as important as in a later situation where more review stations and data collecting stations are connected. That is why the upgradebility of a PACS is such an important factor. The highest cost reduction will be gained when the total hospital is PACSed. Therefore, at a later stage more funds will be available for implementation of higher speeds. A step by step routine if planned well, can offer a gradually implementation of PACS while during this implementation the economical aspects are still favourable.

Mass storage is also extremely important with every PACS. Non erasable optical disks are the ideal storage material as there is no reduction in quality over tens of years. Moreover the size of the disks is extremely small in comparison with the size of the photographic images with their holders. On one single optical disc with a diameter of 15 cm and a thickness of 5 mm one can store up to 60.000 CT images in compressed form. The larger optical disks of a diameter of around 35 cm can store a few hundred thousand of these images. The disks can be put in a 'juke-box' and these juke boxes can be put into the network so that all images generated in many years can be hold on-line. The size of this storage media is only a very small fraction of what is now in use for the archive. Moreover one can save tremendous amounts of time in the retrieval process. With a PACS system one can retrieve an image by typing in the patient name or ID number. When compared with the manual footware system one starts to realize the charm of PACS.

A total PACS system will involve the digitization of all medical image generating devices. In the selection of the PACS one should realize that this is the ultimate goal. Flexibility of the PACS to connect to all these different devices is therefore a matter to discuss while selecting a PACS.

The most difficult matter is the implementation of the logistics. The introduction of PACS will incorporate changes in the way the clinician is working. Instead of sticking an image to the lightbox, interaction with a keyboard will be required. It will take time for all clinicians to adapt to this situation but everyone sees the spread of computers on the desks of the clinicians expand and it will not take long for having the idea of the computer being a clinical tool generally accepted. Implementation of a PACS is a giant operation which cannot be done overnight for a hospital. One should start somewhere and expand from there using the experience built up in the pilot site. The sooner one starts with the implementation, the sooner one will reach the level of maximal cost reduction and maximal quality improvement, and consequentually, the more the hospital, the patients and the staff will profit from it.

"Personal PACS: its Effects on Imaging People and Machinery" or "PACS is worthwhile, but what if you haven't got it?"

Sven Ploem

Radiologist
Canisius Wilhelmina Ziekenhuis
Postbus 9015
6500 GS Nijmegen, Holland

summary

X-Ray Departments, in our hunt for disease, are important. They also are expensive. It pays to try to make them more efficient and raise the quality.

PACS at the moment can't do this: its too complex and too expensive.

Personal-PACS is a low cost Image and Speech/Text Processing System based on the **Personal computer.**

With the best Price-Quality Ratio of the computer bussiness, the Personal Computer gives Patient, Technician and Radiologist benefits that can't be overlooked any longer.

Since there are more X-Ray Departments without PACS than with it, we need such a low cost system to convince Policy Makers.

Radiologists are responsable for part of the diagnostic imaging when looking for disease. Radiodiagnostic tools are important as well as expensive. Two valid reasons to try to improve their efficiency and raise their quality.

PACS stands for Picture Archiving and Communication Systems.
By handling images electronically, PACS promises to better things for us. But does it ?
The tragedy is that yes indeed PACS could better things for us. However: it is prevented from doing so by beeing too complex and too expensive.
Thus it is up to us to create a PACS-like system that is cheap enough to fit ailing budgets and powerful enough to convince Policy Makers.
PACS is very much alive, not in its technical realisation but in its goals.

Personal-PACS is such a low cost system.

Personal points at the Personal Computer (PC), the heart of the system. It is made suitable for Image and Speech/Text Processing.

Personal also points at us, Imaging Professionals. We know our current tools intimately. Can we really expect from others to understand where they fail us? Surely it can be left to us to decide how to improve and to what extent.

My configuration for an X-Ray Department would be: a **Worksite-PC** at every data acquisition machine. A **Hard Copy-PC** at the only Imager of the Department. And a **Reporting-PC** wherever reporting is done.
The CT-PC is connected to the Reporting-PC by a fibre cable for Parallel Viewing.

Simple Things First: MACHINERY

DATA ACQUISITION.

I restrict myself to what I call: **Monitor Diagnostics.**
Ultrasound, Computed Tomography, Digital Subtraction Angiography Magnetic Resonance Imaging and Digital Radiography.
Look at one of the images.
What do you do if you see something interesting?
You annotate.
You annotate for the Clinician or the Teaching File.
And by annotating you block the machinery.

The Ultrsound Machine, in the annotating mode, can't be used to collect ultrsound images (to examine patients). It is used as a word-processor. And a very poor one indeed.
And so are the other, more expensive, modalities.
Even if you have a stand alone viewing console: isn't it a pity to use a $ 100.000 machine to annotate?

The same holds true for measurements (wether geometrical or density) and digital image processing.
With **Personal-PACS** you can do this work at ease, on the Reporting-PC, away from the data acquisition hectic.

PARALLEL VIEWING.

Traditionally we view CT examinations after all the images are made. With personal-PACS we can view the examination while it is going on. I call this parallel viewing.
The CT-PC, in the pauses between image acquisition, sends the last image by fibre cable to the Reporting-PC.
Here a temporary Viewing File is created. We now can make up our minds and we can shorten Patient Examination Time. Thus expensive data acquisition machinery is made much more efficient.

HARD-COPY MACHINERY.

In our department we have a hard-copy machine at every worksite.
This means that we can't buy sophisticated imagers.
And the technicians have to run to and fro with cassettes.

The **Worksite-PC** makes patient image files on Disc Cartridges. These cartridges then are removed and physically (no cables) brought to the **Imager-PC.** If we want we now can make hard-copies.
But do we want to do so?
No, and we have good reasons to want to stop the practice of hard-copying before reporting.

YE OLD CHAIN.

We all know it: first **image acquisition**, then **hard-copying** and finally **reporting.**

To better clinical decision making and to reduce film costs we must change the well known sequence.
We must change it to: **image acquisition, (soft) reporting** and only then **hard copying.**
Only after the reporting is done we know the important images (to be hard copied large), and the unimportant ones (to be hard copied small) and the irrelevant ones (to be discarded).
And we have the report to be hard copied with the images.

Only after reporting we can hard-copy complete examinations and

reduce film costs up to 75%.
Not to mention the fact that even large departments thus run smoothly on only **one** Imager.

And now **PEOPLE.**

THE PATIENT.

In the beginning there is the patient.

The patient wants access to all the diagnostic tools he needs.
No waiting lists.
Therefore we must economise as much as we can (one imager per department).
Therefore we must speed up our machinery and not block it (annotation, etc).

THE TECHNICIAN.

In the beginning there is the technician.

The technician, in the majority of examinations, is with the patient.
She/He is our eyes and ears.
We must free her/him from doing overtime because data acquisition machinery is used in an improper way (annotation,etc).
And we must free her/him from secondary activities like hard-copying.
The technicians must keep an eye on the patient and not on the cassettes.

THE RADIOLOGIST.

In the beginning there is the radiologist.

The radiologist makes the diagnosis the patient seeks.
The patient does not come to the radiology department to enjoy the machinery, wether technical or human.
It is the radiologist who must free the data acquisition- and hard-copy machinery.
Then he must free personnel.
Both are described above.

Finally there are two tasks left for the radiologist.

digital image processing.

At the moment we don't do enough in this field.
For instance: we seldom go beyond window and level for CT images.

No, our endeavours in digital image processing are not comparable to what other imaging professionals do.
Think for instance of people working in Particle Analysis.
Or the Space Industry.
Why are we radiologists behind?
Because from Nine to Five we can't block our machinery when trying our wings on digital image processing. And after Five we can't take the CT with us!
We must have a cheap portable computer to practice: the PC!

reporting.

Our report is the heart of the matter.
In it our expertise is condensed.
And what do we do with it? We glue it on the folder.
Separated from the images inside.

Its crazy, but there it is: we still don't connect images and their interpretation.
With Personal-PACS and changing Ye Old Chain we can include our report in or next to the images. I call this **Biphasic Image Text.**
Biphasic because there is a hard phase and a soft one.
In the hard phase we copy our report on transparant film (Central Imager).
On conventional, developed, film (eg Plain Chest) we can use a plotter and write the report in a non-diagnostic area.

But it is the soft phase that is really important.
Why?
Not because report and images intrinsically belong together.
But foremost because of the nature of PACS itself.
With PACS images are flashed around the hospital. Images.
Nobody talks about the report.
And that is where the danger lies.

If we, radiologists, support systems that flash only images, what we do is not being modern or nice.
What we do is undermining our very profession.
By distributing images without our report, we force all clinicians to interpret the images themselves.
And if we make a good job of PACS without **Biphasic Image Text** we soon will not be needed any more.
From 1985 on, the first CAR Conference, I am intimately involved in PACS. To the best of my knowledge, in Europe, there is not one PACS situation where the report routinely is included in the image.

CONCLUSION.

At the moment there are more radiologists without PACS than with it.
PACS is very much alive, not in its technical realisation but in its goals.
We first have to convince policy makers.
With Personal-PACS we have a splendid tool to do so.
And it should not be perfect, it should be there!

Until now you missed something.
I did not give you what you are given in every lecture on PACS.
And what you read in every article.
I did not give you technicalities.
I did not give you wiring diagrams.
Why?
Because I don't want to lead you astray.
Its not diagrams that count, but markets.
Only markets (What do we gain by it?) convince policy makers.

acknowledgements

The University X-Ray Department, St. Radboud Ziekenhuis (Nijmegen, Holland), has given invaluable help.
Its head is Prof.Dr. J.H.J. Ruys.

Keyboards by: Leon van Erning and Martin Thijssen, Physicists.

Without the help of DIFA Measuring Systems (Breda, Holland) I could not have functioned.
Its head is E. America.
Keyboards by: Ton Schless, Imaging Consultant.

They all touch my heart: the People who created You,

Personal Computer

references

1. Castleman KR, Digital Image Processing. Prentice-Hall Inc., 1979
2. Image Analysis. A Technical Handbook by Joyce Loebl. Short Run Press, Exeter, Great Britain. 1985
3. CAR'85 and CAR'87 Conferences
4. Personal Communications from those the help of whom I acknowledged

Siemens PACS Installations – Practical Experience

C.F.C. GREINACHER, M. HERFORTH, B. LUETKE, G. SEUFERT
Siemens Medical Group, Henkestraße 127, D 8520 Erlangen

Summary:

Siemens PACS is available as a product and several installations are in operation based on the system architecture shown at CAR`87 [1]. Experience gained from these installations since 1986 has been of great value and has led to a continuous upgrading culminating in our products of today. Valuable information not only on technical aspects such as image communication, archiving and display techniques but also on PACS project planning, project engineering and on running a PACS has been acquired.

Whereas many specific requirements of any given radiology department can be met by different combinations of Siemens PACS components, there are many questions we still can not answer today. Systematic analyses concerning efficiency, economics and legal aspects of PACS have to be worked out with the next PACS projects.

1. Introduction

Siemens Medical Group started planning its first PACS pilot project in Victoria General Hospital, Victoria BC, Canada, in 1981. Today 16 Siemens PAC systems are installed worldwide.

Each of these PACS consists of storage nodes, image interpretation and reporting nodes, documentation nodes, and several diagnostic imaging modalities integrated via a digital communication network. Configurations with either one of these four basic components missing (e.g configurations with several CT`s, communicating images via a network to a dedicated remote CT-console but without a shared database on the network), do not meet the definition of a PACS (Fig. 1) and are not included here.

This paper reports some experience in PACS project planning, PACS project engineering and operating a PACS from a manufacturer`s point of view. It might supplement what the PACS users are reporting on their individual results [2 - 11].

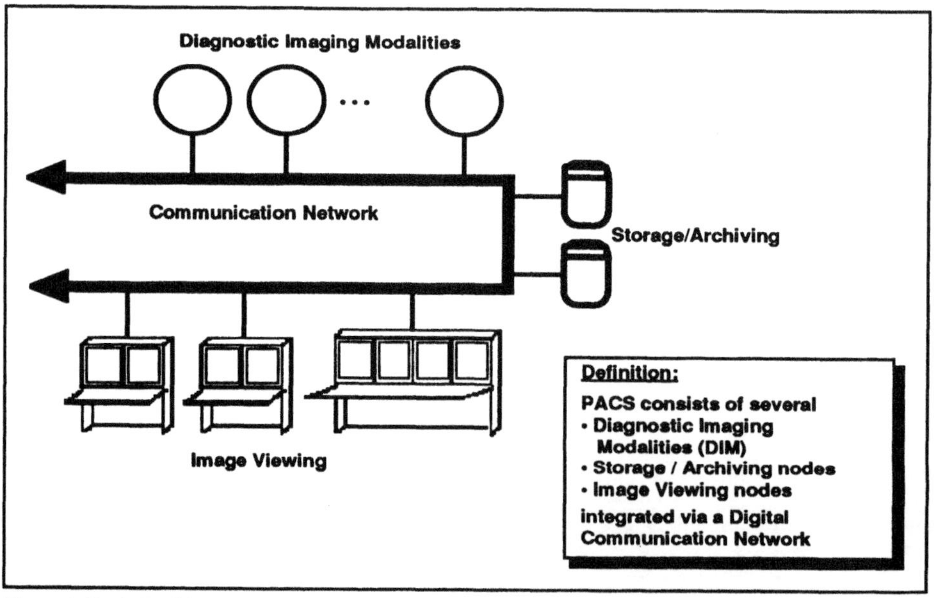

Figure 1: Basic PACS configuration

2. PACS Project Planning

Hospitals, planning to purchase any complex technical system are supported normally by a neutral consultant. The general knowledge on hospital organization, techniques, work-flow, and economics, gathered from multiple similar projects, enables the consultant to make a detailed requirement specification.

We experience a very different situation today when planning a PACS. The personnel in the Hospital responsible for the project have very firm ideas about the performance of the PACS and they have picked up all the fascinating features such as high speed fiber optic networking, optical tape archiving, ACR/NEMA and SPI [12-15] standard interfacing and interactive real-time 3-D imaging workstations. These are the features that scientists and marketing people are discussing and publishing today. What they are not aware of is that many of these features are not yet available, at least not as affordable standard products. When they look for a consultant, they find it very difficult to find someone with the required qualifications in diagnostic imaging equipment, work-flow in radiology departments and hospitals, image interpretation and reporting, digital networking, digital storage and archiving, Health Care economics and legal issues.

The result is that PACS specialists from Industry are asked to act as consultants to Hospitals. No doubt, these specialists have excellent knowledge about the broad variety of PACS related techniques, but do they know enough about medical imaging, work-flow, economics, legal issues etc.? Even if these Supermen did exist could they then work neutrally as Hospital Consultants?

We advise Hospitals planning a PACS project to set up a PACS Consulting Group consisting of different people with expertise in all the relevant fields including Radiology Information Systems (RIS). This PACS Consulting Group should acquire additional expertise from different PACS manufacturers specialists and then work out the specific requirement as described above.

The project planning phase ends with ordering a PACS from a specific manufacturer, based on a contract which has to define

- technical components (hard- and software)
- components and systems functionality
- components and systems performance
- site preparation
- introduction of the system into the routine work
- teaching of personnel
- project time schedule (milestones)
- project costs.

3. PACS Project Engineering

The main objectives of PACS project engineering are project management, site preparation, customizing (customer specific solutions) and bringing the system into routine operation.

3.1 Project Management

Very close cooperation between the manufacturer and the user is a must. A good solution to perform the complex task of project engineering is to install two project managers: one on the customers side, the other on the manufacturers side. Each project manager has to have the expertise and competence to decide which other experts are needed to work on certain specific tasks for the project. (If the head of the radiology department believes he could minimize the risk of the project by performing the task of the project manager himself, this may indicate that he underestimates the time and effort required for that task.)

The customers project manager should report to a decision committee, members of which could be the director of radiology and director of administration (data processing and organization) and - depending on the hospital organization - other high level executives.

3.2 Site Preparation

The project managers will be confronted with severe problems if the following topics are not well understood during the planning phase and not well defined within the contract.

- PACS equipment needs space and air conditioning. On exhibitions or in the manufacturers show rooms only the PACS terminals such as image work stations, film scanners or hardcopy cameras are shown normally. The user might not expect to need additional space for the equipment "behind the wall" such as host computers, storage devices, gateway or database machines. All these equipment have to provide very high performance. It is a technical law that "speed heats". Required space and power consumption are often underestimated!

- Installing equipment with high power consumption in a room where radiographers or radiologists have to do their routine work is not a technical problem at all. But the required air conditioning might generate an unacceptable noise level for the people working in that room.

- Harcopy units, optical disk drives, image work stations and other equipment have to be installed at certain locations within the department to allow easy control and use by the people working there. But cable lengths between these units and host computers are limited, i.e. the host computers can not be installed anywhere far away in the building.

3.3 Customizing

Customers should know that individual hardware or software development is expensive and long-term support of customer specific components can hardly be guaranteed by manufacturers.

If possible in any way it would be much more effective (and cheaper) to adopt the work-flow to standard systems components than to invest in individual development and maintenance.

In this context it becomes obvious that systems flexibility is a very important criteria for selecting a certain product.

3.4 Bringing the system into operation

To move from the conventional to the PACS supported department some changes concerning organization and flow of work will be necessary. This is one of the most critical issues when introducing a PACS. Because across the board experience about PACS is not yet available, we can not provide very PACS specific rules in this context. We suggest to follow the general rule for complex projects:

- Changes in organization and flow of work have to be investigated and specified in close cooperation with the responsible experts. The specified changes have to be introduced and tested carefully step by step. If everybody understands why and how the modifications will improve the departmental efficiency, he or she will support their introduction.

4. Operating a PACS

Experience from operating Siemens PAC systems has been reported by several users [6-11]. Some of the more general results are shown as follows.

4.1 System Performance

The overall system performance is primarily determined by the input/output channels (I/O protocols) of the computers interfaced to the communication network and not by the network speed (10 Mbit/s) itself.

Real-time manipulation of images can be performed nowadays only if image communication via the network is not involved in this manipulation. It is easy to prove that the same restriction will be valid with the next generation of very fast networks (FDDI = 100 Mbit/s), especially for digital radiography images with 2048^2 image matrices.

Real-time reaction of the system at the users interface level can be guaranteed only by using well defined organizational rules, patient scheduling and related information, to pre-fetch images from remote storage devices and to control image communication to dedicated local storages in background-mode.

Not every diagnostic imaging modality needs a high-speed network to get rid of its small amount of image data. Therefore and for economical reasons the high-speed network will function as a backbone to interconnect multiple departmental subsystems, and to provide high-speed communication for high performance nodes such as image workstations or storage devices.

4.2 Image Interpretation

Image interpretation and reporting on Siemens' high-performance Diagnostic Reporting Consoles (DRC) is very much appreciated by radiologists. Even the first prototype image workstations PDM with somewhat reduced performance have been accepted very well.

To define all general rules on how to arrange a series of images on the DRC's monitors is not yet possible. But some rules are well accepted for CT and MRI images:

- Images of CT (MRI) examinations displayed for the first time, should show up in the "overview mode", i.e. a high number of token images (16, 64 or 256 images on one 1024 by 1024 display monitor) are shown on one monitor of the DRC to provide an overview on the examination.. Selected images can be switched from the "overview" to the original image matrix display which we call "diagnostic view".

- The display order should meet the table positions (not the acquisition order!), eaxamination date, type of examination, and others.

- Textual information should not be displayed automatically super-imposed onto the images but optionally.

- Textual information related to all images of a series should be displayed as a "first image".

- It is widely accepted that images are windowed by the technician during image generation at the CT (MRI) scanner and then sent to the PACS as prewindowed.

- A series of images, once manipulated at the DRC, has to show up next time exactly as they were finished during the last display session.

4.3 Improvement of Efficiency

Improvement of efficiency is especially related to image communication, data base management and archiving, multy-modality image interpretation (CT and MRI) and film savings. Unfortunately no systematic analysis and quantitative results have been published up to now.

Based on general results from the operating projects, preliminary results of project requirement specifications and some additional assumptions, the authors have published a quantitative cost efficiency analysis recently [16].

Under the direction of the Deutsche Forschungs und Versuchsanstalt für Luft- und Raumfahrt (DFVLR) the "Consulting Group for Health Care Economics" (Beraterkreis für Gesundheitsökonomie) is working on a study which will suggest to the German Ministry of Research and Technology (BMFT) how such a cost efficiency analysis of PACS should be performed in pilot projects.

5. Conclusion

We have accumulated experience from installing and operating 16 Siemens PAC systems in Canada, Japan, Europe and USA. Digital modalities as CT, MRI, Digital Radiography and DSA are interfaced to these PAC systems. These systems support routine work partially and are well accepted concerning their performance. Results from these projects proved, how such systems can be planned, installed and operated and what has to be improved.

Systematic analysis concerning efficiency and economics of PACS should have highest priority in the near future.

References

1. Greinacher, C.F.C.: Systemarchitektur und Bausteine eines digitalen Bildinformationssystems für die Radiologie. Proc. of the intern. Symposium CAR`87, H.U. Lemke, M.L. Rhodes, C.C. Jaffee, R. Felix (eds.) Springer Verlag, 1987, 544-557

2. Medical Imaging II, R.H. Schneider, S.J. Dwyer III (eds.) Proc. SPIE 914, 1988

3. Medical Imaging III, R.H. Schneider, S.J. Dwyer III (eds.) Proc. SPIE 1093, 1989

4. The UCLA PACS modules and related projects - A progress report - Medical Imaging Division, Department of Radiological Sciences, University of California, Los Angeles, CA 90024, Febr. 1988

5. Barneveld Binkhuysen F.H., et al.: The setup of a clinical evaluation of the Dutch PAC system. Proc. of the fifth intern. symposium on the planning of radiological departments ISPRAD V, A. Chiesa, G. Gasparotti, R. Maroldi (eds.) 1988, 239-241

6. Gell, G., et al.: PACS Erfahrungen beim Einsatz in der Klinik. Proc. 5. Grazer Radiologisches Symposium, G.H. Schneider, E. Vogler (eds.) Springer Berlin, Heidelberg, New York 1988, 594-600

7. Schneider, G.H.,et al.: Practical experience with a PACS in a CT station. Proc. of the fifth intern. symposium on the planning of radiological departments ISPRAD V, A. Chiesa, G. Gasparotti, R. Maroldi (eds.) 1988, 242-245

8. Lamarque, J.L.: PACS. Journè de Prèsentation. Hôpital de Lapeyronie, Montpellier, Octobre 1988

9. Fisher, P. et al.: Clinically based archive model for a PACS. Medical Imaging II, R.H. Schneider, S.J. Dwyer III (eds.) Proc. SPIE 914, 1988, 1097-1104

10. Nosil, J., et al.: A prototype multy-modality PACS at Victoria General Hospital. Medical Imaging II, R.H. Schneider, S.J. Dwyer III (eds.) Proc. SPIE 914, 1988, 1362-1378

11. Mattheus, R., et al.: An experimental and clinically useful PACS. EuroPACS Newsletter, vol. 3, no..2, 1988, 13-14

12. Digital Imaging and communications. ACR-NEMA publication no.300, 1985, Washington D.C.

13. Tesche, G.: A PACS interface specification. EuroPACS Newsletter, vol. 3, no. 2, 1988, 15

14. Prior, F.W.: Standardizing interfaces in radiology. Lecture Notes in Medical Informatics, Vol.37, A general PACS-RIS interface. O.Rienhoff, C.F.C. Greinacher (eds.) Springer Berlin, Heidelberg,New York, 1988, 19-23

15. Herforth, M.: What is the ACR-NEMA standard and the SPI specification? Proceedings of the intern. Symposium CAR`89, H.U. Lemke, M.L. Rhodes, C.C. Jaffee, R. Felix (eds.) Springer Verlag, 1987

16. Greinacher, C.F.C., B. Luetke: Aufbau von PACS-Systemen unter wirtschaftlichen Gesichtspunkten. Standpunkte aus dem Bereich Medizinische Technik der Siemens AG Erlangen. Proc. 3. Frankfurter Gespräche über Digitale Radiographie, Bad Nauheim Okt.1988, H.E. Riemann, J. Kollath, O. Rienhoff (eds.)

Low Cost Workstation for Medical Image Processing

+TORRESIN A., +TOSI G., +BURATTI A., +COLOMBO P., +ZANNI D., *BADI C., ++ CASTELLANI G., ^DELLA VENTURA A.

+ Servizio di Fisica Sanitaria, Ospedale Niguarda Cà Granda, P.za Ospedale Maggiore 3, 20162 Milan, ITALY
* Impianti Medico Scientifici, Ospedale Niguarda Cà Granda;
++ Casa di Cura S.Maria S.p.A., Castellanza - Gruppo Ospedale San Raffaele, Istituto Scientifico, Milano.
^ Consiglio Nazionale delle Ricerche, IFCTR - SIAM, Milano.

SUMMARY

A gradual approach to the development of an inexpensive multipurpose workstation for medical images post processing is described, based on standard hardware components and software easy to be up-dated and improved. The workstation is not dedicated and can easily be adapted to new tasks.

INTRODUCTION

The scientific and clinical usefulness of medical imaging is often limited both by the accessibility to raw data files and by the lack of workstations, separate from the principal ones, able to display and elaborate images. Biomedical images are now produced at a higher rate than the capability of interpreting and "using" them. Besides this, the economical costs of image departments are growing higher and higher, becoming a significant part of hospital costs.

Commercial PACS (Picture Archiving and Communication Systems) are now available on the market, but they are very expensive and demanding in terms of cultural and organization resources. At present time, commercial PACS are justified only in hospitals with a high level of central organization and with an advanced Hospital Information System (HIS). An independent system dedicated to the solution of specific problems met in every day's work has therefore been developed. This allows, for instance, to perform image quality controls (1), quantitative CT analysis (2), (3), (4) and quantitative analysis of film-dosimetry of high energy electron and X-ray beams. This system, however, is not dedicated to a special task and therefore no efforts have been done to optimize any of the hardware components, since the system itself has to be easily adapted to new tasks.

One important problem is to choose how to transfer digital data from existing medical imaging systems to the workstation, without significant loss of information. Different solutions may be used for this aim, depending on the hardware available: the physical nature of the signal must always be taken into account as well as the specific target of the application. For instance, acquisition from video signal may be insufficient for clinical evaluation of CT images, but sufficiently good for rough morphological evaluations. Film's digitalization is a very complex problem which cannot easily be solved with inexpensive systems; however digital images sufficiently good for specific purposes can be obtained by means of the same graphic board used by the workstation: for instance, for the analyses of the blackening of films produced by electron beams of a linear accelerator.

MATERIAL AND METHODS

Software components of the project are based on MS-DOS operative system (80386 microprocessor, 1 MByte RAM, 80 MByte Hard Disk), FORTRAN-77 and C programming languages; ASSEMBLER subroutines, employed by FORTRAN-77 programs were also used in some cases. A first test was performed using a PC with a standard "high resolution" color board (320 x 200 pixels, 256 color levels), with which a display system with many of the standard display functions of CT o MR workstations was implemented. The corresponding "package" included software for "importing" data from systems other than MS-DOS, through an off-line digital data transfer: no attempt was performed at this stage of directly interfacing, because of the lack of standards. The first test was greatly encouraging, but put in evidence the need of choosing a better graphic board. The particular kind of graphic board used at first time was not able to digitalize external data and was rather complicated to use.

This graphic board was therefore substituted by a PCVISION-plus FRAME GRABBER of Imaging Technology (5). The frame memory size (1024x512 pixels; 8 bits per pixel) allows storage of two 512x512 images. A library of routines which can be employed by C and Fortran, can be used in connection with the board. It is also possible to use a ready interactive applicative software. External Professional Monitor RGB, 14", 512x512 is used. Of course, this hardware solution, does not allow to digitalize X-ray radiograms for diagnostic purposes. The system was also rested in a small local area network (LAN), foreseen both for stand-alone and network capabilities.

Software development started with these objectives:

- to set up an environment favourable to increase medical knowledge and to inject this knowledge in the diagnosis process;
- to give to physicians understandable tools to performs image analysis in a controlled way;
- to progressively automatize the interpretation tasks from the clinical ones until complex diagnosis procedures by an increasing awareness and formalization of the visual interpretation criteria. The process automatization not only permits to save time, but forces the explicit statement of interpretative rules making the procedures more objective and repeatable.

Design and test of software modules follows a precise methodology aimed to obtain reliable and portable products. In particular, automated interpretation procedures are introduced in the operational context after repeated cycles of definition/confirmation and satisfactory accuracy evaluation.

RESULTS

Other systems for interactive display and analysis of medical images based on the use of high level computers (6), (7) have been proposed. The low cost workstation described in this work, allows to perform analyses of similar nature on a lot of medical images (CT, NMR, nuclear medicine, radiotherapy) in an "open" and inexpensive environment.

This low cost workstation has been used for an off-line analysis of CT images in order to determine the bone mineral

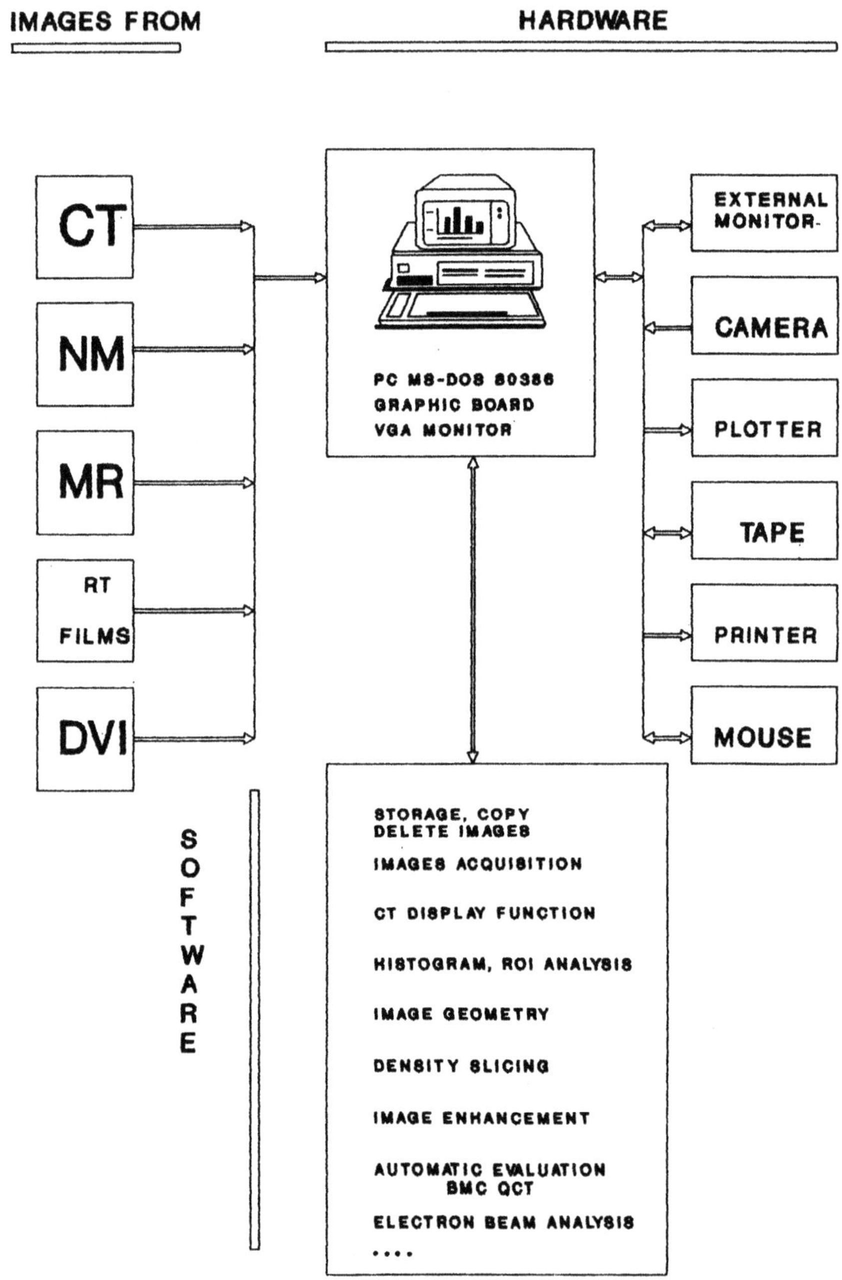

Diagram of major components (hardware and software) of low cost workstation for image processing.

content of the vertebrae, and in many other cases, where a quick and accurate image analysis was required. In particular, it seemed very interesting to set up a software instrument capable to perform comparisons between different interpretation procedures. The instrument compares their results and displaies matchings and discrepancies in a pictorial way. The usefulness of this "comparator" became evident in those cases (very common) in which a substantial lack of interpretative standards exists. The comparator can be considered a working instrument during the definition of an interpretation procedure which can collect a general agreement.

The possibility of analyzing dosimetric films exposed to electron beam of different energies, from a linear accelerator, was also tested.

ACKNOWLEDGMENT

We thank Atel, Milano, for his support for this research.

REFERENCES

(1) Borasi G., Castellani G., Domenichini M., Franchini M., Granata M., Torresin A. and Tosi G. Image quality and dose in computerized tomography: Evaluation of four CT scanners. Med Phys 11:321-325,1984.

(2) Castellani G.C., Liverani B., Mussio P., Pelizzari P., Torresin A. Steps toward an automatic CT images interpretation. Proceeding of the International Symposium CAR'87 Computer Assisted Radiology, pagg. 724-728, Springer-Verlag ed., Berlin, July 1-4, 1987.

(3) Gattinoni L., Pesenti A., Torresin A., Baglioni S., Rivolta M., Rossi F., Scarani F., Marcolin R., Cappelletti G. ARDS profiles by computerized tomography. J Thorac Imag 1(3):25-30,1986.

(4) Tamponi M., Campoleoni M., Zanni D. Critical analysis of dual-energy CT in bone mineral contents assessment. Phys Med III(2):135-138, 1987.

(5) Frei W. Digital Image Processing Software. SPIE Vol.528 Digital Image Processing, 88-94,1985.

(6) Orphanoudakis S.C. Supercomputing in Medical Imaging. IEEE Eng Med and Biol Magazine: 7(4),16-20,1988

(7) Robb R.A. A workstation for Interactive Display and Analysis of Multidimensional Biomedical Images. Proceeding of the International Symposium CAR'87 Computer Assisted Radiology, pagg. 642-656, Springer-Verlag ed., Berlin, July 1-4, 1987.

High Speed Networks for PACS – Introduction and Comparison

B. Fasel, F. Vossebürger, D. Meyer-Ebrecht
Lehrstuhl für Meßtechnik
Aachen University of Technology, West Germany

Summary

PACS systems will only be accepted in clinical environments if they actually allow *interactive* work with digital images. Random access to remotely stored images must be performed within seconds. With regard to the quantities of images to be transferred and the volumes of data per image, *net* transfer data rates far beyond 1MByte/s are required. None of the currently available standardized networks will meet the requirements of a PACS which includes the entire range of image modalities. On the other hand no relevant PACS evaluation can be performed with pilot systems which are based on insufficient communication backbones. This motivated the design of *ImageNet*, a novel LAN which is conceptually optimized for PACS applications. Since *ImageNet* was developed from scratch, a considerable amount of typical medical requirements could be incorporated into the network concept. Beside *ImageNet* another fast fiber optic network called *FDDI* will briefly be introduced.

Speed requirements for PACS networks

As long as we do not have practical experience with full PACS systems which replace every physical image transport by digital transmission, we can only make very simplistic assumptions about the required data flow. The following calculations are based on the current film transport activities in the 1500-bed University Hospital at Aachen. The annual image production amounts to 300.000 X-ray images and 280.000 digital images.

1st assumption
The data equivalent of all images produced during a work day has to be transferred only twice within 8 hours (once for archival and once for evaluation). This leads to an *average* data flow of roughly 2Mbit/s.

2nd assumption
One reporting session of about 20 minutes duration will require access to about 100 images (4MByte/image). To transfer the data equivalent of images required for one session within a period of the duration of that session leads to an average data flow of 3Mbit/s.

3rd assumption
A LAN is used which performs only one data transfer between two partners at a time. Then the *net* data transfer rate must be above (2+3*N)Mbit/s where N is the number of image workstations for reporting.

A data communication specialist will immediately see that the assumptions made above are very naive. Peak data rates will be much higher than the theoretical average data flow. On the other hand the number of transfers per image will be much higher than the minimal value assumed above. Finally, we often mistake the physical bandwidth of the transmission media and the net data rates between two partners, which usually is a factor of five to twenty less.

Why are standard networks much slower than the physical data rate?

It is evident that in standard networks there is a wide gap between the physical data rate on the transmission medium and the data rate which is available for the user. This is mainly due to the methods of media access (MAC, media access control, ISO/OSI sublayer in layer 2) and the high complexity of the used protocols. The dominating MAC procedures are *CSMA/CD* (for *bus* topologies), *Single Token* (for *bus* and *ring* topologies) and *Multiple Token* (*ring* topology). Common to these methods is that the overall dataflow can not exceed the physical data rate on the transmission medium. When the time needed for media dispatching is neglected, this means for example: when on average there are twenty transfers active at the same time, only 5% of the maximum data rate is available for each transfer. These kinds of LANs are not applicable for image transfer in PACS even when the physical data rate seems to be sufficient.

Moreover, the high complexity of protocols on all layers of the ISO/OSI reference model increases the inefficiency furthermore: the maximum data rate can by far not reach the physical data rate even with low work load on the network. The complexity is a result of the demand for maximum universality of the standard networks. Networks become either very slow when you implement the lion´s share of protocols in software or very expensive when using mainly specific hardware.

ImageNet

When designing *ImageNet* [1] [3] we consequently applied decomposition strategies. The basic idea was to separate the world of digital image data strictly from the world of alphanumeric data. This results in several advantages:

- existing systems for managing alphanumeric patient data can be used with almost no changes. (e.g.: RIS/HIS, Radiological/Hospital Information System)
- the dataflows of image data and alphanumeric data do not interfere with each other.
- the existence of two independent communication systems can be used to increase the reliability of the overall system.
- the dedicated application of the *ImageNet* to transfer image data leads to particularly simple and quick implementations of protocols.

Thus *ImageNet* is part of an overlay network consisting of a standard LAN for alphanumeric data (=standard application) and *ImageNet* itself, a novel LAN for the very fast transfer of digital images. The fundamentals of *ImageNet* are:

- *Line Switching*
  In order to transfer images, *ImageNet* builds up a direct link between the transfer partners and maintains the link until the transfer is completed. To the transfer partners the network looks like a transparent point to point link.
- *Asynchronous Transmission*
  Data words are asynchronously transmitted. This gives the advantage of combining transmission lines with different speeds, media, etc. in one and the same network.
- *Decentralized Network Control*
  Routing is performed by independent switching modules. Those modules need information only about devices they are directly connected to. This leads to a very high modularity, the network can very easy be expanded or modified.
- *Hyperstar Topology*
  The *Hyperstar Topology* is highly structurized and modularized. Therefore *ImageNet* can easily be adapted to organizational requirements with respect to optimizing the dataflow. An important feature of the *Hyperstar Topology* is that multiple data transmissions can be performed simultaneously at full network speed. Without restrictions redundant network branches are allowed.

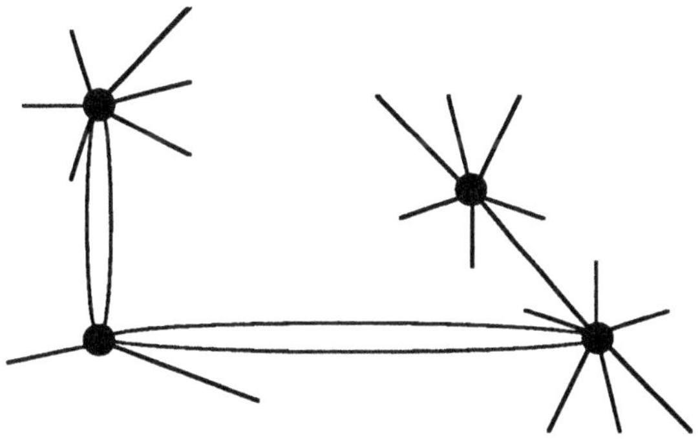

Fig. 1: *Hyperstar Topology* of *ImageNet*

- *Simple DTE-Protocol*
  Especially for the task to transfer images a simple and very efficient DTE-protocol has been developed. The DTEs are not burdened with network internal tasks. Since the network is independent of the DTEs, the overall reliability of the whole system is significantly improved. (DTE=Data Terminal Equipment)

All hard- and software components of *ImageNet* are available. Transmission links are mainly based on fiber optics operating in a speed range between 16Mbit/s and 200Mbit/s. In addition to that, telephone wires have been successfully used (144Kbit/s over a distance of 5km). In several pilot installations it has been demonstrated, that even the transmission of very high resolution images (2048*2048*8bit) takes less than a second.

FDDI

Among the standard systems only the fiber optic LAN *FDDI* is being regarded as a candidate for PACS applications. *FDDI* (Fiber Distributed Data Interface) is a proposed *ANSI* standard. It features:

- 100Mbit/s physical data rate
- fiber optic double-ring topology, total length <100km
- up to 500 network nodes
- timed token passing protocol

FDDI has some drawbacks when used in PACS: Due to the ring architecture the total data throughput is always less than 100Mbit/s. This data rate is shared via the *FDDI*-MAC-procedures among all transferring DTEs. Thus the remaining data rate for the DTEs quickly reaches unacceptable low values. On the contrary a 200Mbit/s-*ImageNet* provides an overall dataflow, which is always above 200Mbit/s. In practical use lots of transfers will be active at the same time, resulting in an overall dataflow which is much higher than the physical data rate. Another disadvantage of *FDDI* is that every single data transfer needs *interaction* with each and every network participant. This is a typical disadvantage of all ring and bus architectures. Thus the probability for all kinds of errors (even for total breakdown) is increased because the exact state of a DTE is often unpredictable. Due to the ring topology of *FDDI* another disadvantage has to be considered: Any modification of the network causes a complete shut down of the whole system. The network speed is strictly fixed and can not be increased in the future neither step by step nor for the whole network at once. Moreover the complexity of the *FDDI*-protocols on all ISO/OSI-layers is evident. This is a result of a basic design goal of *FDDI*: the general purpose application. It must be taken into account that this complexity will reduce the overall performance of the system significantly. Summarizing all *pros* and *cons*, *FDDI* is not the optimum choice for PACS installations. *ImageNet* is currently being tested in several PACS pilot projects with promising results.

Final Clause

We hope that the results of our work will not be misused for military purposes. To the best of our knowledge we will not participate in any activities concerning military applications.

References

[1] B. Fasel *Fibre Optic Transmission Systems for PACS: Network Considerations and Component Design*. Proc. CAR 85, Springer, Berlin, pp. 303

[2] *IMNET Preliminary Specifications* IMTEC Image Technology AB, Uppsala 1986

[3] B. Fasel, F. Vossebürger, D. Meyer-Ebrecht *Image-Network for PACS* Proc. CAR 87, Springer, Berlin, pp. 558

Design for a High-Tech Imaging Center

RAY L. BROVOLD, AIA
Senior Medical Planner
Ellerbe Becket, Inc.
Minneapolis, Minnesota

FRANK D. NEMETH, AIA
Senior Vice President
Design Principal
Ellerbe Becket, Inc.
Minneapolis, Minnesota

JALIL FARAH, M.D.
Chief of Diagnostic Radiology
William Beaumont Hospital
Royal Oak, Michigan

PRESENTATION SPONSORED BY
SIEMENS MEDICAL SYSTEM
Guenter Westermann
Senior Vice President
Iselin, New Jersey

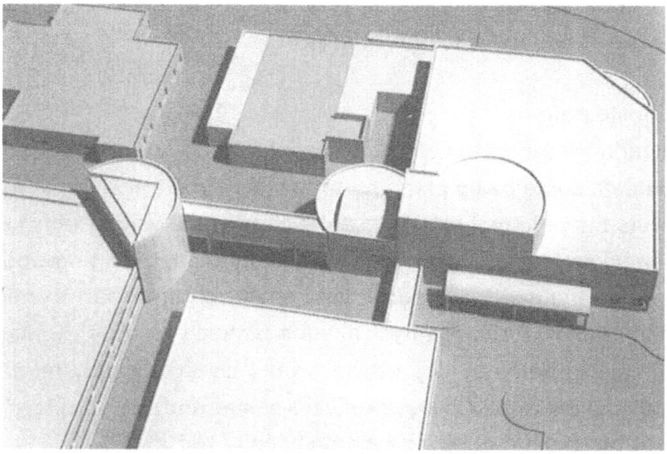

Introduction

The William Beaumont Imaging Center is a freestanding outpatient facility located on the campus of the William Beaumont Medical Center in Royal Oak, Michigan, USA. An MRI and CT outpatient facility exists on the site and will be integrated into a comprehensive new imaging facility. When the first phase of the project is complete, it will include expansion of the existing MRI and CT space, plus four ultrasound rooms, four mammography rooms, eight nuclear medicine rooms, eleven general imaging rooms, and a lithotripsy unit. The design provides for expansion of all these imaging modalities and the introduction of new technology in future phases.

Design Criteria

The owner's requirements for the Imaging Center are:

1. to provide a highly functional facility compatible with existing state-of-the-art technology and designed to easily accommodate transition to new high-tech imaging systems as they are developed;

2. to provide a facility sensitive to the needs of the patient, in an aesthetically pleasing environment;

3. to provide a functionally efficient work environment for the staff; and

4. to provide for expansion and change so that they occur in a logical and orderly manner, with a minimum of disruption.

When the William Beaumont project began in early 1988, it was clear that digital technology would have a significant impact on imaging and the kind of space that imaging would require in the future. Designers of the Beaumont Imaging Center understand that, in response to rapidly changing tèchnology, flexibility has to be given a high priority.

Digital Considerations

The transition from conventional to computer-assisted imaging requires that we re-evaluate some basic planning principles. The transition to totally digital imaging has the potential to significantly change the way in which radiologists practice their profession. Theoretically, totally digital imaging would no longer require the need to process film; physically transport examination requests, film, and reports; or physically file patients records. Consequently, traffic within the radiology department would be significantly reduced. Distribution of reports would be quicker and easier and the problems of lost film and repeat examinations would be significantly reduced.

This transition will take time. Although we have been designing and building facilities for totally digital imaging since 1983, the full impact of this new technology may not be realized for some time. Estimates range from ten years to a generation and beyond. The success of a transitional department will depend on how well the architect and the planner satisfy the needs of both conventional and new technology.

Design

The William Beaumont Imaging Center consists of two stories above grade and one below grade. General imaging, ultrasound, MRI, and CT are located on the main floor. Mammography, nuclear medicine, lithotripsy, and research space are located on the second floor. High-activity, nonpatient or staff areas are separated from support spaces and patient circulation. Separate male and female dressing and gowned waiting rooms are provided for the patients. The film file is on the lower level below grade. Images are processed at seven locations conveniently located to serve groups of imaging rooms.

The Picture Archiving and Communication System (PACS) will link all locations where images are generated with reading locations, radiologists' offices, the archive, and a few key locations outside the Imaging Center. The existing Rapid Telephone Access System (RTAS) will be extended to this facility and will allow radiologists to dictate digitized voice reports from any touchtone telephone. The addition of a Radiology Information System (RIS) is being considered. RIS, PACS, and RTAS will also interface directly with the existing Hospital Information System (HIS).

Because equipment selection typically occurs during the construction phase, the key to smooth and trouble-free installation and operation of imaging systems depends on initially designing for flexibility, with careful attention to detail. Imaging rooms should be sized to accommodate the "worst case scenario" so that similar equipment from any manufacturer can be installed without moving the walls. Reserve capacity should be designed into building systems to ensure that new technology can be accommodated with little or no modification to the building or the building systems. An extensive network of electrical trench duct and cable trays will allow easy access for installation and service of image transmission and communication systems. Mechanical, electrical, and communication systems should be designed with a minimum reserve capacity of twenty percent.

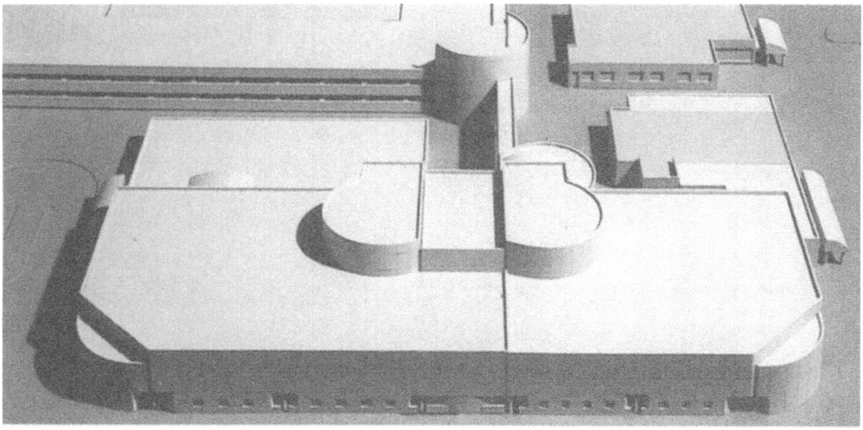

The Future

The William Beaumont Imaging Center is designed to be expanded both horizontally and vertically, and will ultimately be more than double its initial size. By virtue of its size and capacity, its comprehensive "state-of-the-art" technologies, and its growth potential, this project is unique in the United States and represents an important concept and direction in imaging facilities design.

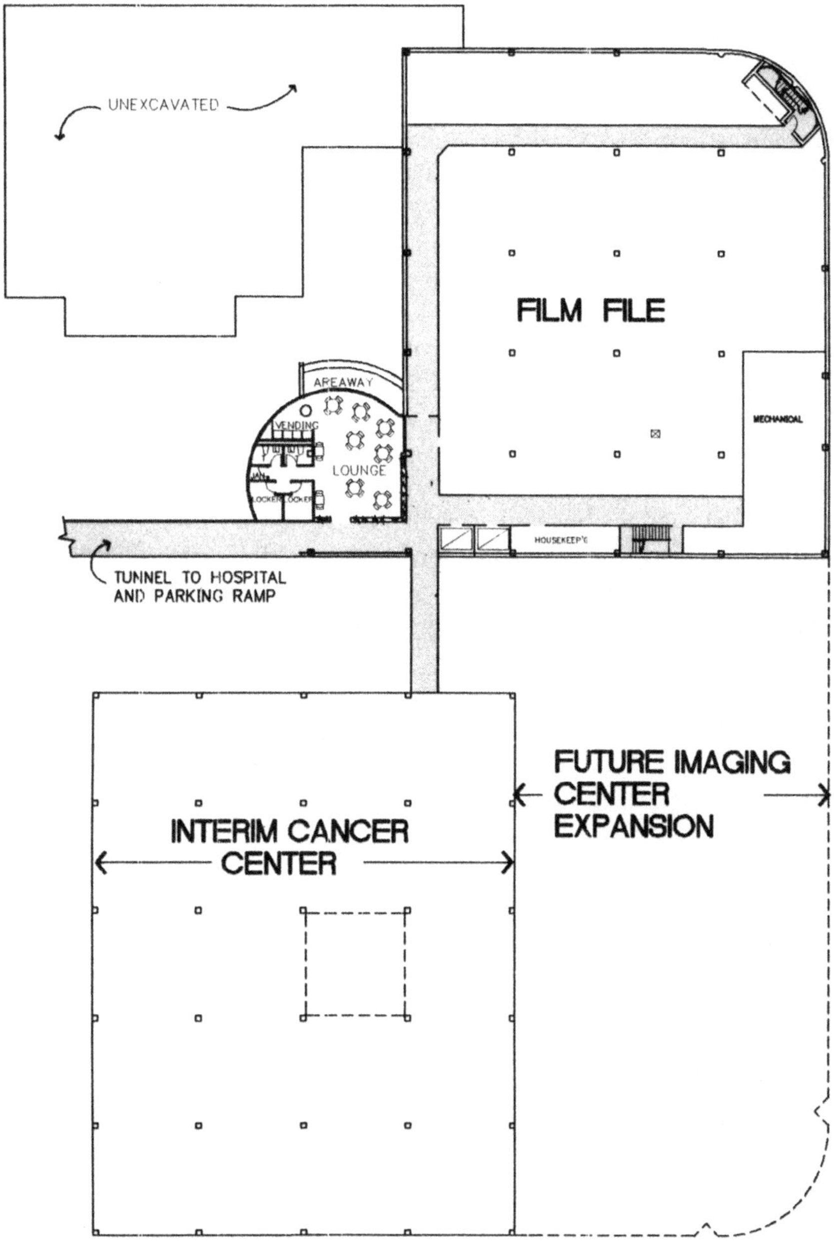

LOWER LEVEL FLOOR PLAN

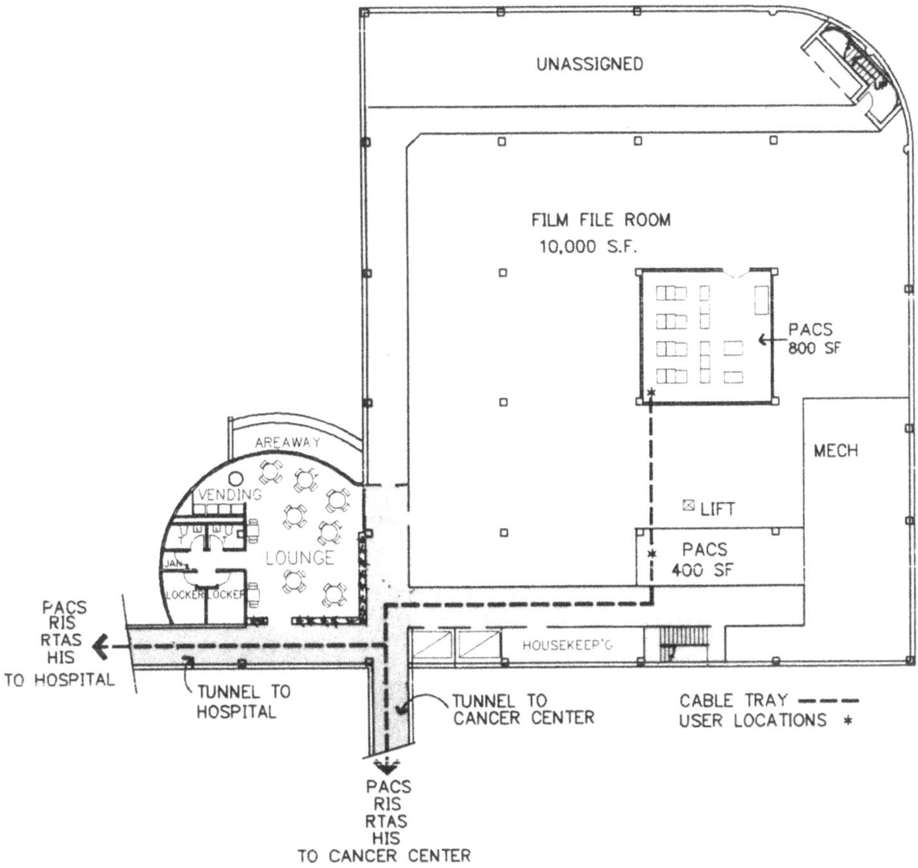

LOWER LEVEL FLOOR PLAN-IMAGING CENTER

The buildings on the William Beaumont Hospital campus are linked together by an extensive system of tunnels, providing a conduit for both building services and pedestrian traffic.

PACS and other communication cables will run through these tunnels, connecting the Imaging Center with several key locations on the hospital campus, including Inpatient Radiology in the hospital.

The film file room and PACS occupy most of the lower level of this Imaging Center. When the transition to electronic imaging is complete, the 10,000 square feet assigned to film files will be available for re-assignment to other functions, representing an 88% reduction in filing space. The 1,200 square feet assigned to PACS includes approximately 25% reserve capacity for future developments.

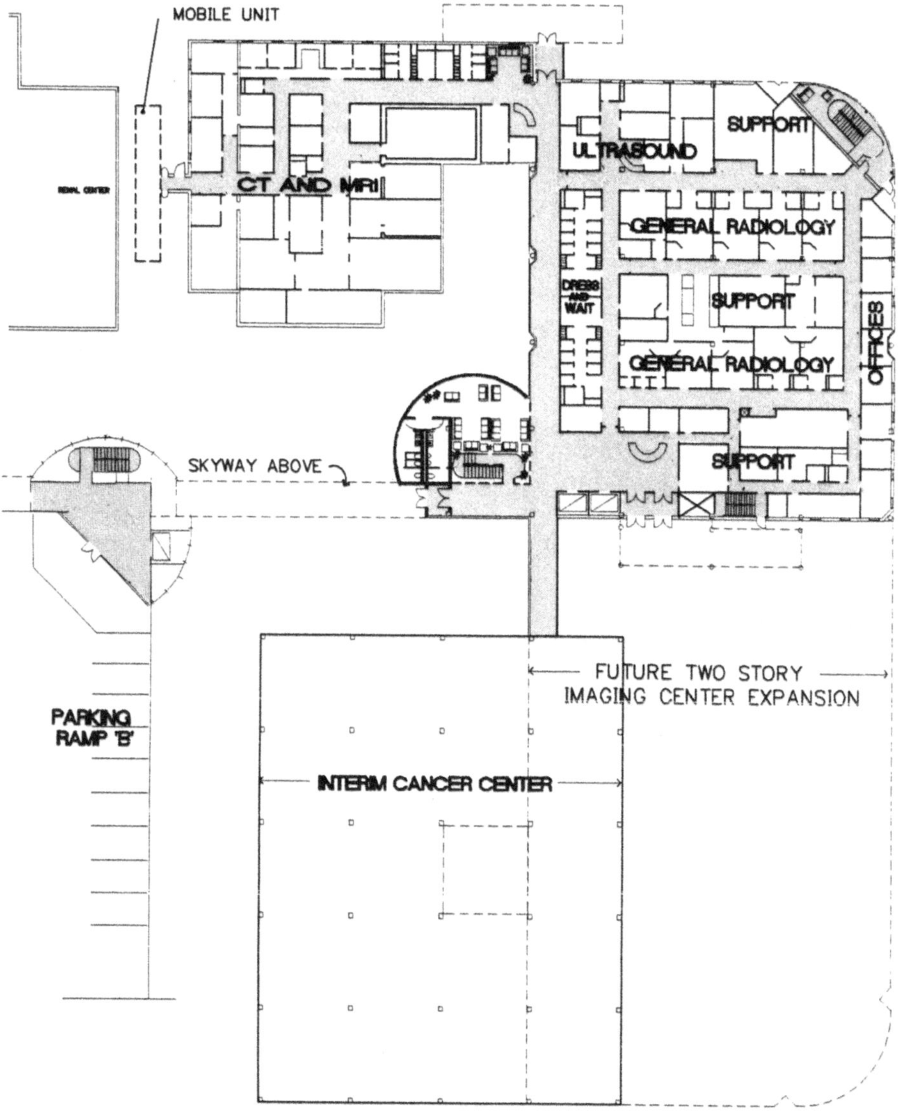

FIRST FLOOR PLAN-IMAGING CENTER

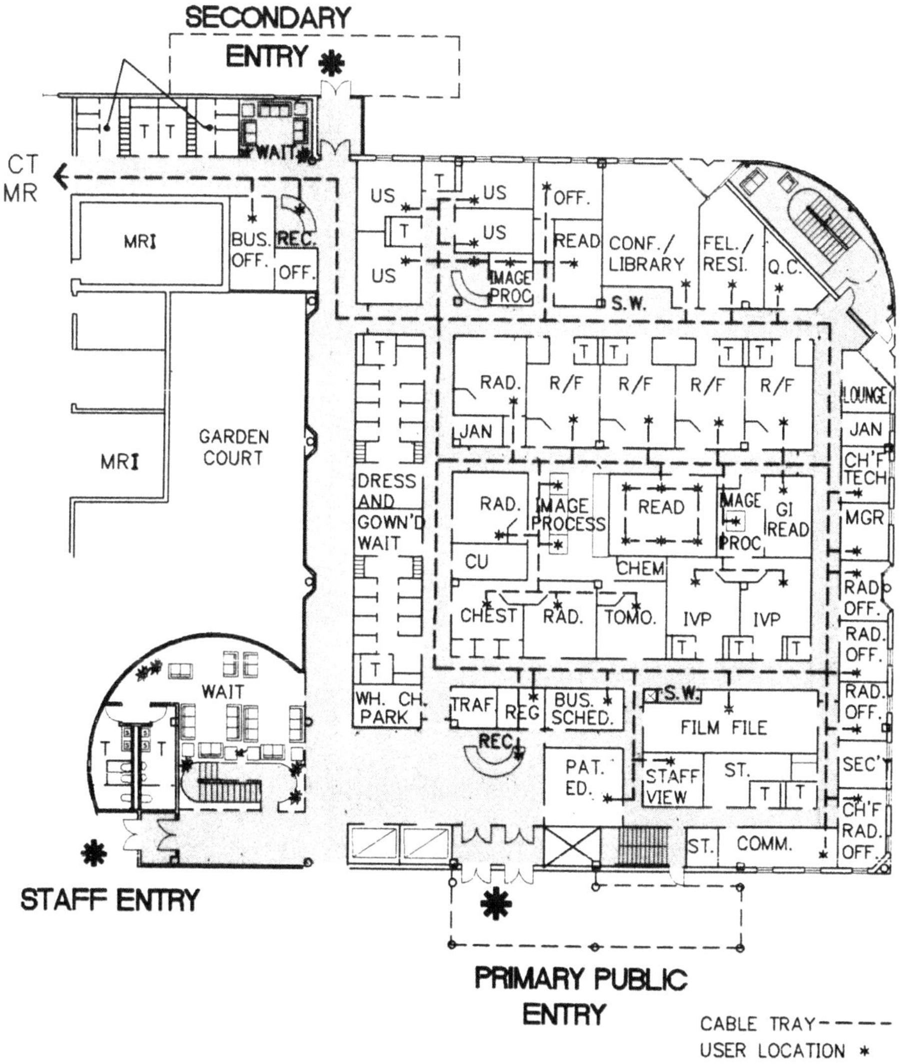

FIRST FLOOR PLAN-IMAGING CENTER

The first floor of the Imaging Center includes CT, MRI, ultrasound, and general diagnostic imaging. MRI and CT currently are housed in a temporary building that will be demolished after a future expansion in which both departments will be relocated into the new Imaging Center. The garden courtyard between the buildings serves to contain the magnetic field and to provide appropriate protection for magnetically sensitive people and equipment.

PACS, RIS, RTAS, HIS, and other communication cables are installed in a network of cable trays located above the ceilings of most corridors in the Imaging Center. Conduits, trench ducts, and secondary cable trays branch off to user locations. This flexible system provides easy access for installation, modification, and service. The transition to totally digital technology can be gradual, keeping pace with new technology as it develops. Imaging modalities that are not presently digitized can easily be brought on line in the future at minimum cost and disruption to the department.

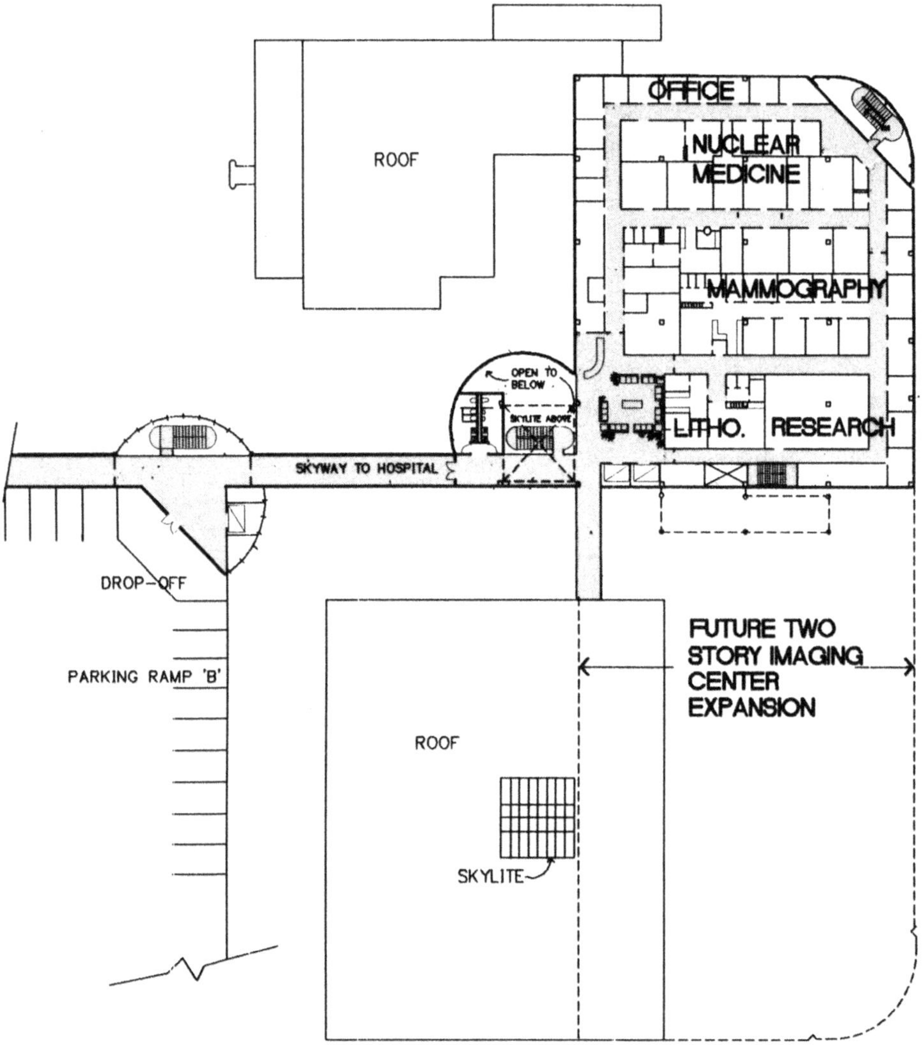

SECOND FLOOR PLAN-IMAGING CENTER

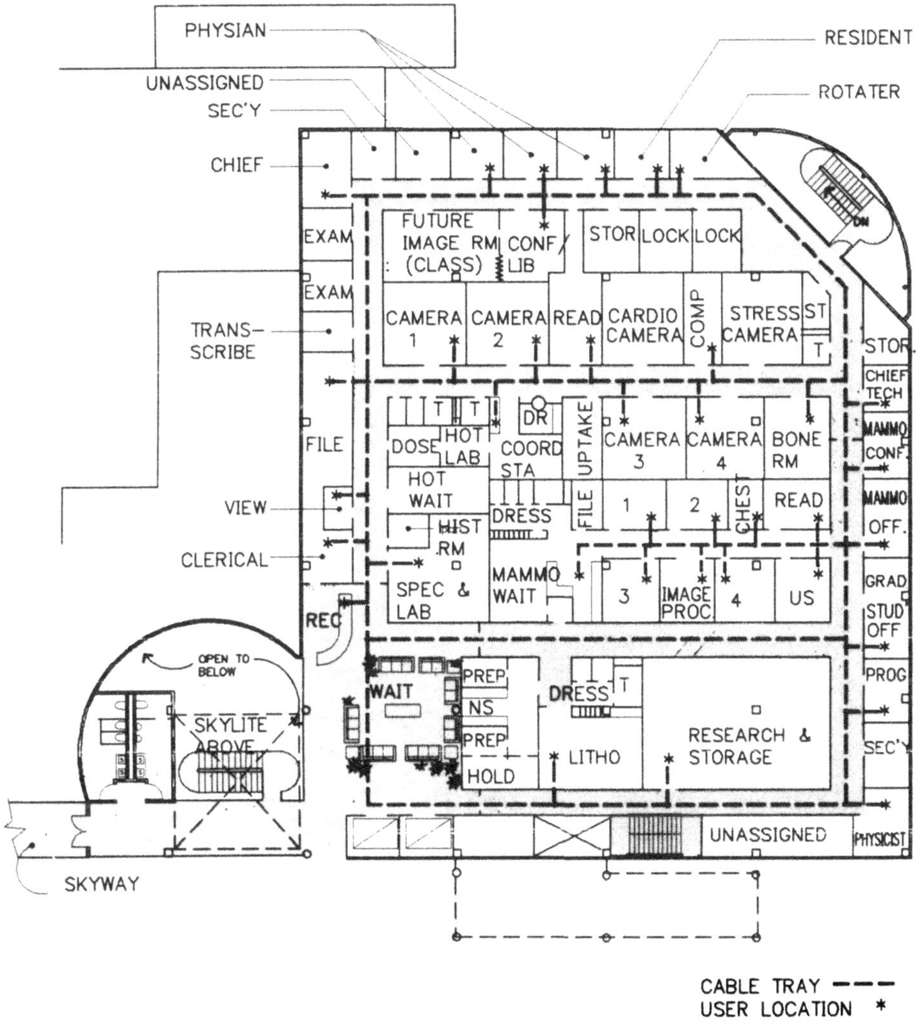

SECOND FLOOR PLAN-IMAGING CENTER

The second floor of the Imaging Center includes nuclear medicine, mammography, lithotripsy, and research. A network of cable trays located above the corridor ceilings serves all locations where images are generated, read, sorted, or otherwise required for normal operations. Because the decision has been made not to utilize PACS in the nuclear medicine department at this time, images on film will be processed here in a conventional darkroom (DR). The cable tray system will allow the change to electronic imaging to take place any time in the future. The space presently occupied by the film processing equipment will then be re-assigned and nuclear medicine images will be stored in the centrally located PACS room at the lower level.

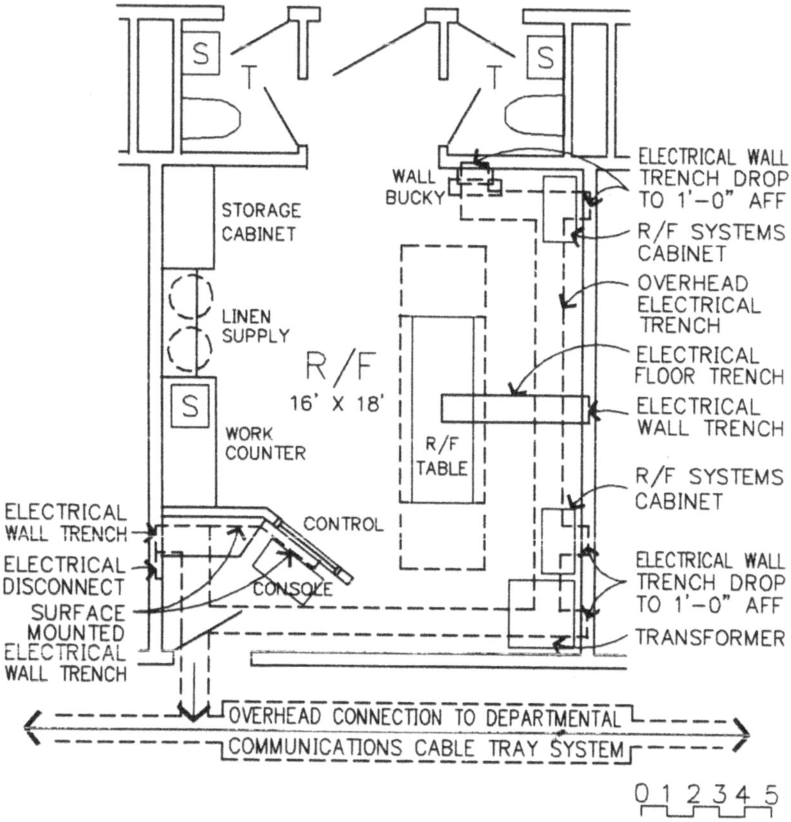

FLUOROSCOPY ROOM TYPICAL TRENCH DUCT LAYOUT

The trench duct layout for the fluoroscopy room is typical of the general diagnostic imaging rooms. A connection to the Imaging Center cable tray system is provided for each imaging room. The connection is inexpensive and is easily justified on the basis of the flexibility it provides. These connections may remain empty in imaging rooms in which the equipment selected and installed is not digitized. In fact, the equipment in some rooms may never be digitized. However, the connection to the Imaging Center may facilitate installation of other communication and data-gathering devices.

Summary
The William Beaumont Imaging Center is designed to bridge the gap between conventional and totally digital imaging technology. Reserve capacity and flexibility are designed into the building and the building systems to facilitate the transition at minimal cost and minimal disruption to daily operations.

Developing an Effective Optical Archive for Medical Images

Marcus W. Hedgcock, Jr., MD[1,2,3], Tod S. Levitt, PhD[1], Suzy Smith[1,2], Louis Humphrey[4]

[1]San Francisco VA Medical Center; Computed Imaging Center; Radiology Service (114) San Francisco, CA 94114
[2]San Francisco VA Medical Center; Research Service (151); San Francisco, CA 94114
[3]Department of Radiology; University of California, San Francisco San Francisco, California 94117
[4]AT&T/Bell Laboratories

Summary

An effective image archive should minimize access time and errors. The archive should provide at least two years of images on line. The size of archive needed to provide two years of storage is more accurately estimated from an imaging department's film use than from case load, since cases may vary in number of films used. Critical to archive performance are protocols which minimize errors and access time. Correct identification of image data is critical as images are moved electronically from institution to institution. A random numerical archive key assures that images will be distributed equally between optical disks in the archive. The national identification number (in the US, this is the social security number) provides both a number key for archiving and an identity check.

INTRODUCTION

Our institution (San Francisco VA Medical Center) is the VA pilot center for total digital imaging and PACS. The image archive is a critical element in system functionality. The system design considerations must address both the size of archive which is required and the archive scheme (the protocols for archiving and retrieval of images.) A flexible archive scheme (protocols for archiving and retrieval) for optical disk storage is imperative to assure efficient image retrieval as archive volume builds.

HOW LARGE MUST THE ARCHIVE BE?

Analysis of our image use[1] shows that active image files which contain the last two years of images will address the vast majority of our clinical image needs. Our radiology service performs about 50,000 examinations per year. Extrapolating from film use, we estimated our yearly storage needs for imaging (excluding nuclear medicine) at 876 Gigabytes(Gb) per year.

ARCHIVE SIZE AND EQUIPMENT

The media currently (or soon) available for an electronic image archive include magnetic tape (.166 Gb), magnetic disk (.7 to 1.4 Gb), single platter optical disks (2 or 6.8 Gb), optical tape (200 Gb) and multiple platter optical disk jukebox system (200 to 1000 Gb, depending on the number and size of disks). Optical tape offers high density storage, but has linear access, and multiple tapes cannot be put on line simultaneously. The best current solution appears to be optical disks in a jukebox arrangement. Table 1 shows the number of years of data we would have on line with various systems, using different data compression ratios, assuming that we add 876 Gbytes of images per year to the archive[2]. The storage capacities two years or greater are shown in bold type.

Table 1

Years of Storage Provided by Various Optical Disk Systems, Assuming Data Accrual of 876 Gigabytes per Year

Disk size Gb	# Disks	Total Gb	Compression Ratios.		
			1:1	2:1	4:1
			---------Years of Storage---------		
2.0	50	100	0.114	0.228	0.456
2.0	100	200	0.228	0.457	0.913
2.0	150	300	0.343	0.685	1.370
6.8	50	340	0.388	0.776	1.553
6.8	100	680	0.776	1.553	**3.105**
6.8	150	1020	1.164	**2.329**	**4.657**

There is no 2.0 Gb disk system which will fulfill our two year requirement. We will require 6.8 Gbyte disks with a 100 platter system using a compression ratio of 4:1, or a 150 platter system at 2:1 compression.

REQUIREMENTS FOR AN EFFECTIVE ARCHIVE SCHEME

Once an archive of appropriate size has been selected, there are several requirements for optimal archive: 1) patient records are uniquely identified (even between multiple institutions); 2) patient records are functionally grouped; 3) optical disk identification codes are meaningful to a human operator; 4) system checks exist for ID correctness; 5) minimal access time for current imagery.

CHOICES FOR AN ARCHIVE KEY

There are several ways in which we can archive our images to optical disk, including temporal sequence, alphabetical, or by a number associated with the patient, either one which is randomly assigned for the archive, or a number which has been previously assigned to the patient for some other activity.

Of these, only a random number seems suitable. If we archive by temporal sequence, the search process will become progressively longer as the number of images in the archive increases, because the images are not in close physical proximity. An alphabetical scheme will group a given patient's images together, however disks will fill unevenly, with disks assigned to common letters accumulating images faster. If a random number is associated with the patient, the disks in the archive will fill at the same rate, so that we can adjust the number of years worth of images on line by changing the number of disks in the optical jukebox, as in Table 1 above.

An ideal number to use as our archive key appears to be the national identification number, since the majority of patients will have such as number, and these numbers all have a random component which can be used as the key. In anticipation of international image exchange, the national identification number should be followed by the abbreviation of the country of origin.

The United States national ID number is the social security number (SSN). The SSN has nine digits, in the form 000-00-0000. Theoretically all US citizens possess this number. The first five digits of the social security number encode geographic data, however the last four digits are random. As the keys for our archive scheme we use SSN digits 8-9 ,then digits 6-7, name, country code (US), and image date. (An alternate method would be to use digits 6-9, and divide this set of 1,000 numbers into appropriate subsets.)

One of the investigators (Humphrey) has developed a program for archiving medical images to the optical disk of our CommView (AT&T) Image Data Management System. This program permits images for archival to be grouped in a set of the last two digits of the SSN. Since we currently have a single platter optical system which uses 2 Gbyte disks, we are beginning with ten 2 Gigabyte optical disks. In this initial arrangement, the program selects all patient images with the terminal SSN digits 00-09. These images are then archived to the disk, subsorted by penultimate two SSN digits, patient name, and date of exam.

Using the terminal four digits of the US national identification number (the SSN) each patient's images are grouped, each disk has same number of years of images, and the disk number corresponds to a subset of the last four digits of the SSN. The yearly imaging volume and number of disks available in the jukebox (or in the set of "current" disks, if a single platter system is used) determine the years of images in the current archive. As the digital imaging volume grows, we will expand the archive to one hundred disks, each corresponding to one of set of terminal SSN digits, e.g. 08, 09, etc. We can switch (at the time we begin a new set of disks) to using the last four SSN digits in sequence, if an on line archive larger than 100 disks is needed. A numeric archive key based on the national identification number is equally useful for single platter or multiple platter "jukebox" archives, and provides the inexpensive option of beginning with a small number of optical disks, and an easy transition to a "jukebox" operation.

Since a single patient's images are physically close together, retrieval time to locate comparison images is decreased. The disk containing a specific patient's images is easily located without having to consult a database, since the disk labels indicate patient image location by SSN. Knowing the rate of image acquisition, we can plan for the bulk purchase (which should reduce the cost per disk) of additional optical disks when the current disk library is filled.

RETRIEVAL PROTOCOLS

Eventually the electronic archive will have several retrieval functions automated, however this automation requires a jukebox. We are developing these retrieval protocols, in preparation for service through an optical jukebox to multiple interpretation and clinical workstations. Retrieval of images from the optical disk is relatively slow. Fortunately the great majority of image requests from an archive can be anticipated. Requests for in-patient images, clinic patients, and patients on the operating room schedule are known at least one day in advance. The optical disk system is instructed to retrieve the appropriate images on these patients the night before. The list of names of these patients to be sent directly from the Hospital Information System to the optical archive. The images retrieved from the optical disk are sent to a large Winchester device (a "gateway") which serves several workstations. As image traffic and number of workstations grows, additional gateways are added, and the images routed by the data management system.

We are developing algorithms for the specific images which will be retrieved when a patient's "image file" is requested. These algorithms will anticipate the need for given requestor groups. Our tentative plan is to retrieve all images done in the past week, and then the most current examination of each type with the previous comparison (if not included in the past week images). The effectiveness of this method will be the subject of clinical trials at our institution.

Retrieval protocols which allow the bulk of images to be recalled from the archive during the low activity night hours will decrease traffic on the optical archive, and allow faster access to images which are requested on an as needed basis (for example, the emergency room).

CONCLUSION

A effective image archive will require analysis of the projected image load in order to select the appropriate size of archive. Calculation of storage requirements is best done using film use, since estimation by case load may give an artificially low figure. Images are most effectively archived using a random number as a key, and the national identification number is well suited to this purpose, since it is already assigned and prevents confusion of images for patients with common names. Clinically tested algorithms for the retrieval of images will be needed to optimize access time and automate archive functions.

[1] Hedgcock MW, Kehr C: Quantitative assessment of the impact of computed radiography and PACS on imaging service delivery. Presented at the 1988 Radiologic Society of North America Meeting, Chicago, IL.

[2] Hedgcock MW, Levitt TS, Smith S: Calculating storage needs for an optical archive. SPIE Medical Imaging 1989 (in press).

Communications: PACS/RIS/HIS

PACS – RIS – HIS
The Demand for Communication Management

O. RIENHOFF

Institute of Medical Informatics
Philipps-University Marburg

1. Introduction

During the last 30 years Medical Informatics specialists have prophesised that computer systems will have a major impact on medicine and may even alter clinical medicine to a large extent. The medical community has treated many of our early pioneers as witchdoctors, comparing their systems approach to the throwing of bones. In some universities where Medical Informatics peaked very early (eg Hannover/Reichertz; Heidelberg/Wagner; University of California/Blois) the negative reaction within the medical community has nearly suffocated the field after the pioneers passed away.

But, as expected - computer systems are now everywhere. No research field can survive in Medicine without them. Hospital Information Systems (HIS) have become a standard equipment to manage hospitals. Radiology Information Systems (RIS) have become a common tool to organize radiology departments. Is there any need to think about Picture Archiving and Communication Systems (PACS) to become the next self-evedent step on the way to total computer support?

Radiologists tend to perceive PACS as something like another digital modality to be plugged into the department. Let us throw the bones again: how do PACS, RIS and HIS relate to each other?

2. Relevant Factors of Development

2.1 Planning PACS-RIS-HIS

The life cycle of HIS seems to be about 15 years and the implementation of a HIS (with its main administrative functions) takes about two years. The life cycle of RIS may be 10 years and it takes about one year to install a RIS. The life cycle of PACS cannot be estimated so far, but in relation to system costs one can expect 3 to 5 years for pilot versions and about 10 years for the commercial products which hopefully will be offered by a few vendors in the early nineties.

To run HIS one needs a staff of highly experienced senior Medical Informatics specialists including specialists on databases, communication and the like. You also need clerical as well as system managers to run the hospital installation in accordance with the hospital development policy. On a much lower scale the same applies to RIS: at least one well-trained person is needed to get all the support a RIS can offer to the management of a radiology department. We do not so-far know what type of staff we need to run PACS.

A well designed strategy is necessary to install HIS and it will cost several man years ranging from clerks to high ranking specialists. The same is true for RIS on a lower scale. With PACS we are still gathering experiences.

It is not possible to plan, install and run HIS successfully without participation of all users, may those be the managing director or the data typists. The same applies for RIS and PACS.

Regarding involvement of staff, importance and long-term-implications this means that setting up of information systems in the hospital environment can be compared with setting up hospital budgets. If this fact is not recognized in the medical departments and dealt with appropriately it will directly lead to more centralised responsibilities managed by the hospital administration (as it happened in many hospitals with budgeting).

2.2 Budgeting PACS - RIS - HIS

In the seventies the overall costs of data-processing in a teaching hospital were estimated to 1-4% of its overall budget.Facing digital communication systems (ISDN) and Local - as well as Wide Area Networks (LAN, WAN) and widely distributed powerful work-stations one can easily expect that these figures have to be doubled in the nineties. Only few administrators and departmental medical heads are aware of this development. No exact figures have been published on detailed cost analyses or implications of this shift of resources on the institutional set up of the hospital nor on its impact on various health related curricula.

The technical linking of systems never linked before will require a different management approach on the administrative side as well as on the medical side. The type of knowledge necessary to be able to harmonise business-, medical-, reserch-, teaching and technical goals is only scarcely available in the existing hospital environment. Right now many institutions believe that they can survive by buying consultants from outside. As if the problems to be tackled could be solved by outsiders in a few months. In addition, well-paid medical departmental heads often seem to be surprised if those outside consultants charge daily rates which reach their own level of turnover and therefore tend not to hire them.

Regarding the life cycles of systems and the costs of equipment for tertiary care in general budgeting for information systems has to be part of a long-term budgeting policy of each hospital. Information systems are the only means available to keep complex hospitals cost-transparent.

Not to invest in them limits the flexibility of hospital management in general but especially in terms of controlling economic performance of the various hospital units.

These aspects apply expecially to radiology/imaging departments with their high annual turnover and with their high investment costs.

2.3 Using PACS - RIS - HIS

The move from traditional hospital administration to a modern hospital management based on various kinds of information systems takes many years and often the exchange of a whole generation of employees. Even with dynamic force in a private environment 3 to 5 years is a minimum. State bureaucracies may happily achieve the same development within 10 to 15 years.

Facing RIS and PACS the radiology departments will undergo a similar move. Depending on the management skills of the heading physician the change will happen in less than a decade or last forever. It will either be an actively prepared, well calculated set of steps or a reacting and painful stumbling.

The main feature of change will be that the diagnostic process may change and consequently: the patient flow, the teaching content and, likely, the kind of institutionalisation of imaging departments.

The driving force behind this will not be computers and their processing power, thus allowing new types of images to be created, manipulated, stored and evaluated for research. The driving force will be in the change in communication possible through installing PACS - RIS - HIS. Therefore the management of communication tools of a hospital or of its departments will be a key issue in health care management, beginning in the nineties.

3. Conclusion

Looking into the near future of PACS - HIS - RIS as communication tools, one can conclude that a new kind of know-how is urgently required on departmental as well as hospital level: planning, using and budgeting of communication systems. Recognizing the importance of this fact, hospital and medical management have to become literate with these terms. Further, highly skilled experts have to be employed and trained, because outside consultants can only be of limited value in the long run.

CARO (Computer Assisted Radiology Organization) – Mangetic Resonance Version

K.Küper, J.Ottinger, U.Klose

Abteilung für radiologische Diagnostik, Radiologische Klinik, Universität Tübingen

System Philosophy

The need for patient data is increasing in all radiological departments. Commercial information systems are not only expensive but mostly written for the necessaries of administation. Informations to support the radiologist in his routine duties or in scientific work are often not recorded to the full extend.

Data base systems that run on personal computers (PC) are single user systems and therefore not suitable for the demands of a radiological department.

We developed a program for a computer assisted radiology (CARO) that runs for four years without problems. CARO manages patient data sets containing all important informations on personal and account administration. Furthermore informations on examine technique are stored.

Because of the VAX/VMS architecture CARO is a multi user system. Up to 164 terminals are supported. All data files are sharable, i.d. patient data can be accessed by sevaral users simultaneously.

Different CARO systems that run in different radiological departments can be linked together by commercial network software (DECNET) which cover up the necessaries of medical data protection.

CARO is software written for the radiologist. It's an easy-to-use program with menus, help functions and virtual screen management (windows).

Hardware

CARO runs exclusivly on the DEC VAX 11 computer family which are available in many radiological departments. For a CARO Add-On version only a multiplexer with eight or sixteen channels for terminals or printers is necessary.

For a CARO Stand-Alone version there a reasonable systems available. Figure 1 shows a CARO hardware configuration based on a MivroVax with system disk (70 - 800 MB) and load and backup medium (streamer tape). In addition a multiplexer for terminals and printers is required.

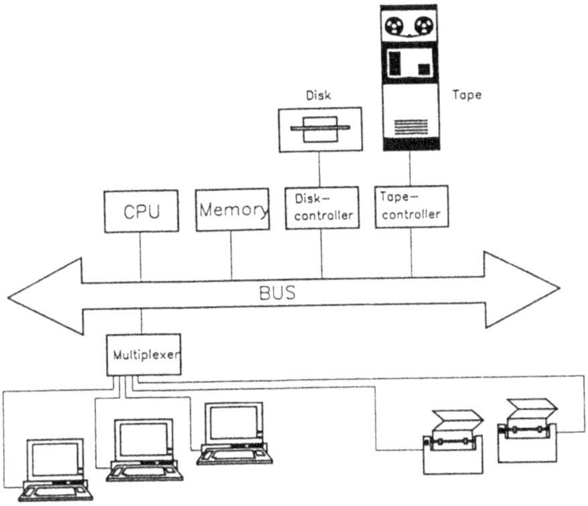

Figure 1: Caro Hardware Configuration

Software

CARO is written in VAX/VMS-FORTRAN. It is a real multi user system. Up to 164 Terminals are supported.

CARO handles an index sequential data base with a record length of about 1500 byte. The data base is accessed via RMS - the Record Management System of the operating system VMS - that is used by all data base system. RMS garantuees high performance. It uses data compression for empty fields in the records. Every record consists of the patient's data and department specific data. All parts of the record which have to be accessed directly are keys in the index sequential data base. The data base files are "sharable", i.d. records can be accessed by several users simultaneously. All print jobs are queued which means that the user does not have to wait until the printer is ready. CARO's characteristic features are virtual display managment (windows), help functions and menus.

Network

Caro can be run as single system for several departments. The user must specify the abbreviation of his department together with the name of the examined organ (e.g. MR/PELVIS, CT/THORAX, DSA/KIDNEY, NUK/SKELETON, ROE/SKULL, US/UPPER ABDOMEN etc.)

Data Retrieval

A record in the data base can be accessed directly by the keyed fields of the record. It takes about two seconds depending on the CPU and the number of records in the data base. The user can retrieve records interactively. He only has to specify a fragment of the desired item. If for example the user wants to look up all patients who have a keyword beginning with "PROS", the program lists all patients with the keyword "PROSTATA", "PROSTATA-CA", "PROSTATITIS" and so on.

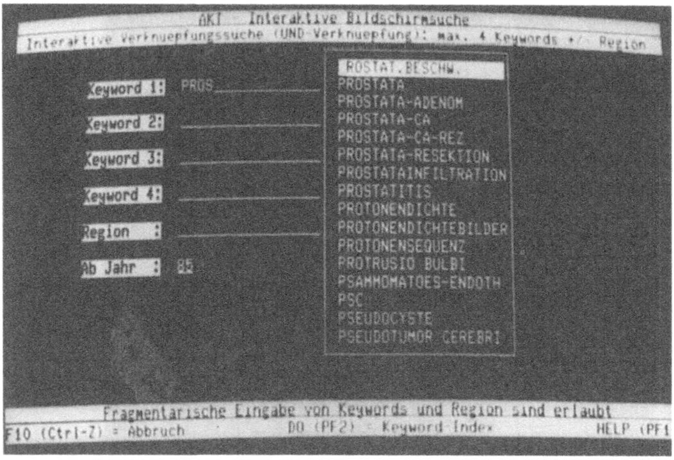

Figure 5: Pull Down Menu (Diagnosis)

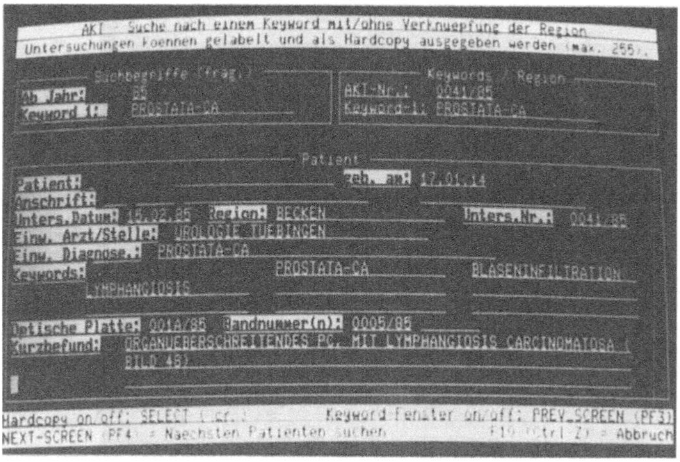

Figure 6: Sample Output

The user can access records by a combination of keywords. Another possibility to retrieve data is what we call the "Multiparameter-Suche". The user can specify a fragment or in some cases even a substring of any field in the record. This program runs on the batch queue and the results are outputted on a printer. Either the full record or only some of the data fields appear on the paper. On a VAX 730 the program processes 3000 records per minute or 180.000 records per hour. CARO supports up to ten different printers.

A Flexible Network Solution for an Extended Information System in a Radiological Department

P.Mildenberger, K.-J.Klose, M.Thelen, W.Offenmüller[2]

Radiological Department, Uniklinik Mainz, Langenbeckstr.1, 6500 Mainz
[2]Siemens AG, UB MED, 8520 Erlangen

The organisational structure of a distributed radiological department has grown over many decades. The main interest of the department is centered around two important factors: The patient and the special medical subjects. If computer-based information systems are introduced, one of the first goals should be to minimize changes arising from this tool. Architectural aspects, different educational levels of the staff, traditional structures and of course future plans must be considered. But not only the internal departmental needs should be covered sufficiently, also future requirements from other locations in the hospital play an increasing role in the forthcoming years.

Adaequate hardware-units and software-modules for information systems are now available, to make sure, that radiological departments of different sizes, location and user-expectations can be given the best solution.

One major planning aspect is the selection of the right network-topology. In teamwork with physicists and the administrative staff one should model the daily work pattern properly, but should not prevent or block the integration of progressive hard- and software concepts.

The RIS "SIMEDOS" as an example of using a distributed network structure was installed in the University-hospital in Mainz/Germany during July 1988. The radiological department is divided into three sub-departments, which are separated by a few hundred metres. Some complex organisational processes like central stand-by services for all locations or overlapping appointments for various special examinations on one patients must be handled as well. Additional the neuroradiology, radiotherapy and orthopedic department should be integrated in the communication-link. These local pre-conditions forced to the use of a broad net based on Ethernet. The average throughput in the diagnostic department alone differs between 300 and 350 patients per day, that means 30 - 35 MB/month patient data. Therefore the storage media and the processors are configured fail-safe to reduce the probability of technical downtime to a minimum. Intelligent and unintelligent terminal servers can be tailored to the specific workloads in the department, six microcomputers control and support the whole system.

The availibility of basic department information make complex correlations more transparent. The increasing practice of all users and the existence of new technologies leads to more applications (i.e. in the medical and administrative field). With the underlying standardized network structure we are very confident that such a solution will face nearly all future requirements.

The Added Value of a RIS Integrated in a HIS

MAARTEN L. KOENS

BAZIS, Central Development and Support Group Hospital Information Systems, P.O. Box 901, 2300 AX Leiden, The Netherlands

Summary
Most Radiology Information Systems (RIS) have only a meagre coupling with the Hospital Information System (HIS) operational in the surrounding hospital. Such a coupling is usually restricted to patient identification from the hospital's patient database and the transfer of examinations performed for billing purposes. Further integration of these system is needed however to optimize the communication between the radiology department and the in- and out-patient departments in the hospital.

1. Introduction

Radiology departments try to solve their management, coordination, communication and information problems with the aid of a RIS.
The different processes within the department, such as:
- the production of images
- the diagnostic process
- the reporting process
- the archiving process

have to be supported in itself and have to be tuned with each other. To realize this the RIS supports the planning and operation of the flows of
- patients
- personnel
- materials
- reports
- film folders.

Although this might lead to an adequate operation of the radiology department this might not be true for the surrounding hospital. It is important to realize that a radiology department is part of a hospital and has many lines of communication to and from the various in- and out-patient departments. However most RISes have only a meagre coupling with the HIS.

The import of patient identification and the export of data on procedures performed for billing purposes are quite common. In a few cases order entry from the in- and out-patient departments is possible [1].

In which way can a RIS, integrated in a HIS, support the communication between departments thus preventing a suboptimal solution for the hospital as a whole?

In the next paragraph we will focus on the possibilities to answer this question.

Integration of the BAZIS RIS into the BAZIS HIS has made it possible to realize a number of these solutions.

Experiences with this integration will be subject of the third paragraph.

2. Improved communications by means of a RIS integrated in a HIS

The communication between the radiology department and the in- and out-patient departments can be improved in several ways. In order to structure the various possibilities and to detect the missing ones we try to categorize these possibilities according to two axes (see figure 1):

1. planning versus operational support
2. the support of various border-crossing flows, such as patients, reports and film folders.

Each cell in figure 1 refers to one of the next subparagraphs.

cat. 1 cat.2	operational planning	operational support
patients	2.1 appointment scheduling	2.2 progress control
reports	2.3 scheduling succeeding consult	2.4 decentralized printing and patient's history
film folders	2.5 planning film folder transfer	2.6 actual location of film folders

Figure 1: Categories of communication possibilities.

2.1 Planning of the patient flow: appointment scheduling

The limited capacity of the radiology department requires a fine tuning in the demand and supply for examinations.
For the radiology department it is important that the flow of patients is as evenly distributed as possible. For the in- and out-patient departments it is profitable to have access to radiology examinations whenever they need these.

To solve these conflicting requirements the radiology department makes use of an appointment scheduling system being a part of the RIS.
Integration of this RIS into the HIS realizes major advantages:

- While making an appointment for X-ray examinations the user has insight into other appointments of the patient in the hospital. E.g. out-patient visits. This gives the user and the patient the opportunity to:
  . prevent the exposure of the patient to more than one heavy examination in one day
  . combine the various appointments of a patient to avoid waste of time
  . to schedule the X-ray appointment a predefined period(e.g. 24 hours) preceding an out-patient visit in order to guarantee the availability of the report before patient visit to the out-patient's department (see also 2.3)
- The same advantages apply for the user making an appointment for an out-patient visit.
- The appointment scheduling and the registration of the request for X-ray examinations can be decentralized to the requesting physician or department. This makes the request notice redundant and will in many cases give the requesting physician and immediate answer to the request for an appointment.

2.2 Operational support of the patient flow: Progress control

In many RISes patient tracking is a standard feature. For each patient visit events are registered:

. the arrival of the patient
. the start of the examination
. the end of the examination
. the departure of the patient

Making this information available for the in- and out-patient departments realizes the possibility to gain a quick insight into the actual progress of the patient's visit to the radiology department. This information might be useful in a situation where the patient is awaited for a next examination or treatment or patient's return is awaited in the clinic.

2.3 Planning of reports: Scheduling on a succeeding consult

Radiology departments try to reduce the leadtime between the departure of the patient and the availability of the report. They realize e.g. a service level of 95% within 24 hours. So 95% of the authorized reports is available within 24 hours from the departure of the patient from the radiology department.
Therefore it is recommended to schedule X-ray examinations at least 24 hours preceding an out-patient visit.

2.4 Operational support in reporting: fast availability in the department

As soon as the report is authorized it should be available and sent to the requesting physician.
There are three ways in which this can be supported by the integration of RIS and HIS:

- The authorized report is send immediately via the HIS terminal network to a printing device in the ward or out-patient department.
- The requesting physician can ask the system for all the available reports with a selection on a certain period.
- The existence of an authorized X-ray report can be reported to the out-patient information system that keeps track of the progress of all patient's examinations, lab tests, etc.
  The requesting physician can, at the moment of his own choice, ask the system for all the available results.

2.5 Planning of transfer of film folders

The availability of old and new X-ray film folders is essential for:

- the radiologist while reporting new images
- the specialist preparing patient's out-patient visit
- the specialist preceding and during the admission of a patient.

In order to realize both a better planning of archive personnel workload and high service levels to the requesting specialists, requests for film folders have to be registered at the moment an appointment for X-ray examinations or an out-patient visit or the admission is scheduled.
With this information the archive management system can generate information which folders have to be transferred when to which departments.
In- and out-patient departments send requests to the different file rooms in the hospital. In daily practice this is found to be an unstructured and unreliable way of communication:
- requests arrive written on a variety of notes and forms, either by telephone or personally
- requests are sent to the wrong file room
- requests are sent for patients that never visisted one of the X-ray departments before
- requests arrive too late.

Integration of the request film folder function of the RIS in the HIS realizes a major improvement of the communication between the requesting departments and the file rooms, and an improved operation of the file rooms.

2.6 Operational support of the search for film folders

Users throughout the hospital need the possibility to find out the actual location of film folders or send an urgent request for film folders via each HIS terminal, in case of an unscheduled or emergency patient visit.

3. Experiences with HIS/RIS integration and future development

The BAZIS HIS is built from about 70 different subsystems around a central hospital database [2]. The BAZIS RIS, called RADI, is one of these subsystems with functions for:
- patient registration
- examination scheduling
- archive management and film folder tracking
- report generation

In development are functions for patient tracking. The scheduling functions are integrated with the subsystem for out-patient visits.

In het next subparagraphs we will describe the HIS/RIS integration from experiences in the Leiden University Hospital [3,4] and the developments in HIS/RIS integration we expect in the near future.

3.1 Appointment scheduling

Both the radiology and out-patient users make use of the appointment scheduling functions and take the information about other scheduled visits of the patient into account.

As an experiment we authorized a small group of users of one of the internal medicine wards to make use of the radiology appointment scheduling functions for a limited set of examinations. This experiment was done to find out the user requirements for this decentralized function which are expected to be different from the requirements of the radiology personnel. Some of our experiences were:

- The possibility to choose an appointment date and time that best suites the patient is found to be an advantage.
- Date and time of the appointment are registered immediately. Therefore duplicated appointments occur less frequently.
- For some examinations the access time for an appointment is too long. In this case the radiology department is called to schedule the patient in the overflow area of the agenda.
- 45% of the requests for an examination is scheduled by the users in the ward. The rest of the examinations require consultation with the radiologist or chief technician.
- The radiology department experiences a decrease in contacts by telephone.
- For the radiology department it sometimes is more difficult to realize an evenly distributed workload.

Future developments will be focussed by the development of the BAZIS nursing information system.

In this system an appointment for e.g. a colon X-ray 'explodes behind the screen' into a series of actions and registrations [5]:

- check on possible contra-indications and conflicts with other appointments for this patient
- check on desired date and time in the agendas of the radiology department and registration of the appointment

- identification of all nursing activities related to a colon X-ray and registration of these activities with date an time in the nursing workplan
- registration of medication orders and initiation of transport of the patient to the radiology department
- if necessary the addition of problems from the standard care plan for a colon X-ray to the individual care plan of the patient

3.2 Decentralized reporting

Reports are printed in the in- and out-patient departments immediately after authorization of the report.
The printing of authorized report in the typing area of the radiology department will become almost redundant:
- Historical reports will be on-line available for a period of more than 5 years.
- Users throughout the hospital have access to the on-line medical record system, called PATIENT, to consult information about:
  . results of lab test
  . radiology examinations and reports
  . admission and discharge diagnosis
  . surgery information
  . radiotherapy treatments
  . blood transfusions
  . medication prescriptions
  . etc.

  This PATIENT system is not only used by in- and out-patient users but also by radiologists to support their diagnostic process.
- The development of a system called MIRAGE [6] with which admission, transfer and admission data, radiology reports and lab results are sent via a public data network to the information system of patient's general practitioner.

  This system will become available in the course of 1989.

In the long term the request for and the reporting of X-ray examinations will become part of an out-patient information system with which the requesting user can keep control of the progress of lab tests and exams, and has access to the available results and reports.

3.3 Archive management

Most of the in- and out-patient departments request film folders via their HIS display terminals. For each archive location the archive management system collects all the requests that have to be dealt with within a certain period. Only requests for film folders that are really available in the archive location are presented on a pull list in archive order. Requests for lend out folders can be printed on a list at the borrowers location.

The number of requests from the departments increased with the introduction of the decentralized request function. The users were interviewed [4] and were found to have positive experiences with the system:

- The overview of all the available film folders with their actual location with the possibility to point out and request the wanted folders encouraged the user to issue a request.
- The required time for requesting has been reduced considerably:
  . no waiting before the telephone is answered
  . no engaged lines
  . no waiting before the required information is found.
- On the other hand users became aware that they sometimes submitted a request for folders that should be already in their possession!
- The users experienced the archive to be a reliable and fast supplier of the requested folders: the leadtime was reduced considerably. This improved the relations between borrowers and archive staff.
- The ratio of honoured/requested folders has been improved, due to the more accurate registration of the actual location of the folders.

The archive personnel experienced:

- a fast decrease in request forms, telephone calls and urgent requests
- a more structured method of working and a more spread out work load
- the requested folders are exactly identified and no request for folders not present in the file room occurred
- less time-consuming searches for lost folders
- the number of lost folders reduced considerably
- high service levels towards the borrowers
- better motivated for their work.

Although the number of requests increased with about 10% and the number of examinations increased with about 18% from 1982 until 1988, the permanent archive staff remained constant at 7½ fte and the number of temporary staff was reduced from 0.2 to zero.

4. Conclusion

Integration of a RIS in a HIS environment realizes a splendid opportunity to contribute to the optimizing of the patient care process in the hospital as a whole in stead of the radiology department only.
Decentralized functions for scheduling of X-ray exams, requests for film folders and insight into patient authorized reports

- improve the communication with the radiology department,
- improve the planning of the workload in the radiology department,
- realize higher service levels towards the in- and out-patient departments,
- realize a better service to the patient.

The direct availability of other patient related HIS data is profitable in the diagnostic process of the radiologist.

References

[1] Cannavo, M.J., Radiology Information Systems can improve department efficiency, Diagnostic Imaging, December 1988, pp. 92-95.

[2] Bakker, A.R., Integrated Hospital Information Systems, Pictorial Information Systems in Medicine, Springer-Verlag Berlin, Heidelberg, New York, Tokyo, 1986, pp. 105-148.

[3] Elsakkers, P. et al, User experiences with the implementation of an extended radiology system within a new hospital organization, proceedings of MIE 1987, European Federation for Medical Informatics, Rome, 1987, pp. 78-85.

[4] Op het Veld, W.A. and Koens, M.L., Experiences with an archive management and lending administration system for X-ray folders and patient medical records, Proceedings of MIE 1987, European Federation for Medical Informatics, Rome 1987, pp. 86-94.

[5] Gondelach, J., Bik, S., Prototype developments for a nursing information system, proceedings of MIE'88, Springer-Verlag, Berlin, Heidelberg, New-York, London, Paris, Tokyo, 1988, pp. 147-151.

[6] Arnou, P.G., Bouman, P.C., Hasman, A., Ferwerda, P.H., A computer communication network for hospitals and general practice, 1988, to be published.

A Medical Image Data Base System for Tomographic Images

S. C. Orphanoudakis †‡, *E. G. Petrakis* †, *and P. Kofakis* †
†Image Analysis and Computer Vision Laboratory
Institute of Computer Science, FORTH, Heraklion, Crete, Greece
and
‡ Departments of Diagnostic Imaging and Electrical Engineering
Yale University, New Haven, CT

Abstract

In this paper we present initial work on the design and implementation of a prototype Medical Image Database system for tomographic images. Our attention has been focused on techniques for the automated description of anatomical crossections in terms of geometrical features which facilitate matching operations and can be used to access tomographic images by content. These techniques are computationally efficient and the resulting descriptions compact in order to meet the requirements of the fast-paced clinical environment and to minimize storage space. The organization of the image database and possible strategies for image retrieval by content are also described and discussed.

Introduction

It is possible to extend the capabilities of diagnostic medical imaging techniques by developing Medical Image Database (MIDB) systems, which support the storage, retrieval, presentation, and interpretation of medical images. MIDB systems are emerging as an important component of Picture Archiving and Communications Systems (PACS), which in turn must become fully integrated with other components of a Hospital Information System (HIS) in order to support administrative, clinical, teaching, and research activities. Retrieval strategies and user interface designs must take into account the requirements of HIS and PACS as well as the multimedia nature of medical records [1,2]. General issues related to pictorial information systems have been raised previously by other investigators [3,4]. Important considerations in the design and implementation of image database systems are: image feature extraction, image content representation and organization of stored information, search and retrieval strategies, and user interface design. To date, a general methodology for the design of such databases has not been developed. Once the requirements of a particular application have been determined, techniques of image analysis and description, with known database design methods, are adopted to develop an image database which satisfies these requirements [5]. The content of particular images can be determined, based on the correspondence between the derived description of a particular image and some appropriate model(s) of an image class [6]. So far, most of the methods which have been developed are knowledge-based, making use of application and domain-specific knowledge, and are not particularly well suited for image retrieval by content and image database work. This is primarily due to problems with computational efficiency, uncertainty, and knowledge representation [7]. Nevertheless, for certain applications within well specified domains, it is possible to develop and implement techniques of image analysis, symbolic content representation and modeling which can be used to manipulate images in such systems efficiently.

Image data can be viewed as a component of a multimedia document, i.e. a structured collection of image, voice, text, and attribute data [8]. Images together with their descriptions and other document components are stored in a *document base*. Queries on a MIDB must be expressed via a well defined and powerful query language capable of handling different image representations and attributes. To date our attention has been focused on the development of techniques for the automated description of anatomical crossections in terms of geometrical features which facilitate matching operations and can be used to access tomographic images by content. These techniques are computationally efficient and the resulting descriptions compact in order to meet the requirements of the fast-paced clinical environment and to minimize the required storage space. Furthermore, the system which is currently being developed provides a flexible and simple storage scheme which facilitates multimedia document retrieval based on image content and minimizes the dependence on the application domain.

Image Analysis and Representation

Picture descriptions are generally given in terms of properties of objects contained within the picture and relationships among such objects. A picture description may also include properties or attributes of the picture as a whole. The specific properties which are used in picture descriptions can be geometric, statistical or textural, or properties specified in a transform domain. The effectiveness of any scheme for retrieving multimedia documents by pictorial content will ultimately depend on the complexity of picture descriptions and picture class models, used in a particular application, as well as the extent to which such descriptions are integrated with representations of other document components. In developing a prototype MIDB system which permits image retrieval by content, we have chosen to work with tomographic images of high tissue contrast because of their relatively well-structured content and their widely acknowledged clinical significance.

Tomographic images are first segmented using lowpass filtering followed by thresholding, with appropriate thresholds determined on the basis of corresponding histograms. A polygonal approximation to segment edges is then obtained and used to compute a set of geometrical features for each segment (Figure 1). Image analysis can be performed online or interactively under the supervision of an expert. Furthermore, the set of geometrical features can be augmented with modality-dependent characteristic tissue parameters. A processing step prior to any further analysis places each image in standard position and orientation so that subsequent descriptions can be rotation and translation invariant.

The description of each segment in the image is given in terms of properties such as: average pixel value, texture, area, perimeter, roundness, orientation with respect to a reference direction, center of mass, etc. In addition to these properties of individual segments, the following relationships between segments are obtained and used in queries: 1) Relative position: takes the values 1,2, or 3 corresponding to a segment being outside, intersecting, or falling inside another segment respectively. 2) Relative orientation: defined as the orientation of the line connecting the centers of mass of two line segments, relative to a reference direction. 3) Minimum distance: defined as the minimum distance between the boundaries of two segments.

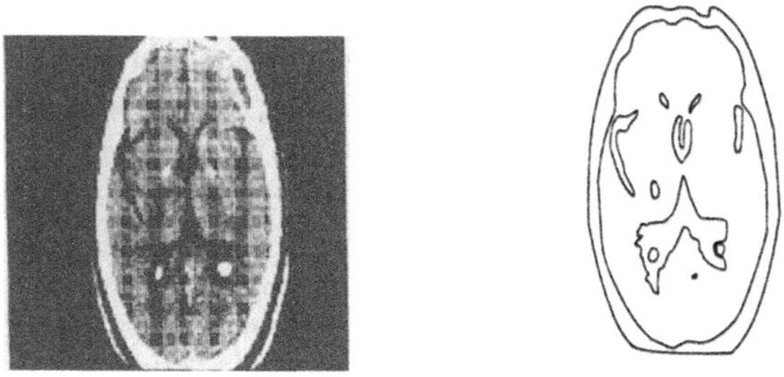

(a) (b)

Figure 1. a) Original image. b) Segmented image.

Image Storage

Images together with their descriptions and other document components are stored in the *"document base"*, which consists of a *"logical"* and a *"physical"* database [9]:

1. The *"logical database"*, is used to store representations of the original images in terms of a set of features extracted by automated or computer assisted image analysis and other attributes. The relational model has been widely used to implement a logical database [10]. An alternative method of storing and manipulating images is based on the object-oriented approach which permits image related data, as well as properties and rules to be represented and manipulated in a uniform way [11]. This method is better suited to handling the semantic content of images with its inherent hierarchical structure.

Our work to date has been based on the relational model and the MIDB system described in this paper has been implemented in a SUN enviroment and using SYBASE. The logical database is organized into the following three tables: a) Table IMAGES contains general information about each image as a whole and has the following attributes: image index, image file name, number of segments, date of image acquisition, patient name, symptoms, diagnosis, physician's comments etc. b) Table SEGMENTS contains descriptions of all segments contained in each image and has the following attributes: image index, segment index, average pixel value, texture parameters, area, perimeter, roundness, orientation etc. It can be easily extended to store other features that are shown to be effective in quickly narrowing down the search space for the purpose of document retrieval by image content in a specific application domain. c) Table RELATIONS contains relations between segments of each image and has the following attributes: image index, first segment index, second segment index, relative position, relative orientation, minimum distance.

A number of different *"views"* can be defined in terms of the above tables, corresponding to classes of specific image/segment properties and relations.

2. The *"physical database"* contains the original image files. Pointers from the logical to the physical database are handled by the MIDB management system. If necessary, image compression techniques may be applied to reduce storage requirements. To date, no image file structure has been introduced in our system and image compression is not implemented. However, images will eventually be stored in clusters based on the likelihood of being retrieved together in responce to a particular query.

Image Retrieval

All queries to the MIDB address the logical database rather than the raw image data stored in the physical database. All documents including images that satisfy the search criteria should be retrieved and displayed for viewing. Thus, an efficient method for browsing documents must be provided. Queries may be executed using a command-oriented query language extended to handle image data. There may be substantial uncertainty as to what a user is looking for, whereas a query answer set may be enlarged due to inexact matching techniques. Query formulation needs to be iterative and flexible, enabling gradual resolution of user uncertainty; it must also be associative, exploiting all possible associations between document components. A query based on pictorial content may specify:

1. The semantic content of an image: an image or part of an image is specified by pictorial example. In this case, a sample image or sketch is provided and the system must analyze it, extract its semantic content, create an appropriate representation, and match this representation against representations of images stored in the database. Alternatively the semantic content can be specified directly by a conditional statement involving features and other attributes similar to those stored in the logical database.

2. Various document features/primitives, either general or particular to objects contained in the image: these are specified as image attribute values and relevant images can be better accessed using indexing techniques.

3. Combinations of (1) and (2).

The response to retrievals depends highly on query type, specificity, complexity, completeness of data stored in the database, amount of online image analysis required and the size of the search. Queries must address properly defined classes of stored image information. It is generally useful to define classes corresponding to what queries usually address. Query search is currently performed on the main database tables. Performance can be substantially improved by limiting the search space with the definition of classes. In the most complicated type of query which has been implemented, a sample image or sketch is given and all images that are similar to it are returned. (Figure 2). Similarity is expressed as a weighted sum of differences of specific attributes. Appropriate thresholds are used to define the required degree of similarity.

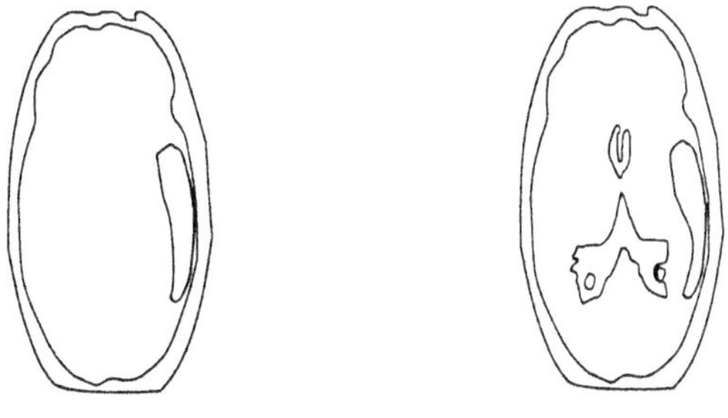

(a) (b)

Figure 2. a) Query image. b) Example of retrieved image.

Conclusions and Future Directions

An approach has been presented for the organization of a prototype MIDB system. The performance of image analysis and retrieval techniques has been evaluated on the basis of a limited number of images and the results are encouraging. Our immediate target is to test more sophisticated image description and retrieval methodologies and to investigate object-oriented techniques. An intelligent user interface must be developed for handling user uncertainty by iterative and flexible query formulation. A repertoire of alternative image analysis and modeling techniques to choose from must be provided in an attempt to minimize the complexity of picture descriptions and modeling in a particular application domain.

References

[1] V.S. Moustakis and S.C. Orphanoudakis. Requirements Definition for an Integrated Hospital Information System. In *EURINFO '88, Proceedings of the First European Conference for Organizational Systems*, pages 852–858, Athens, Greece, May 1988.

[2] S.C. Orphanoudakis. Supercomputing in Medical Imaging. *IEEE Engineering in Medicine and Biology*, 7(4):16–20, 1988.

[3] S.K. Chang and K.S. Fu, editors. *Pictorial Information Systems*. Springer-Verlag, Berlin, 1980.

[4] H. Tamura and N. Yokya. Image Database Systems: a Survey. *Pattern Recognition*, 17(1):29–49, 1984.

[5] S.K. Chang and S.H. Liu. Picture Indexing and Abstraction Techniques for Pictorial Databases. *IEEE Trans. on Patt. Anal. and Mach. Intell.*, 6(4):475–484, July 1984.

[6] M. Nagao. Control Strategies in Pattern Analysis. *Pattern Recognition*, 17(1):45–56, 1984.

[7] J.K. Tsotsos. Knowledge and the Visual Process: Content, Form and Use. *Pattern Recognition*, 17(1):13–27, 1984.

[8] P. Constantopoulos. MUSE: A Multimedia Document System. *IEEE Office Knowledge Engineering*, 1(1):28–38, February 1987.

[9] S.K. Chang, B.S. Lin, and R. Walser. A Generalized Zooming Technique for Pictorial Database System. In S.K. Chang and K.S. Fu, editors, *Pictorial Information Systems*, pages 257–287, Springer-Verlag, Berlin, 1980.

[10] N.S. Chang and K.S. Fu. A Relational Database System for Images. In S.K. Chang and K.S. Fu, editors, *Pictorial Information Systems*, pages 288–321, Springer-Verlag, Berlin, 1980.

[11] L. Mohan and R.L. Kashyap. An Object-Oriented Knowledge Representation for Spatial Information. *IEEE Trans. on Software Eng.*, 14(5):675–681, 1988.

PACS-RIS Interfacing: Experiences and Problems

G. GELL, G.H. SCHNEIDER, M. WILTGEN
Universitätsklinik für Radiologie, A-8036 Graz, Austria

Introduction

A Radiology Information and Documentation System (AURA) based on free text radiological reports has been developed and used since 1970 at the Department of Radiology of the University of Graz /1,2/. Since 1985 we developed and installed in cooperation with Siemens Medical Systems in Erlangen a PACS pilot project /3/. One of the interesting problems during the evolution of this project was the interfacing between PACS and RIS to ensure a unified and consistent database for images and reports.

AURA

Although AURA was originally planned as a documentation system to support radiological research it has grown into an information system which serves patient care and patient management, provides easy access to the stored data for scientific evaluations (case finding) and performs a variety of administrative tasks (billing, workload, statistics etc.). The center of the system is a database consisting of all the diagnostic reports of our department (including nuclear medicine) and of all the reports of the Department of Pathology. If a patient enters the department, his identification data are typed on a terminal - if previous examinations are found, they are immediately displayed and the relevant ones printed. Other terminals are also in the reporting areas where radiologists may request more detailed information on previous examinations. The reports are dictated as usual and typed on terminals and stored in the database. The system is also used for "feedback" informing radiologists if subsequent examinations or histological results confirm or contradict their diagnoses.

PACS

Our PACS serves a CT-station with three CT-scanners (Siemens Somatom, DR2, DR3, DRH). The configuration is given in fig. 1. A LAN (Ethernet) connects the three scanners with a diagnostic console and a digital archive with two optical disk drives. The archive computer (IMS-Image Management System) serves as a link to a hospital-wide second network with the computers of the AURA-System, the computers of the central administration etc. An MR-machine is also linked over the hospital-network to the PACS and images can be exchanged between the CT and the MR station. All the images produced by the CT-scanners are sent over the network to the IMS and are eventually archived on optical disk. A part of the workload is routed through to the diagnostic console (PDM) for reporting.

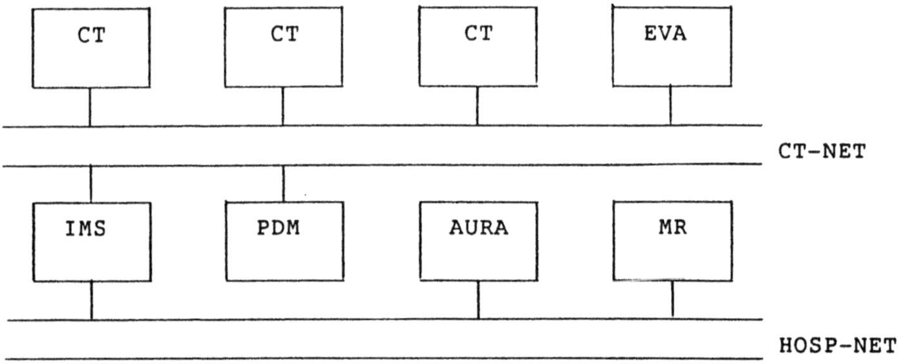

Figure 1.: Configuration of the PACS-network EVA: independent evoluation unit, IMS: image management system and archive (with two optical disk drives), PDM: diagnostic console, AURA: RIS-radiology information system

The transmission of images from the CT-scanners over the network is initiated by a command given by the radiographer. The rest - reformatting of images, grouping the images to form examination folders and patient folders, insertion into the image data base, archiving on optical disk, routing to the PDM, generation of overview images etc. - is done automatically. Independent software processes residing in the different computers communicate over defined interfaces to perform the different tasks mentioned above.

PACS-RIS connection

Our RIS (AURA) and our PACS have been developped independently and may function autonomously. But of course it is desirable and necessary to integrate the two systems if the PACS contains a long term archive. Without links the archive is either useless (because one cannot find the images) or it must duplicate a part of the RIS database.

We decided to put the administration of the image data into the RIS and to treat PACS with respect to long term archival and retrieval of images as a somewhat autonomous subsystem of AURA. There are two connections or interfaces between the two systems: one to inform AURA where the images are stored and one to inform PACS which images should be retrieved.

After the archiving of the images the PACS system writes the information about the location of the images, i. e. the number of the optical disk and the identification of the files on the disk into a file of the AURA system. This information is then added to the patient record within AURA. Retrieval requests for images on optical disks are always initiated from the AURA system. There are basically two kinds of retrievals: the routine retrieval of images from previous examinations if the patient is admitted for a new examination and the retrieval of the images of preselected patients for scientific studies. The former retrieval is started during the admission of the patient into the CT-station (cf. the section on AURA) when the previous reports are displayed. If there are images on optical disk a request is sent to the PACS which retrieves the images (either directly if the disk is still online or by asking a radiographer to insert the disk manually into the drive) and the images are routed to the diagnostic console where they are available for reporting together with the images of the actual examination (e.g. to compare the evolution of a tumor).

AURA allows for selective retrieval of patients depending on specified criteria, e.g. a logical combination of diagnostic terms, age group, sex etc. This feature can also be used to retrieve all the images pertaining to such a patient group and combining them as a special folder in the PACS system.

Experiences and Problems

A complete PACS would mean a filmless operation of the CT-station, where all the diagnostic work is done interactively on diagnostic consoles. Therefore our original idea was to let the radiologist select the images for long term storage during the diagnostic process: it was planned to archive only the most relevant images in full size and reduce the rest (to a 64 x 64 matrix) to save space. This turned out to be a problem for several reasons; the selection process could not be integrated so smoothly into the reporting of the examination because the decision to "throw away" images and information is not an easy one; on the other hand reporting on the diagnostic console is still not done really routinely. So we decided to separate the two functions, reporting and archiving. All the images are archived automatically on optical disk. 512 x 512 matrices are reduced to 256 x 256. No data compression is made as yet (plans are underway). Sending one image (512 x 512) over the net takes about 5 sec. (as a mean value) - a whole examination (20 - 30 images) needs 2 - 2,5 minutes. Routing an examination from a CT over the IMS to the diagnostic console takes about 10 - 15 min. This includes reformatting, insertion into the database etc.

In contrast to some predictions the speed of the Ethernet LAN does not restrict our work so far. Even if all three CT-scanners send their images concurrently immediately after the end of each examination the overall waiting time does not increase significantly.

There exist some problems in our PACS resulting from the fact that we have no access to the software of CT-scanners. They have some fixed software for sending and receiving images over the network but we would like to be able to have more communication - to transmit for example the names of the patients that are waiting for an examination on a certain scanner. As yet, the radiographer has to type the patient data again on the CRT of the CT. If she makes an error we have difficulties to correctly correlate the images with the reports in AURA. Our programs check for errors and correct them, if possible, but sometimes we need human intervention to decide about the correct matching.

We cannot yet give a complete comparison between archival on optical disk and on magnetic tape.

The original prices for the optical disk media was three times the price of magnetic tape. We did not start routine archival until we could bring down that price to the price of magnetic tape (by searching for other suppliers and by hard negotiations). At the moment, the cost for storing one 256 x 256 image on optical disk is approximately DM 0.1. We store 10.000 images on one disk (two sides) and therefore, with our workload of ca. 1.000 CT-images per day will need about 26 disks per year (instead of 1.000 magnetic tapes!).

Acknowledgements

We thank the Bundesministerium für Wissenschaft und Forschung and Siemens Medical Systems, Erlangen for supporting this work.

Literature

/1/ Gell G.: AURA: Routine Documentation of Medical Texts Meth. Inform. Med. 22 (1983) 63-68

/2/ Gell G. AURA: A Radiological Information System In: A. Chiesa, R. Gasparotti, R. Maroldi (Hrsgb.): Planning Considerations in Diagnostic Imaging and Radiation Therapy Clas Int., Brescia (1988) 65-68

/3/ Gell G., G.H. Schneider, M. Wiltgen, M. Becker, G. Seufert, C. Greinacher, H. Gnoyke: PACS: Erfahrungen beim Einsatz in der Klinik In: G.H. Schneider, E. Vogler (Hrsgb.): Digitale bildgebende Verfahren - Integrierte digitale Radiologie Springer, Berlin - Heidelberg - New York, (1988) 594-600

Benefits of HIS-PACS Coupling in Clinical Practice

B. M. van Poppel (1), H. Lodder (1), J. B. M. Wilmink (2), C. Ising (3)

1. BAZIS, Central Development and Support Group Hospital Information System, P. O. Box 901, 2300 AX Leiden, The Netherlands
2. Utrecht University Hospital, Central Computer Department
3. Philips Medical Systems, Hamburg, Federal Republic Germany

Summary

An HIS-PACS coupling, an essential PACS component, has been designed, implemented and tested. Subsequently it has been utilized during clinical evaluation of the HIS/PACS sytem.
This research has been performed within the scope of the Dutch PACS project, a joint effort in the field of PACS by BAZIS, the Utrecht University Hospital and Philips Medical Systems, financially supported by the Dutch Department of Health.
A prototype PACS (Philips) has been installed at the radiology department and at three clinical wards (internal medicine, cardiosurgery and cardiology). The data transferred from the HIS tot the PACS was limited to information about the inpatients of these wards. Only data concerning the internal medicine ward is discussed in this paper. The data flow is one directionally.
The HIS (BAZIS/HIS) is an integrated hospital information system, consisting of a central database and more than 70 subsystems (including a RIS, an ADT-system and a system for registration and identification).

This HIS-PACS coupling was designed and implemented in late 1987 and early 1988. It has been demonstrated during the EuroPACS meeting April 1988 in Utrecht.

The clinical evaluation of the prototype PACS, including the HIS-PACS coupling has been started in October 1988. Daily usage has its effects on working methods within the radiology department and the clinical departments involved.

An overview of the coupling experiment will be presented. The first results of evaluation, including influences on the daily working methods, will be described. The need for future developments of the HIS-PACS coupling will be discussed based on the results gathered thusfar.

Introduction

Benefits of PACS only can be considered when the functionality of PACS is observed in practical use. In this paper it is demonstrated that integration with an HIS is an essential prerequisite for efficient use of a PACS [1]. Text data, originating from the HIS, can be displayed in an integrated fashion with the images. Furthermore, coupling will eliminate the need for retyping work, and thus avoid the possibility of mistakes, especially when image identifications are involved. Triggering the reallocation of images by HIS data can improve image management, resulting in acceptable waiting times [4].

After a short characterization of the BAZIS/ZIS and the Philips-/PACS the operation of a non-integrated PACS, not coupled with a HIS, will be described. Improvements, which can be reached by integration, are proposed. In the next section the realization of these improvements is described. The integrated system is evaluated as far as coupling is concerned.

THE HOSPITAL INFORMATION SYSTEM OF BAZIS

This fully integrated hospital information system, contains over 70 subsystems. Subsystems which are directly involved in our experiment are:

- the radiology subsystem RADI, a RIS integrated within the HIS
- a subsystem for registration and identification of general patient data, PATREG
- a subsystem that keeps track of the location and status of an inpatient within the hospital, the ADT system LOCATI.

THE PROTOTYPE PACS

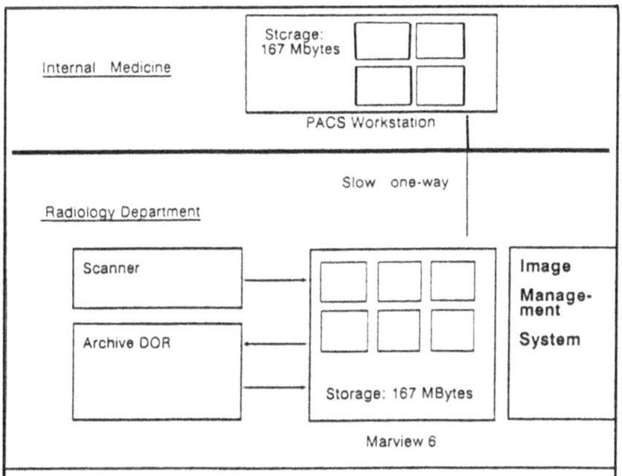

Fig 1. Simplified Architecture

The prototype PACS, Philips MARCOM, has been designed around a central workstation for the radiologists MARVIEW-6. Note that all image transfers, indicated by arrows, have to pass through MARVIEW-6. At the department of Internal Medicine a separate MARVIEW-4 station, with 4 monitors has been installed. This remote workstation can receive image files from the central node via a telephone link (9600 baud). Around 2000 compressed (20:1) images can be stored in the magnetic disk memory of the workstations. From the workstation in the clinic it is not possible to retrieve images directly from MARVIEW. Image transfer commands are given manually by the PACS oparator to the Image Management System (IMS). Retrieval of images can take place because of the unique image identifications provided by the IMS.

OPERATION OF THE PACS WITHOUT HIS/PACS COUPLING

The clinician on the ward can consult only images which are located on his workstation. If it appears that images are not stored in his local equipment he has to notify the PACS operator who will transfer the desired images to his workstation.
The text data of the patient is available on a separate, HIS terminal.

IMPROVEMENT OF FUNCTIONALITY BY A HIS/PACS COUPLING

As outlined in the previous section the retrieval of an image is a rather cumbersome process.
The presence of patients is registrated by the HIS, so the moment of admission (or discharge) indicates the moment certain images have a large (resp. very small) probability to be retrieved. When thus HIS data is transferred to the PACS the IMS can initiate image transport automatically. The retyping of identifications (in case of new images) into the PACS can be avoided when the HIS is able to send this information.

In the above described stand-alone operation no patient-related data such as radiological reports is available on the same device to the clinician. Integration of reports and images is done by sending the text data from the HIS to the PACS. The PACS stores the text data in an additional memory from which it can be retrieved.

HOW THE PROPOSED COUPLING WAS REALIZED

Sending of data will occur after the occurrence of one of the following events:

- admission to ward
- discharge from ward
- radiology department appointment
- radiology department appointment changed
- radiology department appointment cancelled
- radiology report authorized.

Only data is sent concerning inpatients of the Internal Medicine ward, for which the evaluation takes place.

The software module responsible for detecting events, collecting the appropriate data and sending to the PACS is located in the BAZIS/ZIS. This software runs completely automatically and in background. An extensive description of the technical realization can be found in [4].
The admission (or discharge) as registrated by the ADT subsystem LOCATI is recognized by the coupling software, resulting in the sending of among others the data which can be used by IMS to initiate image transfers.
Appointment data, registrated by the subsytem RADI is transferred to provide the IMS with the identifications for new images (event: radiology department appointment).
After recording the spoken report, the report is entered by a typist in the BAZIS/ZIS subsystem RADI. When no modifications are needed the report is authorized (event: report authorized). The occurrence of this event will be detected by the coupling software and in the end the report is available to the PACS user.

THE EXPERIMENT: INTEGRATION IN CLINICAL USE

The coupling as demonstrated during the 6th EuroPACS Meeting, held April 1988, is operational from October 1988 [2]. The coupling software, working reliable on routine basis, was studied by among others the use of logfiles [3]. Working procedures were monitored. The integrated system is clinically in use for referencing. Because of image quality, the PACS is only used for reference and not for primary diagnosis [7].

EVALUATION: IN WHAT WAY WAS THE REALIZED COUPLING EFFECTIVE IN IMPROVING THE FUNCTIONALITY ?

The PACS has the information to do so but didn't yet start the image transfers automatically. Image transfers still had to be initiated by hand.

Admissions were based upon the registration with the subsystem LOCATI, i.e. when the patient actually visits the clinic. There appeared two problems: digitizing loads the other activities in the PACS so heavyly that errors occurred, secondly the communication channel between MARVIEW and the ward is so slow that it takes several hours for a decent image history to transmit. The problem has temporarily been solved by manually forcing the dispatch.

Although the experiment was only meant for inpatients there are sometimes outpatients visits. This category of patients is not detected by the subsystem LOCATI. The switching on of another suitable BAZIS/ZIS subsystem will solve the problem automatically.

Although several shortcomings and a number of severe problems in the PACS arose even for this prototype PACS it was not possible to make adaptions. This problem of modification will be more pronounced in commercially bought PAC(S) systems. Most problems were solved in our case by modifying the coupling software.

As a final remark we can state that radiologists and referring clinicians are becoming more and more enthousiastic.

CONCLUSIONS

- the developed strategy has proven to work and offers good prospects for a hospital wide PACS
- integration, not only with a RIS, is essential for PACS functionality, among others
- old and new data, as well image as text data available
- reduction of waiting times by better image management
- avoiding of retyping image identifications into the PACS
- the more complete the HIS the more subsystems can be used
- a standard should be available for the coupling of the Image Management System with the Hospital Information System. A difficulty is that a lot of integration aspects can only be realized by event driven communication
- when PACS is used on routine basis image transfers must be initiated automatically

FUTURE DEVELOPMENTS

Although images and text data are available on the same unit, the PACS-user interface can be improved further. Technical improvements of the coupling will provide better functionality to the PACS user. The use of more subsystems will cover more needs for example the support of the outpatient visits to the clinic. Research to describe a general applicable coupling including event driven communication is another important topic.

References

[1] Bijl, K., Koens, M.L., Bakker, A.R., De Valk, J.P.J., 'Medical PACS and HIS: Integration Needed', Proc. Medical Imaging Conference, R.H. Dwyer III, S.J., Editors, SPIE Vol. 767, pp. 765-770, 1987.

[2] Lodder, H., van Poppel, B.M., Wilmink, H.B.M., Scharnberg, B., de Valk, J.P.J., Bakker, A.R., HIS-PACS coupling: BAZIS/ZIS and Philips/MARCOM on speaking terms, Medical Informatics, Vol. 13,4, 1988, pp. 361-367.

[3] Lodder, H., van Poppel, B.M., de Valk, J.P.J., Wilmink, H.B.M., Ising, C., Bakker, A.R., HIS-PACS coupling in practice, Medical Imaging III Conference, Jan. 1989, Newport Beach.

[4] Lodder, H., van Weperen, J.H., de Valk, J.P.J., Bijl, K., Bakker, A.R., Helder, J.C., Scharnberg, B., HIS-PACS Coupling First Experiences, In: Proceedings Medical Imaging II Conference, R.H. Schneider, S.J. Dwyer III, Editors, SPIE Vol. 914, 1988, pp. 1041-1046.

[5] Haar Romeny, B.M. ter, Wielen, J. van der, Achterberg, A.J., Barneveld Binkhuysen, F.H., Andriessen, J.H.T.H., PACS efficiency: a detailed quantitative study, Medical Imaging III Conference, Jan. 1989, Newport Beach.

[6] Rijnsoever, R.C. van, Valk, J.P.J. de, Bakker, A.R., A layered storage structure for images confronted with the use of X-ray images in a hospital, Proceedings Applications of Optical Digital Data Disk Storage Systems Conference, W.M. Deese, M. Carasso, Editors, SPIE Vol. 490, 1984, pp. 49-52.

[7] Barneveld Binkhuysen, F.H., Ottes, F.P., Haar Romeney, B.M. ter, Klessens, P.L.M., Vos, C.G., Winter, L.H.L., Calcoen, P.T., Andriessen, J.H.T.H., First results of the diagnostic evaluation studies and the clinical efficacy evaluation in the Dutch PACS project, SPIE Mediacal Imaging III, Proceedings 1989, in press.

Medical Workstation

Computer Vision

arteries. In addition we demonstrate the ability of this technique to enhance magnetic resonance images of cerebral blood vessels.

Our algorithm, which is based in part on the multiwindow concept described in [2], is a simplification and improvement of an earlier algorithm reported in [3].

2. Algorithm

Our algorithm corrects a) mean gray-level shifts and b) motion artifacts. The Steps of our algorithm are as follows.

Step 1: Finding the exclusion template

Threshold the difference between the live image L(i,j) and the mask M(i,j), yielding the binary-valued digital image T*(i,j). The pixels where T*(i,j) = 1 are excluded from the calculations in subsequent Steps.

Step 2: Correction of gray level variations

Two types of corrections in gray level are carried out:
a) correction of variation caused by fluctuations of the power of the x-ray source (global correction); and b) correction of variation caused by the diffusion of contrast medium (local correction).
These corrections produce a gray-level-corrected mask $\tilde{M}(i,j)$.

Step 3: Select the control points

The control points on the live image are chosen every c pixels along the rows and columns of the image matrix. The number c is determined by the user. In our study, c was empirically chosen to be 10. For each control point, a square window of a specified size is selected, with the control point at the center. This window is used to find the best matched point on the mask for the control point in question.

Step 4: Rough translation

Keeping the live image window fixed, shift the mask under the window until a best match is found; then record the location of the the best match on $\tilde{M}(i,j)$. The shifting is done horizontally and vertically with an increment of one pixel at a time.

Step 5: Fine translation

Shift the mask horizontally and vertically around the location found in Step 4, with increments of 0.1 pixel. Record the location of the best-match control point in the mask image $\tilde{M}(i,j)$.

Step 6: Rotation

Rotate the best match window on the mask $\tilde{M}(i,j)$ around its control point in specified increments to find a best match.

Step 7:

Repeat Steps 5 and 6 until no changes take place.

Step 8:

Go back to Step 4 for each control point.

Step 8:

All points on the mask corresponding to the points on the live image are found through bilinear interpolation. Using these interpolated points, the corrected mask $M^*(i,j)$ is computed.

Step 9:

Perform pixel-by-pixel subtraction between the live image $L(i,j)$ and the corrected mask $M^*(i,j)$. Store the results.

In the above algorithm, the user must choose the following parameters: image window size, the translation limits, the rotation angle limits and the angle increments. This algorithm has been implemented in the C language, and executed on a Digital Equipment Corporation MicroVAX II computer at the Image Modeling Laboratory of the Focused Research Program in Image Engineering at the University of California, Irvine.

3. Experimental Results

We tested our flexible registration technique on two types of angiograms: x-ray and magnetic resonance (MR). The results are reported below.

3.1 X-ray angiography

The x-ray angiograms in our experiments were obtained at the Cardiac Catherization Laboratory of the University of California Irvine Medical Center. These images are part of a series of cinematic frames of an *in vivo* human heart shortly before and after the injection of iodine into the coronary arteries. The images were acquired by an x-ray generator with a tungsten target operated at 64 mA and 96kVp, a Siemens photo intensifier/video camera that produced 30 frames per second, and a Fischer Model DA-100 computer system that digitized each frame into a 512 x 512 x 8-bit array, processed the image, and stored the processed image on 1/4-inch tape in a Scotch DC

600 HC tape cartridge. The images were taken at LAO (Left Anterior Oblique) 60 degrees. On our experiments, we chose the following parameters. Window size = 31 x 31 pixels, translation limit = ± 21 pixels horizontally and vertically, rotation limit = ± 10 degrees in increments of 1 degree.

Figure 1 illustrates the subtraction of two x-ray cinematic frames occurring after injection of iodine. Both images include a pacemaker wire. Figure 1a also includes a software phantom of a stenotic artery. Figure 1b is obtained at a later time, at the same cardiac phase (end diastole), without the software phantom. The first image served as the "live" image, while the second, without the phantom, served as the "mask". Our intent was to segment (i.e. isolate) and measure the phantom with the aid of subtraction of one of the two images from the other. Figure 1c is the result of a direct pixel-to-pixel subtraction. The result of our subpixel flexible-mask subtraction algorithm is shown in Figure 1d. Note that in Figure 1d the artifacts in Figure 1c have been almost completely suppressed.

We carried out an experiment to evaluate the usefulness of our algorithm in quantitative angiography. We applied a parametric estimation method similar to that in [4] to estimate the diameters and areas of the phantom in three distinct situations: no subtraction, direct subtraction, and subpixel flexible mask subtraction (our algorithm). The results are shown in Figures 2 and 3. Here we see the substantial improvement in the accuracy of both diameter and area estimates that our subpixel flexible registration made possible.

Subtraction of real angiographic images with a distorted grid:

In this experiment we demonstrate the ability of our technique to correct geometric distortions. Figure 1b is an x-ray image of contrast-enhanced coronary arteries along with a pacemaker wire. This served as a live image. Call it A. Figure 4a, called B, was obtained as follows: we superimposed on image A a perfectly square grid to create a new image A'. We distorted A' globally in a manner similar to pincushion distortion. We also distorted A' locally in a few small regions about the size of a square of the grid. The resulting distorted image is B. The amount of distortion is visible in Figure 4b, which is a direct subtraction of images A and B.

In our experiment, image B served as the mask. Our subpixel flexible registration technique was applied to correct the amount of misregistration between A and B. If the correction were perfect, the subtraction would yield the original perfect square grid. Figure 4c shows the result of our flexible registration algorithm. We see that our algorithm corrected most of the distortion artifacts in image B.

3.2 Magnetic resonance (MR) angiography

The MR images in this study were acquired on a Magnetom 63, manufactured by Siemens. This equipment is part of the MR imaging facility provided to this project by the Radiology Department of the Long Béach Memorial Hospital. The raw images were acquired in the form of 256 x 256 x 12-bit digital arrays, each array representing a coronal MR slice. Each slice is 2 mm thick. Each slice was acquired by a form of partial-flip-angle time-of-flight imaging sequence.

The set of raw images consisted of 48 pairs. In one image of each pair the blood was brightened and consequently enhanced; in the other the blood was dephased and consequently not enhanced. We modified our flexible registration technique, which was previously developed for x-ray angiography, to make it suitable for MR angiography, restricting it initially to 8 bits per pixel.

Figure 5a shows the sum of the unenhanced images among the 48 pairs. Figure 5b shows the sum of the 48 direct subtraction images. Figure 5c shows the sum of the 48 flexible subtraction images.

4. Discussion

Our experiments demonstrate the effectiveness of subpixel flexible registration for estimating the transverse diameters and areas of coronary arterial lumens from digital subtracted x-ray angiograms and for enhancing the visibility of cerebral blood vessels in MR angiography. Our experiments also demonstrate that our algorithm suppresses geometric discontinuities that occur in earlier "flexible-mask" subtraction techniques [3].

We are evaluating the efffectiveness of our subpixel flexible registration technique for 3-D modeling of the coronary arteries in both x-ray and MR angiography.

Acknowledgments

We thank Dr. J. Tobis, Prof. Werner Roeck and Dr. K. Kitamura for providing the angiograms used in this study. We are grateful to Dr. Richard Ong and Mr. Dennis Atkinson of Siemens Medical Systems for providing the magnetic resonance image data for this study, and to Drs. Y.S. Kwoh, P. Burger, and W. J. Wilson of the Long Beach Memorial Hospital for their cooperation. We thank Dr. T. C. Mills, Mr. C. E. Smith, and the members of the Focused Research Program on Image Engineering at the University of California, Irvine for their comments and assistance.

References

1. J. H. C. Reiber, et al, "Quantitative Coronary Angiography with Automated Contour Detection and Densitometry: Technical Aspects," Angiography, edited by H. Just, P. H. Heintzen, Springer-Verlag, Berlin, 1986, pp. 311-331.

2. M. Yanagisiwa, S. Shigemitsu, T. Akatsuka, "Registration of Locally Distorted Images by Multiwindow Pattern Maching and Displacement Interpolation: the Proposal of an Algorithm and its Application to Digital Subtraction Angiography," Proc. of the Seventh International Conference on Pattern Recognition,. 1984, pp. 1288-1291.

3. L. V. Tran, J. Sklansky, "Flexible Mask Subtraction for Digital Angiography," Hybrid Image and Signal Processing, by D. P. Casasent and A.G. Tescher, Proc. of SPIE, Vol. 939, April 1988, pp. 203-211.

4. K. Kitamura, J. M. Tobis, J. Sklansky, "Estimating the 3-D Skeletons and Transverse Areas of Coronary Arteries from Biplane Angiograms," IEEE Trans. on Medical Imaging, Vol. 7, No. 3, Sept. 1988, pp. 173-187.

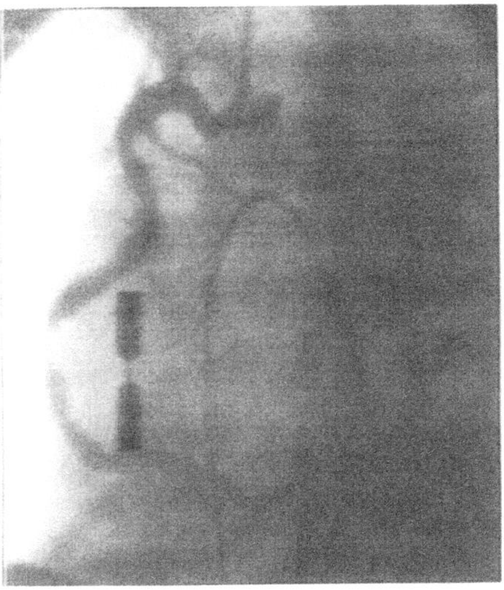

Figure 1a

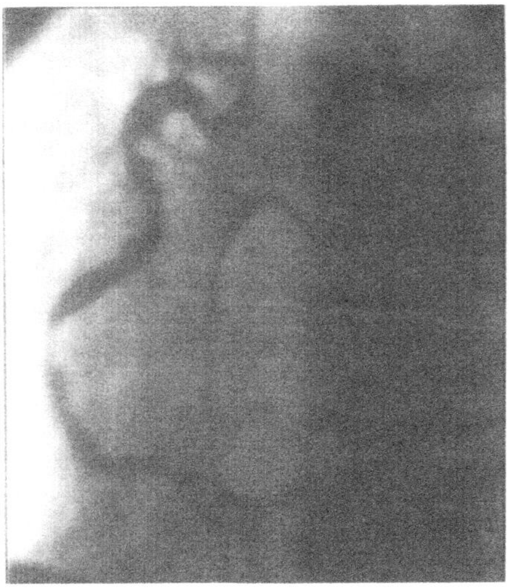

Figure 1b

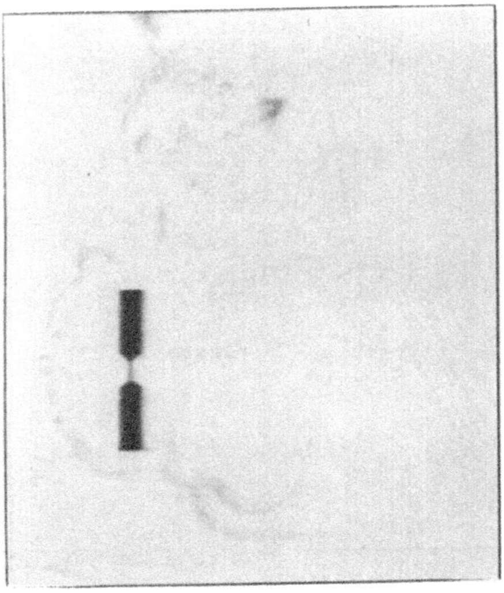

Figure 1c

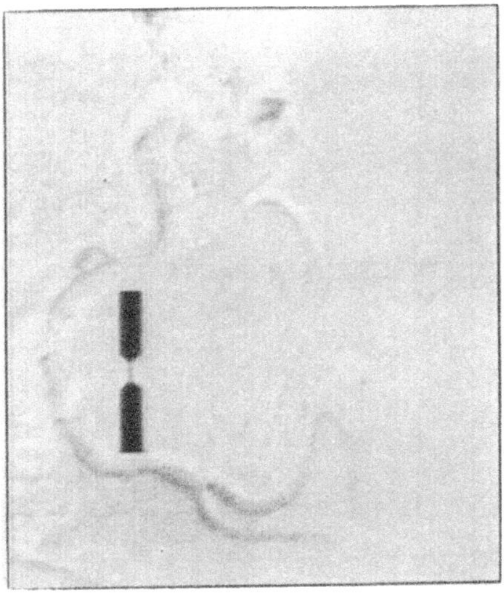

Figure 1d

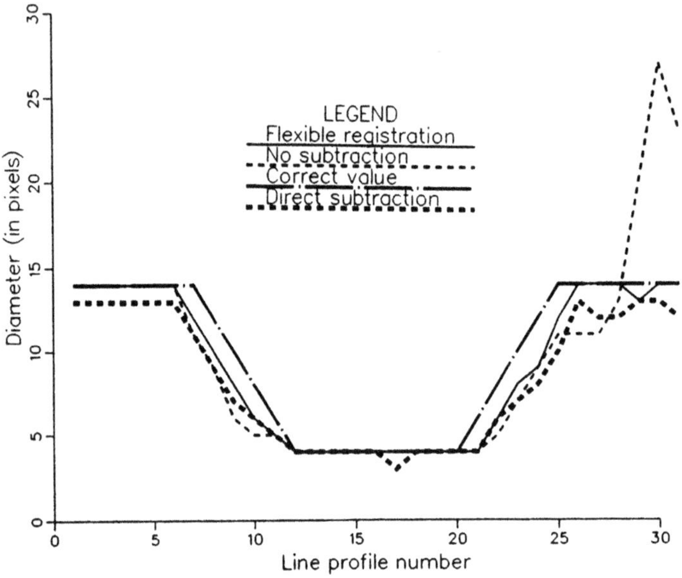

Figure 2

Diameter estimation

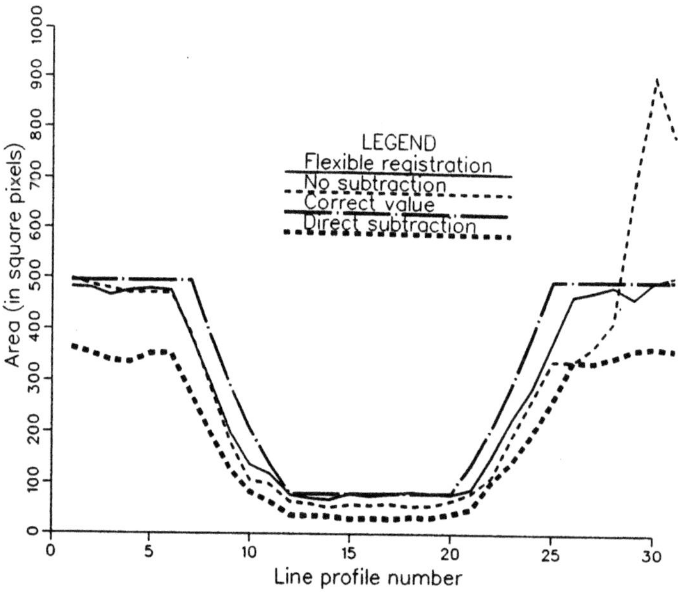

Figure 3

Area estimation

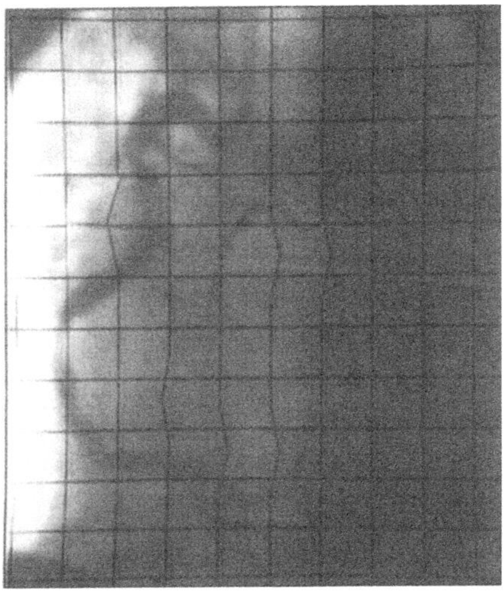

Figure 4a

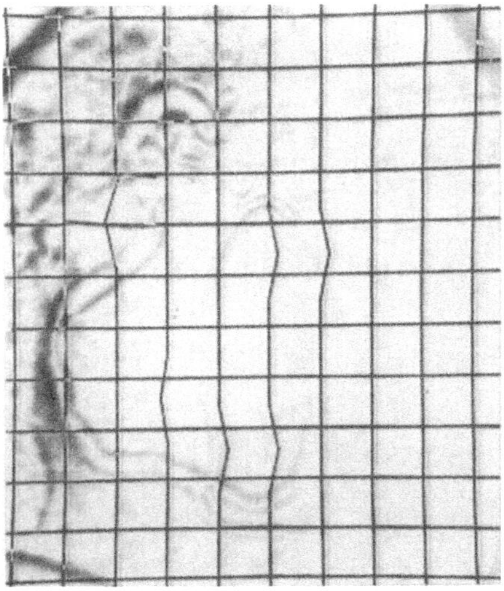

Figure 4b

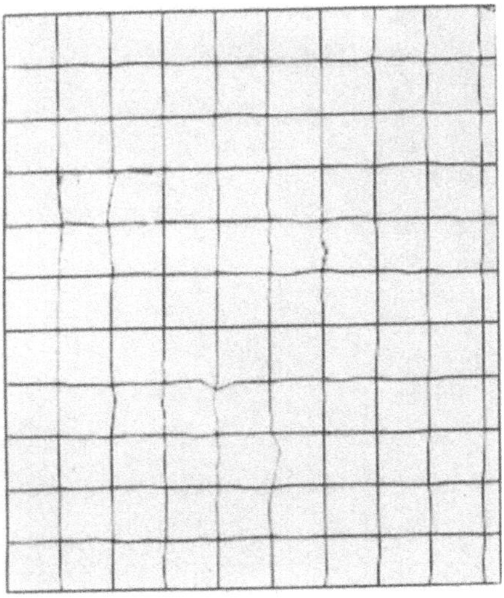

Figure 4c

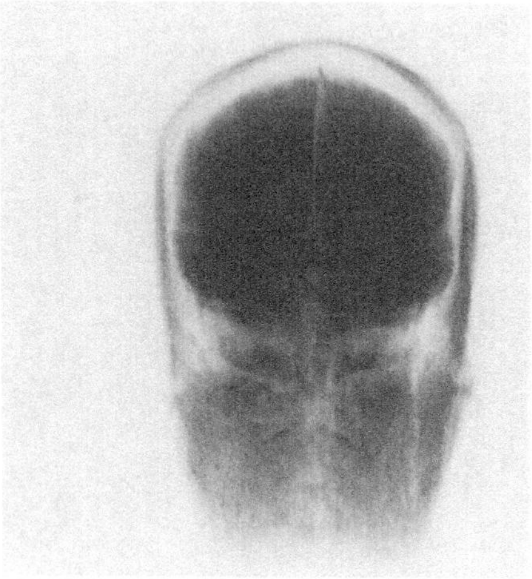

Figure 5a

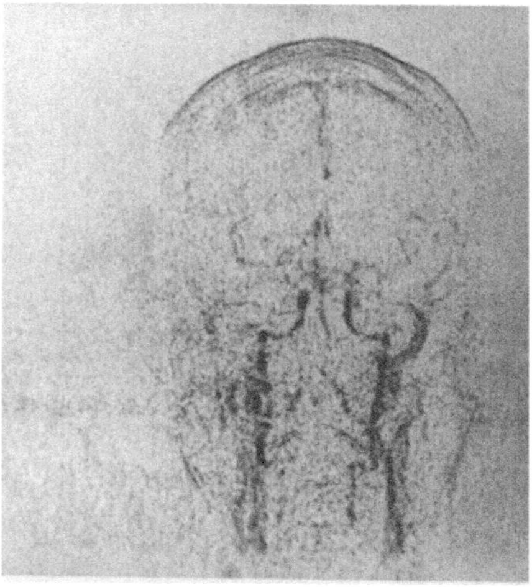

Figure 5b

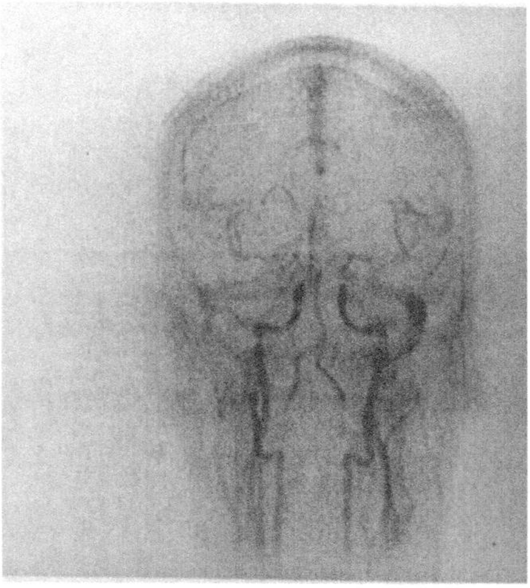

Figure 5c

Delineation of Neuroanatomical Objects Using Deformable Models

D. Vandermeulen, P. Suetens[1], J. Gybels, G. Marchal, A. Oosterlinck.

Interdisciplinary Research Unit Radiological Imaging (ESAT/MI2 & Dept. of Radiology), Dept. of Neurology and Neurosurgery, Catholic University of Leuven (Belgium).

Summary
In this paper, we present work in progress on the semi-automatic delineation of neuroanatomical objects on 3D images (MRI, CT, US). The method we propose starts from an initial, incomplete and/or incorrect approximation of the object and iteratively deforms it in congruence with both the multidimensional image data and generic model knowledge. Some initial results on the 2D segmentation of Doppler US images are shown and an outline of the procedures for the 3D segmentation of other structures is given.

Introduction
During the past decade a number of diagnostic imaging techniques have emerged that generate three-dimensional matrices of grey values. These volumes are pictorial representations of intrinsic scalar tissue properties, such as X-ray absorption coefficients in CT, radio-active tracer concentrations in SPECT and PET, proton concentrations and magnetic spin relaxation parameters in MRI, acoustic parameters (impedance, density, compressibility,...) in Ultrasound.

Unfortunately, 3-D visual examination of the volume is not straightforward.
Originally, images were viewed *slice-by-slice*, leaving the three-dimensional reconstruction as a difficult and subjective mental exercise.
Current 3D display methods apply conventional computer graphics techniques to display *surface* morphology. This implies a preliminary surface detection step, which is done by manually tracing contours in each slice or by simple automatic thresholding schemes. Manual tracing is too tedious, too time consuming and too subjective to be clinically acceptable. Furthermore, the use of thresholding confines the segmentation to relatively high contrast images such as is the case in CT for bone/soft-tissue and soft-tissue/air surfaces. This explains the abundance of literature on 3D display of medical images for

[1]P. Suetens is research associate with the National Fund for Scientific Research (NFWO)

thopedics and plastic surgery.
Recent advances in computational power and especially in larger and cheaper memories have spurred the development of *volumetric rendering* [1,2,5,6], whose major characteristic is the incorporation of segmentation *probabilities* into the display process. This results in 3D pictures with less perceivable artefacts than were previously introduced by the simple segmentation schemes. This display method is particularly appealing for the visualization of images of soft tissue structures as are generated by current MR scanners.

Since both surface and volume displays are critically dependant on the segmentation results, we have begun to develop methods for multidimensional segmentation. In this paper, we present a method for an almost automatic *delineation* of neuroanatomical objects, in casu intracranial tumors on CT and MRI images and cerebral blood vessels on MRI and Doppler Ultrasound flow images. It starts from an initial, incomplete and/or incorrect approximation of the object and iteratively deforms it in congruence with both the multidimensional image data and generic model knowledge until a generalized cost or energy function is minimized.
This method is described for two specific cases, i.e. the delineation of blood vessels and tumor lesions on 3D MR images. Initial results on a 2D delineation of blood vessel cross-sections on Doppler Ultrasound images are presented. Finally, we briefly discuss the extension of deterministic segmentation to probabilistic segmentation schemes.

Three-dimensional visualization of neuroanatomical objects

The goal of three-dimensional visualization is the perception of semantically meaningful patterns (including pathologies), e.g. all kinds of vascular structures. Simple projection procedures, such as the integration of intensity data in ray casting may result in obscuration of these structures. A better perception or an improved visibility amounts to image enhancement, by accentuating or highlighting the characteristics of the particular patterns. Some characteristics of meaningful patterns are its shape, its local feature (e.g. grey value) variations and contrast with background and eventually its context (e.g. continuation of linear structures). Highlighting these structures always implies the application of model knowledge. This can be subdivided in generic model knowledge and specific model knowledge. Generic model knowledge is e.g. linearity of blood vessel center line and circularity of cross-sectional shape. Specific model knowledge is e.g. the particular geometric model of the carotid arteries. The latter model gives the best results for a well restricted application for which generic model knowledge can be easily formulated (e.g. description of man-made regular objects). However, in the case of medical images and particularly images of complex and highly variable vascular structures, this method may be unrealistic.

The method described below applies generic model knowledge to the extraction of anatomical objects of interest. This knowledge is stored in the form of a reference shape model (e.g. tubular form for blood vessels) and shape deformation parameters (e.g. elasticity, rigidity).

Delineation of blood vessels on 3D MR angiograms

We first filter the 3D MR angiograms with a 3D line detector to extract bloodvessel centerlines [7]. This method enhances local intensity maxima, and discriminates between line-like or cylindrical structures and other structures. It is even possible to look for "negative vessels", i.e. the intensity of the blood vessel lumen is lower than the surrounding tissue. This is particularly interesting since existing acquisition procedures in MRI angiography have not found a satisfactory way to solve this clearly diagnostic important problem.

In a second step, we threshold, thin, link and prune the preprocessed images. This results in an initial segmentation of the centerline of the blood vessels. The medium-level processing will have to correct the low-level filter results in different ways :

- Fine tuning of centerline positions.
- Bridging of small gaps generated by imperfect filtering if sufficient evidence exists.
- Elimination of non-cylindrical structures, that may still result after the initial filtering.

In our current implementation, however, this general scheme is broken down in two steps. First, we deform the original centerlines, resulting from the low-level segmentation to lie on intensity maxima, while at the same time imposing smoothness constraints on the form of the centerlines. We use a *deformable model*, called "snakes" [3], to match it with the image data by means of *energy minimization*. Basically, the "snake" is an elastic wire that deforms under the influence of external and internal forces. The internal forces impose smoothness constraints, whereas the image forces push the "snake" towards salient features (viz. lines and edges). Instead of just deforming centerlines, one can also try to find cylindrical structures. Symmetry-seeking models [8] may do the job. These models consist of a centerline and a sheet in the form of a tube. The centerline is forced to lie on intensity extrema, whereas the sheet is attracted to gradient extrema. The two components are coupled by elastic spring forces, with additional forces to impose radial symmetry. Furthermore, expansion or contraction forces act on the tube. This last feature is particularly appealing in our case, since we can start from an infinitely thin cylinder, which we subsequently expand to match the blood vessel edges.

Second, grouping of individual "snakesegments" into perceptual entities has to be performed. We are currently investigating alternatives for this step. An interesting solution, found in [4], is also based on the minimization of an energy functional.

A problem we are facing at the moment is how to incorporate both steps simultaneously, i.e. deforming the snakesegments not only by image forces, but also by interaction forces between the snakesegments.

Delineation of tumor lesions on 3D MR or CT images

In the case of tumor lesions, the starting approximation is entered manually by simply pointing at the interior of the lesion. This point, seen as an infinitesimal small sphere can subsequently be expanded in an edge-limited-region-growing way, until both interior homogeneity and edge strength criteria are optimally met, based on the same energy minimization formulation as discussed in the previous section.

Two-dimensional delineation of blood vessel cross-sections on Doppler Ultrasound Images

Initial values for the delineation of blood vessel cross-sections on Ultrasound slices can be extracted from the blood flow information present in the Color Doppler images. Figure 1 illustrates the procedure when starting from a rectangle surrounding the blood flow patterns in the color images and iteratively deforming this contour, using the same energy minimization framework as discussed above.

Discussion

As already stated in the introduction, volumetric rendering is based on probabilistic segmentation results. Indeed, visualization based on binary segmentation clearly suffers from a number of visually perceivable artefacts. Intrinsic intensity variations within the object, along with partial volume phenomena and inhomogeneities during image acquisition and reconstruction can make the results of intensity thresholding useless. More complicated segmentation procedures are also deterministic in the sense that they partition the scene by all-or-none decisions, introducing visually perceivable artefacts. In many situations, lack of sufficient information may not justify such a drastical categorization. This limitation introduces biases and aliasing and interferes with a correct interpretation. Hence, there is a need for *probabilistic segmentation* techniques, which perform an object intensification instead of object partitioning. We are currently exploring this avenue.

Conclusion

Our initial experience with deformable models and energy minimization for the extraction of vascular structures is promising. Problems we are dealing with now are the integration of perceptual grouping rules in the optimization procedure and the extension towards a probabilistic segmentation for subsequent visualization.

Acknowledgments

The authors would like to thank H.Bosmans and E. Plessers who have contributed to the implementation and testing of the above procedures. This research is supported by the F.G.W.O. (Belgium) under grant number 3.0009.85.

References

1. Drebin R.A. et al., "Volume rendering", Computer Graphics (Proc. SIGGRAPH 1988), Vol.22, No.4, pp.65-74, (1988).
2. Hoehne K.H., "3D bildverarbeitung und Computer Graphik in der Medizin", Informatik-Spektrum, Vol.10, pp.192-204, (1987).
3. Kass M., Witkin A., Terzopoulos D., "Snakes : active contour models", Int'l J. Comp. Vision, Vol.1, pp.321-331, (1987).
4. Lawton D.T. et al., "Knowledge-based vision for terrestrial robots", Proc. Image Understanding Workshop, pp.103-110, (1988).
5. Lenz R. et al., "Presentation and perception of 3-D images", NATO Advanced series on 'Pictorial information systems in Medicine', (1986).
6. Levoy M., "Display of surfaces from volume data", IEEE Computer Graphics and applications , pp.29-37, may (1988).
7. Vandermeulen D., Smets C., Suetens P., Gybels J., Marchal G., Oosterlinck A., "Knowledge-based 3-D segmentation of bloodvessels on a spatial sequence of MRI and Ultrasound images", to be published in *Medical Imaging III*, Proc. SPIE 1092 (1989).
8. Terzopoulos D., "Matching deformable models to images : direct and iterative solutions", in Topical Meeting on Machine Vision, Technical Digest Series, Vol.12, pp.160-167, (1987)
9. Terzopoulos D. et al., "Symmetry-seeking models and 3D object reconstruction", Int'l J. Comp. Vision, Vol.1, pp.211-221, (1987)

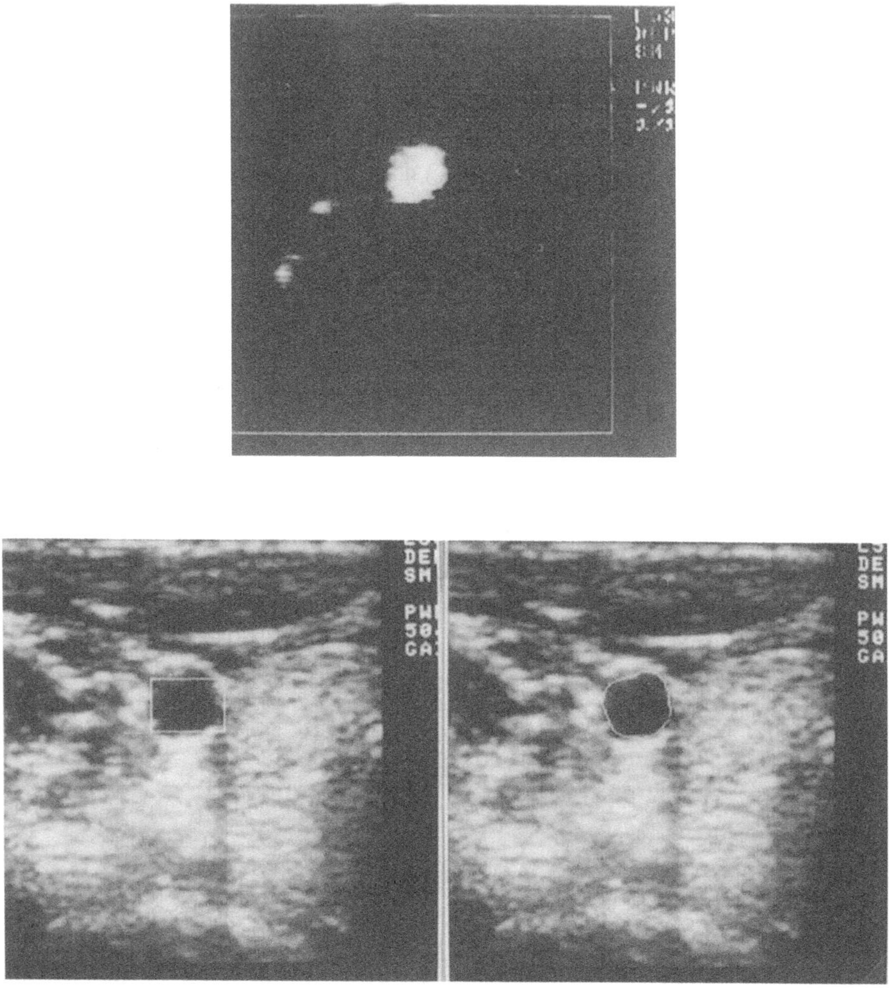

Figure 1: Delineation of blood vessel cross-sections on Doppler Ultrasound Images. *Top* : Colored Doppler blood flow image. *Downleft* : starting contour, derived from color info. *Downright* : final contour, using deformable models.

Comparison Between Several Approaches of Vascular Network Reconstruction from a Limited Number of Projections

Christian HAMON * ‡, Christian ROUX*, René COLLOREC ‡, Jean-Louis COATRIEUX ‡

* Groupe Traitement d'Images. Ecole Nationale Supérieure des Télécommunications de Bretagne.
B.P. 832 - 29285 BREST CEDEX - FRANCE

‡ Signaux et Images en Médecine. Université de Rennes I
35042 RENNES CEDEX - FRANCE

Summary

This paper presents several approaches to tomographic vascular network reconstruction from a limited number of projections. Three methods are tried in 2D parallel geometry : filtered-backprojection method, subtractive reconstruction, detection of objects via maximum likelihood. Then another procedure in 3D is proposed for vascular network reconstruction from a dozen conical projections : simulation results show a reduction in reconstruction artifacts.

I - Introduction

The constraints on taking measurements during angiographic examinations, especially the continuous flow of blood (made opaque for X-rays by a contrast medium) do not enable more than a few tens of views of the vascular network to be obtained by X-ray conic projection. These constraints will always be present in the new generation of 3D machines [1] which will supplant the 2D scanner and the current digital angiographic devices. In order to determine a realistic, efficient approach to 3D v.n.r. (vascular network reconstruction), we first tried in 2D (in parallel geometry) the performances of several methods already proposed in the literature i.e. filtered back-projection method (f-b-p), substractive reconstruction (s-r) [3], detection of vessels via maximum likelihood (d-v-m-l) [4].

These algorithms were chosen mainly for their low computational cost, taking into account the objective of the reconstruction of a volume of 512 x 512 x 512 voxels. They have in common an analytical approach to the problem, unlike the iterative algebraic methods [5] which we dismissed for the above-mentioned reason. For this test we have defined a phantom comprised of 4 disks of different diameter : the methods proposed do not reconstruct the disk with the smallest diameter. We then show that the detection of the disks at each measurement projection enables the reconstruction by f-b-p to be improved, especially for the disks of small diameter. Finally, we applied this approach in 3D, using conical geometry : the detection of vessels is carried out by a robust method of segmentation associated with minimal reconstruction [8] [9], then the 3D conical reconstruction estimates the value of the density function inside the reconstructed support of the object.

II - 2D Reconstruction from a limited number of projections

II.1 - Reconstruction by Filtered Backprojection (f-b-p)

We have from one to several tens 3D views in real experimental conditions, but first we thought it of interest to compare several reconstruction methods for only 4 different projection directions (0, 45, 90, 180 degrees), in order to put us in a configuration similar to that already used to test the algorithms presented. The phantom of 64 x 64 pixels (figure 1.a) is made up of 4 disks of 3 to 9 pixels in diameter. The f-b-p reconstruction of the measurement projections from the 4 simulated projections, allow large streak artifacts to be shown (figure 1.b) : the intersection of the back projection streaks can have a greater density than that of the smallest disk. The negative reconstruction values at the boundaries of the high density objects are due to the increased contrast imposed by the high-pass filtering of the projections [2].

II.2 - Subtractive reconstruction (s-r)

This method [3] enables 3D vascular reconstruction from a few angiographic projections to be done. s-r isolates areas of great intensity both better and more rapidly when these areas are smaller than those of low density. This is an iterative procedure applied to an original volume constituted from the f-b-p reconstruction of the measurement projections (these being few in number, this first reconstruction is contaminated by back projection artifacts). The following process is applied to each 2D plane, then these are "piled up" one upon the other to create the 3D reconstruction volume :

- initialization of the final image,
- computation of the first image by f-b-p measurement projections (parallel geometry),
- localization of the maximum P of the image,
- if P is inferior to the value τ, which represents a criterion of convergence, the reconstruction stops. If not, the processing continues,
- the contribution of a point of amplitude αP is subtracted from the analytic reconstruction : this contribution is obtained by projecting point αP according to measurement geometry, then by reconstructing an image from the projections thus obtained,
- a point of amplitude αP, is added to the final image, with the same coordinates as the maximum P,
- the process starts again with the localization of a new maximum.

The computation of the contribution of point αP is equivalent to the calculation of the impulse response of the whole chain comprising the operation of f-b-p reconstruction. The impulse response obtained in parallel geometry from the 4 projections previously defined can be seen in figure 2.a. The subtractive reconstruction from 4 simulated projections, using a value of 0.5 for the stopping process criterion τ, and the same value for the convergence speed α, can be seen in figure 2.b : "undesirable" points appear in the final reconstruction image whilst the smallest vessel is not reconstructed. Several causes of error could explain this :
- the impulse response contains a negative lobe which does not favour the convergence of the process : the subtraction operation of the analytical reconstruction results locally in an increase in the density function.
- the elimination of the contribution of a point not being complete, some of the streaks of this point will remain : the density function of vessels of small diameter could be inferior to this backprojection residue, in which case these vessels will never be detected.

In conclusion, subtractive reconstruction is not very well suited to the reconstruction of vessels of small diameter when the phantom image contains vessels of greater diameter, because s-r favours the maxima of the density function.

II.3 - Detection of vessels by Maximum Likelihood (m-l)

Another approach to the reconstruction of an image from these projections begins with the following consideration : the image is made up of a background plus objects whose characteristics we know. The projections therefore enable the parameters describing the object to be quantified : its position, its dimensions. The image is given by

$$f(x) = f_b(x) + d.\, f_0(x - c, \text{gamma}) \qquad (1)$$

$f_b(x)$ represents the background of the image, f_0 is an object located in the image, at position x - c, having the gamma geometric characteristics (size, shape, orientation).
Modelization of the object : in 3D, the vascular network can be modelled as a sum of cylinders of fixed height and of variable diameter and orientation.
Reconstruction of the object : the geometric features of the objects are detected and the corresponding parameters are estimated by m-l. [4].

The likelihood of the measurement Y for a disk with centre C and diameter Γ is given by the conditional probability P [Y/C, Γ]. The function of likelihood is given by :

$$L(C, \Gamma) = -\log(P[Y / C, \Gamma]) \qquad (2)$$

It can be shown that, if the projection $Y(\sigma,\theta)$ of the disk is given by

$$Y(\sigma, \theta) = g_0(\sigma - M(\theta)C; \theta; \Gamma) + v(\sigma, \theta) \qquad (3)$$

where $M(\theta)$ is the eccentricity function and $v(\sigma,\theta)$ is a Gaussian noise of variance N_0, then

$$L(C, \Gamma) = \frac{2}{N_0} \sum_{\theta} \int_{-\infty}^{+\infty} Y(\sigma, \theta) g_0(\sigma - M(\theta)C; \theta; \Gamma) d\sigma - \frac{1}{N_0} \sum_{\theta} \int_{-\infty}^{+\infty} g_0(\sigma; \theta; \Gamma) d\sigma \qquad (4)$$

The first term in (4) enables the most likely position of the disk to be determined; the second term (a normalization one) enables a choice of disk diameter detected among the diameters proposed.
The dimension of the disk to be searched for is chosen among the 4 disk diameters of the phantom image.The image is scanned in order to find the most likely position of each of the 4 disks. Figure 3 gives the image obtained for each reference circle from the greatest one (3.a) to the smallest one (3.d). It is clear in this figure that detection by m-l is a sub-optimal method for seeking several objects [6]. In the particular

case where the reference circle is replaced by a Dirac function, the first part of the expression of L (C, Γ) in (4) corresponds to the backprojection of the measures without high-pass filtering, therefore with a reduction in contrast.
The detection of the maximum due to the smallest disk is made more difficult than in the f-b-p method, for the following reasons :
- processing in order to find m-1 uses low-pass filtering of the measurement projections and therefore reduces the contrast in the "likelihood" image for each reference disk.
-the streak artifacts in the "likelihood images" create numerous maxima which do not correspond to a real object (sub-optimality of this processing for several disks present in the phantom image) (figure 3 d).
It therefore does not seem realistic to use this method to reconstruct the vascular network, even if associated with a system eliminating false alarms [6].

II.4 - Reconstruction of the object support

The noise generated by reconstruction from a small number of projections is, as we have seen, very great. It is higher than the measurement noise of a value generally admitted to be around 20 dB S/N in real experimental conditions. We therefore propose a method based on the detection of the disks on each projection. The comparison between the positions estimated on each view of the projection of an object will testify whether a disk is present in the reconstructed image. By giving a minimum diameter to the reconstructed disk, a mask in thus formed (figure 4.a). This mask allows a good reduction in the back-projection streaks on the f-b-p reconstruction (figure 4.b).

III - 3D EXTENSION OF TOMOGRAPHIC RECONSTRUCTION OF THE SUPPORT SURROUNDING THE VASCULAR NETWORK

The 3D method of reconstruction of a vascular network from a low number of projections comprises 2 steps.(i)The reconstruction of the support of the vascular network, then (ii)the estimation of the density function within this support [7].

III.1 - Reconstruction of the support of the vascular network

There are 3 steps in the reconstruction of the support of the object.
a) The extraction of vessel contours on the measurement projections by p.c.a (principal component analysis) of the density function of the projections on small surfaces. The greatest eigenvalue Λ of the measurement matrix measures the data inertia in relation to the first axis of factorial analysis, and therefore reflects the variation in the density function within the window. This value is directly determined from the pixel gray levels in the window of size n x n [8] :

$$\Lambda = (\alpha + \delta) + \sqrt{(\alpha - \delta)^2 + 4\left(\beta^2 + \sigma^2\right)} \qquad (5)$$

$$\alpha = n.\sum_{x=-a}^{a} x^2 \ , \ \beta = \sum_{x,y=-a}^{a} x.\,Z(x,y) \ , \ \sigma = \sum_{x,y=-a}^{a} y.\,Z(x,y) \ , \ \delta = \sum_{x,y=-a}^{a} Z^2(x,y) \ , \ a = \frac{n-1}{2}$$

where x,y represent the centered coordinates of the pixels in the window, and Z(x,y) the reduced density function of the pixel of the x and y coordinates.
b) Filling the space bounded by the contours of the vessels, by a morphological opening from a structuring element made up of a segment of d points, d representing the maximum diameter of a vessel.
c) Minimal reconstruction [9] from segmented projections : each point in the volume is given the minimal value taken by the projection of this point on each view. Thus the support incorporating the vascular network is obtained.

III.2 - Analytical reconstruction of the vascular network

We use an extension of the algorithm for 2D tomographic reconstruction by f-b-p filtering in conical geometry [10]. This extension corresponds to an approximation of the contribution of the projection points located outside the rotation plane of the X-ray source, the processing being only realistic for a small opening of the cone beam.

III.3 - Results

In 3D, simulation of the projections is done from the digital projection of a vascular network, automatically generated in a volume of 64 x 64 x 64 voxels (figure 5.a) [1]. The analytical reconstruction of the volume from 13 noisy conical projections (S/N at 20 dB), spread over 180 degrees, shows streaks of back-projection (figure 5.b). The determining of the support from measurement projections enables these reconstruction artifacts to be reduced (figure 5.c). The first step in detecting the support is no longer worthwhile when the number of projections is superior to several dozen views : in this case the whole volume can be reconstructed. Figure 5.d shows a reconstruction from 33 measured projections of an aluminium phantom obtained on the GE/CGR 3D X-ray experimental system.

IV - CONCLUSION

A preliminary approach to the problem of the reconstruction of a vascular network is proposed. This comprises 2 steps : (i) detection of the support of the object and (ii) reconstruction of the density function of the object within this support. It enables a reduction in streaks artifacts due to angular undersampling to be made, for a relatively low computation cost. It can be pointed out that the first step in detecting the support is not worthwhile when the number of projections is superior to several dozen views. The validation and the implementation of this algorithm in real measuring conditions, using 3D apparatus, is yet to be undertaken.

ACKNOWLEDGMENTS
The help of Janet ORMROD in preparing this manuscript is gratefully acknowledged.

REFERENCES

[1] A. ROUGEE, K.M. HANSON, D. SAINT-FELIX : "Comparison of 3-D tomographic algorithms for vascular reconstruction", Proc. Medical Imaging, SPIE, Newport Beach,California, 1988.
[2] C. ROUX : Contribution à l'étude d'un système d'imagerie cardiaque en tomographie axiale transverse par rayons X, Thèse de troisième cycle, Institut National Polytechnique de Grenoble (France), Dec. 1980.
[3] G.M. MAWKO, T.M. PETERS : "Iterative 3-D reconstruction of vascular images form few views : phantom study results", International Workshop on Phys. and Eng. of Comp. Mult. Imaging and Processing, Newport Beach, California, 1986.
[4] D.J. ROSSI, A.S. WILLSKY : "Reconstruction from projections based on detection and estimation of objects", IEEE ASSP, 1983.
[5] R. GORDON, R. BENDER, G.T. HERMAN, "Algebraic reconstruction technique for three dimensional electron microscopy and X-ray photography", J. Theo. Biol., 29., 1970.
[6] Y. BRESLER, A. MACOVSKI, "A hierarchical bayesian approach to reconstruction from projections of a multiple object", Proc. 7th ICPR, Montréal, 1984.
[7] C. HAMON, C. ROUX, J.L. COATRIEUX, R. COLLOREC : "Tomographic reconstruction of a vascular tree from few projections", Proc. Conf. IEEE EMBS, New Orleans, 1988.
[8] A. BENOIT, "Analyse et segmentation d'images angiographiques", Internal report, SIM, Université de Rennes I, 1986.
[9] R.A KRUGER, D.R. HEINECKE, S.W. SMITH, R. NING : "Reconstruction of blood vessels from X-ray substraction projections : limited angle geometry", Med. Phys. 14, 1987.
[10] L.A. FELDKAMP, L.C DAVIS, J.W. KRESS : "Practical cone-beam algorithm, J. Opt. Soc. Am., 1(6), 612-619, 1984.

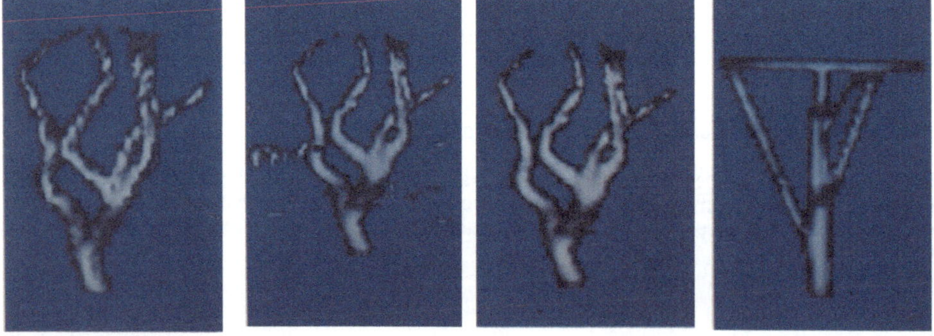

Figure 5. (a) (b) (c) (d)

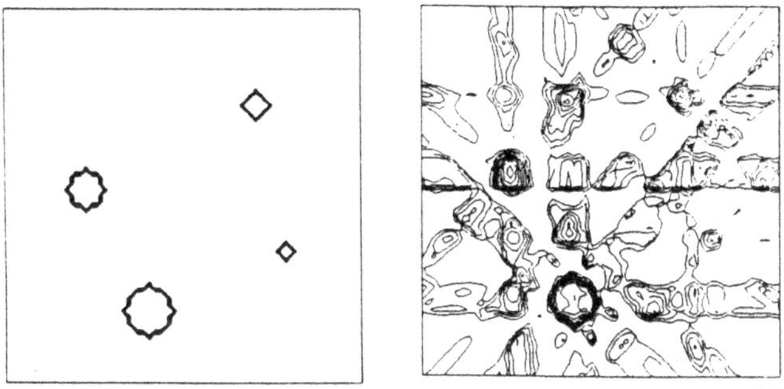

Figure 1. (a) (b)

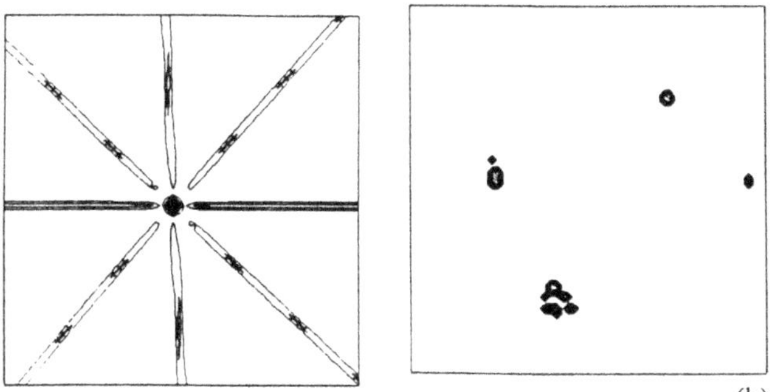

Figure 2. (a) (b)

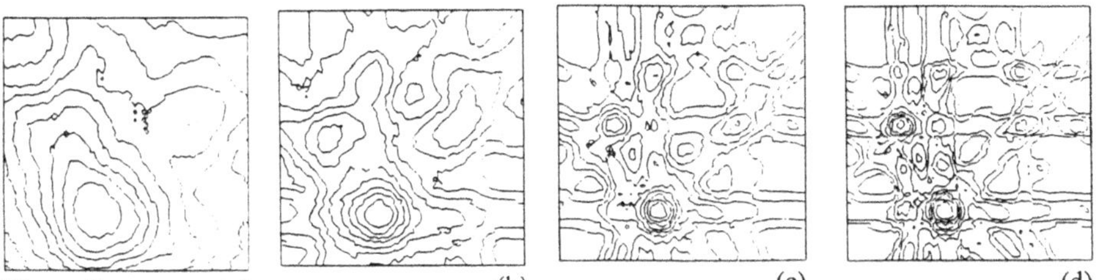

Figure 3. (a) (b) (c) (d)

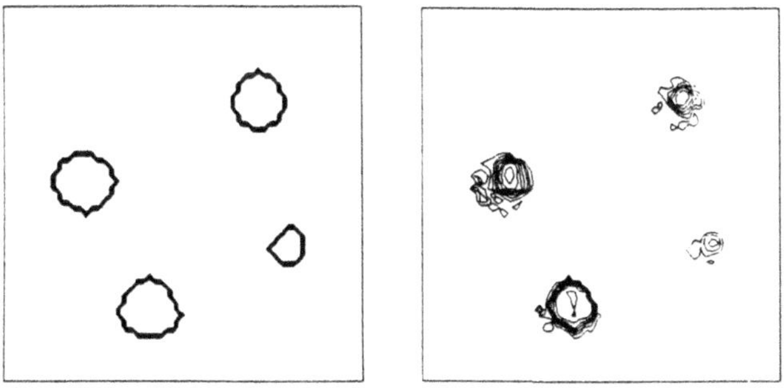

Figure 4. (a) (b)

The Determination of Object Coordinate Systems for Longitudinal Three-Dimensional Quantitative Image Processing

P. Steiger, C.C. Glueer, S. Steiger and H.K. Genant

University of California
Department of Radiology
San Francisco, CA 94143, USA

Summary

High reproducibility and exact repositioning of volumes of interest are primary concerns in quantitative medical image evaluation. We are using quantitative computed tomography (QCT) for the three-dimensional in-vivo assessment of bone density in the proximal femur and lumbar spine. Rather than using computationally intensive image correlation techniques based on the gray-levels of volume elements, we decided to use algorithms that determine object dependent coordinate systems. With our method, the object coordinate system (OCS) is determined at base line and follow-up, and the equations to transform the follow-up measurement into the base line coordinate system are computed. The transformed follow-up image is then used for the longitudinal evaluation of arbitrary sub-volumes of the object as defined at base line. This article focuses on the methods used to define object coordinate systems for the lumbar spine and proximal femur. Evaluation of our object coordinate systems on in-vivo and in-vitro data suggests that residual repositioning errors are in the order of 1mm, based on image acquisition with 3mm slice spacing.

Introduction

Traditional techniques for measuring bone mineral density by QCT employ single, 8-10mm thick slices through the center of the vertebral body or the femoral neck. Regions of Interest (ROIs) are then typically placed over areas containing trabecular bone and evaluated longitudinally in studies spaced 6 to 12 months apart. Specifying these ROIs reproducibly is of major concern in such studies. Automated techniques [1,2] reduce operator dependence and increase reproducibility. However, single slice techniques remain subject to errors induced by slice positioning and allow the assessment of only a limited number of in the bone compartments. We propose a volumetric technique in which the entire object of interest is imaged with contiguous low-dose (300 mrem), 3mm thick CT slices, resulting in a stack of slices aligned with the scanner coordinate system. We use a coordinate system tied to the object of interest, in our case the vertebral body or the proximal femur, to correct for repositioning errors within the scanner coordinate system. The object coordinate system (OCS) can then be used as a reference for arbitrarily defined volumes of interest or can serve as the basis for the computation of the equations to transform the follow-up image into the baseline coordinate system. The images used in this study were generated on a GE 9800 scanner (80 kVp, 140 mAs, 3 mm contiguous slices).

Object Coordinate System for the Spine

Figure 1 illustrates the vertebral OCS. It is determined as follows:

1. The operator specifies the location of true sagittal and coronal planes by placing boxes over the anterior vertebral body of axial slices located at the inferior and superior end-plates (Figure 2). The equations of the planes are given by the axes of the superior box and the center of the inferior box. True coronal and axial images are then reformatted. Generally, the planes of the true coronal and axial images are not aligned with any of the scanner axes.
2. The operator specifies the location of the inferior and superior end-plates on the sagittal and coronal reformations (Figure 3). Using a box rather than a line to do so optimizes the fitting of a line through the curved shape of the end-plates. Two boxes are generated for each end-plate. The mean of the centers defines a point in the end-plate plane and the axes of the boxes define the vectors spanning the plane. The mean of the surface normals of the two end-plates defines the z-unit vector of the OCS.
3. A true mid-vertebral slice, centered between the end-plates and perpendicular to the z-unit vector, is reconstructed. The symmetry axis of the vertebra in that plane defines the y unit vector; the cross product of z- and y-unit vectors yields the x-unit vector of the right-handed coordinate system. The origin of the OCS is the centroid of the spinal canal on the true mid-vertebral slice.

Object Coordinate System for the Proximal Femur

Figure 4 illustrates the OCS for the proximal femur. It is determined as follows:

1. The operator specifies on all axial slices that include the femoral head a number of points on the cortex. A total of approximately 100 points are picked in this fashion. From these, multiple quartets are formed by randomly picking four points. Each quartet

uniquely defines a sphere with a center and radius associated with it. A statistical analysis of all centers and radii yields the coordinates of the center of the femoral head, which is the origin O of the coordinate system.

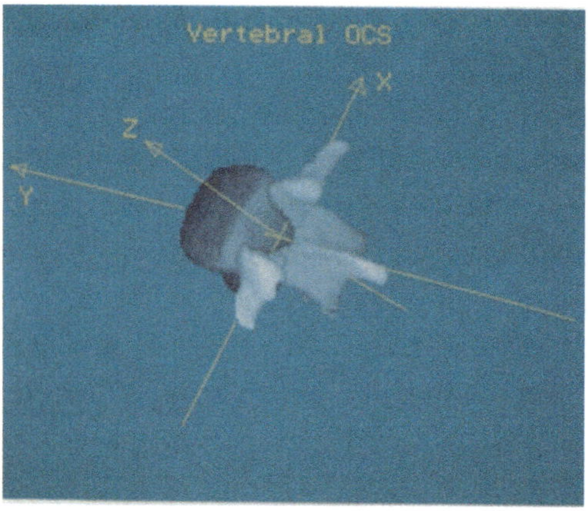

Figure 1. The vertebral object coordinate system. The x- and y-unit vectors lie in the midvertebral plane, the origin is the centroid of the spinal canal and the z-unit vector points in a cranial direction.

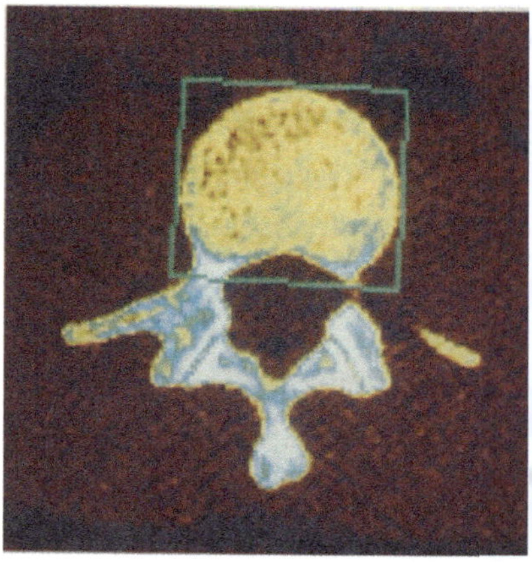

Figure 2. Using a box to specify the orientation of true coronal and sagittal planes. Using a box rather than lines improves reproducibility.

2. The femoral shaft axis is determined based on three CT slices taken more distally. The bone contours are automatically tracked and the centroids computed. The least squares approximation through these three points yield the equation of the shaft axis. The z-unit vector of the OCS is aligned with the shaft axis and points in a cranial direction.
3. Point S is defined as the point on the shaft axis closest to O. The x-unit vector is defined by $\overline{OS}$. The y-unit vector is defined by the cross product of the x and y unit vectors.

Results

To determine intra-observer dependence, the determination of the OCS of a lumbar vertebra was repeated six times by one operator. We found that the standard deviation of the OCS origin location was 0.44 mm, and the standard deviation of the orientation was 0.83 degrees. When four different operators assessed the OCS of the same lumbar vertebra we found that the inter-observer variation of center and orientation were 0.76 mm and 1.08 degrees, respectively. The procedure was performed independently and blinded by all operators after a brief training session that involved the evaluation of another vertebra. Performing the procedure takes a trained operator approximately five minutes.

The hip OCS was tested on in-vivo patient data. Two sets of hip scans were obtained from the same individual with interim repositioning. The variation of OS showed a residual spatial repositioning error of less than 1mm.

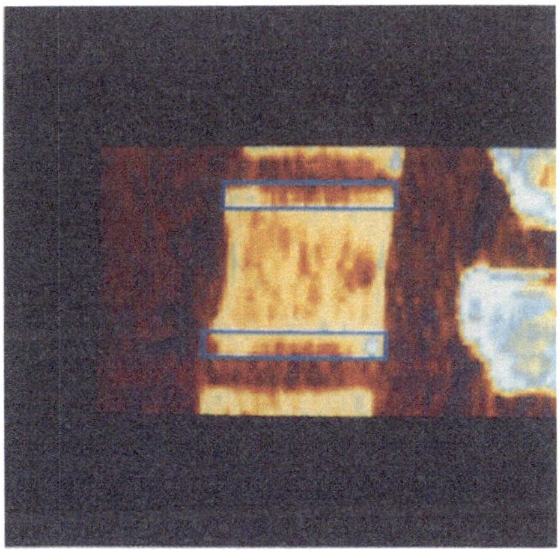

Figure 3. Using a box to specify inferior and superior end-plates. The box is placed such that the end-plate is tangential to it centrally and peripherally.

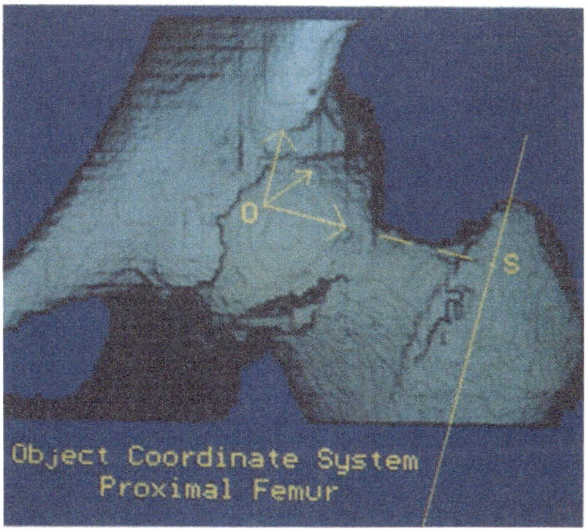

Figure 4. The femur object coordinate system. The origin is located in the femoral head. Unit vectors are determined by the orientation of the shaft axis and the location of point S (see text).

Discussion

We have developed simple techniques to determine object coordinate systems for lumbar vertebrae and proximal femur. Although they still require considerable operator interaction, our methods were designed to minimize observer impact. Furthermore, many of the tasks currently performed manually lend themselves to automation. Our results suggest that object coordinate systems can be computed with minimal inter- and intra-observer variation and little computational effort. Object coordinate systems thus

provide the basis for precise longitudinal and volumetric quantitative image evaluation.

References

1. Kalender WA, Klotz E, Suess C: Vertebral bone mineral analysis: An integrated approach with CT. Radiology 1987;164:419-423.
2. Steiger P: Computer graphics for highly reproducible quantitative medical image processing. Proceedings of the ninth annual conference of the National Computer Graphics Association, Anaheim, CA, March 20-24, 1988.

Application of Image Processing Techniques to the Determination of the Insertion Point for Intramedullary Nailing of the Tibia

M. Köppel and A. Wallin

M. E. M. Institute for Biomechanics
University of Berne, Switzerland

Summary

CT-Scanning of 5 human cadaver tibiae was performed. Using image processing and three-dimensional fitting techniques a new insertion point for a recently developed tibial intramedullary interlocking nail was determined.

Introduction

Intramedullary nailing is a common procedure in the treatment of fractures of the tibia shaft. In this technique of osteosynthesis a metal rod is inserted into the medullary cavity. Biomechanically this procedure can be regarded as internal splinting of the fractured bone. For the treatment of fractures with multiple fragments the interlocking intramedullary nail was developed. Interlocking a nail means passing a screw or bolt in a horizontal direction trough the bone and the nail in the main proximal and distal fragment, allowing to initially transmit all loads through the nail. Possible complications of intramedullary nailing are axial and rotational malalignments. They are either due to incorrect fracture reduction during the operation or to rotation and/or deformation of the nail during implantation. The key parameters responsible for these complications are the geometric misfit between the nail and the medullary cavity and the localisation of the point of nail insertion.
Based on a morphological study a new interlocking tibial nail with better correspondance to the medullary cavity has been developed. It consits in essence of two straight segments with an angle of 11 degrees between them. The intersection of the two segments is at 33% of the nail length. It is intended that after implantation this intersection is situated at about

33% of the length, measured from the tibial plateau [1]. With the material properties and geometry of the nail given, the localisation of the point of nail insertion in the proximal tibia is the only parameter which can be varied.
The aim of this study was to determine the localisation of the insertion point of the newly developed interlocking intramedullary nail by means of three-dimensional image processing techniques based on CT data from human cadaver tibiae.

Material and Method

Based on the quantitative evaluation of CT cross sectional images acquired from cadaver tibiae the shapes of the medullary cavities can be determined. Knowing the interior geometry of the tibiae, and the geometry of the intramedullary nail, it is possible to determine its optimal position after implanatation. This optimality is achieved when the nail is inserted in a plane inducing the smallest rotation and deformation on the nail. It is assumed that this plane is defined by the largest cross sectional diameter of the proximal part of the cavity in addition to the axis described by the distal part.
Transverse sections of 5 fresh human cadaver tibiae with fibulae were aquired by means of a GE 9800 CT scanner. An average of 44 scans were taken from every bone: at intervals of 1% of the tibia length in the proximal 16%, at 2% from 16 to 50% and thereafter at steps of 5% of the tibia length. All bones were scanned with identical scan parameters. Scanning of the bones was performed in a special device to improve accuracy of geometric measurements. For area and density calibration, a calibration procedure was used developed by Zimmerli [2].
The CT data were transferred to a VAX 11/780 computer (Digital Equipment Corporation, Maynard, Massachusetts) for processing. The CT scans from each tibia were divided into two separate groups depending on their distances from the proximal end of the tibia. The proximal group contained all scans down to 33% of the length of the tibia, and the distal group contained the rest. The density threshold value to discriminate between

cortical and cancellous bone was determined by the calibration procedure. This value was then used as an upper threshold for each image since the cavity and not the cortical bone was of interest. In order to avoid the influence of pixels outside the bone, an area of interest (of arbitrarily complex shape) was manually drawn on each image (Fig. 1). The pixels within this area and with density values below the threshold were used to compute the centroid as well as the principal axes for the images in the proximal group.

The centroids of the distal group were fitted to a line utilizing a linear regression technique. The projections of the centroids on the x-z plane and on the y-z plane were used respectively to find the constants for the parameterized line:

$$x = a_1 + b_1 * z$$
$$y = a_2 + b_2 * z.$$

This line defines a vector and a point laying in the plane of insertion. The point is positioned at 33% from the proximal end of the tibia and is the point were the bend in the nail should be. By averaging the unit vectors describing the direction of the main principal axes for each of the scans in the proximal group, a second vector to describe the plane of insertion is computed.

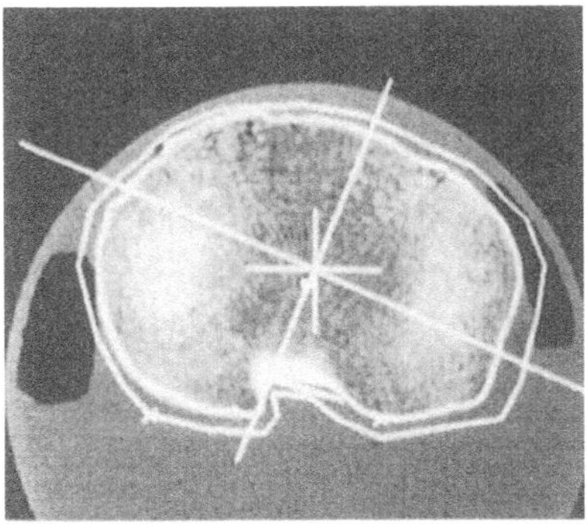

Fig. 1

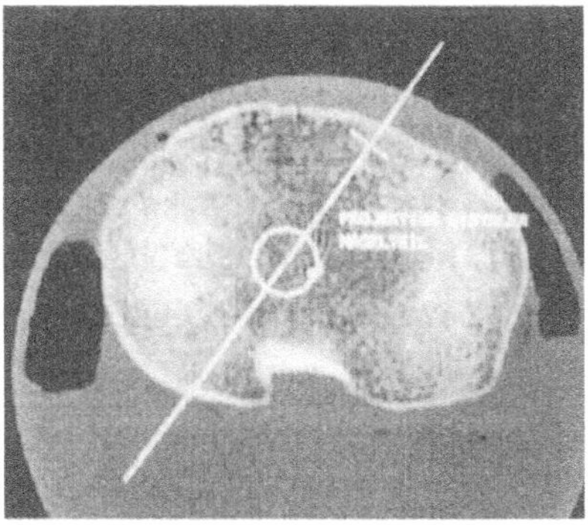

Fig. 2

Fig. 1: Cross-section of a tibia with centroid principal axes and circumferential line for thresholding.

Fig. 2: Cross section with plane of nail implantation, nail insertion point and distal nail axis.

Using the equation of the plane of the nail

$$A*x + B*y + C*z + D = 0$$

and the planes of the scans (z = constant), the intersecting lines between these planes were calculated:

$$y = (-A*x - C*const - D) / B$$

By knowing the bend of the nail (11 degrees), the intersecting ellipses between the nail and the scan, planes were computed as follows. The vector describing the center of the ellipses was solved by the following equations:

$$(v|v_1) = \cos(11)$$

$$(v|nv) = 0$$

$$(v*v_1) = \sin(11)*nv$$

With v being the vector to solve for, v1 the vector computed by the the centroids, and nv the normal vector to the plane. (n|nv) is the dot product, and (v*v1) is the vector product. The equation system has four solutions, with the one corresponding to a proximally-frontally inserted nail being the solution used.

The actual insertion point, laying on the surface of the cortical bone of the tibia, was determined by inspecting the CT images with graphical overlays of the intersecting lines and ellipses (Fig. 2).

Results

Examination of the five human tibiae revealed a new nail insertion point (Fig. 3) on the antero-lateral surface of the bone at 2.3 +/- 0.2% of the bone length distal to the articular surface of the tibial plateau. It is at a distance of 30.3 +/- 2.3% from the tibial plateau width medially to the lateral border of the bone in a plane rotated 35 +/- 5 degrees antero-laterally to the sagittal plane.

Discussion

Image processing techniques based on CT cross sections are well suited for orthopedic problem areas, where true three-dimensional geometry cannot be extracted from conventional radiographs. The CT data can be regarded as an approximate measure of the density of the material in the specimen and thus be used for quantitative examinations.

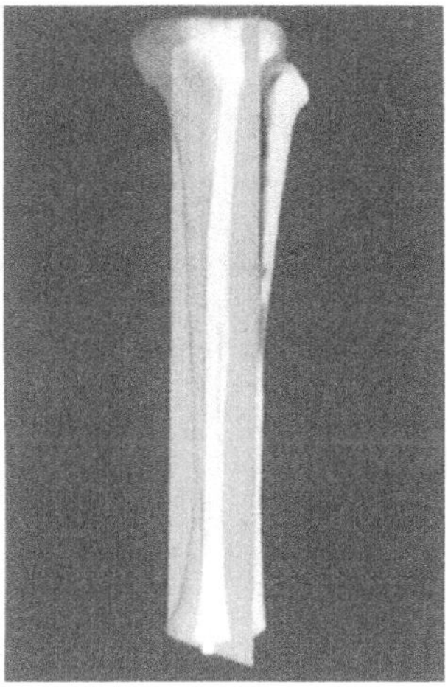

Fig. 3a

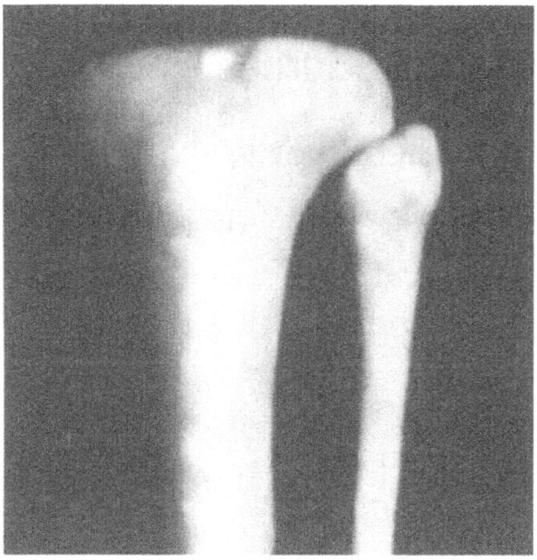

Fig. 3b

Fig. 3:a) 3-D-Image with implanted nail and plane of nail insertion
b) 3-D-Image with nail at the insertion point

Though having data from only five specimens the results suggest that the insertion point of the new tibial interlocking nail may be situated cranial and lateral to the traditional one. Due to the fact that the method used does not take into account the effect of reaming and the insertion process itself, test on a large sample of human cadaver tibiae are required to determine the practical conclusions of the present study.

References

1. Heini, P.; Eulenberger, J.; Schneider, E.: Determination of the inner shape of the tibia. Internal report 24-85, M.E.M.-Institute for Biomechanics, University of Bern, Switzerland. (1985)

2. Zimmerli, S.: Bestimmung der Knochenqualität der menschlichen Tibia mittels Computertomographie nach Entfernung von Osteosynthesenmaterial. Internal report 88, M.E.M.-Institute for Biomechanics, University of Bern, Switzerland. (1988)

Interactive 3D Registration of Image Volumes from Different Sources

C. SCHIERS, U. TIEDE, K.H. HÖHNE

Institute of Mathematics and Computer Science in Medicine (IMDM)
University Hospital Eppendorf
Hamburg, F. R. G.

Summary

Matching of different 3D image volumes is a problem that is encountered in several applications of medical 3D imaging. Using interactively specified 3D landmarks on 3D surfaces, linear and non-linear polynomial matching of arbitrary image volumes is performed. This is shown with the example of MR and CT data.

1 Introduction

Matching of different 3D image volumes is a problem that is encountered in several applications of medical 3D imaging.

- 3D views that show the combined information of different modalities (e.g. CT and MR, MR and PET) can be produced if we match data of the same object acquired with different modalities
- Quantitative volumetric comparison can be achieved if we match data from the same object at different times (e.g. the volume and the morphology of a tumor)
- Simulation of surgery can be done if we match the image volume to a target model (e.g. in craniofacial surgery)

The problem of matching 3D data is not as trivial as it looks at first glance. In most cases the problem cannot be reduced to the 2D case by just matching the individual slices, because the slices are not in the same plane. So real 3D procedures are necessary.

Another problem concerns the matching criteria. We can on the one hand aim at completely automatic methods based on anatomical similarities [1]. Apart from the principal difficulties of pattern matching in the 3D case computation times become prohibitive. The other extreme is the manual specification of corresponding landmarks on the respective image volumes on the base of which an appropriate distortion can be computed.

The interactive specification of large numbers of landmarks is, of course, time consuming. Since a tool for manual interaction is necessary anyway, we have decided to start with interactive landmark specification, and then gradually proceed to semi-automatic and possibly to automatic matching. For the first experiments, we have restricted the matching problem to ridgid objects, so that only linear distortion (scaling, translation, and rotation) can occur.

2.1 Specification of Landmarks

In order to specify arbitrary 3D coordinates on a 2D screen, it is necessary to provide the third dimension. We chose to specify the landmarks on the projection of the outer skin surface of the objects, which can be easily obtained from CT and MR by simple thresholding. For this purpose, we have developed a tool for the interactive specification of 3D landmarks.

In the first step several projections from different points of view of the volumes are calculated to get a realistic impression of the object, which simplifies the detection of remarkable points. The images, the projection parameters, and the distance of the visible surfaces to the viewing plane are stored.

In the next step two arbitrary views are displayed. Now the user can specify the corresponding landmarks on the projections interactivly with a mouse (Fig. 1). Since projection parameters and distances to the surfaces are known, it is possible to determine the 3D coordinates of the specified landmarks.

Landmarks may be altered or deleted, when they seem to be dislocated. Since the projections have already been calculated, it is possible to turn each of the presented views into another direction. The already specified landmarks would turn then, too.

2.2 Mapping Function

Several methods have been described to fit linear and even higher dimensional curves through given points, e.g. polynomials, splines and B-splines [2], or Aikima interpolation [3]. We have decided to use polynomial interpolation and approximation because they are easy in handling. A disadvantage of polynomials of higher degree is that they start oscillating.

The mapping of a point $(x_{in} \mid y_{in} \mid z_{in})$ on $(x_{out} \mid y_{out} \mid z_{out})$ is given by a triple of tri-variate polynomial functions of degree n:

$$x_{out} = \mathcal{F}(x_{in}, y_{in}, z_{in}) = \sum_{k=0}^{n} \sum_{j=0}^{(n-k)} \sum_{i=0}^{(n-k-j)} a_{[ijk]} x_{in}^{i} y_{in}^{j} z_{in}^{k}$$

$$y_{out} = \mathcal{G}(x_{in}, y_{in}, z_{in}) = \sum_{k=0}^{n} \sum_{j=0}^{(n-k)} \sum_{i=0}^{(n-k-j)} b_{[ijk]} x_{in}^{i} y_{in}^{j} z_{in}^{k}$$

$$z_{out} = \mathcal{H}(x_{in}, y_{in}, z_{in}) = \sum_{k=0}^{n} \sum_{j=0}^{(n-k)} \sum_{i=0}^{(n-k-j)} c_{[ijk]} x_{in}^{i} y_{in}^{j} z_{in}^{k}$$

Coefficients $a_{[ijk]}$, $b_{[ijk]}$, and $c_{[ijk]}$ are calculated from the corresponding landmarks of the form $[(x_i \mid y_i \mid z_i) \Leftrightarrow (x_o \mid y_o \mid z_o)]$, which preserve a least square error algorithm with the needed number of equations for solving the following linear equation systems:

$$x_o=\mathcal{F}(x_i,y_i,z_i)=a_{[000]}+a_{[100]}x_i+a_{[010]}y_i+\cdots+a_{[201]}x_i^2z_i+a_{[120]}x_iy_i^2+\cdots$$
$$y_o=\mathcal{G}(x_i,y_i,z_i)=b_{[000]}+b_{[100]}x_i+b_{[010]}y_i+\cdots+b_{[201]}x_i^2z_i+b_{[120]}x_iy_i^2+\cdots$$
$$z_o=\mathcal{H}(x_i,y_i,z_i)=c_{[000]}+c_{[100]}x_i+c_{[010]}y_i+\cdots+c_{[201]}x_i^2z_i+c_{[120]}x_iy_i^2+\cdots$$

The degree of the polynomials is depending on the image material. In many cases, there is only a linear distortion, due to different scaling factors, translation, or rotation of the image volumes. For linear tri-variate functions, there have to be at least four equations (and therefore four landmarks) to determine the four coefficients of the three polynomials. If, however, the distortion is of higher degree, e.g. when a patient's joint has to be matched in different positions, or when pre- and postoperative examinations of a tumor are to be matched, more landmarks are necessary.

3 Results

We have applied the described method to different data sets. The first one was an artificial 3D scene with known geometry, consisting of a cone intersecting with a pyramid. This object was linearly distorted and then used as a second input. Up to fourteen landmarks were defined on the projected 3D surfaces and the registration functions were calculated. In the case of an exact specification, the mapping function is the inverse of the distortion function. For several reasons, this is hardly to achieve, for example sampling errors may occur during surface projection or the user cannot determine the corresponding points exactly.

Fig. 2 shows a projected error image, comparing the original and the distorted-redistorted object volume, using the volume error function described by Pommert [4]. The maximum of the resulting error two voxels, which seems to be tolerable.

Secondly, we applied the method to volume data of a cadaver, consisting of a CT sequence (107 slices, 512^2 pixels) and a MR data set (128 slices, 256^2 pixels). Fig. 1 shows the interactive specification of the landmarks on the 3D skin surfaces of the CT and MR sequences respectivly. Although the skin surfaces vary strongly between CT and MR, the resulting linear registration function matches the data sets sufficiently well. Fig. 3 shows examples of 3D projections, where now different aspects from both data sets are combined into a single 3D view using the program VOXEL-MAN[5]. The skin surface is calculated from MR values and the bone surface is calculated from CT data, using gray level gradient shading [6,7], while the cut planes display the original MR and CT gray values.

4 Conclusion

We have shown that the interactive specification of 3D landmarks on 3D surface projections leads to sufficient results in case of ridgid objects. For non-linear distortion, where higher degree mapping functions are necessary, problems with oscillation occur, unless enough landmarks are specified, or if they are not well-distributed over the whole surface. However, the interactive specification of a large number of landmarks may become difficult. For that reason, our further developement will concentrate on the improvement

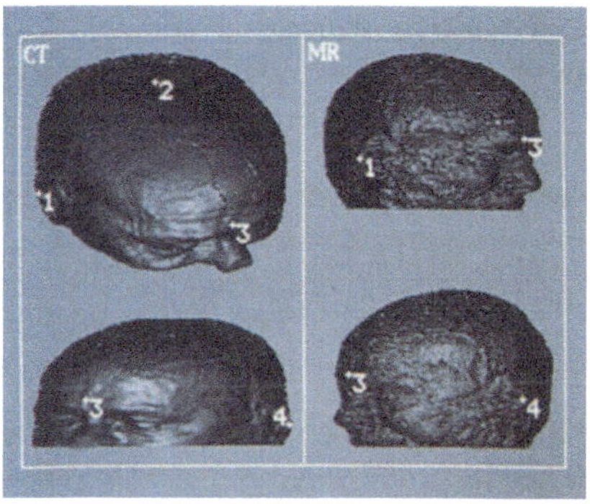

Fig. 1: Example for the interactive specification of landmarks, used for the registration of the CT and MR volumes of a cadaver.

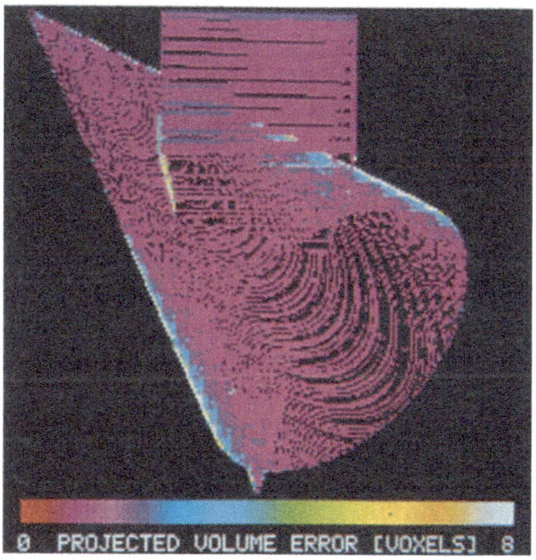

Fig. 2: Error image for the matching of a linearly dislocated artificial object.

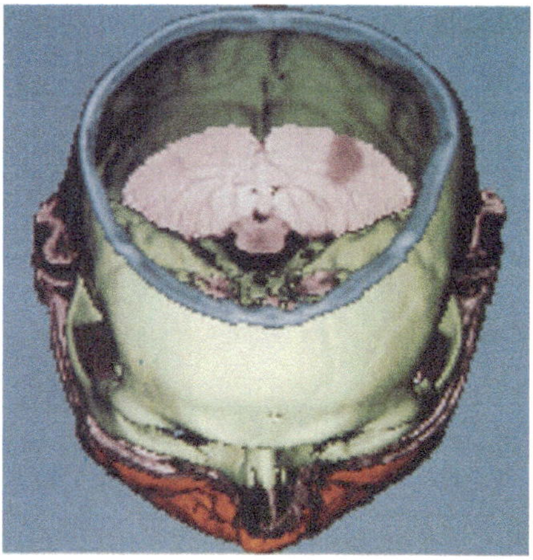

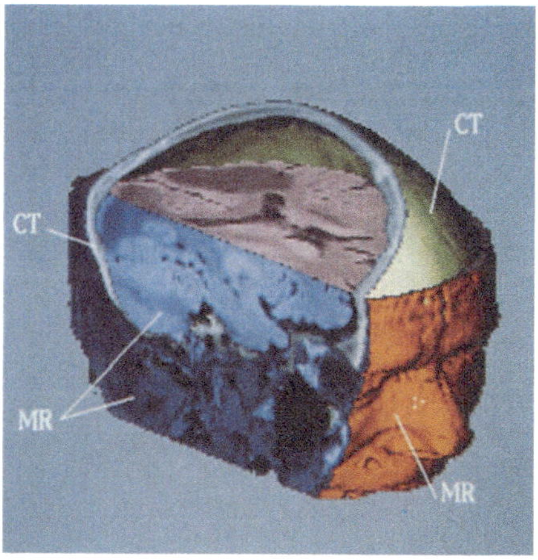

Fig. 3: Multimodality 3D views of a cadaver after registration, using linear mapping functions. Bone surface has been calculated from CT, skin and brain from MR.

of semi-automatic or automatic detection of remarkable points on 3D surfaces and on the problems, arising with higher degree distortion functions.

Acknowledgement

We are grateful to the Gertrud und Erich Roggenbuck-Stiftung for financial support and to the Department of Neuroanatomy, Medical University Hannover for providing the MR and CT data sets of the cadaver. Furthermore, we want to thank A. Pommert for helping us with his volume error function.

References

[1] Pelizzari, C.A.; Chen, G.T.Y: Registration of multiple diagnostic imaging scans using surface fitting. In: Bruinvis, I. et al.: The use of computers in radiation therapy. Elsevier, (1987)437–440.

[2] Newman, W.M..; Sproull, R.F.: Principles of interactive computer graphics. McGraw-Hill, (1984).

[3] Aikima, H.: A method of interpolation and smooth curve fitting, based on local procedures. Comm. ACM, 17(1984)18–20.

[4] Pommert, A.; Bomans, M.; Tiede, U. Höhne, K.H.: Simulation studies for quality assurance of 3D images from computed tomograms. In: Todd-Pokropek A.; Viergever, M.A. (eds.): The formation, handling and evaluation of medical images. NATO ASI Series F, Computer and system sciences, Springer Verlag, in press.

[5] Höhne K.H.; Bomans, M.; Pommert, A.; Riemer, M.; Schiers, C.; Tiede, U.; Wiebecke, G.: 3D visualisation of tomographic volume data using the generalized voxel-model. Visual Computer 5(1989), in press.

[6] Höhne, K.H.; Bernstein, R.: Shading 3D-images from CT using gray level gradients. IEEE Transactions on medical imaging 5(1986)45–47.

[7] Tiede, U.; Riemer, M.; Bomans, M.; Höhne, K.H.: Display techniques for 3D tomographic volume data. In: Proc. NCGA '88, Vol. III, Anaheim, (1988)188–197.

SUNRISE – A Software System for Medical Imaging Analysis

M.STAEMMLER, R.BRILL, K.BECKER, K.H.FOLKERTS, K.GERSONDE

Hauptabteilung Medizintechnik, Fraunhofer-Institut für zerstörungsfreie Prüfverfahren, D-6670 St. Ingbert, Ensheimerstr. 48 und Fachrichtung Medizintechnik der Universität des Saarlandes, D-6650 Homburg/Saar (Germany)

Summary

Modern health care employs different imaging modalities (CT, MR, Ultrasound (US)) for diagnostics. Typically, medical images are processed, for each modality separately, using a specific software and hardware environment. The objective of the software system SUNRISE is to provide an integrated environment for processing and evaluating the image data. SUNRISE is implemented as an additional layer on top of the UNIX operating system featuring a user interface, standard facilities for image processing, modality-dependent evaluations of functional states and an intrinsic database for image data and patient record handling.

Introduction

Medical images originating from different modalities (such as CT, MR, Ultrasound, Emission Tomography) are handled in similar ways, i.e., greyscale or colour display, image processing, object-oriented representation of organs, patient record and image data archiving. Differences are obvious when using algorithms for the evaluation of functional states (e.g., tissue characterization based on parameter-selective MR imaging[1], blood flow determination by Doppler measurements in US or XENON-CT[2]). These modality-dependent and common needs initiated the development of the Saarbrücken Ultrasound, Nuclear Magnetic Resonance and X-Ray Intelligent Software Environment (SUNRISE).

SUNRISE is implemented as an additional layer on top of the UNIX operating system (Fig. 1). This layer is installed during the start up of SUNRISE providing UNIX-like features, i.e., user login, password, user-specific environment set-up, access and permission control for running applications within SUNRISE. Each application results in a process within the UNIX system, thus allowing the exploitation of all UNIX facilities (batch jobs, shell programming, redirection, piping, protocol generation etc.) for the implementation of SUNRISE. Executing each application as a single process passes the effort for command parsing and command argument handling to the UNIX operating system. Even more, the organization of the applications in a tree-like structure can be directly mapped to the UNIX file system[3,4].

This concept is supported by strategies for the management of SUNRISE, for the user interface, for data handling and storage. These strategies are described as follows, beginning with a detailed view on the structure of SUNRISE.

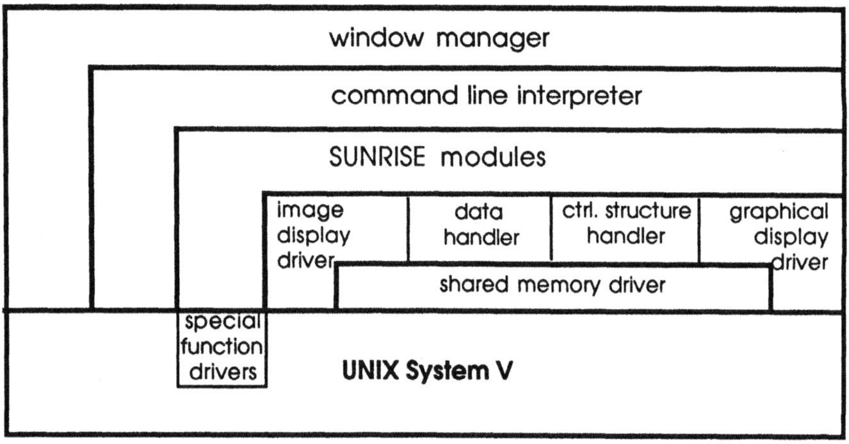

Figure 1: Implementation of SUNRISE as a layer on top of the UNIX operating system.

The Structure of SUNRISE

As shown in Fig. 2 the highest level of SUNRISE provides all means to start the software system, i.e., to provide login, password and access control for the users and to handle the user interface. The next lower level provides the selection of different imaging modalities. On the level of SUNRISE modules the module 'Image Display' is responsible for standard image display operations, whereas the module 'Parameter-Selective Image Display' provides all means for the generation and display of functional images. The module 'Parameter-Selective Analysis' contains modality-dependent analysis methods, which allow for example the calculation of relaxation times or blood flows . The incorporation of data from different imaging devices and the adaptation to a common data format within SUNRISE is performed by the module 'I/O Image Data'.

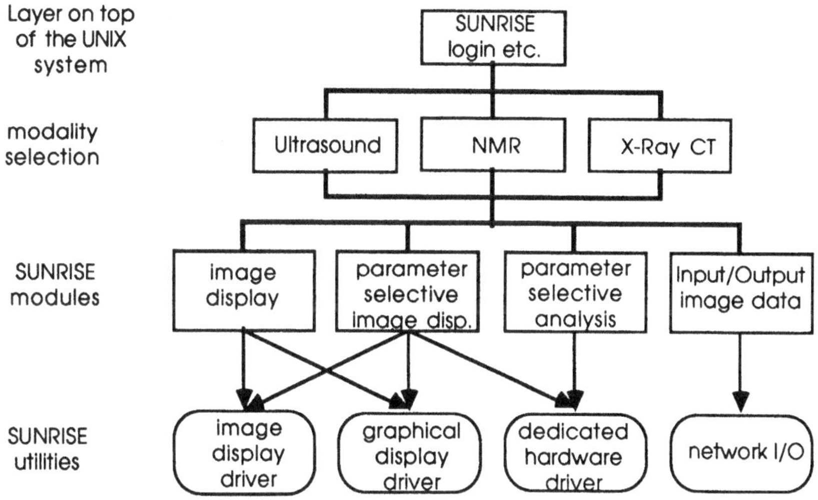

Figure 2: The structure of SUNRISE.

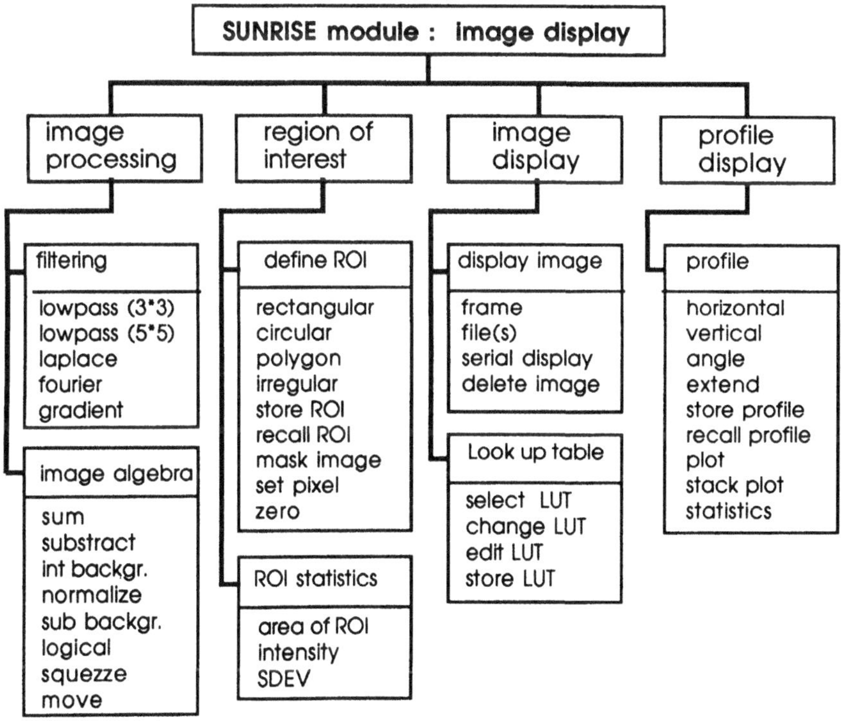

Figure 3: Tree-like structure within the 'Image Display' module.

Each module represents a tree-like structure of applications (Fig. 3). For example, the various applications within the module 'Image Display' include image processing routines, such as filtering and image algebra, region of interest definitions and statistics, image display options and look-up tables, image profile generation and display. Each of these routines has access (if necessary) to the utilities which are available on the level of SUNRISE utilities. These include all input and output routines and drivers for the devices (graphical, image display, LAN, dedicated hardware).

User Interface

Two levels of user interfaces exist in SUNRISE. The higher level is built as a window system. Characteristic for this type of environment are multiple windows, menu trees and sliders which can be manipulated using a trackball or a mouse. All commands received by this higher interface are translated into command strings and passed to the lower level user interface. This level may be directly accessed by the user to allow control of SUNRISE by means of a simple ASCII terminal. All commands have to be in form of strings. The syntax is almost the same as for the common UNIX shell to allow short teach-in times. For graphical or image output a graphic screen is needed in this configuration.

Management of SUNRISE

The continuous development of analysis and evaluation algorithms and the future aspects of multi-modality applications require a method for dynamic upgrading and extending the features of SUNRISE. Therefore, the information exchanged between different applications uses a fixed protocol based on predefined control structures. Thus, the execution of an application does not only require the data for processing, but also at least one control structure, which describes the origin of data and guides the processing step itself. For example, when displaying an image the image display process employs two control structures: (i) for specifying the kind of display (look up table, position on the screen, locations for displaying text and sliders, etc.) and (ii) for identifying the image (patient ID, processing steps, etc.). Furthermore, each user may define specific control structures to configurate the presentation to his own needs.

The generation and updating of the control structure definitions is supported by the control structure handler (Fig. 1). During reading of a control structure from the disk, the handler is also responsible for the updating of formerly stored control structures, which is automatically done at runtime without any user interaction. Thus, the effort to add a feature to SUNRISE consists of two steps: (i) Updating of the control structure definition with the handler and (ii) inserting the application for the new feature in the tree-like structure of the corresponding SUNRISE module. No conversion of 'old' data or control structures has to be performed by the user of SUNRISE.

Data Management

To allow short response times for the execution of an application it is important to reduce file input/output. Therefore, a dedicated structure of shared memories is available for each application within SUNRISE. This shared memory concept supports the complete access to data and control structures. It is supported by SUNRISE library functions building a software interface which is easy to handle for the programmer. Hence for example, an image display module may access data calculated by an analysis module avoiding time-consuming conversion and store/load processes. Only when switching to a completely new job, file input/output is needed to update control and data files and to load the new sets.

Hierarchy of Data Storage

The hierarchy of the data is based on six different levels giving a tree-like structure, which is applicable to all different imaging modalities (Fig. 4). This hierarchy is directly mapped to the UNIX file system. According to the concept of control structures, the data and the corresponding control structure are stored in two separate files in one directory on the appropriate level. For example, the raw data acquired in an examination (EX) of a patient are stored on the measurement acquisition (MA) level plus a control structure file, which describes the measurement procedure (adjustments of the imager, imaging method, etc.). The reconstruction of these data is guided by a control structure describing reconstruction methods and leads to processed measurement data (MP). In the case of CT, reconstructed images may directly be fed to the MP level. The analysis of the functional states is sufficiently specified by the control structure on the level of the analysis (AR). The evaluation of the result itself also uses a control structure for producing a parameter-selective image (PS), which may show a functional or an anatomical image. Each step of processing is documented by the control structures, thus providing a kind of 'intrinsic' database. This procedure is also used for patient records, which are stored on the level of examinations (EX).

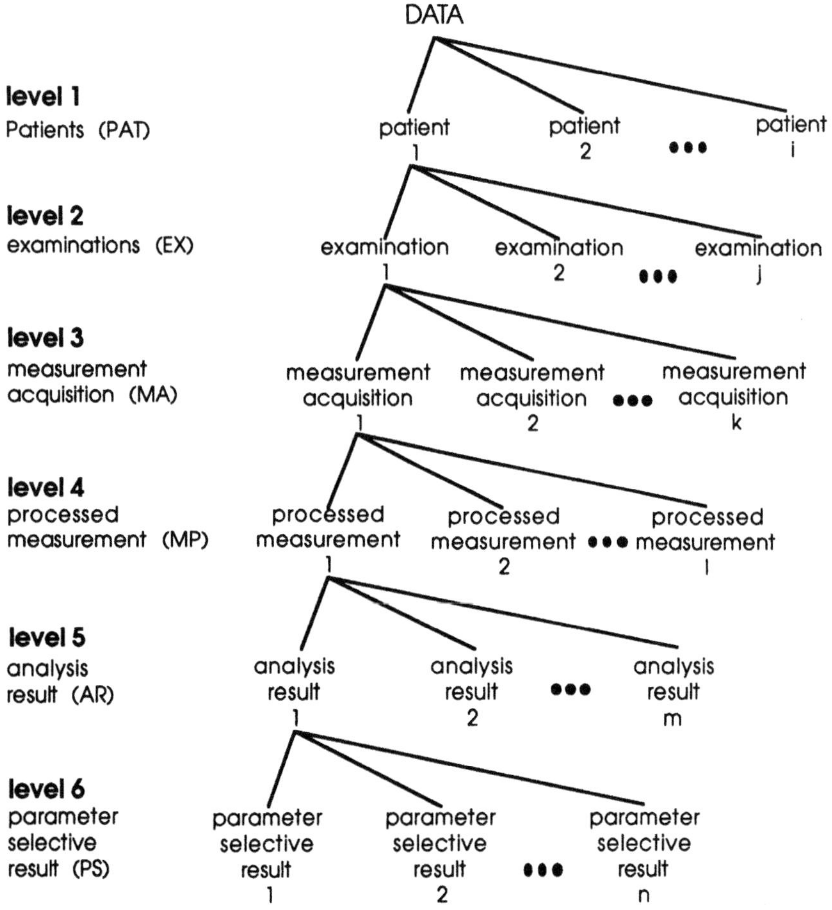

Figure 4: Hierarchy of Data Storage.

Transferability and Communication

SUNRISE can be transferred to UNIX Systems of different manufacturers. The hierarchical file structure of SUNRISE modules, sources and data files enables an easy installation, which is performed within six steps: (i) Copying source and data files to the new UNIX system, (ii) changing system-dependent paths in only one header file, (iii) recompiling SUNRISE, (iv) eventually adding routines to support special hardware, (v) installing users and (vi) running SUNRISE. Due to the modularity of SUNRISE dedicated coprocessors of a specific system (display, pipeline or array processors) can be integrated using the existing fixed interface definitions. Routines for these special devices are optional, as they only help to improve computing power. However, they do not add new features to SUNRISE.

SUNRISE may be run on different systems in a LAN environment. The system-dependent storage format of binary data files is specified by the control structures, thus allowing an automated conversion by the input/output utilities of SUNRISE.

Conclusion

The current state of implementation of SUNRISE comprises of all strategies building the SUNRISE kernel, whose functions are described. This kernel was for example applied to implement parameter-selective MR imaging analysis methods. A first experience confirmes that the SUNRISE concept is a flexible approach for medical imaging analysis.

References

1. K. Gersonde, T. Tolxdorff, L. Felsberg, Identification and Characterization of Tissues by T_2-selective Whole-Body Proton NMR Imaging, Magn. Reson. Med., 2, 390-401 (1985).

2. U. Hübner-Löffler, F.G. Kaup, F.J. Schuier, A.Spitz, Human cerebal blood flow measured by stable xenon, J CBF Metab. (1989).

3. B. Kernigham, R. Pike, "Der UNIX Werkzeugkasten", Hanser Verlag, München, (1987).

4. M.J. Bach, "The design of the UNIX Operating System", Prentice Hall International Editions, New Jersey, (1986).

Automated Identification of the Spine in Magnetic Resonance Images: A Reference Point for Automated Processing

P. E. SHILE*, M. P. CHWIALKOWSKI***, D. PFEIFER***, R. W. PARKEY*, and R. M. PESHOCK*,**

The University of Texas Southwestern Medical Center at Dallas, Department of Radiology* and Medicine**, and the University of Texas at Arlington, Department of Electrical Engineering***.

NMR Imaging Center
1311 Record Crossing
Dallas, Texas 75235 USA
214-630-0674

ABSTRACT

Simple reproducible reference points would be of substantial use in the automated segmentation of magnetic resonance (MR) images. The spine is widely visible in MR images and could thus serve as a suitable reference point for processing in images from the neck to the pelvis. Regional and morphological techniques were examined for their ability to automatically locate the spine. Regional techniques were unsuccessful. Morphological techniques were used to identify features of vertebral bodies and the spinal canal in vertebral-edge enhanced images. This enabled automated identification of the lumbar spine.

INTRODUCTION

Magnetic resonance imaging (MRI) is an extraordinarily flexible technique for the evaluation of pathologic processes. Recent research has examined techniques for enhanced image analysis (1,2,3,4) and improved image display (5,6,7,8). Many existing techniques, however, require user input to define object boundaries or image reference points (1,2,6,7,8). The development of automated techniques for the recognition of reference points within MR images would be of substantial benefit (9).

Given that the spine is widely visible, it can serve as a reference point in MR images from the neck to the pelvis. As the first step in using the spine to establish a frame of reference, we evaluated techniques based on both regional and morphological information for their ability to locate the spine in MR images.

METHODS/RESULTS

Sagittal images from routine clinical studies of the lumbar spine were utilized. Images were transferred to personal computers, where image analysis techniques, based on either regional or morphological information, were developed and tested for their ability to identify the spine. Results of a program for the fully automated identification of the spinal canal were compared with the identification of this structure by two radiologists who had no knowledge of the results of the computer analysis.

A. Regional Analysis

1) Intensity thresholding

Intensity histograms of several anatomic structures typically present in MR images of the spine were examined. Image thresholding was then performed on the basis of these histograms to determine the suitability of thresholding as a method for segmenting image structures of interest.

Intensities associated with any one anatomic structure in an image of the spine overlapped considerably with those associated with many others. Figure 1 demonstrates this using the vertebral bodies as an example. As a result, image thresholding could not be used to segment the spine.

2) Cluster analysis of image textures

Textural properties of structures typical present in MR images of the lumbar spine were generated from an index image. These included the spinal canal, vertebral bodies, intervertebral discs and subcutaneous adipose tissue. The average local intensity and its standard deviation; the maximum and minimum local intensity; and the average and range of local image intensity gradients were calculated for 10 to 20 regions of interest within each structure. Discriminant and principal component analyses were subsequently performed to determine which of these properties could uniquely identify any of the designated anatomic structures (10). Using the linear discriminant functions from the discriminant analysis of the index image, region classification was performed in additional images. The results of these classifications were then analyzed for their ability to segment anatomic structures.

The spinal canal proved to be the most texturally unique category when this type of analysis was employed (Figure 2). With additional processing this method would have enabled automated detection of the spinal canal. However, the substantial processing time required to produce this image rendered this method unacceptable.

B. Morphological Analysis

These techniques focused on the identification of morphological features associated with vertebral bodies and the spinal canal in order to precisely locate the spine in a series of images.

1) Vertebral-edge enhancement and spine detection

Vertebral-edge enhancement was achieved via a multiple step process of first identifying gradient edges within an image and then removing as many edges which were not vertebral-edges as was possible. One of the techniques developed in this procedure enabled identification of images which contained the spine. A description of these techniques follows.

Image edges were first identified using a modified Prewitt gradient-based operator (11) on low-passed filtered images. In the resultant edge-enhanced images it was desirable to eliminate edge artifacts resulting from image noise. This was accomplished with recursive application of an expanding digital erosion mask. The erosion procedure was based on an assumption of an approximately square shape for individual vertebral bodies. In this process, an N x N square mask (initial mask size: 5 x 5 pixels) was passed in a systematic fashion over images. Any edges which fit completely within the mask were eliminated. Those which crossed the mask borders were preserved.

Recursive expansion and application of the erosion mask eliminated progressively larger amounts of noise-related artifacts from images. In order to avoid the eventual elimination of vertebral-edges, however, it was necessary to limit the expansion of the mask in images which contained vertebral bodies. Images which contained vertebral bodies were found by scanning digitally eroded images for the borders of roughly square elements of different size. The height and width dimensions were then used to create two histograms for each image. Images containing vertebral bodies were characterized by a high frequency of occurrence of objects of one size. This was not trūe for images in which vertebral bodies were not present. An examination of peak frequencies enabled identification of images in which the spine was present (Figure 3).

The dimensions corresponding to the peak frequencies in the histograms associated with each image were used to estimate the average respective height and width of vertebral bodies. These dimensions were used to limit the size of the digital erosion mask and thus preserve vertebral borders.

The entire procedure for vertebral-edge enhancement transformed vertebral bodies into relatively signal-free, rectangular structures that were enclosed by nearly contiguous borders. Similarly, the spinal canal was transformed into a long curvilinear, roughly vertical, relatively signal-free strip.

The procedure for vertebral-edge enhancement was examined on images obtained from a single subject with several different pulse sequences. Good delineation of vertebral bodies and the spinal canal was obtained in all cases. This suggests that methods for locating vertebral bodies and the spinal canal, based on their morphological features, may be applicable to images obtained with a wide variety of commonly used pulse sequences.

2) Spinal canal identification

Two scores based on the morphological features of vertebral bodies and the spinal canal were calculated for each image and then used in linear discriminant functions for identifying the spinal canal.

a) Likelihood score

Images identified as containing the spine were scanned for the locations of image elements of the same dimensions as the vertebral bodies. The distribution of these image elements along the horizontal image axis was plotted for each image. The distribution in one of the images was then used to locate the anterior and posterior borders of the spine (Figure 4). The number of occurrences of image elements with the dimensions of vertebral bodies between these borders was then calculated for each of the images. This number tended toward a maximum at or near the midsagittal image (Figure 5). Given that the spinal canal is a midsagittal structure, comparison of these areas could be used to estimate the relative likelihood (**L**) that the spinal canal was present in an image.

b) Ray-path-length score

A technique was developed which aided the identification of structures with morphologies similar to that of the spinal canal following vertebral-edge enhancement. Rays were drawn from a line that horizontally bisected each image. Ray paths were altered by reflections at vertically oriented edges and terminated when the geometry of edges in a neighborhood prevented further vertical travel. Ray-path-lengths were relatively long when they originated in the spinal canal (Figure 6). Given that the spinal canal is located immediately posterior to vertebral bodies within the spine, comparison of the maximum ray-path-length (**maxRPL**) in this region of the images aided identification of the spinal canal (Figure 7).

c) Linear discriminant functions

Using the likelihood and maximum ray-path-length scores, linear discriminant functions (10) were developed for identifying an image in which the spinal canal would be found. Two categories, the set of midsagittal images and all other images, were created for a set of six spinal studies. Values of **L** and **maxRPL** were calculated for each image and a discriminant analysis was performed to define linear discriminant functions which would separate midsagittal images from all other images. These functions were used to prospectively identify the spinal canal in sagittal studies of the lumbar spine.

C. Program Evaluation

The accuracy of locating the spinal canal from a linear discriminant analysis of the two morphologically based image scores, **L** and **maxRPL**, was compared with localization of the spinal canal by two radiologists. Both radiologists were blinded to the purpose of the study and one anothers findings. The results of the program were in agreement with the findings of the two radiologists in all cases.

CONCLUSION

This work demonstrates that morphologically based image processing techniques are capable of automatically identifying the spinal canal and vertebral bodies. Furthermore, preliminary results suggest that this identification process may be useful in images obtained with many commonly used pulse sequences.

REFERENCES

1. Young, SW, Ballerio, C, Carrol, CL. Visual fuzzy cluster analysis of MR images. AJR **1989**;152:19-25.
2. Merickel, MB, Carman, CS, Brookeman, JR, Mugler, JP, Brown MF, Ayers, CR. Identification and 3-D quantification of atherosclerosis using magnetic resonance imaging. Computers in Biology and Medicine **1988**;18:89-102.
3. Jungke, M, Bielke, G, Grigat, M, Meindl, S, Pfannenstiel, P, Seelen, WV. Feature extraction and data preprocessing applied to tissue characterization in NMR-tomography. In: Lemke, HU, Rhodes, ML, Jaffee, CC, Felix, R, eds. Computer Assisted Radiology (CAR'87). New York: Springer-Verlag, **1987**:48-52.
4. Leonardi, M, Della Riccia, G, Lavaroni, A, Battiston, C, Elefante, R, Simonetti, L. Expert systems possibilities in brain NMR imaging. Presented at the 7th Annual Meeting of the Society of Magnetic Resonance in Medicine, San Francisco, CA, August **1988**.
5. Weiss, KL, Stiving, SO, Herderick, EE, Cornhill, JF, Chakeres, DW. Hybrid color MR imaging display. AJR **1987**;149:825-829.
6. McNeal, GR, Maynard, WH, Branch, RA, Powers, TA, Arns, PA, Gunter, K, Fitzpatrick, JM, Partain, CL. Liver volume measurements and three-dimensional display from MR images. Radiology **1988**;169:851-854.
7. Hu, X, Levin, DN, Herrmann, A, Tan, KK, Galhotra, S. Continuous tissue characterization algorithm for 3D volume-rendering of MR images. Presented at the 7th Annual Meeting of the Society of Magnetic Resonance in Medicine, San Francisco, CA, August **1988**.
8. Pizer, SM, Fuchs, H. Three dimensional image presentation techniques in medical imaging. In: Lemke, HU, Rhodes, ML, Jaffee, CC, Felix, R, eds. Computer Assisted Radiology (CAR'87). New York: Springer-Verlag, **1987**:589-598.
9. Stiehl, HS, Jackel D. On a framework for processing and visualizing spatial images. In: Lemke, HU, Rhodes, ML, Jaffee, CC, Felix, R, eds. Computer Assisted Radiology (CAR'87). New York: Springer-Verlag, **1987**:665-670.
10. Ray, AA, ed. SAS User's Guide: Statistics. SAS Institute, Inc. **1982**:295-416.
11. Rosenfeld, A. Image analysis. In: Ekstrom, MP, ed. Digital Image Processing Techniques. Orlando: Academic Press, Inc., **1984**: 257-287.

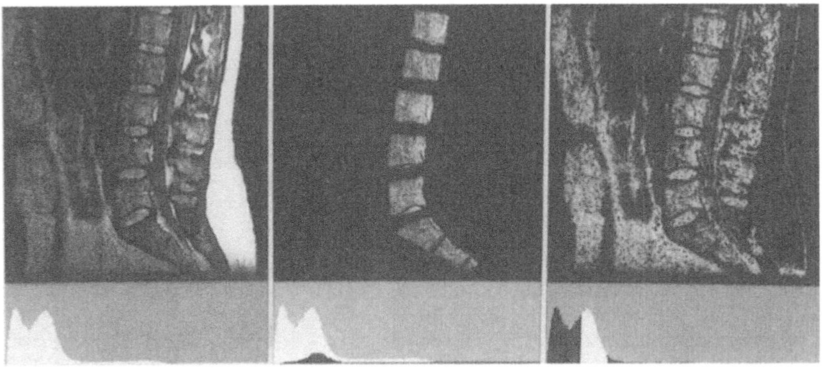

Figure 1: Intensity thresholding. **Left:** Unprocessed 1500/40 (TR/TE) midsagittal image of the spine. Image intensity histogram is plotted below the image (Increasing image intensity plotted horizontally, intensity frequency plotted vertically). **Middle:** All elements of the image except the vertebral bodies have been manually removed from the image. Intensity histogram of the vertebral bodies is plotted below the image in black. **Right:** Image after thresholding for vertebral body intensities (white area in the histogram). Note that many pixels outside of the vertebral bodies have similar intensities.

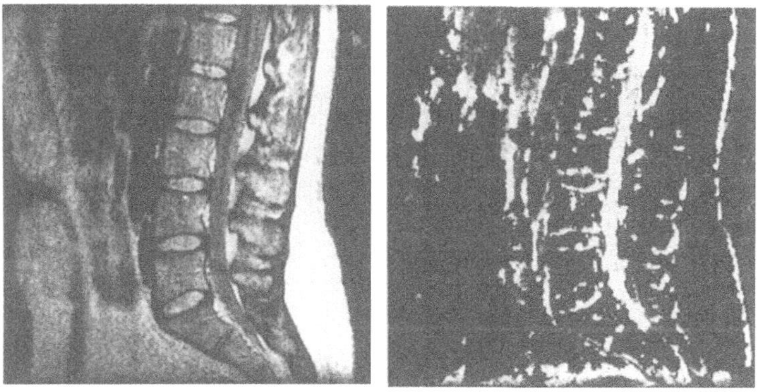

Figure 2: Cluster analysis of image texture. **Left:** Unprocessed 1500/40 (TR/TE) midsagittal image of the spine. A 5 x 5 pixel ROI was passed across this image and parameters of image texture were calculated at each location. Linear discriminant functions developed for an index were used to classify the ROI at each location. **Right:** Those ROI's sharing the textural properties identified for the spinal canal in the index image were then displayed.

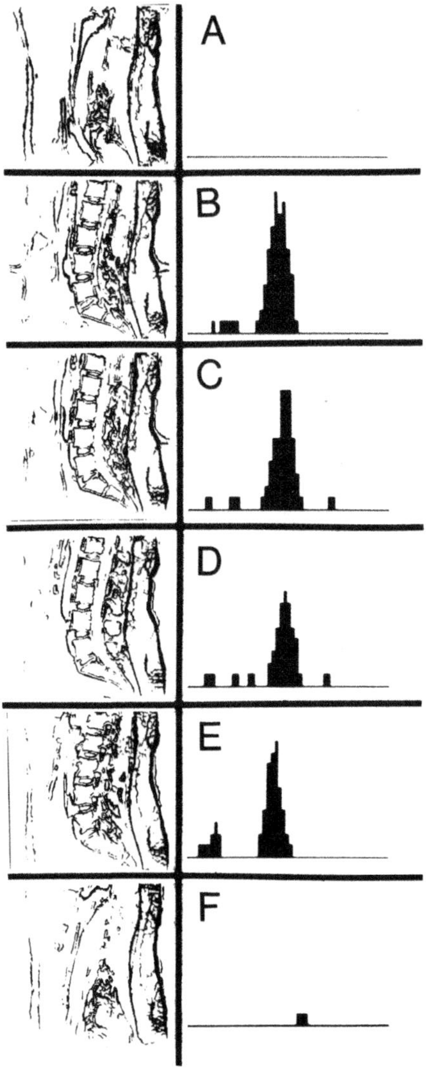

Figure 3: Spine detection. To the right of each digitally eroded, edge-enhanced image is the histogram of heights of roughly square objects within the image. The highest frequencies are associated with images in which the spine was present (B, C, D and E).

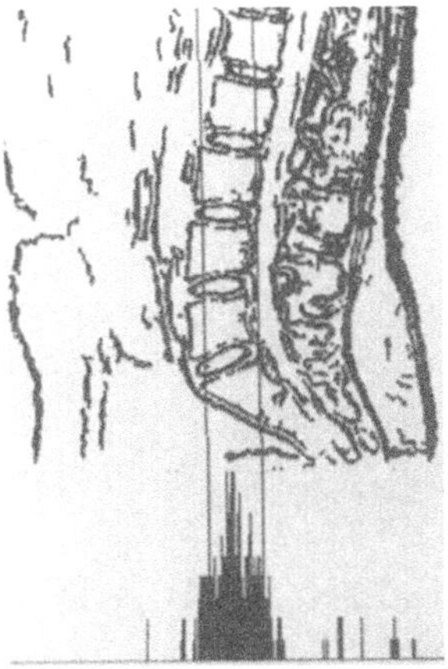

Figure 4: Estimated location of the anterior and posterior borders of the spine. The frequency of image elements with the same dimensions as the vertebral bodies is plotted versus its x-coordinate. The peak frequency was used to roughly locate a region within the image in which the spine could be found. A region one vertebra width wide, centered between the leading and trailing edges of this distribution, provided a estimate of the anterior and posterior boundaries of the spine (black vertical lines).

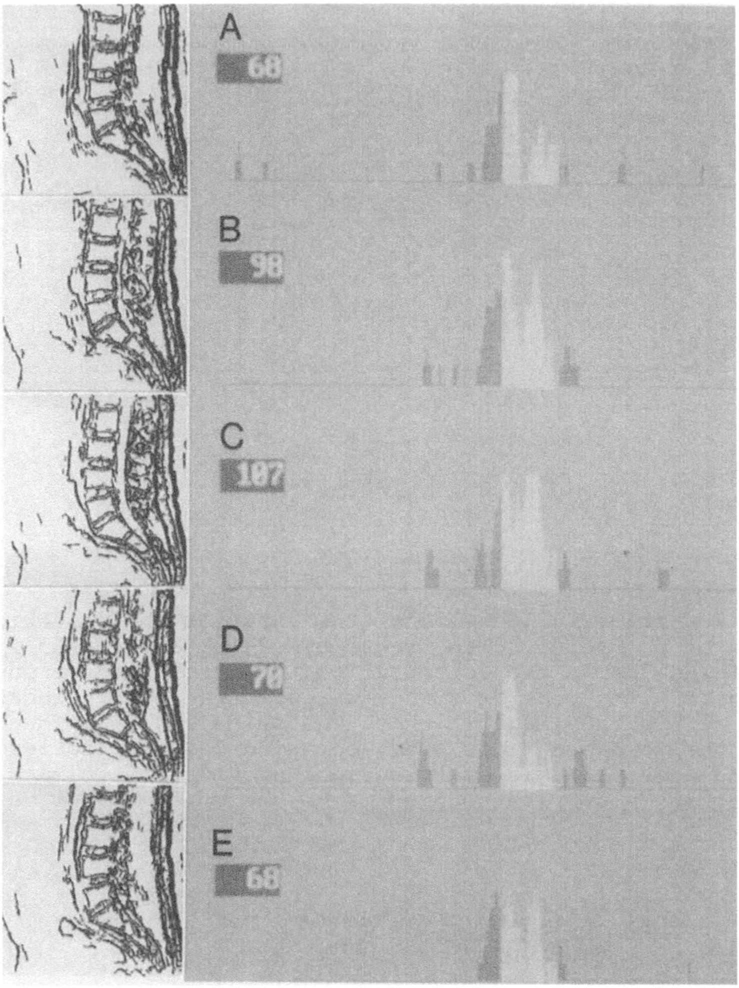

Figure 5: Likelihood score for the presence of the spinal canal. To the right of each image is plotted the frequency (vertical axis), at each x-coordinate, of image elements with the same dimensions as the vertebral bodies. The white area in each distribution identifies the number (white number in black box) of these image elements between the anterior and posterior borders of the spine. As the midsagittal image (C) is approached, this area tends toward a maximum.

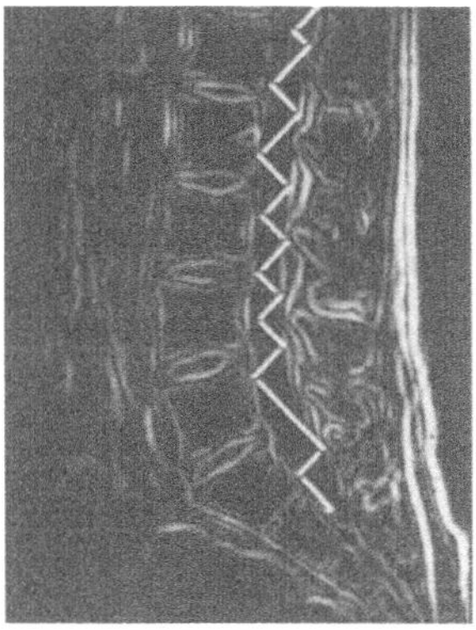

Figure 6: Ray-path-lengths within the spinal canal. Rays were drawn from a line which horizontally bisected each vertebral-edge enhanced image. Rays bounced between vertically oriented edges; their paths were terminated when edges in a neighborhood prevented further vertical travel. Ray path length was relatively long for any ray which originated within the spinal canal, as is shown here.

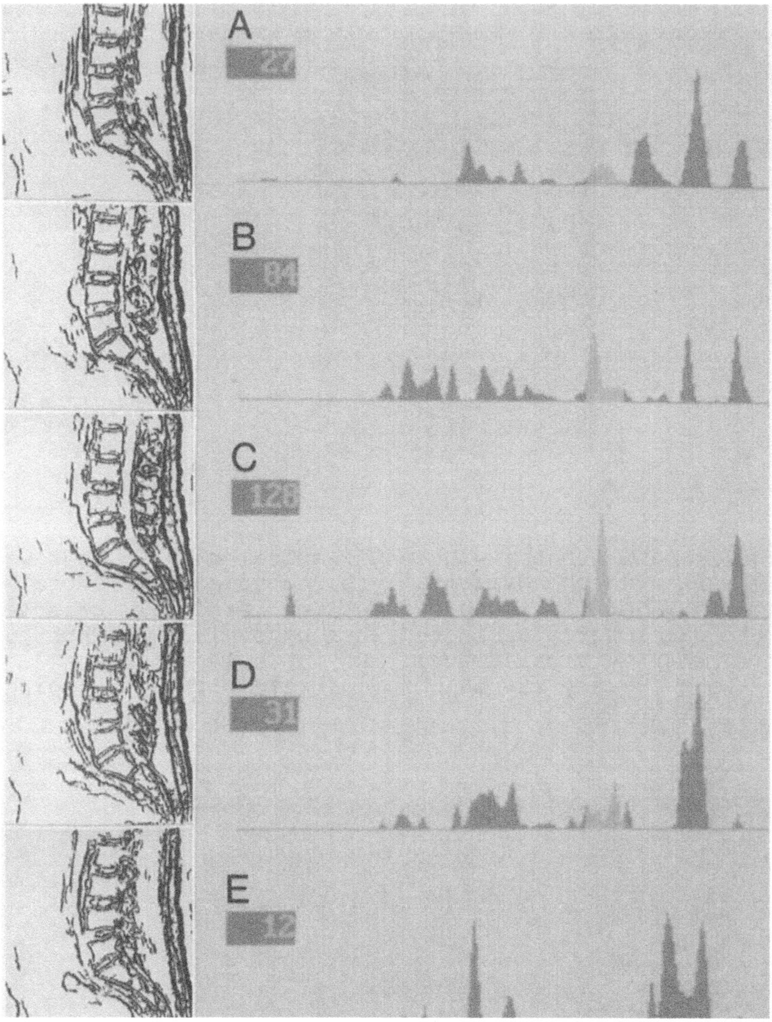

Figure 7: Maximum ray-path-length scores for identification the spinal canal. Ray-path-length as a function of x coordin is plotted to the right of each image. Within a region just posterior to the spine (green area), the maximum ray path len (**maxRPL**) was identified (green number in black box). **maxRPL** tended toward a maximum as the midsagittal image was approach (C).

On Segmenting Computed Tomograms

S. Back, H. Neumann, H.S. Stiehl[1]

Universität Hamburg, Fachbereich Informatik
Bodenstedtstr.16, D-2000 Hamburg 50

Summary

Analyzing the problem of segmenting discrete computed tomograms easily reveals three determinants: First, the specifity of the CT imaging modality (e.g. X-ray vs. MR and associated noise characteristics); second, the slice thickness in the case of anisotropic resolution (e.g. varying from slice to slice even within one sequence) and associated spatially anisotropic partial volume effects; and, third, image structure (e.g. discrete discontinuities and junction/vertice types of both prior unknown different spatial scale and at unpredictable spatial loci). We describe initial research toward a coherent multi-scale edge segmentation scheme independent of modality, noise, slice thickness, and partial-volume induced edge profiles of discontinuities.

The Medico-Technical Problems

Computational segmentation of primitives (e.g. edge points) as well as subsequent grouping of primitives (to e.g. smooth contours) has to be considered the starting point of a processing cascade which maps raw image data into geometric, iconic, or symbolic representations of quantifiable body properties. Such representations are in turn the methodological prerequisite to specific classes of computer-assisted tools useful for e.g. quantitative radiation therapy planning. At present no parameter-free or even self-adaptive (and thus generally applicable in the sense of being independent of e.g. image and/or object class) segmentation scheme does exist. Consequently, locally consistent, accurate, complete, unambigious, reliable, and reproducible results cannot be assumed or even guaranteed per se - except for either appropriately restricted domains or tasks of minor theoretical and computational complexity (for instance, bone segmentation in thin CT slices). The presence of segmentation errors has to invoke error recovery mechanisms via, to name a few, time-consuming user-driven interaction for re-contouring, local/global optimization schemes based upon relaxation to allow for maximum error-tolerance, or exploitation of anatomical domain knowledge by so-called intelligent image understanding processes. On the other hand unrevealed errors in segmentation may

[1]This research has been carried out while the authors have been with the Fachbereich Informatik at the Technische Universität Berlin. One of the authors (S. Back) is now with SUN Microsystem GmbH.

cause severe errors in subsequent processing steps (e.g. for contour-based surface reconstruction and quantitative morphometry). However, one must be also aware of the fact that sources of errors principally inhere in numerical segmentation of discrete images. Hence error sources as well as both the type and quantitative properties of resulting errors have to be carefully identified and analyzed.

In the context of computed tomograms, the quest for image- as well as domain-independent, slice-thickness- as well as noise-insensitive, and object-invariant segmentation to (in the ideal case automatically) detect and locate organs and lesions with utmost accuracy, reliability, efficiency, and generality can be straightforwardly motivated. The need for automated or at least semi-automated segmentation particularly arises from requirements from e.g. *quantitative* stereotaxy and radiation therapy planning: Accurate definition of anatomical landmarks, of the target volume under consideration, and of the organs at risk in terms of 2-D as well as 3-D location and shape is of paramount importance for computer-assisted planning of safe therapeutical procedures minimizing patient risk. Ideally, one single computational approach - being as automated as possible - should cope with all degrees of freedom of the problem of segmenting CT images. In contrast, commonly one has *i*) to select a particular approach from a plethora of methods, *ii*) to tune (e.g. via heuristic/empiric/ad hoc parametrization) to a specific problem and *iii*) to experimentally validate prior defined performance criteria. In operational systems for processing CT images, either interactive contour extraction for arbitrary objects, semi-automated region/volume binarization (via user-driven setting of partly locally erroneous global thresholds) and subsequent simplistic contour following, or automated segmentation for specific high-contrast objects in mainly very thin slices (such as to exclude segmentation errors due to the partial-volume phenomenon in CT images) have been implemented. Hence (semi-) automated segmentation of ***arbitrary*** objects has been identified as being a severe bottleneck of such operational systems: Tedious, fatigueing, and time-consuming as well as error-prone interactive contouring of organs and lesions for a large set of slices per patient sequence - as a prerequisite for generation of a 3-D body geometry representation for advance stereotaxy and/or radiation treatment planning - may even take up to one hour. Since the status quo in operational segmentation schemes prevents from (semi-) automation of clinically useful tools, research has to be directed towards (more) general approaches to segmentation which allow for rapid and less cumbersome extraction of ***arbitrary*** objects. Clearly, the level and type of interaction for semi-automation may only be precisely defined after having investigated in detail how far to a significant reduction of interaction one can get on with a general approach. In addition, incorporation of domain knowledge about e.g. anatomy should not be an essential at the early level of computational segmentation but could play a crucial role for e.g. intelligently supported interaction for error recovery on a later processing stage.

Computed Tomograms and Image Structure

A spatial sequence of computed tomograms (or respectively an either isotropic or anisotropic voxel cube) is a three-dimensional discrete noisy estimate of continuous physical properties of the human body, e.g. smooth morphology as in the case of X-ray computed tomography. Each computed tomogram is defined by a discrete 2-D intensity function $f(x, y)$ (assuming additive noise) which again represents an estimate only due to integration e.g. of absorption coefficients of a continous 3-D distribution of different materials within a slice of bounded thickness (and therefore, of anisotropic resolution). Discontinuities of the intensity function only indicate both the presence and the locus of an intersection of a 3-D object surface with the slice but do *not* allow for direct determination of geometrical properties of the intersecting object surface patch itself. Hence the type (e.g. step or ramp edge) and the local characteristics (e.g. slope and magnitude) of a discontinuity directly depend on both the slice thickness and the local differential geometry of the intersecting surface patch.

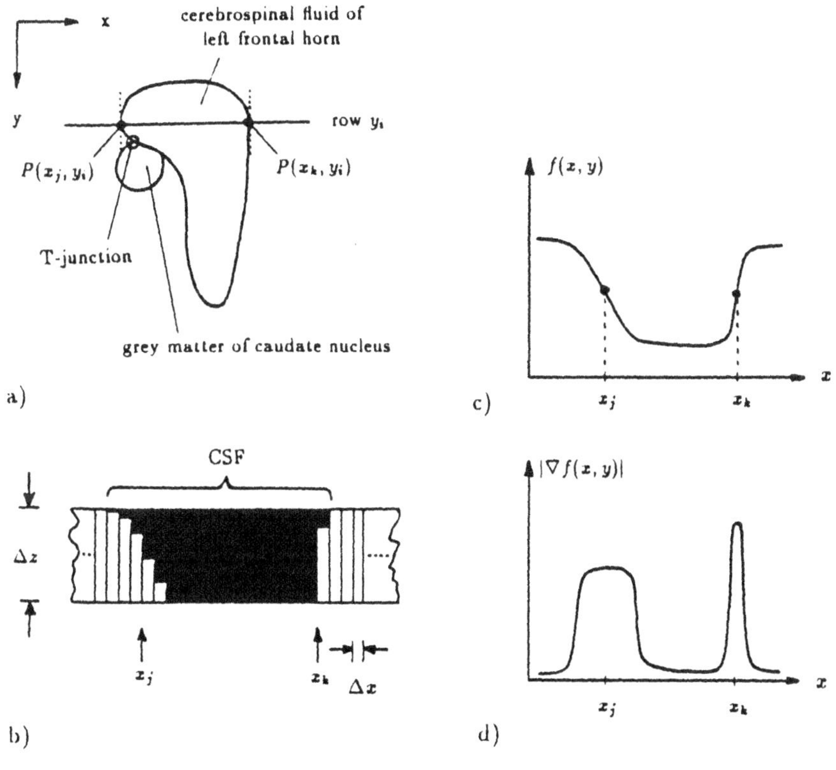

Fig. 1:

a) Schematic depiction of morphology in a CT (contours are assumed to be detected by gradient-based techniques)

b) Idealized discrete 3-D distribution of CSF in row y_i of the anisotropic slice in a) ($\Delta x = \Delta y < \Delta z$)

c) Intensity discontinuities (edge profiles) along row y_i with different slopes indicating different partial volume phenomena according to local 3-D anatomical morphology

d) Magnitude of gradient vector along row y_i in c)

Fig.1 illustrates the problem of discontinuities and associated local anatomical morphology. Since the morphology, the number of slices of a sequence, the slice thickness, the relative orientation as well as the spatial offset of a set of abutting slices are not constant from one examination to another, loci and local geometric properties of the object surfaces intersecting a slice cannot be assumed to be predictable per se. Consequently, the locus, the type and the local characteristics of a discontinuity of a 2-D intensity function representing a slice cannot be expected to be known a priori. For the same reasons, a variety of junction types (e.g. the T-junction in Fig.1 a) where the contours of two different anatomical objects meet) may appear in the 2-D intensity function at any location. Additionally, the constituents of a junction type in turn may have differently scaled edge profiles due to the partial volume phenomenon.

As a conclusion, discontinuities, smooth contours, *and* junction types (aside from e.g. textured or homogeneous regions) determine the basic image structure of a computed tomogram. Instead of applying conventional operators for numerical differentiation (e.g. kernels with a predefined fixed local support which might result in locally suboptimal operator responses), we propose a scale-space segmentation scheme independent of the above explicated boundary conditions. Whereas we are concerned with multi-scale *edge* detection, Pizer and coworkers investigate advanced multi-scale *region* segmentation for computed tomograms ([1]). The results of our pursued initial research will be briefly described below (for details see [2]).

Towards a General Approach

Scale-space or multi-resolution edge detection has been mainly motivated by knowledge about mammalian visual information processing in specific classes of retinal and cortical cells. As explicated in the preceding paragraph, prior unknown different types of both discontinuities and junctions may appear at any discrete location of the intensity function thus making it necessary to *a*) locally apply a ***set*** of specific detection operators at ***each*** location and *b*) to decide upon optimum operator response once one single operator from the set detects a type of image structure it has been designed to be sensitive for.

We adopted some basic principles of a computational approach which at first had been proposed in [3] and recently has been further investigated and developed in [4] and [5]. Multi-scale edge detection is here based upon the first order partial derivatives G_x and G_y of a normalized Gaussian distribution $G(x,y,\sigma_i)$ which are the components of the gradient vector $\nabla G = (\partial G/\partial x, \partial G/\partial y)$. The normalized partial derivatives $(\hat{G}_x, \hat{G}_y)$ have to be sampled and serve as linear shift-invariant discrete convolution kernels to determine $|\nabla G(x,y,\sigma_i) * f(x,y)|$.

Parametrization of the Gaussian by appropriately varying the space constant σ_i (e.g. $\sigma_1 = 0.75$, $\sigma_2 = 1.0$, $\sigma_3 = 1.5$, $\sigma_4 = 2.0$, ... in our discrete case) yields a family of scaled operators such that any operator has a different local support but a similar shape of the constituting impulse response functions $\hat{G}_x$ and $\hat{G}_y$. Convolving the image intensity function $f(x,y)$ with differently scaled $\hat{G}_x(x,y,\sigma_i)$, $\hat{G}_y(x,y,\sigma_i)$, respectively, for $i = 1,2,...,n$, and subsequently computing the gradient magnitude $|\nabla G(x,y,\sigma_i) * f(x,y)|$ at each spatial location (x,y), one yields a three-dimensional scale-space composed of filtered intensity functions. By analyzing the operator response at each location (x,y) through the scale space along the σ_i-axis (i.e. tracking a local maximum in operator response through scale space), a particular scale σ_{opt} can be determined for which the operator response is optimal. For the optimal scale both the local support and the impulse response function of the operator at (x, y) are perfectly adapted (however only in the case of the 1-D continuum) to the spatial extent as well as to the type of the underlying discontinuity. Since the problem in the 2-D case is nontrivial (e.g. in the presence of arbitrarily curved contours), the current implementation is restricted to scaled operators based upon the norm $\| \,.\, \|$ of $\nabla G = (\hat{G}_x, \hat{G}_y)$. Therefore the output of such an operator is locally maximal with respect to the direction of steepest ascent of $f(x, y)$. Starting with the response of a small operator scaled by σ_1, the operator response must increase steadily until - in scale σ_{opt} - it asymptotically converges to a maximum value. The asymptotic approaching of a maximum has been heuristically quantified via e.g. a relative scale-to-scale increase of the operator response less than 4% ([4]). In a later paper, Korn favoured a scale space search for the absolute maximum ([5]).

For practical reasons, the initial set of edge points per scale σ_i is reduced to contour candidates prior to scale space processing by a local non-maximum suppression method. Moreover due to the numerical nature of the discrete 2-D case, scale-to-scale search for increase of operator response has to extend from a candidate location (x, y) in scale σ_i to a local 1-D neighborhood (centered at (x, y) and oriented along gradient direction) in scale σ_{i+1} (see [5] for details).

In contrast to Korn's approach, we assume an error integral curve $\Phi(x, {}^1\sigma)$ as a valid local 1-D edge model for capturing the characteristics of discontinuities (step edges and differently scaled ramp edges due to spatially anisotropic partial volume phenomena) in computed tomograms. The theoretical reasons for choosing the error integral curve are twofold (see [2] for details): *a*) the complete range of differently scaled discontinuities (from step to any ramp edge) can be derived by $\Phi(x, {}^1\sigma) = H(x) * G(x, {}^1\sigma)$; ${}^1\sigma = 0, \ldots, \infty$ where H(x) is the Heaviside function and $\lim_{\sigma \to 0} G(x, {}^1\sigma) = \delta(x)$, and *b*) the optimum scale ${}^2\sigma_{opt} = {}^1\sigma$ of an operator (normalized partial derivative $\hat{G}_x(x, {}^2\sigma)$ can be explicitely determined by comparing operator responses from two subsequent scales ${}^2\sigma_i$ and ${}^2\sigma_{i+1}$ during scale-space processing

$$ {}^2\sigma_{opt} = {}^2\sigma_i = {}^1\sigma \quad if \quad \frac{M_g(x, {}^1\sigma, {}^2\sigma_{i+1}) - M_g(x, {}^1\sigma, {}^2\sigma_i)}{M_g(x, {}^1\sigma, {}^2\sigma_i)} < \frac{{}^2\sigma_{i+1} - {}^2\sigma_i}{2\, {}^2\sigma_{i+1}} $$

where $M_g(x, {}^1\sigma, {}^2\sigma) = \Phi_g(x, {}^1\sigma) * \hat{G}_x(x, {}^2\sigma)$ and Φ_g is the generalized error integral curve. For the 2-D case, we simply substituted $M_g(.)$ by $|\nabla(G * f)|$. Moreover for the purpose of accurate candidate selection per scale, we investigated the noise characteristics in X-ray computed tomograms to allow for exclusion of operator responses due to noise from scale-space processing. As a first attempt, we assumed the noise in the intensity function $f(x, y)$ to be isotropic and to obey a Gaussian distribution $N_f(\mu_f, \sigma_f)$. This hypothesis has been verified via a χ^2 test based upon the statistical distribution of the intensity function in different interactively selected homogeneous local neighborhoods. Since the transformation of $f(x, y)$ into e.g. $G_x * f(x, y)$ is linear, the distribution of operator responses in homogeneous regions of computed tomograms is again a normal distribution, $N_{G_x}(\mu_{G_x}, \sigma_{G_x})$, where σ_{G_x} decreases with increasing scale of the operator G_x. Depending on a particular significance level for a Gaussian distribution, e.g. $\alpha = 4.55$, one can - in statistical terms - precisely define a scale-dependent threshold to exclude operator responses with high statistical evidence of being caused by noise from further scale-space processing.

Lastly, we performed experiments on detecting partial-volume strips (or ribbons) along organ/lesion contours detected by operators based upon first order derivatives of a Gaussian. As it can be seen from Fig.1, only the extrema of the curvature of the discontinuities - and *not* the loci of steepest ascent - are precisely related to contours of e.g. anatomical objects within an anisotropic slice. The width of the discontinuity is a measure of the spatial extent of the partial volume phenomenon. Hence one has to apply an operator based upon the second directional derivative along the gradient direction $\partial^2 f/\partial n^2$ (for $|\nabla f| \neq 0$). Consequently, local extrema of $\partial^2 f/\partial n^2$ are in particular partial-volume invariant precursors for organ contour detection *and* organ surface reconstruction.

Conclusion

We have thoroughly investigated the problem of segmenting discrete noisy computed tomograms. A general framework has been proposed - in the main following Korn's multi-scale approach - to detect and locate image structure independent of modality, slice thickness, and noise. Initial experiments have been carried out on X-ray computed tomograms and the performance of detecting and locating edges has been judged to be clearly superior to conventional operators with fixed single-valued local support. Open questions, directly related to our own further research, are: Use of directionally sensitive scaled operators to improve location of edge points, precise noise estimation, massively parallel grouping (incl. gap filling) and junction detection, contour-based region reconstruction, and extension to the discrete three-dimensional case.

Literature

[1] Pizer, S.M.; Gauch, J.M.; Lifshitz, L.M.: Interactive 2D and 3D Object Definition in Medical Images. Proc. SPIE 914 "Medical Images II" (1988).

[2] Back, S.; Neumann, H.; Stiehl, H.S.: On Scale-Space Edgel Detection in Computed Tomograms (submitted for publication, 1989).

[3] Marr, D.: Early Processing of Visual Information. Phil. Trans. of the Royal Society of London 276 (B) (1976) 483-519.

[4] Korn, A.: Das visuelle System als Merkmalsfilter. in: Syrbe, M.; Thoma, M. (eds.): Fachberichte Messen, Steuern, Regeln 13. Berlin: Springer-Verlag 1985.

[5] Korn, A.: Towards a Symbolic Representation of Intensity Changes in Images. IEEE Trans. on Pattern Analysis on Machine Intelligence 10 (5) (1988) 610-625.

Interpretation of Tomographic Images via Virtual-Data Fusion

S.DELLEPIANE, D.D.GIUSTO, C.REGAZZONI, S.B.SERPICO, AND G.VERNAZZA

Department of Biophysical and Electronic Engineering
University of Genoa, Italy

Summary

The interpretation process of medical tomographic slices (CT or MR) is aimed to recognize different tissues up to obtain an anatomical map and features quantitative measurements of the recognized regions. The data fusion (performed at the various processing levels) of multiparametric, multisensor and transformed ("virtual") images allows to obtain more complete and reliable results than those obtained from the single channel analysis.

Introduction

Automatic interpretation of tomographic images represents one of the necessary preliminary steps towards a full exploitation of the possibilities provided by computerized systems, such as organ reconstruction, therapy and surgery planning, etc. In fact, programs for automatic segmentation and automatic extraction of organs are powerful tools that allow the user to handle a larger amount of data as compared with the interactive techniques used so far.

In this context, the evolution undergone by knowledge-based systems has already yielded some promising results on automatic segmentation and interpretation of complex images. However, such systems still exhibit many drawbacks, and their performances have to be improved by adding new capabilities, such as the use of multisensor or multiparametric inputs, or the exploitation of all possible information extracted by means of processing algorithms.

At the same time, several efficient algorithms have been developed for low-level image processing. We recall, for instance, various types of filters, grey-level scale modifications, segmentations, and so on. Each operation generates a new image that can be regarded as a particular transformation of the original one; we label such transformed images as "virtual images".

In order to achieve a correct automatic interpretation, we are proposing to use all the information derived from various acquisitions of the same tissue or organ, and from as many virtual images as possible obtained from the original ones. From a practical point of view, it seems reasonable to think of a knowledge-based structure with different information channels, where each input is a differently acquired image (or "virtual image"). The data-fusion process allows one to use, at the same time, the information provided by all these kinds of image, thus fully exploiting the potentiality of the image content [1].

Data-Fusion Levels

The data-fusion process can be performed at different levels, since the information contained in virtual images can be merged so as to obtain better results than those yielded by traditional techniques in any step of the recognition process.
Our approach to image interpretation is based on a blackboard system [2], whose knowledge sources correspond to low, middle and high levels. Symbolic primitives are regions obtained by a region-growing algorithm; the actual interpretation step is performed through a comparison of a parametric organ description with organ features extracted from symbolic primitives.
Starting from an original image, the first version of this interpretation system is able to generate a final map where each symbolic primitive (in this case, each region generated by the segmentation process) is labelled according to the organ (or tissue) to which it corresponds. Moreover, each match of a region with an organ is characterized by a fuzzy membership value which defines the match accuracy.
The evolution of IBIS, which is a single-channel system, will involve an extension of its capabilities in order to deal with more than one input, with more than one output, and with more than one intermediate symbolic data set.
The recognition system can be applied to both multiparametric and multisensor inputs, considering each input as a single channel. Recognition results are fused to derive a more complete map from the ones obtained from the separate input channels. During this process, only one segmented image is extracted for each channel on the basis of the pixel grey-level distribution and the data-fusion process is performed at the highest level, on organs and tissues.
An alternative approach is to consider not a single image feature in order to perform the segmentation process, but some other properties besides grey-level distribution. For instance, segmentation tasks can be accomplished that are based on both textural properties and grey levels, thus obtaining more significant symbolic primitives, which are more consistent with the actual image content and which allow the system to discriminate between various tissues or organs.
This approach allows the fusion process to be performed at the pixel level, since more images (both acquired and virtual) are used to produce a unique set of symbolic primitives.
A third kind of approach is based on the use of more segmentation results. In particular, different symbolic primitives can be extracted from various images, or by applying different methods to a unique image (e. g., region- or line-finding methods).
In any case, the use of uncertainty factors allows the system to manage noisy, incomplete and uncertain situations, thus taking also into account doubtful results.
In order to demonstrate the feasibility and capabilities of these kinds of approach, in the following practical implementations of such methods are briefly described, and some preliminary results are reported.

Methods

In MR (Magnetic Resonance) applications, in order to merge various recognition results obtained by applying the system to separate input channels, the same anatomical slice can be acquired by using different pulse sequences and different acquisition parameters.
Up to now, we have been able to deal with T1- and T2-weighted images, by introducing specific information into the knowledge-base of IBIS to tailor

it to the different features of the organs and tissues in the two types of image.
Since some tissues or organs can be better recognized in one channel than in the other, recognition results can be merged appropriately by taking into account the different capability of each channel for discriminating a tissue or organ. This kind of information is embedded in the fuzzy values, defined a-priori in the knowledge-base for each channel, and called "credibility values".
The fusion process is then performed, on the basis of the membership values obtained by the recognition process and on the credibility values associated with the two channels. In particular, a weighted average is computed, thus confirming the fuzzy values of the organs recognized in both channels, and disgregarding those organs which have not been located in a channel, due to its high credibility value.
Since the symbolic primitives in the two channels are not topologically equal, during the fusion process, the two sets of elementary region are combined when such regions are overlapping.
As a result of this process, more detailed maps can be obtained, where it is easy to locate small organs or organ subparts. This is of high practical interest, especially when dealing with biomedical images, where the differences between normal and pathological tissues are not well-defined.
As an example of a data fusion process performed at the lowest level, an image-segmentation method is based on multiparametric inputs and on virtual textural images [3]. Fractal dimension values and histogram parameters have been used to generate a segmented image that better preserves the whole original information.
A more complex fusion process, at the low level, can be performed by merging the information coming from multisensor inputs. A new image is extracted, starting from an MR and a CT scan acquired for the same patient position. In this case, a geometric correction is required, since it is impossible to obtain exactly the same patient position and orientation when using the two techniques. When some errors remain in the fused image, after the preliminary correction, additional knowledge about the model can be used to perform better local adjustments.

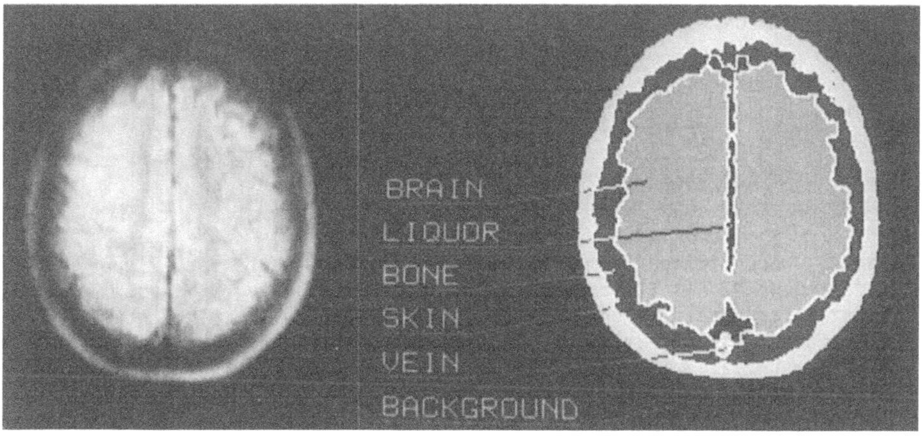

Fig.1. Original high transversal MR slice (left portion); organs recognized by the single-channel system (right portion).

Results

To point out the capabilities of the data-fusion approach, a comparison between the single-channel and the multichannel recognition results is made. Fig. 1 shows the recognition results on a high-transversal slice. The main organs have been located, while some details or organ subparts are missing. Through two acquisitions of the same anatomical slice, more detailed results can be obtained. The original images (Fig. 2) are spin-echo sequences with parameter values equal to TR=200 msec and TE=30 msec for the T1-weighted image (left part), and to TR=2000 msec and TE=120 msec for the T2-weighted one (right part) (acquired by the tomograph EsaOte Biomedica ESATOM MR5000).

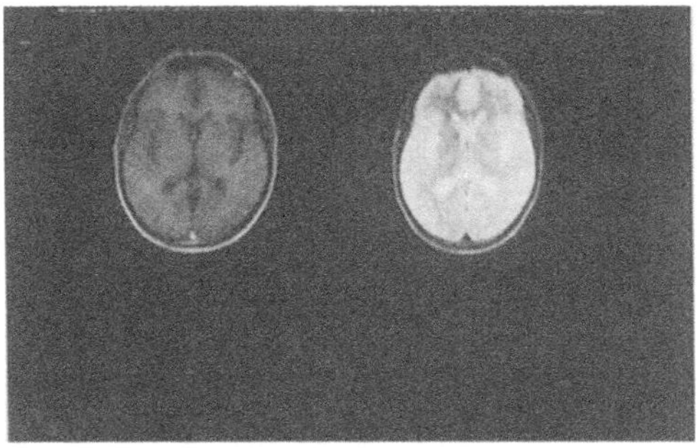

Fig.2. Original T1- and T2-weighted images (left: SE200/30; right: SE2000/120).

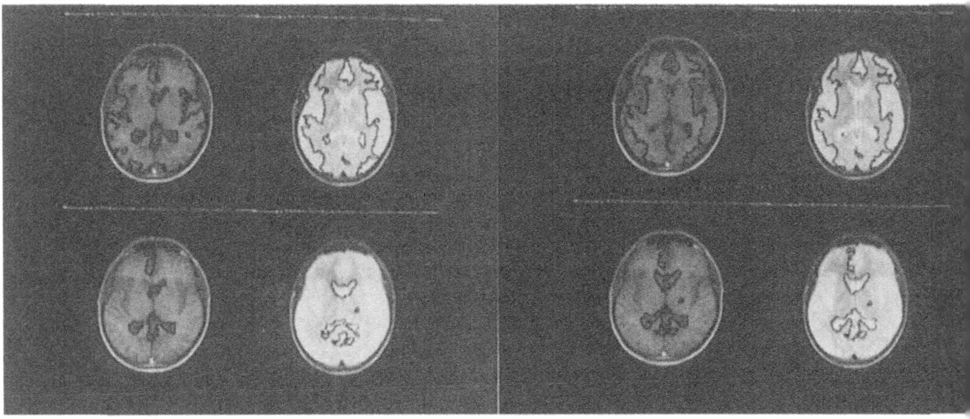

Fig.3. Recognition from the independent T1 and T2 channels. Upper-left: brain recognized in T1-image; right: grey matter in T2; lower-left: liquor in T1; right: liquor in T2.

Fig.4: Grey matter and liquor after data-fusion process. Overlapped to T1-image in the left portion; to T2-image in the right portion.

Fig. 3 shows the brain as recognized in the T1-weighted image (upper-left portion), grey matter as recognized in the T2-weighted one (upper-right portion), and liquor space, as detected in the T1- and T2-weighted images. As a result of the fusion process (Fig. 4), fewer errors remain than in the single-channel case, and a more complete result is obtained (especially for liquor) by merging the two partial results.

In order to show an example of the low-level data fusion performed, Fig.5 presents a CT scan (lower-left) and the corresponding MR slice, beforè (upper-left) and after (upper-right) the geometric correction.
By combining the information derived from the two images, a new virtual image is generated, which make it possible to have, at the same time, a sufficient contrast to distinguish both soft tissues (as in MR images), and bone tissues (as in CT images).
Some local errors, due to a non-rigid deformation and to uncertainty in the patient position, are present in the lower portion of the figure.

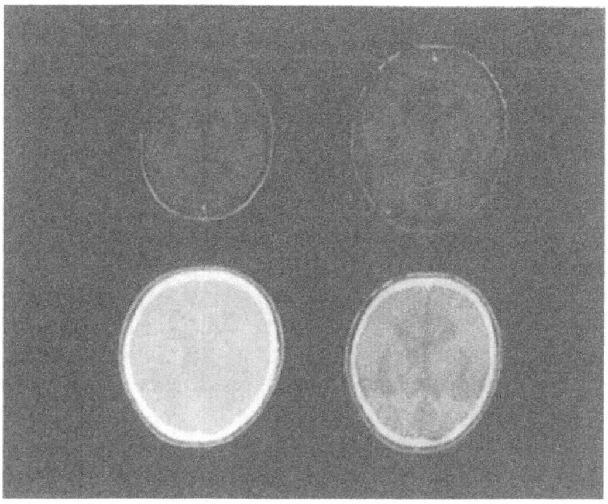

Fig.5. Fusion of CT and MR slices (lower-right portion). Original CT scan (lower-left), original MR slice (upper-left), MR after geometric correction (upper-right).

Conclusions

The methods and results presented in the paper constitute a preliminary approach to medical data fusion. It seems feasible to obtain better results by a contemporary analysis of more than one image, as medical practice suggests.
By integrating data fusion, at various levels, into the recognition system, it will be possible to combine information and partial results yielded by numeric, symbolic and semantic kinds of representation, thus obtaining a more complete and reliable interpretation in a shorter computation time.

Acknowledgments

The authors wish to thank EsaOte Biomedica for supplying the MR images.

References

[1] A.Mitiche and J.K.Aggarwal - "Multiple sensor integration-fusion through image processing: a review" - Optical Engineering, Vol. 25, No. 3, pp. 380-386, March 1986.

[2] G.Vernazza, S.B.Serpico, and S.Dellepiane - "A knowledge-base system for biomedical image processing and recognition" - IEEE Trans. on Circuits and Systems, Vol. CAS-34, No. 11, pp. 1399-1416, November 1987.

[3] D.D.Giusto, L.Parodi, G.Perrone, and G.Vernazza - "An adaptive technique for the analysis of SAR images" - Proc. SPIE, Vol. 1129, April 1989.

Improvement on Dynamic Elastic Interpolation Technique for Reconstructing 3-D Objects from Serial Cross-Sections

Shiuh-Yung Chen[1], Wei-Chung Lin[2], Cheng-Chung Liang[3], and Chin-Tu Chen[4]

Department of Electrical Engineering and Computer Science[123]
Northwestern University
Evanston, IL. 60208

Department of Radiology[4]
The University of Chicago
Chicago, IL. 60637

Abstract

An improved method for automatically reconstructing a three-dimensional object from serial cross-sections is presented in this paper. The method improves the dynamic elastic contour interpolation technique [1]. There are two major improvements : (1) in case of branching that involve concave contours, instead of pairwise interpolation between the start contour and each goal contour, the goal image is considered globally and local constraints are imposed on the forces exerting upon the vertices; (2) it takes the continuity of high order derivatives into consideration and incorporates the schemes of spline theory, quadratic-variation-based surface interpolation, and finite-element-based multilevel surface interpolation to create smoother surfaces of the reconstructed object.

1. Introduction

Recently, a dynamic elastic surface interpolation scheme which reconstructs 3-D objects in volumetric representation has been proposed in [1]. The approach provides a mechanism to generate iteratively a series of intermediate contours for filling the gaps between one start and (one or more) goal contours. However, in case of branching, it works well when the contours have no "significantly concave" portions but its performance is sometimes peculiar when the contours possess concave parts. Moreover, since the object is also formed by stacking up the contours together, even with a fine interpolation the resulting surface might be coarse if the degree of dissimilarity among these contours is high.

To enhance the dynamic elastic interpolation method in case of branching, instead of pairwise interpolation between the start contour and each goal contour, the goal image is considered globally and local constraints are imposed on the forces exerting upon the vertices. More sophisticated schemes beyond simple stack-up are also employed to improve the smoothness of the surfaces reconstructed by using the method in [1]. In this paper, a method which combines the modified elastic interpolation algorithm, the spline theory, and the surface consistency theorem is proposed for reconstructing a smooth 3-D object from serial cross-sections.

2. Contour Interpolation between Cross-Sections

The first step in the proposed method is to generate a series of intermediate contours based on the elastic matching method [2]. During this step, a sequence of intermediate contours (approximated by polygons) between a *start contour* and a *goal contour* are created iteratively. The basic concept behind the method is to identify a "force field" acting on the start contour $C_s = \{(x_i, y_i), 0 \le i < N_s\}$ and trying to distort it to be similar to the goal contour $C_g = \{(x'_j, y'_j), 0 \le j < N_g\}$, where $N_s \ge N_g$. To achieve this, a family of *force vectors* (or *force field*), is determined first. A force vector is defined to be a function of "pushing" and "pulling" forces :

$$DS(x,y) = \gamma^{-1} \left[\frac{\sum_{i=0}^{N_s-1} G_s^i D_s(i,J_i)}{\sum_{i=0}^{N_s-1} G_s^i} - \frac{\sum_{j=0}^{N_g-1} G_g^j D_g(j,I_j)}{\sum_{j=0}^{N_g-1} G_g^j} \right], \quad (1)$$

where

$$G_s^i = \begin{cases} 0 & , \text{ if } m_1 > t \\ exp\frac{-(x-x_i)^2-(y-y_i)^2}{\sigma_k^2} & , \text{ if } m_1 \le t \end{cases} \quad (2)$$

where

$$m_1 = \frac{| D_s(i, J_i) |}{\left[\frac{\sum_{i=0}^{N_s-1} |D_s(i,J_i)|}{N_s} \right]},$$

and

$$G_g^j = \begin{cases} 0 & , \text{ if } m_2 > t \\ exp\frac{-(x-x'_j-D_{gx}(j,I_j))^2-(y-y'_j-D_{gy}(j,I_j))^2}{\sigma_k^2} & , \text{ if } m_2 \le t \end{cases} \quad (3)$$

where

$$m_2 = \frac{| D_g(j, I_j) |}{\left[\frac{\sum_{j=0}^{N_g-1} |D_g(j,I_j)|}{N_g} \right]},$$

are the weights designed to provide the effect that near neighbors have more influence than far neighbors. $D_s(i, J_i)$ is the feature displacement vector defined as the shortest distance from a vertex $p_s(i) \in C_s$ to the line segment formed by two vertices $p_g(J_i), p_g(J_{i+1}) \in C_g$. $D_g(j, I_j)$ is similarly defined by reversing the roles of the start and the goal contours. The parameters t, γ and $\sigma_k(= \sigma_0 f^{-k})$ denote the force constraint threshold, the damping factor and the stiffness measurement (at the kth iteration), respectively.

According to the above equations, all the intermediate contours, indexed from 1 to k, can be iteratively generated with the initial condition $W^0 = C_s$. The iteration stops when the following criterion is satisfied :

$$\frac{1}{N_s} \sum_{i=0}^{N_s-1} | DS^k(x_i, y_i) | < \epsilon,$$

where ϵ is a predefined small number.

As far as branching is concerned, a general approach can be devised slightly by modifying the above algorithm. Let C represent the start polygon and its interior. Let $C_p, 1 \le p \le n$, be the n goal polygons (each includes its interior) in one slice. If I_p^k

is the result obtained after the kth iteration using the pair (C, C_p), the result of the interpolation after the kth iteration is $\bigcup_{p=1}^{n} I_p^k$.

3. Contours-to-Grid Transformation

After the contour interpolation is performed, a sequence of intermediate contours approximated by polygons are obtained. Let's denote these contours by $C^k = \{(x_i^k, y_i^k), 0 \le i < N_k\}$, where $0 < k < n$, and N_k is the number of vertices on the kth contour. Let's denote the start and the goal contours by C^0 and C^n, respectively. For simplicity, assume that the start and the goal contours lie in the two parallel planes $z = z^0$, and $z = z^n$, respectively. Each of the intermediate contours, C^k, $0 < k < n$, generated by the interpolation algorithm lies in a plane $z = z^k$, where the value of z^k is between z^0 and z^n.

The next reconstruction process is to compute the complete surface representation by interpolating an initial representation consisting of sparse surface values which are the mapping from the contours. To facilitate the interpolation process, a transformation is performed which converts the set of contours to a grid. In this method, the contours are treated as a family of parametric curves which are represented by using two parameters u and v as

$$C^v = \{(v,u) \mid \omega(x_u^v, y_u^v, z^v) = (v,u), 0 \le u < N_v\}, 0 \le v \le n,$$

where ω is a function which assigns coordinates (v, u) to the vertex (x_u^v, y_u^v, z^v) of a contour. Without loss of generality, we assume that the grid of size $m \times m$ represents the domain of the surface function. The grid can be conceived as the result of "unfolding" a cylinder. The principal axis and the perimeter of the cylinder (after it is unfolded) become the i and j axes of the grid, respectively.

The transformation first maps x and z coordinates of each vertex to a point (i, j) on the grid. If there exists a correspondence, the coordinates (z^v, x_u^v) of the vertex is stored in $c(i, j)$ and the y_u^v coordinate of the vertex is used as the initial surface value $s(i, j)$. After the transformation is performed, a set of points on the grid which contain the coordinates mapped from the vertices of the contours are obtained. For the remaining points on the grid, the *B-spline cubic* function is adopted to interpolate all the x coordinates on the grid. Since the points on the contour are coplanar, a linear interpolation is used for the z coordinates of $c(i, j)$.

4. Surface Reconstruction

After the preliminary processing which creates the surface function domain and the initial conditions, the problem of reconstructing a 3-D object becomes one of determining the unknown values of the fitting surface at the remaining grid points, subject to the condition that the resulting surface is most consistent with the information provided by the contours. Two alternatives, quadratic-variation-based and finite-element-based surface interpolation methods, are adopted to obtain final surface representation.

4.1 Quadratic-variation-based Interpolation

In [3,4], the best approximation to a discrete surface s is given by the minimal solution to the following objective function :

$$\sum_{i=1}^{m-2}\sum_{j=0}^{m-1}(s(i-1,j)-2s(i,j)+s(i+1,j))^2+\sum_{i=0}^{m-1}\sum_{j=1}^{m-2}(s(i,j-1)-2s(i,j)+s(i,j+1))^2$$

$$+2\sum_{i=0}^{m-2}\sum_{j=0}^{m-2}(s(i,j)-s(i+1,j)-s(i,j+1)+s(i+1,j+1))^2, \tag{4}$$

subject to $s(i,j)-s'(i,j)=0, \forall(i,j)\in\varphi$, where
$\varphi=\{(i,j)\mid \text{a known value mapped from a vertex of a contour to the grid point}(i,j)\}$, and the $s'(i,j)$'s are the initial conditions calculated from the procedures described in Section 3. The approach used to solve the problem of surface minimization is the *gradient projection method.*

4.2 Finite-element-based Multilevel Interpolation

In [5], the surface reconstruction problem is formulated as a variational principle describing the equilibria of a thin flexible "free plate" as shown in the following:

$$\mathcal{E}(v)=\iint_\Omega[\frac{1}{2}(\Delta v)^2-(v_{xx}v_{yy}-v_{xy}^2)]dx\,dy+\frac{\beta}{2}\sum_{(x_i,y_i)\in C}[v(x_i,y_i)-c_{(x_i,y_i)}]^2, \tag{5}$$

where Ω is the domain of the plate and $\Delta\equiv\partial^2/\partial x^2+\partial^2/\partial y^2$ is the Laplacian operator. The first term gives the strain energy within the interior of the plate. The second term measures the potential energy due to deformation of the springs acting on the plate, where C denotes the set of points in Ω at which the imaginary pins are located. The scalars $c_{(x_i,y_i)}$'s denote the heights of the pins (i.e., the depth constraints), and the stiffness of the associated springs are given by positive constants β for all $(x_i,y_i)\in C$.

According to the variational principle, the reconstructed surface is the function u in the ***Hilber space*** which minimizes the potential energy $\mathcal{E}(v)$. To facilitate the computation, the ***finite element method*** [6] is applied which constructs the approximating function u^h of u over a finite element space S^h, a subspace of V. With the finite element scheme, the multilevel reconstruction algorithm can be developed by introducing a hierarchy of finite element spaces $S^{h_1},\ldots,S^{h_L}$, where L is the number of levels and $h_1>\ldots>h_L$ are the fundamental lengths of elements at each level.

5. Experimental Results

To demonstrate the capability of our reconstructing scheme, the real medical image is applied in the experiments. In Figure 1, two consecutive slices of magnetic resonance (MR) images revealing the branching of artery in the pelvis area of human body are shown. After contour and surface interpolations, the reconstructed junction of blood vessels is obtained which is shown in Figure 2, Figures 3(a), and (b), respectively.

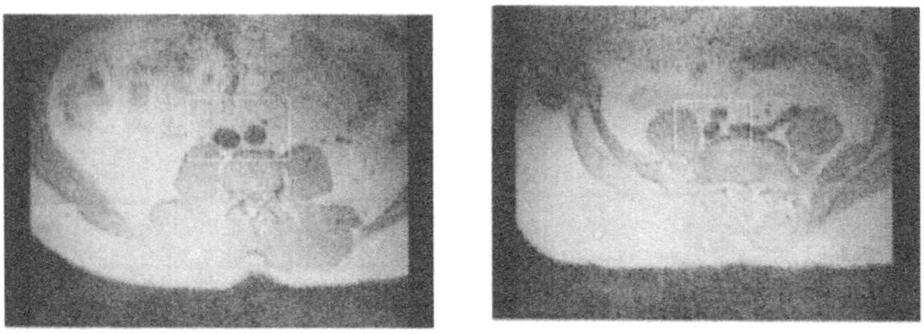

Figure 1: Two digital magnetic resonance (MR) images.

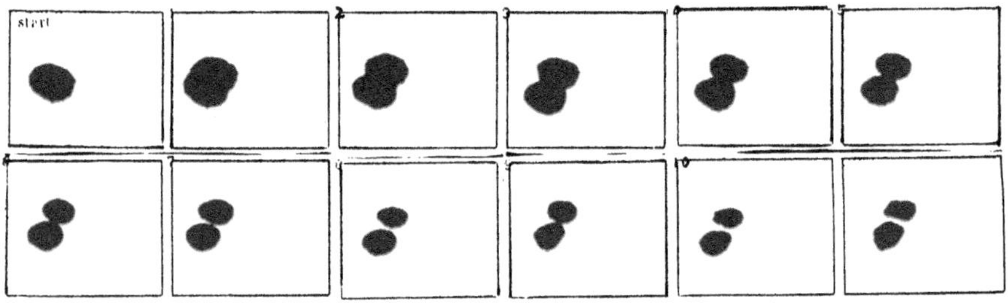

Figure 2: The intermediate contours obtained by applying dynamic elastic interpolation to the contours extracted from the MR images.

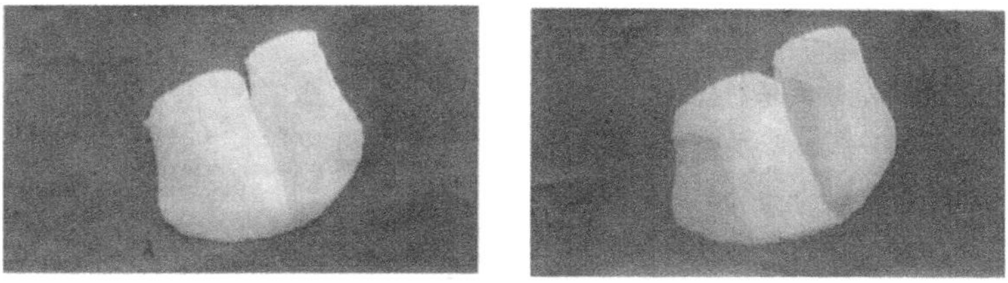

Figure 3: The reconstructed blood vessel junction based on (a) quadratic-variation-based interpolation method and (b) finite-element-based multilevel interpolation method.

Reference

[1] W.-C. Lin and C. C. Liang, and C. T. Chen, "Dynamic Elastic Interpolation for 3-D Object Reconstruction From Serial Cross-Sectional Images," *IEEE Transactions on Medical Imaging,* Vol. 3, No. 3, September, 1988, pp. 225-232.

[2] D. J. Burr, "Elastic Matching of Line Drawings", *IEEE Transactions on Pattern Analysis and Machine Intelligence,* Vol. PAMI-3, No. 6, Nov. 1981, pp. 708-713.

[3] W. E. L. Grimson, "An Implementation of Computational Theory of Visual Surface Interpolation", *Computer Vision, Graphics, and Image Processing,* Vol. 22, 1983, pp. 39-69.

[4] W. E. L. Grimson, "Surface Consistency Constraints in Vision", *Computer Vision, Graphics, and Image Processing,* Vol. 24, 1983, pp. 28-51.

[5] D. Terzopoulos, "Multilevel Computational Processes for Visual Surface Reconstruction," *Computer Vision, Graphics, and Image Processing,* Vol. 24, 1983, pp. 52-96.

[6] E. B. Beckeer, G. F. Carey and J. T. Oden, *Finite Elements: An Introduction,* Prentice-Hall, 1981.

Evaluation of Digital Image Processing in Angiography for the Diagnosis of Renal Neoplasms

K.Jarnicki,J.Jarnicki,R.Marciniak

From the Department of Radiology of the Medical Academy and Institute of Technical Cybernetics of the Technical University Wrocław,Poland

Summary

The paper reports the concept of the quantitative analysis of X-ray angiographic patterns of renal tumores.The features of the x-ray neoplasms images essential in the qualitative analysis have been presented.The quantitative representation of the features applied in the interactive computer aided image processing system have been described.An image processor based on the IBM microcomputer has been applied to analyze the angiographic patterns.

Introduction

Typical procedures in the angiographic diagnosis of renal neplasms is based on the qualitative analysis of following x-ray pattern:[1]

1.Width of renal artery lumen

2.Angulation of renal artery origin from aorta

3.Early apperrance of venous vessels in arterio-venous fistulas

4.Change in the shape and dimensions of kidneys in the parenchytous phase of filling

5.Degree of sharpness demarcation between the tumorous tissue and healthy renal parenchyma

6.Appearance of pathologic vessels which characterize in changing of lumen width and winding course

7.Appearance of avascular zones

8.Appearance of ill-sharpe demarcated patchy contrast conglomerates of sun-ray appearance

9.Appearance of venous collateral circulation in cases of renal vein inpatency

The aim of the performed studies with the aid of the computer was the quantitative evaluation of the characteristics mentioned above.The goal of the quantitative evaluation was not the application of the automation of decision making process but it was elaborated in purpose for establishing the effective[2] algorithms in searching of computer base data containg collection of angiographic pattern [3] for finding of similar cases of renal neoplasms and performace of comparative analysis. Besides,methods of computer aided data-processing proved to be helpful in decision making of dubious cases.

Data Acqusition and Image Processing

The major components of the image processor include the MC9000 series Reticon camera (model MC 9265),the MB 9000 Reticon video data formatter and custon designed IBM-AT microcomputer interface card.The MC 9000 series camera makes used of a two-dimensioned image sensor consisting of individual photodiodes arranged in a 256x256 square array.Analogue video signal is provided as output and only a clock input to the camera is required to scan the image sensor.The MB 9000 video data formatter contains a 6-bit analog-to-digital converter with digital threshold which digitions the signal to a 6-bit gray scale format and 1-bit black-white value.RS-170 timing module is also provided for standard video monitor.An IBM-AT microcomputer interface card allow continuous image data transfer from formatter to the microcomputer memory.Remote programming of the analog-to digitalconverter thresholds using serial RS 232 interface is also realized.The software part of the image processor has been developed as a fully interactive menu oriented program, written using Turbo Pascal and Assembler languages.A mouse and EGA or VGA graphics card with suitable screen monitors has been chosen for standard input-output.The program includes the following options:

-display of gray scale or pseudocolor images using arbitrarily selected colour table(palette)

-remote programming of the analog-to-digital converter in formatter manipulation

-disk operations and directories manipulation

-image processing(full image or its part in the arbitriarily selected window) using standard processing algorithms(various types of filtration and contouring) or special custom created algorithms.

The following special algorithms have been implamented:
-measurement of distance between two points
-measurement of the angle between two stright lines
-measurement of the surface enclosed by a cursor selected contour, and the lenght of the contour
-approximation of the interactivaly drawn curve by spline polynom.[4]

Quantitative analysis

The properties of angiographic images,mentioned in the introduction,are very complex in nature.The width of the renal artery and its angulation origin from the aorta are directly measured after pointing out of the characteristic points by a graphic cursor.The time of exposition of veins can also be measured simply by analysis of sequence of stored images. Other characteristics are much difficult to quantity.Shape of kidneys is obtained by interactive fitting of a curve to the observed renal border.This is realized by manipulating the coefficients of the polynomial describing the curve and observing the results on an optimally chosen filterred image (the filtering algorithms is chosen from the avialable by trail).Degree of sharpness demarcation is characterized by the ratio of the longest and the shortest distances between observed borders.Pathologic vessels and patchy contrast conglomerates are directly counted.In case of avascular zones the measurement are realized in the same way as described for shape of kidney and the number is counted as well.Inpatency of renal vein is registered by binary value.The proposed method of analysis of angiographic images of renal tumors is presentaptly being tested the as to its sensivity to the personal aptitude of the radiologist.

References

1. Abrams H.L:Angiography,2nd ed.Boston:Little Brown 1971
2. Menhardt,W,Schmidt,K-H: Automated Interpretation of Transaxial MR-images of the Brain.Proceedings of the International Symposium CAR 87,Springer Verlag,Berlin,pp.386-390 (1987).
3. Hachimura,K:A Prototype PACS with the Capability of Retrie-

val by Image Data. Proceedings of the International Symposium CAR 87, Springer Verlag, Berlin, pp.507-524(1987).

4. Pawlidis, T: Graphics and Image Processing. Springer Verlag, Berlin-Heidelberg(1982).

Computer-based System Analysis of Radioisotope Scan Image

Abdalla S.A.Mohamed

Department of Computer Science

University of Regina,Sask. S4S 0A2

CANADA

ABSTRACT

Most medical radioisotope images are characterized by rather poor spatial resolution and high noise level.If they are to be improved by manipulation of image data,then a flexible mathematical approach is required to predict effects of different processing techniques on image resolution and noise. A multivariate random field model is introduced for describing the nonhomogeneous images as a two-dimensional nonstationary stochastic processes. Using the statistical decision theory we indicate the development of decision rules for the choice of the neighbors,the homogeneity of the data. This approach is efficient for maximizing the signal-to noise-ratio,and can be used for data compression without losing any information.

1- INTRODUCTION

Radioisotope scan images[RSI's]are inherently noisy because of the random nature of radioactive decay.The number of counts detected in any region of the scan image fluctuates about a true or"expected"value,exhibiting some statistical distributed properties of independent quantum events.

Processing RSI's presents a particular challenging problem.Medical radioisotope scans are characterized by poor resolution and high noise level; most scanning systems have point response widths on the order of one or two centimeters,and most clinical scans are composed of a few hundred to a few thousand recorded quanta per square centimeter.Efforts to enhance resolution tend to increase an already serious noise level,while efforts to reduce noise can further degrade already poor resolution.A good understanding of the influence of various processing techniques upon both resolution and noise is vital if manipulation of scan data is to improve image quality. In any real processing problem both expected count density(resolution,or fidelity of the image to the object protrayed,aside from statistical fluctuations) and variance(noise) are important parameters of image quality.Image noise can be reduced sharply by using low broad weight functions,which average count density over a large region and correspond to very low pass filters,yet resolution in the expected(and observed) image is then seriously degraded.On the other hand,resolution in the expected image can be enhanced, correcting for imperfect transfer characteristics of the primary transfer system,by passing the image through a processing stage with frequency response inversely proportional to the primary system frequency response[1]. If one is interested in detecting a small image area superimposed on an assumed flat or linearly varying background,the signal-to noise-ratio [SNR] may be of interest,since it is related to the statistical significance of the observed image[2] .It should be clear that the SNR of energy and relative root mean square[RMS] deviation of the image spectrum are not changed by filtering[3] .Therefore,if a reasonable mathematical criteria can be established for scan image quality,then calculation of optimal processing techniques may be possible.These optimal techniques depend upon the type of information to be represented,the resolving properties of the scanning system,and the degree of statistical fluctuation in the scan image. This paper records an attempt to develop a unified implementable mathematical approach to analyze nonstationary RSI.The main features of a selected random fields are extracted to identify the shape region.In section-2 a procedure to define the discrete random Markovian field is given.The features of this field are extracted by a multivariate parametric model.Some results of cardiac RSI's are shown in section-3,and discussions in sec.-5.

2-IMAGE ANALYSIS

We have to start with the general assumption that the RSI is characterized by nonstationary statistics(mean and autocorrelation).For the mathematical tractability,it is desirable to transform the nonstationary image into stationary one.The transform

$$f_m(x,y) = f(x,y) - \mu_m(x,y) \qquad (1)$$

will create zero mean image data which has a stationary mean.Also,the transform

$$f_m(x,y) = \frac{f(x,y)}{R_m(x,y,0,0)} \qquad (2)$$

will lead to an image with unit energy to produce a stationary behaviour of the energy.Then,depending on the resulting amplitude distribution,we can define the image model that ensures an optimal solution for the criteria of interest.

Fortunately,considerable studies of imagery for the purpose of data compression has shown that the"hot"areas of RSI can be successfully treated as a Markovian random field.The noise source is equivalent to the error in a best linear mean square error MSE estimate of the field at a point in terms of its nearst neighbors.A P-order homogeneous Markov random field is characterized by the correlations among the elements out to distance P.Also,,this field leads to an analogous discrete representation of homogeneous Gaussian field defined by [4] :

$$f_m(x_n,y_m) = \sum_{\mathscr{D}} h(k,l).f(x_n-k,y_m-l)+ e(x_n,y_m) \quad (3)$$

where

$$E[\, f(x_n,y_m)\cdot e^*(x_k,y_l)\,] = c\ \delta_{nk}\ \delta_{ml} \quad (4)$$

$$\mathscr{D} = D_p - \{\vec{0}\} \qquad D_p \subset \mathscr{D} \quad (5)$$

$$D_p = \{\ (k,l) \quad \text{such that} \quad k^2+ l^2 \leqslant p^2\ \} \quad (6)$$

$$E[\, e(x_n,y_m)\ .\ e^*(x_0,y_0)\,] = \begin{cases} c & m=n=0 \\ -h_{mn}\ c & (m,n)\in \mathscr{D} \\ 0 & \text{elsewhere} \end{cases} \quad (7)$$

$$f(x_i,y_j) = \frac{\beta\ (x_i,y_j)}{M} \quad (8)$$

$h(k,l)$ are the coefficients of the linear minimum MSE [MMSE] estimator of $f(x_n,y_m)$ given its neighbors.
$e(x_n,y_m)$ is the error in the estimate of $f(x_n,y_m)$
c is the error in the MMSE.
$\beta(x_i,y_j)$ represents counts density due to point or line source.
M is the total number of counts in the scan image.

The larger question concerning the kind of non Markov process $\{f(x_n,y_m)\}$ needs to be clarified.It is clear that if $c^{-1}\ R_{ee}(m,n)$ in(7) is positive definite,it can be shown that the solution to(3) driven by $\{e_{nm}\}$ with correlation $R_{ee}(n,m)$ given by(7) is Markov-P in $\mathscr{D}$ as shown in Fig.1.To generate a Markov-P field in some finite region,it is necessary to generate$\{e_{nm}\}$as a finite linear combination of white noise.However,any field generated by (3)driven by a source $\{e_{nm}\}$ of independent non-Gaussian variables will be a Markov field.

The RSI in the selected field is unlikely to be homogeneous and hence cannot be adequately represented by a univariate autoregressive [AR]model[5]. Therefore,the entire nonhomogeneous image is divided into(LxL)blocks labeled(u,v),u=1,...,L;v=1,...,L such that

$$(u,v)\text{block} = \{f(i,j)\ ,\ 1\leqslant i,j \leqslant N\ \} \quad (9)$$

Each block is assumed to be sufficiently small so that it is homogeneous.A separate univariate AR model is fitted for each block.It is clear that the AR's coefficients are not completely independent of one another.Therefore, solving the eigen-value problem of the covariance matrix of AR's coefficients will lead to a solution for this problem.

3-RESULTS

Unprocessed RSI of anterior wall myocardial infarction is given in Fig.2.

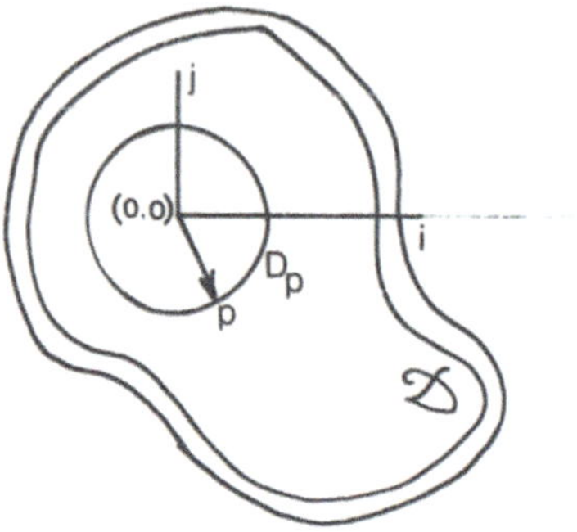

Fig.1 Region for Markov-P field.

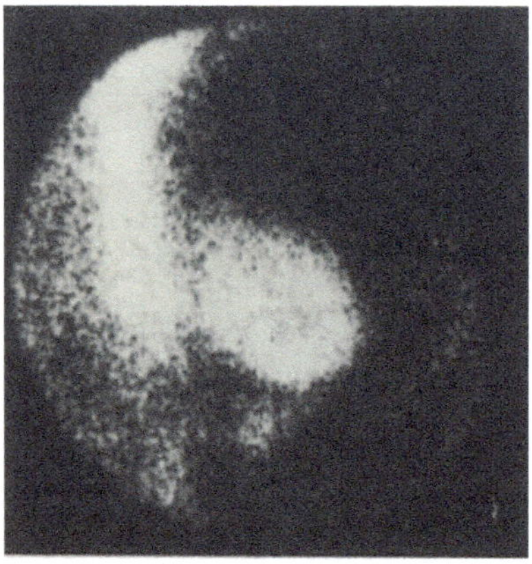

Fig.2 Radioisotope image of cardiac anterior wall with myocardial infarction.

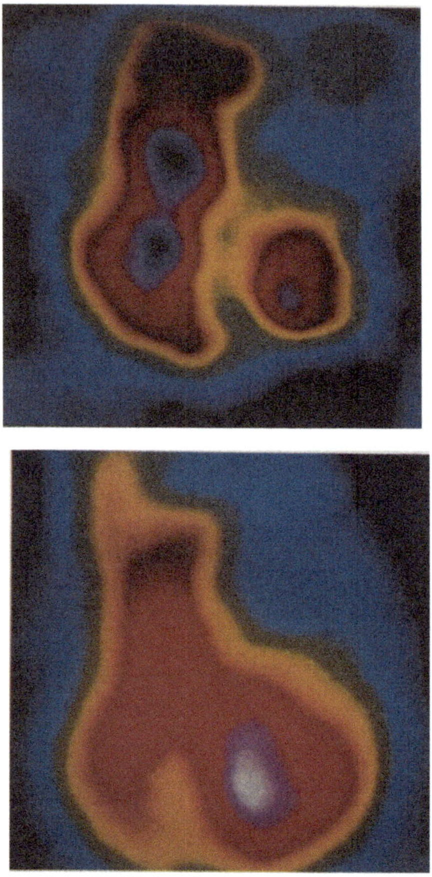

Fig.3 Reconstructed field from Fig.2 with different block size; (a)(16X16) block size; (b)(128X128) block size.

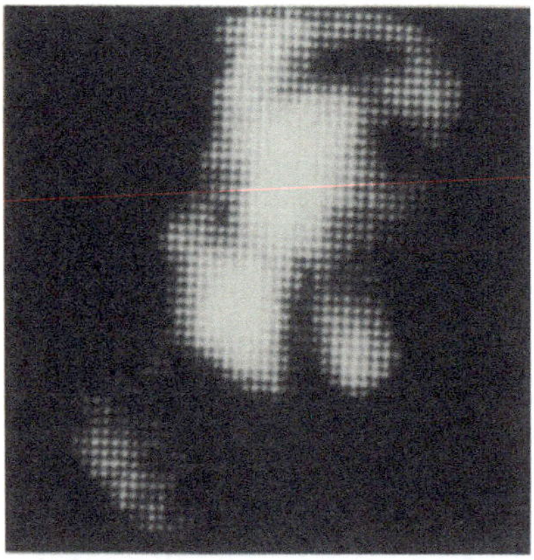

Fig.4 End-systolic gated pool image.

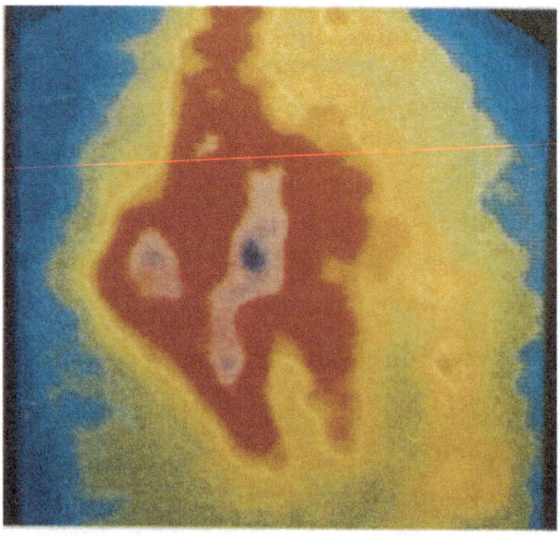

Fig.5 Reconstructed field from Fig.4 with (16X16) block size.

The available computer system used for processing the RSI contains a micro-computer Zenith 286 equipped with a Winchester drive(40M),two floppy drives (3.5"and 5.25"),color display type 8513,640K RAM,Matrox frame grabber(512X 512,8-bit gray level),and RCA 16mm camera.The procedure of Markovian field generation was used to predict the count density in the "hot"area.The influence of number of blocks on reconstructing the selected fields is shown in Fig.3a for(128X128)block size and in Fig.3b for (16X16)block size with respect to the original RSI in Fig.2.The same procedure was followed to reconstruct end-systolic gated blood pool image given in Fig.4,The resulting image is shown in Fig.5.

4-DISCUSSION

The difficulties of practical image resolution enhancement lie in the problems of:(1)finding ways to implement these general forms of frequency response;and (2)choosing the filter response or the linear/ nonlinear transforms useful for a particular primary transfer system,image noise,and object distribution.The formulation of the discrete-space Markov field can be considered as a generalization of field dependence without any constraints on the direction of propagation of the process.Consequently,it helps us to calculate the high-order joint probability densities and design the optimum processors.Moreover,the expected image can be easily predicted.The selection of AR model helps us to test the field hemogeneity and the choice of the relevant set of neighbors,compress the image data,and image recognition.

5-REFERENCES

[1] Witting,T.A.,and Mallard,J.R.,"Theoretical considerations for high resolution radioisotope scanning",Phys.Med.Biol.,Vol.13,pp.657,(1968).

[2] Kuhl,D.E.,and Edwards,R.Q.,"A hybrid processor for modifying and rearranging radionuclide scan data under direct observation",Radiology, Vol.92,pp.558,(1969).

[3] Hunt,B.R.,and Cannon,T.M.,"Nonstationary assumptions for Gaussian models of images",IEEE Trans.System,Man,and Cybernetics,Vol.SMC-6, pp.876-882,(1976).

[4] Yaglom.A M.,"An introduction to the theory of stationary random functions",Englewood Cliff,N.J.:Prentice-Hall,pp.81-84,(1962).

[5] Delp,E.J.,Kashyap,R.L.,and Mitchell,O.R.,"Image data compression using autoregressive time series models",Pattern Recognition,Vol.11, pp.313-323,(1979).

The Iris Image Analysis for Medical Diagnosis

R. ANDONIE, D. WANIEK[1], I. GARBACEA, S. CISMAS

University of Braşov, ITCI, 2200 Braşov, Romania

[1]Perşani Medical Unit, Municipal Hospital, Făgăraş, Braşov, Romania

Summary

Iridology tends to become an independent alternative medical system. We describe a research tool for the computerized iris image evaluation. The iris pictures are acquired by a TV camera and further processed in order to extract significant parameters of the iris pattern. Our approach is based on some new theoretical considerations concerning the iris physiology and the iris pattern significance. Within this framework new studies are made possible in the field of alternative and biomedical systems integration.

Introduction

Iridology is a branch of the alternative medicine which infers statements about one's health and specific organ disorders from diagnostic work confined to iris examination. There are several "iridologic charts" /8/, which may vary in respect with the iris surface areas or "organ zones" (Organfelder) /3/. These organ zones are held to represent (and allow diagnostic inferences about) the respective target organs (e.g., see fig. 1).

This paper applies computer vision expertise /1,2/ to iris image study, being the result of interdisciplinary team work. Throughout this paper we refer to an IRis Image ANalysis system (IRIAN). We require for such a system the following characteristics:

1. The input is the iris picture of a patient (e.g., from a TV camera).
2. The symbolic name of the region of interest or the organ zone (e.g., the heart, the ovary) has to be indicated by the system user.
3. The output consists of computed parameters of the corresponding iris surface areas. These parameters refer to luminosity, texture, and other visual characteristics of the areas.
4. The diagnoser may even not have any knowledge about the iridologic chart, because this chart is a priori considered by the computer.
5. The image analysis should be extended with a connected expert system for medical diagnosis which uses as input image parameters of the organ

zones, and knowledge about the iris signs grammar.
This paper reports on the realization of an IRIAN system as it is designed above, except point five which is still a project for the future.

Looking for the iris pattern significance

The iris pattern may represent a necessity for a basic non-visual function of the eye /9/. Our most important assumption is that the image features of physiologic relevance can be objectively (and thus automatically) extracted. Previous work concerning densitometric values in eye photographs showed that even simple statistics, as the Student **t** test, produce positive results of a computable level of confidence in rejecting the null hypothesis /6/.
The direct observable features of the iris pattern are the color and the iris stroma radial fibers configuration. In a monochrome input picture these correspond to the gray level and the texture, respectively. Another feature should be added if the surface of the iris is viewed in 3D: the relief of the region of interest. The combination of gray level and texture values generates different types of regions. The boundaries of these regions are of two kinds: precisely delimited (step boundaries) and fuzzy (ramp boundaries).
In terms of their average gray level, texture, and boundaries, we observed the following classes of regions which correspond to classical iridologic signs:

1. Light, with different texture (comparing with the background), with step boundary, corresponding to the closed catharral sign.
2. Light, different texture, ramp boundary, corresponding to the open catharral sign.
3. Light, not different texture, ramp boundary, corresponding to the acute inflammatory sign.
4. Dark, not different texture, ramp boundary, corresponding to the chronic inflammatory sign.
5. Dark, different and nearly homogeneous texture, step boundary with well-structured shape, corresponding to the special sign.

Considered in 3D, the catharral signs are concave, the inflammatory signs are convexe, and the special signs are planar areas.
Completing in a combinative way the above classification, three new types of regions may be defined:

6. Light, not different texture, step boundary.
7. Dark, different texture, ramp boundary.

8. Dark, not different texture, step boundary.
There are no corresponding iridologic signs to these three regions and we suppose that for the future they will enrich the classical iris semiology.

Image acquisition

A first difficult problem one is faced with in an IRIAN system is an optical one: there is a superposition over the iris image of the light source image in the convex mirror represented by the cornea. To avoid this, the light source has to be brought in the optical axis of the eye-TV camera system. In this case, the area of the pupil contains the undesired reflexions. We designed a simple mirror device to solve the problem. The image acquisition is performed by a low-cost, one-board microcomputer, IT-88, specifically tailored for the acquisition, processing, and displaying of iris images. In order to increase the processing power, IT-88 was connected to a host minicomputer. IT-88 performs picture acquisition from TV camera (256 x 256 pixels with up to 256 gray levels). The pictures are displayed on a monochrome/color monitor, and/or printed.

Matching the iris

Our aim is to focus attention on an arbitrary organ zone of the input iris, this zone being specified by the iridologic chart. Hence, a matching of the idealized iris (the chart) and the real iris (the image acquired by the TV camera) is needed.
The iris position and dimensions in the input picture are variable. Fortunately, there is no significant retation of the iris around the optic axis when the image is acquired, irrespective of the slightly different subject head positions, because of the postural static and statokinetic correction reflexes. Therefore, we concentrated on matching in terms of translating and shrinking/expanding the input iris image. This basic part of the image analysis is performed on a host minicomputer and has been written in C. An input picture is processed in the following sequence: preprocessing, segmentation, and approximation of the iris frontier with a circle.
First, we tried to smooth the input picture and to reduce the noise. Second, we had to deblur the image and sharpen contours. The combined use of the median smoothing /4,7/ and extremum sharpening /4/ transforms has proven to be sufficient in preprocessing the input pictures (see fig. 2b).

The iris frontier was detected by heuristic thresholding (fig. 2c). Finding the best circle that fits the set of points from the iris frontier is a nonlinear optimization problem. We were satisfied with a suboptimal solution (see /5/, p.293).
The matching of the iris chart and the input iris picture is now obvious: the resulted approximated circle (fig. 2d) has only to be translated and shrinked/expanded to superimpose exactly over the chart circle.

Conclusions

Our system is a research tool for both the standard clinical/biomedical and the alternative medical scientist. It has the following advantages:
1. It extracts from the iris valuable information which could not be obtained with the naked eye. Simple inspection does not process such an amount of refined information as that implied in biological codes, though it eventually leads to rapid recognition of the learned patterns.
2. It archieves the iris patterns and retrieves them by request allowing for further processing.
3. It could lead to a better definition of an iris chart on statistic, syntactic or heuristic grounds.

Presently, we apply the IRIAN system to the solving of several basic problems, such as studies in morphogenesis of the iris pattern, exploring the ovary and uterine zones, in search for changing patterns during biological cycles. According to our current experiments, even simple techniques based on the gray level statistical variance of these organ zones are expected to yield results, possibly to produce biologic fundamental data.

Acknowledgements

We want to thank Prof. M. Olteanu, Head of the Ophthalmologic Department, Medical Institute of Bucharest, for his support. Thanks are also due to Dr. H.S. Stiehl (Univ. of Hamburg) for his helpful suggestions and careful revision of this paper.

References

1. Andonie, R.: A Semiotic Approach to Hierarchical Computer Vision. In: Cybernetics and Systems, J. Rose (ed.), Thales Publications, Lytham St. Annes, England, pp. 930-933, 1987.
2. Andonie, R., Gărbacea, I.: A System for the Analysis of Cranial Computed Tomographies. In: Computer Assisted Radiology (Proceedings CAR' 87, Berlin, July, 1987), H.U. Lemke, M.L. Rhodes, C.C. Jaffee, W.D. Selles (eds.), Springer, Berlin, pp. 749-753, 1987.
3. Jaensch, P.A., Rohen, J., Schreck, E., Huerkamp, B.: Irisdiagnostik. Eine augenärztliche Kritik. Beiheft der Klin. Mbl. für Augenheilkunde, 22, 1955.
4. Lester, J.M., Brenner, J.F., Selles, W.D.: Local Transforms for Biomedical Image Analysis. Comp. Graph. and Image Process., 13, 17-30, 1980.

5. Pavlidis, T.: Algorithms for Graphics and Image Processing. Comp. Sci. Press, Rockville, 1982.
6. Popescu, M.P., Waniek, D.: Improvement of the Iridodiagnostic Method. Rev. Chir. Ophthalmol., 30,29-35, 1986.
7. Rosenfeld, A., Kak, A.C.: Digital Picture Processing. Academic Press, New York, 1982.
8. Simon, A., Worthen, D., Mithas, J.A.: An Evaluation of Iridology. J. of the American Medical Assoc., 247, 1385-1389, 1979.
9. Waniek, D.: A Model for a Possible Non-Visual Function of the Iris, Lens and Peripheral Retina. Medical Hypotheses, 23,309-312, 1987.

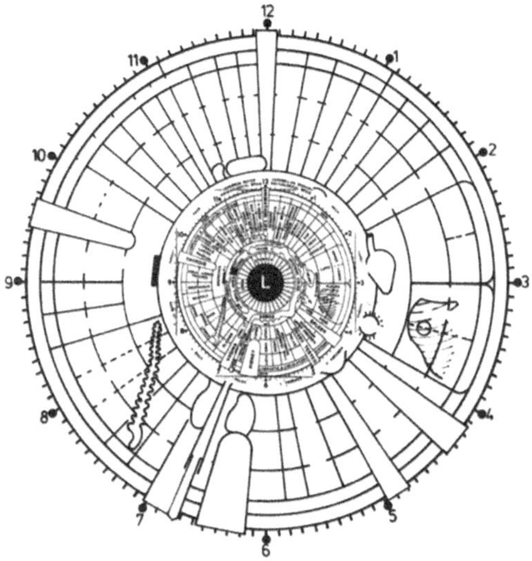

Figure 1.: Typical iridologic chart for the left eye.

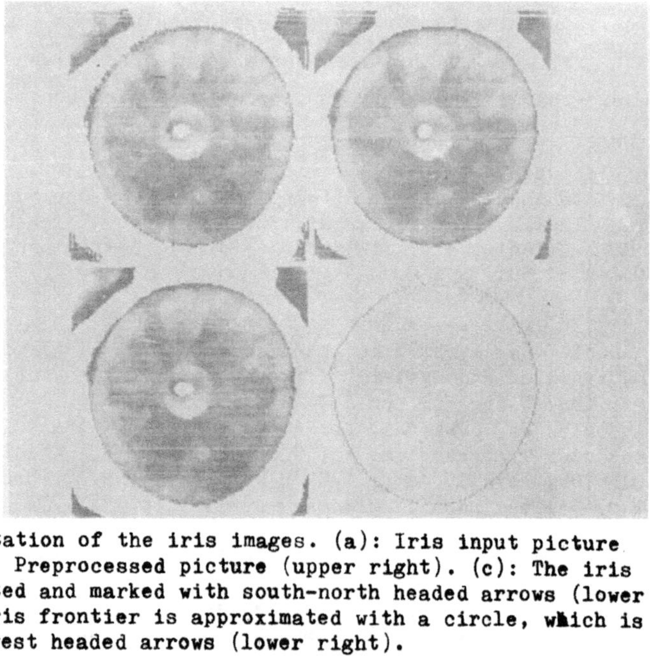

Figure 2.: Segmentation of the iris images. (a): Iris input picture (upper left). (b): Preprocessed picture (upper right). (c): The iris frontier is detected and marked with south-north headed arrows (lower left). (d): The iris frontier is approximated with a circle, which is marked with east-west headed arrows (lower right).

Computer Graphics

3D Display with Minimal Predefinition

Stephen M. Pizer, Henry Fuchs, Marc Levoy,
Julian G. Rosenman, Richard E. Davis, Jordan B. Renner

Medical Image Display Research Group
University of North Carolina, Chapel Hill, NC, USA

Abstract

The portrayal of medical images in three dimensions can be strikingly improved in presentational usefulness and in user time requirements with methods based on volume rendering rather than surface rendering. We describe methods that lead to such improvement. These involve, in order of application, region-of-interest definition, as opposed to object definition; fuzzy surface specification; rendering not only fuzzy surfaces but together with them, surface models and grey-level slices; and fast display. These techniques lead to improved usefulness in many areas of medicine, including radiation oncology, surgery, multimodality image display, and diagnostic imaging from CT, MRI, SPECT, PET, and ultrasound.

Introduction

3D display of medical images has traditionally involved the predefinition of regions of the image volume corresponding to objects to be displayed, followed by the portrayal of the explicitly located surfaces of these regions. This surface rendering approach is flawed by requiring the user fully to comprehend and define the 3D objects to be portrayed before they are ever seen in 3D. The result is that object features can be missed and that only the highest contrast objects can be defined without hours of user time per 3D image.

Recently new methods for more directly displaying 3D data have been developed. Among the techniques developed have been the calculation of shading based on local intensity gradient [Höhne, 1986], the use of ray tracing to generate images [Schlusselberg, 1986], and improved algorithms for detecting surface voxels [Trousset, 1987]. More recently, our group [Levoy, 1988a] and others [Drebin, 1988; Sabella, 1988; Upson, 1988] have developed so-called volume rendering techniques, in which all voxels in the dataset are assigned a fuzzy surface likelihood and participate in the generation of each image. Such an approach has two major advantages over ones based on explicit (binary) surface specification. First, minimal predefinition is required: only the carving out of a region excluding surrounding or adjacent objects that are likely to obscure or confuse the portrayal of the objects of interest. The surface likelihoods are directly calculated from the 3D intensity values in the original image data in the non-excluded region (call this the region-of-interest or ROI). Second, the presentation of the objects is improved since the method takes into account all the image intensities in a surface region rather than limiting the basis to binarily chosen surface points.

Surface rendering and grey-scale display techniques are not made obsolete by volume rendering. Surface rendering remains of significant use, in combination with volume rendering, for the portrayal of objects where binary specification is necessary: e.g., for the portrayal of prostheses, radiation treatment beams, isodose surfaces, surgical hazards, and other objects defined by the therapy planner, as well as anatomic objects whose volumes or surface properties must be measured. In addition, surface rendering may improve the appreciability of multiple, nested complicated objects that appear in human anatomy.

Grey-level display of the original slice intensities is not made obsolete because it portrays subtle intensity differences in a way impossible for any 3D volume display modality. Therefore, we have developed methods that allow the combined display of binary surfaces, surface likelihood, and superimposed grey-scale slices.

Volume rendering is most effective with fast display. Immediate display allows interactive selection of viewing parameters such as point of view, and allows ROIs and parameters affecting surface appearance to be chosen not when the image structures have been appreciated only in 2D slices but when this tuning seems necessary according to structure already portrayed in 3D. The effect of this tuning can be immediately appreciated in 3D, allowing real-time feedback to the parameter setting.

In the following we describe in order of their application each of the required steps, shown in Figure 1.

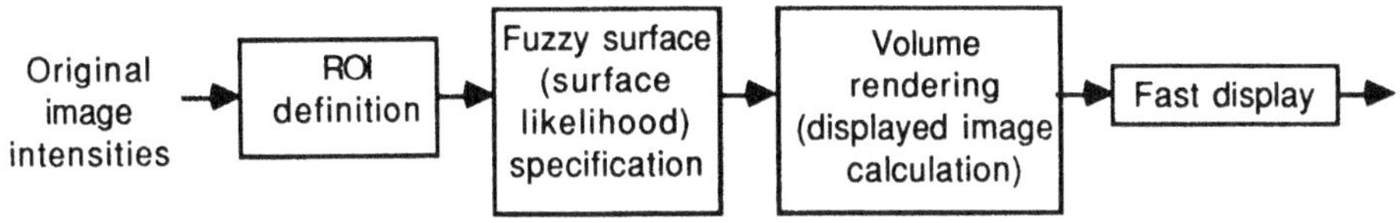

Figure 1. Display steps for volume rendering

Region-of-Interest Definition

Good display via volume rendering requires that surrounding and confusing regions be excluded before display. For example, to present the cortex of the brain from MRI (see Figure 2), the skull and its skin must be removed.

This selection of regions for exclusion, or for inclusion, is a far less demanding objective than the definition of the surface of the object to be displayed. Sometimes the ROI can be specified by a simple geometric construct such as a rectangular prism or an ellipsoid. In other cases simple sculpting may be necessary. Both of these approaches require the visualization of the selected region relative to anatomy during the ROI selection, either in 3D or on an array of grey-scale slices of the original image data. Interactive display is thus necessary.

An alternative to specification of the ROI via simple geometric shapes is based on automatically precomputed image descriptions that divide the image into "sensible", coherent regions, probably hierarchically by scale [Pizer, 1989]. The user can then point to a voxel (perhaps displayed as a pixel on a grey-scale slice) and cause a whole image region to be selected and displayed [Coggins, 1989]. This region may frequently be used without editing in selecting or excluding voxels for volume rendering. Gauch [1988] has recently demonstrated the feasibility of computing such regions in 2D images based on a construct called the *Intensity Axis of Symmetry* that defines regions in terms of symmetry relations on the image intensities in an intensity "ridge" or "course" (see Figure 3). Cullip [1989] is now extending this approach to 3D.

Fuzzy Surface Specification

Volume rendering computes opacities and shades from a surface likelihood value at each voxel. Such a value is far easier to specify with adequate accuracy than binary surface locations, because the display method is more forgiving to errors in this non-binary value.

Drebin [1988] and Levoy [1988a] have described methods for using intensity value and intensity gradient to calculate a surface likelihood or "strength" for every voxel in the original image data. The basic idea is to let the surface strength increase both with the closeness of the voxel's intensity to some intensity range associated with the object to be displayed, and with the magnitude of the intensity gradient at the voxel. It seems possible to have the surface strength depend also on higher order local geometric properties such as curvature, global geometric properties such as symmetry or surface continuity, or surface probabilities calculated using knowledge of the imaging process [Lin, 1989].

Binary surface specification is handled by volume rendering as a special case, where the probabilities are zero or one. This is useful in the portrayal of treatment objects (see Figure 4) and of scalar intensity distributions without edges, such as radiation dose distributions that need to be visualized relative to anatomy in radiotherapy treatment planning. Surface likelihoods of unity can be set at locations having selected levels of the dose. Binary surface specification will be also be useful for portraying anatomic objects whose edge contrast is too low or diffuse to allow surfaces to be portrayed by direct volume rendering. In this case, the user can define object surfaces by the standard manual techniques, and the resulting surface tiles can be inserted into the volume rendering.

Volume rendering, as described by Levoy and Drebin, also requires a surface normal for each voxel. This is normally taken to be the direction of the intensity gradient, though higher order measures of local shape could also have an effect.

Volume Rendering

In volume rendering opacities and shades are computed from the surface likelihood at each voxel, and rays are cast to calculate the effect of the shades as attenuated by the opacities on the display pixel through which the ray passes. Our results show the importance of careful anti-aliasing in the interpolations involved in computing shades and opacities along the rays from their values at the discrete voxel centers (see Figures 5 and 6). They also show the usefulness of specular highlights, fixed light positions and rotating objects, and color use (see Figure 4) to distinguish object types.

Levoy has developed techniques [Levoy, 1988b] for converting surfaces defined by polygonal tiles into the opacities and shades used by volume rendering and doing so in a way that avoids aliasing artifacts. Multiple surfaces, some rendered directly by volume rendering and others by polygonally modeled surfaces rendered into the surface likelihood values, can be used with color and transparency to allow the understanding of multiple, complicated objects, as, for example, are found in the inner ear (see Figure 6). It is frequently useful to portray grey values superimposed on 3D images, with the 3D objects providing context by which to understand the grey values. Höhne [1987] has shown the effectiveness of this approach with grey-scale values coming from the original intensity data on which the 3D display is based. In our group Tsui [1988] has shown that this idea is also effective for showing physiological function from one 3D image intensity set in the context of anatomy coming from another. Our approach is to render such intensities onto planes which are combined with volume rendering via texture mapping onto planes that clip the volume rendered 3D region [Levoy, 1988b].

Fast Display of Volume-rendered Images

Volume rendering of 256 $\times$ 256 $\times$ 256 image intensities can be accomplished in approximately 10 minutes of computer time on a Sun 4 workstation, using techniques developed by Levoy [1988c, 1989]. These techniques allow the re-rendering from a new point of view in only 2 minutes. Furthermore, progressive refinements are possible which allow a coarse rendering in only a fraction of this time, with improvement to full quality occurring over a few minutes when interaction stops.

The techniques described above lead to an even greater promise for volume rendering: allowing real time exploration, in 3D, of 3D intensity data. On machines such as Pixel-Planes 5, a parallel graphics and image processing engine being developed in our laboratory [Fuchs, 1989] we calculate that it will be possible to produce coarse renderings from 256^3 voxels at the rate of 20 per second. Thus it should be possible to vary parameters of the voxel classifications (such as ROI or the opacity function's slope vs. gradient) or viewing parameters (such as point of view, color, transparency, or texture-mapping clipping plane) and see the result fast enough for dynamic adjustment. This approach promises to allow the same sort of tuning of the image to optimally portray an anatomic region that is presently achieved by interactive intensity windowing with 2D images. The power of this ability to explore directly in 3D will not be fully known until this ability is realized, but we think it may qualitatively increase the usefulness of 3D medical image display for diagnostic purposes and improve the already accepted uses of 3D display in planning for therapy in radiation oncology and surgery.

Clinical Results

This section characterizes the clinical effectiveness of 3D display based on volume rendering, as illustrated by the images already shown and one new image. Illustrated properties of volume rendering are also discussed.

Figure 2, showing a volume rendering of the cortex from MRI, illustrates that a complicated surface can be well portrayed without much manipulation. Only the image regions corresponding to the skin on the skull were excluded. It was not necessary to remove bony skull since its intensity is very different from the cortex. Also, since a single layer of surface elements can be made to produce only partial transparency, it was not necessary to remove the closely fitting but thin arachnoid mater to visualize the cortex, which is thicker and thus builds up the opacity that makes it appear as a surface. With such an image of a complex surface, we have found improved visualization by putting the brain into rotation via a cine loop, especially if the light source has a fixed position.

Figure 4 illustrates the potential of volume rendering in radiation treatment planning. It shows the relation between a radiation beam, in blue, and a patient's brain tumor and head. It is our experience that without high quality 3D graphics it is difficult for a clinician to understand the complex spatial relationship between the radiation beam positions with their associated dose distribution, the target treatment region, and radiosensitive normal tissue in a fully 3D radiation treatment plan. The figure illustrates how combination of objects represented by surface likelihood and by polygonal tiles may increase the clinician's comprehension of a given treatment plan. Because of its inherent detail, volume rendering provides an excellent understanding of the patient's anatomy, represented by surface likelihoods computed from CT image data; also, the method's ability to use color allows the clinician to distinguish different tissue types, such as the flesh in red and the bone in yellow. At the same time the method allows the appreciation of the user-defined (and thus tiled) radiation beam and target region, the latter consisting of both the imaged tumor and surrounding tissues into which the tumor may have spread. The result makes it easy to see, for example, if the entire tumor target is covered by the beam or if the beam is unnecessarily impacting upon normal tissues such as the patient's eye.

The clinician typically must interactively select from or vary radiation beam arrangements. Choosing arrangements that correctly intersect or avoid anatomic regions requires volume renderings that are produced fast enough for the viewpoint to be dynamically varied. The final evaluation, requiring the clinician to understand the relation of the radiation dose distribution and the patient's anatomy, can be made from the combination of transparent, colored isodose surfaces, volume rendered directly from the dose distribution, and the anatomy and tumor target, volume rendered directly from the CT image

Summary and Conclusions

Volume rendering can provide a number of benefits that will strongly aid clinical treatment planning, diagnosis, and followup. Volume renderings of objects that appear in image data with high contrast show more detail and more accurate detail than surface renderings. Objects that appear less sharply in image data can provide adequate 3D comprehension while avoiding the enormous time consumption of object definition that is the rate-limiting process with clinical application of surface rendering techniques. In imaging areas like nuclear medicine and ultrasound, in which surfaces are inherently fuzzy, volume rendering may be useful where surface rendering was close to useless. Especially important for diagnosis, volume rendering may allow exploration in 3D of the image data to find out what 3D structures and relationships the data show. Clinically important uses of this technique in surgery planning, radiotherapy treatment planning, and medical imaging diagnosis abound.

However, to obtain these benefits, much work is still needed. Better classification techniques for defining ROIs and the surface likelihood of selected organs are necessary. Improvements in mapping surface likelihood to opacity will improve image quality when surfaces are especially fuzzy. Methods for avoiding the effects of motion or correcting for motion are very important to avoid the major artifacts that motion now causes for volume rendering. Methods for real-time rendering of precollected images must be developed not only to add convenience to preparation of images for therapy planning but to allow at all the explorational forms of 3D display that most diagnostic uses await. And methods of real-time rendering of real-time image data will allow volume rendering to be used to visualize dynamic processes in the body while they happen.

Acknowledgements

Dr. Mark Schiebler has consulted on clinical aspects of MRI imaging and display, and Mr. George Sherouse and Dr. Edward Chaney have consulted on and provided images from radiotherapy treatment planning. Images have been prepared and processing and display programs written by Mr. Andrew Skinner, Ms. Jiuqi Tang, Ms. J. Lynne Hendricks, and Mr. Timothy Cullip. Dr. Benjamin Tsui's project on multimodality 3D display, the project of Dr. James Coggins, Mr. Timothy Cullip and Mr. Eric Fredericksen on interactive display of image region hierarchies, and the project on image description of Mr. John Gauch and Mr. Timothy Cullip have contributed significantly to the reported results. We are also indebted to NeuroCT technicians Bonnie Gordon and Pamela Buckingham, who have spent considerable time archiving CT data sets to be used for 3D display. Finally, all of this work is made possible by our laboratories manager Dr. Graham Gash and by our administrative assistant, Ms. Carolyn Din, for whose help with this manuscript we are grateful.

We much appreciate 3D MRI images provided by Mr. Hartmut König of Siemens, Erlangen. The work leading to this paper was carried out with the support of NCI grant # P01 CA47982 and NSF grant # CDR 8622201.

References

Coggins, J., Cullip, T., Fredericksen, E., Pizer, S., "A Data Structure for Image Region Hierarchies," Internal Report, Department of Computer Science, University of North Carolina, 1989.

Cullip, T., "Intensity Axis of Symmetry and its Application to Image Segmentation and Hierachy Manipulation", *Radiology Research Review,* Univ. of NC Dept. of Radiology, 1989.

Drebin, R.A., Carpenter, L., and Hanrahan, P., "Volume Rendering," *Computer Graphics,* **22**(4):65-74,1988.

Fuchs, H., Poulton, J., Eyles, J., Greer, T., Goldfeather, J., Ellsworth,D., Molnar, S., Turk., Tebbs, & Israel, L., "A Heterogeneous Multiprocessor Graphics System Using Processor-Enhanced Memories," Technical Report 89-005, Computer Science Department, University of North Carolina at Chapel Hill (submitted for publication), 1989.

Fishman, E.K., Magid, D. Ney, D.R., Kuhlmand, J.E., Brooker, A.F., "Three Dimensional Imaging In Orthopedics: State of the Art 1988", *Orthopedics,* **11**(7):1021-1026, 1988.

Gauch, J., "Image Description via the Multiresolution Intensity Axis of Symmetry," *ICCV Proc. IEEE* Catalog #88CH2664-1:269-274, 1988.

Höhne, K.H. and Bernstein, R.,"Shading 3D-Images from CT Using Gray-Level Gradients," *IEEE Transactions on Medical Imaging,* **5**(1):45-47, 1986.

Höhne, K.H., DeLapaz, R.L., Bernstein, R., Taylor, R.C., "Combined Surface Display and Reformatting for the 3D-Analysis on Tomographic Data," *Investigative Radiology,* **22**: 658-665, 1987.

Hottier, F., Personal Communication, 1988.

Levoy, M, "Display of Surfaces from Volume Data," *IEEE Computer Graphics and Applications,* **8**(3):29-37, 1988a.

Levoy, M., "Rendering Mixtures of Geometric and Volumetric Data," Technical Report 88-052, Computer Science Department, University of North Carolina at Chapel Hill (submitted for publication), 1988b.

Levoy, M., "Efficient Ray Tracing of Volume Data", Technical Report 88-029, Computer Science Department, University of North Carolina at Chapel Hill(in review, *ACM Transactions on Graphics)*, 1988c.

Levoy, M., "Volume Rendering by Adaptive Refinement," *The Visual Computer,* **5**(3), 1989a.

Lin, W., *Probablistic Surface Definition in 3D Ultrasound Images,* Ph.D. Dissertation Proposal, UNC Department of Computer Science, 1989.

McCann, H.A., Sharp, J.C., Kinter, P.M., McEwan, C.N., Barillot, C., Greenleaf, J., "Multidimension in Ultrasound Imaging for Cardiology," *Proc. IEEE,* **76**(9):1063-1073, 1988.

Moaddab, M.B., Dumas, A.L., Chavoor, A.G., Neff, P.A., Homayoun, N., "Temperomandibular Joint: Computed Tomographic Three-Dimensional Reconstruction," *Am. J. Orthod. Dentofacial Orthop.*, **88**(4):342-352, 1985.

Pizer, S.M., "Multiscale Methods and the Segmentation of Medical Images," to appear in *The Formation, Handling and Evaluation of Medical Images*, Springer-Verlag, Berlin, 1989.

Sabella, P., "A Rendering Algorithm for Visualizing 3D Scalar Fields," *Computer Graphics*, **22**(4):1-58, 1988.

Schlusselberg, D.S. and Smith, W.K., "Three-Dimensional Display of Medical Image Volumes," *Proc. NCGA* '86, Anaheim, CA, **3**:114-123, 1986.

Trousset, Y. and Schmitt, F.," Active-Ray Tracing for 3D Medical Imaging," *Proc. EUROGRAPHICS '87*: 139-149, 1987.

Tsui, B.M.W., Perry, J.R., Gilland, D.R., Interrante, V.L., Fuchs, H., & McCartney, W.H., "Preliminary Study of Simultaneous Three-Dimensional Display of Anatomical Functional Information Obtained with SPECT," *Proc. Soc. Nuclear Medicine*, 1988.

Upson, C. and Keeler, M.,"VBUFFER: Visible Volume Rendering," *Computer Graphics*, **22**(4):59-64, 1988.

Vannier, M.W., Conroy, G.C., Marsh, J.L., Knapp, R.H., "Three - Dimensional Cranial Surface Reconstruction Using High-Resolution Computed Tomography," *Am. J. Phys. Anthropol.*, **67**:299-311, 1985.

von Ramm, O., Smith, S. W., Sheikh, K.H., Kisslo, J., "Real-Time 3-Dimensional Echo Cardiography", presented in *Circulation,* Supplement II, **78**:35, 1988.

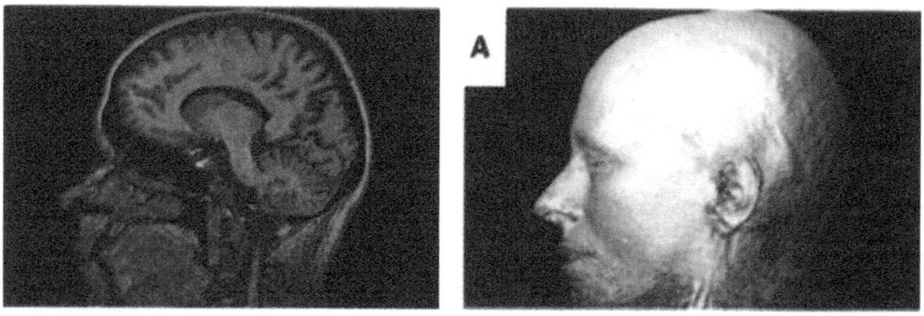

Fig.2a. MRI Sample slice and volume rendering of skin surface (MRI data courtesy of Siemans).

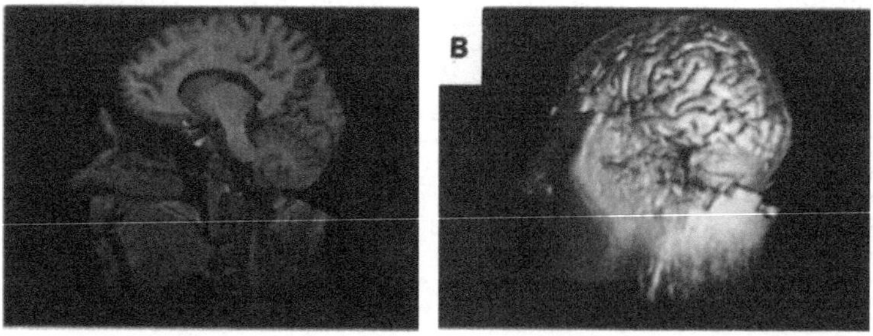

Fig.2b. MRI sample slice edited to exclude skin region, and subsequent volume rendering of exposed cerebral cortex.

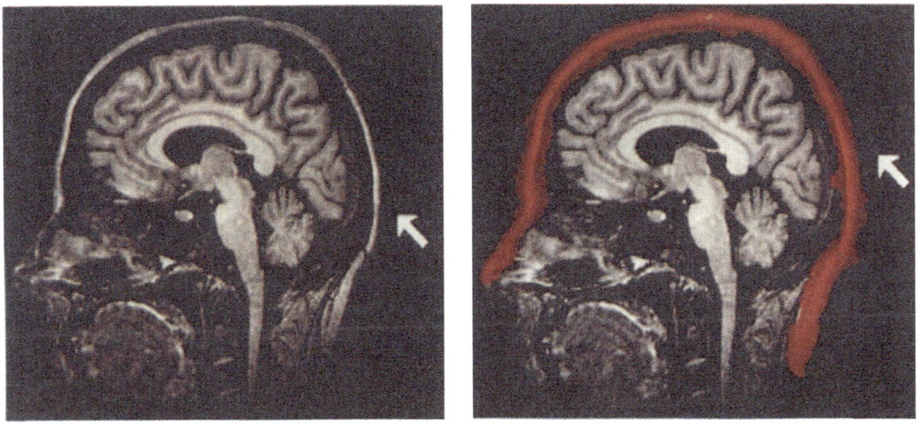

Fig.3. Sensible region computed via Intensity Axis of Symmetry of 3D MRI data (MRI data courtesy of Siemens).

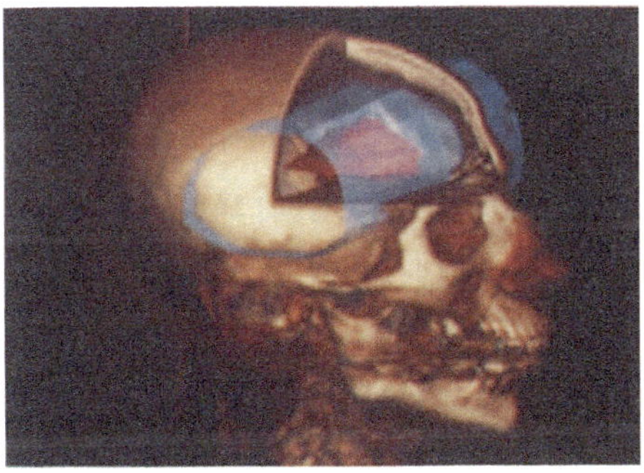

Fig.4. Polygonally tiled radiation treatment beam and tumor target in volume rendered head from CT.

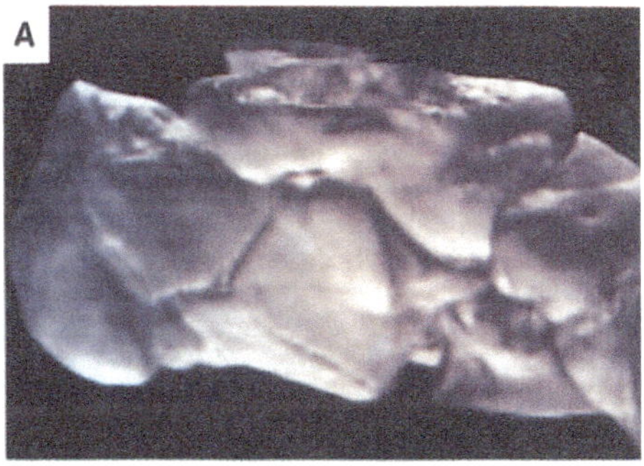

Fig.5a.

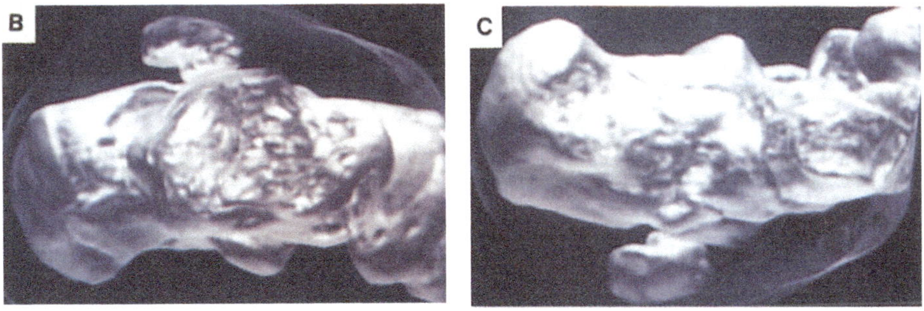

Fig.5. Volume rendered medial (5a), superior (5b), and inferior views (5c) of fractured human heel from CT.

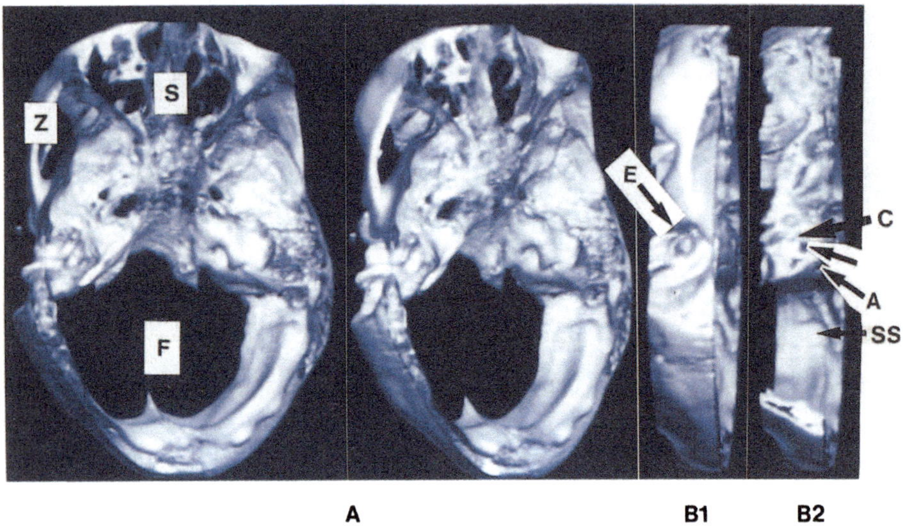

Fig.6. Volume renderings for surgical planning of temporal bone from CT data. Fig.6a shows representative frames from a rotational cine loop. A 3.0 cm slice of the skull is seen from a left superior oblique viewpoint. The left zygomatic arch(z), posterior nasal septum(s) and posterior cranial fossa(f) are indicated. Fig.6b1 shows the left temporal bone from the surgeon's perspective. The patient is lying face up and is viewed down the external ear canal(e). In fig. 6b2, the temporal bone has been "sanded away" to reveal the cochlea(c), the internal auditory canal(i), the contralateral internal auditory canal(a) and the contralateral sigmoid sinus (ss).

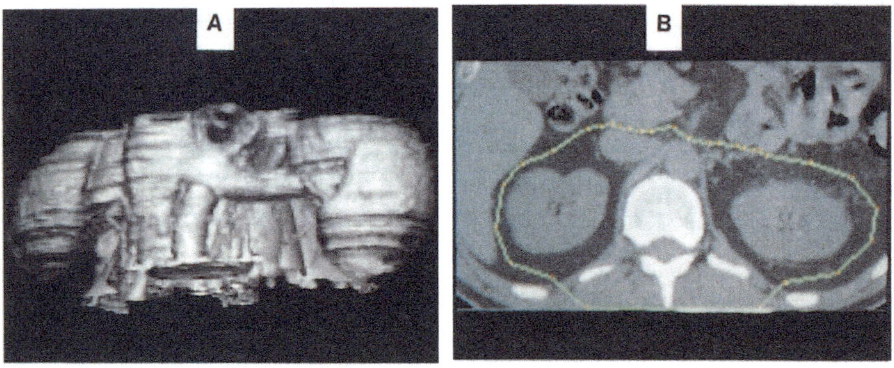

Fig.7. Volume rendering of kidneys from CT DATA (7a), with one slice (7b) showing included ROI.

Image Quality in Voxel-Based Surface Shading

A. POMMERT, U. TIEDE, G. WIEBECKE, K.H. HÖHNE

Institute of Mathematics and Computer Science in Medicine (IMDM)
University-Hospital Eppendorf
Hamburg, F. R. G.

Summary

Three-dimensional (3D) display of medical objects is an essential aid for the interpretation of tomographic volume data. For clinical applications, it is important to ask whether the resulting images show the true anatomical situation. In this paper, we investigate the effects of different shading algorithms—especially of the gray level gradient method—on the quality of 3D-imaging. As a result, an algorithm with an improved spatial resolution is presented.

Introduction

Radiological examinations are increasingly based on tomographic volume data, as obtained in x-ray computed tomography (CT) or magnetic resonance imaging (MRI). In current practice, these data have to be viewed slice by slice in order to get an impression of the three-dimensional (3D) structures which they represent—a very difficult task even for an experienced observer.

A much more natural 3D-perception is obtained if the objects are presented like a surgeon or an anatomist would see them. For this purpose, various methods have been developed during the last few years [4]. Roughly, after image acquisition and some preprocessing, the surfaces of the objects of interest are defined and shaded. The fidelity of the resulting 3D-images thus depends on many variables like slice acquisition parameters and processing algorithms used.

In a previous paper, we presented some first results of our investigations on quality assurance [6]. We are now concentrating on the effects of different shading algorithms on the fidelity of 3D-imaging, with special regard to the gray level gradient method.

Method

For clinical data, it is often difficult to compare the fidelity of different 3D-images since we usually do not know how the reality looks like. Therefore, we are using simulated test data for our investigations. The whole procedure consists of the following steps:

1. Simulated tomographic acquisition of a formally specified test object. Like in CT and MRI, sampling is done in co-planar slices with adjustable distance and thickness. Test objects may be any combinations of geometric primitives like ball, cone, cube etc. with specified gray values. At the object surfaces, the partial volume effect occurs.

2. 3D-image processing as for real clinical data (Höhne et al. [4]): for this investigation, no interpolations and rotations are performed, segmentation is done with thresholding. For shading, we apply z-buffer and gray level gradient methods, which provide the surface normal vectors used for surface rendering with the Phong illumination model [1].

3. Comparison of the results with the original object specification by visual inspection and/or quantitative quality measures.

As a quantitative measure for the accuracy of shading algorithms we use the angles between computed and ideal surface normal vectors (in the range of 0 to 90 degrees). They can be visualized in an *error image* thus presenting the local deviations. The sum of all these angles gives an idea of the *global* performance of the algorithm.

Investigated Shading Methods

Z-buffer gradient method: the surface normal vectors are estimated from the depth image of an object, which is usually stored in the so-called z-buffer [2]. For every pixel, depth information is considered in a 3×3 environment.

Gray level gradient method: as described by Höhne and Bernstein [3], this method makes use of the correspondence between surface inclination and gray level gradients in the 3D-neighborhood of a surface voxel which is caused by the partial volume effect. The method can be applied to any number of neighbours. The algorithm proposed by Tiede [7] calculates the three components of a gradient G for a surface voxel at (i, j, k) directly from the gray levels g of its 6 neighbours along the main axis as

$$G_x = g_{(i+1,j,k)} - g_{(i-1,j,k)},\ G_y = g_{(i,j+1,k)} - g_{(i,j-1,k)},\ G_z = g_{(i,j,k+1)} - g_{(i,j,k-1)}.$$

Normalisation of G yields the surface normal. A narrower asymmetric approximation with only 3 neighbours is given by

$$G_x = g_{(i,j,k)} - g_{(i-1,j,k)},\ G_y = g_{(i,j,k)} - g_{(i,j-1,k)},\ G_z = g_{(i,j,k)} - g_{(i,j,k-1)}.$$

The gray level gradient may also be calculated from all 26 voxels in a $3 \times 3 \times 3$ neighbourhood, weighted according to their distance from the surface voxel [7]. The algorithm is then identical to the Zucker-Hummel operator used for edge detection [8].

Adaptive gray level gradient method: for small objects even a 6 voxel neighbourhood may be too large. An algorithm which adapts to the object using only 3–6 meaningful neighbours calculates G_x as follows (G_y, G_z similarly): if the gray value of the surface voxel at (i,j,k) is greater than the gray values of its neighbours at $(i-1,j,k)$ and $(i+1,j,k)$, it takes the gradient between the surface voxel itself and the neighbour with the lower gray value, otherwise the gradient between both neighbours.

Transparent gray level gradient method: in contrast to all other methods previously mentioned, the approach described by Levoy [5] does not require a segmentation step. Instead, a transparency is assigned to every voxel in a volume depending on its gray value and its gray level gradient. It is then shaded with the gray level gradient method (6 neighbours). It is claimed that the presentation of thin objects can thus be improved.

Experimental Results

We have applied the described algorithms to a number of simulated and clinical data. For a solid object as shown in fig. 1, the gray level gradient method results in better images than the the z-buffer gradient method. It can even visualize details which are below the spatial resolution of segmentation, as a gap of only 0.5 voxels width. The corresponding error images show that the z-buffer gradient method gives only a rough estimate of the real surface inclination, whereas the gray level gradient method comes very close to the ideal values. A further improvement can be achieved using 26 instead of 6 neighbours.

For an investigation of the shading problems which occur at thin surfaces, we have applied the gray level gradient method to a ball with a very thin wall (1 voxel). The error images are shown in fig. 2. Deviations are quite high for 6 (fig. 2b) and much better for 26 neighbours (fig. 2d). The asymmetric approach with 3 neighbours is clearly not suitable for real applications (fig. 2a). It shows however that very good results as in the upper left quadrant can be obtained using a narrow neighbourhood. It is thus not surprising that the adaptive method with 3–6 neighbours (fig. 2c) is superior to the 6 neighbours algorithm and shows a globally similar performance to the 26 neighbours algorithm.

Fig. 3a shows an application of the gray level gradient method with 6 neighbours to a clinical case. With 26 neighbours, the aliasing effects are strongly reduced (fig. 3c). The black spot artefacts at thin parts like in the left orbita which occur in both images are almost eliminated with the 3–6 neighbours algorithm (fig. 3b). In fig. 3d which was produced with the transparent gray level gradient method these artefacts are still visible, but veiled from overlaid structures.

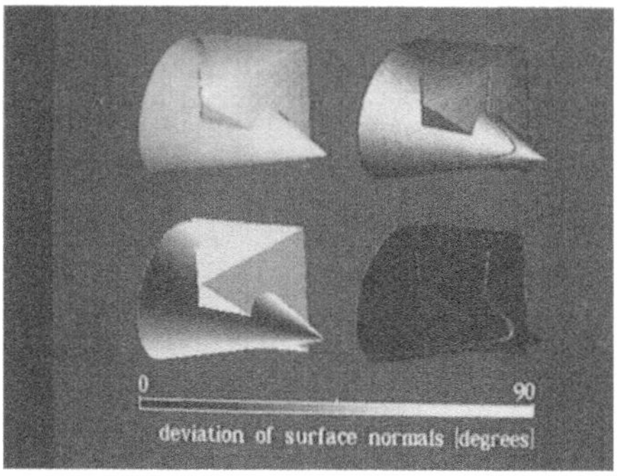

Fig. 1. 3D-images (top) and corresponding error images (bottom) for a solid test object with a small gap. The gray level gradient method (right) calculates the surface normals more accurately than the z-buffer gradient method (left).

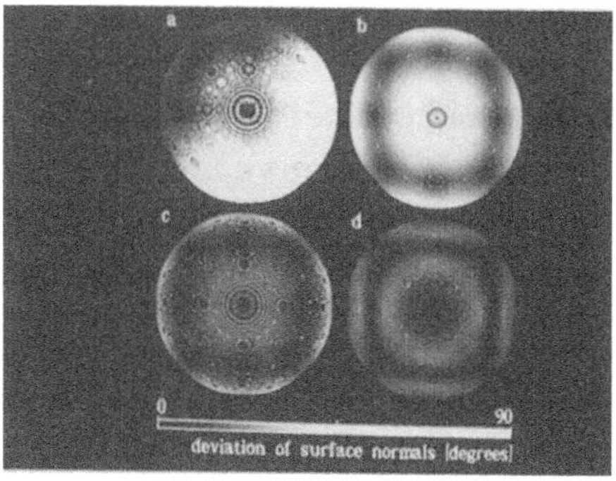

Fig. 2. Error images for the gray level gradient method, applied to a ball with a very thin surface. a: 3 neighbours, b: 6 neighbours, c: adaptive gray level gradient method with 3–6 neighbours, d: 26 neighbours.

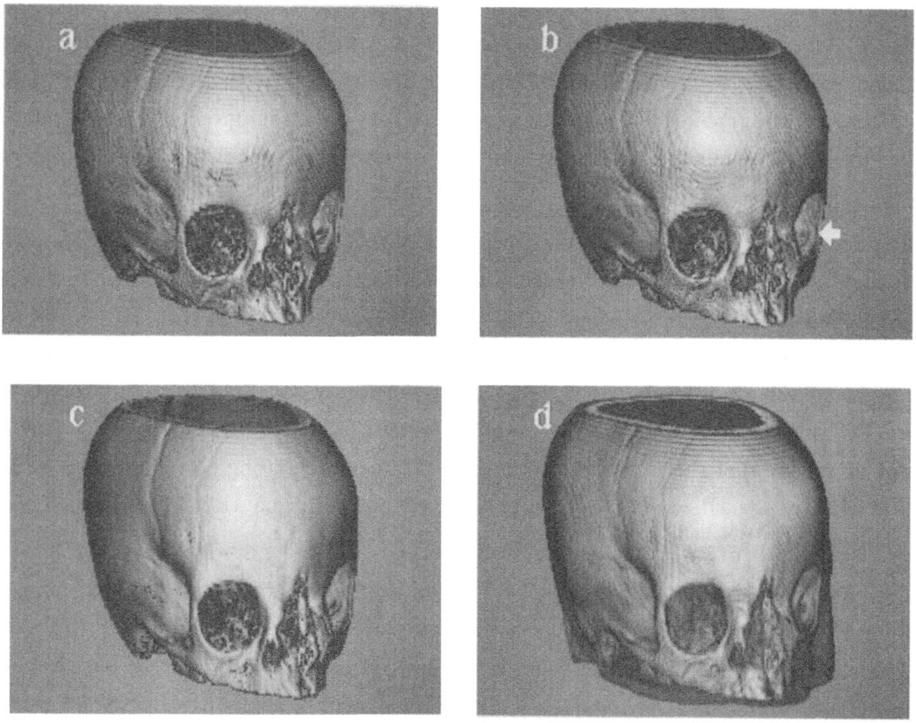

Fig. 3. Skull of a three year old child with a facial cleft, prepared from 61 CT slices. a: gray level gradient method, 6 neighbours. b: adaptive gray level gradient method, 3–6 neighbours. Artefacts at thin surfaces like in the left orbita are reduced (arrow). c: gray level gradient method, 26 neighbours. d: transparent gray level gradient method.

Conclusions

It has been shown that error images are a valuable tool for the assessment of different shading methods. For shading solid objects, the gray level gradient method with 26 neighbours gives the best results. For thin structures however the adaptive gray level gradient method that we have implemented is more accurate. It is expected that an extension which selects the gray level gradient out of all 26 neighbours should be even better.

Compared to the gray level gradient method, the images produced with the transparent gray level gradient method are more blurred and suffer from the same problems at thin surfaces. This approach might benefit from neighbourhood adaption as well.

Acknowledgements

Our thanks go to M. Bomans and M. Riemer for their permanent support and to R. Schubert who helped preparing the images of the skull. We are also grateful to the Werner Otto Foundation, Hamburg, for supporting this project.

References

[1] Bui-Tuong, Phong: Illumination for Computer Generated Images. Comm. ACM 18 (1975), 311–317.

[2] Chen, L.S.; Herman, G.T.; Reynolds, R.A.; Udupa, J.K.: Surface Shading in the Cuberille Environment. IEEE Comput. Graphics Appl. 5, 12 (1985), 33–43.

[3] Höhne, K.H.; Bernstein, R.: Shading 3D Images from CT Using Gray Level Gradients. IEEE Trans. Med. Imaging 5 (1986), 45–47.

[4] Höhne, K.H.; Bomans, M.; Pommert, A.; Riemer, M.; Schiers, C.; Tiede, U.; Wiebecke, G.: 3D-Visualization of Tomographic Volume Data Using the Generalized Voxel-Model. Visual Computer 5, 3 (1989), in press.

[5] Levoy, M.: Display of Surfaces from Volume Data. IEEE Comput. Graphics Appl. 8, 3 (1988), 29–37.

[6] Pommert, A.; Bomans, M.; Tiede, U.; Höhne, K.H.: Simulation Studies for Quality Assurance of 3D-Images from Computed Tomograms. In Todd-Pokropek, A.; Viergever, M.A. (eds.): The Formation, Handling and Evaluation of Medical Images. NATO ASI Series F: Computer and Systems Sciences, Springer, Berlin (1989), in press.

[7] Tiede, U.; Riemer, M.; Bomans, M.; Höhne, K.H.: Display Techniques for 3D-Tomographic Volume Data. In Proc. NCGA '88, Vol. III. Anaheim (1988), 188–197.

[8] Zucker, S.W.; Hummel, R.A.: An optimal three-dimensional edge operator. Report 79-10, McGill University (1979).

Minimizing the Data Preparation for 3D Visualization of Medical Image Sequences

V. Heyers, H.P. Meinzer, F. Saurbier, D. Scheppelmann and R. Schäfer

Department of Medical and Biological Informatics
German Cancer Research Center Heidelberg

Summary

Computed representations of three dimensianal medical structures on a two dimensional screen create the illusion of the third dimension. This kind of data representation facilitates the diagnostic process and enhances the interpretatibility of the images. The observation of 3D relationships is more comfortable than using the image sequence slice by slice.
Volume rendering of unmodified volume data (the greyvalue indicates the opacity of a volume element) performs a proper representation of soft tissue and complex medical scenes without any previous segmentation. In this point the algorithm differs basicly from conventional methods.
However, the visualization won't be suggestive if the opacity of structures fitted into one another doesn't increase between the viewer's eye and the intresting structure in the data space. Therefore the visualization may need data preprocessing which can be minimized.

1. Introduction

Visualization achieves the step from the description of an object to its image [1]. The quality of this description fixes an upperlimit for the quality of the image.
In medicine, the description exists in the form of measuring data (e.g. absorption densities for X-rays with CT images). These data cannot answer medical problems directly. Answers only become feasible by informed analysis and interpretation.
Thanks to technical improvements, the spacial resolution of the measuring data increases. This increase of the quantity of data leads to a growing necessity of a suitable presentation in order to be able to fully exhaust the wealth of information for the purpose of analysis. Visualization procedures achieve this presentation which can become more and more perfect with the growing accuracy of the description.
Classical visualisation procedures present *surfaces*. Therefore, part of the analysis of measuring data will have to be anticipated by the computer, i.e. the

dissection of the volume of data into induvidual components. The calculated image only contains the surfaces estimated by the computer which means a considerable loss as compared to the information content of the original measuring data. Moreover, the necessary dissection can be automatically executed in exceptional cases only.
Of late, *volume elements* have been successfully presented which, next to the definite improvement of the quality of the picture, also has the advantage that the entire information about measuring data can be used to create the picture [2-4]. To generate the pictures, the measuring data need not necessarily be preprocessed; in particular, segmentation or classification is not required [4]. However, covering the structure of interest by an irrelevant one can necessitate preprocessing.

2. Different Levels of Preprocessing

According to data material and interest, the visualization of volume elements requires varying degrees of preprocessing. In the interest of short computing times and images which are as unadulterated as possible, one bewares of unnecessary manipulations of the initial data.

2.1. Visualization of the Original Data

The volume data are interpreted as absorption densities for light.

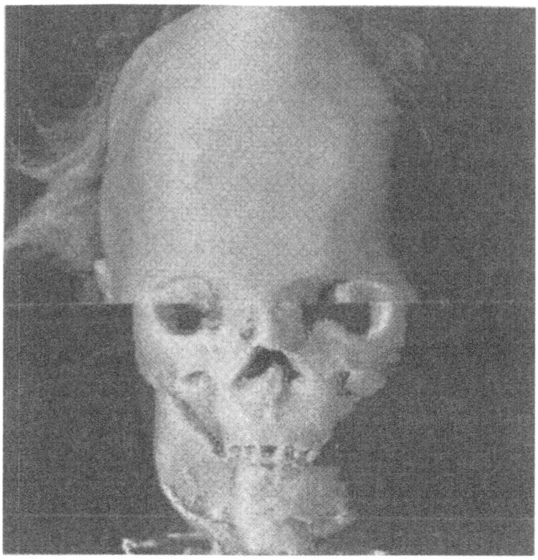

Fig. 1: Example for the presentation of the original data material. By naming a parameter, the transparency of an object can be fixed [4]. Visualization of 130 CT-slices (256×256).

To facilitate the differentiation of the various structures, the picture can be supplemented by colour information.

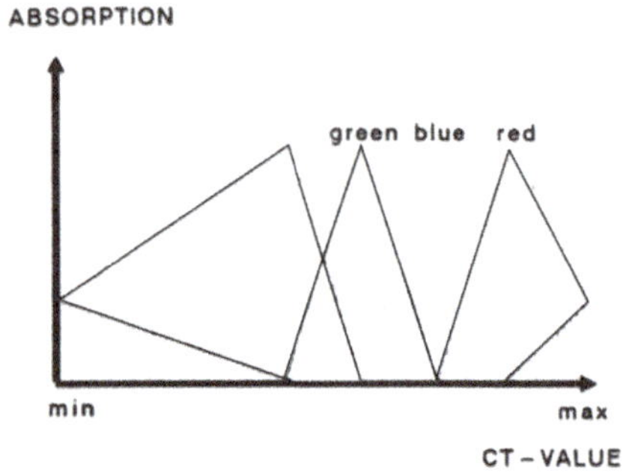

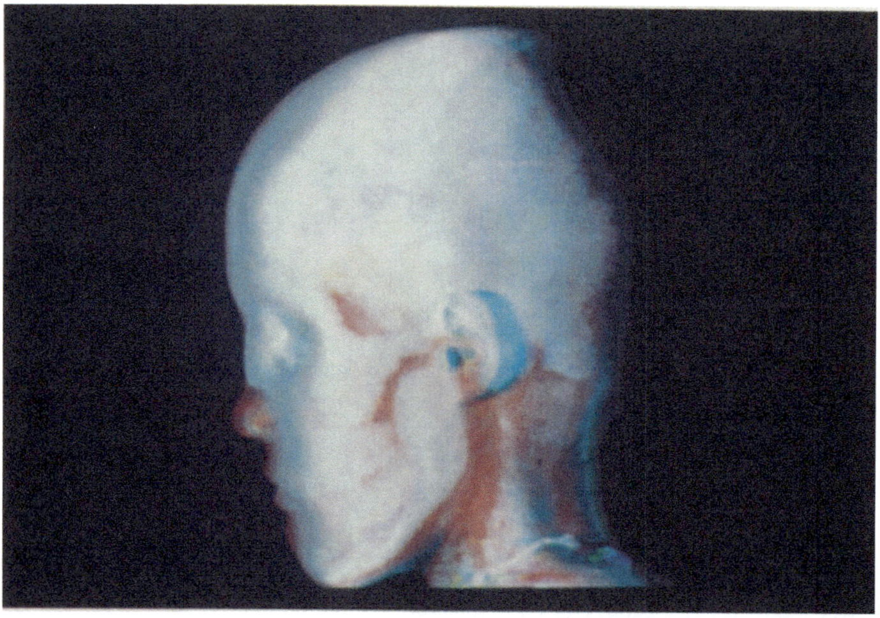

Fig. 2: Colour allocation according to the graph. Visualization of the same dataset.

By varying the colour allocation, individual structures can be differentiated with special accuracy.

2.2 Visualisation of a Reduced Dataset

In the structure of interest is covered, the interfering part of the dataset will have to be removed.

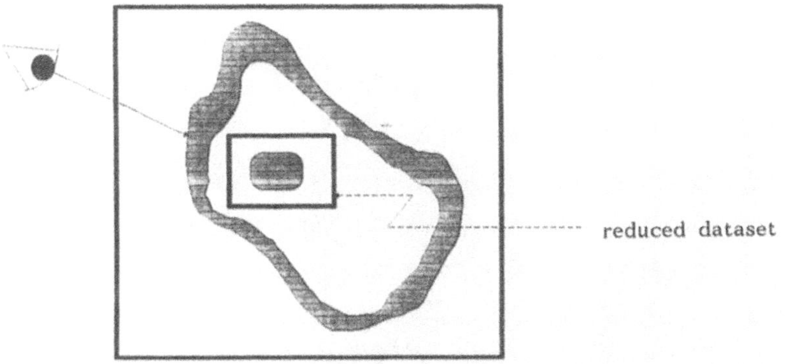

Fig. 3: Schema: Interest is focused on the inner structure. When visualizing the original dataset, only the outer shell is visible. When using the reduced dataset, the inner structure becomes visible (if necessary, the data will have to be inverted).

The problem of suitably minimizing the dataset is disproportionately smaller than the classical segmentation problem, since here the exact shape of interferring and interesting structure may remain unknown for visualization.

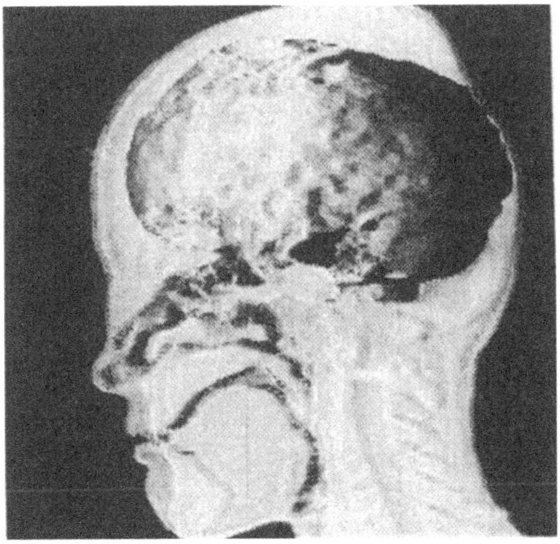

Fig. 4: Here the dataset was reduced with the aid of the topological map [5] and morphological operations [6].

2.3. Visualisation of a Modified Dataset

If one does not restrict oneself to the presentation of measuring values but if the computer is already to achieve a preinterpretation, classifications can be assigned to probabilities. These probabilities can be presented as was previously done with the measuring values.

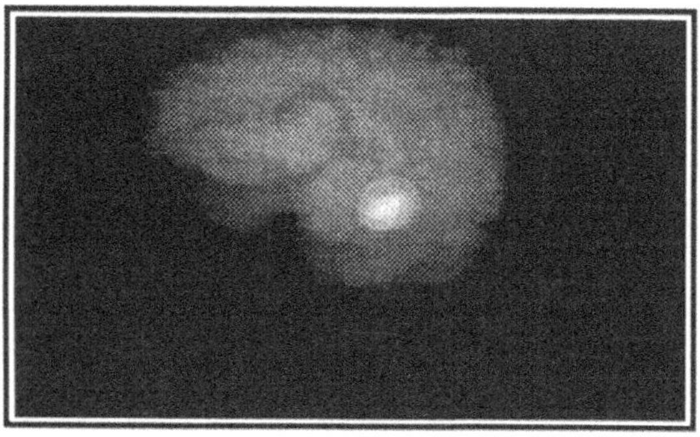

Fig. 5: Brain tumor. 127 MR-slices (256×256). Segmentation and classification with the aid of the topological map.

3. Summary

The visualization of volume elements proves its worth for the 3D presentation of medical image sequences (e.g., CT or MR data). Pictures generated in this manner can support the physician more than those generated with surface-oriented methods. Since data manipulation is not obligatory for the generation of images, great confidence may be placed in the resulting picture. Preprocessing may be reduced to what is absolutely necessary. When using automated segmentation or classification algorithms, a decision security can be presented.

4. References

[1] Kropatsch, W.G.: Wechselwirkung zwischen Visualisierung und Bildanalyse. In Barth W. (Ed.): Visualisierungstechniken und Algorithmen. Proc., Informatik Fachberichte 182, Springer Berlin-Heidelberg-New York-Tokyo 1988

[2] Levoy, M.: Display of Surfaces from Volume Data, IEEE CG&A 5 (1988) 29-37

[3] Drebin, R.A., Carpenter, L., Hanrahan P.: Volume Rendering, Proc. SIGGRAPH 88. Computer Graphics 22, 4 (1988) 65-74

[4] Heyers, V., Dengler, J., Meinzer, H.P.: 3D Visualisierung von Grauwertvoxelräumen. in Bunke, H., Kübler, O., Stucki, P. (Eds.): DAGM 88, Proc., Informatik Fachberichte 180, Springer 1988

[5] Bertsch, H., Dengler, J.: Klassifizierung und Segmentierung medizinischer Bilder mit Hilfe der selbstlernenden topologischen Karte. In Paulus E. (Ed.): Mustererkennung 87, Proc., Informatik Fachberichte 149, Springer 1987

[6] Serra, J.: Image Analysis and Mathematical Morphologie. Academic Press New York-San Francisco-London 1982

Fast Three Dimensional Visualization Using a Parallel Computing System

U.Kliegis, R.Neumann, Th. Kortmann, W.Schwesig, R.Mittelstädt, H.Weigel, W.Zenker[+]

MEK Medizinelektronik GmbH, Düppelstr. 71, D - 2300 Kiel 1, FRG

[+]Abt. Unfallchirurgie, Klinikum der CAU, Hospitalstr., D - 2300 Kiel 1, FRG

Introduction

Three dimensional representations of organs have been proved to be useful in surgery, orthopedics and other disciplines. The shape of an object, its spatial context and correlation to other objects can be optimally judged in a form of display that resembles the natural appearance, as it would look like if the surrounding matter normally objecting the direct view would be subtracted. Naturalistic shading and a dynamic display of different views add to the illusion of seeing an object floating in space, although it is only a calculated reconstruction based on CT or MR slice images.

With the basic problems of the software implementation of the commonly used algorithms performing these tasks solved now, the need for faster systems arises in clinical practice to increase the number of cases to be evaluated with this new technique. The structure of the algorithms used in 3D - visualizations and in the production of organ models leads to the application of parallel computing arrays such as a transputer system.

Sequential versus parallel

Until now, on most 3D - workstations on which artificial images are computed, dynamic sequences of different views of an object from different perspectives are precomputed offline and subsequently displayed as a real time video loop. This is due to the limited computing power present in most workstation computers. Although this technique leads to very useful results already, users ask for more speed now. One way of increasing the update rate of new views could be to decrease the detail resolution, but that would result in a coarser and less usable image. On the other hand, the speed limit of conventional computer architectures is caused by the sequential process that is significant for all von Neumann CPUs. Fortunately, computing the visualisation of 3D objects implies a large amount of tasks that can be executed in parallel. So, the key to more computing power for 3D visualizations lies in the application of parallel processing engines.

There are several types of parallel processors. Essential for all of them is the ability to communicate with each other via fast lines to synchronize processes, exchange data and results and transport them across the array made up by the single computing nodes. Depending on the problem that has to be solved, the processors themselves may range from simple comtrollers capable of performing integer arithmetic to highly sophisticated CPUs including floating point arithmetic units. The latter have the advantage of being prepared for all sorts of problems, they execute compiled high level language programs much faster thus processing more data in shorter time.

Parallel configurations

Not too long ago, building a parallel processing system meant constructing a super computer from ground up. With the appearance of the transputer, however, assembling such a system became very easy. Transputers resemble conventional microprocessors like the Motorola 680x0 series, only augmented by a set of very fast bidirectional I/O channels called *links*. Each transputer has four links connecting it in any imaginable way to other transputers. Regular networks can be configured just as pipeline chains can be built, three dimensional nets can be set up and circular or toroid structures are possible. The structures can even be reallocated dynamically to fulfill a set of tasks with the highest efficiency. The nodes in these arrays are transputers, each with its own private memory, communicating with its neighbors via the link channels. A floating point unit in the transputer features 1.5 MFLOPS, that is 1.5 million floating point operations per second per transputer. The transputers can be programmed in nearly all high level languages like C, Pascal, Fortran and others. The genuine language for parallel programs however is OCCAM, a language that has a set of features that handle the interaction between the single processors. OCCAM programs can be developed in a simple PC environment. These facts made the transputer very attractive for us as a building block for 3D computations.

An accelerator subsystem

We decided to construct an accelerator subsystem for our ENDOPLAN system. ENDOPLAN is the first integrated system not only for the computation of 3D views but predominantly for the production of real, physical organ models.

Our primary goal was to build a system that generates a number of views per second. The data primarily are present as a set of contours derived from CT slice images. These contours are isodensity lines that in most practical cases describe the border between bone and the surrounding soft tissue. The surface between the contour lines is reconstructed in an interpolation process that also benefits from the accelerator. The most computing intensive parts of the existing program, originally written in PASCAL, have been ported to the transputer array. After the surface has been reconstructed as a set of triangles, the corner points of these represent the surface of the object. All possible geometric calculations can be performed on this data set, it can be turned, tangled, zoomed and translated.

Porting code to the accelerator

A number of preprocessing steps has to be performed before the shading for every pixel showing a part of the object's surface can be computed. Simultaneously with the positioning a clipping to the display window takes place. All tasks performed on the data are identical to the well known steps of 3D visualization. The main difference to a sequential program is the solution of the problem of how to distribute the computing power to the different transputers. All separable blocks were checked for their efficiency. The time they consumed was measured, and a raw distribution of the tasks to the processors was made. One must also keep an eye on the time it takes to transport data from one node to another. Since the links are bit serial channels, it takes them some time to send data across the net. This, however, happens synchronously with the execution of program code in the processor, so there is only little idle time overhead in a fine tuned system.

During the implementation of the accelerator, we found that about 80 % of the time were consumed by 5 % of the program, mainly in computation intensive loops. These parts of the program were ported first. A second phase of optimisation brought other parts of the original program down into the accelerator. A dual ported RAM area serves as a letter box for the exchange of data between the host computer and the accelerator. This makes it simple to swap more program parts into the accelerator wherever it makes sense.

Having implemented a large part of an existing program package in a new environment offering a vastly increased operation speed enabled us to bring the generation rate of new 3D views to four frames per second with a transputer array. The bottle neck now is in the I/O path of the display refresh buffer, this will be the next part to be improved for high speed transfers.

Conclusion

Our goal was to add an accelerator for the fast computation of 3D algorithms to our existing ENDOPLAN system. The accelerator was implemented as a parallel computing array of transputers. The resulting data rate was increased until the bottle neck of the I/O path of the image sequence store put an obstacle to further speed improvements. The free programmability of the transputers in high level languages, namely in OCCAM, allowed to port the well tested algorithms in very short time. It could be shown, that the computational power of a parallel computing engine can be employed to full extent in 3D operations. The distribution of the tasks to the single processors, however, has to be done very carefully.

References

1. Stephen M. Pizer and Henry Fuchs, Three Dimensional Image Presentation Techniques in Medical Imaging, Proc. CAR '87, p.589, Berlin 1987

2. K.H.Höhne, M.Riemer, U.Tiede, Viewing Operations for 3 - D - Tomographic Gray Level Data, Proc. CAR '87, p.599, Berlin 1987

3. U.Tiede, K.H.Höhne, M.Riemer, Comparison of Surface Rendering Techniques for 3 D Tomographic Objects, Proc. CAR '87, p. 609, Berlin 1987

4. M.W.Vannier, J.L.Marsh, R.H.Knapp, C.J. Offutt, Recent advances in 3 - D CT Scan Reconstruction for Clinical Applications, Proc. CAR '87, p.657, Berlin 1987

5. U.Kliegis, W.Schwesig, H.Weigel, An Integrated System for the Automatic Manufacturing of Organ Models, Proc. NCGA '88, Vol III, p. 138, Anaheim 1988

6. A.Wallin, K.Klaue, R.Ganz, S.M. Perren, Three Dimensional Evaluation of Pathological Joints and Simulation of Corrective Surgical Procedures, Proc. NCGA '88, Vol III, p. 198, Anaheim 1988

7. E.Fishman, D.R.Ney, D.Magid, J.E.Kuhlmann, Three-dimensional CT Imaging of Soft-Tissue Anatomy, Proc. 74th Scientific Mtg.,p. 256,Chicago 1988

Three-Dimensional Surface Reconstructions Using a General Purpose Image Processing System

J. W. HOFFMEISTER, G. C. RINEHART, M. W. VANNIER

Mallinckrodt Institute of Radiology
Washington University School of Medicine
510 South Kingshighway Blvd.
St. Louis, Missouri 63110 USA

Summary

General purpose image processing software has been employed to produce high quality three-dimensional surface reconstruction images from contiguous CT slices. Surfaces within the data set can be viewed at any orientation with respect to the coordinate axes. Perspective transformations are employed to ensure realistic image appearance during close-up viewing. Images of hidden internal surfaces can be produced by defining any desired "cut plane". Volumetric 3D reconstruction images that simulate transmission radiographs were produced. Three dimensional surface images may be depth-shaded, gradient-shaded, or displayed with any desired combination of depth and surface shading. One millimeter thick contiguous 512 x 512 resolution 12-bit gray scale images were obtained of a dry skull with a Siemens Somatom DR-H scanner. The data set, consisting of 210 axial images, was transferred to a VAX-11/750 operating system for analysis with International Imaging Systems (IIS) System/600 software on an IIS Model 75 image processer. Using these algorithms, any user with a general purpose image processing system can view and analyze serial slice data in 3-D.

Methods

The data set was thresholded to create depth-encoded 3-D surface images. Views along the axis of data acquisition (top and bottom surface views), and surfaces perpendicular to the axis of data acquisition (front, rear, and lateral views) were computed. From these depth-encoded 3-D images, gradient-shaded 3-D surface images were rendered. Depth-encoded and gradient-shaded images were combined to simultaneously display depth and surface characteristics. Volumetric images analogous to transmission radiographs were generated from the original 12-bit CT slice data set by ray-casting.

Orthogonal axis depth-encoded 3-D surface images were computed using 1-D convolution to uniquely mark only the surface pixels from each binary data slice viewed along a selected orthogonal axis. This surface detection kernel has 513 rows X 1 column. The gray levels of the pixels in this kernel were -1 for rows 1 through 512 and +1 for row 513. The image to be convolved with this kernel was created by merging a 512 x 512 blank image (all gray levels = 0) with a 512 x 512 binary image slice such that the blank image occupied the top half and the binary slice the bottom half of a 1024 row x 512 column composite image. Convolution of the 513 x 1 kernel with the 1024 x 512 composite resulted in a 512 x 512 binary image where the first pixels (greater than threshold) convolved in the y-direction have a gray level = 1, and all other output image pixels are less than or equal to zero. This convolution product image was mapped such that all negative pixel intensities were set to zero. These rescaled images represent one-pixel-thick profiles of desired 3D surface. Note that any such surface orthogonal to the axis of data acquisition can be picked by rotating the original data binary images into appropriate alignment with the surface detection convolution kernel.

Convolution with a second kernel depth encodes the desired surface information from each binary image. This second (depth-encoding) kernel was a 511 row and 1 column range, with pixel gray levels descending from 511 to 1. Convolution of the 512 x 512 surface-detected image (gray level = 1 at viewed surface pixels and gray level = 0 elsewhere) with this depth-encoding kernel resulted a 2 row x 512 column image. The first row of this product image contained the information desired — depth-encoded orthogonal surface information from the original data binary slice. This depth-encoded surface information from each binary slice was merged in the proper order to create a depth-encoded 3-D surface image orthogonal to the axis of data acquisition. A linear scale adjustment for anisotropic voxel height was necessary. Cut planes were produced by subselecting the binary slices to include only the area of interest and adjusting the convolution kernel sizes appropriately.

Gradient-shading of surfaces

Gradient-shaded 3-D images were computed from depth-encoded 3-D surface images. A gradient-shaded image has pixel intensities proportional to the cosine of the angle Ø between the vector normal to the surface (the gradient vector, G) and the view direction. For viewed surfaces, Ø varies from 0 to 90°

The value of cos Ø for each pixel in a depth-encoded 3-D surface image can be approximated by analyzing the four nearest neighboring pixels. Consider a cartesian coordinate system such that X and Y are distances along the X and Y axes of the depth-encoded 3-D surface image and Z is the pixel intensity at each image X, Y coordinate. The view direction is parallel to the negative z-axis. For simplicity, first assume that pixel intensities in the depth-encoded 3-D surface image only vary in the X-direction. Note that the inclination angle of this ramp with reference to an isointensity plane is identical to the angle Ø between the ramp's gradient vector and the viewing axis. Therefore:

$$\cos \text{Ø} = \frac{a}{\sqrt{a^2 + b^2}}$$

$$a(x, y) = 2$$

$$b(x, y) = I(x + 1, y) - I(x - 1, y).$$

Now, allowing a depth-encoded 3-D image to vary in both X and Y directions, we have.

$$a_x(x, y) = a_y(x, y) = 2$$

$$b_x(x, y) = I(x+1, y) - I(x-1, y)$$
$$b_y(x, y) = I(x, y+1) - I(x, y-1).$$

as sides of triangles and included angles $Ø_x$ and $Ø_y$ with respect to an isointensity plane. Similar proportional triangles can be defined to relate surface vectors G_x and G_y to a_x, b_x, a_y, and b_y (k is a proportionality constant). When G_x and G_y are the x and y components of the surface gradient vector, G. The angle Ø can be defined in terms of a_x, b_x, a_y, and b_y. Therefore,

$$\cos Ø = \frac{a_x + a_y}{\sqrt{(a_x + a_y)^2 + b_x^2 + b_y^2}}$$

Since $a_x = a_y = 2$ for the "central difference" method of approximation using neighboring pixel intensity values:

$$\cos Ø = \frac{4}{\sqrt{16 + b_x^2 + b_y^2}}$$

This formula is a special case of the general formula for the cosine of the angle between the gradient vector $G=(g_x, g_y, g_z)$ and a view direction along the z-axis:

$$\cos Ø = \left. \frac{g_z}{\sqrt{g_x^2 + g_y^2 + g_z^2}} \right|_{g_z = 4} \qquad \text{(eq. 9)}$$

where

$$g_x(x, y, z) = [f(x+a, y, z) - f(x-a, y, z)]/2a$$
$$g_y(x, y, z) = [f(x, y+b, z) - f(x, y-b, z)]/2b$$
$$g_z(x, y, z) = [f(x, y, z+c) - f(x, y, z-c)]/2c,$$

and f(x, y, z) is the density function (original CT data set); and a, b, and c define the voxel size in the x, y, and z directions (6-7).

Volume images (computed x-rays)

Volume images (analogous to transmission x-rays) were computed by summation of original data in the viewing direction. Top-to-bottom (Z axis) views were made by pixelwise summation of the original slices. Front-to-rear (X axis) and lateral (Y axis) views required use of the convolution operator on original slice data. The convolution kernels required are a 1 x 512 matrix (column kernel) or a 512 x 1 matrix (row kernel) of pixel intensity = 1. Ray sum or line integral images were created by convolution of these kernels with each original data slice. These images were merged appropriately to produce complete volume images.

Discussion

Algorithms were developed to employ existing general purpose image processing software and produce high-quality 3-D surface reconstructions from contiguous CT scan slices. Discontinuous transfer function (threshold), maximum seeking, minimum seeking, algebraic and convolution operators were employed. Linear piecewise intensity scaling and operator-supervised region of interest editing were commonly used. Depth-encoding of 3-D surfaces preserves quantitative information about surface geometric relationships. However, fine surface details (small gray scale differences) are difficult for the human eye to appreciate. Gradient-shaded 3-D surface images demonstrate surface details well, but depth-cues may be misleading. A weighted sum of these two types of images displays surface and depth information simultaneously, and may be most useful for aesthetic enhancements of slowly-varying surfaces. Volume images display features of the entire data set in one view; however, 3-D characteristics of surfaces defined by the data set are lost. Cut-plane images were created from subsets of the data to reveal inner surfaces of hollow data sets.

By implementing the algorithms described in this paper, any user of a general purpose 2-D image processing system can produce high-quality 3-D reconstructions from serial slice data.

References

1. Huijsmans DP, Lamers WH, Los JA: Toward computerized morphometric facilities: A review of 58 software packages for computer-aided three-dimensional reconstruction, quantification, and picture generation from parallel serial sections. Anat Record 216;449-470, 1986.

2. Vannier MW, Marsh JL, Warren JO: Three dimensional computer graphics for craniofacial surgical planning and evaluation. Computer Graphics 17:263-273, 1983.

3. Vannier MW, Marsh JL, Warren JO: Three dimensional CT reconstruction images for craniofacial surgical planning and evaluation. Radiology 150:179-184, 1984.

4. Vannier MW, Marsh JL, Gado MH: Three dimensional display of intracranial soft tissue abnormalities. Amer. J. Neuroradiology 4:520-521, 1983.

5. Totty WG, Vannier MW: Analysis of complex musculoskeletal anatomy using three dimensional surface reconstruction. Radiology 150:173-177, 1984.

6. Geist D, Vannier MW: Medical imaging in an object oriented environment. SPIE Medical Imaging II 914:1299-1306, 1988.

7. Drebin RA, Carpenter L, Hanrahan P: Volume rendering. Computer Graphics 22:65-74, 1988.

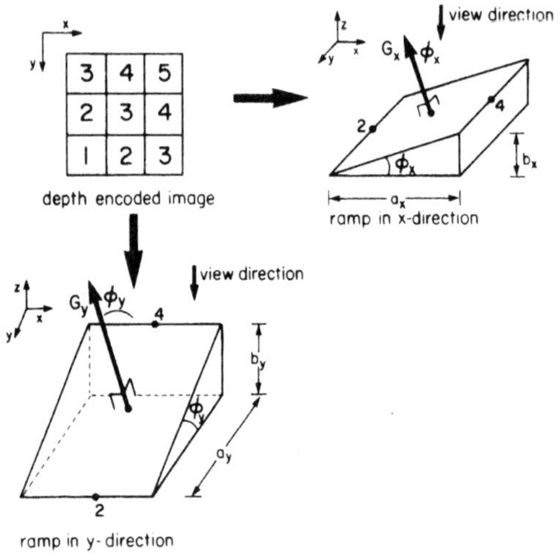

A. Computation of x and y components of surface gradient

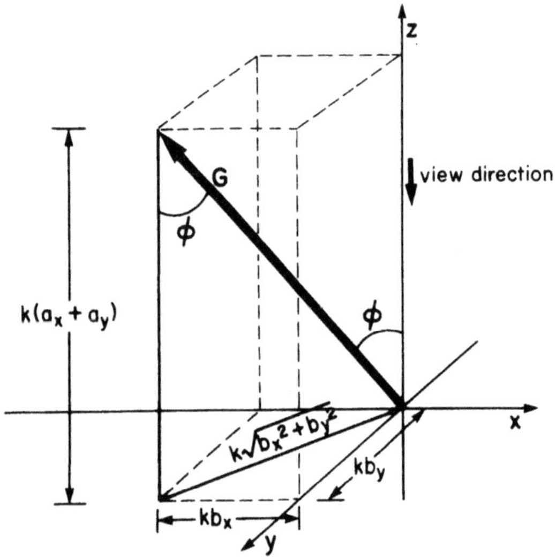

B. Relationship of gradient vector to viewing direction.

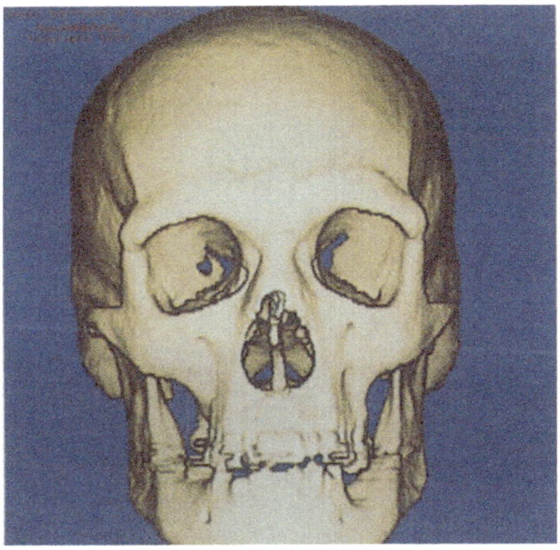

Figure 1 Frontal view of skull.

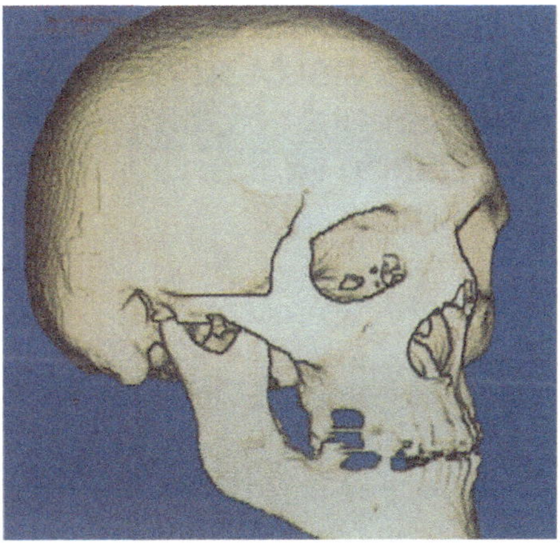

Figure 2 45° Right frontal obliq view of skull.

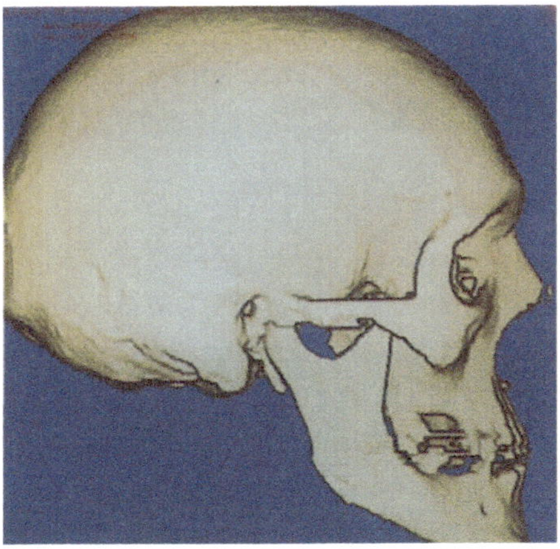

Figure 3 Right lateral view of skull.

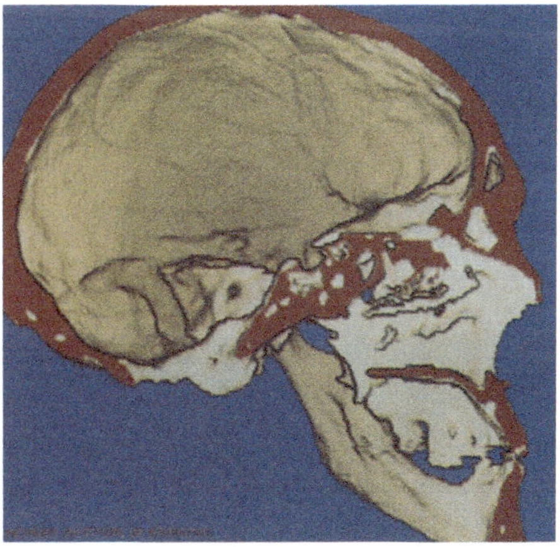

Figure 4 Right midsagittal view of skull.

Improvements to the Cuberille Three-Dimensional Modelling Technique for Assessment of Complex Fractures and Evaluation of Stability in the Musculo-Skeletal System

Å. WALLIN, S. BRESINA, E. SCHNEIDER

M.E.M.-Institute for Biomechanics, University of Bern, Murtenstrasse 35, CH-3008 Bern, Switzerland.

Summary

Conventional x-rays and computed tomography (CT) images are routinely utilized for the diagnosis of complex fractures. The three-dimensional (3-D) shape and relative location of the structures involved are, however, often difficult to assess from these planar views. Standard 3-D reconstructions, as offered by manufacturers of scanner equipment, are in many cases helpful. However, their potential with respect to fracture assessment and evaluation of stability is limited. For the improved display and quantitative evaluation of intact and fractured osseous structures, a number of extensions to the well-known cuberille reconstruction method were implemented. Extensions were introduced to improve and automate object search and detection. The technique allows for quantification of relative object location and orientation. Relative object location and movement has been visualized by kinematic rendering. Rendering enhancements using several display algorithms, utilizing concepts such as transparency and multiple colors were implemented. In order to enhance the impression of fracture lines, an erosion algorithm was added.

Introduction

Studies of the human musculo-skeletal system involve the assessment and quantitative determination of complex three-dimensional structures, e.g. the pelvis or vertebral column. In addition to the capturing and display of such structures, it is essential to describe their shape and location. This permits the comparison between different subjects and/or between different situations of the same individual. Three-dimensional diagnostic imaging has increasingly been utilized in recent years [1-6]. The primary aim has been to improve visualization of fractures or abnormalities.

One popular reconstruction technique is the cuberille method [7]. Its main advantages are algorithmic simplicity, independence from special hardware, and the ability to describe separate objects (e.g. fracture fragments). An important recent development is the determination of the location and orientation of objects [8].

In the first section the cuberille method is briefly described. In the following two sections the main improvements are discussed. These involve the automatic object search and detection, and the determination of object location and orientation. In the last section the improvements in rendering are described and examples are shown.

The cuberille method of three-dimensional reconstruction

The volume described by a stack of CT cross sections contains a large number of parallelepipeds (voxels). Cubical voxels are generated by linear interpolation between the original cross sections. The cubical gray valued voxels are segmented into binary voxels separating tissue of interest (e.g. bone) from background. Starting from a seed point within the volume, a surface of a connected set of equally valued voxels is computed using a surface tracking algorithm. Each surface describes the border of a separate object in the volume. The resulting surfaces can subsequently be displayed using traditional z-buffer and rendering algorithms [9].

Object search and detection

An object is defined as a collection of binary voxels connected over faces and edges. As pointed out above, the surface of an object is computed starting with a seed point. To automatically find seed points for the surface tracking algorithm it is essential to find one point on the surface of each object. This as well as the number of connected voxels (i.e. the volume) are computed using a connected component algorithm. The volume can then be used as a decision criteria for object display. In many cases only the largest objects are of interest. It is not unusual to encounter several hundred objects of differing sizes (mostly very small, i.e. a few tens of voxels) within a volume data set.

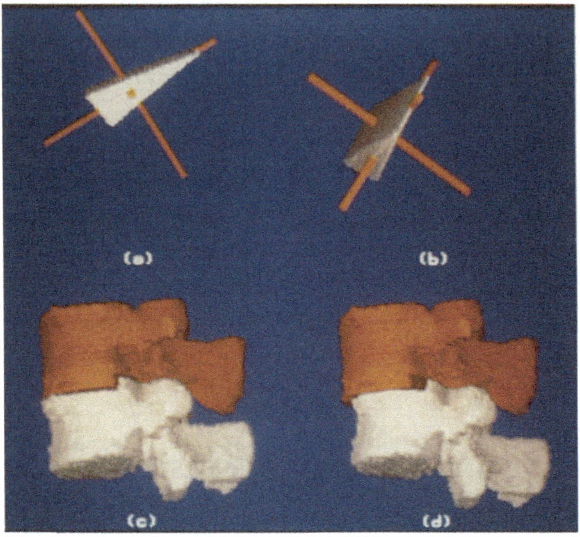

Fig. 1. Two views of a 3-D model of a wedge with principal axes (a, b).
Two instances of a loading experiment on a spine (c, d).

Object location and orientation

In order to describe the location and orientation of a 3-D object by invariants, the centroid and principal axes are used [8]. The relative movement of an object can then be quantified by examining the relative movement of the centroid (translation), and the relative orientation of the principal axes (rotation) (Fig. 1). This requires the three-dimensional reconstruction of two objects (one "fixed", the other one movable) at two different instances in time, e.g. before and after a flexion of the head.

Rendering enhancements

For kinematic rendering of 3-D objects it is necessary to retain the translations, rotations and scaling parameters to obtain the required image on a two-dimensional image plane. Global viewing parameters (defining the "camera") describe the projection of the object coordinate system into the image coordinate system. Local transformations describe separate movements of each object. These are, in most cases, used to visualize kinematics, obtained either using the relative location and orientation, as described above, or by using measurements acquired during experimental measurements. One example is the conversion from nine distance measurements between vertebrae to a rotation around a screw axis and a translation along that screw axis (Fig. 1).

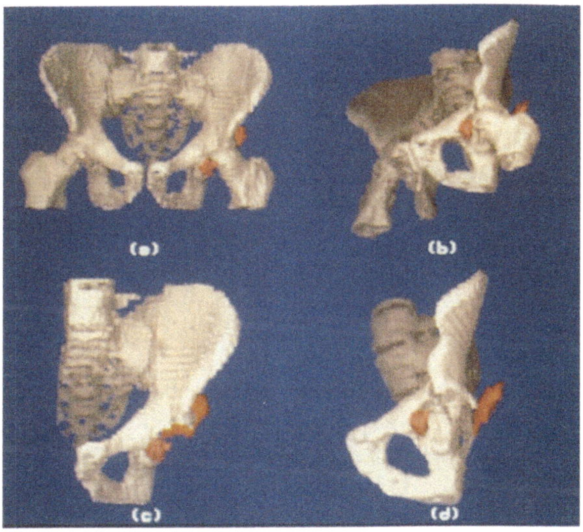

Fig. 2. Four views of a fractured pelvis with (a, b) and without femur (c, d). Fragments displayed in color.

Colors are important cues in enhancing a display. If used with care they can convey the message by immediately drawing the observers attention to a specific part of an image (Fig. 2). If colors are used in conjunction with transparency, it is possible to highlight even hidden structures. Using a combination of colors and transparency factors, several images from the same view point can improve the clinical value of 3-D imaging. This has been used in the display of tumors (Fig. 3).

In the cuberille technique the faces of the voxels are perpendicular, in some cases causing streak artefacts in an image due to sampling effects (aliasing) (Fig. 4). This is especial-

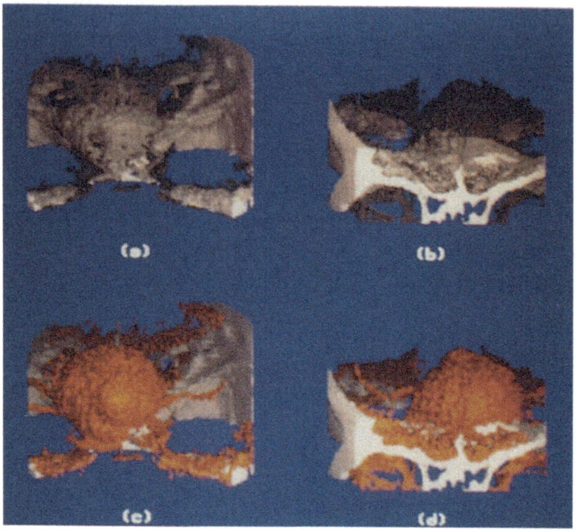

Fig. 3. Two views of a brain tumor: transparency, bone is visible (a, b); color, tumor is visible (c, d).

ly the case in straight views (frontal, side, top etc.). In order to eliminate these effects, smoothing filters can be utilized, or in many cases it is enough to simply choose a slightly oblique view point (Fig. 4).

Images of objects, such as a pelvis in a frontal view, often look flat since only a part of the range of intensity values is utilized for the rendered surface (Fig. 4). The contrast can be improved if the intensity values are stretched and a larger range is present in the output image. This is accomplished by computing a 16 bit value for the intensity at each pixel, finding the minimum and maximum of these pixel values and mapping them into the minimum and maximum values of groups in the output.

Erosion algorithm

The main usage of the cuberille reconstruction method is the visualization of 3-D normal or pathological human anatomy. Often it is used in the display of complex fractures with spatially propagating fracture lines. If these lines are thin, it

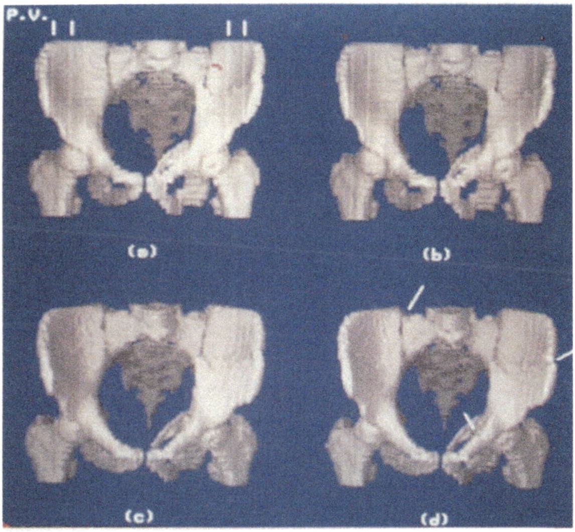

Fig. 4. Frontal view of fractured pelvis.
(a) Rendered normally. Streak artefacts appear.
(b) Rendered "flat", without intensity stretching.
Slightly oblique view of fractured pelvis.
(c) Rendered with additional light source.
(d) Eroded, rendered as in (c).

may be difficult to visualize them. During the segmentation process, it is possible to choose higher threshold values for bone which in some cases widens fracture lines. The drawback is that other parts of the object are thinned as well. Another method is to remove an outer layer of the fractured object to widen the fracture lines (Fig. 4). If a voxel is completely inside an object it is surrounded by voxels with similar density values. If a voxel is on the surface it is connected to some voxels with similar density as well as to others with dissimilar density. A voxel on the surface of a fracture has on the average more neighbours with similar density than not (viz. flat surface: 13 similar and 5 dissimilar). This technique of neighbourhood calculations is not sufficient to detect a fracture line, but it is efficient and can, in some cases, improve the visual impression.

Conclusion

The cuberille three-dimensional reconstruction method has become an important and clinically useful display methods. Its main advantage is the handling of the investigated volume as a collection of distinct objects, consisting of binary voxels. Based on these objects it is possible to describe relative location and movement of parts of the musculo-skeletal system. During the rendering process it is important to understand the limitations regarding the artefacts imposed by the directions of the faces and the possibilities available with respect to display enhancements such as colors, filters, transparency and surface erosion.

References

1 Totty, W.G.; Vannier, M.W.: Complex Musculoskeletal Anatomy: Analysis Using Three Dimensional Surface Reconstruction. Radiology 150 (1984) 173-177.

2 Burk, D.L.; Mears, D.C.; Kennedy, W.H.; Cooperstein, L.A.; Herbert, D.L.: Three-Dimensional Computed Tomography of Acetabular Fractures. Radiology 155 (1985) 183-186.

3 Pozzi Mucelli, R.S.; Muner, G.; Pozzi Mucelli, F.; Pozzi Mucelli, M.; Marotti, F.; Dall Palma, L.: Three-Dimensional Computed Tomography of the Acetabulum. European Journal of Radiology 6 (1986) 168-177.

4 Sartoris, D.J.; Resnick, D.; Bielecki, D.; Andre, M.; Gershuni, D.; Meyers, M.: A Technique for Multiplanar Reformation and Three-Dimensional Analysis of Computed Tomographic Data: Application to Adult Hip Disease. Journal of the Canadian Association of Radiologists 37 (1986) 69-72.

5 Fishman, E.K.; Drebin, B.; Magid, D.; Scott, W.W.; Ney, D.R.; Brooker, A.F.; Riley, L.H.; St.Ville, J.A.; Zerhouni, E.A.; Siegelman, S.S.: Volumetric Rendering Techniques: Applications for Three-dimensional Imaging of the Hip. Radiology 163 (1987) 737-738.

6 Becker, H.; Rolfes. J.; Graul, B.; Giebel, G.: Klinische Anwendung von 3-D Bildverfahren in der Computertomographie. Proc. CAR'87, Springer Verlag, (1987) 630-635.

7 Herman, G.T.; Liu, H.K.: Three-Dimensional Display of Human Organs from Computed Tomograms. Computer Graphics and Image Processing, 9 (1979) 1-21.

8 Bresina, S.J.; Schneider, E.; Wallin, Å.; Perren, S.M.: Determination of Relative Object Position and Orientation from Computed Tomography Data. Journal of Biomechanics (submitted) (1989).

9 Herman, G.T.; Udupa, J.K.: Display of Three-Dimensional Descrete Surfaces. SPIE 3-D Machine Perception 283 (1981) 90-97.

Building Highly Structured Volume Representations in 3D Medical Images

N. Ayache, J.D. Boissonnat, E.Brunet, L. Cohen, J.P. Chièze,
B. Geiger, O. Monga, J.M. Rocchisani, P.Sander

INRIA
BP. 105, F-78153 Le Chesnay
France

Summary

Our purpose is to develop new tools to help the analysis of three-dimensional images. We present a chain of processes which goes from the voxels of a 3D image to a highly structured volume representation of regions of interest. This representation can then be efficiently manipulated for display and analysis.

1 Introduction

Modern medical imaging techniques, such as Nuclear Magnetic Resonance or X-Ray Computed Tomography, provide three dimensional (3D) images of internal structures of the body, usually by means of a stack of tomographic images. In many applications, the physician asks for a segmentation of these 3D images into regions of interest he wants to manipulate, display, and characterize by objective quantitative measurements.

In this paper, we present results of a chain of processes which can help the segmentation into regions of interest, and also their further manipulation for display and analysis.

We present first a new edge detection procedure, which takes into account the 3D nature of the image and provides a good trade-off between localization accuracy and noise immunity. The algorithm implies the use of recursive separable filters which can be implemented very efficiently.

Second, we present a method to fit deformable models of structures to detected edges. Currently the method is implemented in two dimensions only. The idea is to let an elastic curve deform itself from an initial position, to minimize an energy which takes into account internal constraints (curve elasticity) and external constraints (proximity of edges). This method provides a technique to interactively edit the detected edges in any planar cross-section of the studied volume by drawing grossly the shape of the desired region, and let this shape deform itself to fit as closely as possible the nearby detected edges. Moreover, once a fit has been obtained in a given cross-section, it can be used as a starting point in the next cross-section, where it will naturally converge toward the new boundaries of the the currently studied region of interest. Also, when studying moving organs like the beating heart, one can let the shape deform itself in a given cross-section but at different times of observation. We believe that this 2D technique can drastically simplify the interactive segmentation of 3D (possibly dynamic) images.

Third, we present the use of the NUAGES algorithm, designed by J.D. Boissonnat and developped by himself and B. Geiger, which is used to build a volume representation from a set of contours given in a sequence of planar cross-sections. The method handles a variable number of contours in each cross-section and garanties a solid object as a result. This solid object is described by a set of tetraedra, which can be used to compute its volume and its surface. Also, the tetradra structure can be used as a mesh to study mechanical properties of the represented object.

Last, but not least, we describe our current work on surface singularies. We believe that some singularities are stable to small perturbations due to deformations or noise for instance, and can be used to characterize typical points on which registration and/or tracking procedures can robustly rely.

2 Edge Detection

We assume that a 3D region of interest is defined by a set of voxel having an intensity function varying *smoothly* with spatial position. A first stage toward the identification of these regions is to search the surfaces, or edges, which form their boundaries. The points of such surfaces correspond to abrupt changes of the grey level function and are located where the gradient of that function is high. This leads to the classical problem of edge detection widely investigated for 2D images [9], but which had not yet received much attention in 3D.

Basically 2D and 3D edge detection sets the same problem i.e. how to localize the discontinuities of a discrete 2D or 3D function. The classical approach consists in computing the gradient by applying to every point in the image an operator which responds to the discontinuities. However the huge size of data in the 3D case makes algorithmic complexity and storage requirement be key points for 3D edge detection.

Recently, using Canny's design [2] R.Deriche [3] proposed an optimal 1D derivation operator :

$$d(x) = -cxe^{-\alpha|x|}$$

He derives a recursive realization of the optimal filter $d(x)$ by two recursive filters moving in opposite directions. In addition to the noise immunity which results, the recursive nature of the filtering operations leads to a substantial saving in computational effort (12 multiplications and 10 additions for one pixel, independently of the size of the convolution masks). He also shows that $l(x)$ integral of $d(x)$ can also be implemented recursively :

$$l(x) = s(\alpha \mid x \mid +1)e^{-\alpha|x|}$$

The 1D operators (d) and (l) previously defined, have been naturally extended to the 3D case by decomposing the 3D filtering problem into a succession of 1D filtering operations [8]. Figures 1 to 2 present some results provided by the algorithm described in [8].

3 Elastic Deformable Contours

We are then looking for mathematical entities which describe the shapes of objects appearing in images. We thus define a model of an elastic deformable object as in [6]. The model is put on the image by the action of *external forces* which move and deform it from its initial position to best fit to the desired attributes in the image.

The external forces are derived from the image data or imposed as a constraint. Internal forces define the physical properties of the object.

This applies for example to extraction of good edges. We draw a simple curve close to the intended contours and the action of the image forces will push the curve the rest of the way (see figure 3). The final position corresponds to the equilibrium reached at the minimum of energy.

The contour model of deformation is a mapping :

$$\Omega = [0,1] \rightarrow \Re^2$$

$$s \mapsto v(s) = (x(s), y(s))$$

We define a deformable model as a space of admissible deformations Ad and a functional E to minimize which represents an energy of the following form:

$$E : Ad \rightarrow \Re$$

$$v \mapsto E(v) = \int_{\Omega} w_1(s)|v'|^2 + w_2|v''|^2 + P(v)ds$$

where P is the potential associated to the external forces. It is computed as a function of the image data according to the goal aimed. For example, to be attracted by edge points, the potential will depend on the gradient of the image. The mechanical properties of the model is defined by the functions w_j. Their choice determines the elasticity and rigidity of the model.

The energy can be written as the sum of three terms:

$$E = E_{int} + E_{image} + E_{ext}$$

The curve is under control of three forces:

- E_{int} represents internal forces which impose the regularity of the curve.
- E_{image} pushes the curve to the significative lines which correspond to the desired attributes. It is defined by a potential of the shape

  $$P(v) = -|\nabla I(v)|^2.$$

  The curve is then attracted by the local minima of the potential, which means the local maxima of the gradient, that is contours.
- E_{ext} imposes constraints defined by the user.

We minimize E by solving the evolution equation associated by finite differences. We then obtain a linear system and we have to inverse a pentadiagonal symmetric positive matrix.

The first application (shown on figure 3) permits to have precise contours using a simple initial guess. This can be used in a second time to follow contours in a sequence of images. As was pointed out in the introduction, the sequence can be either different slices at the same time or the same slice at successive time steps. By following contours through slices of a same 3D image, it could be possible to build a surface which represents the 3D object. By following contours through time, the deformation of the object could be observed and modelled. A second approach to this last goal is to model the object by a surface which deforms the same way as described for a curve. These are the orientations we are following in our research.

4 Surface and Volume Representation

Once 3D edges have been found (and possibly edited), it is necessary to find a good representation of the volume of the region they enclose. For doing this, one uses NUAGES, a software designed by J.D. Boissonnat [1]. The method makes use of the Delaunay triangulation, one of the most useful data structures in Computational Geometry [12]. One computes successively the Delaunay triangulation of the vertices of the pairs of adjacent cross-sections. The Delaunay triangulation is a set of non-overlapping tetraedra which is isomorphic, in this specific situation (where points lie in two different planes) to a planar graph. This isomorphic correspondance with a planar graph allows an optimal computation of a 3D Delaunay triangulation with only 2D operations.

Unfortunately, the Delaunay triangulation provides a representation of the *convex hull* of a pair of successive contours. It is therefore necessary to prune this triangulation. This is done slice by slice by a technique which handles the case of multiple contours in the cutting planes. The number of contours may be different from one cross-section to the other, and allows multiple inclusions of contours.

Finally, the method garanties that the result is a solid object. As we said in the introduction, this object is described by a set of tetraedra, which can be used to compute its volume and its surface. Also, the tetradra structure can be used as a mesh to study mechanical properties of the represented object.

Figure 4 illustrates a volume reconstruction obtained from the contours of the heart shown in the previous NMR images.

5 Studying Singularities

Recently, P. Sander and S. Zucker introduced the notion of qualitative but *stable* features extracted on 3D shapes, with specific applications to the automatic analysis of 3D medical images. There are three senses in which an image feature can be stable:

1. it must be robust to small perturbations of the underlying physical object;
2. it must not depend on how the object is situated is space relative to the image coordinate axes, that is, it must be *intrinsic* to the object;
3. it must not be sensitive to noise in the image.

It appears that some *curvature* features, i.e. features based on the second derivative, meet these criteria. These features include Gaussian curvature and parabolic points, and also principal directions and umbilic points.

The *Gaussian curvature* is a number which gives the 'total curvature' at a surface point. Its sign allows the classification of surface points into three types, see figure 5(a). Parabolic points, the zeros of the Gaussian curvature, form curves on a smooth surface separating regions of positive Gaussian curvature from negative regions. These singular curves are stable and allow a coarse segmentation of the surface as it is shown in figure 5(b) [5,7];

The *principal directions* associated with each surface point indicate the two directions in which the surface curvature is greatest and least. They form two direction fields which are smooth almost everywhere over the surface. Certain surface points, however, are singularities of these fields, since it is impossible to assign unique directions there. Near such points, termed *umbilic* points, the fields assume characteristic configurations, and umbilic points are generically stable.

Umbilic points can be further classified by considering their *index* (the number of counter-clockwise rotations of a vector aligned with the direction field, considered at points of a small circle around the umbilic). The technical details need not concern us here, but what is extremely important is that the indices of vector fields provide a bridge between local (differential) properties of surfaces and their global (topological) properties [4].

If computation of the first derivative is adversely affected by image noise, then the second derivative is even more so, and it was considered that computation of curvature properties would be extremely delicate. However, the effective use of minimization techniques permits the reliable computation of Gaussian curvatures [10] as well as principal directions [11] and their singularities. Figure 6 shows a direction field on the face region extracted from an MR image, with some of the umbilic points.

Aknowledgements

We thank J. Bittoun from the Kremlin Bicêtre Hospital, Paris, for the acquisition of the NMR images, and R. Di Paola for several helpful discussions and advices. The last NMR image is a courtesy of Dr. T. Peters of the Montreal Neurological Institute.

References

[1] J.D. Boissonnat. Shape reconstruction from planar cross-sections. *Computer Graphics and Image Processing*, 44:1–29, 1988.

[2] J.F. Canny. *Finding edges and lines in images.* Technical Report TR. 720, MIT, June 1983.

[3] R. Deriche. Using canny's criteria to derive a recursively implemented optimal edge detector. *International Journal of Computer Vision*, 1(2), May 1987.

[4] Victor Guillemin and Alan Pollack. *Differential Topology.* Prentice-Hall, Englewood Cliffs, 1974.

[5] David Hilbert and S. Cohn-Vossen. *Geometry and the Imagination.* Chelsea, New York, 1952. Originally published as *Anschauliche Geometrie* (in German).

[6] Michael Kass, Andrew Witkin, and Demetri Terzopoulos. Snakes: active contour models. In *Proceedings of the First International Conference on Computer Vision*, pages 259–268, London, June 1987.

[7] Jan J. Koenderink. The internal representation of solid shape and visual exploration. In Spillman and Wooten, editors, *Sensory Experience, Adaptation, and Perception*, pages 123–142, Lawrence Erlbaum, 1984.

[8] O. Monga and R. Deriche. 3d edge detection using recursive filtering. application to scanner images. In *IEEE Computer Society Conference on Vision and Pattern Recognition*, San Diego, June 1989.

[9] A. Rosenfeld and A. Kak. Digital image processing. *New York: Academic*, 1976.

[10] Peter T. Sander and Steven W. Zucker. Inferring differential structure from 3-D images: smooth cross sections of fibre bundles. *IEEE Transactions on Pattern Analysis and Machine Intelligence*, 1988. To appear. Available as Technical Report CIM-88-6, McGill Research Center for Intelligent Machines, McGill University, Montréal.

[11] Peter T. Sander and Steven W. Zucker. Singularities of principal direction fields from 3-d images. In *Proceedings of the Second International Conference on Computer Vision*, Tarpon Springs, Florida, December 1988.

[12] M.I. Shamos. *Computational Geometry.* PhD thesis, Yale University, 1978.

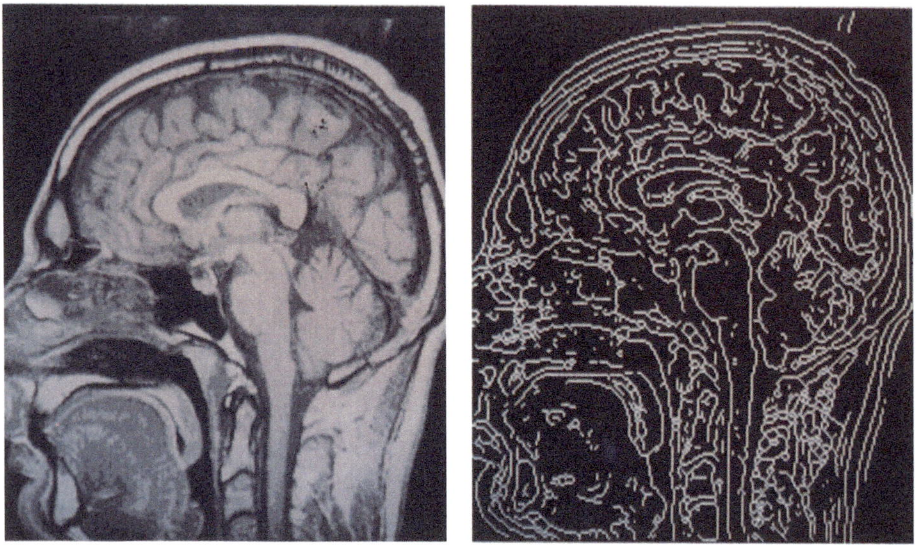

Figure 1: Original NMR image showing a cross-section of a head and corresponding 3D local extrema

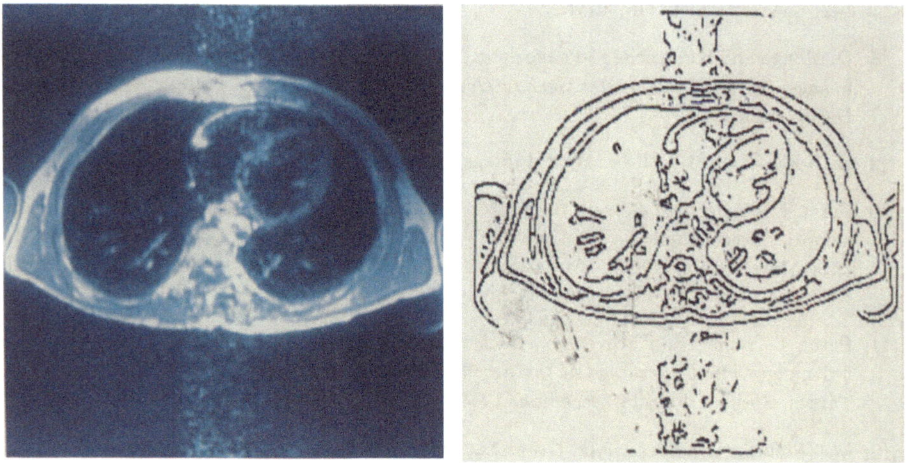

Figure 2: Original NMR image showing a cross-section of the heart and detected 3D edges

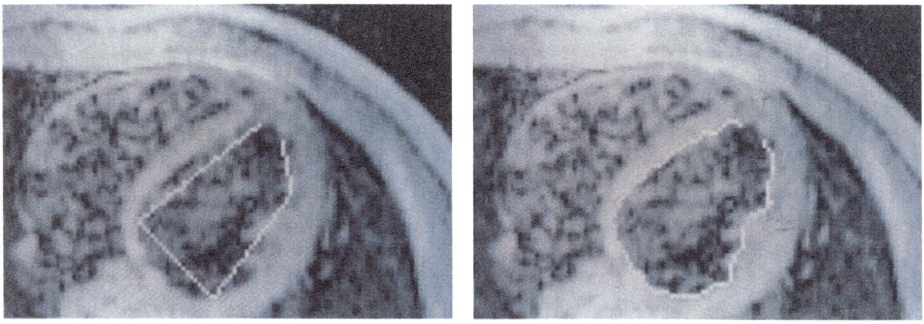

Figure 3: Crude initial estimation of the contour of the left ventriculum and its automatic convergence toward the correct boundary

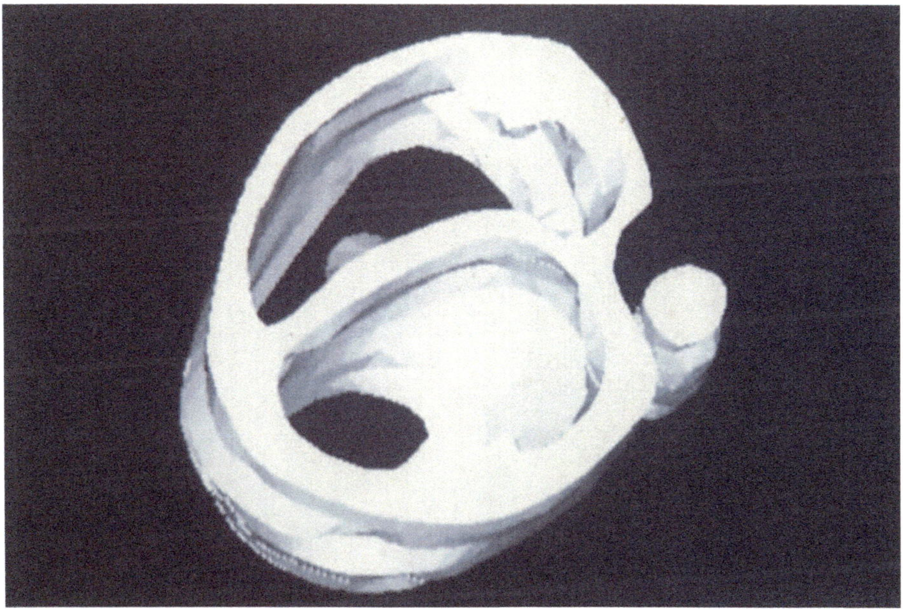

Figure 4: Volume reconstruction of part of a heart obtained from the contours of several cross-sections

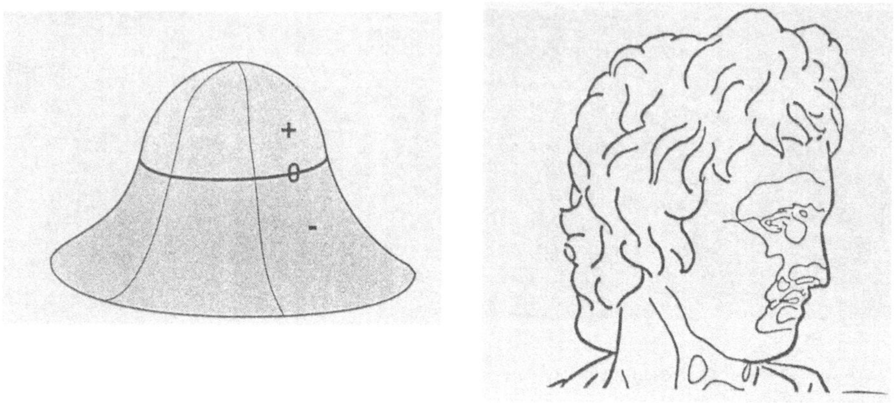

Figure 5: (a) The coarse segmentation of a surface by the sign of the gaussian curvature: positive in the light grey region, negative in the darker region. (b) Parabolic curves traced by hand on a statue [5,7]

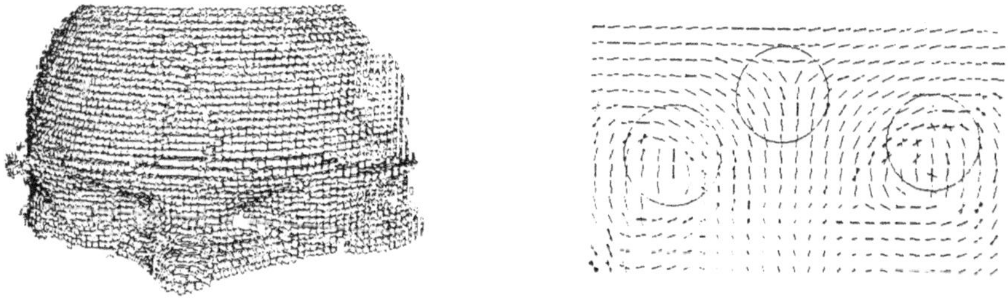

Figure 6: (a) Tangent plane representation of the face region extracted from an MR image. (b) Smoothing of the direction field showing some of its umbilic singularities.

Man Computer Interaction

Cooperative Human-Machine Interfaces for Medical Image Workstations: A Scenario

TH.WENDLER

Philips GmbH, Forschungslaboratorium Hamburg, FRG

Abstract

Adaptive, cooperative systems are designed to take diverging user interface requirements for different users and tasks into account. This is proposed to contribute to the userfriendliness of diagnostic image workstations in digital radiology. The benefits and desirable properties of adaptive user interfaces are discussed. A scenario for adaptive, cooperative diagnostic image workstations is outlined. Principle adaptation mechanisms, knowledge sources and evaluation strategies are described and illustrated by examples.

Introduction

Userfriendly human-computer interfaces have been identified to be an important feature of medical image workstations for fully digital diagnostic imaging departments. Userfriendliness is generally regarded to result from a predominantly user-centered and iterative design process [4][6][7]. However, user-centered design often focusses on some kind of 'average' user and rarely considers changing user interface requirements that arise from different kind of users and the diversity of tasks users have to perform. Today's human-machine interfaces appear to be 'static': They behave uniformly, regardless of what kind of user is working with a system and what task the user is concerned with.

Adaptive user interfaces, in contrast, take into account that users and tasks are variable. In radiology, digital diagnostic display systems will be used by different stereotypic categories of users (e.g.radiologists, physicists, technicians etc.) with diverging user interface requirements. Diverse categories of goal-oriented tasks can be destinguished ranging from simple (routine) to complex (problem-solving) image handling and display procedures. Significant individual differences (e.g.novice/expert, casual/frequent users) are of general importance for user interface design and should be considered as well as personel preferences and styles. Furthermore, user interface requirements are known to change over time. This will be of particular interest in the introduction phase of fully digital systems where probably new image viewing procedures will evolve.

Adaptive user interfaces are designed with the main objectives to increase usability and convenience. From the user's point of view, adaptive systems should appear to be supportive by behaving cooperatively. From the designer's view, user interfaces have to be based on formalized models of clinical users and tasks that define major part of the interaction system.

In this paper some key issues in adaptive user interface design will be addressed. Part of an adaptive user interface for cooperative image workstations is described.

Adaptive user interfaces

Generally speaking, adaptive user interfaces relate their behavior to the environment (users and tasks) they work in. To achieve this, system actions and re-actions have to be based on knowledge about users and tasks and about the general task domain. These formal user- and task-models may be pre- specified (using a priori knowledge of the diagnostic process) or acquired and updated during the users' dialogues. Adaptive systems are designed to realize purposeful changes at the user interface not only as part of the iterative design cycle but as an embedded feature which is part of the functioning of the system. The purpose (objective) of adaptation and the degree of adaptivity may be defined in wide limits. A classification of possible levels of adaptivity in system design has been given in [5]:

adapted	designed (static system that fits to user needs)
adaptable	tailorable by user or system supplier
adaptive (discriminative)	discr. among multiple choises, fixed response
adaptive (self-regulating)	interpret. of environm.,learning,evaluating success
adaptive (self-mediating)	interpret. of environm.,model builder and planner
adaptive (self-modifying)	reasoning about purpose, evaluating the evaluation.

The general discussion of pros and cons of the various kinds of adaptive systems is certainly beyond the scope of this paper. The last two categories mentioned may not be in reach of todays technology and their general desirability for user interfaces in radiology may be questioned. The adaptive diagnostic workstation described in this paper will focus on category 3 and 4.

Some recommendations for a design methodology have been proposed in [1]. For the systematic design of adaptive user interfaces, at least the following questions have to be considered and answered explicitely:

What are the objectives of adaptation ?
What is the adaptation mechanism (theory of how objectives can be reached)?
What are the sources of evidence (data) for adaptation ?
What will adapt (what becomes visible at the user interface) ?
How can the success of adaptation be measured and evaluated ?
What are the implementation strategies for adaptive systems ?

Using this guideline, adaptive user interfaces for image workstations in digital radiology will be discussed.

Objectives for adaptation

A general motivation for adaptive user interfaces in digital radiology has already been given in the introduction of this paper. The global objective is to support multimedia document access in fully digital, hospital integrated picture archiving and communication systems (PACS). In complex information systems, adaptive user interfaces will be supportive by simplifying document access and presentation. The variability of data flow organization in different hospitals and in different countries can be taken into account. Main objectives for system design will be to

increase the effectiveness (by providing the right information),
increase the efficiency (by providing the information timely), and
increase the acceptability (by appropriate information presentation, user guidance and feedback).

Adaptive, cooperative user interfaces will provide active support for the diagnostic process by trying to satisfy multiple user requirements and by taking the initiative to support the user's current task:

Differences among canonical or individual users will be considered. There are a number of stereotypical user categories with diverging user interface requirements. According to their general role in the diagnostic process we can differenciate between e.g.

general or sepecialized radiologists,
referring physicians,
physicists,
technicians or
administrative staff.

Additionally, categories of individual differences among users can be defined according to criteria like skill (naive/novice/experienced/expert users) or usage pattern (frequent/casual users). Personal preferences or styles (e.g.task-dependent personal document arrangement on the display screen) may play an important role for user productivity and satisfaction.

From the procedures of film-based radiology, some categories of goal-oriented tasks with diverging user interface requirements can be derived:

Routine image interpretation (e.g.routine thorax image interpretation).
Problem solving image interpretation (e.g.multimodality image viewing, considering additional non-image information sources).
Pre-examination viewing.
Post-examination image quality check.
Final reporting session.
Demonstration/consultation (involvement of multiple viewers).
Teaching file image access for radiological education.
Image access for research tasks.

Task characterization with respect to users (e.g.casual/frequent task) appears to be of importance.

Adaptation mechanisms

According to user- and task differences, the following principle mechanisms of adaptive, cooperative system behavior are proposed:

Cooperative selection of relevant documents (e.g. model-based selection of the patient's relevant images to be used in a given work session)
Cooperative document display (e.g. model-based arrangement of sets of images on the display screen).

Additionally, adaptation mechanism like adjusting the level of feedback and user guidance or restricting/expanding the user's scope of actions (e.g. reducing the number of functions presented to the user in a routine work situation) may contribute to the adaptation objectives.

Sources of evidence for adaptation

An important consideration for adaptative systems are the information sources which can be utilized as a basis for cooperative actions. There are three principle information sources available: General domain knowledge, domain status data (current object properties) and interactive data captured during the user's work session:

General domain knowledge: This includes general knowledge about the medical diagnostic process (procedures, information needs and data flow models in clinical radiology).

Domain status data: These data describe the properties and status of current domain objects (e.g.the available objects of radiological interest like patients, examinations, images, reports etc.). Also the data base for user-models (characteristics of the current user) and task-models (current goals or sub-goals and status of the user's current task) fall into this category.

As an important pre-condition for cooperative decision processes, the objects of radiological interest have to consist of rich sets of object descriptions from which inferences about adaptive behavior can be made. Example: To enable the system to cooperatively select and display relevant images in a given context, **object descriptions** must provide information like

Patient data consisting of patient demographic data, clinical problem description, examination request, anamnesis.

Diagnostic Code (e.g.using the ACR index code [3]) specifying anatomical and pathological data, localization and savety of diagnosis.

Examination data (imaging modality, examination-type and -parameters).

Geometric data (patient orientation, slice orientation).

Physical image parameters (size, resolution, storage location).

Image display parameters (position, scaling, grayscale mapping).

Clinical requester (institute, department, ward, referring physician).

Responsible staff in radiology department. (radiologist, technician).

Relation to other documents (e.g.part of a study, folder, etc.).

Interactive data: These are data captured during a work session from the user's input. Useful dialogue data to be recorded may be the user's function calls and their sequence. Also direct manipulation data (e.g.the mouse-controlled re-arrangement of objects on the display screen) may be captured and interpreted. The system may gather additional data from meta-dialogues to clarify some of the users goals or needs.

What will adapt ?

This addresses the question how the system's cooperative actions will finally become visible to the user. Some examples of adaptation are given below.

Active parameter setting for functions. Example 1: The user wants to select images for display and looks into the patient's image folder (FOLDER-function). Instead of the total of 80 images available for this patient, the user will be provided with a pictorial index of the 12 images which are relevant in this context. Additional menu fields may also be presented for the access to MORE information. Example 2: After selecting a new patient (PATIENT-function), a meaningful arrangement of objects will automatically appear on the display sceen with appropriate (user- and context-dependent) ordering, size, greyscale mapping etc.)

Actions initiated by the system. Example: Images which are anticipated to be relevant in a given work session will be transported by the system from the archive to fast foreground memories. The user will recognize and appreciate this by consistently fast image retrieval procedures.

Modification of user interface complexity. Example: There will be quite a large number of functions available at digital image workstations which are not needed in all tasks and by all kind of users. Therefore it makes sense to adaptively select functions to be presented to the user (restricting/ expanding the user's scope of actions in routine/problem-solving tasks).

Measuring and evaluating the success of adaptation

For self-regulating adaptive systems there is a need to install mechanisms to evaluate and accordingly modify the adaptation strategy. There are different kind of data sources available that allow to reason about the success of adaptation. The system may (a) monitor the success of cooperative actions by obsering available indicators for mismatched settings, e.g. if the user asks for MORE or different information, UNDOs actions or replaces parameter settings, or re-positions default object arrangements on the screen.

The system may (b) monitor available indicators for user difficulties (e.g. the user's help requests or action retries).

Implementation strategies

A block level architecture of an adaptive, cooperative image workstations which is currently realized as a laboratory prototype is given in [2]. The software structure separates an object-layer (giving full access to complex environmental objects) and a model-layer that represents general domain knowledge and the adaptation strategy. The object-oriented part of the design is described elsewhere in more detail [2]. The model-layer implementation uses a rule-based approach.

Conclusion

Adaptive, cooperative user interfaces offer a promising approach for matching complex information system technology (PACS) with diverging user and task requirements. In adaptive system design, the early and intensive interdisciplinary cooperation between medical and technical disciplines is mandatory for success. Namely the clearly enhances the quality of the user-centered design process and opens the potential for better user interfaces.

References

1. D.P.Browne, R.Trevellyan, P.Totterdell, M.Norman: Metrics for the building, evaluation and comprehension of self-regulating adaptive systems. Proc. Interact'87, North Holland 1987, pp.1081-1087.
2. R.Grewer, K.J.Moennich, J.Schmidt, H.Svensson, Th.Wendler: Object-Oriented Design for Medical Image Workstations. Proc.CAR'89, this issue.
3. Index for Radiological Diagnosies. The American College of Radiology, revised 3rd edition, 1986.
4. D.A.Norman: Cognitive Engineering. In: User-Centered System Design: New Perspectives in Human-Computer Interaction (D.A.Norman, S.Draper, ds.),Lawrence Erlbaum Ass.1986, pp.31-61.
5. P.A.Totterdell, M.A.Norman, D.P.Browne: Levels of adaptivity in interface design. Proc.Interact'87, North Holland 1987, pp.725-721.
6. Th.Wendler, R.Grewer, K.J.Moennich, H.Svensson: Design Considerations for Multi Modality Medical Image Workstations. In: Pictorial Information Systems in Medicine (K.H.Hoehne ed.), NATO-ASI series F, Computer and Systems Sciences, Vol.19, Springer 1986, pp.401-420.
7. Th.Wendler: Human-Computer Interaction in Radiology. Proc.CAR'85, Springer 1985, pp.669-678.

Obejct-Oriented Design for Medical Image Workstations

R.GREWER, K.J.MÖNNICH, J.SCHMIDT, H.SVENSSON, TH.WENDLER

Philips GmbH, Forschungslaboratorium Hamburg, FRG

Abstract

This paper is concerned with the question how to structure different pieces of information in digital image workstations in order to satisfy both user interface and system requirements. It is shown how an object-oriented paradigm can support the design of human-machine interfaces for workstations in radiological information retrieval systems. After a brief introduction to object-oriented design, the system architecture and part of the object structure for a diagnostic image workstation are presented. It is shown how an object-oriented system can be beneficially used for the implementation of adaptive, intelligent workstations that actively support the user's work procedures by integrating rule-based knowledge sources.

Introduction

The transition to fully digital diagnostic imaging departments in radiology will lead to systems that comprise complex information spaces. For clinical staff, medical image workstations have to provide human-machine interfaces that hide this complexity to the user and allow easy and convenient information access and display. Although simple in appearance, advanced diagnostic image workstations themselves will be technical systems which have to create, manage and exploit a large amount of structured information with considerable complexity.

Users are merely concerned with the documents of radiological interest (images, reports, studies, examinations, patient data and other information is routinely used in the diagnostic process). Today, this information is available on different kind of media like films, forms, record cards etc. It is often provided handwritten, informel and uncomplete. In contrast, digital workstations are concerned with technical objects (like files, memories, viewports, user-models etc.) that need to be strictly formalised. Intelligent workstations have to make use of structured knowledge as a basis for user-adaptive, cooperative behavior [5]. Apparently, the information space in a medical image workstation has to be represented on different levels of abstraction. Different kinds of speed/flexibility compromises are imposed by the user-interface requirements and system needs.

Object-oriented design

Object-oriented design means that the system structure is directly derived from natural objects. All real world entities can be described as objects. For example a piece of paper is an object with properties like color, format, contents and so on. More specific to radiology, image type objects have simple properties like size, resolution, identification, imaging modality etc., but also more complex semantic descriptions indicating the anatomical or pathological image content. Most real world objects can be structured into a class hierarchy, for instance the class of images belongs to the class of documents and the class of ultrasound images is a subclass of images. A class can be seen as a prototype for instances (single unique objects) and each instance receives all properties of its class and super classes (inheritance). Describing objects in that way is common to the language tools which can be used to implement objects. The most important advantage of this type of information structuring is that it supports the natural way of human thinking.

In addition to this data structuring, object-oriented programming languages allow the designer to attach a number of methods to each class or prototype. These methods are activated by sending messages to objects. For example: A message 'add image X' may be sent to an image-folder, with the result that the addressed folder now contains also this new image. Using object-oriented programming, large systems can be designed with data (object properties) and processing functions (object methods) being encapsulated in well defined units. Systems which are designed in an object-oriented way are easy to read and therefore easy to maintain.

The information space in an object-oriented system can easily be separated into entities that represent different levels of abstraction (e.g.text strings, viewports, images, user-models etc.). For display systems, the consistency of the user-interface can be largely supported if the internal object implementation directly mirrors the structure of visible objects on the user's display screen. From the system's point of view, the complexity of the information space becomes better manageable by a strict separation and encapsulation of well defined object entities. For user-adaptive systems, object structures represent easily accessible pieces of knowledge which can be used as a rich source of information for high level inferences about users and goal oriented tasks.

The idea of object-orientation as briefly described in this paper is supported by programming languages like for example Simula[2], Smalltalk[3], C++[4], and Objective C [1]. Most of the current expert system shells use objects for their knowledge representation.

Object structure for a diagnostic image workstation

A prototype system of an intelligent diagnostic image workstation with object-oriented control structures is being build in the laboratory. The objects are grouped in three main catagories:

* objects of radiological interest (e.g.images, examinations, reports etc., also object collections like folders)
* objects that support interaction and user guidance (e.g.menu fields, graphical support objects)
* system object that are invisible to the user and help running the workstation (e.g.workfile object)

From the user's point of view, the most important objects in medical image workstations are the documents of radiological interest:

Fig.1a shows the structure of the document class-hierarchy and gives an example for typical object properties. The class DOCUMENT is divided into the subclasses IMAGE, IMAGE_SEQUENCE and TEXT. IMAGE itself is divided into ORIGINAL_IMAGE, DERIVED_IMAGE and RAW_DATA, where a derived image is typically derived from an original image by some type of image processing operation. Original images are divided into modality specific subclasses like CT_IMAGE, DIGITIZED_FILM_IMAGE and so on. Each object class has properties which can be used as facts for intelligent decision processes. The numerous properties defined for e.g. IMAGE type objects are shown in fig.1b.

The workstation user interface is supported by objects for interaction (e.g. mouse, keyboard) and user guidance (e.g. menus, bars, graphical frames or sliders). Interaction objects are the only objects whose properties can be directly changed by the user. For instance moving the mouse updates the mouse-property 'position' (visualized by a cursor on the screen). On special user actions, for instance clicking a mouse button, the mouse object generates messages depending on the current context. For example pressing the mouse button while the graphic cursor is somewhere in the middle of a document, selects that document by changing the mouse property 'move-connected-document'. Thereafter, further movements of the cursor (while the button is still pressed) result in 'move' messages from the mouse object to the selected document. When pressing the mouse button while the graphic cursor is on a menu field, the mouse generates the message 'selected' and sends it to the current menu object. The selected menu decides, depending on its contents, which message has to be generated for other objects involved.

In the background a number of system objects are neccessary to control the system operation. The user does not need to know about their methods, properties or even their existence. For example any graphics appearing on the screen will consist of subobjects like polyline, character, curves and so on. Another example for an administrative object is the 'workfile', which is responsible for managing the storage of documents throughout the various storage facilities of the system.

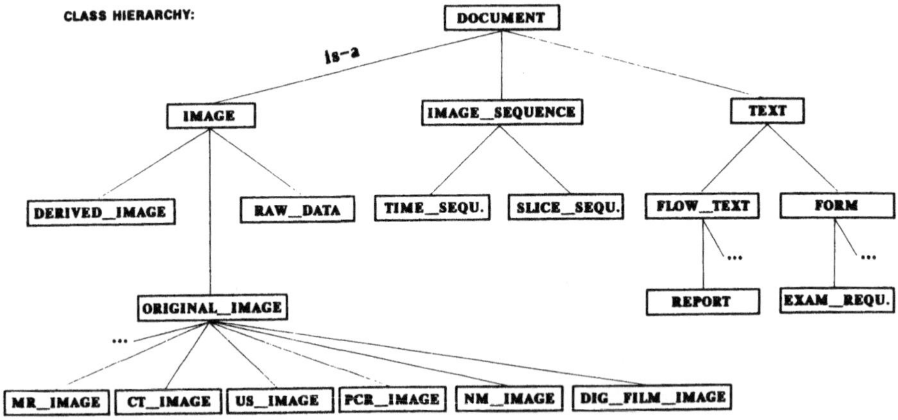

Fig. 1a: Class hierarchy for documents of radiological interest

OBJECT PROPERTIES:

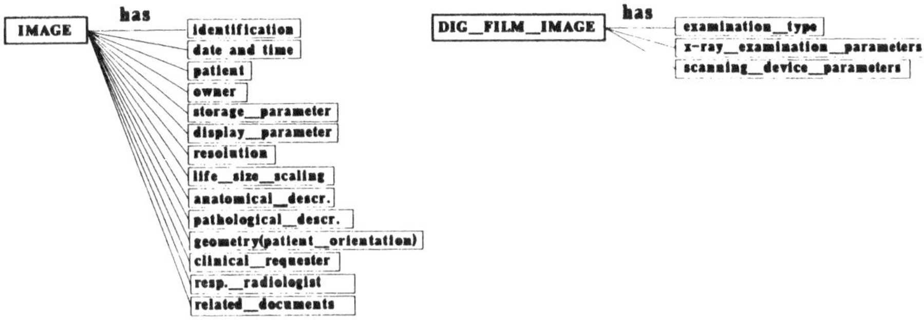

Fig. 1b: Properties of example objects.

The role of knowledge

As a minimum requirement, an image workstation user interface should offer a number of functional tools for fast and convenient document access and display. Advanced image workstations, however, reach far beyond this point and try to make user interfaces more intelligent, aiming to support the user's work procedures by context dependent, cooperative system behavior [5]. Adaptive, intelligent workstations will be able to actively support goal-oriented tasks in radiology and a variety of different users with diverging user interface requirements.

To achieve this, image workstations need not only knowledge about objects as described above, but also knowledge about general information requirements in radiology, clinical work procedures (task-models) and different users (user-models). While the knowledge about objects is represented by the discussed object properties (facts), it turned out that the knowledge about information requirements, user- and task-models can beneficially be stored in terms of rules. For that reason, our general system architecture is divided into different layers (see fig.2):

* The model-layer, containing rules about radiological procedures, tasks and user behaviour.
* The object-layer which represents facts about the environmental objects (object- properties).
* The realtime-layer, logically a subset of the object layer, controls all realtime functions implemented in hardware.

The workstation has different mouldings depending on the point of view: At first, the workstation handles objects (represented by the object-layer) with all their properties which can be retrieved, shown, modified, stored etc. Secondly, the workstation behaves as a functional toolbox, where tools are methods of the various objects. Thirdly, workstations behave like intelligent aids which actively assist users, based on pre-specified or acquired knowledge about the current user and task.

The communication between the model- and object-layer is established by special interface modules. Using an expert system for the model- layer, the messages are generated by 'demons' or 'if-needed meta slots': If the inference machine needs some facts which can be found in some object properties, a special program (the 'demon') is automatically invoked to fetch the required value. Using Prolog for the model-layer, the messages are generated through special predicates. Additionally, the object-methods are able to initiate inference processes in the model-layer by suggesting a hypothesis. The current workstation implementation is based on the languages Objective C and PROLOG for the object- and model-layer respectively.

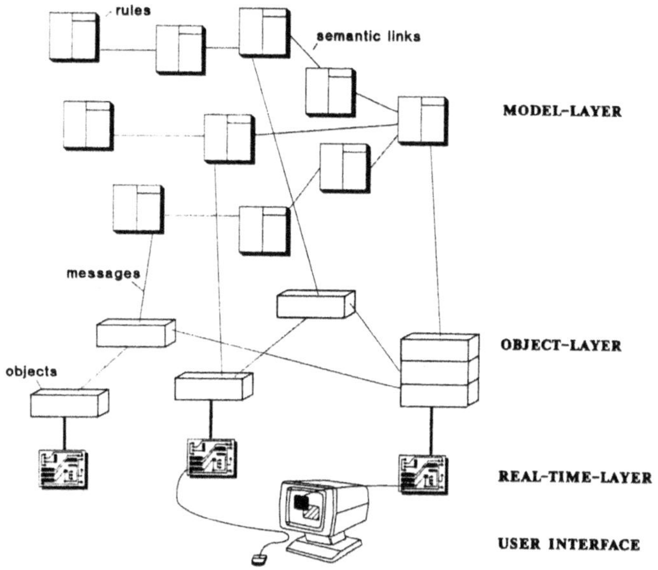

Fig. 2: Workstation architecture

Conclusion

Object-oriented design is a excellent method for the implementation of diagnostic image workstations. The advantage that objects can be derived directly from the existing world can bridge the gap which naturally exists between system designers and the end users. The design of adaptive, intelligent workstation by adding knowledge in terms of rules is straight forward because the object-properties already cover most of the required facts that are needed as a basis for decision processes. As object-oriented design encapsulates properties and methods into well defined units, readability and maintenance is well supported by object-oriented systems.

References

[1] B.J.Cox: Object Oriented Programming, An evolutionary approach. Addison Wesley 1987
[2] O.J.Dahl, B.Myrhaug, K.Nygaard: SIMULA common base language. Norwegian Computing Center S-22, Oslo, Norway 1970
[3] A.Goldberg, D.Robson: Smalltalk-80 The Language and its Implementation. Addison Wesley 1983
[4] B.Stroustrup: The C++ Programming Language. Addison Wesley 1986
[5] Th.Wendler: Cooperative human-machine interfaces for medical image workstations: A scenario. Proc.CAR'89, this issue

Specification and Design of Medical Workstations

Robert Boesecke, Wilfried Müller, Klaus Alandt, Wolfgang Schlegel, W.J.Lorenz

Institut für Radiologie und Pathophysiologie, DKFZ, INF 280, 6900 Heidelberg

Summary

Currently we have a situation where performance and costs of hardware and software products are changing fast and simultaneously the requirements of the users concerning quality of products increase. In the field of medicine, medical physics and biomedical engineering high quality products are required designed to meet modern standards in functionality, speed, userfriendlyness, reliability, cost effectivity, maintainability and portability.

More and more software solutions are used to overcome the problems of the changing hardware market. Software development is extremely man power intensive. Thus both, research and industrial developers have to apply modern software engineering methods to meet the requirements of the user.

Introduction

Medical Workstations

A medical workstation can be defined similar to computer workstations known from technical applications e.g. Computer Aided Design (CAD) workstations in common use for electronical and mechanical engineering and simulation since several years. Typical attributes of these systems are more or less userfriendlyness and interactive response. The workstations are mainly designed as single user systems.
Medical workstations are specified as 2d or 3d image handling systems for support of diagnostics or as therapy planning systems for radiotherapy, neurosurgery and surgery.

User interface

Today userfriendlyness is generally achieved using a mouse/graphic tablet driven graphical user interface with extensive use of windows and hierarchical or parallel menus.
Here the intention is to enable the user to concentrate on his technical know how and minimize the effort for computer specific tasks.

Hardware

Interactive response is mainly due to the used hardware, generally PC's and mini computers. Here two branchs exists, either coming up from the more and more powerful IBM PC compatible systems or coming down from the more and more PC like workstations. The

power of these workstations can be described typically by a CPU speed of 1 to 20 MIPS or up to 4 MFLOPS, one to 64 Mbyte RAM, a theoretical bus speed of up to 10 Mbyte/sec, disk space of some 100 MB. The graphic device is deeply integrated into the computer, thus a special terminal is often not necessary and the graphic screen is used as terminal also.
It uses a single screen with a typical resolution of 1000*1000 pixels and 1 to 24 bit planes. The graphic speed is typically 5000 to 250000 2d or 3d vectors/sec.

Integration
In addition to technical workstations, medical workstations must generally be integrated to some degree into other devices as CT, MR, PET, radiograph and therapy machines. This integration can have the form of a physical integration into the device (e.g. CT), a node in a Local Area Network (LAN) or part of a Picture Archieving and Communication System (PACS).
The minimal form of integration is the property to read and convert medical data from a storage device.

Current situation from the developers point of view
Beginning in the 70's the half-life time for hardware products in the mini/micro computer market began to decrease and has now achieved estimated eighteen month. After this period the computer manufacturers introduce a new generation of computers with substantial improved performance, typically by a factor of two, and often for a reduced price.
Simultaneously graphic userinterfaces developed for non computer experts appeared together with personal computers and brought new standards in userfriendlyness.
To overcome security and maintenance problems several organizations like the FDA (Federal Drug and Health Authority) in USA or the TÜV (Technischer Überwachungsverein) in West Germany try to set quality standards for software, dedicated also for systems used in medicine.
Although these standards appear more or less vague, current software products for use in medicine often do not meet these standards.
More and more software solutions are used to overcome the problems of the changing hardware market. Software development is extremely man power intensive. Here 10 to 50 man years are typical numbers for the amount of work to develop the software for a medical workstation, e.g. a modern Radiotherapy Planning (RTP) system.
Thus both, research and industrial developers have to apply modern software engineering methods for being able to meet the requirements of the users and of their own staying in their frame of costs and manpower.

Methods

Software standards

A very essential point to reduce hardware dependency and development times is to select and choose appropriate software standards. Significant standards exist for:

-operating system
-programming language
-graphic library
-file management library
-data bases
-user interface
-tools
-networks
-documentation

The selection criteria should be oriented according the specification of the system and the quality criterias valid for software and its documentation:

-accuracy
-reliability
-completeness
-performance
-economy
-integration
-maintainability
-portability

Unfortunately some of these criteria are contradictory.

To achieve good portability it is wise to use a high level programming language like C and choose UNIX rather than anything else as operating system. On the other hand will a very speed optimized system often have a poor portability due its hardware oriented code. And, a very complete system will probably not result in a very economic system in terms of development costs.

Software engineering techniques

The extensive use of software engineering techniques [2] and of tools, software by itself, for specification, design, documentation and implementation will not automatically speed up the software development, but will at least help to avoid errors and result in more uniform program and documentation structures.

Tools for development and documentation

Tools for support of the development and maintenance, software by itself, are available in more or less good quality. Project planners, source code and version management systems, structogram editors, automatic documentation systems and performance analyzers are only a small list of available tools. The win in time due to such tools is a trade off with the time needed for training and use of them. In our experience a more flexible, homogenous and error free product will be the result of the use of carefully selected tools.

Specification

The specification is the definition of the product on a level above used hard- and software. It describes the system and is used as working instruction ("Pflichtenheft") for the developers and as purchase information for the customers. The specification of a medical workstation describes all options available, the user interface, the performance, form and range of integration into an environment, used algorithms etc.

From our point of view it is very essential to have the support of experienced "clinical" users for the specification. Very helpful is also an existing prototype workstation or a similar system as a standard to be improved.

We made good experiences with a test installation of our RTP workstation in another clinical oriented site and with using the RTP workstation for scientific programs with users having absolutely no computer experience.

Companies generally require also input from their marketing departments to come in touch with the needs of the market.

In our experience the specification changes dynamically during development. Thus the developers must be aware of this fact and select tools and modularize their system in a way that they can follow changing specifications.

Design

Design describes the method of the implementation of the specification. Here data structures, flow charts, software interfaces etc. are written down, produced with the help of tools.

The design is mainly oriented at the specification, very helpful is here a very experienced development team. The design depends partly on used hardware, thus hardware components should be known when design starts. Design should not be influenced very much by used programming language or operating system.

Unfortunately also design changes dynamically during development, thus modularization and extensive use of tools are the keywords for developers to be able to follow unexpected changes in specification and design.

Quality assurance

Today Quality assurance (QA) is an often used word with a vague meaning. In our understanding quality assurance should join a development project during the total life cycle starting at the selection of software standards. For example the selection of PASCAL rather than FORTRAN will result in a more error free code [3]. Furthermore testing of the software during development and maintenance is essential. Here standardized test procedures are very helpful, preferable is automatic testing with error detection. Other QA tools are inspection of specification, design, documentation and code. Thus QA should be performed by independent people.

Results

Figure 1 and figure 2 show screen layouts of two at the DKFZ in Heidelberg developed RTP systems [1]. Figure 1 shows a system now distributed and maintained by F.L. Fischer dedicated for convergence beam and interstitial therapy planning under stereotactic conditions. Figure 2 shows a 3d RTP system with all features of modern treatment planning. This features are 3d graphics for beam's eye view and dose distribution evaluation, multi-planar reconstruction of image data, ray-tracing dose calculation, dynamic and conformation therapy planning, optimization, dose volume histograms, support of CT-images, MR-images and contours.

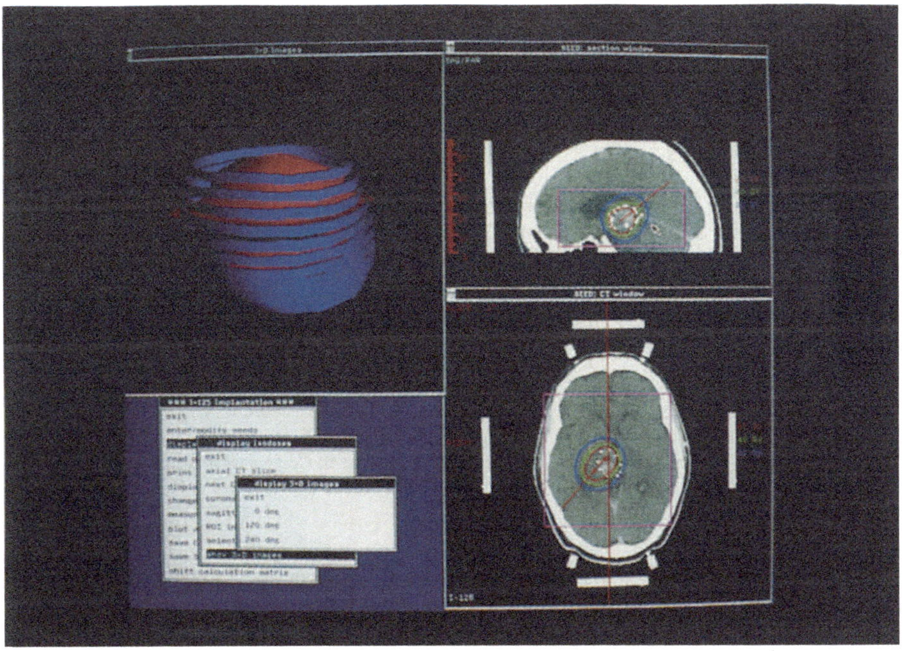

Figure 1: graphic screen of the Fischer stereotactic therapy planning system. The sytem is installed on a VAXstation 3500 with eight bitplanes.

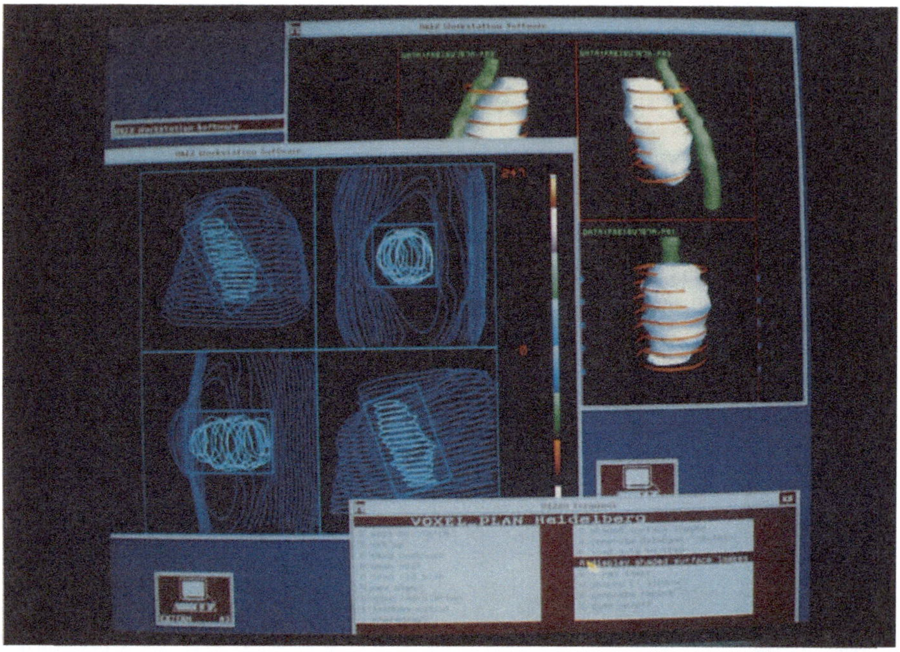

Figure 2: Graphic screen of the threedimensional radiotherapy planning system developed at the DKFZ Heidelberg. Here the flexible window oriented userinterface is demonstrated. The window configuration is up to the user, generally only two windows are necessary for a RTP session. The system is running under VMS on DEC VAXstations.

References

1 R Boesecke, B Bauer, W Schlegel L Schad and W J Lorenz: MEDICAL WORKSTATION FOR RADIATION THERAPY PLANNING; THE USE OF COMPUTERS IN RADIATION THERAPY, Proceedings of the Ninth International Conference on The Use of Computers in Radiation Therapy, held in Scheveningen, The Netherlands, 1987, ELSEVIER SCIENCE PUBLISHERS B.V. (ISBN: 0444702636), 181-183

2 H Balzert: Die Entwicklung von Software-Systemen; Reihe Informatik, B.I. Wissenschaftsverlag 1982 (ISBN 3411016183)

3 F Stetter:Softwaretechnologie; Reihe Informatik, Band 33, B.I. Wissenschaftsverlag, 1987

Social Implications

Imaging, Communication and Perceptions

Heinz U. Lemke

Technische Universität Berlin, Fachbereich Informatik, FR 3-3
Franklinstr. 28/29, 1000 Berlin 10, W. Germany

1. Introduction

In brief, Computer Assisted Radiology (CAR) is the application of computer and communication technology to assist radiological and related medical activities in the generation, processing, management and communication of medical images. Its principle aim is to augment the qualitative impression, the quantitative measurement, and the communication of medical images relating to the state of health of a patient.

At this moment in time it is not much more than a proposition that radiology is the discipline, which should, because of its traditional service function, also provide in the future all medical imaging related services. Other specialities which use extensively medical imaging, such as cardiology, orthopaedics, neurology or radiotherapy may take part in this service too. Modern communication networks allow the corresponding infrastructures for this to be estabished. Computer and communication networks can be adapted to particular needs of hospitals and other regional health care centres. The effectiveness of these infrastructures depends on the creativity with which these new communication networks are being developed, implemented and used.

Several concepts and realisations of Picture Archiving and Communication Systems (PACS) and teleradiology systems demonstrate this point. There exist open and closed systems, single-vendor and multi-vendor systems, star, bus and ring topologies, as well as a multitude of different types of terminals which may start from simple, up to highly sophisticated work stations and optical archiving systems.

At least one medical discipline, such as radiology, should, for the benefit of the patient, obtain familiarity with these concepts and optimise them for patient care.

2. Imaging

Radiology fulfills the commonly accepted definition of a discipline, as defined by a system of

a) objects and definitions, i.e. medical images of patients and the definition of image parts,
b) relationships, i.e. of parts within images and between images,
c) body of propositions and hypotheses, i.e. meaning of particular instantiations of images and their parts,
d) methods and techniques for image generation, modelling, analysis, verification and description, i.e. reporting and consultation on patient specific imaging procedures.

Components a) to c) are the basis for the activities outlined in d). Such activities are part of a diagnostic pathway (Fig. 1). It represents the heart of the radiological activity and computer assistance may be realised at various positions in this pathway.

3. Communication

Implicit in Figure 1 are a number of communication paths for the transfer of information. They may be inter-human communication oriented, such as between referring physician and patient, radiologist and supporting staff, radiologist and medical specialists or radiologist and referring physician and clinician. Traditionally, these communication paths are established by the spoken word, written documents or visits.

Overall support for clerical and administrative activities of communication pathways within the radiological department may be provided by a radiological information system (RIS). Communication to the outside of the radiological department may be realised via an interface of the RIS to a hospital information system (HIS).

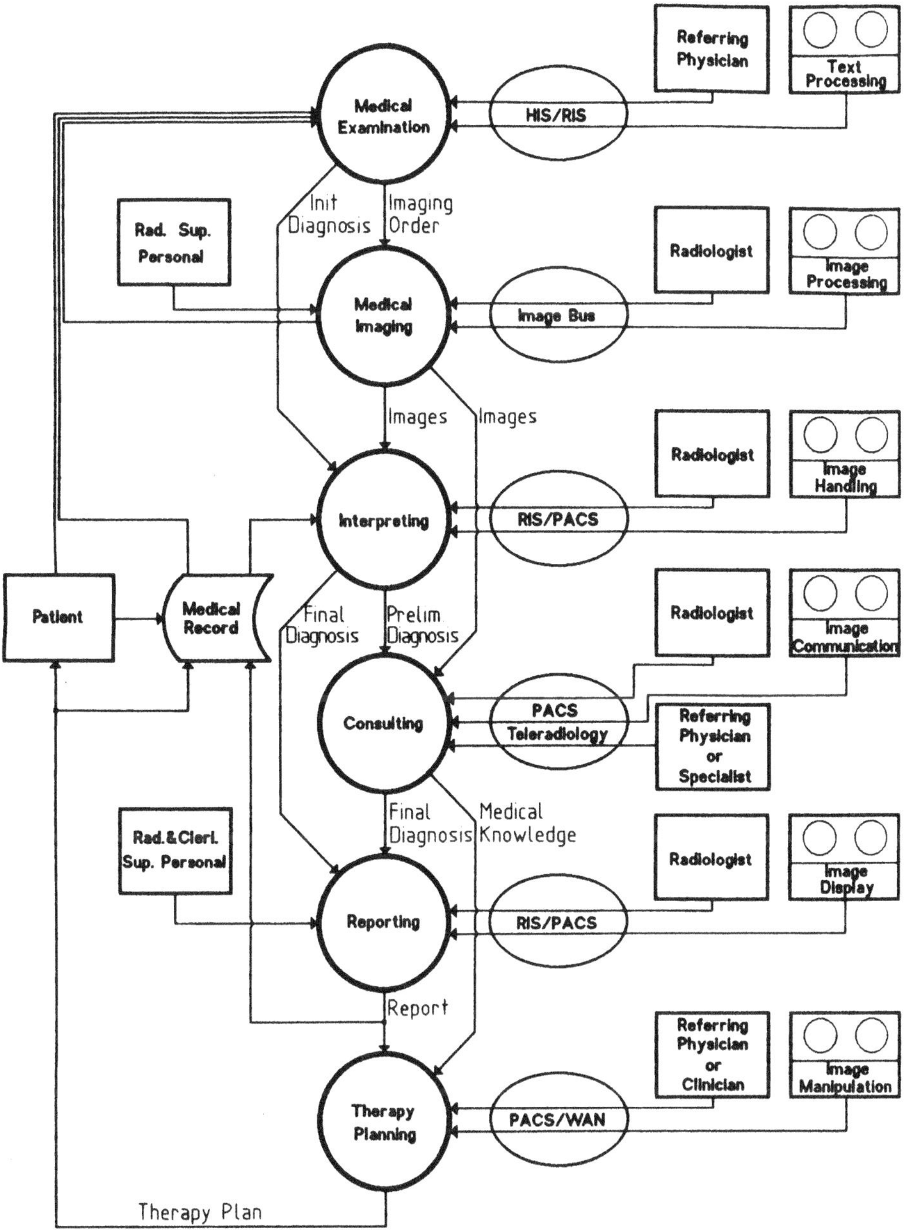

Fig. 1 Diagnostic Pathway Using Images

Medical examination and imaging scheduling

This HIS provides administrative functions supporting entry and management of patient demographic data, order entry, exam scheduling and status data, reporting, billing, etc. RIS includes some HIS-functions but, in addition, some specific radiological functions are available, such as film library management, interpretation, transcription and radiological reporting. Both HIS and RIS are typically realised on traditional hospital and department-wide communication networks for text transmission.

Medical Imaging

Communication within the digital imaging system and local processing system require high bandwidth image transfer channels. Specialised parallel bus-systems with up to 100 megabits per second transfer rate may be required for real-time display of large volumes of image data, as for example in digital angiography or other dynamic imaging procedures.

Interpretation

Interpretation may be supported by PACS, which provides short, medium and long term storage along with various hard and soft copy display facilities. Certain functions are shared with RIS. Some interface problems between RIS and PACS still remain to be solved.

For interpretation, it is necessary to have fast access to new and old images from various imaging modalities held in a patient file. This demands high speed local area networks (LANs) with an open system architecture to allow for the integration of imaging devices and other system components from multiple vendors.

Consulting and Reporting

Local consulting may be supported by PACS. Long distance consulting requires some kind of teleradiology system using wide area network facilities. Teleradiology facilities are being developed in many countries using smallband and broadband ISDN or similar network infrastructures.

The reporting function may be supported by RIS and PACS facilities.

Therapy Planning

Preparatory work for therapy planning may be carried out on the local facility. For certain specialised tasks it is feasible that specialised health care centres can provide this service. The usually required long distance access may be realised via wide area networks (WANs).

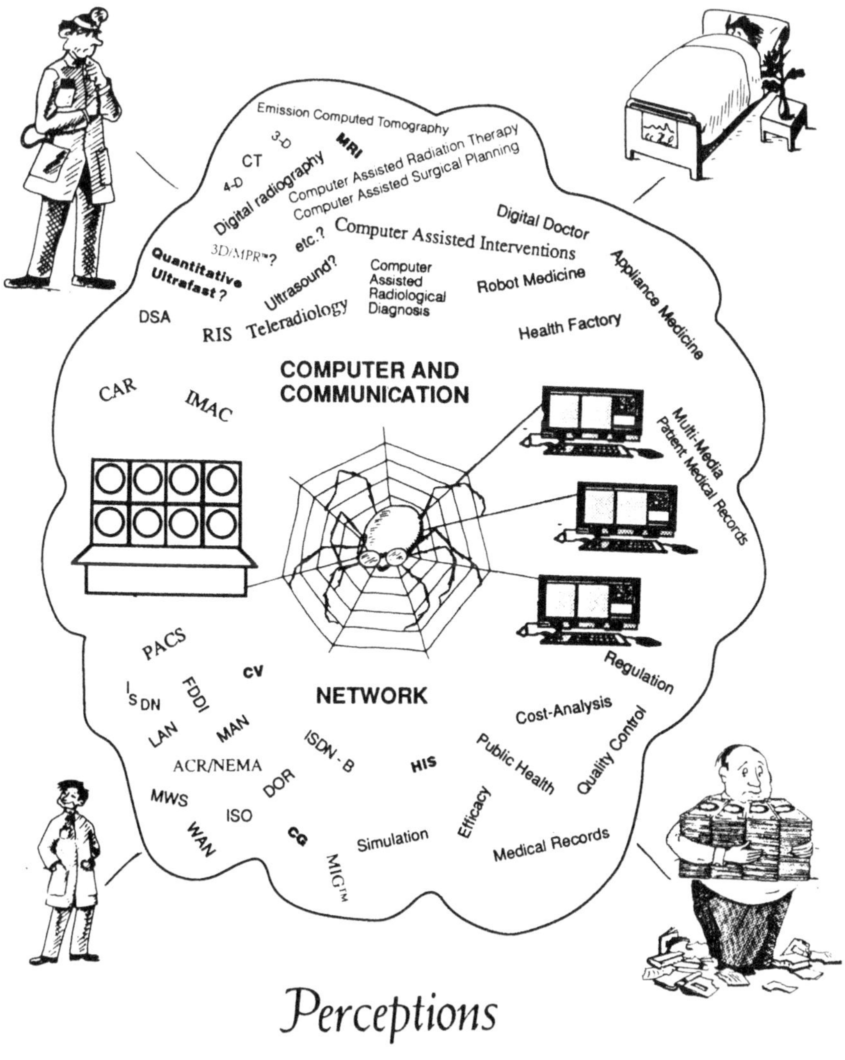

Fig. 2 Group Perceptions on Computer and Communications in Medicine

4. Perceptions

Qualitative and quantitative handling of medical images using modern technology is advancing in an evolutionary manner. Communication of medical images, however, may be realised through entirely new concepts in health care communication infrastructures. This can only be achieved through a new way of thinking by patient, radiologist, physician, engineer, administrator, legislator and even politician. Each group of people involved, however, has established perceptions and patterns of thinking when confronting issues relating to computer and communication technology in medicine, see Figure 2.

These perceptions often reflect the divergence of human activity towards technical progress on one hand and the understanding of social and ethical implications on the other. Group perceptions on the advantages and shortcomings of modern tools and systems in medicine will determine the groups cooperation towards their realisation. Interrelating and collating these perceptions to achieve an improved health care system is a task which must be carried out in parallel to network development. Prototyping of communication systems for medical imaging should therefore be accompanied by investigations into their social implication.

References

J.S. List and K.G. O'Malley: The Impact of PACS Technology on Diagnostic Cycle Time, SPIE Vol.914, pp 961-969, Medical Imaging II, 1988.

The Turbulent Road to a Filmless Hospital St. Mary's Past, Present and Future

Dr. J.O.M.C. Craig
Dr. Richard Dawood

Department of Diagnostic Radiology,
St. Mary's Hospital, London W.2.

The concept to build a new hospital at St.Mary's as a digital filmless hospital began in 1982. At that time there was great expectation that this was feasible by 1987, the date the first phase of the new hospital was to be opened. It is worth recording the years of excited planning, followed by the realisation that technology was unable in 1987 to provide the means whereby a fully operational filmless hospital could be achieved. The problems were both technological and financial. The large cost of such a development in today's climate demands proof of future health service cost savings. The present situation is that St. Mary's Hospital is evaluating the efficacy of VDU reporting compared to film reporting: preliminary studies have used subperiosteal resorption in hyperparathyroidism as a test. At present a wider evaluation is in progress including difficult cases of lung, bowel, breast and bone pathology. The images are obtained from digitised film, at varying matrix sizes and viewed on 1K x 1K monitors.

We feel that before a PACS system is installed in any hospital, it is essential that the medical efficacy of such a system must be proved. Only then can we move to the next phase of archiving and hopefully transmission.

Computer-Controlled Interactive Laser Video-disc as a Medical Encyclopedia

Conrade C. Jaffe[1], M.D., Patrick J. Lynch[2], M.S., Arnold W.M. Smeulders[3], Ph.D.

Depts. of Diagnostic Imaging[1] and Biomedical Communications[2], Yale University School of Medicine, New Haven, Connecticut, and the Dept. of Medical Informatics[3], Erasmus University, Rotterdam, the Netherlands.

Summary

Though prematurely heralded in the past, progress in microcomputers has reached a critical threshold in power, speed, reliability and commercial availability to permit these devices to be used meaningfully as a platform for education and medical diagnostic consultation. Because of visual complexity, radiology training has concentrated on educating limited numbers of individuals since it requires intensive tutorial experience to develop image recognition expertise. Video-disc and CD-ROM image storage media make it possible for microcomputer workstations to provide learning environments substantially equivalent to conventional time-consuming tutorial methods. A "hypermedia" program on echocardiology was constructed which provides a user-controlled learning environment with instant access to 54,000 video frames encompassing 1200 clinical issues. The module is controlled by a microcomputer which electronically links relevant graphics, animations, text, categorized databases and digitized audio. This self-contained system has been successfully used in a clinical environment as a primary instructional tool. This media offers substantial advantages over conventional books as a clinical reference source.

Hypermedia for Computer-Aided Instruction:

Newly available micro-computer controlled video-disc workstations offer sufficient on-line mass storage capacity of images to make possible a totally novel approach to instruction in medical imaging. Central to this is a recently coined term, "hypermedia", which has gained attention for its unique power in managing large repositories of complex information[1]. The term should not be confused with "multimedia" which is merely a mix of audio-visual techniques. "Hypermedia", in contrast, is a concept which connotes a highly integrated electronic environment which allows a user to interactively peruse a very large collection of electronically linked information consisting of real-time moving color video images, sound, text and electronically searchable databanks[2,3]. As a project to demonstrate the impact of this hypermedia concept on medical teaching, we constructed a comprehensive instructional module on echocardiography. This module has been successfully used for the past year in a residency training program as the sole method for giving trainees greater than basic level training in echocardiology.

Hypermedia and HyperCard:

Hypermedia is a general concept ("metaphor") promoted by a several personal-computer visionaries[1,4,5] which employs the organizing capacity of the computer to integrate and control not only large bodies of text and numerical databases but also to mesh them with appropriate sound and moving images (by connections to video-disc players or CD-ROM). A popular software implementation of that concept is offered by Hypercard in the Macintosh. HyperCard software is part graphics program, part database, and includes a built-in programming language[6,7]. HyperCard as a high-level command language, allows users with little or no previous experience in traditional computer programming languages to integrate text, audio and video material, computer graphics and animation into educational and training applications that take full advantage of the computer's easy-to-use graphic interface. The central concept underlying HyperCard applications is hypertext. In existing hypertext applications users browse through a system that is designed to extensively inter-link and cross-reference text and graphics in a non-linear, interactive format that responds to the user's requests for more information on a particular subject. The project described here is one of the first in medicine to offer additional integration with audio and video material. For example, to get further information on a word or graphic the user simply points at the area or text segment and clicks on the mouse button.

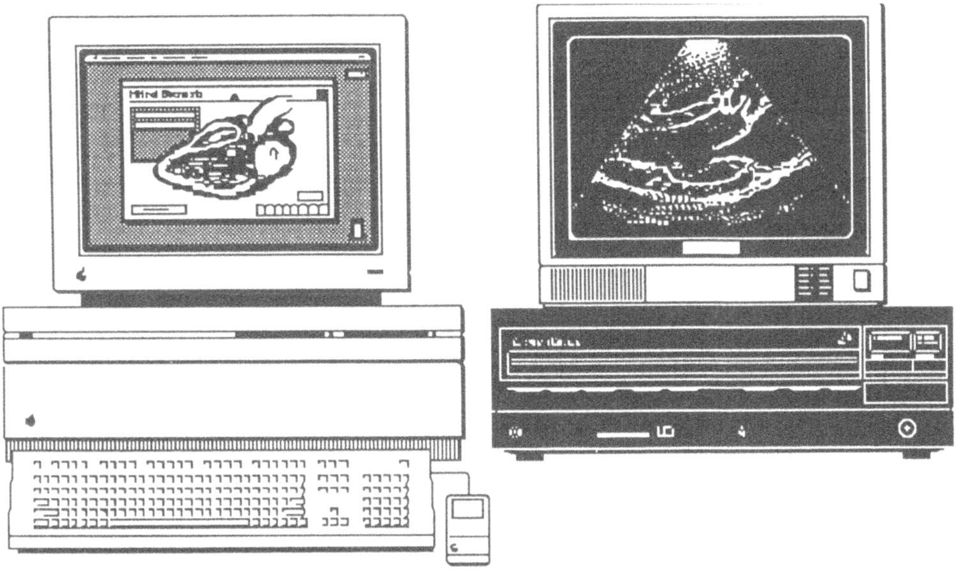

The program then branches to another "card" (individual screen views, composed of multiple fields of information and electronic link points are called "cards"), or plays a video sequence from the laser-disc, or calls up new text from hard-disk or CD-ROM, or in some other way immediately brings forth a new set of related information to the user. Thus the user can explore the program in a heuristic, self-paced manner that takes full advantage of the computer's ability to organize, correlate, store and retrieve information. Hypercard is one of a new family of computer languages: so-called "object oriented" programming. In conventional programming techniques implemented in languages as BASIC, Fortran, Pascal and C, a sharp division is made between programs and data. Programs are abstract, generally applicable recipes which achieve a specific effect by filling in specific data as parameter values. For example, to get the record of a patient Wilson, in the routine: GO_AND_FIND_ME_A_RECORD, the parameter NAME might first receive the value WILSON and then be executed. As the program routines have general value, so the structure of the database and the records must also be fixed (or at least predictable). In object oriented programming, for example the languages LISP, Prolog and to some degree Hypercard, the central issue is a concept where data repesent an instance of the concept and programs perform an action on that concept. In Hypercard, the main concept is the card to which a "script" (i.e., card specific piece of program) and data (in the form of text, or a graphic on the face of the card) are attached. Cards are organized in larger structures called "stacks", as in "stack of cards", and contain smaller structures called button, fields, etc. Once a card is displayed (which is equivalent to activated) other actions may be initiated by selecting buttons (i.e. activating the script of a button), or fields with the mouse. HyperCard allows only one card to be visible at a time regardless of screen size. Users browse through HyperCard applications by moving between cards much as a stack of index cards can be flipped through, one at a time. Cards are composed of two independent layers, a transparent foreground layer containing graphics, buttons and text fields unique to a particular card, and a background layer of graphics, buttons and text fields which may be shared by many cards.

HyperTalk as an Authoring Language:

In an object-oriented language environment nothing happens until the user (or the program somewhere) actively initiates an action on the part of the program, usually by clicking on a screen button. "Mouse button down", "mouse button up", "new card" and "open stack" are all examples of common events occurring as a user navigates through a HyperCard application. Each of these events sends a message through the hierarchical structure of HyperCard objects. An object with an event handler for that event takes action when the event message is received. Otherwise the event is passed on to the next layer, from button to card to stack etc. This capacity to pass on an action automatically implies the inheritance of actions shared by all elements in one concept. For example, if it is desirable to create the button GET_OUT_OF_PROGRAM, it is most efficient, even desirable from the point of compatibility, to create the action of the button at the most general of all levels: the stack.

Objects created within HyperCard exist in a hierarchy, and messages pass through the hierarchy seeking handlers aimed at controlling that event. If the user clicks down on a screen button that does not contain a handler with instruction to act on the "mouse button is down" event, it passes the message along to the card, the card background, the stack, and finally to HyperCard itself. Any of the objects deeper in the hierarchy may be programmed to respond to "mouse button is down". Thus the foremost objects in the hierarchy (like buttons) can "inherit" capabilities from objects deeper along the chain (like card backgrounds), and large numbers of buttons on different cards can act on HyperTalk instructions attached to the card background (or stack) they all share. In addition to the programming possibilities offered by HyperTalk's standard 250-word command vocabulary, HyperCard's native capabilities may be extended and customized through the addition of HyperCard words to which special-purpose segments of Pascal or C-language code are attached. These so-called XCMD or XFCN resources allow the HyperTalk language to control virtually any device with a serial interface, such as videodisc players, CD-ROM drives and other audiovisual and computer equipment[6,7].

REFERENCES

1. Nelson, T. Computer Lib/Dream Machines. Bellevue, Wa.: Microsoft Press, 1987

2. Brunsman S, Messerly J, Lammers S., Publishers, Multimedia, and Interactivity. In: Ambron S., Hooper K, eds. Interactive Multimedia Bellevue, Wa.: Microsoft Press, 1988; 273-281

3. Cook, P., Multimedia Technology. In: Ambron S., Hooper K, eds. Interactive Multimedia Bellevue, Wa.: Microsoft Press, 1988; 217-240

4. Kay AC. Computer Software. Scientific American 1984; 251: 53-59

5. Engelbart DC., A Conceptual Framework for the Augmentation of Man's Intellect. In: Vistas in Information Handling Vol. 1 Spartan Books, London 1963

6. Goodman, D. The Complete HyperCard Handbook. New York: Bantam Books, 1987.

7. Williams, G. HyperCard. Byte. 1987, 12(14):109-117

DARE – Radiology Data Bank, Essen

E.Löhr T.Philippi

University - GHS Essen, FRG

Universitätsklinikum Essen
Hufelandstrasse 55
43 Essen 1

Tel: 49 (0201) 7991 2800 / 3640

Summary:

Five years of experiance in designing and running an image tone and text data base on public networks.

Despite the progress in computing, networking and storaging in recent years, the realisation of a image tone and text database, suitable for the daily background use, is still a hard cumbersome work. Approaches have been made, mainly on the basis of a PC stand alone solution or a LAN with an integrated server.

The database Radiology < DARE > is one of the first European image tone text database on the combination of a central mainframe and a local situated mass storage for images. The necessary link between the two devices, is physically the telephone network, logically the access of images by text. Both, as concerning the inevitable questions of international standardization of different protocols, will be fulfilled by the media of videotex. Briefly, the approach is based on the use of a mainframe for the retrival of the text, the telephon network for transmission of text information, which controls the image storage, situated at the site of the end user, by transmitted pointers.

The use of an analogue mass storage for images with a capacity of 54.000 images of PAL TV quality offers the benefit of movies, analog sound, digital sound and digital data and still frames, which can be digitized at reasonable costs.

Since more than five years we are experienced in processing radiological and neuroradiological topics for learning and instructive programs.

So we collected more than 20 programs, such as:

cerebrovascular malformations
aneurysms
vascular diseases
orbital lesions
brain tumors
tumors of the maxilla
CT diagnosis of the petrous bone
CT of the basal structures
cranio-cervical junction
renal hypertension and PTA
pancreatic diseases
tumors of the esophagus
liver disease
colonic tumors
abdominal lymphomas
fractures of the spine, etc.
CT - findings and liver transplantants
CT - examination of abdominal lesions in infants

etc.

In consideration of the aim of our application, to give reference to the external radiologist and to perform an individual concept of the structure of a image text tone document, this one is basically designed for the use in videotex networks, which will have a mayor impact by the introduction of ISDN. Using the network architecture of ISDN there are the necessary flexibility to install different VAS (Value Added Services), one kind may be the transmission of still video frames.

Today, the database DARE can be accessed by a simple telephone line, even in a bad network enviroment like the Chinese one, only using some cheap basic equipment. Transmission of text in both directions, text retrival with small response times and an enviroment to have access to further 3500 information providers, including 400 different external computers gives a considerable edge to the telephone subscriber, still in an analog transmission mode with a very low baudrate.

At the present time the generation of data, both text and images is to be done centrally in Essen out of technical and economical reasons.

Therefore only a small part of the inherent possibilities of an image based communication, which does not include the task of daily archieving, is part of our realisation. Of the long term image data only a small fraction were used for referencing purposes, the greater volume of the normal use has to be archieved in a different manner.

From the point of view of an.external information provider, there are essential the following characteristica of access:

Time of system login
Time of application login
Time of access to a subject
Time of access to the content
Time of access to the image
Time of image built up

To optimize the system approach of an image provider as a whole one has to consider end users attitude quite fairly. In his position of a decision maker he is not interested to invest time and personal efforts to get familiar with sophisticated technical structures, either hardware or software oriented. He just wants a minimized approach to get access to the content he needs.

Due to the lack of personal front end processors, like an intelligent check card, which covers the personal enviroment of the user, - therefore quite suitable to reduce his effort of system approach - , we have chosen a structured free text database.
In this case, we suppose, there is a quite convenient manner of distributing medical or other knowledge.

To archieve further progress in the development of text, sound and image databases one has to consider three different areas very carefully such as:

Information Provider

Network

End User

In each area there are still quite a lot of problems, which are unseen or not yet a point of discussion. After a couple of years of discussion, PACS is still in progress. Questions are open in storaging, processing and distribution of image data locally or by a WAN.

The transfer of images across the border of a department of radiology will still be a problem, there is no cost effective solution for image transfer between different departments of a hospital. The approach to transfer the image data on existing networks is not to recommend, especially if a lot of short online services have to be done.

Some network considerations were made in the project BERKOM, but there is in fact no solution, which covers the wide spread area, national as international.
In recent years we have made some experience with video conferencing, which is quite suitable for exchanging information or medical approach.

Despite the fact that there is some loss of bandwidth in the video signal this media is quite convenient to exchange images and data. The lack of a missing transparent mode during the conference can be reduced by the preparation of the source material. In our case we are using the video conference channel of 2 Megabyte in two ways, one of them conservational mode and as the second one the graphical, but switched to our database. The quality of the transmitted images was quite resonable although there is the lack of no original image data of full bandwidth.

The transparent use of the 1.55, 2 or 140 MByte channel will solve this problem, which incoperates an suitable interface to the existing computer equipment.

From the point of view of an end user, in local or remote position, there is still no sufficient cheap equipment available, which includes the necessary I/O rate between storage and mass storage, the necessary interfaces to different networks, the logical interfaces are in the beginning, briefly there is still a lot to do.

PACS and the Need for a New Legal Structure

B. LUTTERBECK, R. WILHELM, S. HADJEROUIT

Technical University of Berlin, Institute for Applied Computer Science

1. Temporary Ban on Digital Archiving

On January 1, 1988, a new Radiography Ordinance came into force in the Federal Republic of Germany (Bundesgesetzblatt I 1987, p. 114). § 28, Section 5 of this ordinance contains the following provision:

> "...the storage [of X-ray photographs produced by direct radiography] on a picture or other data carrier [is] permitted only upon expiry of a term of three years."

The new ordinance, then, evidently reflects doubts about the reliability and efficiency of computer-based picture-generating techniques and opts for a temporary ban on digital archiving.

It is our claim that this ruling is based on an obsolete structural model of legal regulation. The new provision models the risks entailed by modern technology in accordance with the archetypal exploding steam engine. It regulates hardware, where it is software that it should be regulating.

2. Information Technology and Law

Modern legal experts see law as fulfilling the following basic functions:

- Decision of individual conflicts

Legal regulations should only intervene where risks have turned into concrete situations of danger. In such cases, law settles the compensation of damages incurred.

- Law as a means of shaping society, to produce specific social effects

Law should govern the development and application of technology in such a way as to eliminate perceptible risks as far as possible.

In each of these cases, the optimal legal means for attaining the objectives are dependent on the respective cultural setting. In the U.S.A., for instance, liability rules play a crucial role here. By contrast, in the Federal Republic of Germany, law is seen as the best form of regulation. Moreover, in the wake of efforts to establish a single European Common Market, the differing approaches to regulation – at least those of the European countries – are likely to be brought into line with one another. There is ample evidence of this in the new EEC product liability provisions [6], which take the correct but still controversial view that software, too, should be treated as a product [7].

In future, then, the attempt will probably have to be made to combine regulatory and approaches with those based on the decision of individual conflicts. As far as PACS and other medical applications are concerned, however, GODES has quite rightly pointed out: "The law is very much unsettled" [4].

3. Ways of Regulating PACS

Within the range of possibilities between these two approaches, the following options have so far presented themselves as viable alternatives:

(1) Lifting the Ban on PACS Techniques

In instituting its medical device regulations, the FDA has taken a step which may be considered of significance for the international community as a whole:

> "Computer products that have a medical purpose, but *which are intended to involve competent human intervention* before any impact on human health occurs, such as where clinical judgement is used to check and interpret the system's output, are medical devices" [3].

This ruling, which should in principle also apply to PACS, is too imprecise, though, since it leaves the following basic issues open:

- Is it useful to make a distinction between human error and errors of technical origin? Shouldn't regulations go beyond mere technical requirements (e.g. on the correctness of programs) and lay down the requirements enabling the correct interpretation of the picture data presented?
  This is also a question relating to the qualification requirements for PACS users.

- A criterion for assessing the risks associated with PACS must be the question as to what extent
  - incorrect or missing picture data,
  - information losses during the various phases of data processing, and

– poor picture quality causing distortion of the objects represented may lead to wrong diagnoses, faulty planning of operations, etc.

It may be assumed that the FDA regulations are unlikely to be adopted in their present form by the EEC.

(2) Product Liability Rules

"Computer liability is in its infancy" [1], because, so far, there have been no precedents enabling corresponding legal provisions to be established. The following cases, though not of direct relevance, should nevertheless attract the attention of PACS designers, too:

- Printing errors in the section on the Carter-Robbins Test in a standard medical reference work caused incorrect medical treatment to be carried out, very nearly resulting in the death of the patient. The case went up before the German Federal Supreme Court [2].
- The case involving the radiotherapy device "Therac 25", manufactured by the Canadian firm Atomic Energy. As a result of its incorrect programming, patients were exposed to overdoses of radiation.
- A radiation count error at the Royal Devon and Exeter Hospital (UK) went undetected for months, impairing the health of several hundred patients and proving, in some cases, lethal. The London Times called it "...one of the biggest medical compensation claims in the history of the health service" [8].

These cases at least help us to pinpoint the essential issues at stake:

- Following GODES [4], there should be clear demarcation of the interests and responsibilities of all the parties involved: the manufacturer of a software program, manufacturers of equipment, program distributors, programmers and medical consultants, hospitals and other health care providers, specialist departments of hospitals, and medical personnel.
- Error classifications should be drawn up relating to the respective spheres of interest. The classification of product defects provided for by German law might well offer a clue as to how to set about solving this problem. Here, distinctions are made between: 'flaw in a product', 'design defects', 'failure to warn', 'development risks' and 'product observation and duty to inform'.

There is, so far, no clear indication of whether, in the interests of consumer and patient protection, strict product liability rules – such as are proposed by

BRANNIGAN and DAYHOFF – are likely to be introduced or gain acceptance, as the case may be.

4. Outlook

The ruling in § 28 of the Radiography Ordinance – curious as it may appear at first sight – does have, on closer inspection, a certain justification. A temporary ban may, and indeed must, result in minimization of risks and more precise definition of the technical requirements. There is no justification, however, for this sort of caution in the long term. Computer scientists, radiologists and legal experts are called upon to seek a suitable solution to this problem. One thing is clear, though, even now: there are no alternatives *to* the law, only alternatives *within* the law.

References

[1] BRANNIGAN, V.M.; DAYHOFF, R.E.:
Medical Informatics. The Revolution in Law, Technology and Medicine, in: The Journal of Legal Medicine, No. 1 (1986), p. 1 ff.

[2] ENTSCHEIDUNG DES BUNDESGERICHTSHOFS (German Federal Supreme Court Ruling) of 7.7.70 ("Printing Errors"), in: Juristenzeitung 1971, p. 63 ff. with commentary by Deutsch.

[3] FLANNERY, E.J.:
FDA Regulation of New Devices – Expedited Review, in: IEEE Engineering in Medicine and Biology Magazine, June 1988, p. 88 f.

[4] GODES, J.M.:
Developing a New Set of Liability Rules for a New Generation of Technology: Assessing liability for computer-related injuries in the health care field, in: Computer/Law Journal, Vol. VII (1987), p. 517 ff.

[5] KILIAN, W.:
Rechtliche Aspekte der digitalen medizinischen Archivierung von Röntgenunterlagen, in: Neue Juristische Wochenschrift 1987, p. 695 ff.

[6] STÜRMER, U.; KOEPKE, J.E.; REISCHER, B.:
Neue EG-Produkthaftungspflicht – Vergleich USA. New EEC Product Liability. The US in Comparison, Gerling-Konzern Allgemeine Versicherungs-AG, Cologne 1988.

[7] VANDENBERGHE, G.P.:
Software Bugs: A matter of Life and Liability. Computer Software and the European Directive on Product Liability: The First Generation Question,

in: Computer Law Association (ed.), Distribution, Access & Communications, Amsterdam 1988.

[8] WRIGHT, P.:
Radiation error count puts more patients at risk, in: The Times, 6.12.88.

Medical Computer Law: Discussion of Legal Problems by Using Computer Science and Expert Systems: Example Radiology

B. R. BEIER

Center for Medical Informatics, Goethe-University
Frankfurt am Main and IMSDD, University of Bonn, W-Germany

Summary

This paper will give an overwiew about the whole field of the problems in "Medical Computer Law". The first legal paper in this session showed the variety of socio-legal communications in which also "Computer Assisted Radiology" works. PACS is one perhaps t h e example for new thinking needed in legal relations when treating patients. On this background we have to understand that modern technology especially in connection with medicine causes problems and consequences which have been are unknown until now: the use of computers in medicine, a technico-legal revolution. The conditions and the consequences of "modern technology" in medicine are discussed and the new research field "Medical Computer and Technology Law" should try to answer the questions and to solve the new type of problems.

1) The field of "Medical Computer Law"

Medical Computer and Technology Law is the term for a new research field in the "triple-area":medicine, computer science and the law and was based on the need to bring together these three sciences because of the need to cooperate. [1]
Medical Computer Law is also a field with a "transborder international relationship". When we will have a look to the practice of "computerized medicine" we have also to accept a fast development of integration of computer technology in the medical records use and - the most important - in diagnosis, and in therapy (assistance).[2]

Legal problems in medicine are well known, especially in the fields of civil law and in criminal law:the physicians code of secrecy, his duty to keep secret all information and the knowledge about the patient . This includes also the right of the patient to know about himself and - very important - about his illnesses.[3]

Malpractice is another important topic of "classic" legal problems in medicine. But just in connection with the expression "malpractice" we can see a fast change in the connection with the use of computer science: f.ex. is it malpractice not to use computerized assisted diagnosis or also: another example, is it malpractice to refuse medical help in the case of "mental death"? Even thc question is discussed: what is l i f e ? When we think on the development of insemination, experiments with egg cells or genetic engeneering! When does life begins? or when does it ends? We see the fundamental ethical question. [4]

But the total change in my view also t h e q u a l i t y step in the e v o l u t i o n of medicine (diagnosis a n d treatment, and also research) comes from the use of modern technology, e.g. computer science, e.g. genetic engeneering.

This especially also changes the qualitiy of medico-legal problems and it brings up new questions and need answers to problems unknown until now.

So we can see two big areas of M E D I C A L C O M P U T E R A N D T E C H N O L O G Y L A W :

1. The area of "old known" problems (as described in short before) with a new special intensity by using modern technology (f.ex. the duty to inform the patient, the quality standards) and

2. The area of new problems unknown until now, the r i g h t of i n f o r m a t i o n a l and m e d i c a l self determination when using databases and medical expert systems, the "ownership" of electronic files (bits and bytes) and the CAR- resp. P A C S - problem: the legal status of the "digitalized" patients and the networking of all kinds of patient related informations as well as the transborder flow of data and -finally- the medical device-discussion concerning the correct use of software.

As we have learned the development of medical informatics is the same in principle in the US and Western Europe (with some differences in the expanse of computer use in the US, especially in the past). We can also consider a time-oriented "wake-up" of similar legal problems: e.g. the liability question when using medical software and the question of ownership on patients information as well as the ethical problem of genetic experiments and genetic engeneering.
Therefore we can formulate the catalogue of working fields in "Medical Computer and Technology Law" (who is still not yet finished) [5]:

- Documentation Law concerning computerized information
- Physician - Patient Contract Law (the duties, obligations and the rights under the use of modern technology)
- Liability (especially for defective medical software, the "authority" of medical expert systems)
- Ownership on patients data (files)
- Data protection, physicians code of secrecy

- Ethics in medical informatics combined with the question of the right of informational self determination
- diagnosis (and treatment) with genetic testing and genetic engeneering.

This enumeration shows only a draft outline of fields with legal input strictly related to the use of computers in medicine. At this moment we are in the early (spring) phase of this new research field.

2) P A C S - the example for critical topics in Medical Computer and Technology Law

Let us now have a look at the example PACS and see some examples of the new socio-legal impacts for which we have to find solutions, as physicians, as computer scientists, as lawyers and last but not least as sociologist and psychologist. PACS [6] means - as you all know - Picture archiving and Communications system including ISDN,digital optical recording and communication, LAN, RIS etc. At first sight concerning the problem of documentation it seems only to be one more possibility of storing data. But, here is the difference: the folder of a patient is unique with handwritten information on the patient's status and the lab- or other (also dictated) results (X-ray, etc.). The computer file already has the possibility of multiple existence and via electronic mail there is an accessibility for all user of the system.[7]

The patient is no more limited to "his physician" with "his" documentation. He is a small "bit" in the computerized medical documentation and information system.
The question of ownership changes dramatically because there is no real legal enforcement of patients interest in fact. Legal regulations is needed under the new situation either in the US or in W-Germany (Europe) or elsewhere.Just the computer itself and the different possibilties of "treating" and using data justifies and enforces the socio-legal control: who works with - and how - and in which way with the "patient"?
The concept of self determination with the special "right on informational self determination" is the expression of a democratic and human socio-legal principle.[8] It shows very clearly that the capacity of information systems with the answer of e.g. "data protection" means on the other hand also the ownership on all information about the patient, especially when stored in digitalized sytems and picture archiving.
Changes of the picture (bones, lungs, organs etc.)
and the transfer of such data including transborder patients data flow are not possible without the expressed consent of the patient. The right of informational self-determination and the herewith included ownership on himself, including information, data and pictures of all kinds is the socio-legal answer to the so-called "visible" human being". It is a quality step in the evolution of a human society and human dignity.
Also the use of modern technology changes the existing struc-

ture for the regulation of health care, for which the physicians have an effective monopoly on primary medical decision making.[9] In the future the development of medical expert systems has the possibility of providing both closer control over the physician's decision making and possible automated substitutes for the physician's decision making. [10]

This is also the landmark to focus on legal requirements of the practice of medicine. The threat to the physician's control comes if the computer can actually compete with the physician in judgement, the information system can check any decision against the established norms prior to treatment and query the physician to explain any deviation:the system can be programmed to act as a "second opinion" or even also as an expert itself (PACS, ?!) to decide the best solution including a proposal for the patient.

This development can arrive at a stage to use and or rely on the computerized system instead of or in place of the physician, e.g.the situation of the recent "third pilot" in the cockpit of an airplane: the engeneer was replaced by expert systems (Airbus A 320).

It is not the place and the time to focus on all social and legal consequences but one very important topic should be discussed: the question of liability, who is responsible in the future when using software in medical decison-making and/or medical expert systems: the physician,the system, the author of the system or the society through a controlling organisation (e.g. the FDA in the US or the BGA in W-Germany or ethical commissions?).

This brings us to the "medical-device-discussion.

3) The medical device-discussion

As we have seen there is fast development in electronic medical devices for diagnosis; they are much more complicated and sophisticated than "normal" X-rays. In the past the physician was the final authority for medical decisions. All medical decision and the treatment basically and finally was controlled by the physisican under his responsibility. In fact, the legal situaton is the same today. Under German Law the help of machines may only be enlisted when they have been checked and are functioning properly. In the case of a machine being defective or functioning incorrectlly,the principle of personal liability also holds for the assignment of unqualified, uninstructed or wrongly instructed personnel as well as for faults of this personnel in the use of the devices.[11] A concept for strict liability is not yet in force.A court decision has been made to to reverse the question of proof.[12]
This means that not the "victim" has to prove the defect, the physician or the hospital has to prove that the function or the use of the machine was correct.

For machines in medical use there exist regulation in W-Germany ("Medizingeräte-, Röntgenverordnung, Gerätesicherheitsgesetz"). The principle idea is to reduce the risk for the patient.On the other hand there is a postulated need for the socalled strict liability (Gefährdungshaftung).
When we see that up till today anyone has been able to produce and install software even for use in the medical field,and this has not been regulated or restricted too much, then we realise that there is a real need for the introduction of strict liability in any use of software or hardware or any machine.
In the field of products we find now the European guideline for strict liability. It means that every product which causes damage is "liable" on its own. [13]
In the US the Food and Drug Administration (FDA) is involved in the question of standardisation and/or/resp. regulation of medical devices including soft- and hardware. [14]

The introduction of complex medical technology challenges the function of the physician as the only authority.
In the case of medical information systems (PACS) these include the designers, developers and users of these information systems. The key question addressed here is the effect on liability and responsibility.[15]

To solve this problem it would be necessary to put an end to the patchwork with special legislation:It seems to be necessary to introduce in any case the principle of strict liability inthe sense that there is an obligation to produce the best work possible in hard- and software and specialized systems (clinical or other medical expert systems) and to install the pinciple of strict liability in all medical related actions including liability for or from systems. This would also effect a protection for the "treating" physician.

4) Conclusion and Proposal

One possible way for improving good quality of hard- and software would be to install third-party certifications. This would involve having a third party licensed by the government, but independent of the government, to examine the software or the system and mark it as approved or disapproved. It would also be required to estabilsh and to define standards and quality measures (Qualitatessicherung) for medical computer software.[16]
Medical computer software and the use of software regulated medical systems for diagnosis and therapy becomes an extraordinary issue in the question of liability and personnel as well as institutional and systems responsibility.

It is clear that all topics shown and discussed in this paper propose new legal concepts to solve these medico-legal problems.

Medical Computer and Technology Law will try to give the answers.

References:

1) The first paper developing a theory of Medical Computer Law was published in 1986, journal of "Recht der Datenverarbeitung", no. 4, p. 186-190 by B. Beier (in German).
2) See the papers of the CAR-Symposium in 1987; Proceedings of the International Symposium edited by H. Lemke, M.L. Rhodes C.C. Jaffee, and R. Felix, Springer, Berlin et al. 1987, p. 485-586, especially the paper of H.K. Huang et al., Los Angeles, Calif.: Picture Archiving and Communications Systems for Radiology, 487-492 and also C.F. Greinacher, Systemarchitektur und Bausteine eines digitalen Bildinformationssystems fur die Radiologie, 544-557 as well as M.W. Engelhorn, a concept to distribute radiological Images, 579-586.
3) Cf. V.M. Brannigan, Compensation or Regulation: The Problem of Medical Computer Software, Journal of Consumer Policy 6 (1983)475-481; Beier/Brannigan, Informational Self Determination: A Choice Based Analysis, Datenschutz und Datensicherung, 4 (1985) 218-224.
4) Cf. Beier, ref. 1, p. 188 et 189.
5) Cf. Beier, ref. 1, p. 190.
6) See ref. 2.
7) Cf. V.M. Brannigan, Remote Telephone Access: The Critical Issue in Patient Privacy, Proceedings of the Eighth Annual Symposium on Computer Applications in Medical Care, Washington, 1984, nov. 4-7, 575-578.
8) See Beier/Brannigan, ref. 3, p.221.
9) Cf. V.M. Brannigan, The regulation of medical computer software as a device under the Food, Drug and Cosmetic Act, Comput. Methods Programs Biomed. 25 (1987) 219-228.
10) Cf. B.R. Beier, Novellierung des Bundesdatenschutzgesetzes, Neuauflage alter Ideen oder tatsächliche Reform? Datenschutz und Datensicherung, no.10, 1988, 506-511 (509).
11) Cf.A. Laufs, Arztrecht, Beck, München 1988, cit.no.353-355 as well as E. Deutsch in Medical Responsibility in Western Europe, Research Study of the European Science Foundation, edited by E. Deutsch and H.L. Schreiber, Springer, Berlin et al. 1985, D 110, p. 261.
12) Decision of the German Federal Civil Court (Bundesgerichtshof), BGH JZ (Juristen-Zeitung), p. 275.
13) Cf.the new EEC Product Liability Council Directive of july 25, 1985. The German version of the Product Liability legislation,draft from march 12, 1987, still in discussion.
14) See the FDA Policy for the regulation of computer products (version from july 1988).
15) Cf. F.W. Engel, Produzentenhaftung für Software, Computer und Recht (1986, W-Germany) 702-708; J.N. Godes, Developing a new set of liability rules for a new generation of technology: assessing liability for computer-related injuries in the health care field, Computer/Law Journal(USA), Vol. VII (1987) 517-534.
16) B.R. Beier, Liability and responsibility for clinical software in the Federal Republic of Germany, Comput. Methods Programs Biomed. 25 (1987) 237-242.

Liability for the Use of Computers in Radiology

Vincent Brannigan

University of Maryland, College Park Md. 20742 USA

The introduction of computer systems into radiology has created legal debates over the use of such systems. The introduction of any radical technology forces the legal system to go through a process of "coming to grips" with the technological development. Because the legal system works by analogies to existing problems, it attempts to fit new technologies into its older models. A problem occurs when the new technology does not have a fair analogy to an old well-defined situation. The introduction of computers into radiology will result in increased demands for strict liability for injuries caused by defective systems.

LEGAL ISSUES IN COMPUTERIZED RADIOLOGY

Radiology has certain characteristics that make it particularly interesting for legal analysis:

1. Radiology is defined by a specific technology rather than a pathology, a structure of the body, a therapeutic approach or other organizing principle. Since it is tied to a specific technology, it is subject to the changes in that technology.
2. The specific technology, imaging, has changed more through the development of computers than any other area of medicine. Many techniques would be impossible without computers, others have been totally automated using computers. Digitized images can be transmitted on a worldwide basis. Standardization of digitizing technique allows easier comparison of images.
3. Radiology is distinctively an information oriented area of medicine. Radiology itself does not treat patients. The information developed must be provided to an overall decision maker. This creates the problem of shared responsibility, and allows the decision maker to decide whether image analysis should be performed by a radiologist, another physician, a technician or a machine.

The distinctive technology of radiology is the creation of images of the internal structure of a patient. The distinctive skill is the interpretation of the images. The radiologist may be thought of as having special expertise in both these areas, based on long hours spent producing and interpreting images. However, technology has relentlessly automated the creation of the images, and treating specialists who were traditionally reluctant to defer to the radiologist's judgement on interpretation, now have the advantage of computerized assistance in enhancing and interpreting the images. Transmission of images will allow ready access to interpretations by remote consultants. Radiologists will find competition either directly from machines and expert systems or indirectly from networks accessing remote consultants. However, the substitution of machine processing for human judgement may require changes in the legal structure of health care. [2] Liability issues will have to be resolved.[3]

DATA AND INFORMATION

In radiology the physical and conceptual separation of data and information is straightforward. Data is the result of the direct examination of the patient, e.g. a radiograph. The radiologist turns the **data** into **information** by analyzing the radiograph and deriving conclusions based on the combination of the patient data with a store of experience.

The information component of medical care includes a complex data flow. Data flows from the patient to the physician and back to the patient. No a priori determination need be made as to who is the decision maker. There can be a model in which the patient is the decision maker, and the physician is a source of data, or in which the physician is the decision maker and the patient is a source of data. A third model has the patient as the data originator, the physician as a conduit and the information system as the decision maker. At least three separate data components can be identified: (1) patient data; (2) the data base of medical experience in dealing with images; (3) the combination of medical experience with the patient specific data, to recommend a course of action.

Computer systems have already begun to dominate the acquisition of patient data. With the development of computerized equipment, most decisions about the quality of an image and its data are not made by physicians, but by the designers and builders of the computerized machines. The data base of "experience" is becoming rapidly computerized. Digital images are easily stored, transmitted and compared. Automated interpretation of medical images depends on the progress of "knowledge engineering" of medicine. Remote interpretation of digital medical images is already being performed. Computers can be sources of data, conduits of data or decision makers. The legal concern caused by computers in radiology is related to the disruption of information flow. The provision of false or misleading data at a critical point may cause direct injury to the patient. Further, the legal system has traditionally set the rules to determine who could perform various tasks in medical treatment. [1]

TECHNICO-LEGAL REVOLUTIONS

A technico-legal revolution occurs when a technological advance cannot be clearly analogized to past legal structures. A technico-legal revolution consists of a series of stages in the legal response to novel developments in technology. [5]

The legal system always has a special problem when faced with technological change. Often there are several available analogies, and no system for choosing among those analogies. False analogies use incomplete or misleading comparisons to attempt to secure a more favorable legal structure, without disclosing the incomplete nature of the analysis. Understanding the legal system's process of coping with a new technology allows us to anticipate and discount false analogies.

The general introduction of computers into medicine has been described as a technico legal revolution. Computers challenge the traditional individual physician-centered model of medical decision making. Computers allow the participation of many physicians in decision making, both by direct participation and by creation of the data base of shared experience. Further, because the product of analysis of this data base may be incorporated into machines, decisions by the makers of the computer systems may determine the outcome of medical treatment.

The technico legal revolution of computers in radiology comes from the disaggregation of a previously a unitary task and the assignment of portions of that task to machines. At one time the radiologist generated the image and did the analysis. Now issues arise such as: Will a hospital be liable for failing to transmit an image to outside specialists for evaluation? Who sets the standards for resolution of digitized images? Will doctors be liable for failure to consult a data base of similar images? Who is liable when reliance on the defective machine causes injury? Answering these questions requires analysis of the underlying assumptions of the legal system.[4]

MODELS OF THE RADIOLOGY DATA FLOW

Two general models for the use of computers in radiology might be described. In the first, the computer system uses devices to acquire data from the patients. The data is transmitted to a centralized facility. Computerized data handling allows separation of data acquisition from information generation both in time and space. A few specialized physicians can use the system to examine many more patients than they could handle with traditional radiographs. This might be called the **centralized** model, since it allows centralized physicians, using the computer to interpret the data for many more patients.

In the **decentralized** model, the computer system is used to capture the decision making process of many physicians, and transfer the information to the patient or local physician for decision. In the decentralized model the computer is essentially a reference for aid in making decisions. Decentralized systems might include both networks linked to expert systems, and smart machines capable of analysis. The critical distinction is the placement of decision making as close to the patient as possible, but with the support of computerized expertise.

ROLE OF THE LEGAL SYSTEM

Under traditional law the radiologist was individually responsible for negligence in creating or interpreting the image. The future of image processing involves sharing the responsibility among various individuals and computer systems. The centralized and decentralized models represent the opposite poles of approaches to computerization of health care and will constitute a technico legal revolution in liability.

The role of the legal system is to prescribe the socially required steps for processing medical information into action on a particular patient, and to define the responsibility for compliance with these norms. The legal tools of liability and regulation are simply two different methods of achieving social control of a technology.[2] Society can either directly control an act, or encourage proper action by requiring certain parties to be financially responsible for improper action.

Both of these control systems are used in medicine. Often regulation is introduced because of the inability of the liability system to accomplish the appropriate mixture of encouragement of new technology and the prevention of the uses of dangerous and ineffective technology.

The role of civil liability itself in most legal systems is twofold. The first principle is compensation. Liability is imposed to ensure that the injured party does not bear all of the costs of the injury. The second principle is regulatory. By setting standards for those who must pay compensation, the legal system defines the required standard of care.

The standard of liability is a fundamental policy decision by the government. It defines who will bear the costs of injuries, and what behavior could be changed to reduce the overall number of injuries. The primary debate in the compensation area is whether the standard for liability should be based on a fault or a no fault concept. Under a fault based regime, a defendant is only liable if the legal system views the defendant as failing in some way to comply with a required standard of care; the regulatory principle is primary. The alternative is strict liability, where liability is found because of the occurrence of a predetermined event. Strict liability is designed to emphasize compensation. It should be emphasized that both fault and strict liability are found in most legal systems. However, the choice between the regimes is not simple, and has a profound effect on the development of liability.

For a variety of historical reasons, fault has been the predominant measure of liability in the rendering of professional services, while strict liability is the usual standard for defects in devices. The technico legal revolution occurs when actions that were previously performed by physicians are performed by, or with the assistance of, machines. Radiology is on the cutting edge of the man/machine interface. Under current legal principles the computer in a decentralized system is clearly a device, subject to strict liability. However, a centralized system that accurately transmits the opinions of the key decision makers would probably not be a device under current law. In other words judgement that has been incorporated into a device would render the manufacturer strictly liable, but a system for merely transferring human judgement would be charged only under a fault regime. The effect of such a legal structure may be to bias the medical system against decentralized decision making, since it imposes greater liability even at the same level of injury. There is some evidence that machine manufacturers have avoided including expert systems in their machines, despite the potential for benefitting patients. To ensure legal equality between the systems, it can be argued that the entire centralized computer system for acquiring, transmitting and interpreting images should be considered a device, even if some of its actions are still performed by humans.

CONCLUSION

The distinction between decentralized and centralized systems should be legally unimportant. The legal system should not bias the choice of the most efficient medical decision maker. It can be assumed that patient representatives will press the legal system to impose strict liability on both centralized and decentralized systems.

REFERENCES

1. Brannigan, V., and Dayhoff, R., "Medical Informatics: The Revolution in Law, Technology and Medicine," Journal of Legal Medicine, March 1986, pp. 1-53.

2. Brannigan, V., "Medical Informatics and the Regulation of Decision Making: The Challenge of a New Technology", Proceedings of the 5th World Conference on Medical Informatics, North Holland Press, pp. 1064-1067 (Vol. 2).

3. Brannigan, V., "The Regulation of Medical Software as a Device Under The Food, Drug and Cosmetic Act," Jurimetrics, Summer 87, Vol. 27, pp. 370-382.

4. Brannigan, V., "Legal Issues in Medical Computing", Encyclopedia of Medical Devices and Instrumentation, John Wiley & Sons. PP 1742-1749

5. Brannigan V. "Biotechnology, a First Order Technico-Legal Revolution" Hofstra Law Review Sp. 1988 pp 545-581

Appendix

Simulation of Osteotomy Surgery about the Hip Joint

STEPHEN B. MURPHY, M.D.[+*], PETER K. KIJEWSKI, Ph.D.[=], MICHAEL B. MILLIS, M.D.[+], AND ANDREW HARLESS, A.B.[=]

Departments of Orthopaedic Surgery, Children's Hospital Medical Center[+] and Massachusetts General Hospital[*]; Joint Center for Radiation Therapy and Department of Radiation Therapy, Harvard Medical School, Boston, Massachusetts

SUMMARY

Acetabular dysplasia is a developmental disease that results in an underdeveloped and poorly positioned hip socket. In order to perform corrective osteotomies of the hip joint accurately, the surgeon must have a detailed understanding of the pre-operative hip anatomy and how the proposed surgery will correct the deformity.

By developing computer methods to display anatomy in three dimensions using CT, we recently have been able to analyze pre-operative hip geometry, compare the patient's hip joint to normal data, and then simulate the proposed surgery. This preoperative analysis has been performed on over fifty hip joints and analysis with pre-operative simulation has been performed on over thirty hip joints. Since patients' hip deformities are highly variable, we currently perform pre-operative analysis and simulation on all patients undergoing pelvic osteotomies.

METHODS.

GEOMETRIC ANALYSIS: A Computed Tomographic (CT) or Magnetic Resonance (MR) study of the patient's hip joints is performed according to a standard protocol. The data is transferred to our computing facility (Digital Equipment VAX 11/785) where the images are reformatted and displayed on graphics processors. The bony surfaces of the femurs and pelvis are calculated automatically based on radiodensity and stored as contours. The contours from sequential images are connected using triangular surface tiles to create three dimensional models of the bone and joint surfaces (8,9,12,15). The models of the bones can then be displayed in any position or orientation.

Femoral (1,11) and pelvic (6,7,9,10) reference coordinate systems are used to eliminate the effect of patient positioning on the geometric analysis. The overall orientation of each acetabulum is determined by calculating a plane which best describes all of the points on the acetabular rim using a computer optimization algorithm. The vector normal to the opening plane of the acetabulum is calculated and resolved to determine the acetabular abduction and anteversion angles (7).

Analysis of the acetabular and femoral head surfaces is based on the subchondral bone surfaces. On each image, the subchondral bone surface of the acetabulum and femoral head is defined using equally spaced points. Non-articular portions are specifically excluded.

The surface points are used to define spheres of specific radius and position that best describe the surfaces of each acetabulum and femoral head using another computer optimization algorithm. The root mean square standard deviation between the real surface and the sphere quantifies the sphericity of each joint surface (12).

The ability of the acetabulum to contain the femoral head is quantified using two methods. First, lateral, anterior, and posterior center-edge angles are measured to quantify containment relative the pelvis. Second, containment of the femoral head by the acetabular rim is measured by modelling the acetabulum as a portion of a globe.

The lateral center-edge angle is similar to the center-edge angle of Wiberg. The anterior and posterior center-edge angles were defined in the transverse plane in an analogous manner. Containment of the acetabular rim was also measured by modelling the acetabulum as a portion of a globe. The polar axis of the globe coincides with a vector perpendicular to the opening plane of the acetabulum. The globe is divided into longitudes and latitudes with the location of the acetabular rim, in degrees latitude, calculated for each longitude (9,10,12). For example, an acetabulum that forms a complete hemisphere would have latitude angles of ninety degrees.

Once the analysis is performed, the patient's hip joint geometry is compared to normal hip joints of the same age and sex. Differences between the patient's hip joint and normal define the goals for surgical simulation.

SIMULATION: Osteotomies about the hip joint can be performed either on the femur or the pelvis, or both. In general, if the major deformity is in the acetabulum, a pelvic osteotomy is simulated. If the major deformity is in the femoral head or neck, a femoral osteotomy is simulated. Femoral osteotomies are performed by combining any desired degree of flexion-extension, varus-valgus, or rotation. To simulate these osteotomies, the femoral head geometry is simply rotated about its center to incorporate the proposed changes in flexion-extension, varus-valgus, or rotation. Femoral osteotomies can move a more congruous portion of the femoral head into the acetabulum to improve the surface area of contact or can correct abnormalities of the femoral neck to normalize the magnitude and direction of the hip joint forces.

Deformities of the acetabulum can be improved by performing one of many types of pelvic osteotomies. Two of the more useful osteotomies, the innominate osteotomy of Salter (14) and the Dial osteotomy (3,4,16,17), are described here. To simulate an innominate osteotomy, a computer model of the innominate bone is divided by a mathematical plane that extends from just above the anterior inferior iliac spine to the sciatic notch. The distal fragment, which contains the acetabulum, is rotated about an axis defined by two points; the pubic symphysis and the sciatic notch. The degree of correction may be varied by the degree of rotation about this axis (13).

The Dial osteotomy of the pelvis is performed by completely separating the acetabulum as a spherical as a shell of bone and cartilage from the rest of the pelvis. This operation allows a large degree of correction. In general, this operation is simulated if there is significant maldirection of the acetabulum in the presence of reasonably spherical and congruous acetabular and femoral head surfaces. The procedure is simulated by separating a shell of acetabular bone from the rest of the pelvic model using a spherical cut. The acetabular shell is then rotated into any desired position. The center of rotation of the osteotomy is the best fit center of the spherical cut. Like the femoral osteotomy, the Dial osteotomy is simulated by placing the acetabulum in the combination of flexion-extension or abduction-adduction that maximally restores normal hip anatomy.

Any combination of femoral and pelvic osteotomies can be simulated in each patient. The combination of procedures that appears to maximally normalize the patient's hip joint anatomy is then recommended to the surgeon.

DISCUSSION

We have found a very wide degree of variability in our studies of patients with hip joint deformities. It is important that patients are not assumed to all have the same deformity and rather that each patient be considered individually. Failure to recognize specific features of an individual patient's dysplasia will result in performing an inappropriate operation or in performing an appropriate operation in an incorrect manner.

Since there is wide individual variation in acetabular dysplasia, both our group and others (5) have developed the practice of routinely analyzing the geometry of patients with acetabular dysplasia and simulating the proposed surgery pre-operatively. According to our protocol, a three-dimensional geometric analysis of the patient is performed and the anatomic abnormalities are defined. Various osteotomies are simulated depending upon the deformity.

Since the anatomy and mechanics of the normal and dysplastic hip joint are not completely understood, we have decided to begin to investigate hip dysplasia by studying three-dimensional geometry. We hope that continued study of the anatomy and mechanics of hip dysplasia will improve our understanding of the disease and will refine our treatment goals.

Acknowledgments: The authors thank N.T. Griscom, M.D., R.H. Wilkinson, M.D., J.E. Hall, M.D., and C.B. Sledge, M.D. for their continued support and M. Chin and C. Pickett for assistance with CT scanning.

REFERENCES

1. Billing, Lars: Roentgen examination of the proximal femur end in children and adolescents. Acta Radiologica. Supp. 110:1-79, 1954

2. Eppright RH: Dial osteotomy of the acetabulum in the treatment of dysplasia of the hip. Journal of Bone and Joint Surgery. 58A: 726, 1976.

3. Ganz R, Klaue K, Vinh T and Mast J: A new periacetabular osteotomy for the treatment of hip dysplasia. Clinical Orthopaedics and Related Research. 232:26-36.

4. Herman GT, and Udupa JK. Display of three dimensional digital images: Computational foundations and medical applications. IEEE Computer Graphics and Applications. 3(5): 39-45. 1983.

5. Klaue K, Wallin A and Ganz R: CT evaluation of coverage and congruency of the hip prior to osteotomy. Clinical Orthopaedics and Related Research. 232:15-25, 1988.

6. McKibbin B: Anatomical factors in the stability of the hip joint in the newborn. Journal of Bone and Joint Surgery. 52B:148-159, 1970.

7. Murphy SB, Kijewski PK, Simon SR. Normal acetabular anteversion. Orthopaedic Research Society, San Fransisco, January, 1987.

8. Murphy SB, Kijewski PK, Walker PS, Scott RD. Computer-assisted pre-operative planning of orthopedic reconstructive procedures. Computer-assisted Radiology. Berlin: Springer-Verlag, pp. 413-418, 1985.

9. Murphy SB, Kijewski PK, Simon SR, Griffin PP, Chandler HP, Reilly DT, Penenberg BL, Landy ML. Computer-aided Analysis, Simulation, and Design in Orthopedic Surgery. Orthopedic Clinics of North America. 17(4): 637-649, 1986.

10. Murphy SB, Kijewski PK, Simon SR, and Millis MB. Simulation of orthopaedic reconstructive surgery. Computer-assisted Radiology. Berlin: Springer-Verlag (pp. 411-415). July, 1987.

11. Murphy SB, Kijewski PK, Simon SR, Wilkinson RH, Griscom NT. Femoral anteversion. Journal of Bone and Joint Surgery. 69:1169-1186, 1987.

12. Murphy SB, Kijewski PK, Millis MB, Hall JE, Simon, SR, and Chandler, HP. The planning of orthopaedic reconstructive surgery using computer-aided simulation and design. Computerized Radiology. Computerized Medical Imaging and Graphics. 12(1):33, 1988.

13. Rab GT: Biomechanical aspects of salter osteotomy. Clinical Orthopaedics and Related Research 132: 82-87, 1978.

14. Salter RB. Innominate osteotomy in the treatment of congenital dislocation and subluxation of the hip. Journal of Bone and Joint Surgery. 43B: 518-539, 1961.

15. Vannier, M., Totty, W., Stevens, W., Weeks, P., Dye, D., Daum, W., Gilula, L., Murphy, W., and Knapp, R. Musculoskeletal applications of three dimensional surface reconstructions. Orthopedic Clinics of North America. 16(3):543-55, 1985.

16. Wagner H.: Experiences with sperical acetabular osteotomy for the correction of the dysplastic acetabulum. Progress in Orthopaedic Surgery. 2:131-145, 1978.

17. Wagner H: Osteotomies for Congenital Hip Dislocation. The Hip. Proceedings of the fourth open scientific meetin of the Hip Society. St. Louis: CV Mosby, 1976.

Possible Social Implication of PACS on Radiology in the United States

ROGER A. BAUMANN, M.D.
Associate Radiologist-in-Chief
Massachusetts General Hospital
Fruit Street
Boston, Ma 02114

Picture Archiving and Communication Systems (PACS) is a term which is used to describe computer-based systems which handle digital and digitized images. Through digitization of analog images or direct digital acquisition methods, images are entered into the system, transmitted over local area computer networks, displayed for review at workstations and archived on computer media. A full PACS would handle all of the radiology images in a single hospital, including transmitting images between local health care facilities and more distant health care facilities and around the world. In the approximately seven years that such systems have been discussed, much progress has been made. Nevertheless, realization of full PACS will require substantial additional work. This paper tries to speak to the development of PACS today and in the future and of some of the possible social implications for Radiology in the United States that might arise from the introduction of such systems.

Efforts to realize partial or complete PACS systems are underway by a number of vendors and at a number of universities. None of the systems to date has claimed to be fully usable on the department level. The current state of the art is that pieces of PAC Systems have been realized and are being sold and used. These vary from workstations to teleradiology systems to partial PACS systems for a small portion of a department.

Teleradiology systems are a form of PACS, although the concept predates that acronym. An analog image is captured and transmitted either in analog or digitized form over some media, most commonly telephone lines, to a receiving station which recreates the image on a display. A significant number of commercial vendors are active in this area. In the simplest versions the standard telephone line is used to connect a fairly inexpensive receiver unit with the transmission station. Transmission usually requires manual handling of the images and the device at the transmitting end. Usually, little or no storage capability is present on the receiving end of these lower cost devices. More sophisticated units permit the use of faster, dedicated telephone company lines, private lines, microwave or satellite transmission. One of the common realities of teleradiology is the acceptance of lower than ideal resolution and slower than ideal the transmission speed with the idea that the original images will be available for subsequent review. Better units now are beginning to deliver the resolution needed to be more confident of initial diagnoses. Also available now is storage in the receiving workstation which can obviate the long transmission time of images by storing them for subsequent radiologist review.

The social implications of teleradiology have been in two principle areas: doing remote emergency interpretation in off-hour situations and in interpreting images from other locations. The on-call radiologist can take his teleradiology computer/monitor receiver with him along with his beeper. By telephone line, he can access the transmission from the hospital, bring over the emergency images and render his impression by return telephone call. This mode of operation is in use in many departments. One of the driving forces is to provide more timely coverage by radiologist. In addition to the issue of better patient care, there is also sometimes the economic competition with emergency physicians who might argue they should bill for the initial interpretation of an image.

The second major use of teleradiology that has effected radiological practice is the more extensive use of transmission of images form remote hospitals or clinics. Formerly, many of these places were served not at all or only infrequently by a visiting radiologist. Teleradiology has permitted rapid interpretation of more emergent studies and even routine studies with subsequent review of original images in many cases. This method of providing radiological services has been very helpful to small hospitals in the Unites States, especially the more sparsely settled regions.

Systems to handle nuclear medicine divisions are emerging. Nuclear medicine is ideal as a starting point, as the images are low resolution. This not only makes transmission speed and display simpler, but archiving of a significant number of images is more realistic given present technology. A nuclear medicine system could be used primarily for acquisition display and archiving, or, if the additional specialized software was available, the various processing routines and calculations unique to nuclear medicine could also be performed thereon. Some of the systems interface with several nuclear gamma cameras, and therefore present an opportunity to consolidate computers in nuclear medicine as opposed to one computer nuclear gamma camera where digital studies are done.

Systems on the market now which interface with other digital acquisition devices such as CT, MR or DSA units are available but expensive. Often, limited archive is present but archiving on-line even one month or several months can be quite expensive. Various storage technologies such as optical disk, VHS cassette and possibly optical tape archiving are beginning to provide large volume archives, but cost remains a very real consideration. A likely application for such partial PACS systems are workstations with a minimal archive for use as a physician's console to view CT, MR and/or DSA studies. If reasonably priced, radiology departments could find these devices useful to promote throughput from these expensive digital machines and to improve the lifestyle of radiologists who may now read images at unusual hours to avoid decreasing clinical throughput during the more common working hours. This, in turn, will provide more timely reporting of results for patient care.

Another application of a partial PACS approach which is likely to be found in Radiology in the not too distant future are systems wherein digital acquisition or digitization of analog images is used to create an inviolate file. This file could be used in one of two ways to distribute images in high loss areas, such as emergency wards, intensive care units and operating rooms. The first way would be to retain the original radiographs in the radiology department, furnishing displays with acceptable resolution for clinicians to review in the clinical area. The second mode would be to release original images prior to interpretation, and to interpret them if promptly returned or to interpret the computer stored image for those cases not returned by a cut off time. The social and economic advantages of not loosing cases to interpretation are clear.

Notice that the smaller pieces of PACS systems discussed above fit into niches where the time demands upon the radiologists are not pressing, either because of low volume or because of low resolution images. Radiologists are paid on a piece work basis. Although this might change in the future, at present the radiologist does not submit a bill unless he interprets the study. The amount of work a radiologist can do in a day is limited by the speed at which he can display and interpret the cases, one after the other. Indeed, the current manual system in use throughout the country is highly tuned. Although some radiologists hang their own radiographs, in many departments other personnel put these studies on multiviewers to permit the radiologist to press a foot switch and quickly display up to eight panels of images. For the complete PACS system to succeed, computer workstations must present images with a speed, ease of use and information content equal or better to the current system. This is but one of several strenuous requirements that developers are working to overcome to permit the full PACS system to serve an entire department and beyond. But, for our discussion let us assume that such a workstation exists, and that the problems of image transmission speed, acquisition and archiving have all been solved. What then are the social implications for medicine in the United States?

A full PACS system should impact patient care directly by reducing delays. The immediate availability of an image after acquisition by one of the digital acquisition devices will eliminate the time now spent manually collecting, organizing and displaying the images for review. Comparison studies should also be available promptly in the PACS system, and the computer would save the time previously spent in locating them and collating them with the new images. In well designed systems radiologists would be offered only the unread cases of interest to him or her in automatic fashion, one after the other for review, possible image processing, for interpretation or to be put on a temporary hold pending resolution of some question. The possibility of selecting only one or a small number of images from a larger set of images for a clinician to review (or for permanent archiving), the possibility of annotating images easily and/or the inclusion of voice messages about the case all represent possible faster means for communicating the interpretation.

A fully functional PACS should alert the radiologist to other pertinent studies on the same patient. Studies from any modality should be easily displayed and reviewed. This access to any digitized or digital image from any full interpretation station raises the question of who should interpret these images. What impact will this have on subspecialization in radiology? Some subspecialty organizations are on an organ basis, whereas in some departments individual physicians specialize in CT or MR imaging. If a GI radiologist does not interpret MR studies, how will he or she benefit from seeing the MR images pertaining to the abdomen on a patient undergoing a barium study? Will this decrease the specialization in CT or MR? The future is more clear for those subspecialists who on an organ basis now do roentgen, ultrasound, CT and MR imaging pertaining to their organ system. The impact on others is not as clear, but the strong mitigating factor which may in the end result in little change rather than great change, is that full PACS systems should also be able to display either preliminary or final interpretations of a case along with the images from that case. This would permit the GI radiologist unfamiliar with the MR image to make use of it in correlating the new GI study with it.

A full PACS system has the benefit of making available simultaneously coequal versions of an image. These images could well be displayed anywhere in the hospital, transmitted to other local health care facilities over various transmission media or around the country and the world by satellite transmission. One social and economic dilemma that will be addressed in some way in the future is whether the role of the radiologist as the primary interpreter of imaging studies is challenged. Will the emergency room physician or the orthopedic surgeon or others claim that they should now interpret and charge for the interpretation for the study? The response from the radiologist will come from at least three perspectives:

1. Greater training in skill and interpreting the images;

2. No implication of self referral;

3. Rapid interpretation of the studies.

Indeed, one great social change that may occur is much more rapid interpretation of images because of the economic pressure brought to bear by the more wide spread availability of the images. It may or may not be prudent to restrict the distribution of images outside of radiology for a short period of time to permit full quality control and interpretation; long periods prior to release will probably not be well tolerated.

The fully functional PACS system we have assumed for our discussion does have the capability for confident primary interpretation. This plus the archiving capability will certainly enhance teleradiology work in the future. One interesting possibility is that radiologists might compete more sharply with each other for interpretation of images, using PACS systems to transmit images even hundreds of miles. Aside from questions of state licensure which would be fairly easily overcome, this possibility seems improbable to this writer because of the piece work nature of radiology. Most radiologists in the United States carry heavy workloads, and there is little room in fact to add additional cases in significant numbers. Some price competition might have a downward impact on reimbursement in the future. However, it is unlikely that a few remote groups could achieve a price structure, reputation and service record so much better than local service that they could capture a very large referral business. The best response to these challenges will be consonant with the overall goal of medicine, better patient care; these pressures are likely to cause radiologists to improve the speed and quality of the service they render in the future.

Other social changes are also likely to benefit the patient. The integration of the fully functional PACS system with Radiology Information Systems will make more information about the patient available to the radiologist at the time of interpretation. At present the lack of such information retards optimal efforts in helping to understand a patient's study. Interfaces with hospital information systems will extend the information base available to the radiologist even further. One of the most significant factors in speeding patient care will be the far greater availability of all imaging studies and reports to the radiologist and the clinician, making possible faster sequencing of necessary studies and the avoidance of studies no longer necessary because of information already gained. Not only is the result faster diagnosis and treatment, but also dollar savings from fewer studies and/or shorter hospitalizations. The patient will also benefit from other computer interfaces, such as a direct interface to the pathology system. Such an interface should permit accurate and timely feedback to radiologists concerning proven disease conditions on patients they studied which in turn will increase their diagnostic skill.

Predicting the social and economic implications of picture archiving and communications systems is of course difficult. Nevertheless, it is clear that there are many areas in the practice of medicine in the United States which will be impacted. While a few of these impacts are negative and while some of them affect current specialty lines, it seems on balance that it is likely that two positive results will occur. The first is the most significant and that is better care for the patient. The second is that in response to pressures created by these systems, physicians will work even harder to provide more speedy and higher quality service.

Noise-Reduction Algorithm for Dual Energy Radiography

Werner Bautz, Willi A. Kalender

Radiologische Universitätsklinik, Tübingen

Siemens AG Bereich Medizinische Technik, Erlangen

Summary

Dual energy material density images obtained by basis material decomposition techniques offer specific tissue information, but they exhibit relatively high pixel noise. We show that noise in the material density images is negatively correlated and that this can be exploited for noise reduction in the two basis material density images. The algorithm minimizes noise-related differences between pixels and their local mean values adhering to the constraint that monoenergetic attenuation values, which can be calculated from the density images, have to remain unchanged. Applied to the material density images, a noise reduction by factors of 2 to 5 is achieved. Quantitative results remain unchanged, which is important for quantitative digital radiography (QDR) applications. However, edge effects may occur in the processed images. To suppress these, locally adaptive algorithms are presented and discussed. Clinical examples and a discussion of the respective applications are given.

Introduction

In general, dual energy digital radiography (DEDR) offers additional and more specific information on the object to be examined than single energy techniques would allow to obtain [1,2]. While standard DR yields images of the X-ray attenuation, dual energy basis material decomposition techniques provide separate density images of high and low atomic number materials [3-5]. However, such images suffer from relatively high pixel noise levels. A reduction of noise can be achieved by an increase in exposure; but, based on simple counting statistics, any noise reduction by a factor of n requires an exposure increase by a factor of n^2, leading to unacceptable dose levels for most purposes. Therefore, the clinical use of dual energy digital radiography has been limited.

In the present study we have investigated algorithms to reduce noise by image processing, i.e. without an increase in exposure [6]. In contrast to previous efforts [7,8], we decided that the integrity of quantitative information had to be conserved. The concept presented here obeys to the constraint that quantitative values remain unchanged. It is tailored to material density images obtained by basis material decomposition techniques and exploits the negative correlation between pixel noise in the two density images.

Basis material decomposition method and error propagation

X-ray attenuation in the diagnostic energy range is mostly due to Compton effect (incoherent scattering) and photoelectric effect (absorption); these two effects can be viewed as linearly independent functions of energy. Alvarez and Macovski have shown that the mass attenuation coefficient, $\mu/\rho(E)$, of any material can be expressed as a linear combination of these two basis functions to a good approximation and, furtheron, that alternatively the mass attenuation coefficient functions of any two materials, which differ sufficiently in their atomic number Z, can equally be used as basis functions [3]. For any arbitrary object, the line integral over the linear attenuation coefficient, $\int\mu(\vec{r},E)ds$, can be expressed accordingly as

$$\int\mu(\vec{r},E)ds = (\mu/\rho)_1(E)\cdot\delta_1 + (\mu/\rho)_2(E)\cdot\delta_2 \tag{1}$$

$$\text{with} \quad \delta_i = \int\rho_i(\vec{r})ds \qquad i = 1,2 \tag{2}$$

δ_i represents the area density in g/cm^2 and $\rho_i(\vec{r})$ the local density in g/cm^3 of basis material i.

The basis material decomposition method has to solve a pair of equations for the two unknown area densities δ_1 and δ_2. If we define attenuation $S:=\ln(I_0/I)$, it can accordingly be described as a transformation of a pair of attenuation values into a pair of density values:

$$(S_h,S_l) \Rightarrow (\delta_1,\delta_2) \tag{3}$$

The variance and covariance of the resulting δ_i are defined as

$$\sigma_i^2 = (\partial\delta_i/\partial S_h)^2\cdot\sigma_h^2 + (\partial\delta_i/\partial S_n)^2\cdot\sigma_n^2 \tag{4}$$

$$Cov(\delta_1,\delta_2) = (\partial\delta_1/\partial S_h)\cdot(\partial\delta_2/\partial S_h)\cdot\sigma_h^2 + (\partial\delta_1/\partial S_n)\cdot(\partial\delta_2/\partial S_n)\cdot\sigma_n^2 \tag{5}$$

where σ_1 and σ_2 are the standard deviations in the measurements of S_h and S_l, respectively.

It has been shown that the covariance term is negative, $Cov(\delta_1,\delta_2) < 0$;

i.e., the noise in the δ_1- and δ_2-images, respectively, is negatively correlated [6]. This fact has previously been exploited in the generation of monoenergetic attenuation images which constitute a linear combination of the two density images and offer minimal pixel noise [4,9]. It can also be exploited for noise reduction.

Algorithms for noise reduction

Let ρ_i represent pixel values in the original high (i=1) and low (i=2) atomic number basis material density images; $\bar{\rho}_i$ represent the respective mean values over a n x n mask around that pixel. The differences between pixel values and local mean values are denoted $\Delta\rho_i := \rho_i - \bar{\rho}_i$. If nonzero values of $\Delta\rho_1$ and $\Delta\rho_2$ at a given location are caused by noise (and not by structures within the mask), they have to be of opposite sign as has been outlined above. Noise reduction now aims at calculating some filtered version ρ_i' of ρ_i, which exhibits minimal noise without impairing quantitative results. To achieve this, we require that monoenergetic attenuation values $\mu = (\mu/\rho)_1 \cdot \rho_1 + (\mu/\rho)_2 \cdot \rho_2$ remain unchanged, i.e. $\mu' = \mu$ is used as a constraint.

We devised the following algorithms [6]:

$$\rho_1' = \rho_1 - \varepsilon \cdot C/(\mu/\rho)_1 \qquad (6a)$$

$$\rho_2' = \rho_2 + \varepsilon \cdot C/(\mu/\rho)_2 \qquad (6b)$$

$$\text{with } C = [(\mu/\rho)_1 \cdot \Delta\rho_1 - (\mu/\rho)_2 \cdot \Delta\rho_2]/2 \qquad (7)$$

$$\text{and } \varepsilon = \min \begin{cases} |\Delta\mu/\varepsilon_m| \\ 1 \end{cases} \qquad (8)$$

μ and $\bar{\mu}$ denote the pixel value and the local mean in the monoenergetic image, respectively, and $\Delta\mu := \mu - \bar{\mu}$. The control parameter ε is used to adapt the filter strength.

Material and methods

All measurements were carried out on a standard Siemens SOMATOM DR CT scanner equipped with radiographic localizer (Topogram^{TM}) and dual energy option. The parameters were 85 and 125 kVp, 1024 projections, 2mm slice thickness and 8.5 s scan times, respectively. The necessary data for the two X-ray spectra are obtained in one scan since rapid kVp-switching from pulse to pulse is employed [4,5].

The algorithms were implemented on the CT scanner's host computer, a DEC

PDP 11/44, using the CT console's 512x512 pixel image memory. Processsing times were from 1 to 10 minutes depending on the filter mask size which was varied from 3x3 to 7x7 pixels.

Results and discussion

DEDR material density imaging suffers from high pixel noise levels. It therefore has been limited to body regions of low attenuation, e.g. thoracic scans in p.a. projection.

Noise could be reduced by any of the following measures: increase of X-ray exposure, optimization of the X-ray spectra, or image processing. An increase of exposure appears critical because of patient dose considerations; also, noise will only be reduced following a square root relationship to exposure. Optimization of the spectra and exam parameters is considered mandatory, but the gain is limited and not all theoretical considerations can be practically implemented. Reduction of noise by image processing appears attractive since no increase in exposure and no technical requirements would result.

Noise in material density images obtained by basis material decomposition techniques can be reduced by exploiting the negative correlation in pixel noise between the two images. A reduction by factors of 2 to 10 has been achieved in simulated images and in clinical examples; this can be considered dramatic since alternatively an exposure increase by factors of 4 to 100 would be required. At the same time, quantitative results were not impaired [6]. The algorithms apply to DEDR and to dual energy CT in a completely analogous manner.

The price we have to pay for noise reduction is additional processing times and possibly edge artefacts where sharp transitions in density occur in either of the two density images. Edge artefacts are of lesser concern for quantitative tasks since ROIs are usually not chosen across tissue boundaries. They are of high concern, however, for imaging purposes.

The presented algorithms have been applied in a number of clinical applications. Improvements in soft tissue density image quality is most noticeably achieved in the upper thorax, the upper abdomen and the pelvis. Morphological diagnosis of the skeleton is above all improved in lateral projections of the thoracic and lumbar spine. Small tissue calcifications, e.g. kidney stones, are more easily recognized. The possibility for quantification is not impaired by the processing steps; in particular,

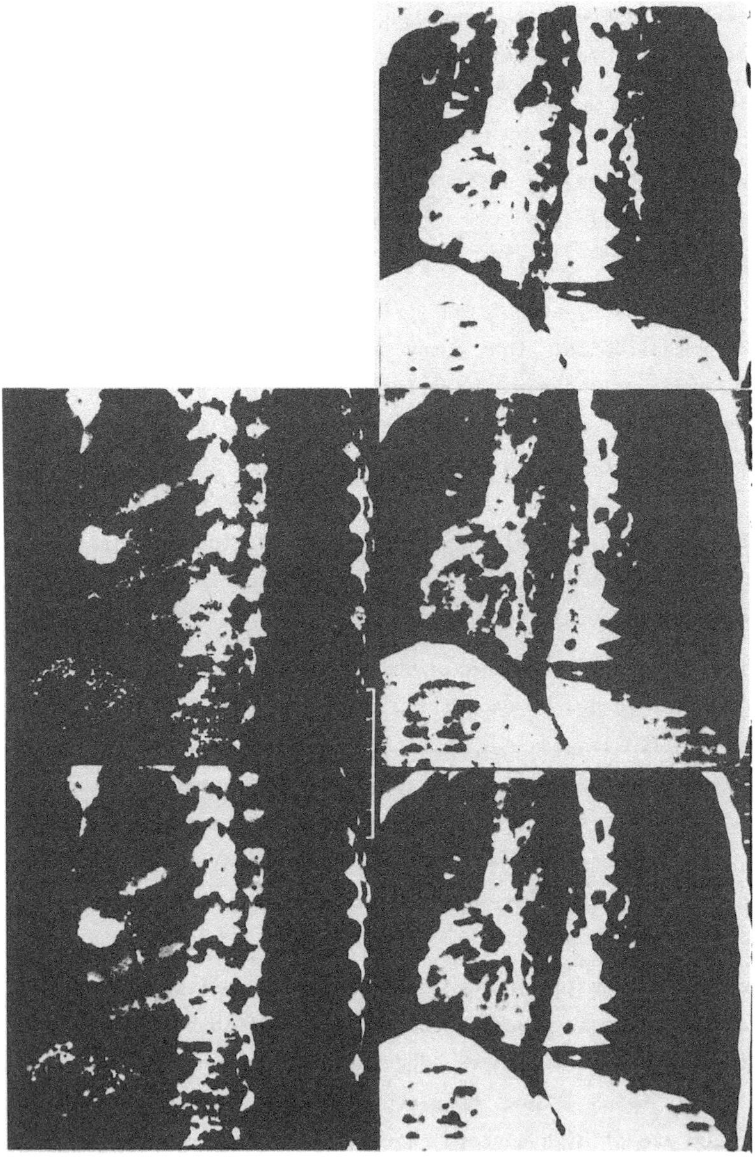

Fig. 1: Dual energy radiography of a calcified myxoma of the right atrium

a. Standard digital radiograph (125 kV),

b,c. Calcium- and soft tissue density image (original),

d,e. Calcium- and soft tissue density image (7x7 Filter).

this applies to bone mineral measurements by arbitrary types of dual energy apparatus. We expect that this type of processing will be adapted to dedicated DEDR systems. Early reports have already confirmed the validity of our approach [10].

References

1. Brody WR, Butt G, Hall A, Macovski A: A method for selective tissue and bone visualizationusing dual energy scanned projection radiography. Med Phys 8 (1981) 353-357
2. Bautz W, Kalender WA: Materialselektive Bildgebung und Dichtemessung mit der Zwei-Spektren-Methode. II. Klinische Anwendung der Zwei-Spektren-Radiographie. Digit Bilddiagn 7 (1987) 95-103
3. Alvarez RE, Macovski A.: Energy-selective Reconstructions in X-ray Computerized Tomography Phys Med Biol 21 (1976) 733-744
4. Kalender WA, Perman WH, Vetter JR, Klotz E: Evaluation of a prototype dual-energy computed tomographic apparatus. I. Phantom studies. Med Phys 13 (1986) 334-339
5. Kalender WA, Bautz W, Felsenberg D, Süß C, Klotz E: Material- selektive Bildgebung und Dichtemessung mit der Zwei-Spektren-Methode. I. Grundlagen und Methodik. Digit Bilddiagn 7 (1987) 66-72
6. Kalender WA, Klotz E, Kostaridou L: An algorithm for noise suppression in dual energy CT material density images. IEEE Trans Med Imag 7 (1988) 218-224
7. Macovski A, Nishimura DG, Doost-Hoseini A, Brody WR: Measurement-dependent filtering: A novel approach to improved SNR. IEEE 2 (1983) 122-127
8. Nishimura DG, Macovski A, Brody WR: Noise reduction methods for hybrid subtraction. Med.Phys. 1984; 11:259-265
9. Alvarez R, Seppi E: A comparison of noise and dose in conventional and energy-selective computed tomography. IEEE Trans Nucl Sc 26 (1979) 2853-2856
10. Ergun DL, Mistretta CA, Brown DE et al.: Dual-energy phosphor cassette for computed radiography. Radiology 169(P) (1988) 89

List of Authors

Glossary

This glossary has been compiled mainly from terms used in the proceedings of CAR. It is divided into medical, medical radiology, and technical terms, each of which will be extended in successive proceedings according to needs and feedback from readers.

Medical

Aneurysm - balloon-like dilation of a blood vessel or portion of the heart

Angina - chest pain caused by narrowing of arteries supplying the heart

Angiography - methods to visualise the blood vessels

Angioplasty - dilation of a narrowed blood vessel

Arteriosclerosis - disease of aging which results in narrowing of the blood vessels

Basal Ganglia - important neural structures at the base for analysis

Cerebrovascular - blood vessels that supply the brain

Congenital Disease - inherited disease

Contrast Material - substance used to increase the intensity difference between tissues. In radiography is usually barium or iodine

Coronary Arteriography - imaging of the vessels that supply the muscle of the heart

Cyst - an abnormal fluid filled mass arising from an organ, commonly kidney or pancreas

Demyelinating - process in which the tissue sheath around the nerves is destroyed

Diastole - the period during which the heart muscle relaxes and the camber fills with blood

DSA - digital subtraction angiography; an imaging method for demonstrating the new appearance of contrast in blood vessels by subtracting a preinjection mask image from image of a later, contrast filled vessel

ECG - electrocardiogram; electrical heart signals

Embolization - blockage of vessels by clot which has traveled from a downstream position within a vessel

Fibroma - abnormal mass of tissue but not cancer

Hemangioma - abnormal tangle of blood vessels

Herniated Nucleus Pulposes - extrusion of tissue between the vertebra which presses on the spinal nerves and causes back pain

Histology - microscopic examination of organ tissues

Hybrid Subtraction - high and low kilovoltage technique subtracting one image from the other to remove representation of the soft tissues

Hypoechogenic - tissue that is free of echoes on ultrasound, usually clear fluid

Infarct - irreversible damage to tissue caused by lack of blood flow or oxygen

Ischemia - tissue suffering from lack of oxygen usually due to restricted blood flow

Laparotomy - surgical opening of the abdomen

Lymphoma - cancer of the lymph tissues

Malignancy - cancer

Mediastinum - organs in the central portion of the chest: primarily the heart and major vessels

Metabolism - cellular chemical activity required to maintain life

Metastasis - cancer which has spread to other organs from its original source

Monoclonal Antibody - biologically produced pure blood protein capable of attaching to a specific target tisue

Myelography - imaging of the spinal cord

Myocardium - muscle of the heart wall

Neoplasm - a new and abnormal growth of tissue

Noninvasive Imaging - imaging achieved without the use of surgery or instruments that penetrate the skin. Examples are ultrasound or nuclear magnetic resonance

Nucleus Pulposes - the spongy tissue between the bony vertebra of the spine

Oncology - the study and knowledge concerning tumors

Percutaneous Transluminal Coronary Angioplasty - procedure by which a ballon catheter is expanded within a stenotic region of the coronary artery to dilate it an open blood flow to the myocardium

Peritoneum - internal contents of the abdomen containing bowel

PET - positrom emission tomography; uses radioactively labelled compounds to selectively label target tissues. Tomographic image results from detection of the paired gamma rays from the positron decay

Retroperitoneum - structures behind the internal contents of the abdomen. This includes organs such as the pancreas, aorta, kidneys and muscular tissues

Scintigraphy - imaging by detecting gamma rays emitted from radioactively labeled tissues

SPET - single photon emission tomography; a tomographic technique using a rotating gamma ray detector to record, then reconstruct radioactive distribution pattern within the body

Supratentorial - above a tissue membrane that separates the cortex of the brain from the cerebellum. Lies in the lower aspects of the skull

Syringomyelia - abnormality in which a fluid cyst forms within the spinal cord

Systole - the period of contraction of the heart muscle in which blood is actively ejected

Therapeutic Radiology - treatment of cancer by targeting portions of the body with ionizing radiation

Thrombosis - clot formation in the arteries or veins resulting in obstruction

Tomography - slice imaging

Venogram - imaging of the veins which return blood to the heart

Ventriculography - display of the left ventricular cavity of the heart ot determine its size and the motion of its muscular walls

Medical Radiology*

Associated Equipment - in a radiological installation, equipment other than those for the production and control of ionizing radiation, but essential for its application

Brachyradiotherapy - intracavitary, interstitial or superficial radiotherapy using one or more sealed radioactive sources

Cineradiography - indirect radiography of moving objects usually in rapid series on cine film

Computed Tomography - reconstructive tomography in which recording and processing is effected by a computing system

Dental Panoramic Radiography - direct radiography of a part of or the complete dentition by the use of an intra-oral X-ray tube

Direct Radiography - radiography in which the recording is effected at an image reception area

Direct Radioscopy - radioscopy in which the visible images are presented at the image reception area, or close to it, in the radiation beam

Direct Tomography - tomography in which recordingof the image of a layer in the object is effected at the image reception area

Display - visual presentation of information

Electro-Optical X-Ray Image Intensifier - X-ray image intensifier incorporating an electro-optical vacuum device

* The medical radiology terminology has been extracted from the 'International Electrotechnical Commission IEC Standard', Publication 788, First Edition, 1984, Geneva, Switzerland.

Electromagnetic Radiation - propagation of emitted energy through space or through a material medium in the form of waves or in the form of kinetic energy of particles

Examination Room - in medical diagnostic radiology, room in which ionizing radiation is applied to the patient or measured on the patient, and which has the required structural means for radiation protection and for mechanical support

Fluorography - indirect radiography by means of a fluorescent screen

Fluoroscopy - technique of radioscopy by means of a fluorescent screen

Gamma Camera - equipment for scintigraphy, which produces an image by simultaneous detection of radiation emitted from the object

...graph, - in radiology, equipment designed to perform radiography preferably by means of a specific technique

Image Reception Area - in radiology, surface on which an X-ray pattern is received

Input Screen - in an electro-optical vacuum device, layer forming the imaging reception area

Intensifying Screen - layer of material used in direct radiography to convert the incident X-radiation or gamma radiation into radiation more suitable for the radiation-sensitive emulsion of the radiographic film

Irradiation - exposing of a living being or matter to radiation. In radiology, exposing of a living being or matter to ionizing radiation

Kymograph - in radiology, equipment designed to perform radiography preferably by means of a specific technique

Kymography - direct radiography in which an image of the displacements of the boundary of an object is obtained

Luminescence - phenomenon in which certain substances, when excited, emit light of wavelength characteristic of the substance

Medical Diagnostic Radiology - medical diagnosis using ionizing radiation

Medical Radiology - branch of radiology applied to human and veterinary medicine as well as to dentistry and chiropractic

Non-Screen Film - in direct radiography, radiographic film with relatively high sensitivity to the radiation emitted by fluorenscence of an intensifying screen

Nuclear Medicine - use of unsealed radionuclides for diagnostic and therapeutic purposes

Operator - person utilizing an equipment individually with or without the aid of an assistant, who controls some or all functions of the equipment in his presence

Output Image - in an electro-optical vacuum device, light image generated in the output screen

Output Screen - in an electro-optical vacuum device, layer in which the electron pattern is converted into a light image

Patient - person or animal undergoing medical examination or treatment. For purposes of radiological protection, a person or animal is a patient only during the intentional application of ionizing radiation to that person or animal

Radiation - propagation of emitted energy through space or through a material medium in the form of waves or in the form of kinetic energy of particles

Radiography - technique for obtaining, recording, and optionally processing directly or after transfer, information contained in an X-ray pattern at an image reception area

Radiological - referring to ionizing radiation, its generation and application for scientific, medical and technical purposes

Radiological Equipment - equipment designed for use in radiology

Radiological Image - information obtained by using ionizing radiation presented as an image suitable for medical diagnosis

Radiological Installation - installed radiological equipment including all means for its intended operation

Radiology - science of ionizing radiation and its application

Radiotherapy - medical therapy essentially consisting of one or more treatments by ionizing radiation

Screen Film - in direct radiography, radiographic film with relatively high sensitivity to the radiation emitted by fluorescence of an intensifying screen

Stereoradiography - radiography in which an object is irradiated from two directions to produce a pair of radiograms which when viewed by suitable optical means give the illusion of a three-dimensional image

Teleradiotherapy - radiotherapy with a large radiation source to skin distance, usually not less than 50 cm

Tomograph - in radiology, equipment designed to perform radiography preferably by means of a specific technique

Tomography - radiography of one or more layers within an object

Transmission - in radiology, passage of ionizing radiation through, and emergence from, matter essentially in the direction of its incidence without or after interaction with that matter

User - when used in an IEC standard on electro-medical equipment, organisation or individual responsible for the use and maintenance of the equipment

Technical

Analog Signal - a s**Analog Signal** - a signal in the form of a continously varying physical quantity

Animation - presentation of a sequence of images on the display screen to create sense of motion

ASCII - American Standard Code for Information Interchange. An 8 bit code for data representation and transfer

Attribute - a characteristic of a display entity, such as, colour, intensity, detectability, etc.

Bandwidth - the difference in cycles per second or hertz between the high and low frequencies of a frequency band

Baud - a unit of signaling speed defining the number of discrete signal conditions

Bit-Rate - the speed at which bits are transmitted over a communication channel, usually expressed in bits per second

Calligraphic Display - a graphic display system for drawing vectors in a refreshed mode

Channel - a facility for a one-way transmission of data

Character - alpha-numerical symbol or other signs and symbols contained in a message

Circuit - a facility for a both-way transmission of data between two or more points

Computer Aided Design - the application of computers to the design process

Computer Assisted Orthopaedics - computer based system to assist in orthopaedic surgical planning, e.g. by means of 3D modelling and computer graphic display of patient specific anatomy and pathology. It may also refer to computer aided design (CAD) and computer aided manufacturing (CAM) methods of patient specific structure models and prothesis

Computer Assisted Radiation Therapy (CART) - radiation dose calculating models and procedures provided by a computerised dose planning system

Computer Assisted Radiology (CAR) - computer based system to assist in generating digital images, storing and transferring images, viewing, analysing and interpreting of images

Computer Assisted Surgical Planning (CASP) - computer based system to assist in general or stereotactic surgical planning

Computer Graphics (CG) - a set of computerised pictoral/graphical procedures for displaying 2D, 3D and 4D data

Computer Vision (CV) - models and procedures provided by a computerised system to support digital image segmentation, analysis and understanding

Data - encoded information that can be stored, manipulated and transmitted by digital devices

Data Communication - the transmission of data from one point to another by means of communication media

Data Tablet - a graphic input device which encodes X and Y data from a hand held stylus on a tablet surface

Data Terminal Equipment - equipment attached to a data network to send or receive data or both

Digitiser - a device that codes images or shapes into digital data

Display Console - a hardware complex consisting of at least one display unit and usually an alpha-numeric keyboard, function keys and/or other input devices

DMA - Direct Memory Access. A technique for transferring data to and from a host memory without interrupting the host processing unit

Frame Buffer - a dedicated memory for a raster display unit used to store image data

GKS - Graphical Kernel System. A standard for programming computer graphic purposes

Graphic Input - a process to define graphical data for modelling or display purposes

Hardware - a generic term to include all equipment, both computer and communication, contained in a system

Hidden Lines, Hidden Surfaces - removal of line and surface segments obscured from view in a projected image of a 3D object

Highlight - a display characteristic, intended to draw the viewers attention to a display entity. It is also used to describe a local maximum in rendering processes using illumination models

Image Transformation - algorithms applied to images to produce change in appearance

Light Pen - an hand held stylus which detects light within a limited area on a refreshed display unit

Look-Up Table - a special memory array containing grey or colour values for pixel display

Man Computer Interaction (MCI) - procedures and devices optimised towards realising a 'natural' interface between man and computer

Menu - a list of options shown on a display unit

Message - a sequence of characters used to convey information or data

Modelling - procedures and devices to represent 2D, 3D and 4D general or patient specific anatomic and clinical data in digital memories

Network - a system consisting of a number of points connected by communication channels

Picture Archiving and Communication Systems (PACS) - computer based system to assist in storage and archiving of images as well as soft and hard copy image generation

Pixel - an individual picture element of a digital image

Point - a coordinate position in 2D or 3D space. Also an addressable position on the display screen

Protocol - a formal procedure required for synchronising data exchange, e.g. to initiate and maintain data communication

Raster - a rectangular matrix of pixels

Raster Scan - a technique for generating an image in a line-by-line sweep

Run-Length Encoding - a data compression technique based on describing the number of pixels along a scan line that have the same intensity value

Scaling - a transformation functions to change the size of a display entity

Shading - computation of the intensities and colours of the surface pixels of 3D objects

Software - a generic term for computer programs and documentation

Spatial Resolution - the smallest distance between two pixels

Translation - movement of a display entity from one position to another

Vector - a line segment drawn between two points. It possesses the attributes of magnitude and direction

Window - a bounded area that contains a subset of the data to be shown on the display unit

H. U. Lemke, M. L. Rhodes,
C. C. Jaffe, R. Felix, (Eds./Hrsg.)

Computer Assisted Radiology/ Computergestützte Radiologie

Proceedings of the International Symposium/

Vorträge des Internationalen Symposiums

CAR '85 Computer Assisted Radiology

Organizer / Organisator: AMK Berlin

1985. XVIII, 724 pp. (122 pages in German).
322 figs. ISBN 3-540-15520-1

Contents: Digital Image Generation: Magnetic Resonance Imaging. Computer Tomography. Nuclear Medicine. Digital Subtraction Radiography. Digital Radiography. – Picture Archiving & Communication Systems. – Application Programming: Computer Assisted Radiation Therapy. Computer Assisted Orthopaedics. Computer Assisted Radiological Diagnosis. – Computer Vision. – Computer Graphics. – Computer Assisted Radiology. – Workshop-Modelling. – Workshop-Man Computer Interaction. – Appendix. – List of Authors.

Springer-Verlag Berlin
Heidelberg New York London
Paris Tokyo Hong Kong

Springer

GPSR Compliance

The European Union's (EU) General Product Safety Regulation (GPSR) is a set of rules that requires consumer products to be safe and our obligations to ensure this.

If you have any concerns about our products, you can contact us on ProductSafety@springernature.com

In case Publisher is established outside the EU, the EU authorized representative is:

Springer Nature Customer Service Center GmbH
Europaplatz 3
69115 Heidelberg, Germany

Batch number: 10844661

Printed by Printforce, the Netherlands